Successful Strategies for the Discovery of Antiviral Drugs

RSC Drug Discovery Series

Editor-in-Chief:
Professor David E. Thurston, *King's College, London, UK*

Series Editors:
Dr David Fox, *Vulpine Science and Learning, UK*
Professor Ana Martinez, *Instituto de Quimica Medica-CSIC, Spain*
Professor David Rotella, *Montclair State University, USA*

Advisor to the Board:
Professor Robin Ganellin, *University College London, UK*

Titles in the Series:

1: Metabolism, Pharmacokinetics and Toxicity of Functional Groups
2: Emerging Drugs and Targets for Alzheimer's Disease; Volume 1
3: Emerging Drugs and Targets for Alzheimer's Disease; Volume 2
4: Accounts in Drug Discovery
5: New Frontiers in Chemical Biology
6: Animal Models for Neurodegenerative Disease
7: Neurodegeneration
8: G Protein-Coupled Receptors
9: Pharmaceutical Process Development
10: Extracellular and Intracellular Signaling
11: New Synthetic Technologies in Medicinal Chemistry
12: New Horizons in Predictive Toxicology
13: Drug Design Strategies: Quantitative Approaches
14: Neglected Diseases and Drug Discovery
15: Biomedical Imaging
16: Pharmaceutical Salts and Cocrystals
17: Polyamine Drug Discovery
18: Proteinases as Drug Targets
19: Kinase Drug Discovery
20: Drug Design Strategies: Computational Techniques and Applications
21: Designing Multi-Target Drugs
22: Nanostructured Biomaterials for Overcoming Biological Barriers
23: Physico-Chemical and Computational Approaches to Drug Discovery
24: Biomarkers for Traumatic Brain Injury
25: Drug Discovery from Natural Products
26: Anti-Inflammatory Drug Discovery
27: New Therapeutic Strategies for Type 2 Diabetes: Small Molecules
28: Drug Discovery for Psychiatric Disorders
29: Organic Chemistry of Drug Degradation
30: Computational Approaches to Nuclear Receptors
31: Traditional Chinese Medicine
32: Successful Strategies for the Discovery of Antiviral Drugs

How to obtain future titles on publication:
A standing order plan is available for this series. A standing order will bring delivery of each new volume immediately on publication.

For further information please contact:
Book Sales Department, Royal Society of Chemistry, Thomas Graham House, Science Park, Milton Road, Cambridge, CB4 0WF, UK
Telephone: +44 (0)1223 420066, Fax: +44 (0)1223 420247,
Email: booksales@rsc.org
Visit our website at www.rsc.org/books

Successful Strategies for the Discovery of Antiviral Drugs

Edited by

Manoj C. Desai
Gilead Science Inc., Foster City, California, USA
Email: manoj.desai@gilead.com

Nicholas A. Meanwell
Bristol-Myers Squibb Research, Wallingford, Connecticut, USA
Email: nicholas.meanwell@bms.com

RSC Publishing

RSC Drug Discovery Series No. 32

ISBN: 978-1-84973-657-2
ISSN: 2041-3203

A catalogue record for this book is available from the British Library

Published by The Royal Society of Chemistry,
Thomas Graham House, Science Park, Milton Road,
Cambridge CB4 0WF, UK

Registered Charity Number 207890

For further information see our web site at www.rsc.org

Printed in the United Kingdom by CPI Group (UK) Ltd, Croydon, CR0 4YY

Preface

Viruses are obligate parasites that enter and reproduce within the cells of their host, so their life cycle relies upon forming an intimate partnership and dependency. For many viruses, such as influenza and respiratory syncytial virus (RSV), this relationship is temporary in nature and self-limiting. However, some viruses, including human immunodeficiency virus (HIV, >35 million people infected worldwide), hepatitis B virus (HBV, >400 million people infected) and hepatitis C virus (HCV, >150 million people infected), cause persistent infections and the virus never leaves the host. Such a long-term association is detrimental to the wellbeing of infected cells, the functions of the organ they infect and, ultimately, the overall health of the host.

The central hypothesis to medical intervention for the treatment of viral diseases is to prevent the spread of a virus to uninfected cells by blocking viral replication, helping the host to control infection. The approval of the nucleoside analog acyclovir for the treatment of herpes simplex virus (HSV) infection in 1978 was a key event in the history of antiviral drug discovery and development, setting the standard for a selective and effective therapeutic that is now available over-the-counter. Oral antiviral agents of this type offer a practical, convenient and rapid means of intervening with virus replication, and over the last 30 years over 50 drugs have been licensed for marketing. In addition, in 2012 alone there were >100 industry-sponsored Phase 3 trials currently registered at ClinicalTrials.gov (www.clinicaltrials.gov).

The demand for antiviral agents has been driven by the HIV-1 epidemic, a virus that continues to present a significant challenge, and the large number of worldwide HBV and HCV infections, all of which contribute to an ever-increasing morbidity and mortality in those unfortunate enough to host these infectious agents. Although HIV-1 and HBV infections have proven extremely difficult to cure, the RNA-based life cycle of HCV is currently susceptible to curative intervention with combinations of pegylated interferon-α, ribavirin

RSC Drug Discovery Series No. 32
Successful Strategies for the Discovery of Antiviral Drugs
Edited by Manoj C. Desai and Nicholas A. Meanwell
© The Royal Society of Chemistry 2013
Published by the Royal Society of Chemistry, www.rsc.org

and a protease inhibitor. However, recent clinical studies have clearly indicated the potential for combinations of small-molecule, direct-acting antiviral agents that selectively and effectively target viral proteins to effect cures with as little as 12 weeks of therapy. Consequently, there is a growing belief that HCV infection will be the first chronic viral infection to be cured by small molecules. For HIV-1 infection, the launch of Atripla®, a combination of the nucleoside phosphonate prodrug tenofovir disoproxil, the nucleoside analog emtricitabine and the non-nucleoside reverse transcriptase inhibitor efavirenz, by the collaborative effort of Gilead Sciences and Bristol-Myers Squibb, represented a watershed in the treatment of this disease by providing a convenient, fixed-dose combination taken once a day that effectively controls viral replication.

The focus of this book is to summarize successful strategies for the discovery and development of antiviral agents into clinically relevant therapeutic agents. The book is organized according to the strategies deployed both to discover and to optimize lead compounds. Section I provides an overview of drug discovery programs that span HCV, RSV, dengue virus and pox viruses and that owe their origin to a robust *in vitro* cell culture system used both to identify lead inhibitors using high-throughput screens and to optimize molecules for potency and selectivity. This kind of chemical genomics screening paradigm has proven to be a highly successful strategy for the discovery of mechanistically interesting antiviral agents, many of which could not be discovered using biochemical assays. By applying selective pressure to viruses grown in cell culture with repeated rounds of replication (passaging) in the presence of increasing concentrations of lead inhibitors, resistant viruses can be isolated and their genomes sequenced for mutations that usually afford insight into the mode of action of a lead inhibitor.

Biochemical screens are an equally important source of leads, a strategy of particular importance in the early days of HCV drug discovery where the enzymatic activities of the NS3 protease and NS5B polymerase could be recapitulated *in vitro* and used to assay compound collections using high-throughput screening methodology. Lead optimization campaigns were subsequently facilitated by structure-based drug design since these proteins were crystallized with inhibitors bound. Section II provides examples of the application of these contemporary technologies that rely upon biochemical screening and structure-based optimization strategies for the discovery of potent and selective antiviral agents for the treatment of HIV-1 and HCV and nicely illustrate the evolution of modern medicinal chemistry technology. Section III includes some of the recent mechanistic approaches that take advantage of host–viral interactions for the treatment of HCV that could be complementary to direct-acting oral antivirals. Finally, the delivery of antiviral agents can present significant challenges and Section IV highlights the development and application of strategies that can be deployed to facilitate oral absorption of nucleosides or the systemic delivery of an entire therapeutic regimen that improves compliance.

The objective of this book is to capture tactical aspects of problem solving in antiviral drug design and development, an approach that not only holds special appeal for those engaged in the antiviral research, but will also be instructive to the broader medicinal chemistry community.

As we compose this Preface in January 2013, we note the passing in 2012 of two chemists who made seminal contributions to antiviral drug discovery. Professor Antonín Holý, a pioneer of the nucleoside phosphonate chemotype that led to the discovery of cidofovir, adefovir and tenofovir, passed away on 16 July 2012 in his native Prague. Jerome P. Horwitz, who synthesized azidothymidine, the first drug approved to treat HIV-1 infection and whose research led to the development of dideoxycytidine, passed away on 6 September 2012 in West Bloomfield, Michigan, close to his native Detroit. We recognize the critically important contributions that these scientists made to the therapy of HIV-1 infection.

We would like to thank the authors of the chapters in this volume for their hard work, patience, dedication and scholarship in the lengthy process of writing, editing and making last-minute revisions to their contributions.

Manoj C. Desai
Nicholas A. Meanwell

Contents

RSC Drug Discovery Series No. 32
Successful Strategies for the Discovery of Antiviral Drugs
Edited by Manoj C. Desai and Nicholas A. Meanwell
© The Royal Society of Chemistry 2013
Published by the Royal Society of Chemistry, www.rsc.org

Section III Host Targets

Chapter 10 TLR-7 Agonists for the Treatment of Viral Hepatitis   **365**
Randall L. Halcomb

Section I
Phenotypic Screening to Discover Antiviral Agents

CHAPTER 1

Discovery and Clinical Validation of HCV Inhibitors Targeting the NS5A Protein

MAKONEN BELEMA,*[a] NICHOLAS A. MEANWELL,[a] JOHN A. BENDER,[a] OMAR D. LOPEZ,[a] PIYASENA HEWAWASAM[a] AND DAVID R. LANGLEY[b]

[a] Department of Medicinal Chemistry and [b] Department of Computer-Assisted Drug Design, Bristol-Myers Squibb Research and Development, 5 Research Parkway, Wallingford, CT 06492, USA
*Email: makonen.belema@bms.com

1.1   Introduction

Significant effort has been invested in elucidating the exact role and function of the NS5A protein in the hepatitis C virus (HCV) replication cycle. Although, unlike the NS3 and NS5B proteins, no enzymatic function has been identified thus far for NS5A, it has become apparent that this protein plays a diverse and critical set of roles both in the replication of the virus and in the mediation of host–virus interactions. Despite its multifunctional role, the lack of a well-characterized function coupled with the limited availability of structural information, compared with the NS3 protease and NS5B polymerase, initially made the NS5A protein a less compelling target for therapeutic intervention. That changed, however, with the validation of NS5A as a clinically relevant target by daclatasvir (**1**), where single doses effected pronounced and rapid declines in viral RNA in HCV-infected subjects (Figure 1.1).[1] Highlights of the

RSC Drug Discovery Series No. 32
Successful Strategies for the Discovery of Antiviral Drugs
Edited by Manoj C. Desai and Nicholas A. Meanwell
© The Royal Society of Chemistry 2013
Published by the Royal Society of Chemistry, www.rsc.org

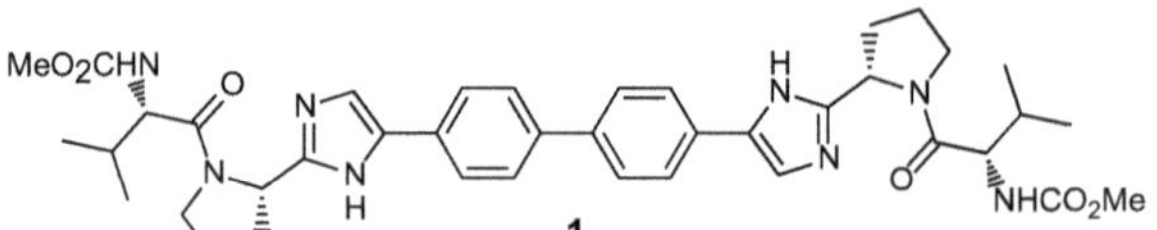

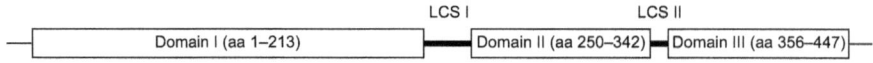

Figure 1.1   Daclatasvir.

	LCS I		LCS II	
Domain I (aa 1–213)		Domain II (aa 250–342)	Domain III (aa 356–447)	

Figure 1.2   NS5A protein organization.

biochemical pharmacology of the NS5A protein, along with the discovery, the mode of action and the clinical characterization of a potent class of NS5A inhibitors, are discussed in this chapter.

1.2   The HCV NS5A Protein

The HCV RNA genome encodes a $\sim$3000 amino acid polypeptide that is processed by both viral and cellular proteases into structural proteins (Core, E1 and E2), an ion channel (p7) and non-structural proteins (NS2, NS3, NS4A, NS4B, NS5A and NS5B).[2] The non-structural proteins are responsible for replication of the viral genome and for the assembly of the viral particle from the structural proteins, with the assistance of host factors. HCV NS5A is a 447 residue peptide that is comprised of three domains, which are interlinked with short fragments designated as low-complexity sequences (LCSs) (Figure 1.2).[3] Various studies have demonstrated that NS5A is an RNA-binding protein, although the specific elements of the protein that are establishing the biologically relevant interactions with ribonucleic acid still need to be identified.[4–6] For example, one study has indicated that all three domains of NS5A exhibit RNA-binding properties, albeit with differential affinities, whereas a different study showed that the Domain I/LCS I peptide fragment exhibited RNA-binding affinity that is comparable to that of the full-length NS5A protein, supported by the observation that the RNA-binding property of NS5A is abolished if Domain I is deleted.[4,5] Whatever their specific RNA-binding properties may be, all three domains contribute to genome replication, while Domain III plays a key role in viral particle assembly.[7,8] In addition, all three domains play a diverse set of regulatory roles in modulating host–virus interactions so as to facilitate the establishment of an environment conducive to successful viral replication.

Domain I of NS5A is a Zn^{2+}-binding moiety with an amphipathic α-helix at its N-terminal that is believed to anchor the protein to cellular membranes. X-ray structural studies by two independent groups on similar amino acid constructs of Domain I, both lacking the amphipathic α-helix motif, revealed that the protein crystallizes as a homodimer (Figure 1.3).[9,10] Interestingly, although the monomeric units in the two X-ray studies were highly structurally

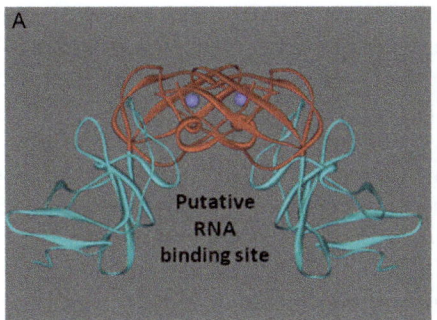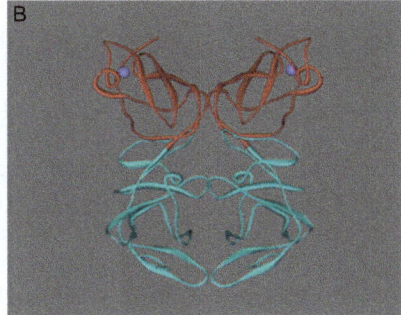

Figure 1.3   Rice dimer (A) and Love dimer (B). The first 100 amino acids (putative inhibitors binding site) are colored red. The Zn^{2+} ions are shown as purple balls.

conserved, their modes of dimerization were different and involved non-overlapping contact surfaces. A positively charged groove created by the dimeric interface of the X-ray structure reported by Rice's group, which had the appropriate dimensions to support the hypothesis that it could be an RNA-binding site, is fully exposed in Love's X-ray structure. The reason for the differing modes of dimerization and how well either one may reflect a biochemically relevant structure of the NS5A protein, especially since about half of the protein is missing from the structural analyses, is not apparent at this stage. Some have postulated that the two dimeric modes may represent snapshots of an oligomeric state, the functional significance of which has yet to be revealed. In general agreement with the X-ray structural findings, a glut-athione-*S*-transferase (GST)-tagged NS5A Domain I was able to pull down a His-tagged NS5A protein, presumably through a dimeric interaction, whereas this was not possible with GST alone. This interaction does not appear to be mediated by the presence of nucleic acids and yet, interestingly, there is a similarity between the minimal NS5A fragment required to effect this pull-down and the minimal fragment required to maintain the RNA-binding affinity of the full-length NS5A protein. It is also noteworthy that the minimal peptide fragment required to effect the dimerization in the pull-down study was longer than the peptide constructs used in the X-ray studies (amino acids 1–240 *versus* 25–215 for the Rice dimer and 33–202 for the Love dimer), and although the reason for this disparity is not apparent, it could be a result of the distinct physical states that the two studies are dealing with and of differences in experimental parameters, such as protein concentration.[11]

In another study, a glutaraldehyde cross-linking experiment demonstrated that either Domain I or the full-length version of NS5A (but not Domain II–III) dimerize in solution, that the dimer is in equilibrium with the monomer and that the presence of uracil-rich RNA, which is known to bind to NS5A, shifts the equilibrium in favor of the dimer.[5] Interestingly, in the same study, NS5A–RNA cross-linking followed by the mapping of the amino acids involved in the cross-linking on to either of the two X-ray structures indicated that, for the Rice dimer, the amino acids decorate the positively charged groove of the protein and not the similarly charged back side of the dimer, whereas for the Love dimer, a ribbon pattern surrounding the structure is observed. It was claimed that this cross-linking result is more consistent with Rice's dimer. Moreover, others have hypothesized that the groove in the Rice dimer may serve as an 'RNA railway system' that connects different functional states that the RNA has to traverse, along with providing a role of protecting the viral RNA from cellular factors that degrade exogenous RNA.[12] Whatever the case may be, the fact that highly potent NS5A inhibitors with resistance mutations that map to Domain I constitute a dimeric pharmacophore that complements the symmetrical features revealed by the X-ray studies and supported by biochemical studies, is unlikely to be coincidental (see below).

Unlike Domain I, Domains II and III are disordered proteins that lack secondary structural elements.[13,14] It is hypothesized that disordered proteins have an extended surface area that promotes simultaneous interactions with multiple proteins and/or an interaction with RNA, which could be a reflection of the multifunctional nature of these domains, the details of which still need to be delineated.[15]

1.3   The Discovery of HCV NS5A Replication Complex Inhibitors

HCV replicons, cell-based assay systems that support the autonomous replication of the subgenomic and genomic HCV, have played a central role in the HCV drug discovery field since the introduction of the first genotype 1b (G-1b) system in 1999, most notably in creating opportunities to exploit the potential of viral targets devoid of enzymatic functions.[16,17] As part of a campaign directed at identifying inhibitors of HCV that act by novel mechanisms to disrupt replication, scientists at Bristol-Myers Squibb (BMS) devised a unique, dual-replicon assay system that was used to conduct a phenotype-based, high-throughput screening (HTS) campaign.[18] Specifically, this assay system utilized a mixture of a G-1b HCV replicon and a replicon of a closely related virus, bovine viral diarrhea virus (BVDV), in the same well. The two replicons had the same Huh-7 cellular background but orthogonal activity reporters – a FRET assay based on NS3 protease activity for HCV and a luciferase expression assay for BDVD. In addition, cell toxicity was assessed in the same well using a standard Alamar Blue assay. It is noteworthy that since a luciferase enzyme assay is more sensitive than a FRET assay, this reporter combination placed a stringent criterion for the identification of HCV-specific inhibitors. The BMS compound collection was screened with this dual replicon assay system and initial hits that had either cytotoxic properties or poor HCV specificity, as reflected by a <10-fold potency spread between HCV and BVDV inhibitory activities, were discarded. Counter-screening of the resultant hit set with NS3 protease, NS3 helicase and NS5B polymerase enzymatic assays afforded a thiazolidinone chemotype, exemplified by carbamate **2**, as a novel class of HCV

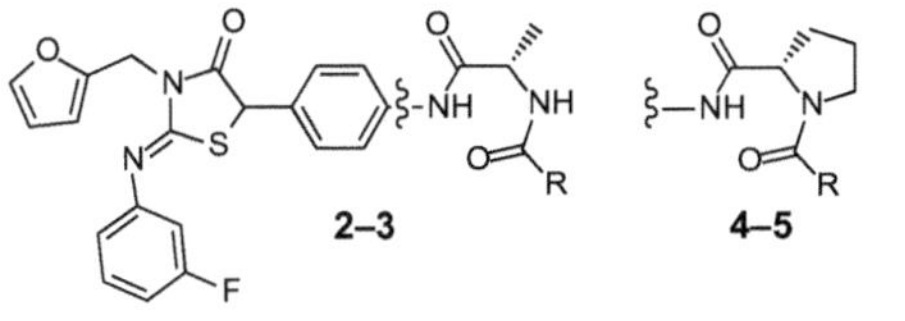

Compd.	R	G-1b EC$_{50}$ (nM)
2	OCH$_2$Ph	580
3	CH$_2$Ph	6.0
4	OCH$_2$Ph	110
5	CH$_2$Ph	2.3

Figure 1.4   SAR highlights of thiazolidinone series.

inhibitor (Figure 1.4). It is noteworthy that over one million compounds were assayed in the HTS effort and only this single chemotype class met the stringent screening criteria to qualify as a suitable lead structure. Carbamate **2** exhibited moderate HCV potency (G-1b EC$_{50}$ = 0.58 μM), very good HCV specificity (BVDV EC$_{50}$ >50 μM) and a CC$_{50}$ of >100 μM.[19,20]

In order to identify the HCV protein that carbamate **2** might be targeting, passage of a G-1b replicon system through increasing concentrations of the compound resulted in a resistant phenotype that was >10-fold less sensitive to the inhibitor. After confirming that the mutation that caused the resistance phenotype was associated with viral RNA and not cellular RNA, sequence analysis of viral RNA from resistant cell lines was conducted and two dominant mutations were identified in the Domain I region of NS5A (Y93H and Y93C). Either mutation, when introduced individually into a G-1b replicon, was sufficient to confer the observed resistant phenotype and no cross-resistance was observed with inhibitors targeting alternative HCV mechanisms. This resistance analysis was the first indication that the thiazolidinone chemotype might be engaging the NS5A protein.

Preliminary structure–activity relationship (SAR) studies directed at establishing the fidelity of the lead revealed that there was a preference for the *S* stereochemistry at the amino acid moiety and that changing the benzyl carbamate to phenylacetamide, as in amide **3**, effected a ~100-fold potency enhancement, an SAR observation that was recapitulated in the analogous proline series (see amides **4** and **5**). The SAR survey of the iminothiazolidinone region of the lead molecule revealed that variation of the substituent pattern also modulated potency; a >10-fold dynamic potency range that was dependent on structure was noted. However, the patterns of SAR associated with this region were less discrete than that of the amino acid moiety. Resistance selection with amide **3** yielded additional mutations in NS5A Domain I (L31V and Q54L) that resulted in a 9–60-fold potency loss and were cross-resistant to amide **2**, suggesting commonality of the inhibitory mechanism between these two molecules, despite the difference in their resistance mutations.

At this juncture, it became apparent that this thiazolidinone chemotype was exhibiting chemical instability in certain organic solvents and in the replicon medium.[21] Careful analysis of degradation products revealed that when **3** is stored in dimethyl sulfoxide (DMSO) under ambient conditions, it undergoes an oxidative rearrangement to afford the thiohydantoin **8**, which was inactive in the replicon assay, EC$_{50}$ >20 μM (Scheme 1.1). Incubation of **3** in the replicon

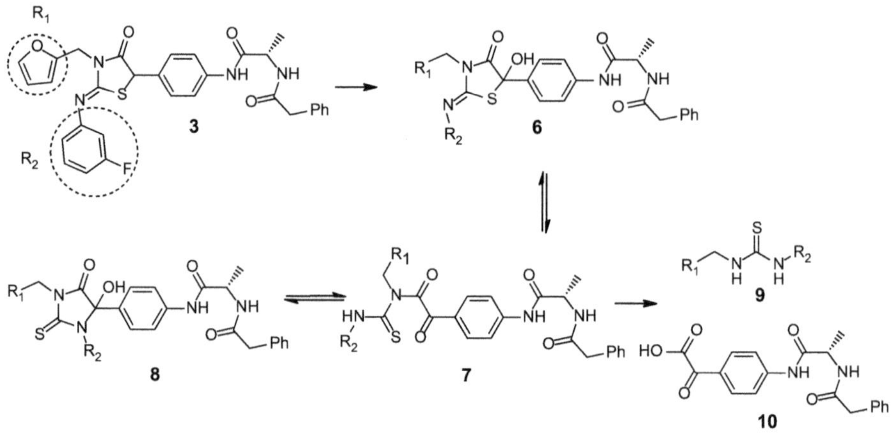

Scheme 1.1   Thiazolidinone oxidative degradation pathway.

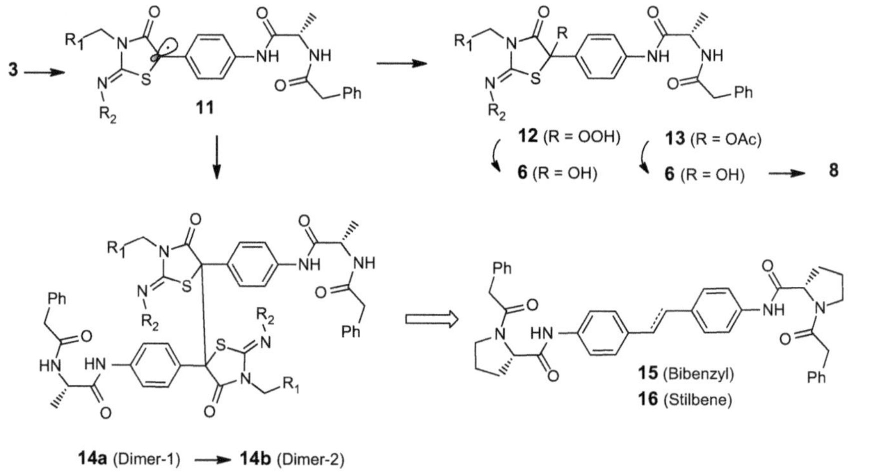

Scheme 1.2   The central role of captodative radical **11** in chemotype degradation and the discovery of stilbene lead **16**.

assay medium initially afforded thiohydantoin **8**, which degraded to thiourea **9**, which also lacked replicon inhibitory activity. A critical and enlightening experiment in which **3** was pre-incubated in the assay medium until complete degradation had occurred, followed by assessment in the replicon, revealed that the HCV inhibitory activity was maintained, clearly indicating that some chemical entity other than the parental analog was likely responsible for the observed effect. A careful HPLC biofractionation study conducted on **3** after incubation in assay medium coupled with detailed spectroscopic analyses of the degradation products revealed the presence of two dimeric derivatives (see **14** in Scheme 1.2), both of which demonstrated inhibitory activity in the G-1b replicon with EC_{50}s of 0.6 and 43 nM. Although the precise stereochemical relationship between these two dimers has not been established, the more potent dimer converts to the weaker dimer when heated at 55 °C in CD_3CN, which is suggestive of either a rotameric or a stereoisomeric relationship. It is hypothesized that the formation of the dimeric species from **3** arises from its susceptibility to form a captodative radical (**11**) in the presence of a radical initiator such as molecular oxygen, which has a diradical ground state. Since **11** is a stabilized radical, it persists such that it can undergo either dimerization to afford **14** or combine with molecular oxygen to afford the peroxy intermediate **12**, which would be susceptible to reduction to alcohol **6** followed by rearrangement to afford thiohydantoin **8** *via* ketoamide **7**.[22] Transfer of the ketoacyl moiety of intermediate **7** to a nucleophilic species in the replicon medium would afford thiourea **9**. Although a simple hydrolysis of ketoamide **7** in assay medium is also possible, the byproduct of such a hydrolytic process, keto acid **10**, was not identified. It is noteworthy that acetate **13**, which can be prepared from **3** in 78% yield by oxidation with $Mn(OAc)_3$–$Cu(OAc)_2$–AcOH, afforded thiohydantoin **8** when treated with MeOH–K_2CO_3, providing supportive evidence for a key step of the proposed mechanism.[23]

Although the identification of the dimeric derivatives represented marked progress for the medicinal chemistry effort, optimizing these architecturally complex leads to a drug candidate appeared to be a challenging task, given that their physical properties fall far outside conventional drug space.[24–26] However, based on insights gleaned from the preliminary SAR investigation, a significant simplification of the dimeric species was achieved when the key pharmacophoric elements were successfully captured in the bibenzyl **15**, which exhibited a G-1b EC_{50} of 30 nM, potency that was improved further with the structurally more rigid stilbene analog **16**, which displayed an EC_{50} of 0.086 nM in the G-1b replicon assay.[20] This new lead molecule was relatively stable when incubated in replicon medium for the length of the assay period and exhibited a resistance profile that mirrored that of **3**, supportive of a similar mode of inhibitory effect and confirming that this molecule contains the key pharmacophore within dimers **14**. With its impressive potency and simplified structure compared with **14**, stilbene **16** served as the starting point for the next phase of the medicinal chemistry campaign. This enterprise focused on expanding genotype coverage, since the EC_{50} of **16** in a G-1a replicon was >10 μM, and optimizing ADME properties. The effort involved significant chemotype evolution based on the application of bioisostere concepts and ultimately culminated in the discovery of the highly potent, first-in-class NS5A replication complex inhibitor daclatasvir (**1**).[1] Daclatasvir inhibits G-1b and G-1a replicons with EC_{50}s of 0.009 and 0.05 nM, respectively. In addition, it inhibits G-2a to G-5a replicons with EC_{50}s ranging from 0.033 to 0.146 nM. This unprecedented *in vitro* potency spectrum established a new benchmark for the HCV field.

A similar cell-based screening of compound libraries conducted by scientists at Arrow Pharmaceuticals (subsequently acquired by AstraZeneca) led to the identification of two distinct hits (**17** and **19**) with HCV inhibitory activity that

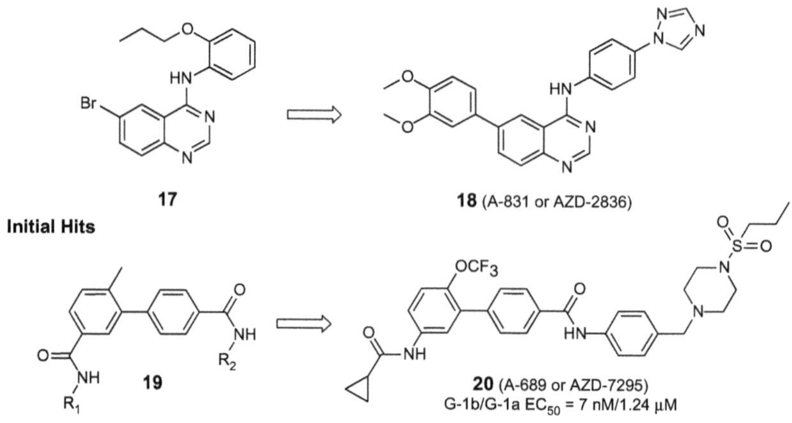

Figure 1.5 AstraZeneca's HCV clinical candidates exhibiting NS5A mutations.

also appeared to target the NS5A protein and which were optimized to the two clinical candidates AZD-2836 (**18**) and AZD-7295 (**20**) (Figure 1.5).[27] Interestingly, although resistance associated with the quinazoline series mapped to the NS5A protein, along with some accompanying mutations in the NS4B and NS5B regions, reverse genetic engineering of the mutations into a G-1b replicon, either alone or in combination, failed to recapitulate the resistant phenotype. On the other hand, the biphenyl carboxamide series afforded mutations in Domain I, the Y93H/C change being the hallmark and in the C-terminal region of the NS5A protein, for which additional details were not provided.

1.4 Highlights of Recent Literature Disclosures

The high *in vitro* inhibitory potency associated with **1** and its clinical validation of NS5A as a target for therapeutic intervention in HCV infection have generated considerable interest in this class of HCV inhibitor. Over the past 4 years, more than 100 patent applications have been published claiming various NS5A inhibitors, the majority of which are based on structural variation of the dimeric pharmacophore element pioneered by **1**.[28] Comprehensive and insightful overviews of the NS5A patent literature that provide distinct perspectives have been published, and in the next section highlights of more recent developments in the field are provided.[27,29,30]

A dimeric pharmacophore that does not necessarily embrace chemical symmetry is a common theme throughout the majority of the published patent applications. Each pharmacophore unit typically contains a spacer element that projects hydrogen bond-donating and -accepting properties attached to a pyrrolidine-like fragment that is derivatized with an amino acid moiety (Figure 1.6). Most of the molecules disclosed maintain some variation of either

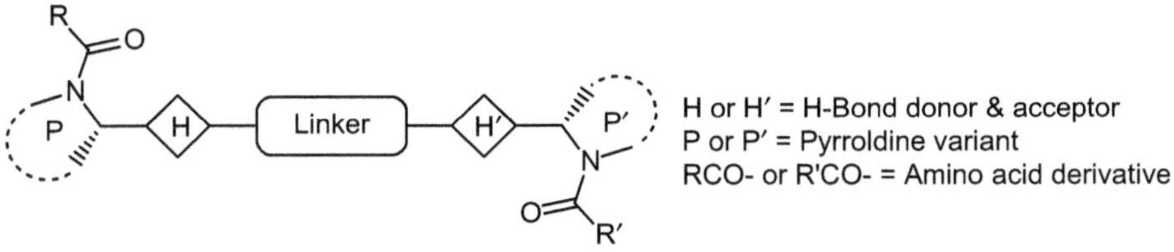
R
O
N
P
H
Linker
H′
P′
N
O
R′
H or H′ = H-Bond donor & acceptor
P or P′ = Pyrroldine variant
RCO- or R′CO- = Amino acid derivative

Figure 1.6 Topology of key pharmacophores needed to effect potent and pan-genotypic NS5A inhibition.

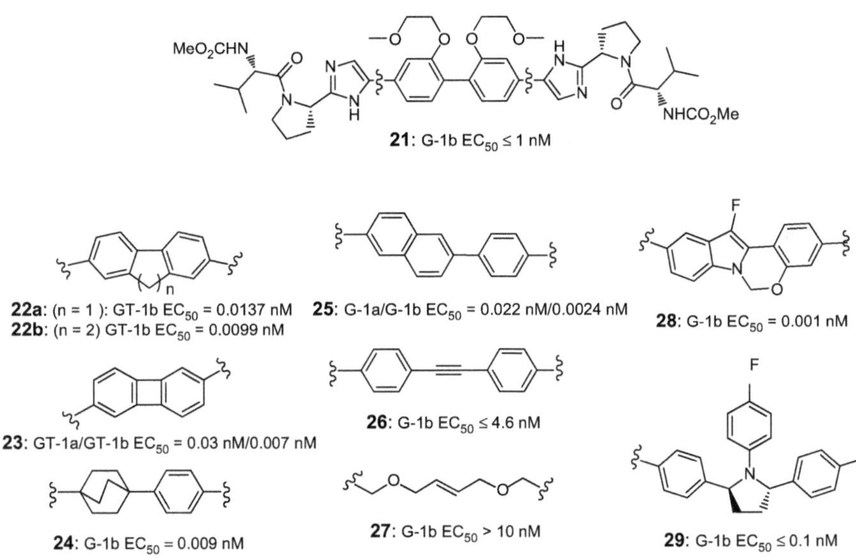
MeO2CHN
NHCO2Me
21: G-1b EC50 ≤ 1 nM
22a: (n = 1): GT-1b EC50 = 0.0137 nM
22b: (n = 2) GT-1b EC50 = 0.0099 nM
25: G-1a/G-1b EC50 = 0.022 nM/0.0024 nM
28: G-1b EC50 = 0.001 nM
23: GT-1a/GT-1b EC50 = 0.03 nM/0.007 nM
26: G-1b EC50 ≤ 4.6 nM
24: G-1b EC50 = 0.009 nM
27: G-1b EC50 > 10 nM
29: G-1b EC50 ≤ 0.1 nM

Figure 1.7 Bis-imidazole core analogs.

a methyl carbamate of an alkyl- or an arylglycine or *N*,*N*-dialkyl derivatives of an arylglycine for the amino acid fragment, and have primarily focused on modifications to the central core and the pyrrolidine regions.

The survey of central core elements has largely been directed towards uncovering alternate scaffolds that maintain the topological disposition of the key peripheral pharmacophoric moieties of **1** (Figure 1.7). Conceptually, the least disruptive strategy involves utilizing the biphenyl core and examining the effect of substitution at every position of the biphenyl moiety, including the installation of bridging elements, as exemplified by compounds **21**–**23**.[31–33] In a case where the bridging element is a single bond, as in **23**, the point of attachment of one of the imidazole moieties is changed to a *meta* position, presumably to reestablish a more linear disposition of the peripheral entities.[34]

Replacement of one of the phenyl groups with a bicylooctane group, as in compound **24** and central scaffold elongations (see **25**, **26** and **29** in Figure 1.7) resulted in compounds claimed to exhibit sub-nanomolar inhibitory potencies in a G-1b replicon.[35–38] Numerous combinations of bridging and elongation strategies that have resulted in tri- and tetracyclic scaffolds claiming to possess potent G-1b inhibitory activity have also been reported (see **28**).[39] Scaffolds with increased flexibility (see **27**) suffered a loss in inhibitory activity.[40]

Another strategy explored the utility of benzimidazole and its aza variants as bioisosteric replacements for the aryl imidazole moiety (Figure 1.8). Here again, analogs with a wide range of core lengths have exhibited G-1b EC_{50}s of <1 nM.[33,41–43] Although detailed data are not available to make a comparative assessment regarding inhibitory activity towards other genotypes, it is noteworthy that **33b** was reported to exhibit G-1a and G-1b inhibitory potencies of 50 and 3 pM, respectively, along with 23% oral bioavailability in monkeys when administered as a dihydrochloride salt.[44]

Numerous hybrid scaffolds have been explored that lie outside the bis-imidazole or bis-benzimidazole scaffolds (see Figure 1.9). The most common structural elements include combinations of imidazole and benzimidazole moieties, as in **34**, or cases where one of these two moieties is hybridized with other potential bioisosteres, including a primary amide (**35**), a thienoimidazole (**36**) or a quinazolinone (**37**).[45–48]

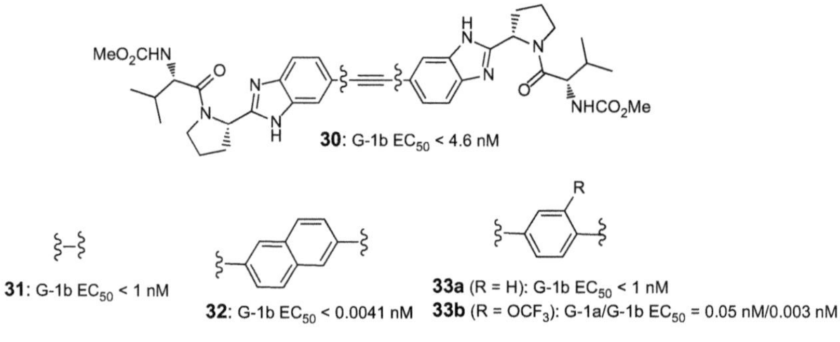

30: G-1b EC_{50} < 4.6 nM

31: G-1b EC_{50} < 1 nM

32: G-1b EC_{50} < 0.0041 nM

33a (R = H): G-1b EC_{50} < 1 nM
33b (R = OCF_3): G-1a/G-1b EC_{50} = 0.05 nM/0.003 nM

Figure 1.8   Bis-benzimidazole core analogs.

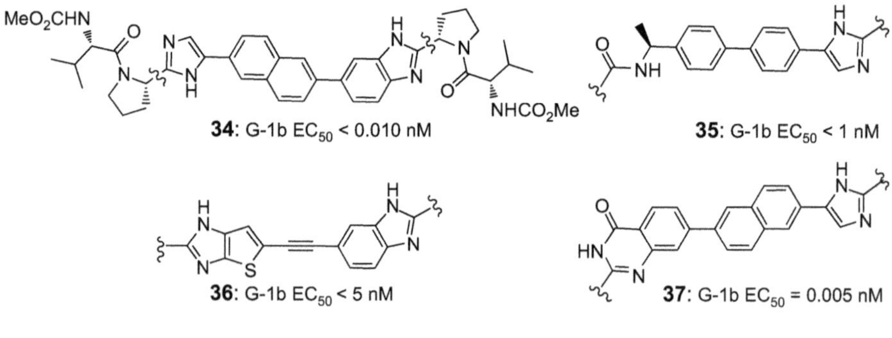

34: G-1b EC_{50} < 0.010 nM

35: G-1b EC_{50} < 1 nM

36: G-1b EC_{50} < 5 nM

37: G-1b EC_{50} = 0.005 nM

Figure 1.9   Hybrid core analogs.

Although the pyrrolidine moiety has been replaced by acyclic amines or by homologous heterocycles, the majority of the reported analogs incorporate functionalized pyrrolidines or five-membered heterocyclic variants in this region (Figure 1.10). Examples include the spirocyclic analog **38** and the bridged analog **39**, where the latter molecule also incorporates a number of the modifications described earlier and has featured prominently in a recent patent application disclosing combinations of advanced HCV therapeutic agents that encompass a range of alternate mechanisms.[49,50] In another example, a dimethylsilane moiety was incorporated into the 3-position of the pyrrolidine ring to afford **40**, claimed to be a potent G-1a/1b inhibitor.[51] Compound **41**, which is a peptidomimetic variant of **1**, exhibited potent G-1b inhibitory activity but is significantly weaker towards a G-1a replicon.[52]

Finally, additional distinct chemotypes with moderate levels of inhibitory activity in replicon systems and resistance mutations that map to HCV NS5A have been reported (see **42–44** in Figure 1.11).[53–56] Although some of the resistance mutations overlap with those observed for **1**, it is not apparent at this stage if they share a similar mode of inhibitory mechanism(s). It is noteworthy that a hybrid chemotype containing pharmacophore elements derived from **1** and **43** has been claimed to exhibit potent inhibitor activity towards both G-1a and G-1b replicons (see **45**).[57]

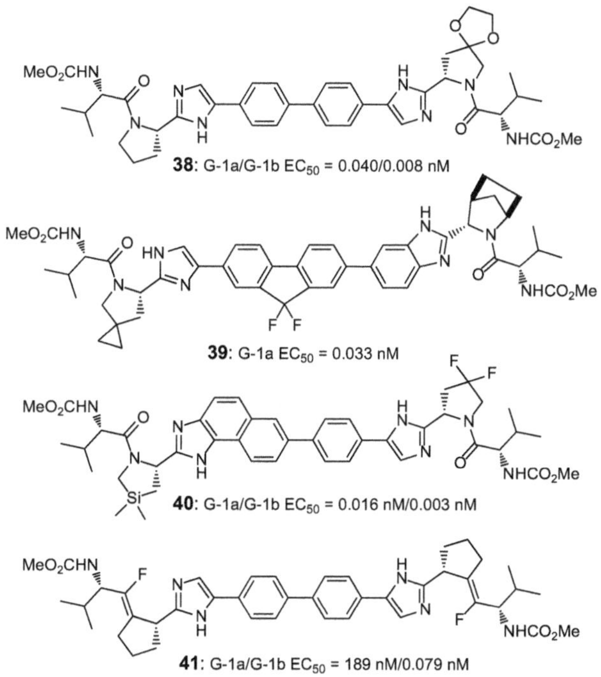

38: G-1a/G-1b EC$_{50}$ = 0.040/0.008 nM

39: G-1a EC$_{50}$ = 0.033 nM

40: G-1a/G-1b EC$_{50}$ = 0.016 nM/0.003 nM

41: G-1a/G-1b EC$_{50}$ = 189 nM/0.079 nM

Figure 1.10 Analogs with peripheral modifications.

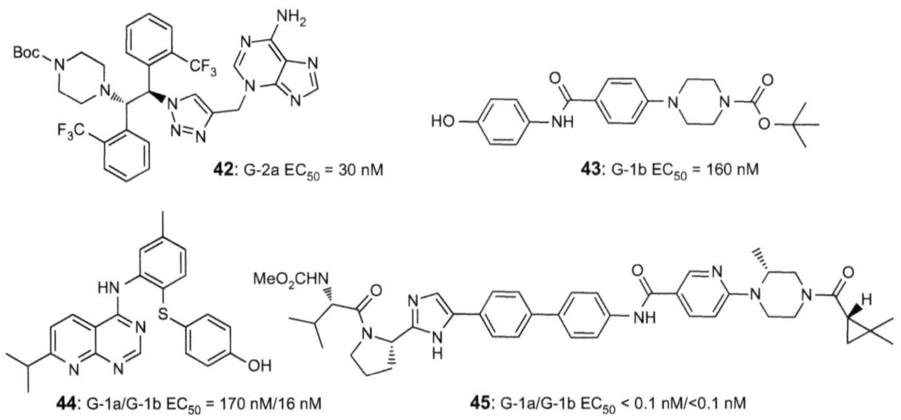

42: G-2a EC_{50} = 30 nM

43: G-1b EC_{50} = 160 nM

44: G-1a/G-1b EC_{50} = 170 nM/16 nM

45: G-1a/G-1b EC_{50} < 0.1 nM/<0.1 nM

Figure 1.11 Miscellaneous chemotypes with NS5A resistance mutations.

1.5 Clinical Trials with HCV NS5A Replication Complex Inhibitors

Clinical validation that inhibitors of NS5A represented an effective approach to the control of HCV replication was obtained in Phase 1 studies with **1**.[1] A single ascending dose (SAD) study conducted in normal healthy volunteers with doses of **1** ranging from 1 to 200 mg established dose-proportional exposure of the drug, with plasma concentrations at 24 h post-dose in all subjects significantly exceeding that required to express antiviral activity in replicons. Administration of an oral solution of **1** to subjects chronically infected with G-1a and G-1b HCV at doses of 1, 10 and 100 mg produced a rapid and dose-related reduction of virus RNA levels. Viral RNA was reduced by an average of 1.8 $\log_{10}$ measured 24 h following a single 1 mg dose of **1**, whereas the 10 and 100 mg doses exerted more profound effects on viral load, with reductions of 3.2 $\log_{10}$ and 3.3 $\log_{10}$, respectively, at 24 h post-dose. The single 100 mg dose was associated with a maximum 3.6 $\log_{10}$ decline in viral RNA levels that was maintained for 144 h following drug administration, with one G-1b virus-infected subject in this cohort achieving RNA levels below the 25 IU mL^{-1} level of quantification at the 144 h time point, and the viral load for a second subject was measured at 35 IU mL^{-1}.

Virus sequencing, conducted prior to dosing and at 24 and 144 h post-dosing, revealed that when reduction in viremia was significant, HCV variants were detectable, with M28T, Q30H/R and L31M observed in G-1a-infected subjects and Y93H in the G-1b patients. These observations presumably reflect potent and effective restriction of wild-type virus replication by **1** that reveals virus-possessing resistant mutations as the thriving species. This scenario is anticipated based on the replication rate of the virus, estimated to be 10^{12} virions per day, combined with the low fidelity of the polymerase, the error rate of which is estimated to range from 0.1 to 1 nucleotide per RNA synthesized.[58–63]

These statistics contribute to the significant population of viruses harboring single (87×10^9 virions per day), double (4.2×10^9 virions per day) and triple (0.13×10^9 virions per day) mutations produced in an infected individual every day.[58–63] Indeed, in a multiple ascending dose study with **1** administered at doses of 1, 10, 30, 60 and 100 mg qd and 30 mg bid for 14 days to chronically infected G-1-infected subjects, the viral load was rapidly reduced by 2.8–4.1 $\log_{10}$ IU mL^{-1} across the cohorts, but most patients experienced rebound on or before day 7 of the treatment period.[64] Viral rebound was associated with the emergence of resistant variants with major substitutions identified at residues M28T/A/V, Q30H/R/K/E, L31M/V and Y93H/C/N for G-1a-infected subjects and L31M/V and Y93H/C for G-1b.[65] These mutant viruses were also observed *in vitro* in G-1a and G-1b replicons placed under selective pressure from **1**. One patient in the 60 mg qd cohort, all of whom were infected with G-1a, had a Q30R mutation detectable at baseline and experienced initial viral suppression, but viral breakthrough occurred by day 14 with a Q30H and Y93H linkage detected that *in vitro* exhibited high resistance to **1**.[65] For two additional patients, Q30E and Y93N variants were detected at day 14, a double mutant associated with high resistance to the drug. The final patient in this cohort experienced failure of therapy with a Q30R virus that emerged within 12 h of drug dosing, despite the fact that the plasma exposure of **1** at day 14 (117 nM) substantially exceeded the *in vitro* replicon EC$_{50}$ of 7 nM.[65] A closer analysis revealed a baseline E62D polymorphism that by itself did not confer resistance to **1** but, when linked to Q30R, conferred high levels of resistance *in vitro*.[66]

Analysis of a cohort of 78 HIV-HCV co-infected subjects and 635 NS5A sequences deposited in the Los Alamos database for the occurrence of baseline resistant mutations to **1** revealed an absence in G-1a and G-3 whereas all G-4 sequences had L31M; the double mutant L31M+Y93H occurred in 7% of G-1b and 13% of G-4 sequences.[67] In a cohort of Japanese subjects infected with G-1b, overwhelmingly the most prevalent in that population, resistance-conferring amino acid substitutions were detected in 11.2% of 294 patients, with Y93H (8.2%) predominating over L31M (2.7%).[68–70]

Taken together, these observations emphasize the anticipation based on virus replication kinetics that combination therapy will be required, either by adding a direct-acting antiviral agent (DAA) to interferon-α/ribavirin therapy or by combinations of DAAs with orthogonal patterns of resistance, to suppress the virus effectively and durably.[61–63,71–75]

In a Phase 2a clinical trial of **1** in conjunction with PEG interferon-α/ribavirin (PEG-IFN/RBV), doses of 3, 10 and 60 mg of the drug were compared with a placebo control arm over a 48 week time span.[76] The primary efficacy endpoint focused on an extended virological response (eRVR), which is defined as undetectable levels of viral RNA at both weeks 4 and 12 after initiation of therapy. Secondary endpoints were rapid virological response (RVR; HCV RNA undetectable at 4 weeks), complete early virological response (cEVR; HCV RNA undetectable at 12 weeks) and sustained virological response at 12

Table 1.1   Clinical results in a Phase 2a study of daclatasvir (**1**) combined with
PEG-IFN/RBV.

Assessments[a]	Placebo (n = 12)[b]	3 mg (n = 12)	10 mg (n = 12)	60 mg (n = 12)
RVR	1 (8%)	5 (42%)	11 (92%)	10 (83%)
eRVR	1 (8%)	5 (42%)	10 (83%)	9 (75%)
cEVR	5 (42%)	7 (58%)	10 (83%)	10 (83%)
SVR_{12}	3 (25%)	5 (42%)	11 (92%)	10 (83%)
SVR_{24}	3 (25%)	5 (42%)	10 (83%)	10 (83%)
Virological failure	9 (75%)	7 (58%)	2 (17%)	2 (17%)

[a]RVR, viral RNA undetectable at 4 weeks; eRVR, viral RNA undetectable at both 4 and 12 weeks; cEVR, RNA undetectable at 12 weeks; SVR_x, sustained virological response at x weeks after end of therapy.
[b]Treated with PEG-IFN/RBV.

and 24 weeks after the end of the dosing period (SVR_{12} and SVR_{24}). The results of this trial are compiled in Table 1.1 and indicate that the two higher doses of **1** are associated with greater efficacy, with 5 of 12 (42%) patients in the 3 mg group achieving eRVR compared with 10 of 12 (83%) and 9 of 12 (75%) of those receiving 10 and 60 mg of drug, respectively, whereas only 1 of the 12 (8%) administered with just PEG-IFN/RBV achieved this endpoint.[76] Based on results from subsequent Phase 2b trials, the 60 mg dose of **1** was selected for Phase 3 studies.

The first step towards a PEG-IFN/ribavirin-free therapy was a short clinical trial in which 74 treatment-naive and null-responding HCV-infected patients – the latter are the most difficult to treat patient subgroup – were administered a combination of the HCV NS3 protease inhibitor danoprevir (**46**) at doses of 100 or 200 mg tid or 600 or 900 mg bid and the nucleoside prodrug mericitabine (**47**) at doses of 500 or 1000 mg bid, for up to 13 days against a placebo control arm ($n = 14$) (Figure 1.12).[77] In the treated subjects who completed the 13 days of therapy, the viral load declined by a median 3.7–5.2 $\log_{10}$ IU mL^{-1}, which compared with an increase of 0.1 $\log_{10}$ IU mL^{-1} observed in the placebo arm.

A more recent clinical trial examined the potential of a combination of daclatasvir (**1**) (60 mg qd) and the NS3 protease inhibitor asunaprevir (**48**) (600 mg bid) to cure a small cohort of G-1-infected null responders who were administered the drugs for 24 weeks with and without PEG-IFN/RBV (**49**).[78,79] In this trial, all 10 patients administered the quadruple drug regimen experienced SVR_{12} and 9 of the 10 maintained this status to 24 weeks post-dosing, while the remaining patient had detectable HCV RNA at week 24 after cessation of drug but had undetectable viral RNA 35 days later. Nine of the 10 patients had an SVR at 48 weeks, with the remaining patient having measurable RNA at less than 25 IU mL^{-1} at week 48, but undetectable 13 days later. The cohort receiving only the combination of the two DAAs comprised nine G-1a and two G-1b infections. Of these, two G-1b- and two G-1a-infected subjects experienced an SVR at 12 and 24 weeks after drug therapy was completed and three of these maintained SVR at week 48, while the fourth one had detectable

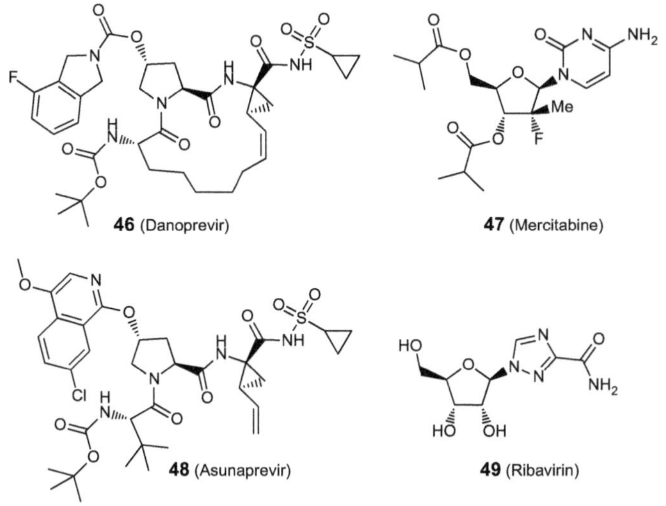

46 (Danoprevir)   **47** (Mercitabine)

48 (Asunaprevir)   **49** (Ribavirin)

Figure 1.12   A subset of compounds investigated in seminal IFN-free HCV trials.

viral RNA that became undetectable upon retest 43 days later.[78,79] One G-1a subject had undetectable viral RNA at the end of treatment, but relapsed 4 weeks later. However, six of the G-1a-infected subjects experienced viral breakthrough, which occurred as early as week 3 and as late as week 12 of therapy, attributed to the selection of resistant variants which included Q30R, L31M/V and Y93C/N in the NS5A protein and R155K and D168A/E/T/V/Y in the NS3 protease domain. This study established for the first time that a chronic HCV infection could be cured without the use of interferon-α and/or ribavirin and is recognized as a watershed event in the therapy of the disease.[79]

Although the results suggest that more difficult to treat G-1a patients will require additional DAAs or the use of HCV inhibitors with a higher genetic barrier to resistance, the successful treatment of G-1b-infected subjects prompted a similar Phase 2a trial in a Japanese cohort with established null response to PEG-IFN/RBV.[80] In this study, daclatasvir (**1**) (60 mg qd) and asunaprevir (**48**) (initially 600 mg bid but subsequently reduced to 200 mg bid) were administered to 10 patients for 24 weeks, with nine subjects completing therapy and one patient discontinuing after 2 weeks. HCV RNA was undetectable in those who completed the course of therapy by 8 weeks and all nine achieved SVR_{12} and SVR_{24}, with no evidence of viral breakthrough during treatment or relapse post-treatment. Most interestingly, the patient who stopped therapy after 2 weeks and who had detectable HCV RNA ($1.8 \log_{10}$ $IU\,mL^{-1}$) at the time that therapy was discontinued, had undetectable levels of viremia measured on follow-up visits at weeks 2, 3, 4, 13 and 24 after discontinuation.[80]

GS-5885 is an HCV NS5A inhibitor for which early clinical data have been reported.[81] In a Phase 1 trial conducted in G-1-infected subjects, GS-5885 was administered qd for 3 days at doses of 1, 3, 10, 30 and 90 mg and the effect on viral load was compared with placebo. Plasma exposure of GS-5885 was close to linear with respect to dose, with C_{max} occurring at 4–6 h after drug administration. Quantifiable concentrations were detectable at 24 h after all doses, exceeding the protein binding-adjusted EC_{90} for the less sensitive G-1a at doses above 10 mg. Once-daily dosing of GS-5885 was supported by a long median apparent plasma half-life that ranged from 22 to 50 h. All doses above 3 mg produced a $>3 \log_{10} IU mL^{-1}$ decline in viral load and suppression was more sustained for G-1b-infected patients, although the median maximum reductions were similar for both genotypes 1a and 1b.[81] Of the 72 patients enrolled in this study, four G-1a subjects and one G-1b subject harbored detectable viruses with resistance-associated mutations. Two of the four G-1a-infected patients experienced maximum HCV RNA reductions of $<1.6 \log_{10} IU mL^{-1}$, of whom one was characterized with a Q30E/Q population and the other harbored L31M virus at baseline. However, one subject dosed with 10 mg of GS-5885 had a maximum viral response of $<1.6 \log_{10}$ and 454 pyrosequencing was necessary to determine that 12% of the virus harbored a Y93C change in the NS5A gene. The emergence of resistance mutations was assessed by population sequencing and all patients dosed at 3 mg or more had virus with detectable changes associated with resistance to GS-5885 *in vitro*. M28T, Q30H, L31M, Q30R and Y93C/H were characterized in G-1a-infected subjects, with Q30R being the most frequent, while Y93H was detected in all 10 G-1b-infected individuals receiving the 10 mg dose of drug. The less resistant M28T and Q30H mutations were not detected in G-1a patients at doses ≥ 30 mg, reflective of plasma concentrations at trough that were above the protein binding-adjusted EC_{90} values.

The quinazoline derivative AZD-2836 (**18**) was the first HCV NS5A inhibitor actually to enter clinical trials, in early 2007, but was abandoned, apparently due to inadequate exposure that was not solved by optimization of the formulation.[27] The antiviral activity of AZD-7295 (**20**) was explored in treatment-naive and treatment-experienced patients infected with genotypes 1a, 1b and 3a virus who were administered the drug for 5 days at doses of 90 and 233 mg tid or 350 mg bid.[27,82] The 90 mg tid cohort comprised five each of G-1a- and G-1b-infected subjects, with the G-1b patients receiving active drug experiencing a mean viral load reduction of over $1 \log_{10} IU mL^{-1}$. The higher doses of AZD-7295 (**20**) produced a more pronounced reduction in viral load, up to a maximum of $2.4 \log_{10} IU mL^{-1}$, although 4 G-1b patients across the cohorts showed no response to the drug. In contrast, the viral load in G-1a-infected subjects was not statistically significant from placebo, reflecting the poor *in vitro* potency of the compound towards G-1a and a C_{24} drug concentration that did not surpass the EC_{90} measured in the replicon assay. Similarly, AZD-7295 (**20**) at 233 mg tid exerted no significant effect on viral load in the G-3a-infected subjects.

The early clinical studies with HCV NS5A inhibitors revealed a class of antivirals that is characterized by high *in vitro* potency, inhibition of multiple genotypes and excellent efficacy at low doses. The first-generation

NS5A clinical candidates appear to have a low genetic barrier to resistance. The NS5A inhibitors evaluated to date were generally well tolerated and offer considerable promise for use in combination therapy that has the potential to cure chronic HCV infection in the absence of pegylated interferon-α and/or ribavirin.[83–85]

1.6 Mode of Action Studies with HCV NS5A Replication Complex Inhibitors

Detailed *in vitro* replicon studies have revealed that the mutations that confer resistance to HCV NS5A replication complex inhibitors, such as **1** and **3**, map to the N-terminal region of Domain I, involving amino acids 24–100.[1,65] Resistance analysis of clinical samples obtained from SAD and MAD studies of daclatasvir (**1**) have corroborated these *in vitro* findings.[86] Although there are differences in the exact mutation composition of different resistant genotypes, there is a broad overlap in the specific locale that are hot-spots for the emergence of resistance mutations, which is suggestive of a common mode and, possibly, region of interactions between NS5A and its putative inhibitors. Although residues 28–31, which is a key resistance mutation region for both G-1a and G-1b viruses, was not captured in the X-ray structural studies of Domain I, it is noteworthy that the mutation Y93H, which is a clinically relevant mutation common for G-1a and G-1b, lies in the dimer interface region of both X-ray structures.

To gain additional insight into the mode of interaction between **1** and the protein, two models for the G-1b NS5A Domain I were constructed using the Rice dimer (1ZH1.pdb, model 1) or the Love dimer (3FQM.pdb, model 2) and the NMR structure of the amphipathic N-terminal α-helix, amino acids 1–31.[9,10,87] The respective missing amino acids were modeled in so as to complete each monomer. Mapping of the primary resistance mutations observed in clinical studies for both G-1a and/or G-1b patients on to the model indicates a clustering of mutation sites that is presumed to indicate the putative inhibitor binding site. Daclatasvir (**1**) was hand-docked into the putative binding site symmetrically across the dimer interface of each model. Docking into the model constructed based on the Rice dimer produced the best fit based on the physical properties of both ligand and protein and also the location of resistance mutations. The hydrophobic regions of **1** – the biphenyl core, pyrrolidine and valine side chain moieties – are surrounded by hydrophobic amino acids (residues M28 and L31), while the polar imidazole and carbamate moieties deployed on each end of the inhibitor are in a position to form hydrogen bonds with the monomeric units (Figure 1.13). It is believed that these inhibitors may interact with the NS5A dimer and induce a conformational change that is not functionally viable. Most interestingly, a recent report suggests that disruption of dimerization may not be the mode of action for **1** and related analogs.[11,88]

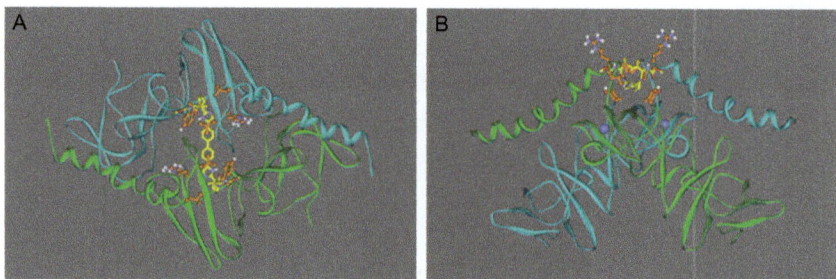

Figure 1.13   Model of daclatasvir (**1**) bound within the NS5A Domain I dimer, with alternative views. The monomers are colored either blue or green and the carbons of daclatasvir are yellow. The position and color of the carbon atoms of the residues associated with primary clinical resistance mutations (amino acids 28, 30, 31, 93) are noted in orange.

Additional evidence that further corroborated the importance of the N-terminal region of Domain I of NS5A as a potential site of interaction for inhibitors of interest was obtained from chimeric replicon studies conducted using compounds with differing inhibitory activities.[19] For example, compound **3**, which inhibits a G-1b replicon with an EC_{50} of 0.018 µM but is devoid of activity in a G-1a replicon ($EC_{50} > 10$ µM), exhibited an enhanced inhibitory activity ($EC_{50} = 0.032$ µM) when tested in a hybrid G-1a replicon in which the first 76 amino acids of its NS5A protein were replaced with the corresponding G-1b sequence. (The difference in the G-1b EC_{50} reported for **3** in different sections (i.e., 18 nM *versus* 6 nM) is a result of assay variations between studies.) Conversely, replacement of the first 76 amino acids of the NS5A region of a G-1b replicon with the corresponding G-1a sequence resulted in a decrease in the inhibitory potency of **3** (EC_{50} of >10 µM). The fact that the inhibitor sensitivity domain of NS5A overlaps with the region that resistance mutations map to is consistent with the direct engagement of NS5A with inhibitors.

A diastereomeric pair of biotin-tagged analogs with differential inhibitory potencies was used in an NS5A pull-down experiment to provide evidence for a direct and specific binding interaction between NS5A and its putative inhibitors.[1] Specifically, NS5A was selectively pulled down when a G-1b replicon was incubated with the biotinylated inhibitor **50a** ($EC_{50} = 33$ nM) (Figure 1.14), lysed and passed over streptavidin–agarose beads, but not when the replicon was lysed prior to treatment with the biotinylated compound. This result signifies that a specific conformation of NS5A is needed for a productive interaction with an inhibitor and is accessible only in a cellular context, presumably in the virus replication complex, which is consistent with a report that NS5A inhibitors failed to bind to the isolated protein.[89] More importantly, in control experiments conducted in parallel, it was observed that little HCV NS5A was pulled down by the inactive stereoisomer **50b** ($EC_{50} > 10$ µM). Moreover, only NS5A, and not NS3 or NS5B, was detected in the bound

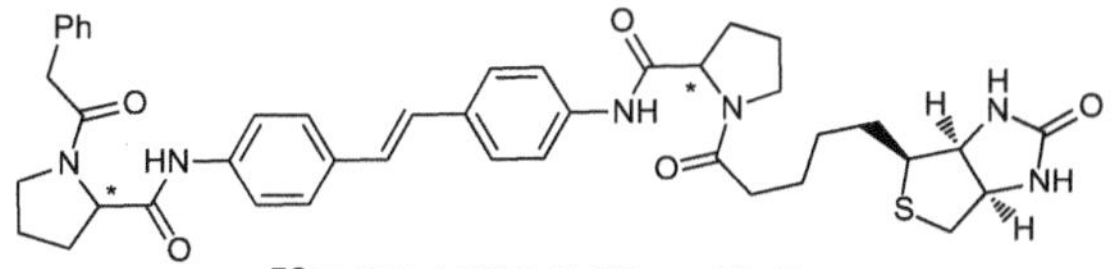

50a: (S,S at C*) G-1b EC$_{50}$ = 33 nM
50b: (R,R at C*) G-1b EC$_{50}$ > 10 μM

Figure 1.14 Biotin-tagged tool compounds.

proteins and **50a** failed to pull down the NS5A protein of BVDV under similar conditions.

The availability of potent NS5A-targeting tool compounds has catalyzed additional efforts that are directed at shedding light on the ramifications of inhibition of the NS5A protein which, in turn, provides additional insights into the mode of action of the inhibitors. NS5A is a phosphoprotein that exists in both basally phosphorylated (p56) and hyper-phosphorylated (p58) forms. It was discovered that NS5A inhibitors such as daclatasvir (**1**) dose-dependently inhibit the formation of the p58 form, without affecting basal phosphorylation, in a manner that correlates with their RNA replication-inhibitory activities.[19,90] Moreover, inhibition of hyperphosphorylation did not depend on the presence of Domains II and III of NS5A. Although the exact significance of phosphorylation of NS5A is still unknown, it is hypothesized that phosphorylation is a regulatory mechanism that toggles the NS5A protein between functional states in the viral replication cycle. Interestingly, it was observed that protease inhibitors could similarly inhibit the formation of p58 in a dose-dependent manner, clearly indicating that the modulation of the phosphorylation state of NS5A is not unique to NS5A inhibitors, albeit this observation may signify the spatial and functional associations of the NS3 and NS5A proteins within the HCV replication complex or polyprotein.[19] Corroborating evidence for a possible interaction of NS5A inhibitors at the polyprotein-processing stage came from a recent study that demonstrated that treatment with **1** results in the accumulation of the NS4B–NS5A polypeptide *in vitro*, an effect that was sensitive to the presence of resistance-conferring mutations.[90]

A combination of morphological and biochemical studies have demonstrated that **1** alters the subcellular distribution of NS5A from that of localized foci to diffuse cytoplasmic patterns.[88] This inhibitor-induced effect on the subcellular disposition was specific to the NS5A protein and, in line with expectations, was minimized in a replicon harboring the Y93H-resistant mutant up to certain concentration ranges. A different set of studies revealed that NS5A inhibitors deregulate the normal distribution of NS5A by relocating the protein from the endoplasmic reticulum to lipid droplets, an effect that is minimized in the context of the Y93H-resistant mutant and which is also specific to the NS5A replication complex inhibitor class.[89] Finally, a cellular imaging study that utilized a click chemistry approach involving an azide-containing NS5A

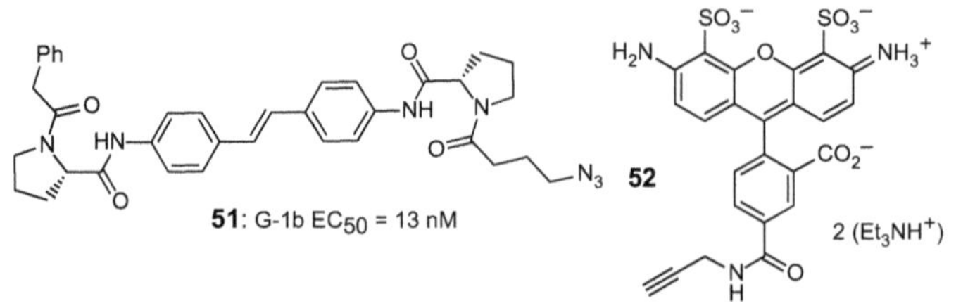

Figure 1.15   Click chemistry substrates for NS5A co-localization study.

inhibitor (**51**) and an alkyne-containing fluorophore (**52**) demonstrated the colocalization of an NS5A replication complex inhibitor with the NS5A protein (Figure 1.15).[91]

The unusually high *in vitro* and *in vivo* inhibitory potency of the NS5A replication complex inhibitor class, coupled with the absence of clearly defined functions of the NS5A protein and of inhibitor binding data for the purified protein, has catalyzed numerous mode of action studies. Although considerable advances have been made, much remains to be discovered in order to illuminate further the complex set of processes that NS5A appears to orchestrate during the HCV replication cycle.

1.7   Conclusion

The complex but ill-defined role of NS5A in the HCV replication cycle continues to drive the curiosity of investigators across academia and industry. Although NS5A was not among the initial targets of choice in the HCV drug discovery campaigns that started over 15 years ago, the pioneering work by Bristol-Myers Squibb scientists in this field has culminated in the discovery of daclatasvir (**1**). The key to success included devising a unique dual-assay screening system that identified a single hit from a collection of over one million compounds; defining aspects of the instability of one of the lead compounds that uncovered significantly active dimeric degradants; simplifying the dimeric pharmacophore into a progressible lead; and successfully optimizing a chemotype with a molecular footprint that is outside of what is considered to be traditional drug-like space. Daclatasvir (**1**) established clinical proof-of-concept for the HCV NS5A protein as a therapeutic target and set a potency benchmark for the HCV field. In addition, as part of a DAA combination, it demonstrated for the first time that a PEG-IFN/RBV-free regimen could cure HCV infection. The clinical validation of NS5A has made it an attractive target for therapeutic intervention, as evidenced by the significant number of patent filings and the growing number of NS5A-targeting compounds entering clinical trials. It is anticipated that NS5A replication complex inhibitors will become integral components of the more effective HCV combination therapies that are expected to emerge in the near future. The genesis of Bristol-Myers Squibb's NS5A drug discovery effort that enabled this successful endeavor is a clear testament to the utility and power of phenotype-based screening approaches in uncovering valuable targets that otherwise would have remained unexplored because of the lack of biochemical information.

References

1. M. Gao, R. E. Nettles, M. Belema, L. B. Snyder, V. N. Nguyen, R. A. Fridell, M. H. Serrano-Wu, D. R. Langley, J.-H. Sun, D. R. O'Boyle II, J. A. Lemm, C. Wang, J. O. Knipe, C. Chien, R. J. Colonno, D. M. Grasela, N. A. Meanwell and L. G. Hamman, *Nature*, 2010, **465**, 96.
2. A. Grakoui, C. Wychowski, C. Lin, S. M. Feinstone and C. M. Rice, *J. Virol.*, 1993, **67**, 1385.
3. T. L. Tellinghuisen, J. Marcotrigiano, A. E. Gorbalenya and C. M. Rice, *J. Biol. Chem.*, 2004, **279**, 48576.
4. T. L. Foster, T. Belyaeva, N. J. Stonehouse, A. R. Pearson and M. Harris, *J. Virol.*, 2010, **84**, 9267.
5. J. Hwang, L. Huang, D. G. Cordek, R. Vaughan, S. L. Reynolds, G. Kihara, K. D. Raney, C. C. Kao and C. E. Cameron, *J. Virol.*, 2010, **84**, 12480.
6. L. Huang, J. Hwang, S. D. Sharma, M. R. Hargittai, Y. Chen, J. J. Arnold, K. D. Raney and C. E. Cameron, *J. Biol. Chem.*, 2005, **280**, 36417.
7. M. Hughes, S. Griffin and M. Harris, *J. Gen. Virol.*, 2009, **90**, 1329.
8. N. Appel, M. Zayas, S. Miller, J. Krijnse-Locker, T. Schaller, P. Friebe, S. Kallis, U. Engel and R. Bartenschlager, *PLoS Pathogens*, 2008, **4**, e1000035.
9. T. L. Tellinghuisen, J. Marcotrigiano and C. M. Rice, *Nature*, 2005, **435**, 374.
10. R. A. Love, O. Brodsky, M. J. Hickey, P. A. Wells and C. N. Cronin, *J. Virol.*, 2009, **83**, 4395.
11. P. J. Lim, U. Chatterji, D. Cordek, S. D. Sharma, J. A. Garcia-Rivera, C. E. Cameron, K. Lin, P. Targett-Adams and P. A. Gallay, *J. Biol. Chem.*, 2012, **287**, 30861.
12. N. Appel, T. Schaller, F. Penin and R. Bartenshclager, *J. Biol. Chem.*, 2006, **281**, 9833.
13. Y. Liang, H. Ye, C. B. Kang and H. S. Yoon, *Biochemistry*, 2007, **46**, 11550.
14. X. Hanoulle, D. Verdegem, A. Badillo, J.-M. Wieruszeski, F. Penin and G. Lippens, *Biochem. Biophys. Res. Commun.*, 2009, **381**, 634.
15. K. Gunasekaran, C.-J. Tsai, S. Kumar, D. Zanuy and R. Nussinov, *Trends Biochem. Sci.*, 2003, **28**, 81.
16. V. Lohmann, F. Korner, J.-O. Koch, U. Herian, L. Theilmann and R. Bartenschlager, *Science*, 1999, **285**, 110.
17. R. Bartenschlager, *Nat. Rev. Drug Discov.*, 2002, **1**, 911.

18. D. R. O'Boyle II, P. T. Nower, J. A. Lemm, L. Valera, J.-H. Sun, K. Rigat, R. Colonno and M. Gao, *Antimicrob. Agents Chemother.*, 2005, **49**, 1346.
19. J. A. Lemm, D. R. O'Boyle II, M. Liu, P. T. Nower, R. Colonno, M. S. Deshpande, L. B. Snyder, S. W. Martin, D. R. St. Laurent, M. H. Serrano-Wu, J. L. Romine, N. A. Meanwell and M. Gao, *J. Virol.*, 2010, **84**, 482.
20. J. L. Romine, D. R. St. Laurent, J. E. Leet, S. W. Martin, M. H. Serrano-Wu, F. Yang, M. Gao, D. R. O'Boyle II, J. A. Lemm, J.-H. Sun, P. T. Nower, X. Huang, M. S. Deshpande, N. A. Meanwell and L. B. Snyder, *ACS Med. Chem. Lett.*, 2011, **2**, 224.
21. J. A. Lemm, J. E. Leet, D. R. O'Boyle II, J. L. Romine, X. S. Huang, D. R. Schroeder, J. Alberts, J. L. Cantone, J.-H. Sun, P. T. Nower, S. W. Martin, M. H. Serrano-Wu, N. A. Meanwell, L. B. Snyder and M. Gao, *Antimicrob. Agents Chemother.*, 2011, **55**, 3795.
22. H. G. Viehe, Z. Janousek, R. Merenyi and L. Stella, *Acc. Chem. Res.*, 1985, **18**, 148 and references cited therein.
23. S. A. Kates, M. A. Dombroski and B. B. Snider, *J. Org. Chem.*, 1990, **55**, 2427.
24. C. A. Lipinski, F. Lombardo, B. W. Dominy and P. J. Feeney, *Adv. Drug Deliv. Rev.*, 2001, **46**, 3.
25. M. C. Wenlock, R. P. Austin, P. Barton, A. M. Davis and P. D. Leeson, *J. Med. Chem.*, 2003, **46**, 1250.
26. N. A. Meanwell, *Chem. Res. Toxicol.*, 2011, **24**, 1420 and references cited therein.
27. P. Najarro, N. Mathews and S. Cockerill, in *Hepatitis C: Antiviral Drug Discovery and Development*, ed. S.-L. Tan and Y. He, Caister Academic Press, Norwich, 2011, p. 271.
28. C. Bachand, M. Belema, D. H. Deon, A. C. Good, J. Goodrich, C. A. James, R. Lavoie, O. D. Lopez, A. Martel, N. A. Meanwell, V. N. Nguyen, J. L. Romine, E. H. Ruediger, L. B. Snyder, D. R. St. Laurent, F. Yang, D. R. Langley, G. Wang and L. G. Hamann, *Patent Application*, WO2008-021927, 2008.
29. U. Schmitz and S.-L. Tan, *Recent Pat. Anti-Infect. Drug Discov.*, 2008, **3**, 77.
30. D. G. Cordek, J.T. Bechtel, A. T. Maynard, W. M. Kazmierski and C. E. Cameron, *Drugs Fut.*, 2011, **36**, 691.
31. M. Zhong and L. Li, *Patent Application*, WO2010-111673, 2010.
32. C. Bachand, M. Belema, D. H. Deon, A.C. Good, J. Goodrich, C. A. James, R. Lavoie, O. D. Lopez, A. Martel, N. A. Meanwell, V. N. Nguyen, J. L. Romine, E. H. Ruediger, L. B. Snyder, D. R. St. Laurent, F. Yang, D. R. Langley, G. Wang and L. G. Hamann, *Patent Application*, WO2009–102568, 2009.
33. H. Guo, D. Kato, T. A. Kirschberg, H. Liu, J. O. Link, M. L. Mitchell, J. P. Parrish, N. Squires, J. Sun, J. Taylor, E. M. Bacon, E. Canales, A. Cho, J. J. Cottell, M. C. Desai, R. L. Halcomb, E. S. Krygowski, S. E. Lazerwith, Q. Liu, R. Mackman, H.-J. Pyun, J. H. Saugier, J. D. Trenkle, W. C. Tse, R. W. Vivian, S. D. Schroeder, W. J. Watkins, L. Xu, Z.-Y.

Yang, T. Kellar, X. Sheng, M. O. H. Clarke, C.-H. Chou, M. Graupe, H. Jin, R. McFadden, M. R. Mish, S. E. Metobo, B. W. Phillips and C. Venkataramani, *Patent Application*, WO2010-132601, 2010.
34. J. A. Kozlowski and B. B. Shankar, *Patent Application*, WO2012-018534, 2012.
35. O. D. Lopez, Q. Chen, M. Belema and L. G. Hamann, *Patent Application*, WO2010-117704, 2010.
36. J. B. J. Milbank, D. C. Pryde and T. D. Tran, *Patent Application*, WO2011-004276, 2011.
37. L. Li and M. Zhong, *Patent Application*, WO2010-065668, 2010.
38. D. A. DeGoey, W. M. Kati, C. W. Hutchins, P. L. Donner, A. C. Krueger, J. T. Randolph, C. E. Motter, L. T. Nelson, S. V. Patel, M. A. Matulenko, R. G. Keddy, T. K. Jinkerson, T. N. Soltwedel, D. K. Hutchinson, C. A. Flentge, R. Wagner, C. J. Maring, M. D. Tufano, D. A. Betebenner, T. W. Rockway, D. Liu, J. K. Pratt, M. J. Lavin, K. Sarris, K. R. Woller, S. H. Wagaw, J. C. Califano and W. Li, *Patent Application*, WO2010-144646, 2010.
39. C. A. Coburn, J. A. McCauley, S. W. Ludmerer, K. Liu, J. P. Vacca, H. Wu, B. Hu, R. Soll, F. Sun, X. Wang, M. Yan, C. Zhang, M. Zheng, B, Zhong and J. Zhu, *Patent Application*, WO2010-111483, 2010.
40. Y. S. Or, X. Peng, L. Ying, W. Ce, D. Tang and Y.-L. Qiu, *Patent Application*, WO2010-096462, 2010.
41. M. Belema, A. C. Good, J. Goodrich, R. Kakarla, G. Li, O. D. Lopez, V. N. Nguyen, J. Kapur, Y. Qiu, J. L. Romine, D. R. St. Laurent, M. Serrano-Wu, L. B. Snyder and F. Yang, *Patent Application*, WO2010-017401, 2010.
42. L. Li and M. Zhong, *Patent Application*, WO2010-065681, 2010.
43. Y.-L. Qiu, C. Wang, L. Ying, X. Peng and Y. S. Or, *Patent Application*, WO2010-091413, 2010.
44. K. X. Chen, G. N. Anilkumar, Q. Zeng, S. B. Rosenblum, J. A. Kozlowski and F. G. Njoroge, *Patent Application*, WO2010-138790, 2010.
45. R. Lavoie, J. A. Bender, Z. Yang, M. Belema, O. D. Lopez, Q. Chen, G. Wang and P. Hewawasam, *Patent Application*, WO2011-059887, 2011.
46. L. Li and M. Zhong, *Patent Application*, WO2010-096777, 2010.
47. K. Vandyck, W. G. Verschueren and P. J.-M. B. Raboisson, *Patent Application*, WO2012-013643, 2012.
48. L. C. C. Kong, S. Giroux, T. J. Reddy and Y. L. Bennani, *Patent Application*, WO2011-119870, 20110.
49. P. Chen, R. Couch, M. Duan, R. M. Grimes; W. M. Kazmierski, B. A. Norton and M. Tallant, *Patent Application*, WO2011-028596, 2011.
50. W. E. Delaney, J. O. Link, H. Mo, D. W. Oldach, A. S. Ray, W. J. Watkins, C. Y. Yang and W. Zhong, *Patent Application*, WO2012-087596, 2012.
51. A. G. Nair, K. M. Keertikar, S. H. Kim, J. A. Kozlowski, S. Rosenblum, O. B. Selyutin, M. Wong, W. Yu and Q. Zeng, *Patent Application*, WO2011-112429, 2011.

52. W. Chang, R. T. Mosley, S. Bansal, M. Keilman, A. M. Lam, P. A. Furman, M. J. Otto and M.J. Sofia, *Bioorg. Med. Chem. Lett.*, 2012, **22**, 2938.
53. A.C. Krueger, D. L. Madigan, D. W. Beno, D. A. Betebenner, R. Carrick, B. E. Green, W. He, D. Liu, C. J. Maring, K. F. McDaniel, H. Mo, A. Molla, C. E. Motter, T. J. Pilot-Matias, M. D. Tufano and D. J. Kempf, *Bioorg. Med. Chem. Lett.*, 2012, **22**, 2212.
54. I. Conte, C. Giuliano, C. Ercolani, F. Narjes, U. Koch, M. Rowley, S. Altamura, R. De Francesco, P. Neddermann, G. Migliaccio and I. Stansfield, *Bioorg. Med. Chem. Lett.*, 2009, **19**, 1779.
55. P. Gastaminza, S. M. Pitram, M. Dreux, L. B. Krasnova, C. Whitten-Bauer, J. Dong, J. Chung, V. V. Fokin, K. B. Sharpless and F. V. Chisari, *J. Virol.*, 2011, **85**, 5513.
56. P. Gastaminza, F. V. Chisari, S. M. Pitram, K. B. Sharpless, V. V. Fokin, L. B. Krasnova and J. Dong, *Patent Application*, WO2011-091152, 2011.
57. R. M. Mckinnell, D. D. Long, L. J. Van Orden, L. Jiang, M. Loo, D. R. Saito, S. Zipfel, E. L. Stangeland, K. Lepack, G. Ogawa, X. Huang and W. Zhang, *Patent Application*, WO2012-061552, 2012.
58. A. U. Neumann, N. P. Lam, H. Dahari, D. R. Gretch, T. E. Wiley, T. J. Layden and A. S. Perelson, *Science*, 1998, **282**, 103.
59. A. S. Perelson, E. Herrmann, F. Micol and S. Zeuzem, *Hepatology*, 2005, **42**, 749.
60. E. Shudo, R. M. Ribeiro and A. S. Perelson, *Expert Opin. Drug Metab. Toxicol.*, 2009, **5**, 321.
61. L. Rong, H. Dahari, R. M. Ribeiro and A. S. Perelson, *Sci. Transl. Med.*, 2010, **2**, 1.
62. L. Rong and A. S. Perelson, *Crit. Rev. Immunol.*, 2010, **30**, 131.
63. J. Guedj, L. Rong, H. Dahari and A. S. Perelson, *J. Viral Hepatitis*, 2010, **17**, 825.
64. R. E. Nettles, M. Gao, M. Bifano, E. Chung, A. Persson, T. C. Marbury, R. Goldwater, M. P. DeMicco, M. Rodriguez-Torres, A. Vutikullird, E. Fuentes, E. Lawitz, J. C. Lopez-Talavera and D. M. Grasela, *Hepatology*, 2011, **54**, 1956.
65. R. A. Fridell, C. Wang, J.-H. Sun, D. R. O'Boyle II, P. Nower, L. Valera, D. Qiu, S. Roberts, X. Huang, B. Kienzle, M. Bifano, R. E. Nettles and M. Gao, *Hepatology*, 2011, **54**, 1924.
66. J.-H. Sun, D. R. O'Boyle II, Y. Zhang, C. Wang, P. Nower, L. Valera, S. Roberts, R. E. Nettles, R. A. Fridell and M. Gao, *Hepatology*, 2012, **55**, 1692.
67. Z. Plaza, V. Soriano, E. Vispo, M. Del Mar Gonzalez, P. Barreiro, E. Seclen and E. Poveda, *Antiviral Ther.*, 2012, **17**, 921.
68. K. Hayashi, Y. Katano, T. Kuzuya, Y. Tachi, T. Honda, M. Ishigami, A. Itoh, Y. Hirooka, T. Ishikawa, I. Nakano, F. Urano, K. Yoshioka, H. Toyoda, T. Kumada and H. Goto, *J. Med. Virol.*, 2012, **84**, 438.
69. F. Suzuki, H. Sezaki, N. Akuta, Y. Suzuki, Y. Seko, Y. Kawamura, T. Hosaka, M. Kobayashi, S. Saito, Y. Arase, K. Ikeda, M. Kobayashi,

R. Mineta, S. Watahiki, Y. Miyakawa and H. Kumada, *J. Clin. Virol.*, 2012, **54**, 352.

70. Y. Tanaka, K. Hanada, M. Mizokami, A. E. T. Yeo, J. W.-K. Shih, T. Gojobori and H. J. Alter, *Proc. Natl. Acad. Sci. U. S. A.*, 2002, **99**, 15584.
71. M. A. Gelman and J. S. Glenn, *Trends Mol. Med.*, 2010, **17**, 34.
72. P. Halfon and C. Sarrazin, *Liver Int.*, 2012, **32**(Suppl. 1), 79.
73. P. Ferenci, *Liver Int.*, 2012, **32**(Suppl. 1), 108.
74. V. K. Rustgi, *Expert Opin. Drug Safety*, 2010, **9**, 883.
75. T. L. Kieffer, A. D. Kwong and G. R. Picchio, *J. Antimicrob. Chemother.*, 2010, **65**, 202.
76. S. Pol, R. H. Ghalib, V. K. Rustgi, C. Martorell, G. T. Everson, H. A. Tatum, C. Hézode, J. K. Lim, J.-P. Bronowicki, G. A. Abrams, N. Bräu, D. W. Morris, P. J. Thuluvath, R. W. Reindollar, P. D. Yin, U. Diva, R. Hindes, F. McPhee, D. Hernandez, M. Wind-Rotolo, E. Hughes and S. Schnittman, *Lancet Infect. Dis.*, 2012, **12**, 671.
77. E. J. Gane, S. K. Roberts, C. A. M. Stedman, P. W. Angus, B. Ritchie, R. Elston, D. Ipe, P. N. Morcos, L. Baher, I. Najera, T. Chu, U. Lopatin, M. M. Berrey, W. Bradford, M. Laughlin, N. S. Shulman and P. F. Smith, *Lancet*, 2010, **376**, 1467.
78. A. S. Lok, D. F. Gardiner, E. Lawitz, C. Martorell, G. T. Everson, R. Ghalib, R. Reindollar, V. Rustgi, F. McPhee, M. Wind-Rotolo, A. Persson, K. Zhu, D. I. Dimitrova, T. Eley, T. Guo, D. M. Grasela and C. Pasquinelli, *N. Engl. J. Med.*, 2012, **366**, 216.
79. C. R. T. Chung, *N. Engl. J. Med.*, 2012, **366**, 273.
80. K. Chayama, S. Takahashi, J. Toyota, Y. Karino, K. Ikeda, H. Ishikawa, H. Watanabe, F. McPhee, E. Hughes and H. Kumada, *Hepatology*, 2012, **55**, 742.
81. E. J. Lawitz, D. Gruener, J. M. Hill, T. Marbury, L. Moorehead, A. Mathias, G. Cheng, J. O. Link, K. A. Wong, H. Mo, J. G. McHutchison and D. M. Brainard, *J. Hepatol.*, 2012, **57**, 24.
82. E. Gane, G. R. Foster, J. Cianciara, C. Stedman, S. Ryder, M. Buti, E. Clark and D. Tait, presented at the 45th Meeting of the EASL, 14–18 April 2010, Vienna, LB Poster 2003.
83. A. N. Fox and I. M. Jacobson, *Clin. Infect. Dis.*, 2012, **55** (Suppl. 1), S16.
84. R. G. Gish and N. A. Meanwell, *Clin. Liver Dis.*, 2011, **15**, 627.
85. F. Poordad and D. Dieterich, *J. Viral Hepatitis*, 2012, **19**, 449.
86. R. A. Fridell, D. Qiu, C. Wang, L. Valera and M. Gao, *Antimicrob. Agents Chemother.*, 2010, **54**, 3641.
87. F. Penin, V. Brass, N. Appel, S. Ramboarina, R. Montserret, D. Ficheux, H. E. Blum, R. Bartenschlager and D. Moradpour, *J. Biol. Chem.*, 2004, **279**, 40835.
88. C. Lee, H. Ma, J. Q. Hang, V. Leveque, E. H. Sklan, M. Elazar, K. Klumpp and J. S. Glenn, *Virology*, 2011, **414**, 10.

89. P. Targett-Adams, E. J. S. Graham, J. Middleton, A. Palmer, S. M. Shaw, H. Lavender, P. Brain, T. D. Tran, L. H. Jones, F. Wakenhut, B. Stammen, D. Pryde, C. Pickford and M. Westby, *J. Virol.*, 2011, **85**, 6353.
90. D. Qiu, J. A. Lemm, D. R. O'Boyle II, J.-H. Sun, P. T. Nower, V. Nguyen, L. G. Hamann, L. B. Snyder, D. H. Deon, E. Ruediger, N. A. Meanwell, M. Belema, M. Gao and R. A. Fridell, *J. Gen. Virol.*, 2011, **92**, 2502.
91. L. H. Jones, D. Beal, M. D. Selby, O. Everson, G. M. Burslem, P. Dodd, J. Millbank, T.-D. Tran, F. Wakenhut, E. J. S. Graham and P. Targett-Adams, *J. Chem. Biol.*, 2011, **4**, 49.

CHAPTER 2

Respiratory Syncytial Virus Fusion Inhibitors

DAVID SPERANDIO* AND RICHARD MACKMAN

Gilead Sciences, Inc., 333 Lakeside Drive, Foster City, CA 94404, USA
*Email: david.sperandio@gilead.com

2.1 Introduction

Respiratory syncytial virus (RSV) is the leading cause of lower respiratory tract infections in infants and the elderly, with an estimated 64 million people infected worldwide, leading to 160 000 deaths each year.[1,2] In the USA alone, 125 000 infants require hospitalization with an estimated economic healthcare burden of $5000–$12 000 per patient.[3–5] Infection is prevalent in the winter months, the virus spreads easily and can stay viable on surfaces for hours.

In the general population, the infection causes limited symptoms comparable to those of the common cold, but in the pediatric population, elderly or immunocompromised, the upper respiratory tract infection can often progress to a lower respiratory tract infection requiring medical attention. In infants, inflammation of the bronchioles (bronchiolitis) is common and characterized by wheezing, labored breathing (dyspnea) and poor feeding. Without proper intervention with i.v. fluids and supplemental oxygen, cyanosis (lack of oxygen) may develop, leading to lung damage. Lung damage, in turn, has been linked to the occurrence of long-term conditions such as asthma.[6]

The frail geriatric population, especially those suffering from COPD (chronic obstructive pulmonary disease), are also vulnerable to RSV infection, with mortality rates of up to 5 in 1000 patients. Infection rates in this patient segment vary significantly, with the highest risk associated with long-term care

RSC Drug Discovery Series No. 32
Successful Strategies for the Discovery of Antiviral Drugs
Edited by Manoj C. Desai and Nicholas A. Meanwell
© The Royal Society of Chemistry 2013
Published by the Royal Society of Chemistry, www.rsc.org

facilities.[7] Immunocompromised patients, especially lung transplant and bone marrow transplant patients, are a third, small, but significant patient population, with high mortality rates due to RSV-related complications.[8]

Despite the burden on the healthcare system in the developed world and the high mortality rates in certain high-risk groups, there is no treatment for RSV infection. Ribavirin (**1**, Scheme 2.1) was approved by the US Food and Drug Administration (FDA) but is rarely used owing to marginal efficacy and side effects for both patient and caregiver. Immunoprophylaxis against RSV infections with palivizumab (Synaggis) is only moderately effective and reserved for the highest risk preterm infants and those with conditions such as chronic lung disease and congenital heart disease due to the high cost of treatment. A cheap, effective and convenient RSV vaccine has not materialized yet and prospects seem rather slim after an early clinical trial with a formalin-inactivated RSV that exacerbated the disease course. Furthermore, the immune response that results after natural infection is modest and not highly protective, so multiple infections of the same individual within the same season have been reported.[9]

RSV is an enveloped virus with a negative single-strand RNA genome. The virus belongs to the family Paramyxoviridae that includes some other important respiratory pathogens such as parainfluenza virus and the closely related human metapneumovirus. RSV encodes for 11 proteins, including three surface glycoproteins (F, G and SH) and several proteins that comprise the viral RNA polymerase complex (N, P, L and M2-1). It replicates effectively in the upper and lower respiratory tract and can cause respiratory symptoms by directly damaging the integrity of the small airway epithelium and indirectly by inducing strong immune responses in lungs that lead to airway obstruction. Two major antigenic subgroups of RSV are known, RSV A and RSV B, that differ primarily in the genetic sequence of the G glycoprotein while maintaining a higher degree of homology across other parts of the genome. Both subgroups show comparable pathogenicity and can co-circulate in the same community during a seasonal epidemic, but their individual prevalence usually varies from season to season. A clinically effective RSV therapeutic therefore requires efficacy across a broad range of diverse isolates from both subgroups. Small molecule inhibitors identified to date, that are active against both subgroups,

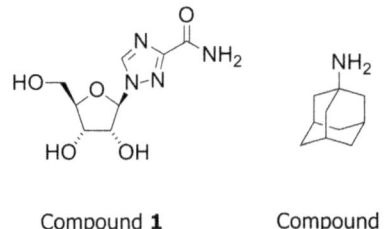

Compound **1**   Compound **2**

Scheme 2.1   Structures of ribavirin (**1**) and amantadine (**2**).

target the viral entry process, particularly fusion or various functions of the RNA transcription/replication complex.

A significant unmet medical need exists for the development of a convenient, safe and effective antiviral therapy for RSV infection. The goal of an antiviral therapeutic would be to block the progression to a lower respiratory tract infection and reduce the number and duration of hospitalizations. The development of direct-acting antivirals for RSV has been slow but a common theme has been the ability to identify inhibitors of the virus fusion process with its target cell. This chapter reviews the challenges, both pre-clinical and clinical, and recent progress that has been made in the development of RSV fusion inhibitors.

2.2 Challenges in the Development of RSV Antivirals

The challenges associated with the discovery and development of RSV fusion (F) inhibitors range from understanding the fundamental interactions of inhibitors with the fusion protein to clinical trial design and demonstration of clinical efficacy. The RSV fusion protein is a critical component of the fusion process along with the attachment protein. The mechanism under which an uninfected cell becomes infected with the virus is poorly understood, in part because it is a dynamic, energy-driven process involving large structural changes of the fusion protein. The F protein is a type I, single-pass integral membrane protein that is embedded in the viral envelope and inserts into the host cell membrane. Until recently, the identity of the target cell surface protein that might be involved in cell recognition and attachment was not known. However, recent reports now suggest that nucleolin, a nuclear receptor that is shuttled from the nucleus to the cell surface, might play a key role in mediating RSV infection.[10,11]

The fact that the fusion protein is a membrane-bound protein with a dynamic mode of action has made access to structural information, especially small-molecule X-ray structure, difficult to obtain. The first report of an X-ray structure of a F protein–inhibitor complex appeared only recently and provides some insight into one potential binding interaction that is consistent with data concerning residues that mutate to reduce inhibitor binding.[12] Other reports have identified sites at which monoclonal antibodies, *e.g.*, motavizumab may also bind to residues of the RSV fusion protein.[13] At present, however, the utility of structural information that can guide structure-based design approaches to inhibitors of the RSV F protein is very limited, which presents a challenge in the design of optimal fusion inhibitors. In addition to the lack of structural information, there is also a limitation in biochemical assays available for evaluation of binding interactions. Therefore, to date, small-molecule inhibitors that have been identified and optimized have been discovered by screening of libraries using a virus replication assay, followed by empirical approaches to optimization. Indeed, compared with many other viruses such as HIV and HCV, the amount of X-ray structural data to drive RSV inhibitor design has been minimal. Given the wide variety of small-molecule structures that have been discovered to interact with the fusion protein and the breadth of the structural chemotypes, it is very possible that multiple different binding modes and interaction points can occur with the fusion protein.

Beyond the biochemical and structural challenges, there is the common challenge of resistance development. Literature reports on the more common fusion inhibitors have identified that resistance to inhibitors can develop rapidly *in vitro* and just single mutations in the fusion protein can effectively abolish antiviral activity.[14] Despite wide differences in the structure of small-molecule inhibitors, the resistance mutations occur in the same region and cross-resistance between fusion inhibitors is common. Given that fusion inhibitors have not progressed clinically, it is not possible to determine whether the *in vitro* results will translate into the clinical setting, especially in the light of the short treatment duration in an infected individual.

Stepping beyond resistance development for a fusion inhibitor, there is also the need for inhibitors to possess broad activity across both subtypes of RSV and the various different strains within those subtypes as discussed above.

Another common preclinical challenge is the availability of relevant animal models to study the effectiveness of RSV inhibitors *in vivo*. The two most common *in vivo* models are the cotton rat model of infection and the BALB/c mouse.[15–17] These models, however, are not ideal since neither represents clinical disease in humans and, as such, the models are of limited use. Other potential animal models include sheep and calves.[15,18–24] In the latter, bovine RSV is a common ailment of calves and livestock during the winter months. The course of infection in calves is very similar to that of infant disease and, therefore, represents a potential model in late-stage optimization of compounds. To date, no small molecules have been reported in the calf model of RSV infection.

A significant challenge for RSV inhibitors, in general, not just limited to fusion inhibitors, is the ability to deliver the drugs to patients effectively. Given the diverse nature of the patient populations, from small infants to the frail elderly, consideration has to be given to the different formulation challenges. In the elderly, it is likely that oral tablets, syrups or even inhaled options for drug delivery could be utilized, whereas infants present a more significant challenge for drug formulation. Pulmonary delivery of a drug *via* inhalation is an attractive option for infants since this directly targets the organ of interest and limits systemic levels of the drug, increasing safety. However, inhaled delivery in infants is challenging since they cannot interact with the device as adults can. Additional complications include the shallow breathing of sick infants that could be inadequate to trigger delivery devices. One solution that is exemplified by viramidine, a prodrug of ribavirin (**1**), involves placing the infant inside a tent-like enclosure and then aerosolizing drug into the atmosphere. This is less than ideal and certainly not practical for a caregiver in the outpatient setting. Therefore, the challenge for infants is to develop safe and, preferably, orally bioavailable drugs that can be administered to even very young neonates. A liquid form or elixir would be ideal; however, the amount of additives needs to be minimal, since the acceptable options for very young infants are limited. Taste is also a critical aspect of drugs for infants when given orally. Other potential routes of administration include suppository or buccal methods involving fast-dissolving minitablets or strips that are placed in the mouth. Although these appear to be attractive options for older infants above 6 months, they have not been explored for neonates.

Even after all the challenges cited above have been negotiated, there are considerable challenges in clinical trial design and demonstration of clinical efficacy for registration. Some of these factors are expanded upon in more detail in the last section of this chapter, but in brief, for adults the main challenges are enrolment of trials to power endpoints and virus detection and diagnostics. Immunosuppressed patients are too few to enroll and power trials easily. Elderly patients with underlying conditions such as COPD are more abundant, but the low level of virus that is typical in these subjects can be difficult to detect. In contrast, all infants by the age of 2 years will have an RSV infection and typically have high viral loads that can be readily detected, thus obviating the diagnostic issue. However, trials in infants require consent from guardians and carry an ethical hurdle in that demonstration of efficacy and safety in other patient groups, *i.e.* adults, is likely required prior to infant trials in order to justify the risk-to-benefit factor.

A final challenge of note clinically is understanding the timing of intervention in the disease. Ideally, any treatment option, regardless of mechanism, is most effective when given as early as possible in the disease course. However, this is not always feasible and it is anticipated that infection would have progressed for several days before symptoms are severe enough that patients or caregivers would seek medical attention. Most importantly, the disease begins in the upper respiratory tract and then progresses to the lower respiratory tract, where the greatest risk and highest morbidity occur. Intervening before lower respiratory tract infection is preferred. Although this logically applies to all RSV inhibitors regardless of mechanism, it is perhaps most relevant for a fusion inhibitor, which by virtue of mechanism is designed to prevent virus spread to uninfected cells. If widespread infection in the upper respiratory tract has occurred, it is unlikely that a significant impact can be made on virus in this cavity. Similarly, for the lower respiratory tract infections, if intervention is too late then minimal benefit may be anticipated for a drug that blocks new infection. In contrast, replication inhibitors might still impact virus production in infected cells in both the upper and lower respiratory tract. It is reasonable on this basis to suggest that fusion inhibitors will be particularly sensitive to the timing of intervention in the disease course. Therefore, the challenge for drug developers of fusion inhibitors would be early detection and early treatment, with early diagnosis of RSV critical. There are several options for the detection of RSV: cell culture, antigen-based tests and PCR, with RT-PCR being the most rapid detection system. Antigen-based tests also offer rapid detection, but lack specificity. Multiplexing several rapid detection systems such as RSV, human metapneumovirus and influenza A and B would be highly desirable, since co-infections often exist and RSV infections could be misdiagnosed as influenza.[25]

Overall, the challenges associated with the development of RSV fusion inhibitors span the whole development process and present a significant hurdle.

It is therefore not surprising that many organizations shy away from tackling this disease based on an apparent high risk-to-reward ratio. Despite these limitations, several groups continue to make inroads and discover fusion inhibitors as explored in subsequent sections.

2.3 Small Molecule RSV Fusion Inhibitor Target Product Profile

In the absence of clinically validated efficacious small-molecule therapies for RSV, proposing a target product profile (TPP) is highly speculative. Nevertheless, some assumptions can be made that may offer some guidance on what a reasonable drug profile should offer.

The safety profile is a function of the severity and time course of the disease, the side effects and risks associated with the current standard of care (SOC) and the specific risk profile of the target population. For RSV in the developed world, mortality in the pediatric population is very rare and the disease is self-limiting. The SOC, although very costly, consists mostly of i.v. fluid/oxygen that does not carry significant risk and, with the target population being premature infants or those <2 years of age, this is arguably one of the least tolerant populations for adverse effects. As a consequence, the standards for the safety profile could not be higher. Physicians will likely not prescribe a drug that carries even a minimal risk to the infant patient. Immunosuppressed patients who contract RSV will likely be given ribavirin in addition to immunosuppressant therapy. Given the fragility of their condition, any prospective treatment will need to be safe. Finally, in the frail elderly, it is likely that these patients are on multiple medications for other disease states and, as such, the new treatment must be safe from drug–drug interactions (DDIs). Regardless of the patient group, the safety bar is high for RSV treatments.

The target exposure levels of small-molecule drugs for clinical efficacy is speculative since there are no direct-acting anti-RSV drug examples to follow. What is the ideal drug exposure in human plasma? How does plasma-adjusted (*i.e.* free fraction) antiviral activity correlate with efficacy for a lung disease? Some answers can be gleaned from other viral diseases such as influenza, HIV and HCV. A significant body of clinical efficacy data with antivirals for HIV and HCV (and also other diseases) correlates plasma-adjusted potency, rather than intrinsic target potency, with efficacy.[26] Furthermore, the large set of clinical data from HCV and HIV therapeutics supports plasma-adjusted trough levels (C_{min}) and not area under the curve (AUC) or C_{max} as the best predictors of a sustained drop in viral load without development of resistance.[27] The inhibitory quotient (IQ_{50}), defined as the ratio of the plasma-adjusted EC_{50} (IQ_{90} if based on IC_{90}) of a drug to the molar concentration of the drug at C_{min} *in vivo*, has been a very useful guide for dose selection that would result in maximum efficacy and minimal potential for resistance development. Oseltamivir (Tamiflu) for influenza, for example, has an IQ_{50} of ~ 20 for its clinically approved dose.[28,29] Recently, HCV protease inhibitors have been described that showed clinical efficacy that exceeded what was expected based in their plasma IQ. This was attributed to accumulation of the inhibitor in the liver relative to the plasma compartment.[30] A similar situation may exist for RSV: fusion inhibitors with an accumulation in the lung have been described (see Section 2.5); however, it is not clear if this afforded improved efficacy, mostly because of limitations on the dynamic range of animal models for RSV.

Overall, although very speculative, to set an exposure bar comparable to oseltamivir for the first RSV fusion inhibitor to reach the clinic seems sensible. As a consequence of this goal, very potent inhibitors with a low human plasma shift stand the best chance of success. Targeting highly potent oral inhibitors requiring a low dose has additional benefits such as minimizing DDIs and idiosyncratic toxicities. This could also be achieved by direct pulmonary delivery to reduce systemic exposure; however, technologies for inhalation dosing to neonates and infants less than 2 years of age have not been clinically validated.

Another lesson from HIV, HCV and other viruses is the requirement of pan-genotype coverage for all strains of virus that are in circulation (2.2 Challenges in the development of RSV antivirals). Both RSV subtypes, A and B, typically alternate in prevalence in subsequent winter months and within each A and B strain there are a large number of variants.[31,32] Lastly, in addition to the issue of natural pre-existing resistance, avoiding the development of resistance acquired under drug pressure is essential. This is especially challenging for HIV antivirals, since therapy has to be maintained for life and could be much less of an issue with the self-limiting and relatively short-duration RSV infecton. The few reported cases of resistance towards oseltamivir are believed to be caused by a natural strain and not a result of drug pressure.[33,34] Other antiviral drugs such as amantadine (**2**, Scheme 2.1), however, are known to develop resistance after a few rounds of replication of the flu virus.[35]

The pharmacokinetics and physicochemical properties should also be considered in the target profile. For example, if oral therapy is targeted, good bioavailability is essential and preferably in multiple non-clinical species. Whereas differences in absorption and elimination pathways in adults can be managed, they can create difficulties in predicting doses for infants, for example. Inhaled therapies would avoid these concerns. The physicochemical properties are also important as, for example, high solubility would allow multiple formulations to be considered for all age categories.[35] These factors as they relate to challenges regarding the clinical development are discussed in Section 2.6.

Overall, the target profile is stringent, and appropriate choice of target properties and goals in the discovery and development process can be a significant factor for clinical success.

2.4   RSV Fusion Inhibitors – Biologics

Preventive biological therapies such as palivizumab (Synagis) and mota-vizumab (Numax) and vaccines are not discussed in detail in this chapter and the reader is referred to more specialized reviews.[36] These monoclonal antibodies, however, have recently been studied as therapy for RSV infection in the pediatric population and, therefore, serve as important benchmarks. Ablynx disclosed a highly potent fusion inhibitor based on their 'Nanobody' technology and this is included in the discussions below and summarized in Table 2.1.

2.4.1   Palivizumab (Synagis)

Palivizumab (Synagis) is a monoclonal antibody (mAB) developed by MedImmune that was approved by the FDA for the prevention of RSV in a fairly narrow patient population.[37] Despite the highly debated cost–benefit ratio,[2,38–40] sales in 2009 and 2010 were above $1 billion; however, sales slipped slightly in 2011 to $975 million. Given prophylactically, palivizumab reduced the rate of hospitalization by 55% in Phase 3 registrational trials. Since the rate of hospitalization in the placebo group was low (10.6%), only a few infants out of many on drug avoided hospitalization (placebo arm, $N = 500$, hospitalizations $= 53$; palivizumab arm, $N = 1002$, hospitalizations $= 48$). Surprisingly, infants on the drug who were hospitalized did not show a less severe disease than hospitalized infants in the placebo arm.[37]

Malley *et al.* studied palivizumab in a therapeutic setting.[41] Infants of 2 years of age or less who where hospitalized with confirmed RSV infection and requiring mechanical ventilation were enrolled. The patients were then randomized and given $15 \, \text{mg kg}^{-1}$ palivizumab i.v. or placebo. Given the potency of $5.6 \, \text{nM}$ (EC_{50}) of palivizumab and the trough levels reported, high plasma-based IQ levels (potency not plasma binding-adjusted) of 360 and 289 at days 1 and 2, respectively, were established. A viral load drop was observed, $-1.7 \log_{10}$ on day 1 and $-2.5 \log_{10}$ on day 2 compared with $-0.6 \log_{10}$ and $-1.0 \log_{10}$ in the placebo arm. The interpretation of these high IQ numbers with antibodies and their relevance to small-molecule efficacy is not straightforward, since distribution into the lungs and lung epithelial lining fluid (ELF) may be very different. No change in disease severity or duration of hospitalization was observed, which may be attributed to the very late initiation of therapy when clinical symptoms required mechanical ventilation.

2.4.2   Motavizumab (Numax)

Motavizumab (Numax) was developed by MedImmune as a second-generation mAB to palivizumab and has improved affinity with an EC_{50} of $\sim 0.3 \, \text{nM}$ compared with $5.6 \, \text{nM}$ for palivizumab. Although early clinical trials suggested improved efficacy of motivizumab compared with palivizumab, a Phase 3 study designed for equivalence but not superiority showed a small but significant increase in adverse events (AE) such as inflammation compared with palivizumab, and motavizumab was ultimately rejected by the FDA.[42] AstraZeneca decided to halt further development. A treatment study reported by Lagos *et al.*[43] reported better treatment results than in the palivizumab study,

Table 2.1 RSV fusion inhibitors – biologics.

Company Development status	Drug	Potency PK	Dosing, patient/species	Efficacy: key observations	Ref.
Ablynx Phase 1	F-VHHb (Nanobody)	$EC_{50} = 0.056$ nM PK, no data published, Ablynx projects qd or bid (i.n.)	60 µg intranasal, $+4$ h or $+24$ h inoculation, BALB/c mouse	Viral load drop (plague reduction) $-2\log_{10}$ in both arms	45
Medimune/AstraZeneca Commercialized	Palivizumab (Synagis)	$EC_{50} = 5.6$ nM or 0.363 µg mL^{-1} PK (trough levels 24 h, 48 h, 15 mg kg^{-1}): 131, 105 µg mL^{-1}	15 mg kg^{-1} i.v., <24 h after initiation of mechanical ventilation, infants <2 years ($N=35$)	Viral load drop (plague and PC-R) -1.7 day 1 (placebo -0.6) -2.5 day 2 (placebo -1.0) No change in disease severity, hospitalization	41 PK: 86
Medimune Development halted	Motavizumab (Numax)	$EC_{50} = 0.3$ nM or 0.02 µg mL^{-1} PK (trough levels, day 2): 62, 171, 333 µg mL^{-1} (5, 15, 30 mg kg^{-1}) PK (trough levels, day 30): 17, 59, 80 µg mL^{-1} (5, 15, 30 mg kg^{-1})	15, 30 mg kg^{-1} i.v., within <24 hospitalization with confirmed lower tracked RSV, infants <2 years ($N=31$)	Total viral load (plague and PC-R) 1.3 day 1 (placebo 3.0) in 15 and 30 mg kg^{-1} arm (data combined) 0.9 day 2 (placebo 3.0) in 15 and 30 mg kg^{-1} arm (data combined) No change in disease severity, hospitalization	43

suggesting that motavizumab may continue development as a therapeutic product.

2.4.3 RSV Nanobody (F-VHHb)

Ablynx developed a functional antibody for RSV that is based on their 'nanobody' technology of heavy chain-only antibodies.[44] These nanobodies, derived from camelids and fish, show improved stability and high affinity and may bind to epitopes that are not accessible to regular monoclonal antibodies. Indeed, the Ablynx fusion inhibitor is the most potent antibody described so far, with an EC_{50} of 0.056 nM, and efficacy in mice ($-2log_{10}$) was observed after intranasal delivery 4 and 24 h post-inoculation.[45] The viral load drop was even higher when dosed 48 h after inoculation; however, the authors cautioned that carry-over of the antibody into the assay may have affected the data, a common issue with plaque-based viral load assays (see Table 2.1). Ablynx projects their RSV nanobody (F-VHHb) to be intranasally qd or bid.[46]

2.5 Small Molecule Fusion Inhibitors

2.5.1 J&J 2408086 and TMC-353121

The discovery of TMC-353121 dates back to a screening campaign in 1990 by Johnson & Johnson (J&J). A cellular assay based on cytopathic effects (CPE) was used and a benzimidazole series moved into a hit-to-lead optimization. Starting with potency in the double-digit micromolar (EC_{50}) range (see **3**, Scheme 2.2), empirical medicinal chemistry optimization afforded significantly more potent compounds with EC_{50}s in the picomolar range that appeared in 2005 in the patent literature.[47–50]

A time-of-addition assay established the entry mechanism of action (MOA) of this class of compounds. Resistant mutations in the F protein further clarified the antiviral target as the fusion protein (S398L, K394R/ S398L, D486N).[12,51]

Early compounds in this series, such as JNJ-2408086 (**4**), had unexpected long tissue retention of 153 h in the lungs, which was considered a safety concern, especially since this was not accompanied by a long plasma half-life. The structural features responsible for the tissue retention were systematically explored and found to be based on the highly polar and cationic ethylenediamine side chain. This observation is in line with the observation that many cationic lipophilic compounds tend to have elevated lung exposure, a property that may seem beneficial for a respiratory disease such as RSV. It is, however, far from clear how well lung tissue exposure correlates with viral load drop for inhibitors that interfere with an extracellular target such as the fusion protein. Replacement of the highly basic ethylenediamine side chain with a morpholinopropylene side chain reduced the lung tissue half life to an acceptable 14 h; however, the potency decreased by over 1000-fold back to high nanomolar

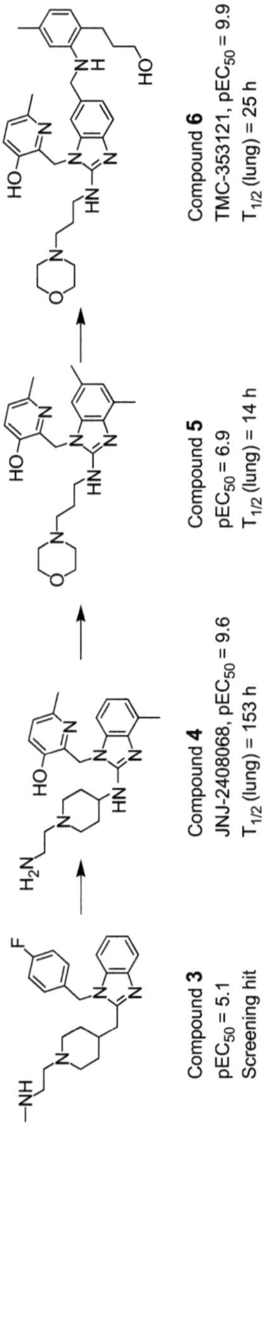

Scheme 2.2 Development of the J&J RSV fusion inhibitor TMC-353121 (**6**).

Compound **3**
pEC$_{50}$ = 5.1
Screening hit

Compound **4**
JNJ-2408068, pEC$_{50}$ = 9.6
T$_{1/2}$ (lung) = 153 h

Compound **5**
pEC$_{50}$ = 6.9
T$_{1/2}$ (lung) = 14 h

Compound **6**
TMC-353121, pEC$_{50}$ = 9.9
T$_{1/2}$ (lung) = 25 h

levels (**5**). A structure-guided approach using modeling and an X-ray structure of the fusion protein guided medicinal chemistry efforts to build back potency. The morpholinopropylene side chain that modulated lung tissue distribution was retained and the benzimidazole core was further elaborated with a substituted aryl group which bound in a sub-pocket and ultimately afforded picomolar fusion inhibitors with the lead compound being TMC-353121 (**6**).[52]

Detailed efficacy studies of TMC-353121 (**6**) in cotton rats and BALB/c mice were published (see Table 2.2).[53] In line with the fusion mechanism and the half-life of the compounds, efficacy in the cotton rat correlated with drug exposure in a relatively short time window after inoculation. Efficacy was observed when dosed i.v. ($10 \, mg \, kg^{-1}$, range from -96 to $+24 \, h$), p.o. ($10–40 \, mg \, kg^{-1}$, $-2 \, h$) and inhaled ($0.25–5 \, mg \, mL^{-1}$, $-1 \, h$), with the best efficacy achieved by the inhalation route. A caveat to inhalation delivery is that the inoculum, which is administered intranasally, will be exposed to some extent to the fusion inhibitor and may block establishment of the infection. The maximum viral load (TaqMan, lung) was $5.67 \log_{10}$ and the observed maximum viral load drop in the bronchoalveolar lavage fluid (BALF) was $1.5 \log_{10}$. These rather modest numbers and the significant variability reported by Olszewska *et al.*[53] reflect the well-known limitations of the cotton rat model. Interestingly, in the BALB/c mouse, significant efficacy was observed in a therapeutic regimen ($+2$ days and $+4$ days, $10 \, mg \, kg^{-1}$, viral load drop $1.0 \log_{10}$), including minimal lung inflammation compared with controls. The pharmacokinetics of TMC-353121 (**6**) in BALB/c mice ($2.5 \, mg \, kg^{-1}$) reported suggests very low drug levels in plasma after 2 days, although in the lung much higher exposure was observed; however, no data were reported on drug exposure after 2 days in the $10 \, mg \, kg^{-1}$ arm.[53] Limited data on further progress into the clinic such as pharmacokinetic (PK) studies in higher animals were published. The development status of TMC-353121 as of 2011 was reported by Bonfanti *et al.*[52] to be undergoing further evaluation.

A recent series of patent publications described the evolution of the SAR around TMC-353121 (**6**) and disclosed an interesting hybrid between **6** and the Bristol-Meyers Squibb (BMS) lead compound **10** described below. The morpholine side chain responsible for the long lung tissue retention times can be deleted without compromising potency (see **7**, Scheme 2.3). Further potency can be built in by retaining the C2 imidazole nitrogen substitution in **8**. A polar and highly charged phosphonic acid can be introduced on the benzimidazole aminomethyl moiety, as shown in **9**. Furthermore, the right-hand benzimidazol-2-one moiety of the BMS lead compound BMS-433771 (**10**) was built into the TMC leads **6** and **8**, affording potent compounds.[48,54]

2.5.2 BMS-433771

A high-throughput screening (HTS) campaign at BMS using a cellular assay with a CPE readout afforded several hits, with a benzimidazole (see **13**, Scheme 2.4) originally designed as an analgesic and antiarrhythmic being the most attractive one.[55] Structurally, the BMS leads differ significantly from the

Table 2.2 RSV fusion inhibitors – small-molecules.

Company Development status	Drug	Structure	Potency PK	Selected animal studies	Ref.
BMS Halted	BMS-433771		$EC_{50} = 10$ nM, $CC_{50} > 218$ μM Plasma binding (human): 92% **Rat PK:** F = 13%, Cl = 61 mL min^{-1} kg^{-1} $V_{ss} = 0.57$ L kg^{-1} **Dog PK:** F = 72%, Cl = 12 mL min^{-1} kg^{-1}, $V_{ss} = 1.3$ L kg^{-1} **Monkey PK:** F = 42%, Cl = 7.3 mL min^{-1} kg^{-1}, $V_{ss} = 2.6$ L kg^{-1}	*Cotton rat:* p.o. 50 mg kg^{-1} 1 h, 0.55log$_{10}$ viral load reduction p.o. 25 mg kg^{-1} 1 h start dose, 4 days bid, 0.59log$_{10}$ viral load reduction *BALB/c mouse:* p.o. 50 mg kg^{-1} 1 h, 1.33log$_{10}$ viral load reduction p.o. 25 mg kg^{-1} 1 h start dose, 4 days bid, 1.14log$_{10}$ viral load reduction	16
J&J Discovery	TMC-353121		$EC_{50} = 0.15$ nM, $CC_{50} = 5$ μM	*Cotton rat:* i.v. 10 mg kg^{-1} 5 min, viral load reduction = 1.2log$_{10}$ Inhaled –1 h, 5 mg mL^{-1}, 1.5log$_{10}$ viral load reduction p.o. 40 mg kg^{-1} 2 h, 0.9log$_{10}$ viral load reduction *BALB/c mouse:* i.v., 10 mg kg^{-1} 1 h, 1.0log$_{10}$ viral load reduction	53
AstraZeneca Unknown	Representative structure		$EC_{50} = 1.7$ nM	N/A	57
Biota Halted	Representative structure		$EC_{50} = 27.5$ nM	N/A	59

Table 2.2 (*Continued*)

Company Development status	Drug	Structure	Potency PK	Selected animal studies	Ref.
Wyeth Halted	RFI-647		$EC_{50} = 6\text{–}109$ nM Inhaled	*African green monkey:* Inhaled, 15 mg mL^{-1} for 20 min–2 h, 0.73–1.34log$_{10}$ viral load reduction	65
Viropharma, MicroDose, Gilead Clinical	VP-14637 MD-637		$EC_{50} = 1$ nM, $CC_{50} > 3$ μM Inhaled	*Cotton rat:* Inhaled on day 0 and day +1, 126 μg kg^{-1} d^{-1}, 2.3log$_{10}$ viral load reduction	87
University of Gothenburg, Sweden Discovery	P13 C15		P13: $EC_{50} = 110$ nM, $CC_{50} = 310$ μM C15: $EC_{50} = 130$ nM, $CC_{50} = 75$ μM	N/A	68

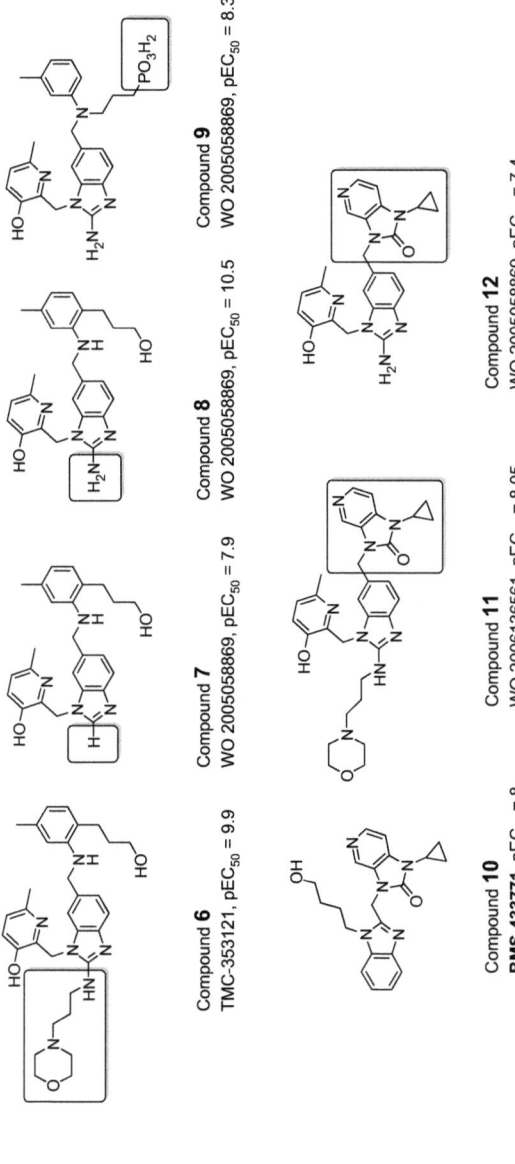

Scheme 2.3 Evolution of TMC-353121 (**6**): examples from WO 2005058869, WO 2006136561 and BMS-433771.

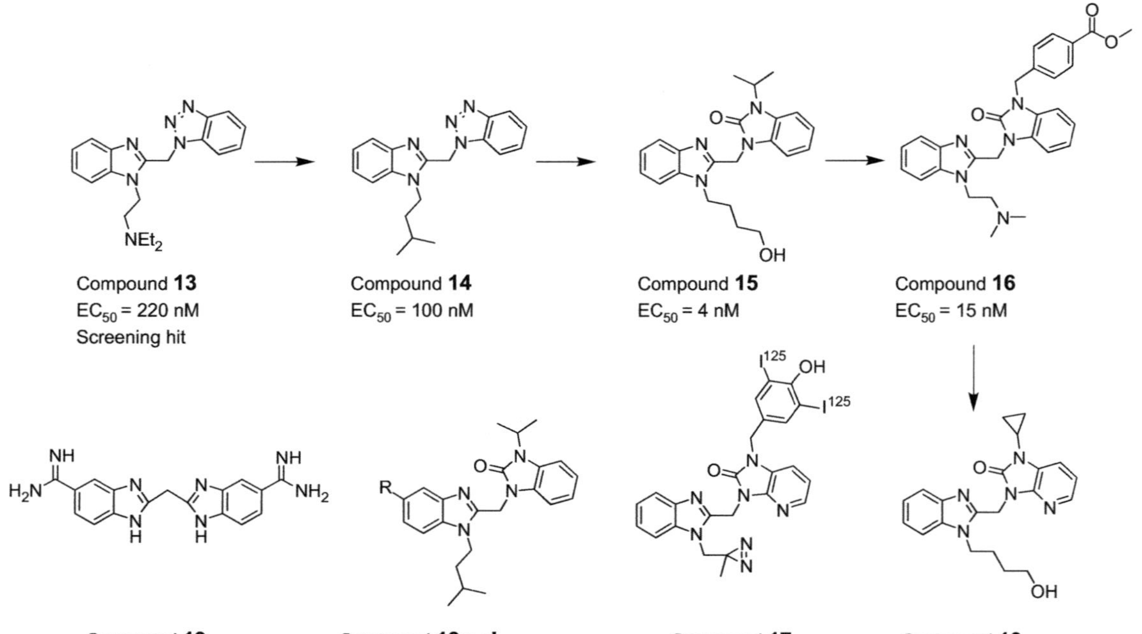

Compound **13**
EC_{50} = 220 nM
Screening hit

Compound **14**
EC_{50} = 100 nM

Compound **15**
EC_{50} = 4 nM

Compound **16**
EC_{50} = 15 nM

Compound **19**
BABIM

Compound **18a–d**
18a: R = CO_2H, EC_{50} = 238,00 nM
18b: R = NH_2, EC_{50} = 125 nM
18c: R = CH_2NH_2, EC_{50} = 2 nM
18d: R = CH_2OH, EC_{50} = 50 nM

Compound **17**

Compound **10**
EC_{50} = 10 nM
BMS-433771

Scheme 2.4 Development of the BMS fusion inhibitor BMS-433771 (**10**).

benzimidazoles pursued by J&J, but shows some similarities with the well-known RSV fusion inhibitor BABIM.[19] The initial lead, which has an excellent potency, an EC_{50} of 220 nM, a high CC_{50} of 83 μM and no cross-reactivity against Sendai, parainfluenza, vesicular stomatitis viruses, HIV and HSV, was advanced into a rigorous medicinal chemistry campaign with the goal of identifying an orally bioavailable inhibitor. Initial SAR exploration focused on the benzimidazole substitution while keeping the parent methylenebenzotriazole unchanged. A wide range of functionalities, including lipophilic, anionic, cationic and uncharged polar groups, were tolerated provided that an ethylene spacer between the core was present. The most potent substitution, *N*-isoamyl (**14**), was moved forward into a second round of optimization interrogating the benzotriazole moiety. In part due to synthetic accessibility, but with the key advantage of having an additional vector for functionalization, the triazole was replaced with a benzimidazol-2-one. In this second round of optimization, potency was further improved from double-digit to single-digit nanomolar EC_{50}. Reoptimization of the benzimidazole *N*-substitution revealed a series of low single-digit nanomolar compounds (**15**). This set of compounds was studied in both the cotton rat and the BALB/c mouse model and, despite exploring multiple routes of delivery (p.o., i.p., s.c.), no efficacy was observed. Proof of concept in an animal model was deemed critical at this point. This was achieved by delivering the compound *via* small-particle aerosol (SPA) delivery in collaboration with Baylor College (Houston, TX). SPA delivery requires highly water-soluble compounds (10 mg mL^{-1}) and the lead molecules were probed for tolerance of polar groups. The benzimidazol-2-one *N*-position was elaborated with a benzyl moiety, which tolerated a wide variety of polar charged functionalities and delivered animal proof of concept with **16** (log$_{10}$ viral load reduction: 2 at 2 mg mL^{-1} concentration in the aerosol).[56] This set the stage for further medicinal chemistry optimization to improve potency and *in vivo* exposure following p.o. delivery. Following SAR for potency, human liver microsome stability and Caco-2 permeability, several key structural changes were introduced, including replacing the isoamyl group with *n*-butanol, installing a nitrogen atom in the benzimidazol-2-one ring and installing a cyclopropyl group as the *N*-substituent, affording the clinical candidate BMS-433771 (**10**). PK (i.v., p.o.) in mouse, rat, dog and monkey was disclosed and the compound showed good absorption, with high to intermediate clearance. Allometric scaling predicted a human half-life of 8 h and, with an estimated 50% oral bioavailability, a human dose of 500 mg was predicted. A scalable and efficient synthesis of **10** was developed to support clinical and preclinical material supply.[55]

Interestingly, resistant mutations induced by **10** map to K394R, which is identical with what was found for TMC-353121 (**6**). Elegant MOA studies using a radiolabeled photoaffinity probe (see **17**, Scheme 2.4,) tagged Y198, which is located in the N-terminal heptad repeat adjacent to the fusion peptide. Y198 lines a hydrophobic pocket inside the deep groove of the HR-N57 heptad repeat trimer. Modeling studies of the BMS series of fusion inhibitors suggested binding in this pocket and was in line with the SAR seen on the right-hand benzimidazole moiety. Modeling proposed that C5 (see **18a–d**, Scheme 2.4) can be further elaborated, possibly forming productive H-bonds with Asp200. Installation of a benzylamine or benzyl alcohol (**18c** and **d**) indeed afforded potent compounds whereas a carboxylate or an aniline were much less active (**18a** and **b**). This SAR is, as mentioned above, reminiscent of pharmacophore of BABIM (**19**), a known RSV fusion inhibitor, and furthermore reflected in the recent AstraZenca series (**20–23**).

In vivo studies with **10** were performed in the cotton rat and the BALB/c mouse model. Oral dosing approaching $50 \, \text{mg} \, \text{kg}^{-1}$ at -1 h with intra-tracheal inoculation showed maximum efficacy ($>1.0\log_{10}$ viral load drop) in the cotton rat model. In the BALB/c mouse model, prophylactic dosing as low as $5 \, \text{mg} \, \text{kg}^{-1}$ was efficacious.[16] Toxicity studies in three animal species were performed, enabling human clinical studies. However, BMS reported that clinical studies were not pursued owing to a change in corporate business strategy.[55]

2.5.3   AstraZeneca WO 2010/103306

AstraZeneca recently disclosed a series of quinazolinones that have striking similarity to the BMS series (see **20–23**, Scheme 2.5).[57] Compounds with high picomolar potency were disclosed, and also extensive, small-molecule X-ray and differential scanning calorimetry (DSC) data for a series of compounds. Interestingly, the installation of a basic moiety on the benzimidazole at the C5 position afforded significant potency and structurally mimics the BMS series of compounds (**18a–f**) and also the BABIM-type RSV fusion inhibitor. Improved solubility of this compound with this basic functionality was claimed. No further data regarding development status are available.

2.5.4   BTA9881

Biota claimed novel RSV fusion inhibitors in a series of patent applications, with potency in the double-digit nanomolar range (see **24**, Scheme 2.6).[58–61] The compound progressed to a Phase 1 clinical study and demonstrated good oral bioavailability and half-life. The compound was subsequently

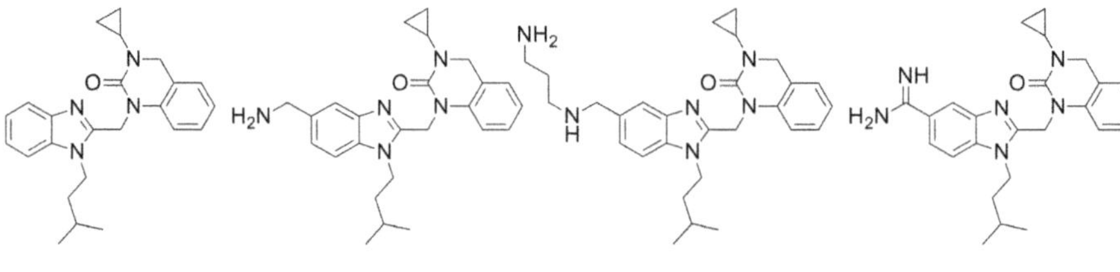

Compound **20**   Compound **21**   Compound **22**   Compound **23**
EC_{50} = 1890 nM EC_{50} = 1.7 nM EC_{50} = 32 nM EC_{50} = 3.2 nM

Scheme 2.5   Selected compounds for AstraZeneca WO 2010/103306.

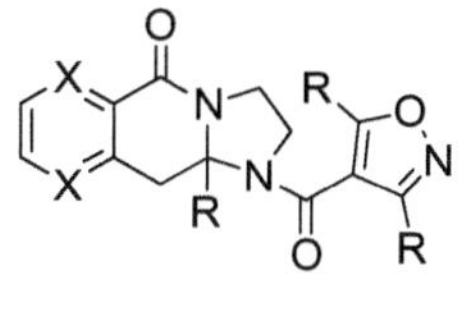

Biota
WO 2012068622

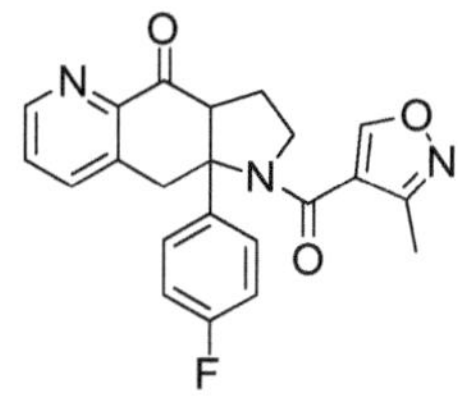

Compound **24**
EC_{50} = 27.5 nM

Scheme 2.6 Biota fusion inhibitors.

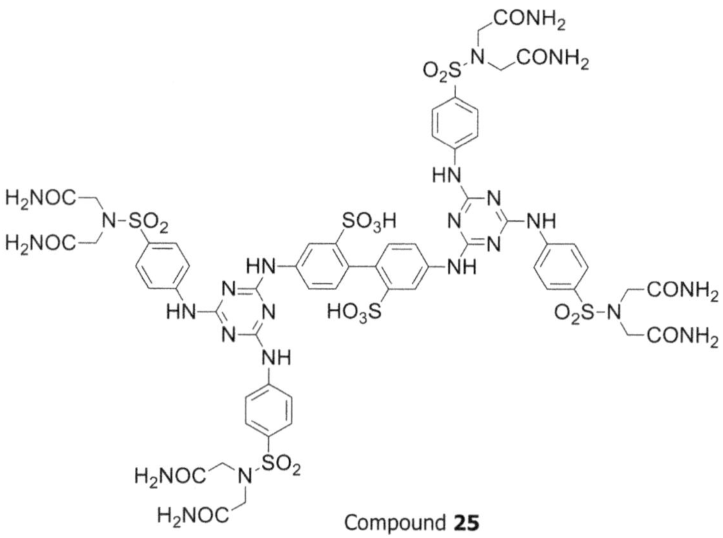

Compound **25**

Scheme 2.7 Wyeth fusion inhibitor RFI-641 (**25**).

discontinued, citing safety concerns in press releases, and the licensing rights were returned to Biota. Biota intends to pursue a series of back-up compounds (according to a press release on 10 August 2009).[62]

2.5.5 RFI-641

Wyeth developed a polyanionic compound that was shown to be an RSV fusion inhibitor but also had some activity against herpes virus (see **25**, Scheme 2.7).[63,64] Intranasal delivery of **25** was efficacious in the RSV murine model and also in the African green monkey. A $2\log_{10}$ viral load drop was observed in BAL samples in the high-dose arm (2 h inhalation *versus* 15 min); no statistical difference was see in nasal washed and throat swap titters.[65] This compound progressed into Phase 2 studies, but development was halted.

2.5.6 VP-14637, MDT-637

VP-14637 (see **26**, Scheme 2.8) was discovered in an HTS campaign at Viropharma. A small impurity in a reagent formed the active component, which was traced down and ultimately afforded **26**. SAR data disclosed so far point to a very sharp loss of antiviral activity when the tetrazole moieties are replaced by diazoles ($-30\times$ potency) or triazoles ($-50\times$ potency). Also, the hydroxy groups adjacent to the tetrazoles are key for potency ($-10,000\times$ potency with OMe), whereas the central hydroxy group has little effect. The potency range of VP-14637 was reported for clinical isolates from Europe and North America to be 0.1–80 nM (EC_{50}). MOA studies suggested an early time point in the viral life cycle and resistant mutations mapped to the F-protein.[66] Good efficacy (-1.3 to $2.8\log_{10}$ reduction in viral titer) was observed in the cotton rat model.[67] In April 2010, Gilead Sciences and MicroDose Therapeutx announced that the companies had entered into an exclusive worldwide license and collaboration agreement for the development and commercialization of **26** using the MicroDose dry powder inhaler technology. Phase 1 studies were concluded in 2012 with high intratracheal levels and low systemic exposure (press release from MicroDose Therapeutx, 24 April 2012).

2.5.7 University of Gothenburg, Sweden

Trybala and co-workers recently identified, from screening a commercial library of small molecules (ChemBioNet), two RSV inhibitors with the structures shown in Scheme 2.9.[68] MOA studies (time of addition, resistant

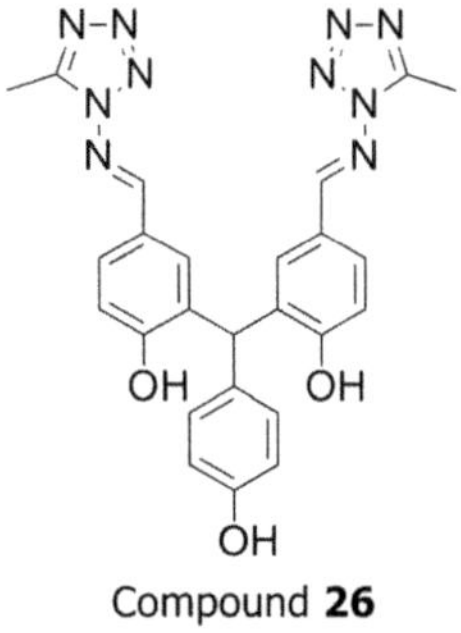

Compound **26**

Scheme 2.8 Viropharma (VP-14637)/MicroDose inhibitor (MDT-637).

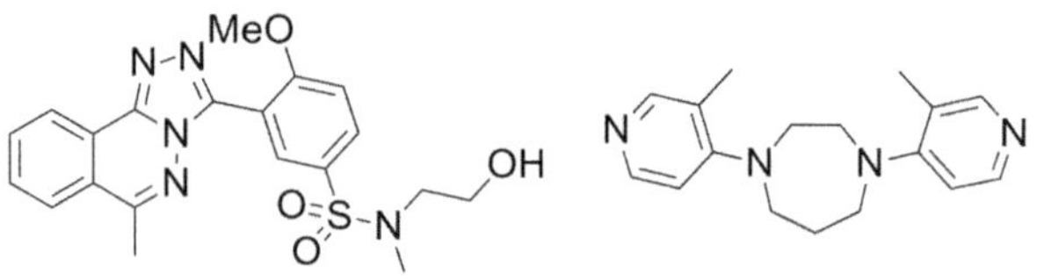

Compound **27**, P13   Compound **28**, C15

Scheme 2.9 University of Gothenburg fusion inhibitors P13 and C15.

mutations) suggested that the F-protein was the likely target. The potency (EC_{50}) of these HTS hits is very high, 110 nM (P13, **27**) and 130 nM (C15, **28**), with negligible toxicity ($CC_{50} = 310$ and 75 μM, respectively).

2.5.8 RSV Inhibitors Targeting Other RSV Genomic Proteins

A series of RSV inhibitors with a post-entry or unknown MOA have been published (Scheme 2.10) that go beyond the scope of this review and are mentioned only briefly. A post-entry MOA seems attractive since the window of intervention in the short-duration RSV infection could be longer, although no animal or clinical data have been reported that support this hypothesis. Unfortunately, the success rate for post-entry RSV inhibitors has been even lower than that for the fusion type inhibitors.

RSV604: The most extensive studies with non-fusion inhibitors were reported by Arrow with RSV604 (**29**) and reviewed by Chapman and Cockerill.[69] Overall, RSV604 has progressed to a Phase 2 study in hematopoietic stem cell transplant patients, which was challenging to enroll. Plasma exposure was variable and drug exposure reached only the EC_{90} range.[70]

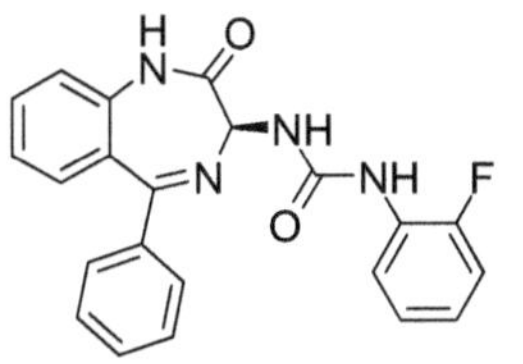

Arrow RSV604
compound **29**

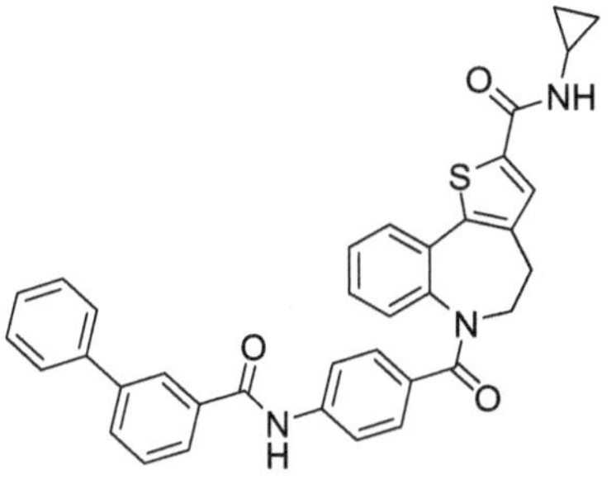

Yamanouchi YM53403
compound **30**

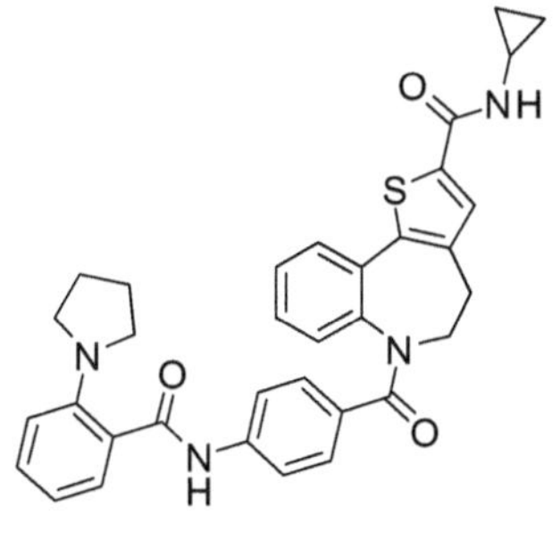

Gilead, WO/2011/005842
compound **31**

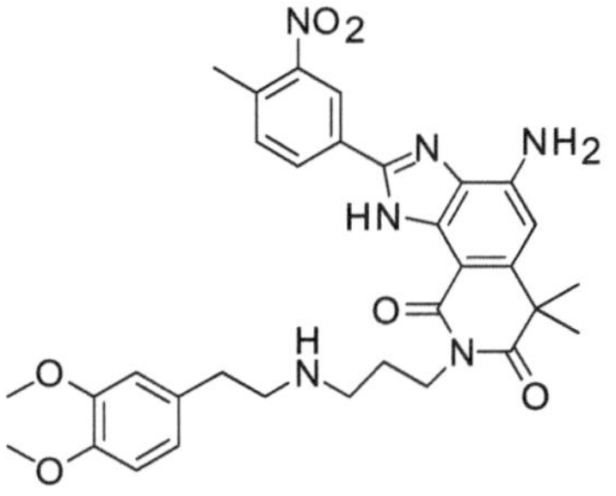

BI WO2005042530A1
compound **32**

Scheme 2.10 Selected RSV inhibitors targeting other RSV genomic proteins.

Nevertheless, the observation that patients with higher drug exposure above the EC_{90} had a larger drop in viral load is very promising. Novartis is continuing clinical trials.[69]

YM53403: Yamanouchi reported RSV inhibitor (**30**) with submicromolar potency and a detailed MOA study suggested a post-entry (L-protein) mechanism.[71]

Gilead WO/2011/005842: Gilead Sciences recently claimed a series of RSV inhibitors that have structural features of the YM series (**31**).[72] No further data regarding MOA and development status were disclosed.

Boehringer-Ingelheim (BI): Novel L-protein inhibitors with activity in the low double-digit nanomolar range were identified by BI (**32**). The L-protein is part of the rather complex RSV polymerase complex. The authors suggested a novel mechanism by which the synthesis of RSV m-RNA is blocked *via* inhibition of guanylation of viral transcripts.[73,74]

ALN-RSV01: Alnylam is developing an RNAi therapeutic targeting the RSV N-protein.[75,76] Phase 1 studies in 2006 and 2007 established the safety and tolerability of ALN-RSV01 and in 2012 results from a Phase 2b trial of ALN-RSV01 in adult lung transplant patients for the treatment of RSV infection were reported. ALN-RSV01 treatment was associated with a clinically meaningful treatment effect, although the primary endpoint of reduced progressive bronchiolitis obliterans syndrome (BOS) in an 'intention-to-treat' (ITTc) analysis was narrowly missed (Alnylam press release, 30 May 2012).

2.6 Options for the Clinical Development of RSV Fusion Inhibitors

Following ribavirin (**1**), only a handful of compounds targeted for the treatment of RSV infection have progressed into clinical trials and none have reached formal Phase 3 registrational trials. This fact alone testifies to the difficulty associated with developing treatments for RSV. A few clinical programs have been reported in Phase 2 and these include the small molecule RSV-604, the siRNA drug ALN-RSV01 and a γ-globulin enriched with RSV neutralizing antibodies, RI-001. None of these products, however, are small-molecule fusion inhibitors, which, despite being more plentiful than other inhibitor classes for RSV, as described earlier, have often stumbled in Phase 1 trials. The early fusion inhibitor VP-14637 (**26**) was discontinued in Phase 1, as was the more recent BTA-9881 fusion inhibitor. Therefore, at the time of writing, only one small-molecule fusion inhibitor is currently in clinical development for RSV, which is the above-mentioned VP-14637 (**26**) in a dry powder formulation using the MicroDose inhaler technology (MDT-637). This inhaler is based on a high-frequency piezo transducer inhaler, avoiding the earlier ethanol based formulation of **26**. Clinical development is sponsored by Gilead (Gilead Science press release, 20 April 2011). Antibody or antibody-based therapeutics such as ADMA-001 are competitors to the approved prophylactic monoclonal antibody Synagis, but can also be considered as potential treatments. Along these lines, the nanobody ALX-0171, the antigen site of which is part of the fusion protein, was reported to be in Phase 1 development in late 2011 (Ablynx press release, 13 December 2011). Both monoclonal antibodies, Synagis and Numax, have also been studied as potential treatments for RSV in hospitalized infant trials.[41,43]

Programs that successfully negotiate Phase 1 safety and PK analyses can initiate treatment-based efficacy studies in one of several naturally infected patient populations, the immunosuppressed, especially hematopoietic stem cell (HSCT) and lung transplant patients, young infants below 2 years of age and the frail elderly with underlying conditions such as COPD or congestive heart failure (CHF). However, each of these diverse patient groups carries significant challenges to prospective pharmaceutical companies for developing RSV therapeutics. An alternative to evaluating efficacy in natural infected populations is to use normal healthy volunteers (NHVs) experimentally infected with virus, which is discussed in Section 2.6.4.

2.6.1 Clinical Trials in Immunosuppressed Patients

RSV infection in immunosuppressed patients is a major concern, since progression to lower respiratory tract infection and pneumonia carries significant risks, including high mortality rates of 70–80%.[77] Most immunosuppressed patients presenting to a healthcare facility having contracted an upper respiratory tract infection, and within 1 week 40–50% will progress to pneumonia. Improvements in managed care and isolation of RSV outbreaks in the main transplant centers over the years have reduced the risk of RSV spreading and also mortality. Therefore, one of the main challenges in conducting efficacy studies in immunosuppressed patients is the difficulty in powering trials without invoking a large geographically diverse set of study sites.[77] For example, aerosolized ribavirin (1) was studied in HSCT patients with the intention of enrolling up to 90 subjects, but the trial was discontinued after 5 years owing to low enrollment numbers.[77] The main reasons cited for this problem were the reduced incidence of RSV as a result of improved care practices as described above and the complexity of the study design, which required hospitalization of patients in the trial and daily blinded analysis by investigators. Despite insufficient patient accrual leading to a lack of statistical significance, subjects who were treated with ribavirin were found to have lower serial viral load in nasal wash samples as detected by qt-PCR (quantitative polymerase chain reaction). Furthermore, a trend towards reduced pneumonia was noted following radiographic analysis. Additional ribavirin-based studies in HSCT patients have been reported in which dosing regimens and alternative routes of administration, oral *versus* inhaled, are under investigation. Once again, these trials appear to recruit only small numbers of subjects over many years. The small-molecule RSV604 was also studied in HSCT patients and, not surprisingly, similar recruitment difficulties complicated the study. Altogether 20 patients were enrolled in 1 year from 10 sites in the USA, Europe and Australia.[69] The primary endpoint for the study was a $2\log_{10}$ viral load reduction in nasopharyngeal swabs analyzed by qt-PCR. Encouragingly, an antiviral effect was noted in patients who were found to have plasma levels of drug that exceeded the EC_{90} for a short period of time. Unfortunately, administration of the compound required a high-fat meal to improve exposure and the resulting drug exposure was too variable across the trial participants for the target endpoint to be met. No further development has been reported for this compound.

Lung transplant patients offer another immunosuppressed patient population for performing RSV treatment trials and have been used in the efficacy trials of aerosolized ALN-RSV01. A total of 521 transplant recipients were followed for RSV infection and 24 were eventually enrolled into the initial study from 22 sites in four countries. Owing to differences in the baseline viral loads of placebo- and drug-treated arms and time from symptom onset, it was not possible to determine a statistically significant antiviral effect. Mean daily symptom scores were lower in the treated group and the cumulative symptom score was significantly lower. At day 90 following treatment, the incidence of new or progressive BOS was significantly reduced compared with placebo. Encouraged by these initial results, a follow-on study of ALN-RSV01 in lung transplant patients was performed. This second study enrolled 87 lung transplant recipients by recruiting from 33 centers in six countries. The study narrowly missed its primary endpoint of reduction in new or progressive BOS at 180 days in the confirmed RSV-infected intention-to-treat group (Alnylam press release, 31 May 2012).

One method for mitigating the difficulty in recruitment of sufficient RSV-infected patients is to utilize the RSV season in both the northern and southern hemispheres. This strategy was used for clinical trial of tamiflu. Many of the trials to date have done this to boost the numbers. One advantage of performing studies in immunosuppressed patients is the ability to monitor disease progression from the upper to the lower respiratory tract. The primary endpoint in the ribavirin trial in HSCT patients was the progression of disease to clinical pneumonia, as determined by chest radiographs and blinded symptom evaluations.[77] Furthermore, in subjects with confirmed pneumonia, bronchoalveolar lavage was performed at the discretion of the patient's physician to provide details of RSV lung infection. This type of information is not readily accessible in other patient groups. Despite the hurdles, it is likely that clinical trials will continue in immunosuppressed patients, but large registrational trials powered for statistical significance will remain logistically challenging. More likely, smaller trials analogous to those described above will provide initial POC promoting drug developers to invest further toward registrational trials in an alternative population, such as infants.

2.6.2   Clinical Studies in COPD or CHF Patients

An alternative adult population for consideration of natural infection RSV trials are the frail elderly, especially those with underlying conditions of COPD and CHF. In the USA and Western Europe, there are more deaths due to RSV infection in the elderly population than in other patient groups and, therefore, there exists a significant unmet need.[78] For patients who go on to respiratory failure, prognosis is particularly poor, with mortality rate as high as 56%.[79] Only 10% of high-risk patients with symptoms compatible with RSV go on to have RSV confirmed by qt-PCR. This is because these patients have low but prolonged viral loads and so current rapid care diagnostics are inadequate to identify these subjects effectively. Although no clinical studies have been attempted to date, improvements in diagnostics may well change this in the future and open up opportunities for clinical trials in this important RSV-susceptible group.[80] An observational study has been reported to be under way to profile RSV disease in the COPD population, the data from which will be useful in further defining the incidence and severity of the disease and identifying potential endpoints for clinical trials. As yet, however, no clinical trials for RSV-related treatments in this patient group have been reported (see ClinicalTrials.gov, Identifier: NCT01455402).

2.6.3 Clinical Studies in Infants

Infants below the age of 2 years can be considered for potential treatment-based fusion inhibitor RSV trials. Unlike the adult natural infection populations, infants offer some advantages, but along with that they also bring some additional hurdles. Almost all infants will be infected with RSV by the age of 24 months and RSV is the leading cause of hospitalization to pediatric general inpatient units in the USA and Western Europe. Of the infants less than 1 year old, >68% will contract RSV.[8] Infections occur primarily in the winter months, but this varies by region. Typically, patients present to physicians or emergency units with symptoms of an upper respiratory tract infection which may progress to a lower respiratory tract infection within days. Children with respiratory distress who experience difficulty with feeding due to tachypnea, or develop hypoxemia, will often be admitted to hospital. The most predictive clinical variable for progression to severe lower respiratory tract disease is the patient's age, with the highest risk for deterioration in children less than 3 months of age and an incremental decline in risk of progression with increasing age.[81,82] Most infants require only supportive care, *e.g.* hydration and oxygen, but ~20% of inpatients are admitted to intensive care units, where some ultimately require mechanical ventilation. Mortality is not high for infants in the developed world, but in the developing or underdeveloped world, where maintenance care is not readily available, mortality rates are significantly higher. In 2005, an estimated 66 000–199 000 infant deaths resulted from RSV infection worldwide, with 99% occurring in the developing world.[43] The viral loads in infants hospitalized for RSV and their relationship to disease severity have been studied.[83,84] These prospective studies have reported that higher viral loads are associated with an increased risk of intensive care, prolonged hospitalization and respiratory failure. Viral loads tend to be higher than in adults with RSV infection, typically reaching beyond 6–$7\log_{10}\mathrm{PFU\,mL^{-1}}$. Measurements of lower respiratory tract virus infection have been obtained from some infants upon mechanical ventilation, which allows deep tracheal aspirate samples to be acquired. A good correlation of virus levels between nasal and tracheal aspirates taken at the same time point has been reported. This correlation is helpful since it allows the nasal aspirate samples to be used as surrogate measurements for lung virus levels in patients who are not on mechanical assistance. Overall, the larger patient numbers, reliable seasonal occurrence of virus and associated infections, high viral loads compared with adult infection and the association of viral load and age to disease severity make the design of clinical trials in infants very attractive. However, since the approval of inhaled ribavirin (**1**) for infants, no small-molecule RSV inhibitor, fusion or otherwise, has been reported to have been studied in an infant setting. The monoclonal antibodies palivizumab and motavizumab, the former approved for prophylaxis in high-risk infants, have been tested as treatments in hospitalized infants.[41,43]

There are several intertwined hurdles to the initiation of infant studies, including ethical considerations, drug delivery and dose prediction considerations and parental or guardian consent. The ethical considerations are the risk–benefit factor to the patient. The risk posed by the treatment can be well understood and studied in adult volunteers first, but results from adults may not necessarily translate to the pediatric population. Therefore, a high level of safety needs to be demonstrated in the early adults to ensure that the initial drug exposures administered to infants are safe. The safety afforded by highly selective monoclonal antibodies is presumably an advantage in this respect over small-molecule fusion inhibitors. In addition to safety, there also needs to be a tangible clinical efficacy benefit to ethically support treatment. For this reason, step-down safety and PK studies in pediatrics under 18 years of age cannot be considered as a means to bridge the age gap between adults and infants <2 years old. This is because RSV is not a significant disease in older children and there would be no clinical benefit. A further unknown is the degree of efficacy data that would be needed for regulatory authorities to approve infant studies based on patient benefit. A low bar might be efficacy in preclinical models such as cotton rat or BALB/c mice, whereas a high bar would be efficacy observation in naturally RSV-infected adults. This, of course, creates a 'catch 22' situation in that demonstrating clinical efficacy in natural RSV-infected adults is not trivial.

A further challenge to infant trials is that of drug delivery and dose selection. The highest risk infants for progression to more severe lower respiratory tract disease are neonates in the 0–6 month age range. Infants of this age cannot be administered solid dosage forms such as capsules and tablets, including chewable solid forms. Therefore, for oral delivery, liquids and solutions are optimal and have the advantage that dosing can be readily adjusted for the sizeable body weight differences between the very young neonates and infants aged 2 years. Taste is also a consideration for solution delivery in infants and flavorings may be needed to ensure patient compliance, which can be poor even without taste issues, as many parents can testify. For the very sick in intensive care, where patients may be on i.v. or mechanical ventilation, delivery and taste are less of a consideration, since solutions can be administered to infants through nasogastric tubes or by i.v. lines, provided that the solubility of the drug is adequate. Liquid formulations often require the use of preservatives and excipients, and some of these can be incompatible with dosing to very young infants and neonates.[85] Altogether, while formulation factors need to be seriously considered as a hurdle, they are not insurmountable provided that they are considered and addressed early in the discovery and development process.

An additional complication for dosing in infants is efficacious dose selection, which is dependent on the absorption, distribution, metabolism and excretion properties of the drug. These characteristics need to be fully understood, preferably through adult Phase 1 studies, in order to predict accurately the appropriate starting dose for trials in infants. Given that newborns and the very young tend to have immature processes for metabolizing drugs due to differences in metabolic enzyme activities, a thorough understanding of hepatic clearance mechanisms is needed. Access to infant hepatocytes can be helpful in this respect for understanding metabolic properties in infants across the 0–2 year age range. Of course, this information is only valuable in the circumstance where hepatic metabolism is a relevant clearance mechanism for the drug in question.

Given all the challenges cited above, it is not surprising that no small-molecule drug has progressed into infant efficacy trials. Indeed, the development of drugs for just young infants below 2 years of age is extremely rare (surfactant being one example). The 'high bar' afforded by infant trials therefore presents a daunting obstacle for many pharmaceutical organizations to undertake. The monoclonal antibody therapeutics solve some of these issues because they are administered intravenously, have long half-lives avoiding some of the adult-infant scaling concerns, and would be considered highly selective and relatively safe, thereby minimizing risk. Palivizumab reduced viral load in the lower respiratory tract of infants, based on results from tracheal aspirates, but did not have an impact on upper respiratory tract virus levels.[41] Motavizumab, however, has been shown to be more potent that palivizumab in the cotton rat model. When tested in infants, a non-statistically significant antiviral impact on cultures from the upper respiratory tract on day 1 post-treatment was observed.[43] In addition, the duration of hospitalization from initiation of drug treatment was shorter for the motavizumab-treated group by almost 2 days but, once again, not statistically significant. These initial studies in infants are encouraging and it is anticipated that for an effective and safe fusion inhibitor in the future, treatment-based infant studies will be a potential option given the substantial benefit potential for these patients.

2.6.4 RSV Challenge Strain (Memphis 37)

Prompted by the challenges of using the age-diverse natural RSV-infected populations for clinical trials, an RSV challenge strain, Memphis 37, has emerged in recent years.[74] Arguably, this is the most significant event in the ability to develop small-molecule antivirals and fusion inhibitors for RSV. The challenge strain is approved for use in healthy adult volunteers by the European Medical Agencies. Exposure of adult volunteers to 3.0–5.4$\log_{10}$PFU mL^{-1} virus produces an infectivity rate ranging from 71 to 86%. Furthermore, the onset of infection measured by viral culture or qt-PCR coincided with the onset of upper respiratory tract disease symptoms such as mucous weight. The challenge virus in healthy adults does not progress to a lower respiratory tract disease, so the study does not provide a means for following progression of upper to lower respiratory tract disease. However, shortly after approval of the challenge strain, the siRNA compound ALN RSV-01 was studied in the experimental model.[74] In this study, ALN RSV-01 or placebo was delivered daily by nasal spray for 5 days starting 2 days before inoculation with virus. The direct impact on viral production was assessed through regular sampling of the nasal cavity using twice daily nasal washes during the study. Infection rate, determined by culture of the nasal wash samples, was reduced by 38% in the treated group ($N = 19$ infected on drug *versus* 30 infected on placebo). A reduction in peak viral load of $\sim 1.0\log_{10}$PFU mL^{-1} and also reduction in viral load area under the curve (AUC) was observed for the treated group, although the study was not powered for statistical significance. As such, the study provided a POC for evaluation of an RSV replication inhibitor targeting the N-protein when administered prophylactically. Prophylactic administration of an RSV fusion inhibitor would also be expected to result in antiviral and symptomatic responses. However, it has not been proven whether treatment-based therapy, for example, treatment upon detection of virus or symptoms, would be effective, in terms reduction of viral load, in this model. In this respect, the fact that disease remains in the upper respiratory tract should be borne in mind since this is a key difference to the natural setting where progression to the lower respiratory tract is the goal of antiviral intervention. Nevertheless, the model provides a significant advance in the ability to test RSV antivirals early in clinical development and avoids some of the hurdles associated with naturally infected patients. Furthermore, demonstration of even a modest antiviral effect in the experimental model may be sufficient to justify a potential benefit for infants and allow initiation of trials in infants infected with RSV.

2.7   Conclusion and Outlook

Significant efforts by BMS, J&J, AstraZeneca, Biota, Wyeth, Viropharma, MicroDose, Gilead and others have afforded potent small-molecule RSV fusion inhibitors with a surprisingly high structurally diversity. Some of these compounds have reached clinical development, but the road to late-stage clinical trials for RSV has been too demanding and to date no Phase 2 study demonstrating efficacy in a therapeutic setting for a fusion inhibitor has been reported. This is not surprising in the light of the challenges both in the preclinical setting for optimization and in the clinical setting for development: unclear MOA of fusion inhibitors, lack of structural information that would guide inhibitor design, development of resistance, pan-genotype potency, lack of good animal models and formulation for certain patient groups, *e.g.* the pediatric population, are just some of the factors. Furthermore, the target product profile calls for a compound which undoubtedly must have an excellent off-target profile in line with the high-risk pediatric and elderly patient groups. Several compounds discussed in this chapter have come close to meeting all of these challenges and a few compounds have progressed into the clinic but, disappointingly, they have been halted or slowed, sometimes for non-scientific reasons such as an apparent lack of urgency or changes in business strategies. The commercial success of palivizumab (Synagis), a monoclonal antibody approved for prophylactic use in high-risk infants, may reignite interest. The experience gained in early clinical trials with ribavirin (**1**) and more recent trials with RSV604 (**29**) and ALN-RSV01 will also inform the design of future clinical trials that will be more successful. To this end, the introduction of an RSV challenge strain (Memphis 37) should also help pave the way for more straightforward clinical trials, at least for POC. Finally, the motivation to pursue the development of an RSV antiviral could not be higher with an estimated annual infant mortality of 66 000–199 000 in the developing world, significant mortality rates in the geriatric population and the huge burden on the healthcare system for pediatrics in the developed world.

References

1. Y. Aujard and B. Fauroux, *Respir. Med.*, 2002, **96** (Suppl B), S9–S14.
2. M.-S. Lee, R. E. Walker and P. M. Mendelman, *Hum. Vaccine*, 2005, **1**, 13–18.
3. B. C. Buckley, D. Roylance, M. P. Mitchell, S. M. Patel, H. E. Cannon and J. D. Dunn, *J. Manage. Care Pharm.*, 2010, **16**, 15–22.
4. C. Hampp, T. L. Kauf, A. S. Saidi and A. G. Winterstein, *Arch. Pediatr. Adolesc. Med.*, 2011, **165**, 498–505.
5. L. Palmer, C. B. Hall, J. P. Katkin, N. Shi, A. S. Masaquel, K. K. McLaurin and P. J. Mahadevia, *Pediatr. Pulmonol.*, 2010, **45**, 772–781.
6. S. P. Brearey and R. L. Smyth, *Perspect. Med. Virol.*, 2007, **14**, 141–162.
7. A. R. Falsey, *Exp. Lung. Res.*, 2005, **31**(Suppl 1), 77.
8. W. P. Glezen, L. H. Taber, A. L. Frank and J. A. Kasel, *Am. J. Dis. Child.*, 1986, **140**, 543–546.
9. J. Schlender, G. Zimmer, G. Herrler and K.-K. Conzelmann, *J. Virol.*, 2003, **77**, 4609–4616.
10. F. Tayyari and R. G. Hegele, *Expert Rev. Respir. Med.*, 2012, **6**, 215–222.
11. F. Tayyari, D. Marchant, T. J. Moraes, W. Duan, P. Mastrangelo and R. G. Hegele, *Nat. Med.*, 2011, **17**, 1132–1135.
12. D. Roymans, H. L. De Bondt, E. Arnoult, P. Geluykens, T. Gevers, M. Van Ginderen, N. Verheyen, H. Kim, R. Willebrords, J.-F. Bonfanti, W. Bruinzeel, M. D. Cummings, H. Van Vlijmen and K. Andries, *Proc. Natl. Acad. Sci. U. S. A.*, 2010, **107**, 308–313.

13. J. S. McLellan, M. Chen, A. Kim, Y. Yang, B. S. Graham and P. D. Kwong, *Nat. Struct. Mol. Biol.*, 2010, **17**, 248–250.
14. J. L. Douglas, M. L. Panis, E. Ho, K.-Y. Lin, S. H. Krawczyk, D. M. Grant, R. Cai, S. Swaminathan, X. Chen and T. Cihlar, *Antimicrob. Agents Chemother.*, 2005, **49**, 2460–2466.
15. L. G. Byrd and G. A. Prince, *Clin. Infect. Dis.*, 1997, **25**, 1363–1368.
16. C. Cianci, E. V. Genovesi, L. Lamb, I. Medina, Z. Yang, L. Zadjura, H. Yang, C. D'Arienzo, N. Sin, K.-L. Yu, K. Combrink, Z. Li, R. Colonno, N. Meanwell, J. Clark and M. Krystal, *Antimicrob. Agents Chemother.*, 2004, **48**, 2448–2454.
17. J. B. Domachowske, C. A. Bonville and H. F. Rosenberg, *Pediatr. Infect. Dis. J.*, 2004, **23**, S228–S234.
18. C. D. Lapin, P. W. Hiatt, C. Langston, E. Mason and P. T. Piedra, *Pediatr. Pulmonol.*, 1993, **15**, 151–156.
19. D. K. Meyerholz, B. Grubor, S. J. Fach, R. E. Sacco, H. D. Lehmkuhl, J. M. Gallup and M. R. Ackermann, *Microbes Infect.*, 2004, **6**, 1312–1319.
20. A. Olivier, J. Gallup, M. M. de Macedo, S. M. Varga and M. Ackermann, *Int. J. Exp. Pathol.*, 2009, **90**, 431–438.
21. F. B. Sow, J. M. Gallup, S. Krishnan, A. C. Patera, J. Suzich and M. R. Ackermann, *Respir. Res.*, 2011, **12**, 106.
22. F. B. Sow, J. M. Gallup, A. Olivier, S. Krishnan, A. C. Patera, J. Suzich and M. R. Ackermann, *Am. J. Physiol. Lung Cell. Mol. Physiol.*, 2011, **300**, L12–L24.
23. M. H. Wagner, J. F. Evermann, J. Gaskin, K. McNicol, P. Small and A. A. Stecenko, *Pediatr. Pulmonol.*, 1991, **11**, 56–64.
24. L. J. Gershwin, E. S. Schelegle, R. A. Gunther, M. L. Anderson, A. R. Woolums, D. R. Larochelle, G. A. Boyle, K. E. Friebertshauser and R. S. Singer, *Vaccine*, 1998, **16**, 1225–1236.
25. W. C. Hymas, A. Mills, S. Ferguson, J. Langer, R. C. She, W. Mahoney and D. R. Hillyard, *J. Virol. Methods*, 2010, **167**, 113–118.
26. G. L. Trainor, *Expert Opin. Drug Discov.*, 2007, **2**, 51–64.
27. A. Barrail-Tran, L. Morand-Joubert, G. Poizat, G. Raguin, C. Le Tiec, F. Clavel, E. Dam, G. Chene, P. M. Girard and A. M. Taburet, *Antimicrob. Agents Chemother.*, 2008, **52**, 1642–1646.
28. H. Lam, J. Jeffery, D. S. Sitar and F. Y. Aoki, *Ther. Drug Monit.*, 2011, **33**, 699–704.
29. D. Morrison, S. Roy, C. Rayner, A. Amer, D. Howard, J. R. Smith and T. G. Evans, *PLoS One*, 2007, **2**, e1305.
30. M. B. Reddy, P. N. Morcos, S. Le Pogam, Y. Ou, K. Frank, T. Lave and P. Smith, *Antimicrob. Agents Chemother.*, 2012, **56**, 3144–3156.
31. P. D. Scott, R. Ochola, M. Ngama, E. A. Okiro, D. J. Nokes, G. F. Medley and P. A. Cane, *J. Infect. Dis.*, 2006, **193**, 59–67.
32. G. A. Storch, C. B. Hall, L. J. Anderson, C. S. Park and D. E. Dohner, *J. Infect. Dis.*, 1993, **167**, 562–566.
33. S. C. Duwe, M. Wedde, P. Birkner and B. Schweiger, *Antiviral Res.*, 2011, **89**, 115–118.

34. A. Lackenby, C. I. Thompson and J. Democratis, *Curr. Opin. Infect. Dis.*, 2008, **21**, 626–638.
35. Y. Suzuki, R. Saito, H. Zaraket, C. Dapat, I. Caperig-Dapat and H. Suzuki, *J. Clin. Microbiol.*, 2010, **48**, 57–63.
36. J. Chang, *BMB Rep.*, 2011, **44**, 232–237.
37. The IMpact-RSV Study Group, *Pediatrics*, 1998, **102**, 531–537.
38. W. Ji, T. Zhang, X. Zhang, L. Jiang, Y. Ding, C. Hao, L. Ju, Y. Wang, Q. Jiang, M. Steinhoff, S. Black and G. Zhao, *BMC Health Serv. Res.*, 2010, **10**, 82.
39. C. B. Hall, G. A. Weinberg, M. K. Iwane, A. K. Blumkin, K. M. Edwards, M. A. Staat, P. Auinger, M. R. Griffin, K. A. Poehling, D. Erdman, C. G. Grijalva, Y. Zhu and P. Szilagyi, *N. Engl. J. Med.*, 2009, **360**, 588–598.
40. L. L. Han, J. P. Alexander and L. J. Anderson, *J. Infect. Dis.*, 1999, **179**, 25–30.
41. R. Malley, J. DeVincenzo, O. Ramilo, P. H. Dennehy, H. C. Meissner, W. C. Gruber, P. J. Sanchez, H. Jafri, J. Balsley, D. Carlin, S. Buckingham, L. Vernacchio and D. M. Ambrosino, *J. Infect. Dis.*, 1998, **178**, 1555–1561.
42. T. F. Feltes, H. M. Sondheimer, R. M. Tulloh, B. S. Harris, K. M. Jensen, G. A. Losonsky and M. P. Griffin, *Pediatr. Res.*, 2011, **70**, 186–191.
43. R. Lagos, J. P. DeVincenzo, A. Munoz, M. Hultquist, J. Suzich, E. M. Connor and G. A. Losonsky, *Pediatr. Infect. Dis. J.*, 2009, **28**, 835–837.
44. P. Vanlandschoot, C. Stortelers, E. Beirnaert, L. I. Ibanez, B. Schepens, E. Depla and X. Saelens, *Antiviral Res.*, 2011, **92**, 389–407.
45. B. Schepens, L. I. Ibanez, S. De Baets, A. Hultberg, P. Bogaert, P. De Bleser, F. Vervalle, T. Verrips, J. Melero, W. Vandevelde, P. Vanlandschoot and X. Saelens, *J. Infect. Dis.*, 2011, **204**, 1692–1701.
46. E. Depla, ALX-0171: a highly potent Nanobody® as inhalation treatment for respiratory syncytialvirus infection, presented at the Next Generation Protein Therapeutics Summit, San Francisco, 25–27 June 2012, http://www.ablynx.com/wp-content/uploads/2012/06/Next-Generation-Protein-Therapeutics-ALX-0171-presentation_final.pdf., Ablynx (last accessed 18 February, 2013).
47. J.-F. Bonfanti, K. J. L. Andries, J. M. C. Fortin, P. Muller, F. M. M. Doublet, C. Meyer, R. E. Willebrords, T. V. J. Gevers and P. M. M. B. Timmerman, to Tibotec Pharmaceuticals, *Patent Application*, WO2005-058874A1, 2005.
48. J.-F. Bonfanti, K. J. L. Andries, J. M. C. Fortin, P. Muller, F. M. M. Doublet, C. Meyer, R. E. Willebrords, T. V. J. Gevers and P. M. M. B. Timmerman, to Tibotec Pharmaceuticals, *Patent Application*, WO2005-058869A1, 2005.
49. J.-F. Bonfanti, K. J. L. Andries, F. E. Janssens, F. M. Sommen, J. E. G. Guillemont and J. F. A. Lacrampe, to Tibotec Pharmaceuticals, *Patent Application*, WO2005-058873A1, 2005.
50. J.-F. Bonfanti, K. J. L. Andries, F. E. Janssens, F. M. Sommen, J. E. G. Guillemont and J. F. A. Lacrampe, to Tibotec Pharmaceuticals, *Patent Application*, WO2005-058870A1, 2005.

51. D. Roymans, H. L. De Bondt, E. Arnoult, P. Geluykens, T. Gevers, M. Van Ginderen, N. Verheyen, H. Kim, R. Willebrords, J.-F. Bonfanti, W. Bruinzeel, M. D. Cummings, H. Van Vlijmen and K. Andries, *Proc. Natl. Acad. Sci. U. S. A., Early Ed.*, 2009, p. 6.

52. J.-F. Bonfanti, G. Ispas, F. Van Velsen, W. Olszewska, T. Gevers and D. Roymans, in *Antiviral Drugs: from Basic Discovery through Clinical Trials*, ed. W. M. Kazmierski, Wiley, New York, 2011, pp. 341–352.

53. W. Olszewska, G. Ispas, C. Schnoeller, D. Sawant, T. Van de Casteele, D. Nauwelaers, B. Van, D. Roymans, M. M. De, M. C. Rouan, R. P. Van, J. F. Bonfanti, V. F. Van Kerckhove, A. Koul, M. Vanstockem, K. Andries, P. Sowinski, B. Wang, P. Openshaw and R. Verloes, *Eur. Respir. J.*, 2011, **38**, 401–408.

54. J.-F. Bonfanti, P. Muller, J. M. C. Fortin and F. M. M. Doublet, to Tibotec Pharmaceuticals, *Patent Application*, WO2006-136561A1, 2006.

55. N. A. Meanwell, C. W. Cianci and M. R. Krystal, in *Antiviral Drugs: from Basic Discovery through Clinical Trials*, ed. W. M. Kazmierski, Wiley, New York, 2011, pp. 353–366.

56. K.-L. Yu, X. A. Wang, R. L. Civiello, A. K. Trehan, B. C. Pearce, Z. Yin, K. D. Combrink, H. B. Gulgeze, Y. Zhang, K. F. Kadow, C. W. Cianci, J. Clarke, E. V. Genovesi, I. Medina, L. Lamb, P. R. Wyde, M. Krystal and N. A. Meanwell, *Bioorg. Med. Chem. Lett.*, 2006, **16**, 1115–1122.

57. H. Blade, E. A. Carron, H. M. Jackson, J. A. Lumley, C. J. Pilkington, G. P. Tomkinson, A. J. F. Thomas and J. Warne, to Astrazeneca UK, *Patent Application*, WO2010-103306A1, 2010.

58. S. Bond, V. A. Sanford, J. N. Lambert, C. Y. Lim, J. P. Mitchell, A. G. Draffan and R. H. Nearn, to Biota Scientific Management, *Patent Application*, WO2005-061513A1, 2005.

59. P. A. Mayes, J. P. Mitchell, A. G. Draffan, G. R. W. Pitt, K. H. Anderson and C. Y. Lim, to Biota Scientific Management, *Patent Application*, WO2012-068622A1, 2012.

60. J. P. Mitchell, A. G. Draffan, V. A. Sanford, S. Bond, C. Y. Lim and P. A. Mayes, to Biota Scientific Management, *Patent Application*, WO2008-037011A1, 2008.

61. J. P. Mitchell, G. Pitt, A. G. Draffan, P. A. Mayes, L. Andrau and K. Anderson, to Biota Scientific Management, *Patent Application*, WO2011-094823A1, 2011.

62. G. Pitt, A. G. Draffan, K. Anderson, S. Bond, J. Fenner, J. Iswarn, J. N. Lambert, A. Luttick, P. Mayes, M. McCarthy, C. Y. Lim, J. D. Mitchell, C. Morton, T. N. Nguyen, J. Ryan and S. Tucker, presented at the International Conference on Antiviral Research (ICAR), Sapporo, 2012.

63. C. C. Huntley, W. J. Weiss, A. Gazumyan, A. Buklan, B. Feld, W. Hu, T. R. Jones, T. Murphy, A. A. Nikitenko, B. O'Hara, G. Prince, S. Quartuccio, Y. E. Raifeld, P. Wyde and J. F. O'Connell, *Antimicrob. Agents Chemother.*, 2002, **46**, 841–847.

64. A. A. Nikitenko, Y. E. Raifeld and T. Z. Wang, *Bioorg. Med. Chem. Lett.*, 2001, **11**, 1041–1044.

65. W. J. Weiss, T. Murphy, M. E. Lynch, J. Frye, A. Buklan, B. Gray, E. Lenoy, S. Mitelman, J. O'Connell, S. Quartuccio and C. Huntley, *J. Med. Primatol.*, 2003, **32**, 82–88.
66. J. L. Douglas, M. L. Panis, E. Ho, K.-Y. Lin, S. H. Krawczyk, D. M. Grant, R. Cai, S. Swaminathan and T. Cihlar, *J. Virol.*, 2003, **77**, 5054–5064.
67. J. McKimm-Breschkin, *Curr. Opin. Invest. Drugs*, 2000, **1**, 425–427.
68. A. Lundin, T. Bergstroem, L. Bendrioua, N. Kann, B. Adamiak and E. Trybala, *Antiviral Res.*, 2010, **88**, 317–324.
69. J. Chapman and G. S. Cockerill, in *Antiviral Drugs: from Basic Discovery through Clinical Trials*, ed. W. M. Kazmierski, Wiley, New York, 2011, pp. 367–382.
70. J. Chapman, E. Abbott, D. G. Alber, R. C. Baxter, S. K. Bithell, E. A. Henderson, M. C. Carter, P. Chambers, A. Chubb, G. S. Cockerill, P. L. Collins, V. C. L. Dowdell, S. J. Keegan, R. D. Kelsey, M. J. Lockyer, C. Luongo, P. Najarro, R. J. Pickles, M. Simmonds, D. Taylor, S. Tyms, L. J. Wilson and K. L. Powell, *Antimicrob. Agents Chemother.*, 2007, **51**, 3346–3353.
71. K. Sudo, Y. Miyazaki, N. Kojima, M. Kobayashi, H. Suzuki, M. Shintani and Y. Shimizu, *Antiviral Res.*, 2005, **65**, 125–131.
72. R. L. Mackman, D. Sperandio and H. Yang, to Gilead Sciences,, *Patent Application*, WO2011-005842A1, 2011.
73. M. Liuzzi, S. W. Mason, M. Cartier, C. Lawetz, R. S. McCollum, N. Dansereau, G. Bolger, N. Lapeyre, Y. Gaudette, L. Lagace, M.-J. Massariol, F. Do, P. Whitehead, L. Lamarre, E. Scouten, J. Bordeleau, S. Landry, J. Rancourt, G. Fazal and B. Simoneau, *J. Virol.*, 2005, **79**, 13105–13115.
74. B. Simoneau, J. Bordeleau, G. Fazal, S. Landry, S. Mason and J. Rancourt, to Boehringer Ingelheim Canada, *Patent Application*, WO2005-042530A1, 2005.
75. J. DeVincenzo, J. E. Cehelsky, R. Alvarez, S. Elbashir, J. Harborth, I. Toudjarska, L. Nechev, V. Murugaiah, A. Van Vliet, A. K. Vaishnaw and R. Meyers, *Antiviral Res.*, 2008, **77**, 225–231.
76. J. DeVincenzo, R. Lambkin-Williams, T. Wilkinson, J. Cehelsky, S. Nochur, E. Walsh, R. Meyers, J. Gollob and A. Vaishnaw, *Proc. Natl. Acad. Sci. U. S. A.*, 2010, **107**, 8800–8805.
77. M. Boeckh, J. Englund, Y. Li, C. Miller, A. Cross, H. Fernandez, J. Kuypers, H. Kim, J. Gnann and R. Whitley, *Clin. Infect. Dis.*, 2007, **44**, 245–249.
78. A. R. Falsey, P. A. Hennessey, M. A. Formica, C. Cox and E. E. Walsh, *N. Engl. J. Med.*, 2005, **352**, 1749–1759.
79. C. B. Duncan, E. E. Walsh, D. R. Peterson, F. E. Lee and A. R. Falsey, *J. Infect. Dis.*, 2009, **200**, 1242–1246.
80. T. Popow-Kraupp and J. H. Aberle, *Open Microbiol. J.*, 2011, **5**, 128–134.
81. R. C. Holman, D. K. Shay, A. T. Curns, J. R. Lingappa and L. J. Anderson, *Pediatr. Infect. Dis. J.*, 2003, **22**, 483–490.

82. L. J. Stockman, A. T. Curns, L. J. Anderson and G. Fischer-Langley, *Pediatr. Infect. Dis. J.*, 2012, **31**, 5–9.

83. J. P. DeVincenzo, *Pediatr. Infect. Dis. J.*, 2005, **24**, S177–S183, discussion, S182.

84. J. P. DeVincenzo, C. M. El Saleeby and A. J. Bush, *J. Infect. Dis.*, 2005, **191**, 1861–1868.

85. R. G. Strickley, Q. Iwata, S. Wu and T. C. Dahl, *J. Pharm. Sci.*, 2008, **97**, 1731–1774.

86. K. N. S. Subramanian, L .E. Weisman, T. Rhodes, R. Ariagno, P. J. Sanchez and J. Steichen, *et al.*, *Pediatr. Infect. Dis. J.*, 1998, **17**, 110–115.

87. P. R. Wyde, S. Laquerre, S. N. Chetty, B. E. Gilbert, T. J. Nitz and D. C. Pevear, *Antiviral Res.*, 2005, **68**, 18–26.

CHAPTER 3

Phenotypic Screening to Discover Inhibitors of Dengue Virus

QING-YIN WANG, BIN ZOU, SIMON J. TEAGUE
AND PEI-YONG SHI*

Novartis Institute for Tropical Diseases, 10 Biopolis Road, 05-01 Chromos,
Singapore 138670
*Email: pei_yong.shi@novartis.com

3.1   Introduction

3.1.1   Disease Burden of Dengue

Dengue is an emerging mosquito-borne viral disease of humans that is endemic in the tropics and subtropics. The spectrum of illness ranges from a mild, non-specific febrile syndrome through classic dengue fever (DF) to the severe forms of the disease, dengue hemorrhagic fever (DHF) and dengue shock syndrome (DSS). Globally, as many as 100 million people are infected yearly, including 500 000 DHF cases and 22 000 deaths, mostly among children.[1] Dengue disease is caused by any one of the four distinct but closely related serotypes of dengue virus (*i.e.*, DENV-1, -2, -3 and -4). There are not yet any clinically approved antivirals or prophylactic vaccines for treating and/or preventing dengue. Currently, the most effective measure for prevention of dengue is vector control to minimize mosquito bites. When infected, early diagnosis and prompt supportive treatment can substantially reduce the risk of developing severe

RSC Drug Discovery Series No. 32
Successful Strategies for the Discovery of Antiviral Drugs
Edited by Manoj C. Desai and Nicholas A. Meanwell
© The Royal Society of Chemistry 2013
Published by the Royal Society of Chemistry, www.rsc.org

disease. Dengue vaccine development has been challenging because of the need simultaneously to immunize and induce a long-lasting protection against all four serotypes of DENV; an incompletely immunized individual may be sensitized to life-threatening DHF or DSS. With the incidence of dengue infection growing dramatically around the world, there is an urgent need for therapies. However, the development of both vaccine and therapy has been hampered by the lack of an animal model that resembles human disease and by the concerns about the role of the immune system in disease pathogenesis.[2]

3.1.2   Antiviral Targets of Dengue Virus

DENV belongs to the genus *Flavivirus* within the family *Flaviviridae*. Many flaviviruses are important human and animal pathogens, including West Nile virus (WNV), yellow fever virus (YFV), Japanese encephalitis virus (JEV), St. Louis encephalitis virus (SLEV) and tick-borne encephalitis virus (TBEV). The flavivirus genome is a single-stranded, positive-sense RNA of about 11 kb in length. The genomic RNA encodes a long polyprotein precursor, which is processed proteolytically upon translation by both cellular and viral proteases to 10 proteins, including three structural proteins [capsid (C), premembrane (prM) and envelope (Env)] and seven nonstructural proteins (NS1, NS2A, NS2B, NS3, NS4A, NS4B and NS5) (Figure 3.1). The structural proteins form the viral particle. The non-structural proteins participate in the replication of the RNA genome, virion assembly and attenuation of host antiviral responses.[3] For viral replication, NS1 participates in replication complex formation,[4] NS3 acts as a viral serine protease with the cofactor NS2B,[5,6] a nucleotide triphosphatase,[7,8] an RNA triphosphatase[9] and an RNA helicase.[10] NS4A induces membrane rearrangements in infected cells,[11,12] whereas NS4B, along

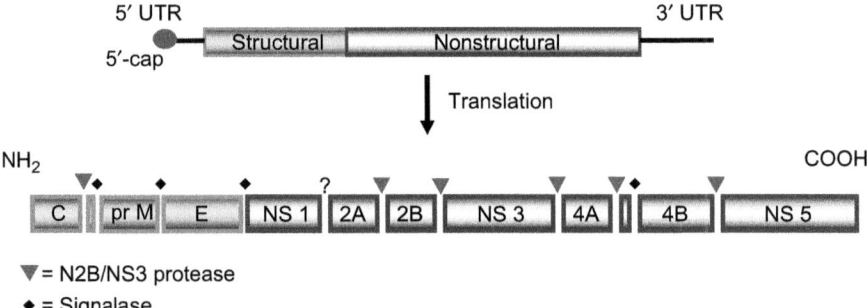

▼ = N2B/NS3 protease

◆ = Signalase

Figure 3.1   Schematic diagram of the flavivirus genome organization and polyprotein processing. At the top is the viral genome with the structural and non-structural protein coding regions, the 5′ cap and the 5′ and 3′ UTRs indicated. Boxes below the genome indicated mature proteins generated by the proteolytic processing. Cleavage sites for host signalases (◆), the viral NS2B/NS3 serine protease (▼) or unknown proteases (?) are indicated.

with NS2A, co-localizes with replication complexes.[11,13] NS5 functions as a methyltransferase[14–16] and an RNA-dependent RNA polymerase (RdRp).[17,18] For attenuation of host antiviral responses, NS1 antagonizes complement activation.[19] NS4B, and to a lesser extent NS2A and NS4A, block type I interferon signaling.[20–23] NS5 also antagonizes type I interferon signaling[24,25] and mediates STAT2 degradation.[26] For antiviral drug discovery, both the structural and NS proteins are targets to block the viral life cycle. In addition, pharmacological blocking of viral protein-mediated antagonism of the host immune response may also bring clinical benefit to infected patients.

Besides viral targets, host proteins could also be targeted for potential antiviral development.[27] A number of host proteins or cellular pathways contributing to DENV replication have been discovered. Furin cleaves the flavivirus prM protein on the virus surface into the mature M protein.[28,29] Glucosidase is responsible for the proper folding and glycosylation of the flaviviral prM, E and NS1.[30] Host kinases have been shown to be involved in DENV and WNV assembly and secretion.[31,32] The cholesterol biosynthesis pathway has been linked to DENV entry and replication and also to the host immune response.[33,34] Host *de novo* pyrimidine biosynthesis pathway has been demonstrated to be essential for DENV replication.[35–37] In addition, with the advent of genome-wide gene knockdown technology, our knowledge of the complex virus–host interaction keeps expanding.[38] The future challenge will be an in-depth understanding of the function of the identified cellular factors/ pathways in the proper host context, which is crucial for developing host-directed antiviral therapy.

3.2 Approaches for Anti-dengue Drug Discovery

3.2.1 Overall Antiviral Approaches

In principle, dengue drug discovery approaches can be grouped into target- and cell-based approaches, both of which have been pursued to identify a lead molecule for development. To date, the structures of five out of the 10 viral proteins – C, prM, E, NS3 (protease, NTPase and helicase) and NS5 (MTase and RdRp) – have been solved by X-ray crystallography or by NMR spectroscopy.[39–45] The structural information has set a solid foundation for target-based drug discovery. Enzyme activity-based high-throughput screening (HTS), fragment-based screening, structure-based rational design and virtual screening have been actively pursued by various academic and industrial groups. This approach has been highly successful for the identification of human immunodeficiency virus (HIV) inhibitors, where more than two dozen antiviral drugs are currently in clinical use.

From the perspective of drug discovery, the strength of the target-based approach is that the mechanism of action is known. Such information typically permits rational drug design and minimizes trial and error if an X-ray structure is available and also allows precise SAR development in the absence of structural information. However, DENV, like all other infectious agents, continues to evolve. The error-prone nature of the viral RdRp is expected to make drug resistance an unavoidable issue. When searching for therapy with novel mechanisms of action, the cell-based phenotypic approach becomes a powerful tool. The replication-based cellular assay could be used to identify inhibitors of either host or viral targets. Host proteins that are critical for viral replication or antiviral activity could potentially be identified by the cell-based approach.

3.2.2   Cell-based Phenotypic Assays

A number of cell-based assays have been developed for DENV HTS. (i) A CPE assay employs a raw DENV to infect susceptible cells and monitors cell viability upon infection.[27,46] (ii) A full-length reporter virus assay uses a complete virus containing a reporter gene (*e.g.*, luciferase gene or GFP) to indicate viral replication.[47] Both CPE and reporter virus assays allow for screening of inhibitors active against any steps of the viral infection cycle, including viral attachment/entry, translation, RNA synthesis, genome encapsidation and virion assembly and egress. (iii) A reporter replicon assay uses a cell line that harbors a DENV replicon in which the viral structural genes are deleted.[48,49] The replicon assay covers the steps of viral translation and RNA synthesis, but not the steps of viral entry and virion assembly. (iv) A virus-like particle (VLP) infection assay uses recombinant VLPs to infect susceptible cells.[50,51] VLPs contain a reporter replicon packaged by viral structural proteins. A VLP infection assay includes a complete viral infection cycle except for the step of virion assembly. The above assays have provided complementary methods to validate HTS 'hits' and to study modes of action.

   One of the major challenges for a cell-based antiviral approach is target deconvolution. With the advent of chemical proteomics, a post-genomic version of classic drug affinity chromatography, target deconvolution is no longer an unattainable mission.[37,52] The following section illustrates a few examples of DENV inhibitors that we have identified in cellular screens. The technical details of the methodology and assays have recently been reviewed elsewhere.[27,46]

3.3   DENV Inhibitors Identified Through Cell-based Screens

3.3.1   Aminothiazole Compound: an Inhibitor Targeting Viral NS4B

We performed an HTS using a cell line harboring a luciferase replicon of DENV-2. After screening a compound library of 1.8 million small molecules, NITD-618 (Scheme 3.1) was identified.[53] It suppressed 85% of the replicon luciferase activity at 5 μM (HTS concentration) in the screen. The antiviral activity and spectrum of NITD-618 were further examined using a viral titer

Table 3.1 Antiviral spectrum of NITD-618.

Virus	Strain	EC_{50} (μM)
DENV-1	West Pacific	1.5
DENV-2	NGC	1.0
DENV-3	H87	1.6
DENV-4	H241	4.1
YFV	17b	>40
WNV	3356	>40

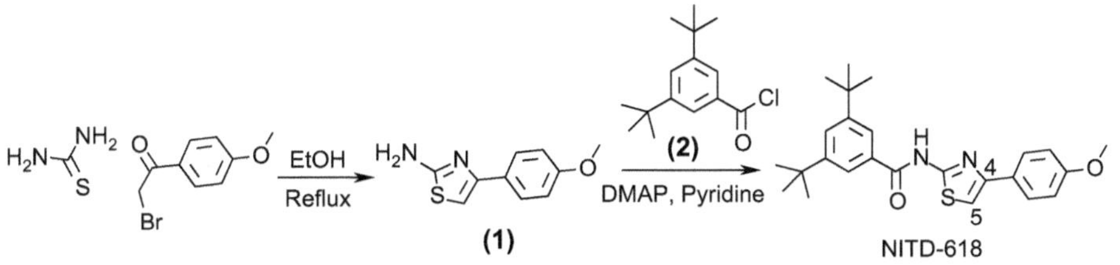

Scheme 3.1

reduction assay. NITD-618 was active against all four serotypes of DENV, with EC_{50}s ranging from 1.0 to 4.1 µM (Table 3.1). When NITD-618 was tested against other RNA viruses, including the two closely related flaviviruses WNV and YFV, two plus-strand RNA alphaviruses [chikungunya virus (CHIKV) and Western equine encephalitis virus (WEEV)] and a negative-strand RNA rhabdovirus [vesicular stomatitis virus (VSV)], it did not reduce viral titers at concentrations up to 40 µM. These results demonstrate that NITD-618 selectively inhibits all four serotypes of DENV.

To determine the target of NITD-618, prolonged culture of replicon cells in the presence of the compound was performed to isolate resistant clones. Mapping of the compound-resistant replicon cell lines revealed a couple of changes in the amino acids of the NS4B protein that, when engineered into the wild-type DENV replicon or the wild-type DENV, conferred the resistance phenotype to NITD-618. Genetic analysis strongly suggested that NS4B might be the target of NITD-618; this, however, does not exclude the possibility of NS4B being an indirect binder. Further biophysical characterization of the interaction between inhibitors and purified NS4B and a co-crystal structure of the compound–NS4B complex are required to define the mechanism of inhibition precisely. However, analyses of biophysical binding and determination of the solid-state structure of the compound–NS4B complex have been hampered by the challenge of expression and purification of recombinant NS4B due to its nature as a membrane protein.

NITD-618 was prepared by acylation of the aminothiazole **1** with the acid chloride **2** (Scheme 3.1). Since NITD-618 is uncharged at physiological pH and has a high calculated lipophilicity (cLog$P$ > 7), it is anticipated that it will have low solubility in aqueous media, high non-specific binding and high metabolic clearance *in vivo*.[54] Although NITD-618 is stable in an *in vitro* mouse liver microsomal stability test, aminothiazoles are known to produce reactive metabolites *in vivo* through epoxidation of the 4,5-bond.[55] Epoxides of this type often result in toxic side effects upon repeated dosing.[56,57] The NS4B protein is membrane bound and is therefore very lipophilic, which results in affinity for lipophilic compounds such as NITD-618. Attempts to reduce the lipophilicity in analogs of NITD-618 may result in the loss of affinity for the NS4B protein and a reduction in its selectivity for the dengue protein with respect to related viral proteins, such as those from HCV.

3.3.2 Benzomorphane Compound: an Inhibitor of Viral Translation

DENV infection of many mammalian cells causes a cytopathic effect (CPE). Compounds that inhibit DENV replication without cytotoxicity reduce or eliminate the CPE of infected cells. A CPE-based HTS assay was developed and used to screen the Novartis compound library. The HTS led to the identification of a class of compounds with a benzomorphane core as inhibitors of DENV.[58] One such compound, NITD-2636 (see Table 3.3, entry **1**), exhibited 76% cell protection at $5 \mu M$ (HTS concentration). Unlike NITD-618, which is DENV specific, NITD-2636 displayed a broad spectrum of antiviral activity (Table 3.2). It not only reduced the titer of DENV, but also the titers of other flaviviruses, including YFV and WNV. However, the compound did not efficiently suppress non-flaviviruses such as alphavirus WEEV or rhabdovirus VSV.

Synthetic chemistry efforts were made to explore the structure–activity relationships (SARs) of the benzomorphan scaffold using NITD-2636 as a starting point. Since NITD-2636 exhibited a short half-life and a high extraction ratio in a rat liver microsomal stability assay (Table 3.3, entry **1**), initial attempts were focused on improving its metabolic stability. Several derivatives were synthesized to replace the pharmaceutically undesirable phenolic –OH group (Table 3.3), which underwent oxidation and/or glucuronidation after intravenous administration of the compound to mice. *Para* fluorine substitution on the 6-phenyl ring of NITD-2636 (**3**) had no negative impact on potency in a CFI assay [an ELISA (enzyme-linked immunosorbent

Table 3.2 Antiviral spectrum of NITD-2636.

Virus	Strain	EC_{50} (μM)
DENV-2	NGC	0.55
YFV	17b	1.1
WNV	3356	1.7
VSV	New Jersey	6.7
WEEV	Cova 746	19

Table 3.3 Structure–activity relationship of the benzomorphane chemotype.

Compound	R	R'	EC_{50} $(\mu M)^a$	CC_{50} $(\mu M)^b$	Half life $(min)^c$	Hepatic extraction ratio^d
NITD-2636	OH	F	0.55	63	9.7	0.82
3	OH	H	0.65	>100	8.5	0.84
4	CH$_3$	F	0.93	69	nd^e	nd
5	N(CH$_3$)$_2$	H	1.4	>100	2.1	0.96
6	4-Morpholinyl	F	6.6	>100	nd	nd
7	CN	H	0.47	57	405	0.1
8	CONH$_2$	H	0.64	>100	10.4	0.81
9	COCH$_3$	F	1.2	>100	nd	nd
10	Cyclopropyl	F	4.5	57	nd	nd

^{a}EC$_{50}$ values were derived from CFI assay ($n \geq 3$).
^{b}CC$_{50}$ values were derived from cell-viability assay using A549 cells.
^{c}Half life was determined in rat liver microsomal stability assay.
^{d}Hepatic extraction ratio were determined in rat liver microsomal stability assay.
^{e}nd: not determined.

assay)-based assay that quantifies viral E protein expression in infected cells[59]]. Substitution of the phenolic –OH group with an electron-donating methyl group (**4**), an electron-withdrawing cyano (**7**) or an acetyl group (**9**) retained antiviral activity. Replacement with a dimethyl amine (**5**) or a polar amide group (**8**) was also well tolerated. In contrast, replacement with a bulky morpholine (**6**) or a cyclopropyl group (**10**) decreased the potency by 12- and 8-fold, respectively.

Active compounds were subjected to an *in vitro* rat liver microsomal stability test to predict their *in vivo* metabolic clearance. This assay measures the depletion of test compounds after incubation with rat liver microsomes and allows the calculation of *in vitro* metabolic parameters, including the half-life (time required to metabolize 50% of input compound) and hepatic extraction ratio (fraction of input compound that has been metabolized). Compound **7** showed an improved metabolic stability, with a half-life of 405 min and a hepatic extraction ratio of 0.10 (Table 3.3). Based on these results, we selected compound **7** for further analysis.

The synthesis of compound **7** is shown in Scheme 3.2. According to the literature,[60] compound **11** can be made in reasonable quantity and yield. With compound **11** in hand, derivatization of the phenolic –OH by triflate activation followed by a Negishi coupling afforded compound **7** in 50% yield for the two

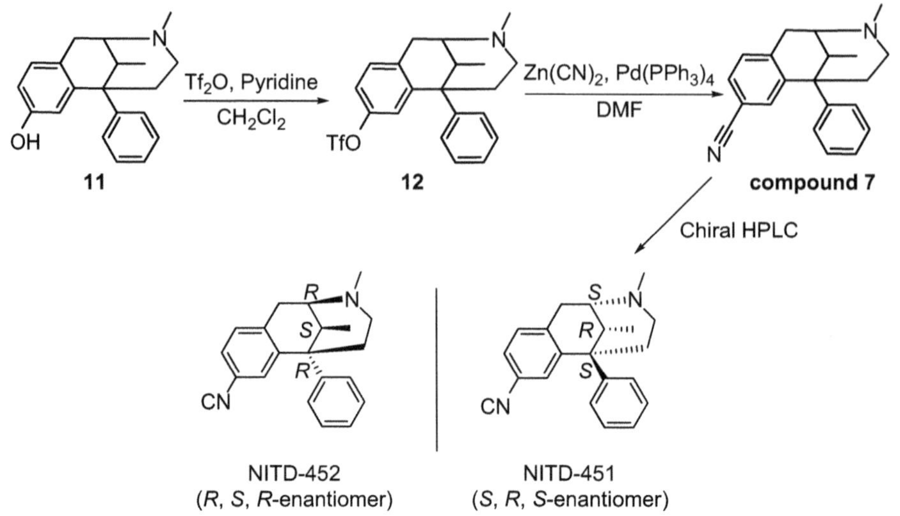

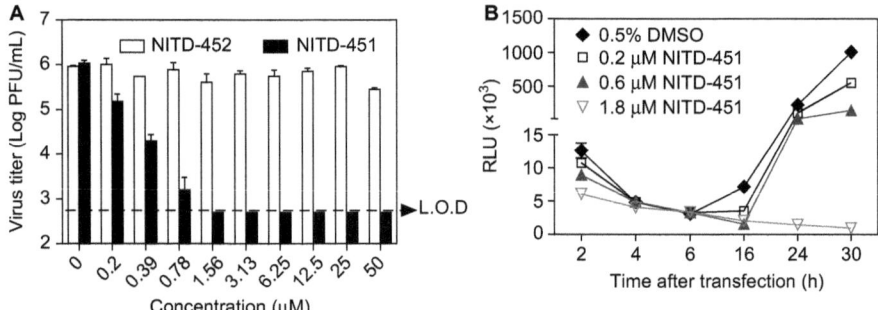

Scheme 3.2

Figure 3.2 (A) Antiviral activities of the enantiomers, NITD-451 and NITD-452. LOD, limit of detection. (B) NITD-451 inhibits viral RNA translation. DENV luciferase-reporting replicon was transiently transfected into target cells followed by NITD-451 treatment at the indicated concentrations. Luciferase activities were measured at 2, 4, 6, 16, 24 and 30 h after transfection.

steps. Since the synthesis of **7** yields a racemic mixture, its two enantiomers were separated by chiral HPLC to afford NITD-451 and NITD-452. The absolute configurations of NITD-451 and NITD-452 were determined to be *S,R,S* and *R,S,R*, respectively, by X-ray crystallography. Remarkably, when tested using a viral titer inhibition assay, only the *S,R,S* enantiomer (NITD-451) showed antiviral activity; treatment with 1.56 µM NITD-451 suppressed viral titers by >103-fold, to an undetectable level (Figure 3.2A). In contrast, the *R,S,R* enantiomer (NITD-452) was inactive; no significant antiviral activity was observed for NITD-452, up to a 50 μM concentration. Similar results were obtained for analogs within the same chemical class. These results demonstrate that the *S,R,S* configuration is the required stereochemistry for the anti-DENV activity.

To elucidate the mechanism of action of NITD-451, a DENV luciferase-reporting replicon was transiently transfected into target cells followed by treatment with the inhibitor and luciferase activity was then measured at various time points after transfection. NITD-451 suppressed luciferase activity as early as 2 h post-transfection (Figure 3.2B), indicating its inhibitory role in RNA translation. The translation inhibition as a mechanism of action of NITD-451 was further demonstrated in a cell-free system, where *in vitro* translation of a reporter RNA was directly suppressed by the compound. As DENV uses host machinery for translating its genome, NITD-451 is most likely targeting a host factor that remains to be determined. The interaction between the compound and its target is specific, as suggested by the observation that only the *S,R,S* enantiomer NITD-451 was active in inhibiting DENV replication and RNA translation, whereas the *R,S,R* enantiomer NITD-452 was inactive in neither assays.

The major concern for targeting host proteins is toxicity. Nevertheless, assuming that upregulation of specific genes in certain cellular pathways is required to ensure productive viral infection, a partial inhibition of the host gene function may be sufficient to disrupt the virus–host interaction while limiting side effects. In agreement with this notion, an *in vivo* efficacy experiment showed that treatment of DENV-infected mice with 25 mg kg^{-1} of NITD-451 reduced peak viremia by about 40-fold without any obvious adverse effects.

3.3.3 Pyrazole Compound: an Inhibitor of Host Pyrimidine Biosynthesis

From the CPE-based HTS campaign mentioned above, another interesting class of compound with an isoxazole-pyrazole core was identified.[35] One such compound, NITD-982 (Scheme 3.3), displayed nanomolar potency against four different RNA virus families, including *Flaviviridae*, *Paramyxoviridae*,

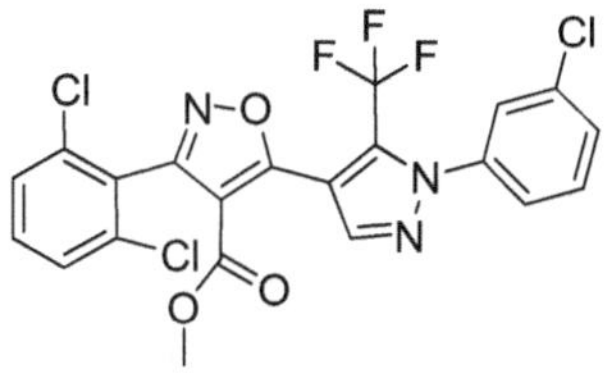

NITD-982 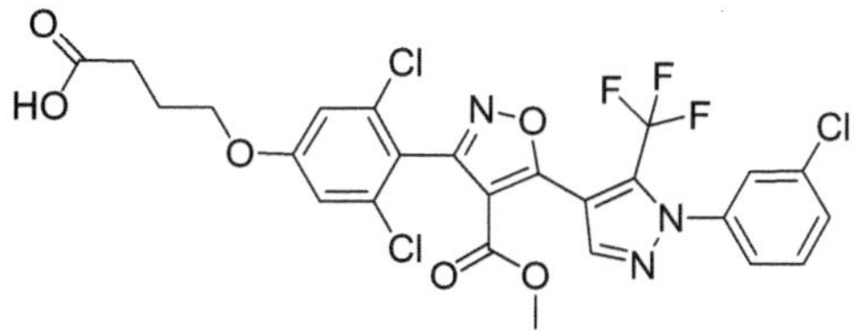   NITD-102

Scheme 3.3

Table 3.4 Antiviral spectrum of NITD-982.

Virus	Strain	EC_{50} (nM)
DENV-2	NGC	2.4
YFV	17b	8.2
HCV	2a	13
RSV	A Long	7.0
HIV	MDR769	0.97
Influenza H1N1	A/PR/8/34	32

Orthomyxoviridae and *Retroviridae* (Table 3.4). The broad-spectrum activity of this compound strongly suggested that a cellular target was involved in the mechanism of action.

The synthesis of NITD-982 starts with the pyrazolecarboxylic acid **13**, which was prepared by condensation of (3-chlorophenyl)hydrazine with ethyl 3-ethoxy-2-(trifluoroacetyl)-2-propenoate followed by hydrolysis under basic conditions.[61] Acid **13** was converted to methyl 3-oxopropenoate **14** in two steps, which was further condensed with 2,6-dichloro-*N*-hydroxybenzenecarboximidoyl chloride in the presence of sodium methoxide to afford NITD-982 (Scheme 3.4).

One of the issues associated with NITD-982 is poor solubility due to its high lipophilicity with a cLog*P* of 7.3. Medicinal chemistry efforts to decrease the lipophilicity or to attach solubilizing groups to the lead compound failed to develop any optimized derivatives that exhibited good potency and desired physicochemical properties. Hence NITD-982 was used as a tool compound to evaluate its biological activities further.

To identify the cellular target(s) through which NITD-982 inhibits viral replication, a three-channel iTRAQ quantitative chemical proteomics technology was implemented.[37] NITD-102 (Scheme 3.3), a bioactive analog of NITD-982, was synthesized and immobilized to affinity capture cellular proteins from lysates of virus-infected cells. The protein identified was dihydroorotate dehydrogenase (DHODH), which is a mitochondrial protein that catalyzes the oxidation of dihydroorotate to orotate, the fourth enzymatic step of the *de novo* pyrimidine biosynthesis. An integral approach of biochemical, biophysical and pharmacological techniques was used for target validation. (i) The compound inhibits the enzymatic activity of the recombinant DHODH. (ii) An analog of NITD-982 directly binds to the recombinant DHODH. (iii) The compound-mediated inhibition of viral replication could be reversed by supplementing the culture medium with uridine. (iv) DENV-2 variants resistant to brequinar (a known DHODH inhibitor) are cross-resistant to NITD-982.

Inhibition of DHODH activity by NITD-982 resulted in depletion of intracellular pyrimidine pools, leading to the suppression of viral RNA synthesis. However, the *in vitro* efficacy did not translate into *in vivo* efficacy. This lack of antiviral efficacy *in vivo* could be attributed to the uridine uptake from diets that replenish/maintain a high concentration of pyrimidine in plasma, therefore counteracting compound-mediated inhibition of viral replication.

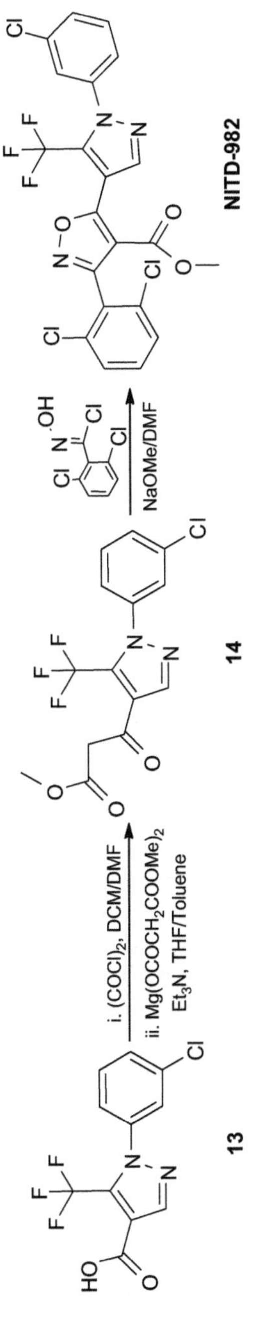
13
i. $(COCl)_2$, DCM/DMF
ii. $Mg(OCOCH_2COOMe)_2$
Et_3N, THF/Toluene
14
NaOMe/DMF
NITD-982

Scheme 3.4

3.4   Discussion

3.4.1   Stratification of Inhibitors of Viral and Cellular Targets

The development of antiviral therapy for DENV represents a major unmet medical need. Cell-based phenotypic screening can be used to identify inhibitors targeting multiple steps and proteins involved in a viral infection cycle. This approach has successfully led to the development of BMS-790052, a potent inhibitor of NS5A that has achieved proof-of-concept in clinical trials for the treatment of HCV infection.[62] A cell-based assay allows for the identification of inhibitors of both viral and cellular targets. Development of viral target inhibitors is relatively more straightforward than that of host target inhibitors. Therefore, it is critical to profile biologically the 'hits' to understand their antiviral mechanisms and targets before a major chemistry resource is committed. Our experience in dengue drug discovery is that the majority of the hits identified from the cellular screens were host target inhibitors; viral target inhibitors were rarely found. This observation could be explained by the fact that the number of host factors required for a productive viral infection cycle is significantly higher than the virally encoded proteins (10 viral proteins in the case of DENV). Consequently, the probability of identifying inhibitors of host targets is much higher than the probability of identifying inhibitors of viral targets. To differentiate viral target inhibitors from cellular target inhibitors, one could filter the primary screening 'hits' using an RNA virus from a different genus (*e.g.*, HCV replicon cell lines) or an RNA virus from a different family (*e.g.*, Venezuelan equine encephalitis virus replicon cell lines). If the compound inhibits both viral replicons, it should be suspected of targeting a host factor that is required for both viruses to replicate. If inhibition of only the DENV replicon is observed, the compound is more likely to be targeting a viral protein or a host protein that is uniquely required for DENV replication.

Targeting a host factor for antiviral development has the advantage of posing a high hurdle for the emergence of resistance. For HCV antiviral development, Debio-025, an inhibitor of host cyclophilin required for HCV replication, showed a much lower rate of emergence of resistance than inhibitors of viral protease and polymerase did in patients and cell culture.[63] Notably, targeting a host factor for antiviral development has two major challenges. The first challenge is potential toxicity. The compound should selectively inhibit the function of the host protein involved in viral replication without affecting its normal cellular function. The second challenge is the feasibility of achieving *in vivo* efficacy under physiological conditions. This challenge is exemplified by the pyrazole compound series, an inhibitor of a host pyrimidine biosynthesis enzyme (Section 3.3.3). Although this inhibitor potently suppressed DENV in cell culture (EC_{50} 2.4 nM), the compound did not show any *in vivo* efficacy in the DENV AG129 (defective in interferon $\alpha/\beta/\gamma$ receptors) mouse model, most likely due to the exogenous uptake of pyrimidine from the diet. The external source of pyrimidine could maintain a high concentration of pyrimidine in plasma, masking the compound-mediated antiviral activity.

Despite the above challenges, maraviroc, an inhibitor of HIV-1 that targets the co-receptor CCR5, has been successfully developed for the treatment of HIV-1-infected patients.[64] Encouragingly, a clinical trial was recently launched in Singapore seeking to establish proof-of-concept for whether celgosivir, an oral prodrug that inhibits host α-glucosidases, therefore blocking DENV morphogenesis,[65–67] could be used as a potential therapy for DENV infection (http://www.celaden.sg/).

3.4.2 Rationale for Dengue Antiviral Therapy

The association between high viremia in the first few days of illness and severe disease outcomes has encouraged an antiviral discovery effort for dengue.[68] These clinical observations led to the current rationale for anti-DENV therapy: a 10-fold reduction of viremia by an antiviral treatment during the early phase of infection would prevent the infected individuals from developing the more severe diseases of DHF/DSS.[27] This hypothesis is frequently challenged by the fact that the duration of viremia is short, just 2–7 days. When patients present at clinics with fever and disease symptoms, they are already in days 3–5 of the infection and viral titers are already beginning to decrease. Unlike chronic HIV and HCV infections, the transient nature of acute viremia associated with DENV makes the window of antiviral treatment extremely narrow. Therefore, the feasibility of antiviral therapy for dengue remains to be clinically tested once a potent clinical candidate becomes available.

Balapiravir, a prodrug of a nucleoside analog (also known as R1479), was the first direct-acting antiviral agent to be tested in clinical trials for the treatment of dengue infection. Treatment of dengue patients recruited within 48 h of the onset of fever did not reduce the kinetics of virological markers, fever clearance time, plasma cytokine concentration or whole blood transcriptional profile.[69] Human pharmacokinetic data showed that the plasma C_{min} barely reached the EC_{50} value derived from DENV cell culture experiments. Although several factors could account for the failure of balapiravir in this dengue trial, one major reason is the weak inhibitory efficacy of balapiravir against DENV. This argument is supported by the clinical data from the balapiravir HCV trial. In the HCV trial, no significant antiviral effect was observed at a dose of 500 mg of balapiravir given BID, where the mean plasma trough was similar to the HCV replicon EC_{50} value, even though C_{max} was more than fivefold above the HCV replicon EC_{50}.[70]

Another direct-acting antiviral candidate for potential dengue clinical trial is a therapeutic antibody. For example, the potent neutralizing, serotype-specific monoclonal antibody HM14c10 was recently reported to protect the AG129 mice from DENV-1 infection at picomolar concentrations.[71] Although this antibody only inhibits DENV-1, it could be used to prove the concept of whether a direct antiviral agent can effectively impact on the disease progression and severity for DENV in a clinical setting.

Besides direct-acting antiviral therapy for dengue, treatments targeting the host immune system have been proposed and pursued. The rationale behind this approach is to suppress the presumed immunological over-activity of the host in response to virus infection. In the 1980s, a few small randomized studies were conducted to evaluate the benefit of corticosteroids in patients (children <15 years old) with DSS. However, the outcomes of these clinical trials were inconclusive, since most of the studies were underpowered and lacked methodological quality.[72] More recently, a randomized, placebo-controlled trail of early corticosteroid therapy in Vietnamese children and young adults was performed.[73] The use of oral prednisolone for 3 days during the early acute phase of virus infection was tolerated, but no reduction in the incidence of severe dengue symptoms was observed.

3.5   Conclusion

With the increasing awareness of the global burden of dengue, the field of dengue antiviral research has been invigorated over the past decade. Phenotypic screening carried out by a number of academic and industrial groups has provided several interesting lead series. Although it is a feasible approach, a sizable investment in medicinal chemistry is often required for hit-to-lead optimization, and also access to a state-of-the-art target identification platform. With patients and physicians waiting for therapies, the effort to develop new antivirals or other therapeutic drugs with activity against dengue virus must continue.

References

1. World Health Organization, *Global Alert and Response (GAR): Impact of Dengue*, http://www.who.int/csr/disease/dengue/impact/en/ (last accessed 19 February 2013).
2. S. J. Thomas and T. P. Endy, *Curr. Opin. Infect. Dis.*, 2011, **24**, 442–450.
3. B. D. Lindenbach, H.-J. Thiel and C. M. Rice, in *Fields Virology*, ed. D. M. Knipe and P.M. Howley, Lippincott William & Wilkins, Philadelphia, PA, 5th edn., 2007, vol. 1, pp. 1101–1152.
4. B. D. Lindenbach and C. M. Rice, *J. Virol.*, 1997, **71**, 9608–9617.
5. B. Falgout, R. H. Miller and C. J. Lai, *J. Virol.*, 1993, **67**, 2034–2042.
6. H. Li, S. Clum, S. You, K. E. Ebner and R. Padmanabhan, *J. Virol.*, 1999, **73**, 3108–3116.
7. P. Warrener, J. K. Tamura and M. S. Collett, *J. Virol.*, 1993, **67**, 989–996.
8. G. Wengler and G. Wengler, *Virology*, 1991, **184**, 707–715.
9. G. Bartelma and R. Padmanabhan, *Virology*, 2002, **299**, 122–132.
10. P. Borowski, A. Niebuhr, O. Mueller, M. Bretner, K. Felczak, T. Kulikowski and H. Schmitz, *J. Virol.*, 2001, **75**, 3220–3229.
11. S. Miller, S. Kastner, J. Krijnse-Locker, S. Buhler and R. Bartenschlager, *J. Biol. Chem.*, 2007, **282**, 8873–8882.
12. J. Roosendaal, E. G. Westaway, A. Khromykh and J. M. Mackenzie, *J. Virol.*, 2006, **80**, 4623–4632.

13. J. M. Mackenzie, A. A. Khromykh, M. K. Jones and E. G. Westaway, *Virology*, 1998, **245**, 203–215.
14. M. P. Egloff, D. Benarroch, B. Selisko, J. L. Romette and B. Canard, *EMBO J.*, 2002, **21**, 2757–2768.
15. D. Ray, A. Shah, M. Tilgner, Y. Guo, Y. Zhao, H. Dong, T. S. Deas, Y. Zhou, H. Li and P. Y. Shi, *J. Virol.*, 2006, **80**, 8362–8370.
16. Y. Zhou, D. Ray, Y. Zhao, H. Dong, S. Ren, Z. Li, Y. Guo, K. A. Bernard, P. Y. Shi and H. Li, *J. Virol.*, 2007, **81**, 3891–3903.
17. K. J. Guyatt, E. G. Westaway and A. A. Khromykh, *J. Virol. Methods*, 2001, **92**, 37–44.
18. M. Ackermann and R. Padmanabhan, *J. Biol. Chem.*, 2001, **276**, 39926–39937.
19. K. M. Chung, M. K. Liszewski, G. Nybakken, A. E. Davis, R. R. Townsend, D. H. Fremont, J. P. Atkinson and M. S. Diamond, *Proc. Natl. Acad. Sci. U. S. A.*, 2006, **103**, 19111–19116.
20. J. T. Guo, J. Hayashi and C. Seeger, *J. Virol.*, 2005, **79**, 1343–1350.
21. W. J. Liu, X. J. Wang, V. V. Mokhonov, P. Y. Shi, R. Randall and A. A. Khromykh, *J. Virol.*, 2005, **79**, 1934–1942.
22. J. L. Munoz-Jordan, M. Laurent-Rolle, J. Ashour, L. Martinez-Sobrido, M. Ashok, W. I. Lipkin and A. Garcia-Sastre, *J. Virol.*, 2005, **79**, 8004–8013.
23. J. L. Munoz-Jordan, G. G. Sanchez-Burgos, M. Laurent-Rolle and A. Garcia-Sastre, *Proc. Natl. Acad. Sci. U. S. A.*, 2003, **100**, 14333–14338.
24. S. M. Best, K. L. Morris, J. G. Shannon, S. J. Robertson, D. N. Mitzel, G. S. Park, E. Boer, J. B. Wolfinbarger and M. E. Bloom, *J. Virol.*, 2005, **79**, 12828–12839.
25. M. Laurent-Rolle, E. F. Boer, K. J. Lubick, J. B. Wolfinbarger, A. B. Carmody, B. Rockx, W. Liu, J. Ashour, W. L. Shupert, M. R. Holbrook, A. D. Barrett, P. W. Mason, M. E. Bloom, A. Garcia-Sastre, A. A. Khromykh and S. M. Best, *J. Virol.*, 2010, **84**, 3503–3515.
26. J. Ashour, M. Laurent-Rolle, P. Y. Shi and A. Garcia-Sastre, *J. Virol.*, 2009, **83**, 5408–5418.
27. C. G. Noble, Y. L. Chen, H. Dong, F. Gu, S. P. Lim, W. Schul, Q. Y. Wang and P. Y. Shi, *Antiviral Res.*, 2010, **85**, 450–462.
28. S. Elshuber, S. L. Allison, F. X. Heinz and C. W. Mandl, *J. Gen. Virol.*, 2003, **84**, 183–191.
29. K. Stadler, S. L. Allison, J. Schalich and F. X. Heinz, *J. Virol.*, 1997, **71**, 8475–8481.
30. M. P. Courageot, M. P. Frenkiel, C. Duarte Dos Santos, V. Deubel and P. Despres, *J. Virol.*, 2000, **74**, 564–572.
31. J. J. H. Chu and P. L. Yang, *Proc. Natl. Acad. Sci. U. S. A.*, 2007, **104**, 3520–3525.
32. A. J. Hirsch, G. R. Medigeshi, H. L. Meyers, V. DeFilippis, K. Fruh, T. Briese, W. I. Lipkin and J. A. Nelson, *J. Virol.*, 2005, **79**, 11943–11951.
33. J. M. Mackenzie, A. A. Khromykh and R. G. Parton, *Cell Host Microbe*, 2007, **2**, 229–239.

34. C. Rothwell, A. LeBreton, C. Young, Ng, J. Y. H. Lim, W. Liu, S. Vasudevan, M. Labow, F. Gu and L. A. Gaither, *Virology*, 1920, **389**, 8–19.
35. Q. Y. Wang, S. Bushell, M. Qing, H. Y. Xu, A. Bonavia, S. Nunes, J. Zhou, M. K. Poh, P. Florez de Sessions, P. Niyomrattanakit, H. Dong, K. Hoffmaster, A. Goh, S. Nilar, W. Schul, S. Jones, L. Kramer, T. Compton and P. Y. Shi, *J. Virol.*, 2011, **85**, 6548–6556.
36. M. Qing, G. Zou, Q. Y. Wang, H. Y. Xu, H. Dong, Z. Yuan and P. Y. Shi, *Antimicrob. Agents Chemother.*, 2010, **54**, 3686–3695.
37. A. Bonavia, M. Franti, E. Pusateri Keaney, K. Kuhen, M. Seepersaud, B. Radetich, J. Shao, A. Honda, J. Dewhurst, K. Balabanis, J. Monroe, K. Wolff, C. Osborne, L. Lanieri, K. Hoffmaster, J. Amin, J. Markovits, M. Broome, E. Skuba, I. Cornella-Taracido, G. Joberty, T. Bouwmeester, L. Hamann, J. A. Tallarico, R. Tommasi, T. Compton and S. M. Bushell, *Proc. Natl. Acad. Sci. U. S. A.*, 2011, **108**, 6739–6744.
38. W. Fischl and R. Bartenschlager, *Curr. Opin. Microbiol.*, 2011, **14**, 470–475.
39. Y. Modis, S. Ogata, D. Clements and S. C. Harrison, *Proc. Natl. Acad. Sci. U. S. A.*, 2003, **100**, 6986–6991.
40. Y. Modis, S. Ogata, D. Clements and S. C. Harrison, *Nature*, 2004, **427**, 313–319.
41. L. Ma, C. T. Jones, T. D. Groesch, R. J. Kuhn and C. B. Post, *Proc. Natl. Acad. Sci. U. S. A.*, 2004, **101**, 3414–3419.
42. L. Li, S. M. Lok, I. M. Yu, Y. Zhang, R. J. Kuhn, J. Chen and M. G. Rossmann, *Science*, 2008, **319**, 1830–1834.
43. D. Luo, T. Xu, C. Hunke, G. Gruber, S. G. Vasudevan and J. Lescar, *J. Virol.*, 2008, **82**, 173–183.
44. T. L. Yap, T. Xu, Y. L. Chen, H. Malet, M. P. Egloff, B. Canard, S. G. Vasudevan and J. Lescar, *J. Virol.*, 2007, **81**, 4753–4765.
45. M. P. Egloff, D. Benarroch, B. Selisko, J. L. Romette and B. Canard, *EMBO J.*, 2002, **21**, 2757–2768.
46. Q.-Y. Wang, Y.-L. Chen, S. P. Lim and P.-Y. Shi, in *Molecular Virology and Control of Flaviviruses*, ed. P.-Y. Shi, Caister Academic Press, Norwich, 2012, pp. 257–270.
47. G. Zou, H. Y. Xu, M. Qing, Q. Y. Wang and P. Y. Shi, *Antiviral Res.*, 2011, **91**, 11–19.
48. C. Y. Ng, F. Gu, W. Y. Phong, Y. L. Chen, S. P. Lim, A. Davidson and S. G. Vasudevan, *Antiviral Res.*, 2007, **76**, 222–231.
49. F. Puig-Basagoiti, M. Tilgner, B. M. Forshey, S. M. Philpott, N. G. Espina, D. E. Wentworth, S. J. Goebel, P. S. Masters, B. Falgout, P. Ren, D. M. Ferguson and P. Y. Shi, *Antimicrob. Agents Chemother.*, 2006, **50**, 1320–1329.
50. M. Qing, W. Liu, Z. Yuan, F. Gu and P. Y. Shi, *Antiviral Res.*, 2010, **86**, 163–171.
51. C. Ansarah-Sobrinho, S. Nelson, C. A. Jost, S. S. Whitehead and T. C. Pierson, *Virology*, 2008, **381**, 67–74.

52. S. M. Huang, Y. M. Mishina, S. Liu, A. Cheung, F. Stegmeier, G. A. Michaud, O. Charlat, E. Wiellette, Y. Zhang, S. Wiessner, M. Hild, X. Shi, C. J. Wilson, C. Mickanin, V. Myer, A. Fazal, R. Tomlinson, F. Serluca, W. Shao, H. Cheng, M. Shultz, C. Rau, M. Schirle, J. Schlegl, S. Ghidelli, S. Fawell, C. Lu, D. Curtis, M. W. Kirschner, C. Lengauer, P. M. Finan, J. A. Tallarico, T. Bouwmeester, J. A. Porter, A. Bauer and F. Cong, *Nature*, 2009, **461**, 614–620.

53. X. Xie, Q. Y. Wang, H. Y. Xu, M. Qing, L. Kramer, Z. Yuan and P. Y. Shi, *J.Virol.*, 2011, **85**, 11183–11195.

54. M. P. Gleeson, *J. Med. Chem.*, 2008, **51**, 817–834.

55. R. Subramanian, M. R. Lee, J. G. Allen, M. P. Bourbeau, C. Fotsch, F. T. Hong, S. Tadesse, G. Yao, C. C. Yuan, S. Surapaneni, G. L. Skiles, X. Wang, G. E. Wohlhieter, Q. Zeng, Y. Zhou, X. Zhu and C. Li, *Chem. Res. Toxicol.*, 2010, **23**, 653–663.

56. C. J. Helal, Z. Kang, J. C. Lucas, T. Gant, M. K. Ahlijanian, J. B. Schachter, K. E. G. Richter, J. M. Cook, F. S. Menniti, K. Kelly, S. Mente, J. Pandit and N. Hosea, *Bioorg. Med. Chem. Lett.*, 2009, **19**, 5703–5707.

57. A. S. Antipas, L. C. Blumberg, W. H. Brissette, M. F. Brown, J. M. Casavant, J. L. Doty, J. Driscoll, T. M. Harris, C. S. Jones, S. P. McCurdy, E. McElroy, M. Mitton-Fry, M. J. Munchhof, D. A. Reim, L. A. Reiter, S. L. Ripp, A. Shavnya, M. I. Smeets and K. A. Trevena, *Bioorg. Med. Chem. Lett.*, 2010, **20**, 4069–4072.

58. Q. Y. Wang, R. R. Kondreddi, X. Xie, R. Rao, S. Nilar, H. Y. Xu, M. Qing, D. Chang, H. Dong, F. Yokokawa, S. B. Lakshminarayana, A. Goh, W. Schul, L. Kramer, T. H. Keller and P. Y. Shi, *Antimicrob. Agents Chemother.*, 2011, **55**, 4072–4080.

59. Q. Y. Wang, S. J. Patel, E. Vangrevelinghe, H. Y. Xu, R. Rao, D. Jaber, W. Schul, F. Gu, O. Heudi, N. L. Ma, M. K. Poh, W. Y. Phong, T. H. Keller, E. Jacoby and S. G. Vasudevan, *Antimicrob. Agents Chemother.*, 2009, **53**, 1823–1831.

60. N. Yokoyama, P. I. Almaula, F. B. Block, F. R. Granat, N. Gottfried, R. T. Hill, E. H. McMahon, W. F. Munch and H. Rachlin, *J. Med. Chem.*, 1979, **22**, 537–553.

61. J. R. Beck and F. L. Wright, *J. Heterocycl. Chem.*, 1987, **24**, 739–740.

62. M. Gao, R. E. Nettles, M. Belema, L. B. Snyder, V. N. Nguyen, R. A. Fridell, M. H. Serrano-Wu, D. R. Langley, J. H. Sun, D. R. Boyle II, J. A. Lemm, C. Wang, J. O. Knipe, C. Chien, R. J. Colonno, D. M. Grasela, N. A. Meanwell and L. G. Hamann, *Nature*, 2010, **465**, 96–100.

63. J. M. Robida, H. B. Nelson, Z. Liu and H. Tang, *J. Virol.*, 2007, **81**, 5829–5840.

64. S. Sayana and H. Khanlou, *Expert Rev. Anti Infect. Ther.*, 2009, **7**, 9–19.

65. A. P. S. Rathore, P. N. Paradkar, S. Watanabe, K. H. Tan, C. Sung, J. E. Connolly, J. Low, E. E. Ooi and S. G. Vasudevan, *Antiviral Res.*, 2011, **92**, 453–460.

66. S. Watanabe, A. P. S. Rathore, C. Sung, F. Lu, Y. M. Khoo, J. Connolly, J. Low, E. E. Ooi, H. S. Lee and S. G. Vasudevan, *Antiviral Res.*, 2012, **96**, 32–35.

67. K. Whitby, T. C. Pierson, B. Geiss, K. Lane, M. Engle, Y. Zhou, R. W. Doms and M. S. Diamond, *J. Virol.*, 2005, **79**, 8698–8706.

68. D. H. Libraty, P. R. Young, D. Pickering, T. P. Endy, S. Kalayanarooj, S. Green, D. W. Vaughn, A. Nisalak, F. A. Ennis and A. L. Rothman, *J. Infect. Dis.*, 2002, **186**, 1165–1168.

69. N. M. Nguyet, T. N. B. Chau, P. K. Lam, D. T. H. Kien, H. L. A. Huy, J. Farrar, T. H. Quyen, T. T. Hien, N. Van Vinh Chau, L. Merson, H. T. Long, M. L. Hibberd, P. P. K. Aw, A. Wilm, N. Nagarajan, N. T. Dung, P. P. Mai, N. T. Truong, H. Javanbaht, K. Klumpp, J. Hammond, R. Petric, M. Wolbers, N. T. Chinh and C. P. Simmons, *J. Infect. Dis.*, 2012, doi: 10.1093/infdis/jis470.

70. K. Klumpp and D. B. Smith, in *Antiviral Drugs*, ed. W. M. Kazmierski, Wiley, New York, 2011, pp. 287–304.

71. E. P. Teoh, P. Kukkaro, E. W. Teo, A. P. C. Lim, T. T. Tan, A. Yip, W. Schul, M. Aung, V. A. Kostyuchenko, Y. S. Leo, S. H. Chan, K. G. C. Smith, A. H. Y. Chan, G. Zou, E. E. Ooi, D. M. Kemeny, G. K. Tan, J. K. W. Ng, M. L. Ng, S. Alonso, D. Fisher, P. Y. Shi, B. J. Hanson, S. M. Lok and P. A. MacAry, *Sci. Transl. Med.*, 2012, **4**, 139ra83.

72. S. Rajapakse, *Trans. R. Soc. Trop. Med. Hyg.*, 2009, **103**, 122–126.

73. D. T. H. Tam, T. V. Ngoc, N. T. H. Tien, N. T. T. Kieu, T. T. T. Thuy, L. T. C. Thanh, C. T. Tam, N. T. Truong, N. T. Dung, P. T. Qui, T. T. Hien, J. J. Farrar, C. P. Simmons, M. Wolbers and B. A. Wills, *Clin. Infect. Dis.*, 2012, **55**, 1216–1224.

CHAPTER 4

Discovery and Development of Antiviral Drugs for Treatment of Pathogenic Human Orthopoxvirus Infections

ROBERT JORDAN

Gilead Sciences, Inc., 333 Lakeside Drive, Foster City, CA 94404, USA
Email: robert.jordan@gilead.com

4.1 Introduction

Smallpox was a devastating disease, with epidemics that swept through nations altering the course of history and affecting the lives of millions. Smallpox caused the death of Queen Mary II of England, Emperor Joseph I of Austria, King Luis I of Spain, Tsar Peter II of Russia, Queen Ulrika Elenora of Sweden and King Louis XV of France.[1] Prior to the discovery of vaccination, it is estimated that smallpox killed nearly 30% of the human population and those who survived were often left disfigured or blind.[2] It is hard to imagine today the effect that this scourge had on humanity.

Smallpox can be characterized as a severe systemic xanthematous disease caused by infection with variola virus, an orthopoxvirus (OPV) that is unique to humans with no known natural animal reservoir.[2] Variola major, the most common form of the virus, was associated with more severe disease and higher mortality rates compared with variola minor.[2] The selective tropism of variola virus proved critical for the eradication campaign sponsored by the World

RSC Drug Discovery Series No. 32
Successful Strategies for the Discovery of Antiviral Drugs
Edited by Manoj C. Desai and Nicholas A. Meanwell
© The Royal Society of Chemistry 2013
Published by the Royal Society of Chemistry, www.rsc.org

Health Organization (WHO) that used a live attenuated OPV (vaccinia virus) to vaccinate the general population and prevent new infections. The original smallpox vaccine was derived from the work of Edward Jenner in 1798, who discovered that scab material from cowpox lesions on dairy cows could be used to inoculate naive individuals who became immune to infection by variola virus.[3] Development of the smallpox vaccine was a major public health triumph, resulting in dramatic reductions in deaths caused by smallpox and ultimately its eradication.[2] The last known case of smallpox occurred in Somalia in 1977, and by 1980 the WHO declared that the virus had been eliminated from the environment.[2]

Although smallpox no longer exists in nature, variola virus isolates obtained during the eradication campaign have been retained and are stored in high-containment facilities located at the Centers for Disease Control and Prevention in Atlanta, GA, USA and at the State Research Center of Virology and Biotechnology in Koltsovo, Russia. The international community has lobbied hard to destroy the remaining virus stocks to safeguard the world from potential release of virus into the environment. However, some nations cite the threat of bioterrorism as a reason to continue research on this virus.[4] This threat has taken on new significance based upon recent allegations from a former deputy director of the Soviet Union's civilian bioweapons program reporting that since 1980, the Soviet government weaponized smallpox for use in bombs and intercontinental ballistic missiles. According to this source, the program was extensive and could produce tons of weaponized smallpox virus annually.[5] Moreover, it is feared that virus stocks produced during this period cannot be reliably accounted for, raising the possibility of their continued existence. Even if live virus stocks are not available, the sequence of the variola virus genome is publicly accessible, raising the possibility that the virus could be reconstructed from parts, as was demonstrated with poliovirus.[6] Based on this information, the US Institute of Medicine issued a report calling for the development of two antiviral compounds that act by different mechanisms for the treatment of OPV infections to complement our existing stocks of smallpox vaccine.[7]

Variola virus is not the only OPV that can infect humans. Three other OPV species (monkeypox, cowpox and vaccinia-like viruses) continue to circulate in the environment and cause sporadic human disease. The most virulent of these is monkeypox virus, which is endemic in some areas of the Democratic Republic of the Congo (DRC).[8] Infection with monkeypox virus causes a systemic disease that resembles a milder version of smallpox, with a mortality rate of 1–5%.[9] Unlike variola virus, monkeypox virus is less transmissible and outbreaks are short-lived and with limited human-to-human spread.[2,9] Cowpox virus and vaccinia-like viruses cause less severe disease, with fewer lesions and with little to no mortality. In certain parts of South America, a vaccinia-like virus termed Cantagalo virus, which shares phenotypic and genotypic features of the smallpox vaccine strain that was used in Brazil during the eradication campaign, causes localized outbreaks in dairy workers.[10] In cattle, the disease is characterized by pustular lesions on the teats and udders and is accompanied by fever and occasionally secondary mastitis. Dairy workers become infected during milking, leading to the appearance of localized lesions at the site of contact, usually on the hands and arms, lymphadenopathy, fever and prostration.[11] Cowpox virus, which is endemic in certain regions of Europe, causes a disease similar to Cantagalo virus, with low mortality rates, fever and localized lesions on the skin.[12] All three OPVs are maintained in the environment through a rodent reservoir and are transmitted to humans by contact with infected animals or through an intermediate species such as cattle or domestic pets.[9,12,13] Exposure rates vary and one study, conducted in Ghana, showed that OPV antibodies could be detected in 53% of the people living in proximity to forest-dwelling rodent populations.[14] Interestingly, variola virus is thought to have evolved from a rodent virus between 16 000 and 68 000 years ago.[15,16]

Other species of OPVs that are genetically related to variola virus, such as camelpox and ectromelia viruses, can cause severe systemic disease in their natural hosts (camels and mice, respectively), but have not been found to infect humans. The genetic susceptibility of OPVs to a particular host has been attributed to the acquisition and adaptive evolution of host response modifier genes.[17,18] These genes are often found to be virulence factors that down-regulate the host immune response and facilitate systemic virus spread.[19] Phylogenic analyses have shown that many of these genes are undergoing positive selection, suggesting that OPVs are continuing to evolve and increasing the likelihood of zoonotic transmission and appearance of variants with altered virulence.[17] Although smallpox is no longer a disease found in humans, the possibility exists that new variants of circulating OPVs may emerge to cause more frequent human disease.

4.2 Natural History of Human OPV Infections

The life cycle of variola virus, like other OPVs, begins with virus replication at the site of entry followed by systemic spread to distal sites.[20] Disease severity is related to the ability of a particular OPV to spread in the host and can be influenced by many factors that include the amount of virus entering the host, route of entry and host response to infection.[20] Clinical observations made during smallpox epidemics have defined four major types of disease associated with variola virus infection characterized by the morphology of the virus-specific lesion and severity of disease symptoms. Ordinary smallpox is characterized by raised pustular skin lesions that can be confluent or discrete, with a mortality rate of $\sim 30\%$. Variola sine eruptione is characterized by fever without rash, requiring serological analysis to confirm diagnosis. Flat-type and hemorrhagic smallpox are the most severe, with mortality rates of 97%, and are characterized by confluent flat pustules and widespread hemorrhages in the skin and mucous membranes, respectively.[2]

Orthopoxviruses are large, double-stranded DNA viruses whose genomes vary in size from 145 to 290 kb.[17] Unlike most DNA viruses that replicate in the nucleus, OPVs replicate exclusively in the cytoplasm of infected cells. Vaccinia virus is the prototype OPV and has been used to study many aspects of the virus replication cycle (reviewed by Fields *et al.*[21]). Vaccinia virus can infect many mammalian cells types, suggesting the presence of a ubiquitous receptor. The virus binds to the cell through interactions with cell surface glycosaminoglycans and enters by fusion of the viral membrane with the host cell membrane to release the viral nucleoprotein core into the cytoplasm. The core contains a virus-specific RNA polymerase that catalyzes expression of early genes encoding proteins required for the establishment of 'virus factories' where viral genome replication takes place. Viral structural proteins are synthesized after genomic replication is completed. Newly replicated virus genomes are condensed into core particles which are formed after a series of temporally regulated proteolytic cleavage events of viral core proteins. The core particles are enveloped by intracellular membranes that appear as crescent-shaped structures in the cytoplasm. The enveloped virus, termed intracellular virus (IV), is infectious and is released upon cell lysis.

Whereas IV particles compose the majority of the virus produced during productive infection, a small fraction of the virus particles, ~10%, are transported to the cell surface and released as extracellular virus (EV).[22] Production of EV particles requires a unique set of virus gene products that wrap IV particles in additional virus-modified membranes derived from the late endosome compartment and the trans Golgi network.[23,24] The wrapped virus is transported to the cell surface on microtubules and released into the extracellular space.[25,26] Newly released EV particles contain an additional envelope with different antigenic properties that makes the virus less susceptible to complement-mediated neutralization.[27,28] Virus variants containing mutations in the genes required for EV formation produce small plaques on cell monolayers and are attenuated for virus spread *in vivo*, suggesting that EV particles are responsible for efficient cell-to-cell spread and long-range dissemination of the virus in the host.[29–32]

The natural history of smallpox has been described from clinical observations in humans and inferred from animal studies, especially those conducted in mice infected with ectromelia virus.[2,33] An in-depth description of clinical disease caused by variola virus infection has been described in the literature.[2] Variola virus enters the respiratory tract *via* aerosolized droplets, seeding mucous membranes and passing rapidly into local lymph nodes where virus replication in the local lymph tissue results in a primary viremia. Virus released during the primary viremia circulates and seeds the spleen, liver and reticuloendothelial system, where continued replication results in a secondary viremia. During this period of intense virus replication, the host response to infection is limited and the patient remains asymptomatic.

Clinical latency ends with the rapid onset of severe headache, backache and fever, termed the prodromal phase. The prodromal phase correlates with a secondary viremia in which infectious virus can be detected in the mucous membranes of the mouth and pharynx. Patients are contagious, transmitting virus *via* aerosolized droplets to new hosts. The virus invades the capillary epithelium of the dermal layer in skin, perivascular cells and epidermis where replication results in necrosis and the formation of a rash which develops into pustules, which then form scabs causing the discrete pocks that are the hallmark of OPV infections. Virus can be cultured from the scab material and the patient remains infectious until the scabs fall off and the skin heals. Mortality in humans has been attributed to toxemia, associated with immune complexes and hypotension. Toxemia is a poorly defined clinical condition thought to be caused by an excessive inflammatory immune response similar to septicemia associated with systemic bacterial infections.

4.3 Antiviral Discovery and Development

4.3.1 Regulatory Path to Developing OPV Therapeutics

The regulatory path for approval of therapeutics to treat smallpox is complicated by the lack of human disease. Since it is not possible to conduct human efficacy trials with new therapeutics that target variola virus, it is difficult to define pharmacokinetic–pharmacodynamic (PK–PD) relationships for human dose selection. Biomarkers based upon virological endpoints are not available for predicting disease outcome since current molecular markers and techniques used to describe human disease had not been developed prior to smallpox eradication.

The US Food and Drug Administration (FDA) proposed guidance (FDA regulation 21 CFR 314 Subpart I) designed to evaluate new drug products when human efficacy studies are not ethical or feasible. The intent of this guidance was to provide a mechanism for approval of drugs to treat diseases caused by pathogens considered to be potential bioweapon threats. Drug candidates would follow the normal regulatory path for approval through Phase 1 and modified Phase 2 safety evaluations in healthy volunteers. However, in place of human efficacy trials, the guidance proposed the use of validated animal models to establish PK–PD relationships to facilitate human dose selection (Table 4.1). Although there are numerous models of OPV infection, none of these models recapitulates all aspects of human disease. Therefore, interpretation of this guidance has been challenging for companies developing therapeutics in this area. A recent FDA advisory committee meeting addressing this topic will hopefully bring clarity to this issue.[34] Alternative regulatory paths, such as conditional approvals or emergency use authorization, may allow the distribution of therapeutics in the event of an outbreak. The development of appropriate validated animal models will be essential for approval of smallpox therapeutics.

4.3.2 Animal Models of OPV Infection

The current CDC clinical case definition for smallpox is given as 'An illness with acute onset of fever >101 °F (38.3 °C) followed by a rash characterized by firm, deep seated vesicles or pustules in the same stage of development without other apparent cause.'[35] Polymerase chain reaction (PCR) using primers specific for variola virus is used to confirm the diagnosis. The CDC case

Table 4.1 Animal and human data that can be used to support human dose selection.[a]

Drug	Route	Animal model	Minimum effective dose (qd) (mg kg⁻¹)	AUC (h ng mL⁻¹)	Active species	Active species concentration[b]	Human dose (mg)	Human AUC (h ng mL⁻¹)	Ref.
CDV	i.v.	NHP/MPX	20	197	CDVpp	$52.9–75.6\ pg/10^6$	150	1740	49
CMX001	p.o.	Rabbit/RPX	20	179	CDVpp	$52.1\ pg/10^6$	150	2650	50
ST-246	p.o.	NHP/MPX	3	4530	ST-246	N.A.[c]	400	21387	103

[a] A comparison of the animal and human PK parameters can be used to link efficacy in animal models of OPV infection with exposure levels in humans that would be predicted to be antiviral. For CDV and CMX001, correlating the active metabolite, CDVpp, in PBMC serves as a biomarker for efficacy. For ST-246 the AUC correlates with efficacy in animal studies and can be used to link drug exposure to predicted outcome. The exposure level for ST-246 in humans is almost five times greater than the exposure level in non-human primates receiving the minimum effective dose that protected >95% of non-human primates from lethal monkeypox infection.

[b] The concentration of CDVpp (active species) for CDV and CMX001 was measured in isolated peripheral blood monocyte cells (PBMC). The concentration of CDVpp has not been measured in human PBMC.

[c] N.A.: not applicable.

definition for smallpox provides a framework for developing animal models of OPV infection that can be used to evaluate effectiveness of new OPV therapeutics. Lesion formation, viremia and mortality are clinically relevant endpoints that can be used to evaluate the efficacy of OPV therapeutics in animal models. Although these endpoints can be used to evaluate the efficacy of therapeutics in animal model systems, their predictive value for disease outcome in humans has not been determined.

The most relevant animal models of OPV infection involve the use of host-adapted virus inoculated at a peripheral site to establish infection that spreads to distal sites in a manner similar to variola virus infection of humans. Host adaptations are defined as genetic alterations that result in increased virulence for a specific host. These changes may include evolutionary adaptations, such as acquisition of host response modifier functions or unknown changes that result in increased virulence. Models that use host-adapted virus in a natural host are appealing because virus replication in target tissue and systemic spread is determined by the host response to infection. Some animal models of orthopoxvirus infection have been developed using non-host-adapted OPVs to establish disease. These models require inoculation of animals with large quantities of virus, sometimes delivered by unnatural routes. Delivering large amounts of virus by unnatural routes alters pathogenesis by allowing virus access to different tissues and stimulating a non-natural host response.

A number of models of OPV infection have been developed in different animal species such as mice, including BALB/c, NMRI, ANC/R and Nu/nu strains, rabbits, prairie dogs and ground squirrels (reviewed by Chapman *et al.*[36] and Smee[37]). These models have been used to evaluate the antiviral activity of OPV therapeutics against multiple species of orthopoxviruses, including vaccinia virus strains IHD-J, Lister and WR, ectromelia virus, strain Moscow, cowpox virus, rabbitpox virus and monkeypox virus.[37] Infections were established by a variety of routes, including intranasal, intravenous, intradermal, subcutaneous and aerosol delivery of virus.

4.3.2.1 Vaccinia and Cowpox Virus Mouse Models

Mouse models using vaccinia virus or cowpox virus have been developed to measure the efficacy of anti-poxvirus compounds.[37] The pathogenesis of infection is dictated by the route of viral entry and models using intracranial, intravenous or intranasal inoculation have been described.[37] Intranasal inoculation of mice with vaccinia virus or cowpox virus produces local replication in the nasal tissue and systemic spread of the virus to distal sites, providing a system capable of assessing antiviral activity of compounds that inhibit multiple steps in the virus life cycle. Infections are established by inoculation of mice with 1×10^4–1×10^6 plaque-forming units (PFUs) of virus in a small volume (10 µL) to each naris. Although a large dose of virus is required to establish lethal infection, replication of the virus starts locally in the nasal tissue and lungs before spreading systemically through the reticuloendothelial system.[38] High levels of virus can be found in liver, spleen, lung and kidney.[38] Mice begin to lose weight and become lethargic by day 4 post-infection. As the disease progresses, mice continue to lose weight and their general appearance declines, with most animals exhibiting signs of severe disease such as ruffled fur, hunched posture and unsteady gait. By day 8 post-inoculation, mice are moribund and have lost as much as 30% of their initial body weight.

Mortality is the primary endpoint in this model, with 100% of mice succumbing to infection by day 10 post-inoculation. Disease progression can be monitored by measuring the change in weight during the course of infection. The change in weight relative to placebo-treated animals is a quantifiable, non-invasive method of measuring disease severity and correlates with systemic replication of virus in mice. The percentage weight change is useful for determining disease severity when treatment protects mice from lethal infection. Thus, the efficacy of compounds that prevent mortality but do not completely inhibit virus replication and all aspects of disease progression can be assessed in this model. To quantify the level of virus spread, animals must be sacrificed at selected time points post-infection and virus titers measured in the liver, spleen, lung, kidney and other organs. Antiviral efficacy is measured by the decreased mortality, inhibition of virus-induced weight loss and reduction in viral titers in liver, spleen, lung, kidney and other tissues.

4.3.2.2   Ectromelia Virus Mouse Model

Ectromelia virus is a laboratory pathogen of mice that has been used as a model system of OPV disease pathogenesis.[33] The pathogenesis of ectromelia virus disease closely resembles human smallpox, with distinct stages of localized replication, systemic virus spread and lesion formation; however, the time course of infection and disease progression is much shorter.[33,39] Unlike vaccinia and cowpox viruses, small amounts of virus delivered to peripheral sites can initiate a lethal infection in susceptible mice.

Ectromelia virus enters through abrasions in the skin and replicates in local lymphoid cells.[33,40] Virus can be detected in these tissues by immunofluorescence within a few hours after inoculation.[40,41] The virus multiplies in the lymphatic endothelial cells, macrophages and lymphocytes within the node over a period of 2–4 days.[33] Following this latency period, the virus spreads through the lymph and enters the bloodstream to cause a primary viremia. The virus is rapidly removed by macrophages lining the sinusoids of the liver, spleen and bone marrow.[41] Infection of the parenchymal cells of liver and lymphoid cells of spleen produces high virus titers that are released into the blood stream to cause a secondary viremia.[40] In highly susceptible animals, replication in liver and spleen produces focal necrotic lesions, acute hepatitis and multi-organ failure. In mice that are less susceptible to infection, a rash develops following the secondary viremia.[39] The rash is caused by virus replication in the perivascular cells and dermal endothelial cells and epidermis.

A lethal ectromelia virus mouse model has been established to evaluate the efficacy of antiviral drugs.[42] In susceptible mice, as little as 1 PFU of ectromelia virus causes lethal infection in >95% of animals.[33] Mice are inoculated with ectromelia virus either by footpad scarification, which is similar to the natural route of infection, or intranasal delivery.[39] The virus multiplies in the lymphatic endothelial cells, macrophages and lymphocytes within the regional node over a period of 2–4 days. By day 4 post-inoculation, animals appear ill, with hunched posture, ruffled coat and increased respiration. Viral replication in liver and spleen and other internal organs causes death in the infected animal between days 6 and 10 post-inoculation.[39] Like the vaccinia virus and cowpox models, antiviral efficacy is measured by decreased mortality, inhibition of virus-induced weight loss and reduction in viral titers in liver, spleen and other tissues.

4.3.2.3 Rabbitpox Virus Model

Rabbitpox virus is genetically related to vaccinia virus but highly adapted to replicate in rabbits, and as little as 15 PFU can establish productive infection in most rabbit species.[43] Rabbits are inoculated with rabbitpox virus in the footpad by intradermal injection or the nasal cavity by aerosol spray. Virus replicates in the mucosa or local lymph tissue to produce a primary viremia which lasts 2–4 days.[44] Animals develop fever within the first few days of infection and focal lesions appear on the ear by day 3 post-infection and can be visualized as small red spots near blood vessels visualized by backlighting.[44] The virus spreads through the lymph to the blood, ultimately seeding lung, liver, spleen and other internal organs. Virus replication at these sites often results in multi-organ failure and fatal disease.[43] Rabbits that survive infection of internal organs develop a secondary viremia where high levels of circulating virus seed the endothelial cells that line the dermal blood vessels to produce a rash on the skin that appears by day 6 post-inoculation. Vertical transmission of virus to susceptible rabbits has been observed, allowing for potential assessment of drug effects on virus spread within a population.[43]

To establish a lethal infection, rabbits are inoculated with 100 PFU of rabbitpox virus by dermal abrasion or intranasal administration through aerosol delivery to the respiratory tract. Dermal lesions appear near blood vessels in the ears between days 3 and 5 post-infection and can be visualized by backlighting. The appearance of these lesions provides a therapeutic trigger for initiation of antiviral treatment. By day 6 post-inoculation, animals develop fever, listlessness and purulent discharges from the eyes and nose.[44] Most animals experience respiratory distress by this time in the infection. A rash develops between days 6 and 8 post-inoculation; however, skin lesions range from a few scattered lesions to confluency. Most animals die without developing a rash and death is accompanied by a fall in body temperature to below normal levels. Thus, quantifying lesion number or severity of the rash is a subjective measure of systemic virus spread and may not be possible in all cases. Since rabbits can tolerate more frequent and larger volume blood draws, blood chemistries can be measured to correlate changes in hematological status with disease progression. To quantify the level of virus spread, animals are sacrificed at selected time points post-infection and virus titers are measured in the liver, spleen, lung, kidney and other organs. Antiviral efficacy is assessed by measuring decreased mortality, reduced virus titers in organs and improvement of hematological status. Histological examination of tissue from sacrificed animals further defines the effects of antiviral compounds on viral pathogenesis.

4.3.2.4   Non-human Primate Models

Primate models of OPV disease have been developed to evaluate the efficacy of new smallpox vaccines and antiviral drugs.[45–48] Inoculation of cynomolgus macaques with monkeypox virus or variola virus, by intravenous injection or intratracheal inoculation, produces a lesional disease that resembles human smallpox. Lesions appear by day 3 post-infection and develop in size and number throughout the course of infection, and ultimately resolve by day 30 in animals that survive the infection. Mortality is observed in >95% of non-human primates infected with monkeypox virus whereas mortality in animals infected with variola virus is highly variable. In fatal cases of monkeypox or variola virus infection, animals die between days 10 and 18 post-infection, usually with over 750 poxvirus lesions, found mostly on the extremities. The severe lesional disease resembles human smallpox following the secondary viremia phase of infection. In addition, mortality, lesion number and viral load are viable endpoints that can be measured in individual animals to quantify antiviral efficacy, making this an ideal model for human smallpox.

Despite the fact that therapeutic efficacy can be measured in non-human primate (NHP) models of monkeypox and variola virus infection, the validity of these models for predicting human disease outcome has not been established. The advantage of using non-human primates as a model of human smallpox is that their physiology more closely resembles humans, providing a mechanism for establishing PK–PD relationships. The disadvantage of these models is that establishment of infection requires intravenous injection of large quantities of virus directly into the bloodstream, eliminating the primary viremia phase and prodrome described for human smallpox. Furthermore, protecting animals from severe disease and death after direct injection of virus into the bloodstream sets a high bar for efficacy for an OPV therapeutic since systemic infection has already been established and disease progression is rapid. Finally, monkeypox and variola viruses are select agents requiring studies to be conducted in high-containment BSL-3 (monkeypox virus) and BSL-4 (variola virus) facilities, limiting the number of animals that can be used in a study and reducing the number of disease endpoints that can be measured. Even with these limitations, randomized, placebo-controlled trials can be conducted in these model systems following GLP guidelines.[47]

Supplementing studies in non-human primates with trials in mice and rabbits can provide additional information of the antiviral efficacy of OPV therapeutics against host-adapted virus in the presence of a more physiological relevant host response. The advantage of these models is that small amounts of virus inoculated at the periphery can establish lethal infection and systemic disease. While mortality and viremia can be used as therapeutic endpoints, the cause of death in these models is likely organ failure caused by replication of the virus at these sites and not 'toxemia' as was described for human smallpox. It is likely that a combination of validated animal models will be required to provide robust data sets for establishing PK–PD relationships that can be used in conjunction with human PK data to provide justification for human dose selection.

4.4 Development of OPV Therapeutics

Three clinical stage compounds, cidofovir (CDV), CMX001 and ST-246, are currently being considered for use in the treatment of OPV infections. Their discovery and development paths are unique and represent different approaches for development of OPV therapeutics.

CDV and CMX001 are acyclic nucleoside phosphonate analogs which belong to a class of compounds that have been developed into potent, clinically significant antiviral drugs.[49,50] The antiviral potency of nucleotide and nucleoside analogs is dependent upon their metabolic activation to the diphosphate and triphosphate form, respectively, and their resistance to metabolizing enzymes (Figure 4.1). For acyclic nucleoside phosphonates, the diphosphate form is the active inhibitor of viral replication typically targeting the viral polymerase. The process of activation through metabolism is determined by cellular permeability, efficiency as a substrate for host and/or viral kinases required for phosphorylation to the diphosphate form, metabolic stability of the diphosphate, efflux from the cell, activity against the viral polymerase and selectivity relative to host polymerases. These factors can determine the efficacy and target tissue specificity of acyclic nucleoside phosphonate analogs.

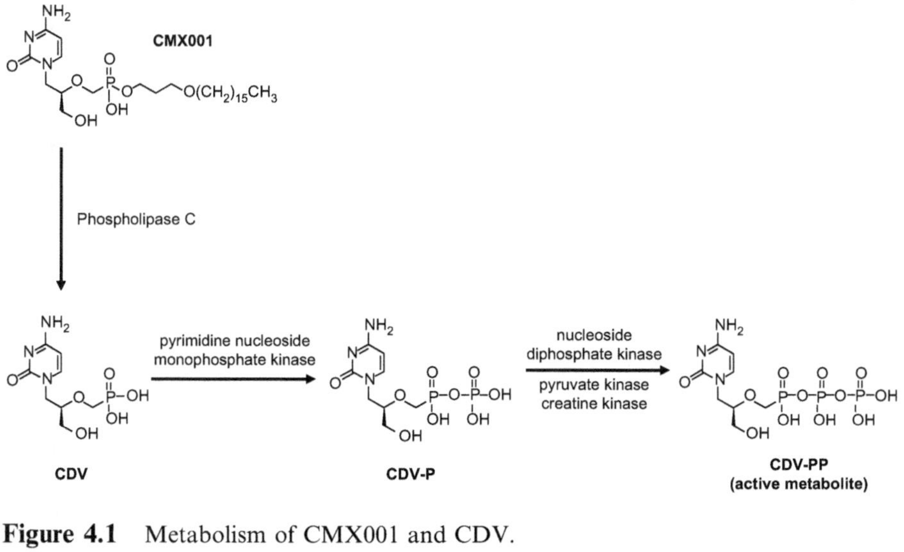

Figure 4.1 Metabolism of CMX001 and CDV.

ST-246 is a new chemical entity discovered by high-throughput screening of chemical libraries to identify compounds the inhibited OPV replication. This unbiased or target neutral approach has proven successful for discovery of novel inhibitors of OPV replication.[51] Cell-based assays have been designed to measure virus replication and can be optimized for high-throughput screening of compound libraries to identify inhibitors of OPV replication. Simple assays have been developed that measure virus-induced cytopathic effects (CPEs) and can be easily conducted in 96- or 384-well formats. Vaccinia virus is often used as a prototype OPV in these assays because it is replicates to high levels in cell culture producing robust CPEs and can be handled safely in biosafety level 2 containment.

4.4.1 Cidofovir

CDV {(S)-1-[3-hydroxy-2-(phosphonomethoxy)propyl]cytosine, Vistide] is an acyclic nucleoside phosphonate analog that exhibits broad-spectrum antiviral activity against DNA-containing viruses (Figure 4.2).[52] The acyclic nucleoside

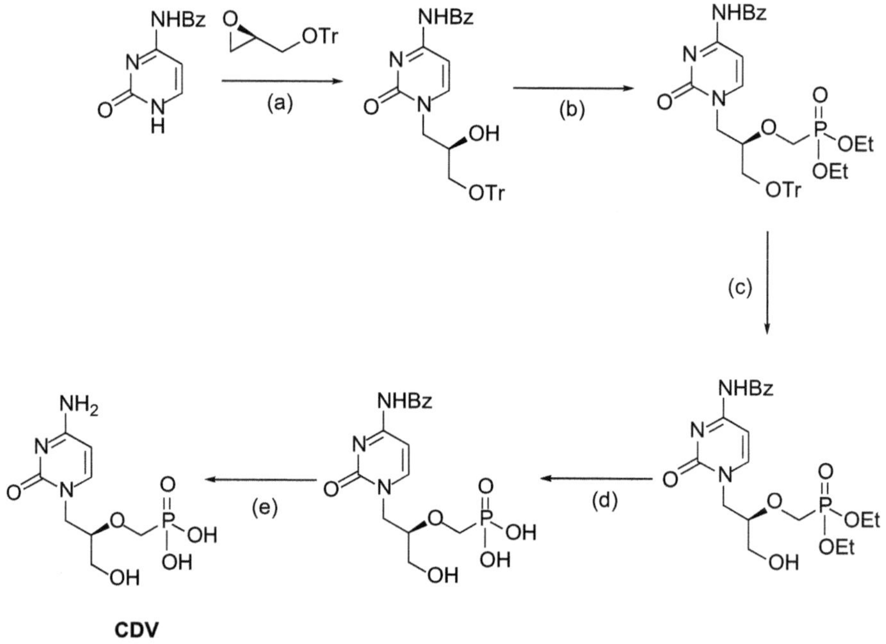

(a) NaH (0.22 equiv.), DMF, 105 °C, 5 h
(b) TsOCH₂P(O)(OEt)₂, NaH (3 equiv.), DMV, 0 °C, 6 h
(c) HCl, CH₂Cl₂, 0–5 °C, 10 min.
(d) TMSBr, CH₂Cl₂, r.t., 18 h
(e) Concentrated NH₄OH, r.t., 4 h

Figure 4.2 Synthesis of cidofovir (CDV). Chemical synthesis adapted from Brod-fuehrer *et al.*[119]

phosphonates have proved most valuable and have provided marketed products for HIV (tenofovir, PMPA), HBV (adefovir, PMEA) and CMV (cidofovir, HPMPC) and derivatives of cidofovir (CMX001) are being evaluating clinically for the treatment of poxvirus, polyomavirus and adenovirus infections (reviewed by Lee and Martin[53] and De Clercq[54]). Acyclic nucleoside phosphonates contain an isosteric O–CH_2–P linkage in place of the labile 5' CH_2–O–P bond.[55] The O–CH_2–P linkage improves the chemical and metabolic stability by making the compounds poor substrates for phosphatases, thereby providing increased stability compared with the typical 5' –CH_2–O–P bond. This increased stability, in turn, provides for long intracellular half-lives and persistent antiviral effects.[56,57] In addition, the intrinsic phosphonate allows bypass of the first phosphorylation step required for activation of nucleotide analog during anabolic phosphorylation to the diphosphate form. However, the high polarity of the phosphonate often necessitates development of a prodrug for optimal delivery (membrane permeability).

CDV is currently licensed for the treatment of human cytomegalovirus retinitis and is administered as a topical ophthalmic solution. Topical application of CDV has been used to treat molluscum contagiosum, a cutaneous skin disease caused by molluscum contagiosum virus, a poxvirus related to OPVs.[58] CDV exhibits antiviral activity against adenoviruses, herpesviruses, iridoviruses, polyomaviruses, hepadnavirus, papillomavirus and poxviruses in viral replication assays in cell culture (reviewed by De Clercq *et al.*[59]). CDV is active against a spectrum of OPV, including variola virus (Table 4.2), and the synthesis of CDV is outlined in Figure 4.2.

Table 4.2 Antiviral activity of CMX001, CDV and ST-246.

Viral class	Virus	EC_{50} (µM) CMX001	CDV	ST-246	Ref.
Orthopoxvirus	VARV	0.1	27.3	0.05	50, 51
	MPXV	0.07	4.6	0.01	50, 51
	CTGV	N.D.[a]	7.7	0.001[b]	117
	VACV	0.8	46	0.01	51, 75
	CPXV	0.5	42	0.05	51, 118
	RPXV	0.5	39	N.D.	50
	ECTV	0.5	12	0.07	42, 51
Herpesvirus	HSV-1	0.06	15	>40	50, 51
	CMV	0.0009	0.38	>40	50, 51
	HHV 6	0.004	0.2	N.D.	50
	VZV	0.0004	0.5		
	EBV	0.04	>170		
Adenovirus	AdV 5	0.02	1.3		
Papillomavirus	HPV	17	200		
Polyomavirus	BKV	0.13	115.1		
	JCV	0.045	N.D.		

[a]N.D.: not determined.
[b]EC_{50} value from Damasso *et al.* (unpublished observations).

Inhibition of viral replication by CDV requires the formation of the diphosphate of CDV (CDVpp). CDV is readily taken up by cells by fluid-phase endocytosis.[60] Pyrimidine nucleoside monophosphate kinase phosphorylates CDV.[60,61] Pyruvate kinase, creatine kinase and nucleoside diphosphate kinase can convert CDV monophosophate to the diphosphate form (Figure 4.1).[61] The half-life for CDV and phosphorylated adducts is 6–48 h, explaining why cells remain refractory to infection even after CDV has been removed from the medium.[60]

CDVpp is a substrate for the viral DNA polymerase and is incorporated into DNA during replication with lower catalytic efficiency compared with dCTP.[62] While CDV-terminated primers can serve as substrates for the next deoxynucleoside monophosphate addition step, additional rounds of deoxynucleoside monophosphate addition are inhibited. Whereas CDV can be excised from the primer 3′ terminus by the 3′-to-5′ proofreading exonuclease activity of vaccinia virus polymerase, DNAs bearing CDV as the penultimate 3′ residue are completely resistant to exonuclease attack.[62] Therefore, CDV not only inhibits the rate of primer extension once incorporated, but also inhibits the activity of the proofreading exonuclease; the misincorporation of CDV can also promote error-prone DNA synthesis during poxvirus replication.[62,63] Whereas CDV is readily incorporated into duplex DNA, the structure of the duplex is less stable compared with native DNA.[64] Drug-resistant variants of vaccinia virus have been isolated with reduced susceptibility to CDV and the resistance determinants map to the viral polymerase.[65] Virus variants resistant to CDV are attenuated for virus replication *in vivo*, suggesting that the compound interacts with amino acid residues critical for polymerase function.[66]

Although CDV has a suitable selectivity index for antiviral treatment, it is cytotoxic at high concentrations. To reduce toxicity associated with CDV, a cyclic prodrug was synthesized that is metabolized to the monophosphate within cells.[67] The cyclic prodrug of CDV is less toxic in certain animal species and is converted to CDV by cellular cyclic CMP phosphodiesterase.[68]

The *in vivo* metabolism of CDV has been extensively studied in multiple animal species. Biodistribution studies using radiolabeled CDV administered to mice, rats, rabbits and monkeys suggest that the compound is concentrated in the kidneys.[69] Intravenous administration of radiolabeled CDV or cCDV shows a multiexponential decay of compound in the plasma with a clearance rate of 0.6 and 1.10 L h^{-1} kg^{-1}, respectively.[70] Concentrations in the kidney were 20-fold higher for CDV than cCDV, consistent with the improved therapeutic index for the cyclic prodrug. In non-human primates, the kidney toxicity associated with intravenous CDV administration can be overcome by co-administration of a competitive inhibitor of organic anion transport in the proximal tubular epithelial cells.[71]

4.4.1.1   CDV Animal Efficacy

CDV has proven effective at inhibiting OPV infections *in vitro* with EC$_{50}$ values in the micromolar range (Table 4.2). The *in vivo* antiviral activity of CDV has been extensively evaluated in multiple animal models of OPV infection (reviewed by Smee[37]). Intravenous administration of CDV protected mice from lethal infection with vaccinia, cowpox and extromelia virus. Administration of CDV as an inhaled powder or by intravenous injection protected rabbits from lethal infection of aerosolized rabbitpox virus.[72] CDV has also been shown to be effective at reducing mortality and lesional disease in non-human primates infected with monkeypox or variola virus.[46,73] While the cCDV is more potent than CDV *in vitro* against poxvirus infections, it is less active than CDV in mice infected with vaccinia virus.[74,75] Despite extensive evaluation of the efficacy of CDV in animals, efficacy studies in humans have been limited to compassionate use cases due to the nephrotoxicity associated with systemic compound delivery. Therefore, for now, the clinical utility of CDV is limited to topical applications.

4.4.2 CMX001

CMX001 is a prodrug of CDV that was designed to improve the compound's pharmacokinetic properties and safety profile (Figure 4.3) (reviewed by Hostetler[76]). Although CDV exhibits broad-spectrum activity and has established clinical effectiveness, it is not readily absorbed and exhibits poor oral

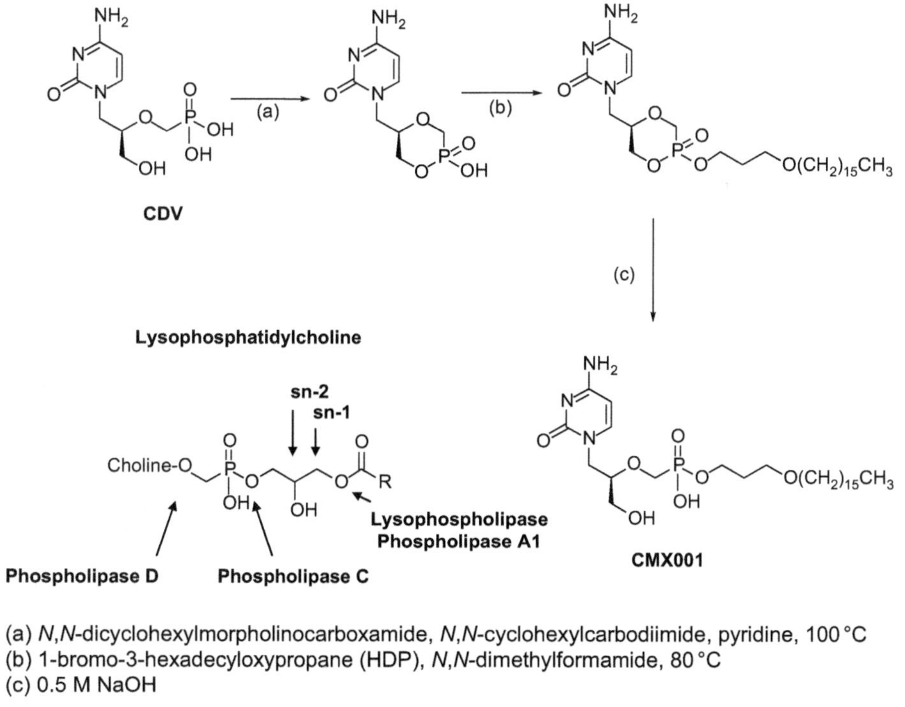
CDV
(a)
(b)
(c)
O(CH2)15CH3
Lysophosphatidylcholine
sn-2
sn-1
Choline-O
R
Lysophospholipase
Phospholipase A1
Phospholipase D
Phospholipase C
CMX001
(a) N,N-dicyclohexylmorpholinocarboxamide, N,N-cyclohexylcarbodiimide, pyridine, 100 °C
(b) 1-bromo-3-hexadecyloxypropane (HDP), N,N-dimethylformamide, 80 °C
(c) 0.5 M NaOH

Figure 4.3 Synthesis of CMX001 and similarity to lysophosphatidylcholine. Chemical synthesis of CMX001 and structure of lysophosphatidylcholine adapted from Kern, Hostetler and co-workers.[75,76]

bioavailability. The compound must be administered by intravenous injection and is associated with toxicity related to concentration in the kidney proximal tubule.[77] Its clinical utility is further limited by the requirement for co-administration with probenecid and hydration therapy to minimize kidney toxicity. To improve the oral pharmacokinetic properties and antiviral activity of CDV, a prodrug strategy was devised to camouflage the acyclic nucleoside phosphonate with a partially metabolized phospholipid moiety.[78] The resulting alkoxyalkyl analogs of CDV had greatly improved antiviral properties, oral bioavailability and decreased accumulation in the kidney.[75,79,80]

Phospholipid prodrugs of acyclic nucleoside phosphonates mimic the metabolism of lysophosphatidylcholine (LPC) in animals to increase absorption (Figure 4.3).[76] Once absorbed, the LPC is reacylated to phosphatidylcholine in enterocytes and incorporated into chylomicrons. Chylomicrons are secreted into intestinal lymphatics, where they enter the circulation through the thoracic duct. Three lipid moieties were evaluated as produgs of CDV, hexadecyloxypropyl (HDP), octadecyloxyethyl (ODE) and oleyloxypropyl (OLP), all of which were shown to improve the antiviral activity relative to CDV.[75,79] Replacing the acyl ester bond at the sn-1 position of LPC with an ether linkage prevents hydrolysis of the acyl group by lysophospholipase during absorption (Figure 4.3).[76] The hydroxyl at the sn-2 position of glycerol in LPC was replaced with a hydrogen atom, which prevented reacylation by lysophosphatidylcholine acyltransferases present in small intestinal enterocytes and other tissues.[76] Finally, it should be noted that compounds such as CDV have a phosphonate ($-P-CH_2-$) linkage to the acyclic nucleoside base that is not subject to cleavage by phospholipase D or phosphodiesterase. Therefore, metabolic cleavage is catalyzed by phospholipase C, which is not present in plasma or pancreatic secretions, providing stability for the lipid prodrug and other compounds of this type during oral absorption and transport in plasma to tissues.

CMX001 is designed to be actively taken up by cells by phospholipid transport pathways and passive diffusion. Indeed, the rate of radiolabeled CMX001 uptake into MRC-5 human lung epithelial cells is $\sim$23-fold greater than CDV, which is taken up by cells *via* fluid-phase endocytosis.[81,82] The enhanced intracellular uptake leads to a significant increase in antiviral potency.[75,81] Plaque reduction assays comparing CDV and cCDV in human foreskin fibroblast cells infected with vaccinia virus or cowpox virus showed that HDP-CDV was 57- and 13-fold more potent than CDV and cCDV, respectively.[75] Although these compounds were more toxic than the parent nucleotide analogs, the selectivity index was increased by 4–13-fold.

Mice administered [^{14}C]CMX001 orally at 5 mg kg^{-1} showed drug-associated radioactivity absorbed from the gastrointestinal tract and wide distribution to tissues by 2 h post-dosing. The highest levels of drug were detected in the small intestine and levels in the lung, liver, kidney and spleen ranged from 0.1 to 10 μg-equivalents per gram of tissue through 24 h post-dosing. Moderate to low levels of drug were detected in remaining tissues and those in which levels were below the limits of quantification included the brain and spinal cord.[83] Thus, CMX001 is readily absorbed following oral administration and widely distributed into tissues.

4.4.2.1   CMX001 Animal Efficacy

The *in vivo* activity of CMX001 has been evaluated in multiple animal models of poxvirus disease. CMX001, administered by oral gavage, protected mice from lethal intranasal challenge with VACV, CPXV and ECTV.[84,85] CMX001 administered by oral gavage protected A/Ncr mice infected with a lethal challenge dose of ECTV (5×10^4 PFU). This amount of virus is 10 000-fold above the LD_{50} delivered by intranasal inoculation or directly to the respiratory tract *via* aerosolized droplets. Moreover, CMX001 protected mice even when the compound was delivered 5 days post-infection.[86] Oral delivery of CMX001 protected rabbits infected with rabbitpox virus when administered prophylactically or at symptom onset.[44,87] These studies were conducted as randomized, double-blind, placebo-controlled trials, providing robust data sets that can be used to establish PK–PD relationships. These data demonstrate that CMX001 is well tolerated in multiple animal species and can protect animals from lethal poxvirus challenge.

Despite showing dramatic antiviral activity in numerous animal species against poxviruses and other DNA-containing viruses, CMX001 was not active in NHPs infected with monkeypox virus when delivered orally. This result was unexpected since CMX001 has been shown to be active against monkeypox virus in cell culture.[50] Moreover, CMX001 was active in a mouse model of monkey pox virus infection established in STAT1-deficient knockout mice.[88] The lack of antiviral activity was likely caused by compound metabolism, since PK assessment following oral administration of CMX001 to NHPs showed low systemic exposure of the parent compound. Intramuscular administration improved systemic exposure but did not protect animals from lethal infection. Whereas CDV and CMX001 produce the same active metabolite (CDVpp) and CDV was effective at protecting NHP from lethal MPX infection when administered by intravenous injection, CDVpp levels were much lower in PBMCs from animals treated with CMX001 compared with intravenous CDV administration. This is consistent with CMX001 being poorly metabolized to CDV in monkeys.[50]

Non-human primate models of OPV infection are considered by some as essential for evaluating antiviral efficacy of smallpox therapeutics. Since CMX001 is not efficiently metabolized in this species, alternative models need to be considered in order to develop PK–PD relationships to establish the human dose. Since CMX001 has broad antiviral activity against DNA viruses, it is possible to establish dosing regimens in humans against other naturally occurring viral infections once the safety and tolerability of CMX001 have been assessed.

4.4.2.2   CMX001 Clinical Studies

CMX001 was evaluated for safety and pharmacokinetics in single and multiple ascending-dose human trials.[89] The compound was found to be well tolerated in humans at all dose levels. Oral administration of single doses of CMX001 of 0.25–2 mg kg^{-1} and multiple doses ranging from 0.1 to 1.0 mg kg^{-1} were well tolerated, with no severe adverse events reported. Moreover, no clinically significant drug-related changes in blood chemistry, hematology, renal function or intraocular pressure were observed. Systemic exposure was dose proportional with no significant accumulation of compound after administration of multiple doses. Importantly, the concentration of CDV in the plasma remained low, indicating that CMX001 was not readily metabolized outside the cell. Hence the lipid prodrug moiety effectively eliminated the dose-limiting toxicity caused by systemic administration of CDV. At doses above $1\,mg\,kg^{-1}$, the plasma concentration at C_{max} was in the range expected to provide antiviral activity against herpes virus, adenovirus and variola virus.[89]

CMX001 was used in several of compassionate use cases for the treatment of disease caused by adenovirus or polyomavirus. The cases provided human data that demonstrated safety and efficacy of CMX001 in the treatment of viral disease.

CMX001 was used to treat a severely immunocompromised pediatric stem cell transplant recipient infected with adenovirus. In normal healthy adults, adenovirus causes a self-limiting infection of the respiratory tract leading to flu-like symptoms that resolve in several days. In severely immunocompromised patients, adenovirus spreads systemically, producing high levels of circulating virus and severe disease, with high levels of associated mortality. In this case, the patient failed to respond to intravenous CDV treatment and CMX001 was administered at $2\,mg\,kg^{-1}$ twice per week. Upon administration of CMX001, viral titers in the blood dropped below the limit of detection after the fifth dose and the clinical outcome significantly improved. These data demonstrate that CMX001 can be used to treat severe infections caused by adenovirus in immunocompromised patients.[90]

The antiviral activity of CMX001 in humans was further demonstrated in a multicenter cohort of 13 severely immunocompromised patients infected with adenovirus who failed CDV therapy. In this study, CMX001, administered at $2\,mg\,kg^{-1}$ per week, reduced the viral load in all patients and significantly improved the clinical outcome. Although the patient population was too small to determine the statistical significance of the data, the results suggest that CMX001 had an antiviral effect in severely immunocompromised humans. Hence CMX001 is safe and well tolerated in severely ill patients.[91]

The utility of CMX001 was further established against polyomaviruses (BK and JC) that are clinically significant DNA viruses associated with distinct human disease (reviewed by Bennett *et al.*[92] and Berger[93]). BK virus infection is a major cause of allograft failure in renal transplant recipients. BK virus is associated with nephropathy and urethral stenosis and in hematopoietic stem-cell transplant recipients with hemorrhagic cystitis. JC virus causes a demyelinating disease of the brain called progressive multifocal leukoencephalopathy, which can be fatal.

In vitro, CMX001 inhibited BK virus in primary human proximal renal tubular cells. CMX001 was shown to inhibit both intracellular and extracellular virus as measured by Q-PCR of viral genomic DNA and Western blotting and immunofluorescence detection of viral proteins.[94] Likewise, CMX001 inhibited

JC virus replication in human fetal brain SVG cells with an EC_{50} value of 0.045 µM. Moreover, intracellular viral DNA levels were reduced by 60%, consistent with the proposed mechanism of action.[95] Furthermore, CMX001 was active against JC virus in progenitor-derived astrocytes and in COS-7 cells. The EC_{50} value was 5.55 nM and the CC_{50} value was 184.6 nM.[96]

CMX001 was used in a recent case of a patient with progressive multifocal leukoencephalopathy and idiopathic CD4 + lymphocytopenia caused by JC virus.[97] At admission, the patient had extremely low CD4 + lymphocyte counts and presented with slurred speech and partial paralysis. An MRI revealed multiple white matter lesions throughout the brain, consistent with a demyelinating process that was confirmed by brain biopsy, establishing progressive multifocal leukoencephalopathy as the diagnosis. *In situ* hybridization for JC viral genome sequences was positive and the JC viral load was reported as 3600 copies mL^{-1}. Intravenous CDV was administered for 2 weeks followed by oral administration of CMX001. In addition, the patient received antibiotic therapy (mefloquine) to prevent opportunistic infection and an experimental interleukin-7 to boost CD4+ cell counts. Administration of CMX001 correlated with reduced JC virus DNA in the blood and clinical improvement.[97]

CMX001 shows promise as an antiviral compound for the treatment of DNA virus infections. The antiviral activity in humans against several viral pathogens provides a path to establish effective dosing strategies that can be extrapolated to infections with OPVs. The broad-spectrum activity and high genetic barrier to resistance provide impetus for continued development of CMX001 for treatment of OPV infections.

4.4.3 ST-246

ST-246 is a tricyclononene carboxamide identified by high-throughput screening of a chemical compound library for inhibitors of OPV replication (Figure 4.4).[51] Over 350 000 compounds were screened for their ability to inhibit vaccinia virus-induced CPE. A total of 759 compounds were identified that inhibited vaccinia virus-induced CPE by >50% and could be grouped into nine distinct chemical series.

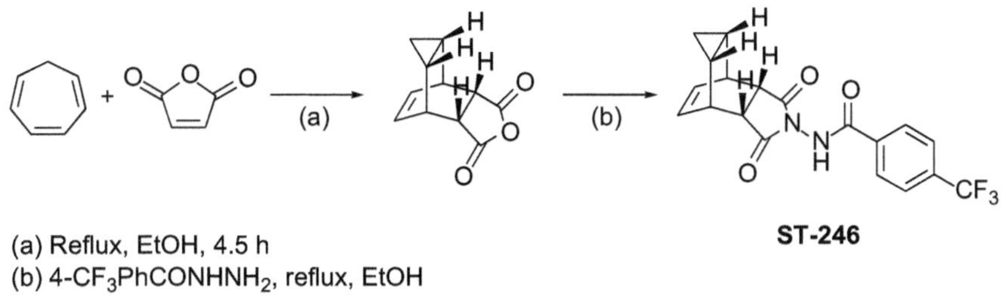

(a) Reflux, EtOH, 4.5 h
(b) 4-CF$_3$PhCONHNH$_2$, reflux, EtOH

ST-246

Figure 4.4 Synthesis of ST-246. Chemical synthesis adapted from Bailey *et al.*[98]

One series of compounds identified from the high-throughput screening was a group of tricyclononene carboxamides that were potent inhibitors of OPV replication. Nascent structure–activity relationships (SARs) from the screening hits indicated that electron-withdrawing substitution on the carboxamide aryl or heteroaryl moiety enhanced the potency of the molecules in the CPE assay (Figure 4.5).[98] To validate this nascent SAR, a series of analogs were prepared and tested for their ability to inhibit both vaccinia and cowpox virus replication in cell-based CPE assays. The antiviral activity of these analogs supported the observation that electron-withdrawing substitution on the carboxamide carbonyl R-group increased the potency of the inhibitors. The 4-nitrophenyl-substituted carboxamide was 100-fold more potent than the electron-donating 4-dimethylaminophenyl analog against both vaccinia and cowpox viruses. Whereas the π-deficient 3- and 4-pyridyl compounds displayed antiviral activity against vaccinia virus, the 2-pyridyl analog displayed a dramatic loss of potency. In all cases, heterocyclic substitution provided modest to weak activity against vaccinia virus and exhibited no activity against cowpox virus up to the highest concentrations tested. For the chloro- and bromo-substituted phenyls, a similar pattern was observed for both viruses, where 3- and 4-

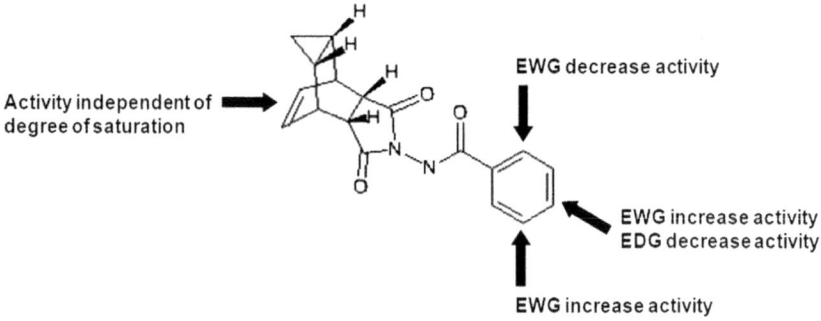

R (Phenyl ring)	Vaccinia EC$_{50}$ (μM)	Cowpox EC$_{50}$ (μM)	CC$_{50}$ (μM)
4-nitrophenyl	0.02	0.15	86 ± 20
4-Me$_2$Nphenyl	2.0	15.5	>100 ± 0
4-aminophenyl	7.7	>20	92 ± 11
2-pyridyl	>20	>20	> 100 ± 0
3-pyridyl	0.74	>20	> 100 ± 0
4-pyridyl	0.5	17.2	> 100 ± 0
2-chlorophenyl	3.0	>20	> 100 ± 0
3-chlorophenyl	0.04	0.6	> 100 ± 0
4-chlorophenyl	0.02	0.8	> 100 ± 0
2-bromophenyl	2.3	>20	> 100 ± 0
3-bromophenyl	0.05	0.6	> 100 ± 0
4-bromophenyl	0.02	1.6	> 100 ± 0
4-CF$_3$phenyl ◄—(ST-246)	0.04	0.6	> 100 ± 0
sat. 4-CF$_3$phenyl	0.02	0.3	> 100 ± 0
4-methoxyphenyl	2.2	>20	> 100 ± 0
2-(1-methyl)pyrrolyl	15.8	>20	> 100 ± 0
5-(3-methyl)pyrazolyl	7.1	>20	> 100 ± 0

Figure 4.5   Structure–activity relationships for ST-246 and related analogs. EWG, electron-withdrawing groups; EDG, electron-donating groups. Adapted from Bailey *et al.*[98] and Jordan *et al.*[103]

substitution were more potent than 2-substitution. Reduction of the olefin had little effect on potency. Examination of selected pairs of compounds in CPE assays against vaccinia, cowpox, monkeypox and camelpox viruses and two strains of variola virus established a broad-spectrum trend in antiviral activity where 4- and 3-substitutions were more potent than 2-substitutions (Figure 4.5).[98] The trifluoromethylbenzyl analog (ST-246) was selected for further development based on compound stability in S9 microsomal fractions from liver extracts.

ST-246 is highly selective for OPVs, with no antiviral activity observed against a panel of RNA- and DNA-containing viruses (Table 4.2). The compound was not cytotoxic up to the limit of solubility in aqueous solvents (50 µM).[51] The EC_{50} values for OPVs ranged from 10 to 50 nM in cell-based CPE assays but no activity was observed against other poxviruses, including myxoma virus (Leporipoxvirus) and ORF (Parapoxvirus). This exquisite specificity suggests that the compound targets a unique OPV gene product.

Genetic mapping studies of drug-resistant virus variants mapped ST-246 resistance to the F13L gene of vaccinia virus.[51] Moreover, virus recombinants containing deletions in F13L are not sensitive to ST-246, suggesting that the F13L gene product is the only viral target. The F13L gene encodes a highly conserved, 37 kDa peripheral membrane protein (p37) required for the production of EV particles.[32] Virus recombinants containing deletions of F13L produce small plaques in cell culture and reduced levels of extracellular virus and are attenuated for replication in mouse models of vaccinia virus infection consistent with a defect in EV particle production.[32,99] ST-246 reduced EV particle production while having little effect on the synthesis of IV particles, as measured by equilibrium centrifugation of radiolabeled virus particles prepared in the presence and absence of ST-246.[100]

The p37 protein catalyzes the formation of a putative wrapping complex derived from virus-modified late endosomal (LE) membranes.[32] The wrapping complex catalyzes the envelopment of IV particles to produce egress-competent forms of the virus.[22,32] The formation of the wrapping complex requires the activities of p37 and other viral proteins that interact with the host Rab-9 protein. Rab-9 is a member of the Ras superfamily of GTPases and is required for the formation of LE transport vesicles.[23,101] Rab9-dependent transport vesicle formation is mediated through interactions with the tail-interacting protein of 47 kDa (TIP47), a Rab9-specific effector protein.[101] Some viruses have evolved mechanisms to circumvent this pathway for the assembly of virus particles. Indeed, RNAi-mediated depletion of Rab9 inhibited replication of human immunodeficiency virus type 1, filoviruses and measles virus, suggesting that Rab9-containing vesicles play a role in virus assembly.[102] Immuno-precipitation of p37 from membrane fractions of vaccinia virus-infected cells coprecipitates Rab-9 and TIP47 proteins, suggesting that these proteins are associated in membrane vesicles. In the presence of ST-246, Rab-9 and TIP47 no longer coprecipitate with p37, an observation consistent with inhibition of a virus-specific wrapping complex.[100]

The non-clinical pharmacokinetic profile of ST-246 demonstrates high levels of oral bioavailability in multiple animal species, which increases when the compound is co-administered with food. The absolute measure of bioavailability in mice and non-human primates was ~40% when administered to animals in the fed state as an oral suspension. The increase in bioavailability was non-linear at high compound concentrations, likely due to decreased absorption. ST-246 is metabolically stable and is widely distributed in tissues. Studies using the radiolabeled compound showed that 72% of the radioactivity is eliminated in the feces and 24% in the urine as intact ST-246 by 96 h post-compound administration. ST-246 was well tolerated in mice and non-human primates in 28 day repeat-dose toxicity studies with a no observable adverse effect level (NOAEL) of 2000 mg kg^{-1} for mice and 300 mg kg^{-1} for non-human primates. Hence ST-246 is well tolerated with very little observable toxicity at doses that far exceed the proposed human dose required for the antiviral activity.[103]

4.4.3.1 ST-246 Animal Efficacy

ST-246 was evaluated in a variety of animals, including mice, rabbits, ground squirrels, prairie dogs and non-human primates.[48,51,104–107] In the mouse model, ST-246 protected animals from lethal OPV infection established by a variety of routes, including intranasal, intravenous, intradermal, subcutaneous and aerosol delivery of vaccinia, cowpox and ectromelia viruses.[104,108] ST-246 treatment has been demonstrated to inhibit poxvirus dissemination, virus shedding and systemic disease in mice.[108] Data from these studies were used to establish the optimal dosing regimen for protective efficacy. These studies demonstrated that once per day oral dosing in mice at 100 mg kg^{-1} for a period of >7 days was optimal for protective efficacy. Treatment can be initiated as late as 72 h post-infection for full protection. Moreover, mice that survive lethal infection due to ST-246 treatment are resistant to subsequent challenge with lethal doses of vaccinia virus due to acquisition of protective immunity during the initial infection.[51] ST-246 has also been shown to protect mice that are deficient in either CD4+ or CD8+ T cells from lethal infection, but not both, regardless of the presence or absence of B-cell deficiency.[109] ST-246 treatment in combination with smallpox vaccination does not appear to diminish the immune response, raising the possibility that ST-246 could be co-administered with the smallpox vaccine to reduce vaccine-related side effects and protect individuals from infection prior to acquisition of protective immunity.[109]

In non-human primates infected with monkeypox virus, ST-246 reduced viral load and lesion formation and protected animals from lethal infection.[47,48] The minimum effective dose was 3 mg kg^{-1}, which produced plasma drug levels below the proposed human dose of 400 mg in the fed state. A randomized, double blind, placebo-controlled study was conducted to evaluate the efficacy of ST-246 in cynomolgous macaques inoculated with a lethal dose of monkeypox virus *via* intravenous injection.[103] Treatment was initiated at 3 and 4 days post-infection at the onset of lesions and ST-246 delivered at 10 mg kg^{-1} or placebo was administered by oral gavage once per day for 14 consecutive days. The results showed that ST-246 protected animals from lethal infection and reduced lesion formation and viral DNA levels in the blood. A compilation of the efficacy data from NHP studies conducted with ST-246 showed a strong correlation between viral DNA levels in the blood, lesion formation and mortality.[103] In all studies, ST-246 protected NHPs from lethal infection with monkeypox virus and reduced lesion formation and viral DNA levels in blood even when compound was administered as late as 5 days post-inoculation. These results demonstrate that ST-246 provides therapeutic efficacy against lethal monkeypox virus infection of NHP.

ST-246 was also shown to be effective at reducing disease in NHP infected with variola virus.[48] A randomized, double-blind, placebo-controlled study was conducted to evaluate the efficacy of ST-246 in cynomolgous macaques infected with variola virus *via* intravenous injection. Treatment was initiated at lesion onset (between days 3 and 4 post-infection). The results showed that ST-246 reduced lesion numbers compared with placebo-treated animals.[103] Moreover, the rate of increase in viral DNA levels in the blood were significantly different between ST-246-treated and placebo-treated animals. The effects of ST-246 on mortality rates could not be evaluated since all of the animals survived the challenge. Taken together, these results suggest that ST-246 administered at lesion onset at $10\,mg\,kg^{-1}$ can protect monkeys from lethal infection with monkeypox virus and reduce variola virus-induced disease.

4.4.3.2 ST-246 Clinical Studies

Single and multiple ascending dose studies in healthy human volunteers established that ST-246 was safe and well tolerated, with plasma drug exposures in the range predicted to be sufficient for inhibiting OPV replication.[110,111] No severe adverse events (SAEs) were observed and no subject was withdrawn due to ST-246. The most commonly reported drug-related adverse effect was neutropenia, which was found, upon further analysis, not to be treatment related. ST-246 was readily absorbed following oral administration with mean times to maximum concentration in plasma of 3–4 h. Absorption was greater in non-fasting than fasting volunteers.

The robust antiviral activity across multiple animal species provided a means to estimate the efficacious human dose. Data from non-human primate models of OPV infection were used as the most relevant model of human smallpox. Using survival as an endpoint, an oral dose of $\sim 3\,mg\,kg^{-1}$ in non-fasted non-human primates confers 100% protection from death following intravenous injection of monkeypox virus. Pharmacological assessment in non-human primates found that a dose of $10\,mg\,kg^{-1}$ results in blood exposure levels comparable to levels attained in humans administered a 400 mg dose in the fed state. Given the variability in exposure levels in monkeys and humans in the fed and fasted states, it was predicted that human doses of 400 mg in the fed state will encompass plasma drug exposure levels comparable to those that provide protective efficacy in the non-human primate model of orthopoxvirus disease.[103]

ST-246 was used as part of the treatment regimen for several clinical cases of orthopoxvirus infection. ST 246 was used for the treatment of a child with eczema vaccinatum.[112] The patient, a 28-month-old male child with a history of eczema and failure to thrive, was exposed to virus through direct contact with a vaccinee. He presented to the emergency room with high fever and severe eczema. Vesicular skin scrapings and viral culture supernatant from vesicles on the child's skin were obtained and determined by PCR to be positive non-variola orthopox virus. The child's condition continued to worsen despite initial treatment with vaccinia immune globulin intravenous (VIG i.v.) and he exhibited progressive metabolic then respiratory acidosis, hypoalbuminemia, hypothermia and hypotension. ST-246 was administered orally *via* a naso-gastric tube. The subject also received one dose of CDV and repeated doses of VIG IV. Clinical signs of the child's improvement were observed within 1 week of the antiviral intervention (ST 246, CDV and VIG IV).

A second case was reported in a 20-year-old male who received the smallpox vaccine and was subsequently diagnosed with acute myeloid leukemia ~ 12 days after vaccination.[113] Chemotherapy was initiated to treat the leukemia, which caused a severe impairment of immune function, resulting in progressive vaccinia 6–7 weeks after vaccination. He received topical imiquimod (5%) at the vaccination site and VIG IV was administered intermittently throughout the course of treatment. ST-246 was administered at 400 mg once per day for 15 days and increased to 800 mg for 5 days and then 1200 mg for ~ 2 months. CMX001 was administered at 200 mg ~ 3.5 weeks after diagnosis or progressive vaccinia and 100 mg every week for 5 weeks. An increase in lymphocyte count correlated with improvement of symptoms and the patient was declared virus free and treatment was discontinued 9 weeks after diagnosis.

4.5   Conclusion

Although smallpox is no longer a disease that affects humans, OPVs continue to circulate in the environment, causing sporadic disease outbreaks in isolated populations. The continued threat of terrorism, coupled with reports of smallpox bioweapons experimentation, provides impetus for the development of new therapeutics that will complement our existing vaccine stocks. Developing two antivirals that work by different mechanisms will increase the genetic barrier to resistance and may be required to prevent disease caused by engineered viruses. Reports in the literature describe the construction of recombinant vaccinia and ectromelia viruses expressing IL-4 that are highly virulent, causing lethal infection in mice that were previously vaccinated against OPV infection.[114] Fortunately, the combined effects of CMX001 and ST-246, which have been shown to be synergistic, could protect mice from lethal infection with these viruses.[115,116] As smallpox fades from our memories, complacency becomes our enemy, making it essential that we continue our efforts to prevent the re-emergence of this devastating disease.

References

1. World Health Orgainization, *Smallpox Facts Sheet*, http://www2a.cdc.gov/nip/isd/spoxclincian/contents/references/factsheet.pdf (last accessed 26 March 2013).
2. F. Fenner, D.A. Henderson, I. Arita, Z. Jezek and I.D. Ladnyi, *Smallpox and Its Eradication*, World Health Organization, Geneva, 1988.
3. E. Jenner, *An Inquiry into the Causes and Effects of the Variolae Vaccinae,* 1798, Dawsons, London, 1966.
4. D. A. Henderson, *Biosecur. Bioterror.*, 2011, **9**, 163.
5. K. Alibek and S. Handelman, *Biohazard: the Chilling True Story of the Largest Covert Biological Weapons Program in the World, Told from the Inside by the Man Who Ran It*, Random House, New York, 1999.
6. J. Cello, A. V. Paul and E. Wimmer, *Science*, 2002, **297**, 1016.
7. Institute of Medicine, Committee on the Assessment of Future Scientific Needs for Live Variola Virus, *Assessment of Future Scientific Needs for Live Variola Virus*, National Academies Press, Washington, DC, 1999.
8. V. B. Mukinda, G. Mwema, M. Kilundu, D. L. Heymann, A. S. Khan and J. J. Esposito, *Lancet*, 1997, **349**, 1449.
9. S. Parker, A. Nuara, R. M. Buller and D. A. Schultz, *Fut. Microbiol.*, 2007, **2**, 17.
10. C. R. Damaso, J. J. Esposito, R. C. Condit and N. Moussatche, *Virology*, 2000, **277**, 439.
11. N. Moussatche, C. R. Damaso and G. McFadden, *J. Infect. Dev. Ctries.*, 2008, **2**, 156.
12. R. M. Vorou, V. G. Papavassiliou and I. N. Pierroutsakos, *Curr. Opin. Infect. Dis.*, 2008, **21**, 153.
13. H. G. Schatzmayr, R. V. Costa, M. C. Goncalves, P. S. D'Andrea and O. M. Barth, *Vaccine*, 2011, **29**(Suppl. 4), D65.
14. M. G. Reynolds, D. S. Carroll, V. A. Olson, C. Hughes, J. Galley, A. Likos, J. M. Montgomery, R. Suu-Ire, M. O. Kwasi, J. Jeffrey Root, Z. Braden, J. Abel, C. Clemmons, R. Regnery, K. Karem and I. K. Damon, *Am. J. Trop. Med. Hyg.*, 2010, **82**, 746.
15. Y. Li, D. S. Carroll, S. N. Gardner, M. C. Walsh, E. A. Vitalis and I. K. Damon, *Proc. Natl. Acad. Sci. U. S. A.*, 2007, **104**, 15787.
16. J. J. Esposito, S. A. Sammons, A. M. Frace, J. D. Osborne, M. Olsen-Rasmussen, M. Zhang, D. Govil, I. K. Damon, R. Kline, M. Laker, Y. Li, G. L. Smith, H. Meyer, J. W. Leduc and R. M. Wohlhueter, *Science*, 2006, **313**, 807.
17. A. McLysaght, P. F. Baldi and B. S. Gaut, *Proc. Natl. Acad. Sci. U. S. A.*, 2003, **100**, 15655.
18. C. Gubser, S. Hue, P. Kellam and G. L. Smith, *J. Gen. Virol.*, 2004, **85**, 105.
19. J. B. Johnston and G. McFadden, *J. Virol.*, 2003, **77**, 6093.
20. R. M. Buller and G. J. Palumbo, *Microbiol. Rev.*, 1991, **55**, 80.
21. B. N. Fields, D. M. Knipe and P. M. Howley, *Fields' Virology*, 5th edn., Wolters Kluwer/Lippincott Williams & Wilkins, Philadelphia, 2007.

22. G. L. Smith, A. Vanderplasschen and M. Law, *J. Gen. Virol.*, 2002, **83**, 2915.
23. M. Schmelz, B. Sodeik, M. Ericsson, E. J. Wolffe, H. Shida, G. Hiller and G. Griffiths, *J. Virol.*, 1994, **68**, 130.
24. J. Tooze, M. Hollinshead, B. Reis, K. Radsak and H. Kern, *Eur. J. Cell Biol.*, 1993, **60**, 163.
25. C. M. Sanderson, M. Hollinshead and G. L. Smith, *J. Gen. Virol.*, 2000, **81**, 47.
26. B. M. Ward, *J. Virol.*, 2005, **79**, 4755.
27. L. G. Payne, *J. Gen. Virol.*, 1980, **50**, 89.
28. A. Vanderplasschen, E. Mathew, M. Hollinshead, R. B. Sim and G. L. Smith, *Proc. Natl. Acad. Sci. U. S. A.*, 1998, **95**, 7544.
29. R. J. Stern, J. P. Thompson and R. W. Moyer, *Virology*, 1997, **233**, 118.
30. W. H. Zhang, D. Wilcock and G. L. Smith, *J. Virol.*, 2000, **74**, 11654.
31. A. A. McIntosh and G. L. Smith, *J. Virol.*, 1996, **70**, 272.
32. R. Blasco and B. Moss, *J. Virol.*, 1991, **65**, 5910.
33. D. J. Esteban and R. M. Buller, *J. Gen. Virol.*, 2005, **86**, 2645.
34. FDA. *Antiviral Drugs Advisory Committee Meeting*, http://www.fda.gov/AdvisoryCommittees/Calendar/ucm274968.htm (last accessed 2011).
35. CDC, *CDC Smallpox Case Definition*, http://www.bt.cdc.gov/agent/smallpox/diagnosis/casedefinition.asp (last accessed 2012).
36. J. L. Chapman, D. K. Nichols, M. J. Martinez and J. W. Raymond, *Vet. Pathol.*, 2010, **47**, 852.
37. D. F. Smee, *Antivir. Chem. Chemother.*, 2008, **19**, 115.
38. M. Zaitseva, S. M. Kapnick, J. Scott, L. R. King, J. Manischewitz, L. Sirota, S. Kodihalli and H. Golding, *J. Virol.*, 2009, **83**, 10437.
39. D. Esteban, S. Parker, J. Schriewer, H. Hartzler and R. M. Buller, *Methods Mol. Biol.*, 2012, **890**, 177.
40. C. A. Mims, *Br. J. Exp. Pathol.*, 1968, **49**, 24.
41. C. A. Mims, *Br. J. Exp. Pathol.*, 1959, **40**, 533.
42. R. M. Buller, G. Owens, J. Schriewer, L. Melman, J. R. Beadle and K. Y. Hostetler, *Virology*, 2004, **318**, 474.
43. M. M. Adams, A. D. Rice and R. W. Moyer, *J. Virol.*, 2007, **81**, 11084.
44. A. D. Rice, M. M. Adams, B. Lampert, S. Foster, A. Robertson, G. Painter and R. W. Moyer, *Viruses*, 2011, **3**, 63.
45. K. J. Stittelaar, G. van Amerongen, I. Kondova, T. Kuiken, R. F. van Lavieren, F. H. Pistoor, H. G. Niesters, G. van Doornum, B. A. van der Zeijst, L. Mateo, P. J. Chaplin and A. D. Osterhaus, *J. Virol.*, 2005, **79**, 7845.
46. K. J. Stittelaar, J. Neyts, L. Naesens, G. van Amerongen, R. F. van Lavieren, A. Holy, E. De Clercq, H. G. Niesters, E. Fries, C. Maas, P. G. Mulder, B. A. van der Zeijst and A. D. Osterhaus, *Nature*, 2006, **439**, 745.
47. R. Jordan, A. Goff, A. Frimm, M. L. Corrado, L. E. Hensley, C. M. Byrd, E. Mucker, J. Shamblin, T. C. Bolken, C. Wlazlowski, W. Johnson, J. Chapman, N. Twenhafel, S. Tyavanagimatt, A. Amantana,

J. Chinsangaram, D. E. Hruby and J. Huggins, *Antimicrob. Agents Chemother.*, 2009, **53**, 1817.

48. J. Huggins, A. Goff, L. Hensley, E. Mucker, J. Shamblin, C. Wlazlowski, W. Johnson, J. Chapman, T. Larsen, N. Twenhafel, K. Karem, I. K. Damon, C. M. Byrd, T. C. Bolken, R. Jordan and D. Hruby, *Antimicrob. Agents Chemother.*, 2009, **53**, 2620.
49. G. Andrei and R. Snoeck, *Viruses*, 2010, **2**, 2803.
50. R. Lanier, L. Trost, T. Tippin, B. Lampert, A. Robertson, S. Foster, M. Rose, W. Painter, R. O'Mahony, M. Almond and G. Painter, *Viruses*, 2010, **2**, 2740.
51. G. Yang, D. C. Pevear, M. H. Davies, M. S. Collett, T. Bailey, S. Rippen, L. Barone, C. Burns, G. Rhodes, S. Tohan, J. W. Huggins, R. O. Baker, R. L. Buller, E. Touchette, K. Waller, J. Schriewer, J. Neyts, E. DeClercq, K. Jones, D. Hruby and R. Jordan, *J. Virol.*, 2005, **79**, 13139.
52. E. De Clercq, *Antiviral Res.*, 2002, **55**, 1.
53. W. A. Lee and J. C. Martin, *Antiviral Res.*, 2006, **71**, 254.
54. E. De Clercq, *Biochem. Pharmacol.*, 2011, **82**, 99.
55. E. De Clercq, A. Holy, I. Rosenberg, T. Sakuma, J. Balzarini and P. C. Maudgal, *Nature*, 1986, **323**, 464.
56. B. L. Robbins, R. V. Srinivas, C. Kim, N. Bischofberger and A. Fridland, *Antimicrob. Agents Chemother.*, 1998, **42**, 612.
57. A. S. Ray, J. E. Vela, C. G. Boojamra, L. Zhang, H. Hui, C. Callebaut, K. Stray, K. Y. Lin, Y. Gao, R. L. Mackman and T. Cihlar, *Antimicrob. Agents Chemother.*, 2008, **52**, 648.
58. K. P. Meadows, S. K. Tyring, A. T. Pavia and T. M. Rallis, *Arch. Dermatol.*, 1997, **133**, 987.
59. E. De Clercq, T. Sakuma, M. Baba, R. Pauwels, J. Balzarini, I. Rosenberg and A. Holy, *Antiviral Res.*, 1987, **8**, 261.
60. H. T. Ho, K. L. Woods, J. J. Bronson, H. De Boeck, J. C. Martin and M. J. Hitchcock, *Mol. Pharmacol.*, 1992, **41**, 197.
61. T. Cihlar and M. S. Chen, *Mol. Pharmacol.*, 1996, **50**, 1502.
62. W. C. Magee, K. Y. Hostetler and D. H. Evans, *Antimicrob. Agents Chemother.*, 2005, **49**, 3153.
63. W. C. Magee, K. A. Aldern, K. Y. Hostetler and D. H. Evans, *Antimicrob. Agents Chemother.*, 2008, **52**, 586.
64. O. Julien, J. R. Beadle, W. C. Magee, S. Chatterjee, K. Y. Hostetler, D. H. Evans and B. D. Sykes, *J. Am. Chem. Soc.*, 2011, **133**, 2264.
65. M. N. Becker, M. Obraztsova, E. R. Kern, D. C. Quenelle, K. A. Keith, M. N. Prichard, M. Luo and R. W. Moyer, *Virol. J.*, 2008, **5**, 58.
66. G. Andrei, D. B. Gammon, P. Fiten, E. De Clercq, G. Opdenakker, R. Snoeck and D. H. Evans, *J. Virol.*, 2006, **80**, 9391.
67. N. Bischofberger, M. J. Hitchcock, M. S. Chen, D. B. Barkhimer, K. C. Cundy, K. M. Kent, S. A. Lacy, W. A. Lee, Z. H. Li and D. B. Mendel, *et al.*, *Antimicrob. Agents Chemother.*, 1994, **38**, 2387.
68. D. B. Mendel, T. Cihlar, K. Moon and M. S. Chen, *Antimicrob. Agents Chemother.*, 1997, **41**, 641.

69. K. C. Cundy, B. G. Petty, J. Flaherty, P. E. Fisher, M. A. Polis, M. Wachsman, P. S. Lietman, J. P. Lalezari, M. J. Hitchcock and H. S. Jaffe, *Antimicrob. Agents Chemother.*, 1995, **39**, 1247.
70. K. C. Cundy, A. M. Bidgood, G. Lynch, J. P. Shaw, L. Griffin and W. A. Lee, *Drug Metab. Dispos.*, 1996, **24**, 745.
71. S. A. Lacy, M. J. Hitchcock, W. A. Lee, P. Tellier and K. C. Cundy, *Toxicol. Sci.*, 1998, **44**, 97.
72. D. Verreault, S. K. Sivasubramani, J. D. Talton, L. A. Doyle, J. D. Reddy, S. Z. Killeen, P. J. Didier, P. A. Marx and C. J. Roy, *Antiviral Res.*, 2012, **93**, 204.
73. R. O. Baker, M. Bray and J. W. Huggins, *Antiviral Res.*, 2003, **57**, 13.
74. D. F. Smee, K. W. Bailey and R. W. Sidwell, *Chemotherapy*, 2003, **49**, 126.
75. E. R. Kern, C. Hartline, E. Harden, K. Keith, N. Rodriguez, J. R. Beadle and K. Y. Hostetler, *Antimicrob. Agents Chemother.*, 2002, **46**, 991.
76. K. Y. Hostetler, *Antiviral Res.*, 2009, **82**, A84.
77. K. C. Cundy, *Clin. Pharmacokinet.*, 1999, **36**, 127.
78. G. R. Painter and K. Y. Hostetler, *Trends Biotechnol.*, 2004, **22**, 423.
79. J. R. Beadle, C. Hartline, K. A. Aldern, N. Rodriguez, E. Harden, E. R. Kern and K. Y. Hostetler, *Antimicrob. Agents Chemother.*, 2002, **46**, 2381.
80. S. L. Ciesla, J. Trahan, W. B. Wan, J. R. Beadle, K. A. Aldern, G. R. Painter and K. Y. Hostetler, *Antiviral Res.*, 2003, **59**, 163.
81. K. A. Aldern, S. L. Ciesla, K. L. Winegarden and K. Y. Hostetler, *Mol. Pharmacol.*, 2003, **63**, 678.
82. M. C. Connelly, B. L. Robbins and A. Fridland, *Biochem. Pharmacol.*, 1993, **46**, 1053.
83. D. C. Quenelle, B. Lampert, D. J. Collins, T. L. Rice, G. R. Painter and E. R. Kern, *J. Infect. Dis.*, 2010, **202**, 1492.
84. K. Y. Hostetler, J. R. Beadle, J. Trahan, K. A. Aldern, G. Owens, J. Schriewer, L. Melman and R. M. Buller, *Antiviral Res.*, 2007, **73**, 212.
85. D. C. Quenelle, D. J. Collins, W. B. Wan, J. R. Beadle, K. Y. Hostetler and E. R. Kern, *Antimicrob. Agents Chemother.*, 2004, **48**, 404.
86. S. Parker, E. Touchette, C. Oberle, M. Almond, A. Robertson, L. C. Trost, B. Lampert, G. Painter and R. M. Buller, *Antiviral Res.*, 2008, **77**, 39.
87. A. D. Rice, M. M. Adams, G. Wallace, A. M. Burrage, S. F. Lindsey, A. J. Smith, D. Swetnam, B. R. Manning, S. A. Gray, B. Lampert, S. Foster, R. Lanier, A. Robertson, G. Painter and R. W. Moyer, *Viruses*, 2011, **3**, 47.
88. J. Stabenow, R. M. Buller, J. Schriewer, C. West, J. E. Sagartz and S. Parker, *J. Virol.*, 2010, **84**, 3909.
89. W. Painter, A. Robertson, L. C. Trost, S. Godkin, B. Lampert and G. Painter, *Antimicrob. Agents Chemother.*, 2012, **56**, 2726.
90. K. Paolino, J. Sande, E. Perez, B. Loechelt, B. Jantausch, W. Painter, M. Anderson, T. Tippin, E. R. Lanier, T. Fry and R. L. DeBiasi, *J. Clin. Virol.*, 2011, **50**, 167.

91. D. F. Florescu, S. A. Pergam, M. N. Neely, F. Qiu, C. Johnston, S. Way, J. Sande, D. A. Lewinsohn, J. A. Guzman-Cottrill, M. L. Graham, G. Papanicolaou, J. Kurtzberg, J. Rigdon, W. Painter, H. Mommeja-Marin, R. Lanier, M. Anderson and C. van der Horst, *Biol. Blood Marrow Transplant.*, 2012, **18**, 731.
92. S. M. Bennett, N. M. Broekema and M. J. Imperiale, *Microbes Infect.*, 2012, **14**, 672.
93. J. R. Berger, *Drug Saf.*, 2010, **33**, 969.
94. C. H. Rinaldo, R. Gosert, E. Bernhoff, S. Finstad and H. H. Hirsch, *Antimicrob. Agents Chemother.*, 2010, **54**, 4714.
95. Z. G. Jiang, J. Cohen, L. J. Marshall and E. O. Major, *Antimicrob. Agents Chemother.*, 2010, **54**, 4723.
96. R. Gosert, C. H. Rinaldo, M. Wernli, E. O. Major and H. H. Hirsch, *Antimicrob. Agents Chemother.*, 2011, **55**, 2129.
97. A. Patel, J. Patel and J. Ikwuagwu, *J. Antimicrob. Chemother.*, 2010, **65**, 2697.
98. T. R. Bailey, S. R. Rippin, E. Opsitnick, C. J. Burns, D. C. Pevear, M. S. Collett, G. Rhodes, S. Tohan, J. W. Huggins, R. O. Baker, E. R. Kern, K. A. Keith, D. Dai, G. Yang, D. Hruby and R. Jordan, *J. Med. Chem.*, 2007, **50**, 1442.
99. D. W. Grosenbach, A. Berhanu, D. S. King, S. Mosier, K. F. Jones, R. A. Jordan, T. C. Bolken and D. E. Hruby, *Proc. Natl. Acad. Sci. U. S. A.*, 2010, **107**, 838.
100. Y. Chen, K. M. Honeychurch, G. Yang, C. M. Byrd, C. Harver, D. E. Hruby and R. Jordan, *Virol. J.*, 2009, **6**, 44.
101. K. S. Carroll, J. Hanna, I. Simon, J. Krise, P. Barbero and S. R. Pfeffer, *Science*, 2001, **292**, 1373.
102. J. L. Murray, M. Mavrakis, N. J. McDonald, M. Yilla, J. Sheng, W. J. Bellini, L. Zhao, J. M. Le Doux, M. W. Shaw, C. C. Luo, J. Lippincott-Schwartz, A. Sanchez, D. H. Rubin and T. W. Hodge, *J. Virol.*, 2005, **79**, 11742.
103. R. Jordan, J. M. Leeds, S. Tyavanagimatt and D. E. Hruby, *Viruses*, 2010, **2**, 2409.
104. D. C. Quenelle, R. M. Buller, S. Parker, K. A. Keith, D. E. Hruby, R. Jordan and E. R. Kern, *Antimicrob. Agents Chemother.*, 2007, **51**, 689.
105. A. Nalca, J. M. Hatkin, N. L. Garza, D. K. Nichols, S. W. Norris, D. E. Hruby and R. Jordan, *Antiviral Res.*, 2008, **79**, 121.
106. E. Sbrana, R. Jordan, D. E. Hruby, R. I. Mateo, S. Y. Xiao, M. Siirin, P. C. Newman, D. A. R. Ap and R. B. Tesh, *Am. J. Trop. Med. Hyg.*, 2007, **76**, 768.
107. S. K. Smith, J. Self, S. Weiss, D. Carroll, Z. Braden, R. L. Regnery, W. Davidson, R. Jordan, D. E. Hruby and I. K. Damon, *J. Virol.*, 2011, **85**, 9176.
108. A. Berhanu, D. S. King, S. Mosier, R. Jordan, K. F. Jones, D. E. Hruby and D. W. Grosenbach, *Antimicrob. Agents Chemother.*, 2009, **53**, 4999.

109. A. Berhanu, D. S. King, S. Mosier, R. Jordan, K. F. Jones, D. E. Hruby and D. W. Grosenbach, *Vaccine*, 2010, **29**, 289.
110. R. Jordan, D. Tien, T. C. Bolken, K. F. Jones, S. R. Tyavanagimatt, J. Strasser, A. Frimm, M. L. Corrado, P. G. Strome and D. E. Hruby, *Antimicrob. Agents Chemother.*, 2008, **52**, 1721.
111. R. Jordan, J. Chinsangaram, T. C. Bolken, S. R. Tyavanagimatt, D. Tien, K. F. Jones, A. Frimm, M. L. Corrado, M. Pickens, P. Landis, J. Clarke, T. C. Marbury and D. E. Hruby, *Antimicrob. Agents Chemother.*, 2010, **54**, 2560.
112. S. Vora, I. Damon, V. Fulginiti, S. G. Weber, M. Kahana, S. L. Stein, S. I. Gerber, S. Garcia-Houchins, E. Lederman, D. Hruby, L. Collins, D. Scott, K. Thompson, J. V. Barson, R. Regnery, C. Hughes, R. S. Daum, Y. Li, H. Zhao, S. Smith, Z. Braden, K. Karem, V. Olson, W. Davidson, G. Trindade, T. Bolken, R. Jordan, D. Tien and J. Marcinak, *Clin. Infect. Dis.*, 2008, **46**, 1555.
113. *MMWR Morb. Mortal. Wkly. Rep.*, 2009, **58**, 532.
114. R. J. Jackson, A. J. Ramsay, C. D. Christensen, S. Beaton, D. F. Hall and I. A. Ramshaw, *J. Virol.*, 2001, **75**, 1205.
115. N. Chen, C. J. Bellone, J. Schriewer, G. Owens, T. Fredrickson, S. Parker and R. M. Buller, *Virology*, 2011, **409**, 328.
116. D. C. Quenelle, M. N. Prichard, K. A. Keith, D. E. Hruby, R. Jordan, G. R. Painter, A. Robertson and E. R. Kern, *Antimicrob. Agents Chemother.*, 2007, **51**, 4118.
117. D. M. Jesus, N. Moussatche and C. R. Damaso, *Int. J. Antimicrob. Agents*, 2009, **33**, 75.
118. J. C. Ruiz, J. R. Beadle, K. A. Aldern, K. A. Keith, C. B. Hartline, E. R. Kern and K. Y. Hostetler, *Antiviral Res.*, 2007, **75**, 87.
119. P. R. Brodfuehrer, H. G. Howell, C. Sapino and P Vemishetti, *Tetrahedron Lett.*, 1994, **35**, 3.

CHAPTER 5

HCV Replication Inhibitors That Interact with NS4B

CHRISTOPHER D. ROBERTS* AND ANDREW J. PEAT

GlaxoSmithKline Research & Development, Infectious Diseases Therapeutic Area Unit, 5 Moore Drive, Research Triangle Park, NC 27709, USA
*Email: christopher.d.roberts@gsk.com

5.1 Introduction

Hepatitis C Virus (HCV) infection is a wide-ranging and significant global health problem. A blood-borne pathogen, HCV develops into a chronic liver infection in $\sim 85\%$ of primary infections. Over a prolonged period, usually decades, this infection causes significant hepatic damage, including cirrhosis and/or hepatocellular carcinoma, and is the leading cause of liver transplantation.[1,2] As there is no vaccine and a substantial infected population, the need for antiviral therapy is pressing. Within the last two decades, a significant amount of research and development has been devoted to identifying anti-HCV agents.[3] Two protease inhibitors have been approved and there are multiple agents with different, and complementary, mechanisms of action under development. The expectation is that ultimately HCV will be successfully treated with a combination regimen containing multiple agents to suppress the emergence of resistance. However with the genetic diversity of HCV and the ever-present risk of resistance, there remains a need for more agents to include in these therapeutic combinations.

RSC Drug Discovery Series No. 32
Successful Strategies for the Discovery of Antiviral Drugs
Edited by Manoj C. Desai and Nicholas A. Meanwell
© The Royal Society of Chemistry 2013
Published by the Royal Society of Chemistry, www.rsc.org

5.2 Identification of NS4B as the Target of Inhibitors Discovered in a Phenotypic HCV Replicon Screen

Beyond the traditional, and successful, strategy of screening viral enzymes for inhibitors, an additional approach to identifying new antiviral inhibitors is to utilize phenotypic screens whereby an HTS is conducted in virally infected cells. The clear advantage of this approach is that inhibitors identified in the screen are, by definition, antiviral. In contrast, compounds identified in biochemical screens may fail to demonstrate antiviral activity due to poor cell penetration, inadequate biophysical properties (*e.g.* high binding to serum in the antiviral assay) or poor correlation of the biochemical assay with the actual, functional form within a virally infected cell. Following the identification of positive hits in a phenotypic screen, the compounds are counter-screened in the same cell line with a general cytotoxicity format to rule out false positives; compounds that are toxic to the cell will also exhibit an antiviral signal in a phenotypic screen. Once the antiviral and cytotoxicity data are obtained for these hits, a therapeutic index (*i.e.* cytotoxic concentration divided by antiviral concentration) can be determined. Typically, compounds with a therapeutic index of <10 are discarded. There are several downsides when running a phenotypic screen. One is the risk of false negatives; the compound screened is an exemplar of chemical space that could deliver an excellent drug but, owing to a feature unique to that exemplar, such as poor cell permeability, is missed in the screen. In practice, there is little one can do to overcome these false negatives. The major disadvantage of phenotypic screening is the challenge of biological deconvolution. Following the initial, and often relatively straightforward, aspect of logistically running a phenotypic screen, one is confronted with many, potent antiviral compounds of unknown mechanism. Although it is theoretically possible to optimize, select and ultimately develop a drug without knowing how it functions, the reality is that to optimize and develop properly a molecule identified in a phenotypic screen usually requires identifying and investigating the compound's antiviral mechanism. There are many techniques that are applied during biological deconvolution, some of the most common being genetic approaches to identifying the molecular target or proteomics, wherein the actual molecular protein that the compound targets is identified or isolated.[4,5] These efforts are usually technically challenging and more laborious and longer than actually running the initial HTS campaign. This 'back-end' commitment likely dissuades the actual running of a phenotypic screen despite the obvious advantages.[6]

In an effort to identify anti-HCV agents with novel mechanisms of action, a phenotypic screen of the HCV genotype 1b replicon was undertaken. Following the running of a high-throughput phenotypic screen *versus* a luciferase reporter gene containing replicon in Huh7.5 cells,[7] multiple, active hits were identified, including compound **1** (Figure 5.1). This compound inhibited both HCV subtypes 1a and 1b replicon with an EC_{50} of $\sim 0.6\,\mu M$ with no observable cytotoxicity up to $25\,\mu M$. A therapeutic index of >10 and the relatively

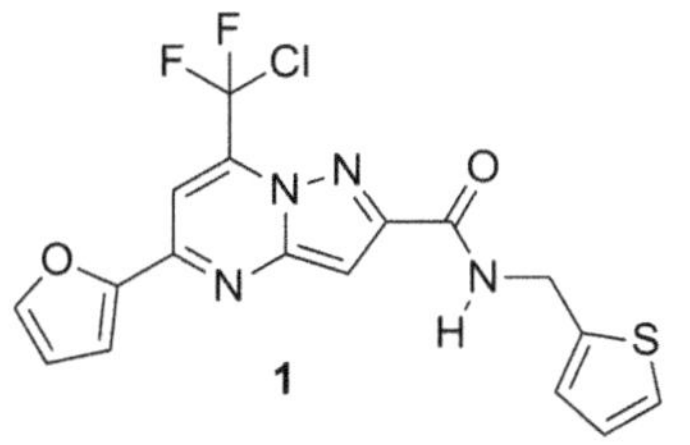

Figure 5.1   Compound **1**, identified *via* high-throughput screening in an HCV replicon phenotypic assay, potently inhibits both genotypes 1a and 1b.

equivalent potency *versus* both the 1b and 1a genotypes of HCV made this compound a potentially attractive starting point.

It was also noted that this compound, along with multiple other chemical series, had previously been disclosed in a patent application as active *versus* the HCV replicon.[8] This patent described genetic resistance and tryptophan fluorescence quenching data to suggest that the chemical series, including **1**, were interacting with the HCV non-structural protein NS4B.

Generation of the genetic data to link a viral target to the antiviral activity of a compound identified in a phenotypic screen involves passaging the HCV-infected replicon cells in the presence of the compound under positive selection pressure. In this case, the HCV replicon contains a gene that confers resistance to the toxic compound G418. Loss of the replicon (and thus G418 resistance) is fatal to the cell; passaging of the replicon in the presence of both the antiviral (fatal to the HCV replicon) and G418 (fatal to the Huh7.5 cell line) requires that the virus mutate to generate resistance to the antiviral so as to retain the ability to continue to produce the G418 resistance (and thus survive). Therefore, application of **1** at different selective pressures, $5 \times EC_{50}$ and $10 \times EC_{50}$, for multiple passages ultimately led to the formation of dividing (*i.e.* surviving) colonies that were then isolated to identify the source of the resistance to **1**. Sequencing of the HCV genome led to the identification of two amino acids, His94, mutated to arginine or asparagine, and Val105, mutated to leucine or methionine, as the potential source of the resistance. Cloning of these mutations back into the replicon conferred resistance to **1**, with a 12-fold shift in EC_{50} against the H94R mutant *versus* wild-type virus.[9] This genetic resistance data suggested that **1** was exerting its antiviral effect *via* an inter-action with NS4B or the interference of NS4B's function.

5.3   NS4B Function and Mechanism of Action of NS4B Inhibitors

The hepatitis C virus is an RNA virus belonging to the Flaviviridae family. Its positive-stranded 9.6 kb genome encodes a polyprotein of ~ 3000 amino acids. Following viral fusion and entry, the HCV RNA is translated in the cytoplasm to the HCV polyprotein containing both structural (C, E1 and E2) and non-structural proteins (p7, NS2, NS3, NS4A, NS4B, NS5A and NS5B). Once translated, the polypeptide is processed by both cellular and viral enzymes into the individual HCV proteins. Most of the non-structural proteins have some enzymatic function; NS2 and NS3/4A are proteases, with NS3 also having a helicase functionality, and NS5B, which is the RNA-dependent RNA polymerase. p7 is an ion channel protein and NS5A is a protein implicated in multiple roles including RNA binding, replication and interferon sensitivity.[3]

NS4B is a highly hydrophobic 27 kDa protein released *in trans* from NS4A after cleavage by the HCV NS3A/4A protease. NS4B has been difficult to study owing to its lack of enzyme function, its highly hydrophobic character and its requisite membrane association. Nonetheless, the function of NS4B has been extensively studied in recent years and its role in HCV infection, replication and pathogenesis is being unraveled.[10,11]

NS4B has been shown in HCV-infected cells to induce the 'membranous web,' a morphological alteration comprised of aggregates of membrane vesicles of the endoplasmic reticulum (ER) membrane. It is thought that this altered membrane forms a protected scaffold for HCV replication. All positive-stranded RNA viruses share the feature of a membrane-associated replication complex composed of viral and host proteins.[12–14] While the HCV NS4B shares no significant sequence homology to other Flaviviridae NS4B sequences, all of these are transmembrane proteins as based on experimental data and/or prediction algorithms.[15–18] Intracellularly, NS4B is localized to the ER membranes, where it is integrally associated. NS4B is postulated to be comprised of three domains, the N-terminal domain composed of two amphipathic helices, a central domain containing four membrane-spanning segments and the C-terminal domain with two helical membrane binding or anchoring regions (Figure 5.2).[19] It is believed that during the processing of the polypeptide by the viral protease both the N- and C-termini are located on the cytoplasmic side of the ER membrane.[15] Some data suggest that the N-terminal region may have a dynamic function wherein the N-terminus can cross the ER to the luminal side with alpha helix 2 forming another membrane-spanning segment.[20,21] The C-terminal helices are thought to anchor membrane association. Additionally, there are palmitoylation sites in this C-terminal domain. NS4B has been shown to oligomerize within the ER membrane, and that this oligomerization may be triggered by palmitoylation of the sites described above.[22]

Once associated in the membrane, it is not clear how NS4B induces the formation of the membranous web. A leading hypothesis is that the

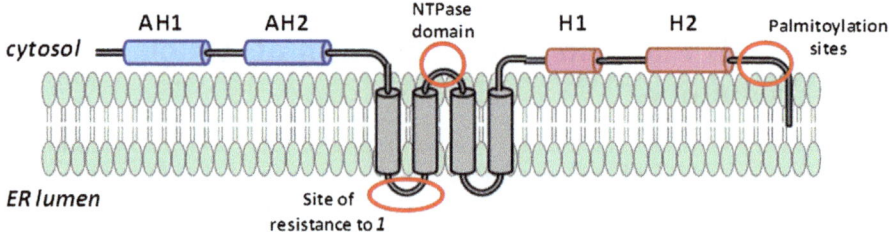

Figure 5.2   Proposed structure of the domains of NS4B in the membrane of the ER.

oligomerization of multiple NS4B molecules induces a curvature of the ER membrane and formation of these membrane vesicles, which then aggregate to alter the morphology of the ER into the observed membranous web and consequent viral replication.[10] It has been observed in cell culture that adaptive mutations in NS4B lead to increased replication.[23–25]

The viral genomes of RNA viruses, including HCV, have limited coding capacity and many virally encoded proteins have multiple functions. In addition to its membrane anchoring and altering role, NS4B has been shown to be involved in HCV RNA replication.[21,26–30] probably *via* protein–protein interactions with other viral and cellular proteins. There has also been a report that NS4B is involved in viral assembly.[15] Finally, NS4B contains an amino acid sequence, on a loop connecting two trans-membrane loops, that is predicted to have NTPase activity. Biochemical studies demonstrated that this loop binds and hydrolyzes purine nucleotide triphosphates, albeit with more affinity for ATP.[31] The role of this NTPase function is not well understood and remains controversial.

Given the resistance data implicating NS4B, we sought to investigate further the mechanism of the NS4B mediated anti-HCV activity of these compounds. Recombinant NS4B from the genotype 1b replicon was cloned into insect S9 cells (with histidine tag added to aid in purification), isolated and purified. In a similar fashion, the NS4B protein containing the H94N resistance mutation was also obtained. Tryptophan fluorescence quenching is a method that aims to take advantage of the propensity for the intrinsic fluorescence of the amino acid tryptophan to be quenched within a protein when another molecule is proximate or when the protein rearranges in response to ligand binding, and is able to absorb emitted fluorescence.[32] This quench can be quantified and is directly proportional to the strength of the binding interaction. Measured across decreasing concentrations of ligand, a binding constant (K_D) can be calculated. NS4B contains six tryptophan residues and several of these amino acids were proximal to the locations that generated the genetic resistance data. Using this assay, the binding constant of **1** was determined as 2.2 µM. Importantly, the binding constant *versus* the mutant H94N NS4B was shifted to 17.6 µM, linking the relative binding potency *in vitro* to the relative antiviral activity. Furthermore, compound **2** (Figure 5.3), a potent inhibitor of genotype

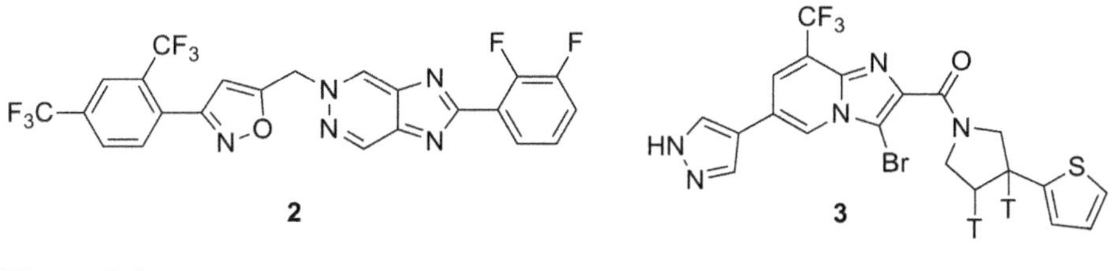

2   **3**

Figure 5.3 Compound **2**, a potent inhibitor of HCV replication, was used as a negative control in the NS4B tryptophan quenching assay. Compound **3**, identified through the optimization of the initial hit **1**, was labeled with tritium and was used to develop the NS4B scintillation proximity assay.

1b replication ($EC_{50} = 0.02\,\mu M$), showed no activity in the tryptophan-quenching assay *versus* wild-type or mutant NS4B, suggesting that the activity of **1** was due to a specific interaction with NS4B rather than a non-specific assay artifact. The almost 10-fold discrepancy between the NS4B binding assay ($2.2\,\mu M$) and the replicon activity ($0.6\,\mu M$) was not worrisome as many factors could explain the measured discrepancy; for example, the relative mimicry of the binding assay to the actual intracellular condition, the amount of 'active' NS4B actually present in the assay or the appropriate concentration of protein to utilize to most closely match the cellular situation are all factors that may account for the difference. To investigate further the interaction between the putative NS4B inhibitors and the NS4B protein, the development of a radioligand binding displacement assay was sought. Taking advantage of emerging structure–activity relationship (SAR) in the 'hit to lead' program (see below), a radiolabeled compound **3** (Figure 5.3) was synthesized and a scintillation proximity assay (SPA) was developed using the recombinant NS4B protein to measure compound affinity and displacement.[33] Compound **3** was found to bind in a dose-dependent manner ($K_D = 12\,nM$). When radioligand **3** was displaced with **1**, the displacement constant was found to be $0.21\,\mu M$, thus further strengthening the link between NS4B binding and HCV inhibition, and also providing an assay with sufficient throughput to allow the binding SAR to be determined, in addition to the anti-HCV activity, for the ongoing optimization program. The development of a biochemical assay is an important milestone for a program arising from a phenotypic screen in that it allows for the potential deconvolution of the multifactorial SAR that is inherent in optimizing from a cellular assay.

Continuing the investigation into the mechanism of these compounds, a plasmid that coded for NS4B linked to green fluorescent protein (GFP) was transfected into the Huh7.5 cell line (the same as used for the 1b replicon) and the NS4B was imaged with fluorescence microscopy in order to observe the localization of NS4B subcellularly.[9] In cells untreated with the compound, NS4B-GFP displayed an ER-associated pattern of localization, whereas treatment of these cells with **1** at $5\,\mu M$ clearly altered the subcellular distribution pattern of NS4B-GFP.

In spite of the complicated nature and relatively poor understanding of the role that NS4B plays in HCV infection, a variety of genetic, biochemical and molecular biological data suggest a reasonable hypothetical answer to the important question of 'how does **1** inhibit HCV replication'? Resistance generated at binding sites situated between AH2 and the transmembrane domain combined with the confirmed binding of anti-HCV compounds to free NS4B protein and their ability to disrupt the subcellular localization of NS4B clearly suggests a hypothesis wherein the compounds described here function by binding to NS4B and preventing the formation of the membranous web (and thus replication). More investigation will be required to test and refine this hypothetical and novel antiviral mechanism.

5.4 Lead Optimization of Series

The lead optimization effort relied heavily on phenotypic screening of new compounds in HCV replicon assays (genotypes 1a and 1b). Although a high-throughput NS4B binding assay was also developed, the upper limit of this assay was ~5 nM and, therefore, was of limited utility as compound potency improved beyond this threshold. In general, the binding data was used to confirm that changes in replicon potency were a direct result of the compound's affinity for NS4B and not due to off-target effects. The screening hit **1** displayed good affinity for NS4B (IC$_{50}$ of 30 nM) and modest potency against HCV genotype 1a and 1b replicons (EC$_{50}$ of 0.6 μM). Our initial program aim was to deliver a clinical candidate with low nanomolar activity (EC$_{50}$ < 10 nM) *versus* the 1a and 1b genotypes and a favorable pharmacokinetic (PK) profile to support qd dosing in humans. Given the likelihood for combination with other HCV drugs, it was critical that the candidate did not show cross-resistance to drugs with other mechanisms of action (*e.g.* protease inhibitors) or exhibit drug–drug interactions that would complicate development.

Although **1** was reasonably active, the molecular weight (MW = 409) and the cLogP (3.9) were higher than we would have preferred at the start of lead optimization. It is well documented that MW and lipophilicity tend to increase during the course of lead optimization, and such trends are cited as underlying causes for the high attrition rates of drugs in clinical development.[34–36] Therefore, a major goal was to increase HCV antiviral activity while maintaining reasonable physicochemical properties. Initially, we sought to avoid increases in MW by examining isosteric replacements of the pyrazolo[1,5-*a*]pyrimidine scaffold of **1**. Several alternative cores were synthesized (**5–8**) and the effects on replicon potency were compared with compound **4** (EC$_{50}$ of 1.5 μM), a derivative of **1** that was synthetically easier to prepare (Figure 5.4). Exchanging nitrogen for carbon at the 3-position of the pyrazolopyrimidine (**4**) proved detrimental to replicon potency. For example, both compounds **5** and **6** were less potent (EC$_{50}$s of 17 and 9 μM, respectively) than the original analog **4**. In contrast, replacing carbon with nitrogen at either of the ring fusion positions was well tolerated, with both the

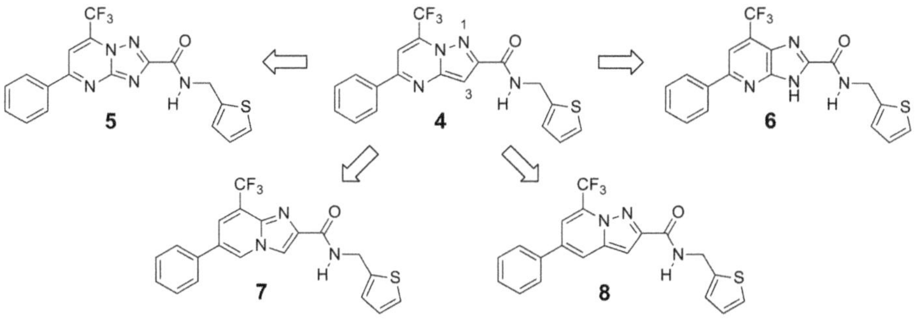

Figure 5.4 Isosteric modifications (**5–8**) to the original pyrazolo[1,5-*a*]pyrimidine core (**4**).

imidazo[1,2-*a*]pyridine (**7**) and pyrazolo[1,5-*a*]pyridine (**8**) analogs displaying similar low micromolar activity (EC$_{50}$s of 2.2 and 1.7 µM, respectively).

Our decision to pursue the imidazopyridine (**7**) core in spite of the fact that it was less active than the corresponding pyrazolopyrimidine (**4**) and pyrazolo-pyridine (**8**) scaffolds was influenced by several factors. First, the key imid-azopyridine intermediate **10** was easily prepared in one step from the commercially available 2-aminopyridine **9** (Figure 5.5).[33] The ability to access large quantities of **10** (>500 g) allowed the rapid development of SARs within the series and represented a significant advantage over the synthetically chall-enging pyrazolopyridine (**8**) scaffold. Second, compounds based on a pyrazo-lopyrimidine core (*i.e.* **1**) have been extensively prepared and profiled for a variety of therapeutic indications, including viral diseases.[8] In comparison, the imidazopyridine scaffold presented an opportunity to explore new chemical space and, therefore, became the primary focus of our efforts.

Compound **10** proved to be a versatile intermediate, allowing for modifi-cations at C3, C5 and the C2-amide (Figure 5.5). For example, **10** underwent electrophilic aromatic substitution at C3 with *N*-chlorosuccinimide (NCS) and was subsequently converted to **11**. The addition of chlorine to the imid-azopyridine core led to a fourfold increase in genotype 1b replicon activity compared with the C3-protio scaffold (compare the EC$_{50}$s of 0.5 and 2.2 µM for analogs **11** and **7**, respectively). Bromine was also well tolerated at this position and provided access to other functional groups such as methyl, cyano, and vinyl *via* Pd(0) cross-coupling reactions. Similarly, a wide variety of substituents could be selectively installed at C5 *via* Pd-catalyzed cross-coupling reactions. In this manner, the 3-furanyl moiety was introduced, thereby affording a compound with a high degree of structural similarity to the original screening hit (compare analogs **12** and **1**). Compound **12** shows a threefold improvement in genotype 1b replicon activity over analog **1** (EC$_{50}$s of 0.2 and 0.6 µM, respectively) with only a modest increase in MW (426 and 409, respectively).

The 3-chloro-5-furanylimidazopyridine core was combined with various amine fragments to probe the SAR at the C2-carboxamide position

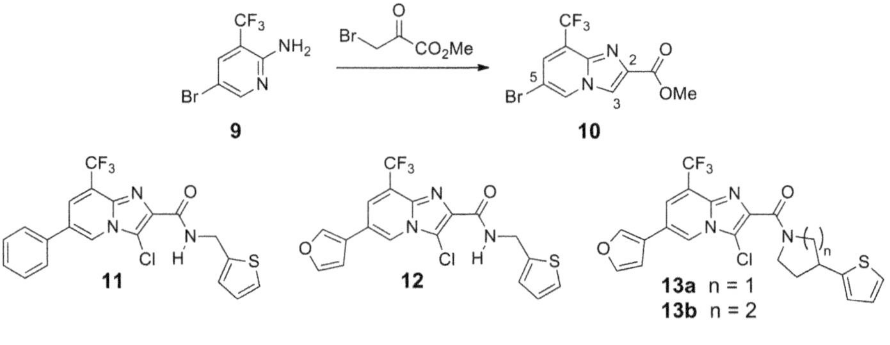

Figure 5.5   One-step synthesis of the versatile imidazopyridine analog **10** that permitted rapid modifications at C3, C5 and the C2-amide (analogs **11**, **12** and **13**, respectively).

(Figure 5.5). Replacing the secondary amide of **12** with a conformationally restrained tertiary amide [(±)-**13a**)] resulted in a threefold improvement in genotype 1b replicon activity (EC$_{50}$ of 0.07 µM). The pyrrolidine tail was subsequently combined with other core modifications (*e.g.* C3 = Br and C5 = 4-pyrazole) to afford the potent analog **3** (EC$_{50}$ of 0.03 µM). The radio-labeled version of **3** was the first compound in this chemical class to demonstrate direct binding to NS4B protein (K_d of 12 nM) and served as the ligand for our initial NS4B scintillation proximity assay. Progress within the pyrrolidine series was often hindered by the need for resolution of enantiomers; fortunately, the pyrrolidine could be replaced with a symmetrical piperidine to afford a derivative (**13b**) that also possessed high affinity for NS4B (IC$_{50}$ of 7 nM). Elimination of the chiral center and access to commercial (or readily prepared) piperidine intermediates accelerated the development of SARs pertaining to the tail fragment.

Our initial optimization efforts within the imidazopyridine series had succeeded in improving replicon activity while avoiding dramatic increases in MW (relative to the screening hit **1**). However, in many cases the increase in potency was associated with an undesirable increase in lipophilicity. The thiophene moiety significantly contributed to the high cLog*P* of the molecules and was also associated with rapid metabolism and irreversible trapping of glutathione in our *in vitro* reactive metabolite assays. We sought to replace the non-polar heterocycle with more water-soluble groups (Figure 5.6). The thiazole analog **14** was equipotent to thiophene **13b** (EC$_{50}$ of 0.12 µM *versus* genotype 1b replicon) and lowered the cLog*P* (3.8 *versus* 5.2, respectively). While heteroaromatic groups in general were well tolerated, the basic pyrrolidine heterocycle **15** was weakly active in both the replicon and NS4B binding assays (EC$_{50}$ of ~4 µM). Given that the basic amine is protonated at physiological pH, the resulting charged molecule is likely not well suited for binding to the hydrophobic NS4B protein. Consistent with this hypothesis, reducing the basicity of the amine by introducing a carbonyl group in the form of a lactam (**16**) restored the 1b replicon activity (EC$_{50}$ of 0.16 µM) while further reducing cLog*P* (2.3).

Modification of the cyclic lactam of **16** was challenging, with most changes decreasing replicon activity. Replacing the α-methylene group with oxygen (Figure 5.7) was tolerated and yielded a compound (**17**) with low nanomolar potency in the genotype 1a replicon assay (EC$_{50}$ of 7.4 nM). In addition, **17**

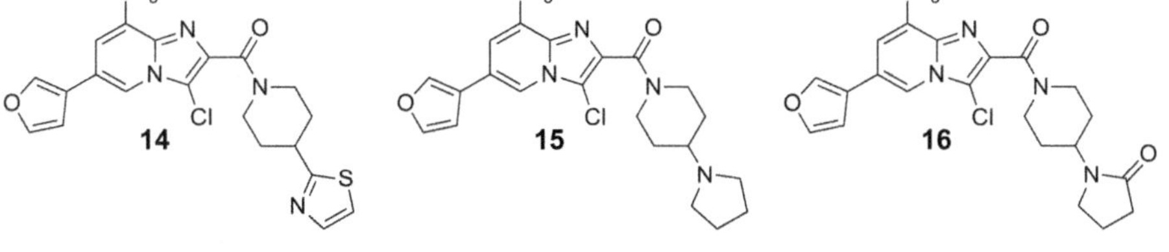

Figure 5.6 Replacement of the lipophilic thiophene moiety with polar substituents (**14–16**) in the tail region to lower cLog*P*.

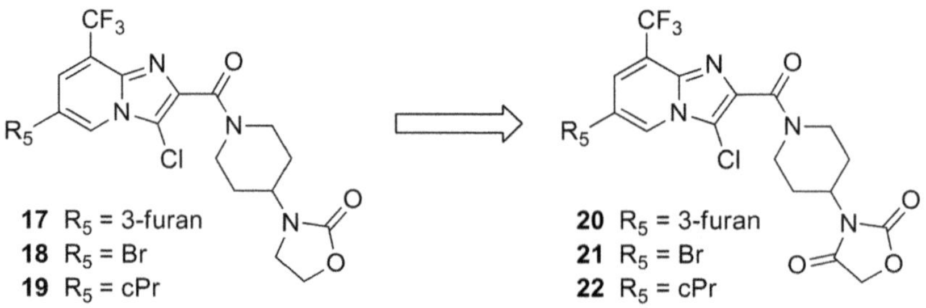

Figure 5.7 A series of oxazolidinones (**17–19**) were identified that potently inhibit HCV replication but undergo rapid metabolism *in vivo* to the corresponding oxazolidinediones (**20–22**). The diones were found to induce mRNA production of cytochrome P450s in rat and human hepatocytes.

displayed an acceptable PK profile in multiple species. Unfortunately, the low cLog$P$ (2.6) associated with **17** did not translate to high aqueous solubility (1.5 µg mL^{-1} in fasted-state simulated intestinal fluid), which ultimately manifested in the inability to obtain high plasma drug exposures in the rat upon escalating the dose (Figure 5.8a). Surprisingly, the bromo-variant **18** was able to achieve much higher plasma drug levels in the rat relative to **17** (Figure 5.8a), albeit with a slight drop in replicon activity (EC$_{50}$ of 31 nM *versus* genotype 1a replicon).

Compound **18** was advanced into a rat 7 day dose-range finding (DRF) study and was well tolerated up to 2000 mg kg^{-1} per day. However, the toxicokinetic (TK) profile showed a dramatic drop ($\sim$80%) in plasma drug exposure from day 1 to day 7 (Figure 5.8b). We also observed high plasma concentrations of a circulating active metabolite (**21**, EC$_{50}$ of 73 nM), which was formed upon oxidation of the oxazolidinone group. The time-dependent decrease in drug exposure and histological evidence of hepatocellular hypertrophy were indicative of drug-induced metabolism. In many such cases, induction of metabolism results from drug activation of nuclear receptors such as pregnane X receptor (PXR), constitutive androstane receptor (CAR), or aromatic hydrocarbon receptor (AhR) that result in upregulation of xenobiotic metabolizing enzymes. Both **18** and its metabolite **21** were profiled for enzyme induction in rat hepatocytes and **21** was shown to induce mRNA production of a number of cytochrome P450s at physiologically relevant concentrations. The two compounds also induced CYP mRNA levels in human hepatocytes, suggesting that drug-induced metabolism would also occur in humans. Although oxazolidinone substitution yielded potent HCV replicon inhibitors, the rapid *in vivo* metabolism to form the oxazolidinedione (*i.e.* **20–22**) caused considerable concern for potential drug–drug interactions. Efforts to block oxidation and formation of the oxazolidinedione metabolite were unsuccessful and the series was ultimately terminated. However, we were successful in replacing the C5 furan (**17**) or bromo (**18**) substituents, both of which significantly increased MW and lipophilicity. Ultimately, the C5-cyclopropyl exhibited the optimum balance of potency *versus* physicochemical properties as

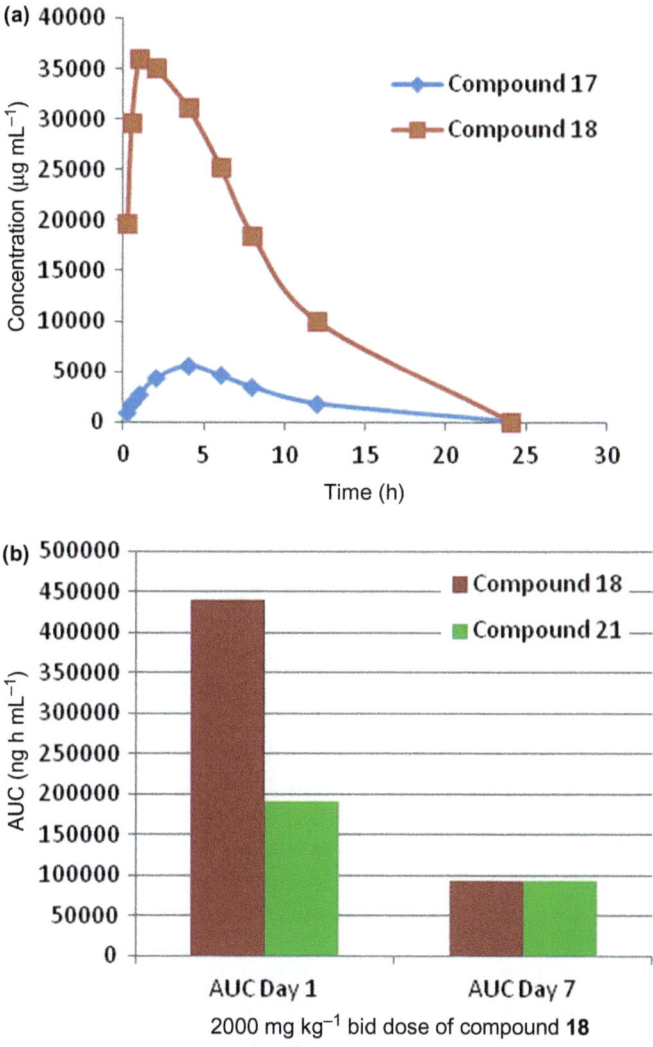

Figure 5.8 (a) Comparison of the drug plasma exposures (AUC_{0-24h}) in rats for compounds **17** and **18**. (b) The drug plasma exposures (AUC_{0-24h}) of compound **18** and its active metabolite **21** in rats on day 1 and day 7 upon bid dosing of **18** at $2000\,mg\,kg^{-1}$.

in the case of analog **19**, which displayed an EC_{50} of 7.8 nM (genotype 1a) with a cLog*P* of 2.5. Therefore, C5-cyclopropyl was introduced on all subsequent analogs.

The impressive replicon activity achieved by the addition of the oxazolidinone substituent (*e.g.* **17**–**19**) suggested that a hydrogen-bond acceptor in this portion of the molecule might serve as an important pharmacophore. *In silico* modeling guided the design of other amine tail fragments aimed at preserving this structural feature. The structurally diverse analogs **23**–**25** were

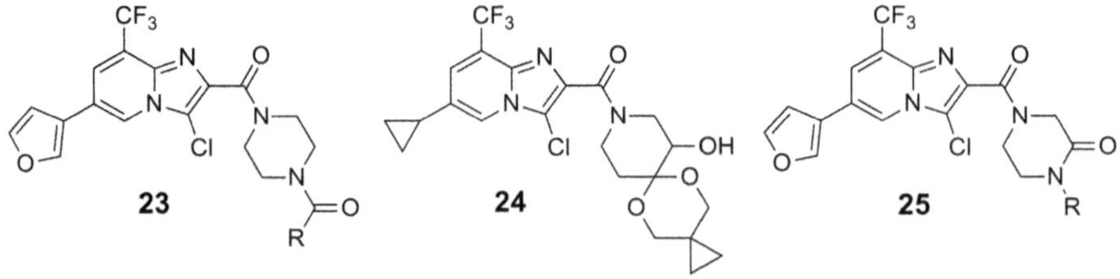

Figure 5.9   A series of analogs (**23–25**) designed to maintain the key hydrogen-bond acceptor (C=O or C–OH) in the tail region of the molecule.

synthesized and found to possess antiviral activity (Figure 5.9). For the piperazine carboxamide derivatives **23**, the plethora of commercially available carboxylic acid reagents permitted the preparation of hundreds of analogs. The SARs revealed that replicon activity was highly dependent on the electronic nature of the side chain (R), as replacing the carboxamide group with isosteric urea or carbamate moieties negatively impacted replicon potency. Within the carboxamide series, relatively small steric modifications in the R group produced significant changes in potency. For example, substitution of the amide (R) with Me and Et resulted in poor activity (EC$_{50}$ of 4 and 1 μM, respectively), while addition of small cycloalkyl substituents proved beneficial. More specifically, increasing the size of the ring from cyclopropyl to cyclobutyl to cyclopentyl led to a progressive improvement in potency against the genotype 1a replicon (EC$_{50}$s of 88, 12 and 5.0 nM, respectively). There were limitations to size of the apparent binding pocket as further ring expansion to the cyclohexyl decreased activity (EC$_{50}$ of 340 nM). Consistent with this finding, substitutions on the cycloalkyl rings (*e.g.* R = cyclobutyl) of **23** were restricted to sterically small groups (*e.g.* methyl, fluoro, hydroxyl). Unfortunately, the *in vivo* rat clearance was uniformly high for this compound class and further efforts were discontinued.

Molecular modeling of **17** suggested that the key carboxamide H-bond acceptor (C=O) could potentially make similar interactions if transposed to the neighboring ring. This element was incorporated into the design of two distinct series, a hydroxyl spirocyclic ketal **24** and the piperazinone **25** (Figure 5.9). The racemic spirocycle **24** exhibited low nanomolar potency against both genotypes 1a and 1b (EC$_{50}$s of 14 and 6.3 nM, respectively) and possessed a cLog*P* (2.7) in the aspirational range (<3). In addition, the ketal was stable under physiologically relevant conditions and was, therefore, explored in combination with other core scaffolds aimed at replacing the imidazopyridine core (see below).

The piperazinone series (**25**) also afforded highly active HCV replication inhibitors.[33] The initial SARs associated with piperazinone substitution (R) reflected the observations made in the piperazine carboxamide series (**23**), with the H and Me analogs inactive and small cycloalkyl groups imparting replicon potency. In the piperazinone series (**25**), the cyclopropyl compound was weakly active (EC$_{50}$ > 2 μM) while the cyclobutyl, cyclopentyl and cyclohexyl analogs potently inhibited 1a and 1b genotypes (EC$_{50}$s < 20 nM). As seen previously, there was a limit to the size of the ring that could be accommodated as the cycloheptyl derivative was ~10-fold less active relative to the cyclohexyl analog. Both the cyclopentyl and cyclohexyl derivatives tested positive for the formation of reactive metabolites (rat and human microsomes) due to apparent oxidation of the C5-furan followed by the addition of glutathione. Replacing the furan with cyclopropyl (**26**, Table 5.1) eliminated this liability and yielded a potent inhibitor of HCV replication.

Although the potency of the piperazinone **26** was attractive, the high rat clearance (Cl_{rat} > hepatic blood flow) precluded further development. An analysis of the metabolites generated *in vitro* from rat and human hepatocytes showed rapid oxidation of the terminal cyclohexyl substituent. Efforts to block the metabolism *via* methylation or fluorination of the ring were unsuccessful. Realizing that the cyclohexyl substituent was cleared *via* an oxidative pathway, we became interested in introducing hydroxyl groups at various positions of the ring as a means to alter metabolism and reduce lipophilicity. The racemic 2-hydroxy (*trans*-**27**) and 3-hydroxy (*cis*-**28** and *trans*-**29**) cyclohexyl analogs were significantly less active than the unsubstituted cyclohexyl analog (**26**) in the replicon and NS4B binding assays (Table 5.1). The 4-hydroxy derivatives proved far more interesting with the *trans*-4-OH analog **31** exhibiting low nanomolar potency against the 1a and 1b replicons. The spatial orientation of the hydroxyl group was important as the corresponding *cis*-isomer **30** was 20–100-fold less potent. As expected, the addition of the hydroxyl group reduced the cLog*P* relative to the unsubstituted analog.

An important benefit of adding the 4-*trans*-OH group to the cyclohexyl substituent (*i.e.* **31**) was the substantial improvement in PK relative to the unsubstituted derivative **26**.[33] The clearance of **31** was low (<10% hepatic blood flow) in mouse, dog and cynomolgus monkey and moderate in rat (Table 5.2). Low oral doses of **31** (5 mg kg^{-1} in a solution formulation) achieved high oral bioavailability (*F* > 80%) in three of the four species. Furthermore, **31** did not inhibit any human cytochrome P450 isozymes (IC$_{50}$ > 30 µM *versus* 3A4, 1A2, 2C9, 2C19 or 2D6) and was negative in the glutathione-trapping reactive metabolite assay (rat and human microsomes). Based on experience with the oxazolidinone compounds (**17–19**, Figure 5.7), the potential for **31** to induce metabolism was also evaluated. Compound **31** did not activate rat or human PXR and did not upregulate the mRNA of the major P450 isozymes in rat, dog, or human hepatocytes.

The high binding affinity to NS4B, attractive antiviral profile, and favorable low-dose PK supported the further preclinical development of **31**. Despite the relatively low cLog*P* (3.3), the FaSSIF solubility of **31** was poor (11 µg mL^{-1}), which negatively impacted the ability to achieve the higher plasma drug exposures needed to support preclinical safety studies. For example, a 300 mg kg^{-1} suspension dose of crystalline **31**, wet-bead milled to maximize the surface area for faster dissolution, achieved low levels of exposure in rat (AUC$_{0-24h}$ = 11.3 µg h mL^{-1}, Figure 5.10). Ultimately, we examined the use of prodrugs to overcome this liability. Two common prodrug classes were selected (phosphate **32** and esters **33–36**) and appended to the 4-hydroxyl group.

Table 5.1 Comparison between a series of hydroxylated cyclohexanol analogs (**27–31**) and the unsubstituted derivative **26** with respect to cLog$P$, binding and replicon potency.

26–31

EC_{50} (nM)

Assay	Genotype	26	(+/−)-27	(+/−)-28	(+/−)-29	30	31
cLog$P$		5.2	4.3	3.3	3.3	3.3	3.3
Replicon	1a	6.0	120	630	96	78	0.8
Replicon	1b	2.1	50	540	42	120	5.4
Binding	1b	8.3	130	580	120	300	53

Table 5.2 Pharmacokinetic parameters of **31** in four species.

	i.v. (1 mg kg^{-1})				p.o. (5 mg kg^{-1})	
Species	Cl (mL min^{-1} kg^{-1})	Cl (% hepatic blood flow)	$T_{\frac{1}{2}}$ (h)	V_{dss} (L kg^{-1})	$AUC_{0-\alpha}$ (µg h mL^{-1})	F (%)
Mouse	4.7	5	1.8	0.5	6.6	37
Rat	20	36	4.2	4.1	3.8	98
Dog	1.1	3	13	0.9	52	65
Monkey	2.1	5	4.9	0.7	50	>100

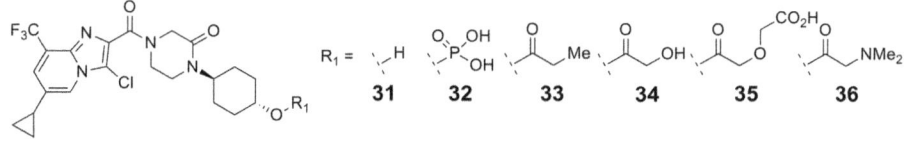

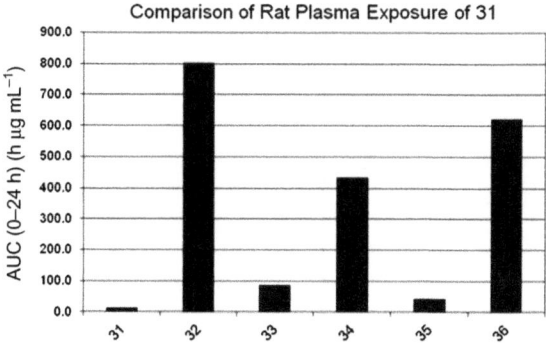

Figure 5.10 Plasma drug exposures of 31 in rat following p.o. administration of 300 mg kg^{-1} of 31 or prodrugs 32–36.

A diverse set of esters were examined to determine the effects of polarity and ionic character on high-dose oral PK. The prodrugs 32–36 were dosed at 300 mg kg^{-1} (p.o.) in rats and the plasma concentrations of 31 were determined (Figure 5.10).[37] The phosphate prodrug 32 outperformed the ester prodrugs, achieving the highest plasma exposures of 31 with no evidence of systemic exposure for the uncleaved prodrug. The phosphate prodrug also formed a stable, crystalline solid and was selected for all forthcoming *in vivo* studies.

The high plasma exposure achieved with prodrug 32 enabled the progression of 31 into rat 7 day safety studies. Unfortunately, an adverse cardiovascular finding was identified that led to its termination. In addition, *in vitro* resistance passaging studies identified several single-point mutations in the NS4B protein (positions 94 and 105) that render the HCV replicons partially resistant to 31.[38] A backup effort was initiated to address these limitations: we aimed to eliminate the need for a prodrug and to improve the viral resistance profile. A significant number of analogs (>1500) had already been prepared and profiled in the imidazopyridine series and, therefore, we chose to pursue new chemical directions.

Recalling that the pyrazolopyridine 8 possessed slightly improved potency relative to the imidazopyridine 7 (Figure 5.4), we re-examined isosteric replacements to the imidazopyridine core in combination with highly optimized tail fragments. Our initial aim was the syntheses of the pyrazolo[1,5-*a*]pyridine 37, *N*-methylbenzimidazole 38, and benzofuran 39 cores in combination with the 4-hydroxycyclohexylpiperazinone (37–39a) and the spirocyclic ketal (37-39b) tail fragments (Table 5.3).[37] Stable genotype 1b replicons bearing the H94N mutation in NS4B were created to allow the development of SARs. The three cores (37–39) with either of the tail subunits (a or b) showed comparable or improved potency relative to the imidazopyridine core (24 and 31). Activity

Table 5.3 cLog*P* and wild-type (WT) and H94N genotype 1b replicon potency for a series of isosteric core modifications (**37–40**) bearing two structurally unique tail fragments (**a** and **b**).

R

24, 31   **37**   **38**   **39**

31, 37–39a   **24, 37–39b**

Compound	cLog*P*	WT (nM)	H94N (nM)	cLog*P*	WT (nM)	H94N (nM)
31/24	3.3	7.0	200	2.7	16	230
37a/37b	3.3	2.0	32	2.7	2.4	39
38a/38b	2.6	3.0	150	2.1	11	>500
39a/39b	4.4	0.8	22	3.7	0.6	25

against the wild-type replicons was similar for all of the analogs but was significantly diminished *versus* the H94N variant. With respect to replicon potency, the pyrazolopyridine **37** and benzofuran **39** cores appeared superior to the benzimidazole **38** scaffold, especially against the H94N mutant replicon. The benzofuran derivatives appeared slightly more potent than the corresponding pyrazolopyridine analogs, but this core was also associated with a greater increase in lipophilicity (cLog*P*). For this reason, we opted to explore the pyrazolopyridine series further.

The development of the SARs within the pyrazolopyridine series required access to large quantities of the intermediate acid **45**. The 10-step synthetic route relied on the *de novo* construction of the pyrazolopyridine core from the commercially available pyridine **40** (Scheme 5.1).[37] Unfortunately, the critical step involving *N*-amination of the pyridine was problematic owing to the low reactivity of the nitrogen, a consequence of the electron-withdrawing CF_3 substituent. Traditional reagents for this transformation, such as hydroxylamine-*O*-sulfonic acid (HOSA), failed to deliver the desired product and a more reactive reagent was required. Treating **40** with *O*-(mesitylene-sulfonyl)hydroxylamine (MSH) afforded **41**,[39,40] which subsequently underwent a 1,3-dipolar cycloaddition reaction with dimethylacetylene dicarboxylate (DMAD) to give the pyrazolopyridine core **42** in 30% yield.[41,42] Heating the diester **42** in the presence of H_2SO_4 led to hydrolysis followed by selective decarboxylation at C3 of the pyrazolopyridine.[43] The resulting intermediate was then converted to the requisite acid **45** in six additional steps.[37]

The initial route provided gram quantities of acid **45**, which was sufficient for developing SARs within the pyrazolopyridine series. However, the progression of lead molecules into *in vivo* animal studies demanded greater compound quantities and necessitated a more efficient synthesis. The low yield for the formation of the pyrazolopyridine ring system and the required use of MSH, a thermally unstable reagent, were major limitations to the existing route. The simplest solution to both issues was to avoid the *N*-amination reaction and begin with the pyrazolopyridine ring system already intact. Unfortunately, no 7-(trifluoromethyl)pyrazolopyridine analogs were available from commercial

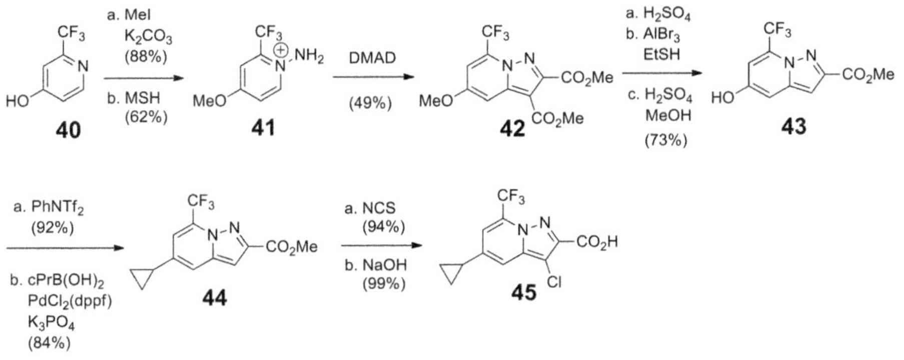

Scheme 5.1 First-generation route to the pyrazolopyridine core acid **45**.

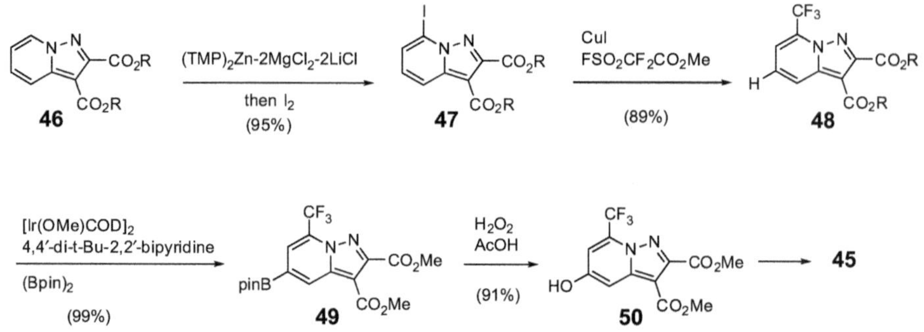

Scheme 5.2   Second-generation synthesis of **45**.

vendors. However, the des-CF$_3$ analog **46** (R = Me or Et) could be purchased and represented a potentially viable starting point (Scheme 5.2).

Multiple attempts to deprotonate C7 of **46** selectively with LDA or TMPMgCl·2LiCl failed to produce any of the desired product. It seemed likely that the two ester moieties could be reactive under such harsh reaction conditions, so a milder, more selective base was employed. Treating **46** (R = Me) with (TMP)$_2$Zn·2MgCl$_2$·2LiCl in THF at room temperature followed by the addition of I$_2$ gave the 7-iodopyrazolopyridine **47** in 60–89% yield.[44,45] Overall, the reaction proceeded cleanly but some hydrolysis of the methyl ester at C3 was also observed. Switching to the less reactive ethyl ester **46** (R = Et) decreased unwanted hydrolysis and afforded the desired 7-iodo analog **47** in 85% yield (~96:4 ratio of **47** to the hydrolyzed side product). The reaction conditions were optimized whereby the addition of 0.7 equiv. of (TMP)$_2$Zn·2MgCl$_2$·2LiCl in THF at −10 °C followed by the addition of I$_2$ gave **47** in 95% isolated yield.[46]

With the functionalized C7-pyrazolopyridine in hand, the conversion of the aryl iodide to the corresponding –CF$_3$ derivative was required. Historically, this has been difficult but within the past 5 years, a number of metal-mediated procedures have been developed to effect this transformation.[47,48] Initial efforts using copper(I) iodide and CF$_3$TMS (Ruppert's reagent) as a trifluoromethyl donor in the presence of CsF gave the desired product **48** along with multiple side products (Scheme 5.2). Other trifluoromethylation reagents were examined and methyl fluorosulfonyldifluoroacetate proved to be superior, affording **48** in 89% yield within 2 h at 80 °C.[46] The reaction is robust and does not require purification of the iodide intermediate **47**, thereby making the overall sequence amenable to scale-up. To date, the zincation/iodination/trifluoromethylation procedure has been carried out successfully on 300 g of starting material **46**, providing **48** in 86% isolated yield.

A consequence of utilizing the commercially available pyrazolopyridine starting material **46** was the need to introduce functionality selectively at the C5 position of **48** (Scheme 5.2). The insertion of transition metals into C–H bonds has emerged as a powerful method for the functionalization of otherwise unactivated aromatic ring systems. In an example involving an iridium-catalyst,[49–51] the selectivity for C–H bond activation is dominated by steric effects which appeared well suited for substrate **48** as the C–H bond at C5 is the least sterically hindered. Treating **48** with [Ir(OMe)COD]$_2$, a bipyridine ligand, and bispinacolborane led to the regioselective borylation at C5 to afford **49** in quantitative yield. Oxidation of the boronate ester with H$_2$O$_2$ gave the phenol **50**, which was converted to **45** in six steps.[37]

The pyrazolopyridine acid **45** was coupled with a small set of structurally diverse amines that were identified during the optimization of the imidazopyridine scaffold. Initially, 56 analogs were prepared and in each case a corresponding imidazopyridine compound existed. Profiling the matched molecular pairs in the NS4B binding assay showed that in most cases (55 of 56), the pyrazolopyridine analog had greater NS4B binding affinity relative to its imidazopyridine counterpart (Figure 5.11). The same trend was observed within the HCV replicon data. On average, the pyrazolopyridine core improved potency by ∼0.5 log units over the imidazopyridine.

The major objective of the lead optimization effort was to improve the viral resistance profile. Therefore, we prepared over 300 pyrazolopyridine analogs and profiled them in the H94N replicon assay. As shown in Figure 5.12, a reasonable correlation existed ($r^2 = 0.68$) between inhibition of wild-type virus and inhibition of the H94N mutant (genotype 1b). In general, the analogs were ∼10-fold (1.0log$_{10}$ unit) less active in the H94N assay. Although 96 compounds were identified with EC$_{50}$s < 10 nM *versus* genotype 1b virus, essentially no compounds exhibited similar activity *versus* H94N. However,

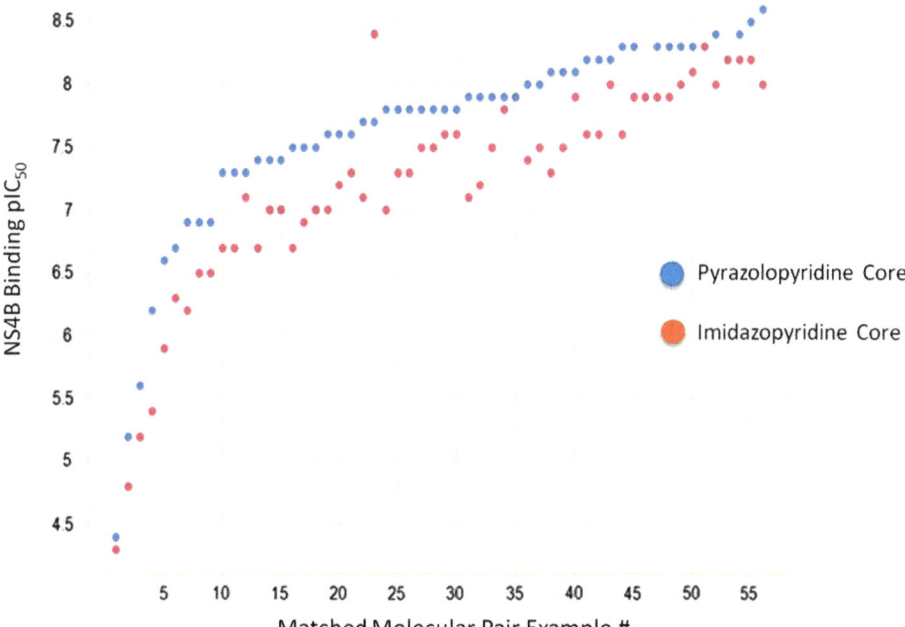

Figure 5.11 Analysis of matched-molecular pairs containing a pyrazolopyridine (blue) or imidazopyridine (red) core with respect to NS4B binding affinity (pIC$_{50}$).

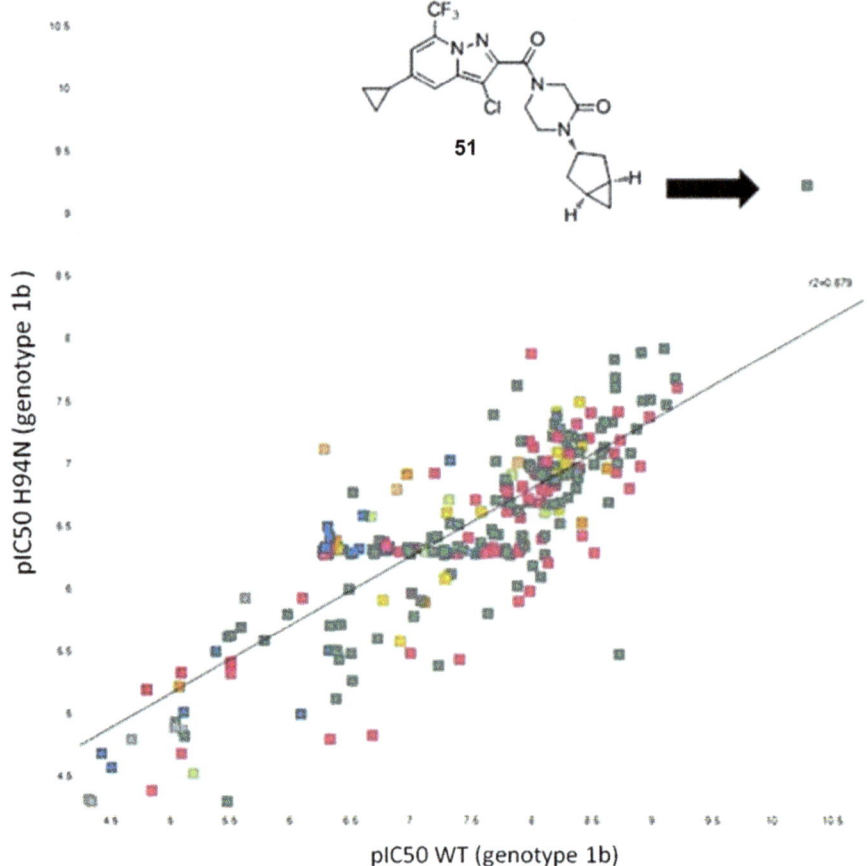

Figure 5.12   Comparison of wild-type (WT) and H94N genotype 1b replicon potency
(pIC$_{50}$) within the pyrazolopyridine series (colors correspond to different
structural sub-classes of tail fragments). The analysis identified one
analog (**51**) with superior wild-type and H94N replicon potency.

there was one notable exception: **51** was remarkably potent against both wild-
type and H94N genotype 1b viruses (EC$_{50}$s of 0.05 and 0.7 nM, respectively).
As far as we are aware, **51** represents the most potent inhibitor of HCV
replication reported to act *via* an NS4B mechanism of action.

Our initial excitement regarding **51** was tempered by the poor PK profile
observed in rats. As seen previously with unsubstituted cycloalkyl substituents
in the imidazopyridine series (*e.g.* **26**), rapid oxidation of the bicyclohexane ring
resulted in high *in vivo* clearance (Cl$_{rat}$ > hepatic blood flow). For **26**, we had
successfully lowered the clearance while maintaining antiviral potency by
introducing a hydroxyl group at the 4-position of the cyclohexyl ring (*i.e.* **31**).
To determine if a similar tactic would improve the metabolic clearance of **51**, a
series of [3.1.0]bicyclohexanol analogs (**52–54**) were envisioned (Figure 5.13).
Initially, analog **52** was deemed most attractive given the lack of stereogenic

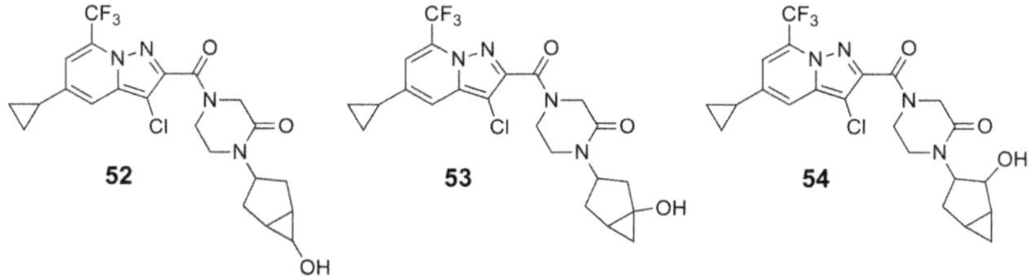

Figure 5.13 A series of envisioned hydroxylated [3.1.0]bicyclohexane derivatives (**52–54**) aimed at reducing metabolic clearance and lipophilicity.

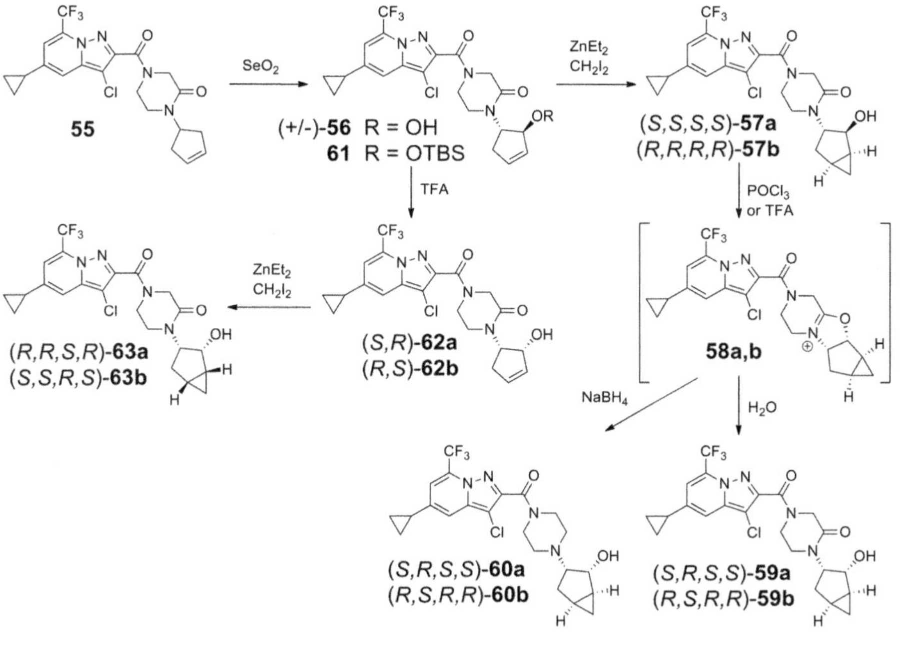

Scheme 5.3 Syntheses of a series of [3.1.0]bicyclohexan-2-ol derivatives.

centers and structural similarity with the 4-cyclohexanol derivatives known to bind NS4B (*e.g.* **31** and **37–39a**). Unfortunately, the desired amine intermediate appeared unstable under acidic conditions and further efforts were abandoned. There were stability concerns and a lack of literature precedent related to analog **53**, so attention shifted to the 2-hydroxybicyclohexane derivative **54**.

The synthesis of **54** began with allylic oxidation of the pendant cyclopentene ring of **55**, which occurred opposite the bulky piperazinone substituent, thereby affording **56** as a racemic mixture (Scheme 5.3).[52] The resulting hydroxyl group directed the cyclopropanation to the *syn* face of the olefin to give (±)-**57**,[53,54] which upon resolution by chiral supercritical fluid chromatography (SFC) yielded enantiomers (+)-**57a** and (−)-**57b**.[37] In the course of exploring the SARs within the bicyclohexan-2-ol series, we found that treatment of **57b** with POCl₃ led to the epimerization of the 2-hydroxyl group. Other strong acids, such as TFA and HCl, also promoted epimerization. The inversion appears to occur *via* the cyclic intermediate **58b**, which upon aqueous workup ring opens to give the diastereomeric alcohol **59b**. The corresponding enantiomer **59a** was prepared in an analogous fashion upon treatment of **57a** with TFA. Separately, the cyclic intermediates **58a,b** could be reduced with NaBH$_4$ to afford the piperazine derivatives **60a,b**.

We were able to exploit this serendipitous finding further in the synthesis of another diastereomer (**63a,b**) by epimerizing the alcohol of **61** before installing the cyclopropyl group (via intermediate **62a,b**). Ultimately, we were successful in preparing six of the eight possible stereoisomers.[37]

The low yield for the allylic oxidation of **55** and the need for chiral resolution of the resulting racemic alcohols (±)-**57** necessitated the development of a more efficient synthesis. Utilizing the Evans oxazolidinone (Scheme 5.4), an asymmetric *anti*-aldol reaction between **64** and cinnamaldehyde installed two of the four desired stereocenters with a high diastereomeric ratio (19 : 1) to give **65**.[55] Ring-closing metathesis of the diene **65** with the second generation Grubb's catalyst afforded the cyclopentenol **66** in 94% yield. The hydroxyl-directed cyclopropanation of the olefin, followed by protection of the alcohol and cleavage of the chiral auxiliary, gave intermediate **67** as a single diastereomer. Curtius rearrangement of the carboxylic acid with DPPA and hydrolysis of the isocyanate with KOTMS afforded the amine **68** in 95% yield.[56] The piperazinone was formed in two steps from **68** *via* amide formation with Nosyl-protected glycine followed by alkylation/ring closure with dibromoethane. Removal of the Nosyl group with thiophenol[57] gave **69**, which was subsequently coupled to the pyrazolopyridine acid **45**. The silyl protecting group was removed with TBAF, thereby affording (+)-**57a** as a single isomer in >95% enantiomeric excess. The absolute stereochemistry of (+)-**57a**, was confirmed by single-crystal X-ray crystallography. The 11-step route (from **64**) provided access to >100 g of (+)-**57a** in 33% overall yield.

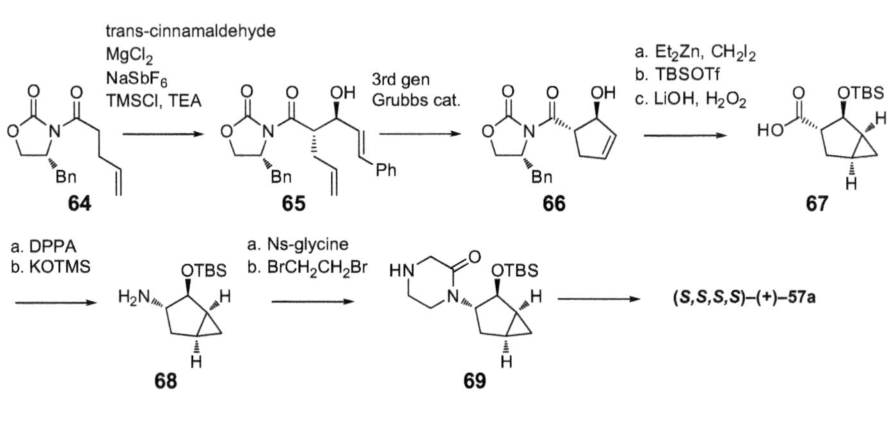

Scheme 5.4   Asymmetric synthesis of the (*S,S,S,S*)-[3.1.0]bicyclohexan-2-olpiperazinone tail fragment of (+)-**57a**.

The bicyclohex-2-ol derivatives **57a,b**, **59a,b** and **63a,b** showed similar binding affinity for NS4B, which correlated well with the genotype 1b replicon potency (Table 5.4). However, much larger differences were observed *versus* the H94N virus. For example, three compounds (**59a**, **63a**, **63b**) were inactive against the H94N variant and (+)-**57a** exhibited low nanomolar activity. The antiviral potency of (+)-**57a** represents a vast improvement over the majority of compounds in the series, despite being about eightfold less potent than the unsubstituted analog **51**. The corresponding enantiomer (−)-**57b** also showed good replicon potency, although it was less active than (+)-isomer. As expected, removing the key H-bond acceptor (C=O) from the piperazinone ring (analogs **60a** and **60b**) resulted in a marked decrease in binding and replicon activity (EC_{50}s of 264 and 50 nM *versus* genotype 1a replicon, respectively).

The initial justification for pursuing the complex bicyclohexan-2-ol tail of was the hypothesis that introducing a hydroxyl group would improve metabolic stability. The PK profile of (+)-**57a** was examined and, consistent with the findings in the cyclohexyl series (*i.e.* **31**), the *in vivo* rat clearance of (+)-**57a** was about threefold lower than that of **51** (Cl_{rat} of 25 *versus* $70 \, \text{mL} \, \text{min}^{-1} \, \text{kg}^{-1}$, respectively) (Table 5.5). Overall, (+)-**57a** exhibited low to moderate clearance across all four species and high bioavailability (>90%) in all species except mouse. In addition, higher doses of (+)-**57a** yielded plasma drug exposures that were sufficient to support animal safety studies without the need for a prodrug.

Using potency, metabolic stability and physicochemical properties to guide the medicinal chemistry direction, we successfully transformed our initial lead **1** into the highly optimized (+)-**57a**. The net result was a >1200-fold increase in replicon 1a and 1b activity and with only a modest inflation of MW (increase of 77 amu). Equally important was the net reduction in lipophilicity that accompanied the optimization of **1** into (+)-**57a** (cLog*P*s of 3.9 and 3.6, respectively). A number of synthetic challenges were overcome to enable the profiling of this molecule, including the development of a novel, scalable route to the pyrazolopyridine core and an asymmetric route to the cyclopentenol tail bearing four contiguous chiral centers. The high binding affinity for NS4B, potent antiviral activity against the HCV replicons, and favorable PK profile in multiple species supported the preclinical development of (+)-**57a**, the details of which will be published later.

5.5   *In Vivo* Proof of Concept

Several small molecules have previously been reported to interact directly with NS4B but no clinical or *in vivo* data have emerged to validate this target as a viable approach for treating HCV infection. Having identified a number of potent antiviral compounds with high affinity for NS4B, we aimed to demonstrate *in vivo* efficacy *via* this novel mechanism of action. The human specificity of the HCV virus has hindered the development of preclinical animal models of infection.[58] Chimpanzees are susceptible to HCV and have been used to study the disease and potential treatments;[59] however, studies in chimpanzees are undesirable for many reasons. Therefore, we focused on a model involving

Table 5.4 cLogP and binding and HCV replicon activity for a series of [3.1.0]bicyclohex-2-ol derivatives.

Assay	Genotype	(+)-57a	(−)-57b	59a	59b	63a	63b
cLogP		3.6	3.6	3.6	3.6	3.6	3.6
Binding	1b	8.0	23	18	18	16	23
Replicon	1b	0.3	2.3	14	4.9	30	19
Replicon	1b H94N	5.0	38	>1000	64	>1000	>1000
Replicon	1a	0.6	4.0	41	7.9	120	95

EC_{50} (nM)

Table 5.5 Pharmacokinetic parameters of (+)-**57a** in four species.

| | *i.v. (1 mg kg^{-1})* | | | | *p.o. (5 mg kg^{-1})* | |
Species	*Cl (mL min^{-1} kg^{-1})*	*Cl (% hepatic blood flow)*	*T$_{\frac{1}{2}}$ (h)*	*V$_{dss}$ (L kg^{-1})*	*AUC$_{0-\alpha}$ (μg h mL^{-1})*	*F (%)*
Mouse	19	22	0.4	0.6	1.0	12
Rat	25	38	2.0	2.0	3.1	92
Dog	6.0	16	2.0	0.7	25	>100
Monkey	3.0	14	4.0	0.7	25	97

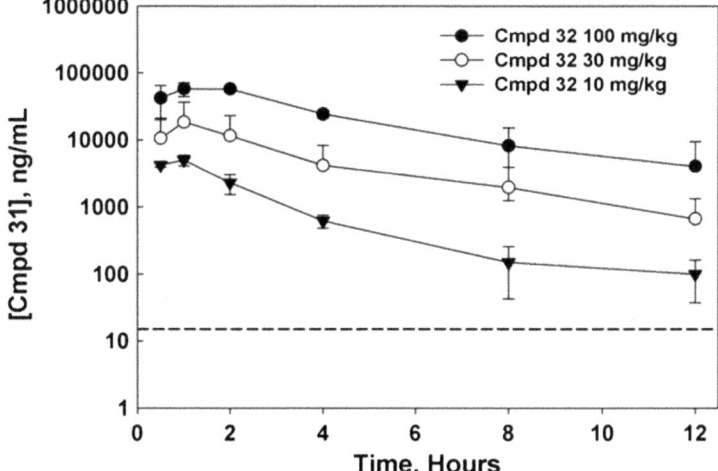

Figure 5.14 Drug plasma concentrations of **31** achieved upon p.o. administration of prodrug **32** to PXB mice prior to infection with HCV. The dashed line represents the serum-shifted EC$_{90}$ (14 ng mL^{-1}) for **31** *versus* the HCV genotype 1a virus used to infect the PXB mice.

chimeric 'humanized' mice. The PXB mice are uPA/SCID mice with livers that have been repopulated with 70–90% human hepatocytes.[60] As a result, the mice can be infected with human liver pathogens such as HCV and HBV. We decided to use the PXB model to test whether an HCV antiviral agent that interacts with NS4B can inhibit viral replication *in vivo*.

Prodrug **32** was selected for use in the efficacy study, in part due to the high plasma drug exposures obtained in the rat PK studies. To ensure that the prodrug would also perform well in the efficacy model, the pharmacokinetic profile of **32** was determined in PXB mice prior to infection with HCV (Figure 5.14).[37] Oral administration of **32** at doses of 10, 30 and 100 mg kg^{-1} resulted in a linear increase in plasma drug exposure of **32** (AUC$_{0-12h}$ = 15, 66, 280 μg h mL^{-1}, respectively). In both dose groups, the 12 h plasma concentrations (C$_{12h}$) of **31** were well above the serum-shifted EC$_{90}$ (14 ng mL^{-1} as determined using a genotype 1a replicon assay). Based on the PK data, doses of

10, 30, and $100\,\mathrm{mg\,kg^{-1}}$ were selected for the efficacy study. In addition, we opted for bid dosing to ensure that plasma drug levels remained above the protein-adjusted EC_{90} throughout the 7 day study, thereby reducing the possibility of viral resistance that can occur with monotherapy treatment regimens.

The PXB mice were infected with HCV genotype 1a virus and after 5–8 weeks were treated with prodrug **32** at 10, 30, or $100\,\mathrm{mg\,kg^{-1}}$ bid. All doses of prodrug **32** produced a rapid and robust drop ($\sim 1\log_{10}$) in HCV viral mRNA titers (viral load) within 12 h of the first dose (Figure 5.15).[38] The maximum viral load reduction of $\sim 4\log_{10}$ units occurred by day 4 for all prodrug dose groups. The two higher doses of **32** (30 and $100\,\mathrm{mg\ kg^{-1}}$) maintained viral suppression throughout the 7 day treatment period, at which point the average viral load reduction was -3.4 and $-3.6\log_{10}$ units, respectively. By day 7, a dose-dependent decrease in viral load was apparent as the low dose of **32** ($10\,\mathrm{mg\,kg^{-1}}$) showed evidence of viral breakthrough (mean log reduction of –2.55). There was no evidence of drug accumulation in the plasma or liver over the 7 day treatment period for the dose groups. Two positive controls were included in the study, a non-nucleoside NS5B polymerase inhibitor (HCV-796) and an NS3/4A protease inhibitor (BILN-2061). Both HCV-796 and BILN-2061 have demonstrated efficacy in human HCV clinical trials and in HCV-infected chimeric mouse models similar to that described here.[61,62] In our study, both positive controls reduced the HCV viral load by $\sim 1.5–1.8\log_{10}$

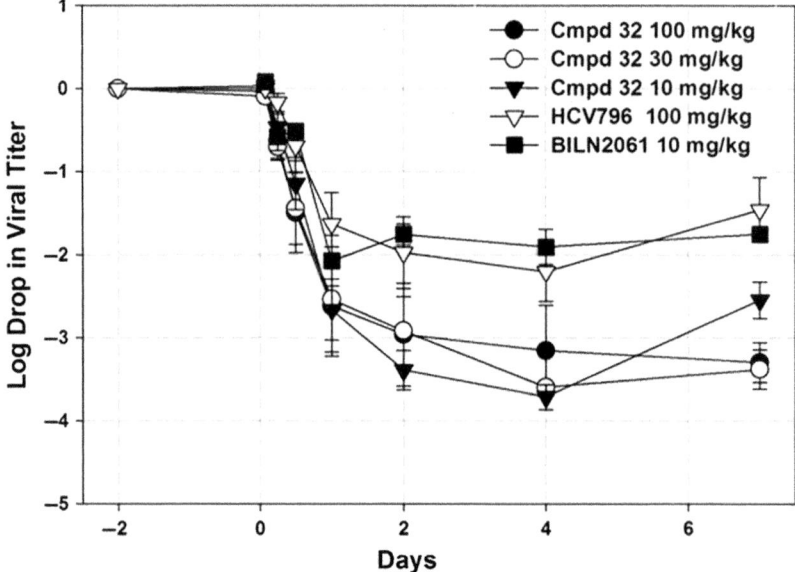

Figure 5.15 Reduction in HCV viral titers in plasma of PXB mice after infection with HCV genotype 1a and treatment with prodrug **32** or positive controls (HCV-796 or BILN-2061).

units, which is in good agreement with published results. In comparison, all doses of prodrug **32** reduced HCV viral titers to a greater extent than did either of the positive controls.

Using deep sequencing technology, the circulating virus on day 7 was analyzed to identify sequence changes in the NS4B region that might lead to drug resistance.[38] Multiple mutations were identified, three of which (N56I, G60V, N99H) predominated in a number of animals dosed with prodrug **32** (Table 5.6). Similar mutations did not arise in animals treated with HCV-796 or BILN-2061. Transient replicons containing the single-point mutations were created and the antiviral activity of **31** was determined. The G60V transient replicon was not viable; however, the N56I and N99H transient replicons showed significant resistance to **31**. There was a >600-fold shift in replicon potency ($EC_{50} > 1000$ nM) between the N99H and wild-type virus.

It is noteworthy that the most prevalent mutations identified in the PXB mice study were not observed in the *in vitro* resistance passaging experiments. Discrepancies between resistance mutations generated *in vivo versus in vitro* have also been reported for HCV NS3/4A protease inhibitors. However, in many instances resistance in the replicon assay is predictive of the *in vivo* outcome, as was the case for HCV-796 in our PXB study (the major mutation was C316Y in the NS5B protein). The reason for the selection of different mutations is unclear, but may result from the presence of adaptive mutations in HCV replicon required for replication or from protein interactions unique to the full viral life cycle of a replicating live virus. Regardless, it serves as a reminder of the potential pitfalls associated with generating resistance in artificial *in vitro* systems and the continued need for developing (and utilizing) *in vivo* models for infectious diseases.

Table 5.6 Mutations in the NS4B protein identified through sequencing of the HCV virus present in PXB mice treated for 7 days with prodrug **32**. Deep sequencing allowed for quantitation and relative percentages of the mutations are reported.

		A21V	I35X	N56I	G60V	F98L	N99H	V105X	A115D	A231V
100	10									
mg kg⁻¹	102			9.60%	17%					
GSK23588	103			1.10%						
53A	104			10.10%	23.90%	3.70%		(M) 2.5%		
30	201			16.0%	11.0%	2.8%	10.3%	(M) 2.6%		
mg kg⁻¹	202	1.2%		30.6%	6.5%		6.8%	(W) 2.1%		
GSK23588	203			8.7%	6.4%		12.5%			1.2%
53A	204			3.1%	7.8%	2.1%	4.3%	(M) 1.7%	1.0%	
10	301	1.0%		1.7%			38.7%			
mg kg⁻¹	302		(V) 1.2%	1.7%			89.8%			
GSK23588	303			1.9%		1.7%	65.7%			
53A	304						81.0%			

5.6 Challenges Ahead

Our efforts towards identifying a clinical candidate that interacts with NS4B produced many exciting results. We demonstrated that small molecules binding to NS4B are capable of inhibiting HCV replication *in vitro* with exceptional potency, on a par with the second-generation NS3/4A protease inhibitors. More importantly, a molecule from this compound class dose-dependently inhibited viral replication *in vivo*, providing a >3$\log_{10}$ drop in viral load over the course of the 7 day treatment (30 and 100 mg kg^{-1} bid). Overall, the intrinsic properties of the NS4B compounds approach the 'desirable' range for physicochemical space, including MW and cLog*P*, which may aid in the development of soluble, low-dose drugs with favorable safety profiles. Furthermore, we have not observed any shift in potency against replicons containing resistant mutations to molecules targeting the NS3/4A, NS5A, or NS5B proteins.[33,38] The apparent lack of cross-resistance to existing drugs provides further support for the unique mechanism of action associated with these compounds and should allow for combination with other agents currently in clinical development.

However, many challenges remain to be solved for this class of compounds. For example, none of the analogs prepared to date have shown any appreciable activity against the genotype 2 virus (Table 5.7). Although the series is active *versus* genotypes 3a, 4a, and 5a, a significant number of patients in Asia are infected with HCV genotype 2 and lack of efficacy in this population represents a serious limitation. It is unclear what sequence variations are responsible for the lack of efficacy against genotype 2; however, utilizing a genotype 1b/2a chimera replicon we believe that the region containing amino acids 89–110 may be responsible for conferring resistance.[63]

The three major resistance mutations (positions 94, 98 and 105) identified through *in vitro* passaging studies with the lead series also remain a concern, especially the significant decrease in efficacy observed with mutations at position 94. Our analysis of reported HCV patient sequences revealed that F98 and V105 are well conserved ($\geq$95%) in both genotypes 1a and 1b.[63] In contrast, H94 is well conserved in genotype 1a but is highly variable in genotype 1b with a large percentage of naturally occurring polymorphisms (Figure 5.16). For this reason, a stable H94N replicon assay was developed and considerable effort was spent improving activity against this polymorphic virus. The optimization of the hit **1** to cyclohexanol **31** and finally to the [3.1.0]bicyclohexan-2-ol (+)-**57a** resulted in substantial potency gains against

Table 5.7 Activity of **31** and **57a** across HCV genotypes.

	EC_{50} *(nM) for genotype*								
	1a	*1b*	*2a*	*2a*	*2b*	*3a*	*4a*	*5a*	*6a*
Backbone	H77	Con1/ET	JFH-1	Con	JFH-1	JFH-1	Con1/ET	JFH-1	JFH-1
31	0.4	1.0	>10000	>10000	3000	41	2.5	2.5	6900
57a	0.6	0.3	>1000	>1000	>1000	0.4	3.8	0.7	ND

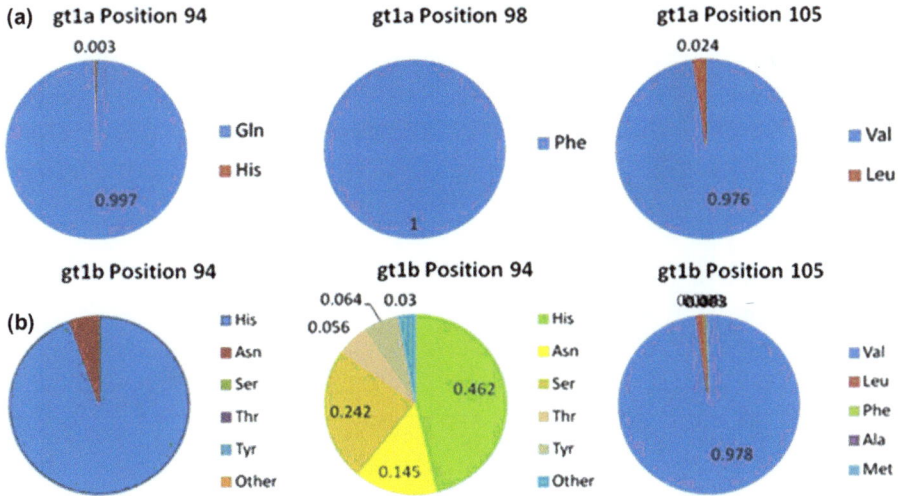

Figure 5.16 Relative sequence homology at three positions (94, 98, 105) of NS4B for genotype 1a (a) and genotype 1b (b) for which *in vitro* resistance mutations arose under selective pressure to compound **31**.

this resistant variant. However, it was critical that the compounds also demonstrate activity against the other naturally occurring variations (T, A, Y, S, Q) at position 94. Using a transient replicon assay (Figure 5.17), compounds **31** and (+)-**57a** were shown to be equally active (relative to H94N) against H94Q/A/Y/S but potency was slightly reduced *versus* H94T. The low nanomolar potency of the lead molecules against these single-point mutations suggested that our optimization efforts had successfully overcome the issues related to the polymorphic nature of position 94. Unfortunately, we observed dramatic shifts in potency (>100-fold) when **31** and (+)-**57a** were evaluated against clones constructed from genotype 1b-infected patient sequences found to harbor mutations at position 94 (Figure 5.18). In three of the four cases (clones 52, 53 and 60), a dramatic decrease in activity was observed. This suggests that other mutations combined with variations at position 94 contribute to the resistance to our lead series. The consequences of these results on compound efficacy, including the impact of the distinct resistance mutations observed *in vivo*, will need to be evaluated in clinical trials.

Perhaps the greatest challenge towards developing a compound with an NS4B mechanism of action is the plethora of HCV drugs in clinical trials. At the last count there were >30 compounds under clinical development acting by ~10 unique mechanisms of action. Although our data suggest that medicines targeting NS4B can provide effective anti-viral therapy that could be combined with other HCV drugs, there is reasonable doubt as to whether additional agents will be needed 10 years from now. Unlike HIV, HCV infection can be cured and does not require life-long therapy. This reduces the potential for the emergence of new resistant viruses and, therefore, the continued need for new

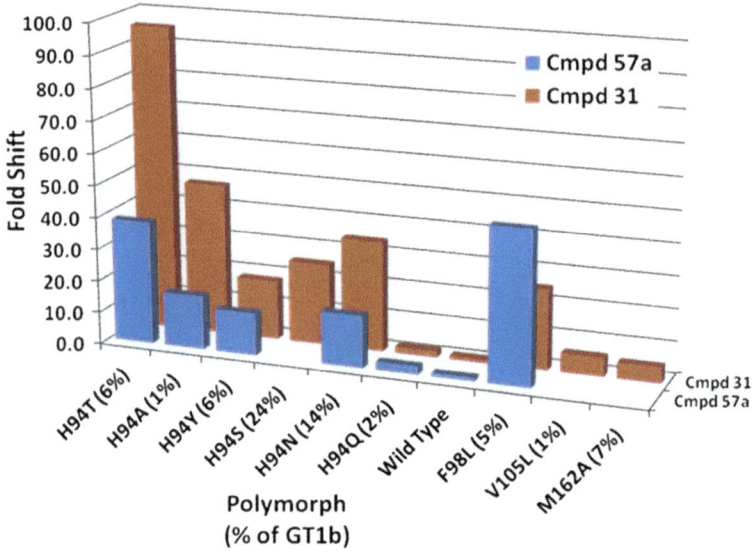

Figure 5.17   The activity (fold-shift relative genotype 1b wild-type replicon) of **31** and (+)-**57a** against known resistance mutations and polymorphic variations at position 94 (genotype 1b) in the transient replicon assay. The prevalence of the mutation is shown as a percentage of the genotype 1b population.

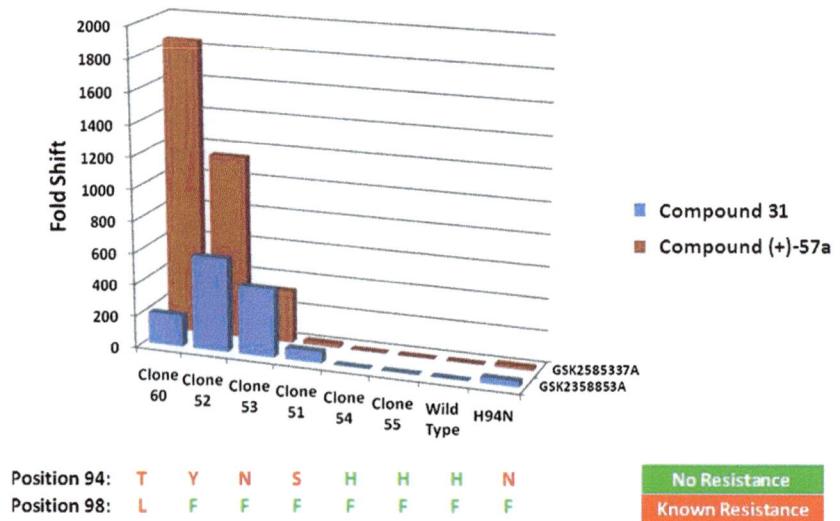

Figure 5.18   The activity of **31** and (+)-**57a** against transient replicons containing individual patient sequences of the NS4B region. Clones 54 and 55 contain the prevailing substitution (H94) while clones 51, 52, 53 and 60 contain polymorphic variation (H94S/N/Y/T) shown to confer resistance to our NS4B compound class.

antivirals. It may well be the case that the current agents are sufficient to deliver all oral regimens with high cure rates. However, the recent termination of several high-profile HCV drugs hints that perhaps it is still premature to declare victory over this notoriously challenging virus. cLogPs were calculated using the Daylight/Biobyte v4.3 algorithm, BioByte Corp., Claremont, CA.

References

1. C. W. Shepard, L. Finelli and M. J. Alter, Global epidemiology of hepatitis C virus infection, *Lancet Infect. Dis.*, 2005, **5**, 558–567.
2. D. Lavanchy, The global burden of hepatitis C, *Liver Int.*, 2009, **29**(Suppl. 1), 74–81.
3. D. N. Fusco and R. T. Chung, Novel therapies for hepatitis C: insights from the structure of the virus, *Annu. Rev. Med.*, 2012, **63**, 373–387.
4. M. Gao, R. E. Nettles, M. Belema, L. B. Snyder, V. N. Nguyen, R. A. Fridell, M. H. Serrano-Wu, D. R. Langley, J.-H. Sun, D. R. O'Boyle II, J. A. Lemm, C. Wang, J. O. Knipe, C. Chien, R. J. Colonno, D. M. Grasela, N. A. Meanwell and L. G. Hamann, Chemical genetics strategy identifies an HCV NS5A inhibitor with a potent clinical effect, *Nature*, 2010, **465**, 96–100.
5. J. A. Lemm, D. O'Boyle II, M. Liu, P. T. Nower, R. Colonno, M. S. Deshpande, L. B. Snyder, S. W. Martin, D. R. St. Laurent, M. H. Serrano-Wu, J. L. Romine, N. A. Meanwell and M. Gao, Identification of hepatitis C virus NS5A inhibitors, *J. Virol.*, 2009, **84**, 482–491.
6. F. Cong, A. K. Cheung and S.-M. A. Huang, Chemical genetics-based target identification in drug discovery, *Annu. Rev. Pharmacol. Toxicol.*, 2012, **52**, 57–78.
7. V. Lohmann, F. Korner, J. Koch, U. Herian, L. Theilmann and R. Bartenschlager, Replication of subgenomic hepatitis C virus RNAs in a hepatoma cell line, *Science*, 1999, **285**, 110–113.
8. S. K. Chunduru, C. A. Benetatos, T. J. Nitz and T. R. Bailey, Methods for treatment and prophylaxis of hepatitis C viral infections and associated diseases using NS4B-inhibiting compounds and compositions, *Patent Application*, WO2005-051318A2, 2005.
9. P. D. Bryson, N.-J. Cho, S. Einav, C. Lee, V. Tai, J. Bechtel, M. Sivaraja, C. Roberts, U. Schmitz and J. S. Glenn, A small molecule inhibits HCV replication and alters NS4B's subcellular distribution, *Antiviral Res.*, 2010, **87**, 1–8.
10. J. Gouttenoire, F. Penin and D. Moradpour, Hepatitis C virus nonstructural protein 4B: a journey into unexplored territory, *Rev. Med. Virol.*, 2010, **20**, 117–129.
11. R. Rai and J. Deval, New opportunities in anti-hepatitis C virus drug discovery: targeting NS4B, *Antiviral Res.*, 2011, **90**, 93–101.
12. A. Salonen, T. Ahola and L. Kaariainen, Viral RNA replication in association with cellular membranes, *Curr. Top. Microbiol. Immunol.*, 2005, **285**, 139–73.

13. J. Mackenzie, Wrapping things up about virus RNA replication, *Traffic*, 2005, **6**, 967–977.
14. S. Miller and J. Krijnse-Locker, Modification of intracellular membrane structures for virus replication, *Nat. Rev. Microbiol.*, 2008, **6**, 363–374.
15. T. Hugle, F. Fehrmann, E. Bieck, M. Kohara, H.-G. Krausslich, C. M. Rice, H. E. Blum and D. Moradpour, The hepatitis C virus nonstructural protein 4B is an integral endoplasmic reticulum membrane protein, *Virology*, 2001, **284**, 70–81.
16. L. Qu, L. K. McMullan and C. M. Rice, Isolation and characterization of noncytopathic pestivirus mutants reveals a role for nonstructural protein NS4B in viral cytopathogenicity, *J. Virol.*, 2001, **75**, 10651–10662.
17. M. Lundin, M. Monné, A. Widell, G. von Heijne and M. A. A. Persson, Topology of the membrane-associated hepatitis C virus protein NS4B, *J. Virol.*, 2003, **77**, 5428–5438.
18. S. Miller, S. Sparacio and R. Bartenschlager, Subcellular localization and membrane topology of the dengue virus type 2 non-structural protein 4Bd2, *J. Biol. Chem.*, 2006, **281**, 8854–8863.
19. C. Welsch, M. Albrecht, J. Maydt, E. Herrmann, M. W. Welker, C. Sarrazin, A. Scheidig, T. Lengauer and S. Zeuzem, Structural and functional comparison of the non-structural protein 4B in Flaviviridae, *J. Mol. Graphics Modell.*, 2007, **26**, 546–557.
20. M. Lundin, H. Lindstroem, C. Groenwall and M. A. A. Persson, Dual topology of the processed hepatitis C virus protein NS4B is influenced by the NS5A protein, *J. Gen. Virol.*, 2006, **87**, 3263–3272.
21. J. Gouttenoire, V. Castet, R. Montserret, N. Arora, V. Raussens, J.-M. Ruysschaert, E. Diesis, H. E. Blum, F. Penin and D. Moradpour, Identification of a novel determinant for membrane association in hepatitis C virus nonstructural protein 4B, *J. Virol.*, 2009, **83**, 6257–6268.
22. G.-Y. Yu, K.-J. Lee, L. Gao and M. M. C. Lai, Palmitoylation and polymerization of hepatitis C virus NS4B protein, *J. Virol.*, 2006, **80**, 6013–6023.
23. M. Elazar, P. Liu, C. M. Rice and J. S. Glenn, An N-terminal amphipathic helix in hepatitis C virus (HCV) NS4B mediates membrane association, correct localization of replication complex proteins, and HCV RNA replication, *J. Virol.*, 2004, **78**, 11393–11400.
24. M. Lundin, M. Monne, A. Widell, G. von Heijne and M. A. A. Persson, Topology of the membrane-associated hepatitis C virus protein NS4B, *J. Virol.*, 2003, **77**, 5428–5438.
25. V. Lohmann, S. Hoffmann, U. Herian, F. Penin and R. Bartenschlager, Viral and cellular determinants of hepatitis C virus RNA replication in cell culture, *J. Virol.*, 2003, **77**, 3007–3019.
26. M. Elazar, P. Liu, C. M. Rice and J. S. Glenn, An N-terminal amphipathic helix in hepatitis C virus (HCV) NS4B mediates membrane association, correct localization of replication complex proteins, and HCV RNA replication, *J. Virol.*, 2004, **78**, 11393–11400.

27. D. M. Jones, A. H. Patel, P. Targett-Adams and J. McLauchlan, The hepatitis C virus NS4B protein can trans-complement viral RNA replication and modulates production of infectious virus, *J. Virol.*, 2009, **83**, 2163–2177.
28. J. Gouttenoire, R. Montserret, A. Kennel, F. Penin and D. Moradpour, An amphipathic α-helix at the C terminus of hepatitis C virus nonstructural protein 4B mediates membrane association, *J. Virol.*, 2009, **83**, 11378–11384.
29. H. Lindstrom, M. Lundin, S. Haggstrom and M. A. A. Persson, Mutations of the hepatitis C virus protein NS4B on either side of the ER membrane affect the efficiency of subgenomic replicons, *Virus Res.*, 2006, **121**, 169–178.
30. K. J. Blight, Allelic variation in the hepatitis C virus NS4B protein dramatically influences RNA replication, *J. Virol.*, 2007, **81**, 5724–5736.
31. A. A. Thompson, A. Zou, J. Yan, R. Duggal, W. Hao, D. Molina, C. N. Cronin and P. A. Wells, Biochemical characterization of recombinant hepatitis C virus nonstructural protein 4B: evidence for ATP/GTP hydrolysis and adenylate kinase activity, *Biochemistry*, 2009, **48**, 906–916.
32. J. Q. Hang, Y. Yang, S. F. Harris, V. Leveque, H. J. Whittington, S. Rajyaguru, G. Ao-Ieong, M. F. McCown, A. Wong, A. M. Giannetti, P. Le, F. Talamas, N. Cammack, I. Najera and K. Klumpp, Slow binding inhibition and mechanism of resistance of non-nucleoside polymerase inhibitors of hepatitis C virus, *J. Biol. Chem.*, 2009, **284**, 15517–15529.
33. J. B. Shotwell, S. Baskaran, P. Chong, K. L. Creech, R. M. Crosby, H. Dickson, J. Fang, D. Garrido, A. Mathis, J. Maung, D. J. Parks, J. J. Pouliot, D. J. Price, R. Rai, J. W. Seal, U. Schmitz, V. W. F. Tai, M. Thomson, M. Xie, Z. Z. Xiong and A. J. Peat, Imidazo[1,2-*a*]pyridines that directly interact with hepatitis C NS4B: initial preclinical characterization, *ACS Med. Chem. Lett.*, 2012, **3**, 565–569.
34. N. A. Meanwell, Improving drug candidates by design: a focus on physicochemical properties as a means of improving compound disposition and safety, *Chem. Res. Toxicol.*, 2011, **24**, 1420–56.
35. P. D. Leeson and J. R. Empfield, Reducing the risk of drug attrition associated with physicochemical properties, *Annu. Rep. Med. Chem.*, 2010, **45**, 393–407.
36. C. A. Lipinski, F. Lombardo, B. W. Dominy and P. J. Feeney, Experimental and computational approaches to estimate solubility and permeability in drug discovery and development settings, *Adv. Drug Deliv. Rev.*, 2001, **46**, 3–26.
37. J. Miller, P. Chong, B. Shotwell, J. Catalano, V. Tai, J. Fang, A. Banka, C. Roberts, M. Youngman, H. Zhang, Z. Xiong, A. Mathis, J. Pouliot, R. Hamatake, D. Price, J. Seal, L. Stroup, K. Creech, L. Carballo, D. Todd, A. Spaltenstein, S. Furst, Z. Hong and A. J. Peat, *J. Med. Chem.*, 2013, DOI: 10.1021/jm400125h.
38. J. Pouliot, M. Thomson, M. Xie, J. Horton, J. Johnson, D. Krull, A. Mathis, Y. Morikawa, D. Parks, R. Peterson, T. Shimada, E. Thomas,

J. Vamathevan, S. Van Horn, Z. Xiong, R. Hamatake and A. J. Peat *J. Virol.*, submitted.

39. L. A. Carpino, *O*-Acylhydroxylamines. II. *O*-Mesitylenesulfonyl-, *O*-(*p*-toluenesulfonyl)-, and *O*-mesitoylhydroxylamine, *J. Am. Chem. Soc.*, 1960, **82**, 3133–3135.
40. Y. Tamura, J. Minamikawa, Y. Miki, S. Matsugashita and M. Ikeda, A novel method for heteroaromatic *N*-imines, *Tetrahedron Lett.*, 1972, **13**, 4133–5.
41. P. L. Anderson, J. P. Hasak, A. D. Kahle, N. A. Paolella and M. J. Shapiro, 1,3-Dipolar addition of pyridine N-imine to acetylenes and the use of carbon-13 NMR in several structural assignments, *J. Heterocycl. Chem.*, 1981, **18**, 1149–1152.
42. J. D. Kendall, Synthesis and reactions of pyrazolo[1,5-*a*]pyridines and related heterocycles, *Curr. Org. Chem.*, 2011, **15**, 2481–2518.
43. S. Loeber, H. Huebner, W. Utz and P. Gmeiner, Rationally based efficacy tuning of selective dopamine D4 receptor ligands leading to the complete antagonist 2-[4-(4-chlorophenyl)piperazin-1-ylmethyl]pyrazolo[1,5-*a*]pyridine (FAUC 213), *J. Med. Chem.*, 2001, **44**, 2691–2694.
44. S. H. Wunderlich and P. Knochel, (tmp)$_2$Zn·2MgCl$_2$·2LiCl: a chemo-selective base for the directed zincation of sensitive arenes and heteroarenes, *Angew. Chem. Int. Ed.*, 2007, **46**, 7685–7688.
45. Z. Dong, G. C. Clososki, S. H. Wunderlich, A. Unsinn, J. Li and P. Knochel, Direct zincation of functionalized aromatics and heterocycles by using a magnesium base in the presence of ZnCl$_2$, *Chem. Eur. J.*, 2009, **15**, 457–468.
46. P. Chong, R. Davis, V. Elitzin, M. Hatcher, B. Liu, M. Salmons and E. Tabet, Synthesis of 7-trifluoromethylpyrazolo[1,5-*a*]pyridinedicarboxylate, *Tetrahedron Lett.*, 2012, **53**, 6786–6788.
47. Q. Chen and S. Wu, Methyl (fluorosulfonyl)difluoroacetate; a new trifluoromethylating agent, *J. Chem. Soc., Chem. Commun.*, 1989, 705–706.
48. S. Roy, B. T. Gregg, G. W. Gribble, V.-D. Le. and S. Roy, Trifluor-omethylation of aryl and heteroaryl halides, *Tetrahedron*, 2011, **67**, 2161–2195.
49. J. F. Hartwig, Borylation and silylation of C–H Bonds: a platform for diverse C–H bond functionalizations, *Acc. Chem. Res.*, 2012, **45**, 864–873.
50. T. Ishiyama, J. Takagi, K. Ishida, N. Miyaura, N. R. Anastasi and J. F. Hartwig, Mild iridium-catalyzed borylation of arenes. High turnover numbers, room temperature reactions, and isolation of a potential inter-mediate, *J. Am. Chem. Soc.*, 2002, **124**, 390–391.
51. T. Ishiyama, Y. Nobuta, J. F. Hartwig and N. Miyaura, Room temperature borylation of arenes and heteroarenes using stoichiometric amounts of pinacolborane catalyzed by iridium complexes in an inert solvent, *Chem. Commun.*, 2003, 2924–2925.
52. E. N. Trachtenberg and J. R. Carver, Stereochemistry of selenium dioxide oxidation of cyclohexenyl systems, *J. Org. Chem.*, 1970, **35**, 1646–1653.

53. A. B. Charette and A. Beauchemin, Simmons–Smith cyclopropanation reaction, *Org. React.*, 2004, 1–415.
54. E. C. Friedrich and G. Biresaw, Zinc dust–cuprous chloride promoted cyclopropanations of allylic alcohols using ethylidene iodide, *J. Org. Chem.*, 1982, **47**, 1615–1618.
55. D. A. Evans, J. S. Tedrow, J. T. Shaw and C. W. Downey, Diastereoselective magnesium halide-catalyzed anti-aldol reactions of chiral *N*-acyloxazolidinones, *J. Am. Chem. Soc.*, 2002, **124**, 392–393.
56. B. Ma and W.-C. Lee, A modified Curtius reaction: an efficient and simple method for direct isolation of free amine, *Tetrahedron Lett.*, 2010, **51**, 385–386.
57. T. Fukuyama, C.-K. Jow and M. Cheung, 2- and 4-nitrobenzenesulfonamides: exceptionally versatile means for preparation of secondary amines and protection of amines, *Tetrahedron Lett.*, 1995, **36**, 6373–6374.
58. P. Meuleman and G. Leroux-Roels, HCV animal models: a journey of more than 30 years, *Viruses*, 2009, **1**, 222–240.
59. D. B. Olsen, M.-E. Davies, L. Handt, K. Koeplinger, N. R. Zhang, S. W. Ludmerer, D. Graham, N. Liverton, M. MacCoss, D. Hazuda and S. S. Carroll, Sustained viral response in a hepatitis C virus-infected chimpanzee via a combination of direct-acting antiviral agents, *Antimicrob. Agents Chemother.*, 2011, **55**, 937–939.
60. R. Kikuchi, M. McCown, P. Olson, C. Tateno, Y. Morikawa, Y. Katoh, D. L. Bourdet, M. Monshouwer and A. J. Fretland, Effect of hepatitis C virus infection on the mRNA expression of drug transporters and cytochrome p450 enzymes in chimeric mice with humanized liver, *Drug. Metab. Dispos.*, 2010, **38**, 1954–1961.
61. N. M. Kneteman, A. Y. M. Howe, T. Gao, J. Lewis, D. Pevear, G. Lund, D. Douglas, D. F. Mercer, D. L. J. Tyrrell, F. Immermann, I. Chaudhary, J. Speth, S. A. Villano, J. O'Connell and M. Collett, HCCV796: a selective nonstructural protein 5B polymerase inhibitor with potent anti-hepatitis C virus activity *in vitro*, in mice with chimeric human livers, and in humans infected with hepatitis C virus, *Hepatology*, 2009, **49**, 745–752.
62. N. M. Kneteman, A. J. Weiner, J. O'Connell, M. Collett, T. Gao, L. Aukerman, R. Kovelsky, Z.-J. Ni, A. Hashash, J. Kline, B. Hsi, D. Schiller, D. Douglas, D. L. J. Tyrrell and D. F. Mercer, Anti-HCV therapies in chimeric scid-Alb/uPA mice parallel outcomes in human clinical application, *Hepatology*, 2006, **43**, 1346–1353.
63. J. J. Pouliot, J. Johnson, M. Thomson, J. Vamathevan, R. Hamatake and A. J. Peat, HCV NS4B inhibitors show reduced potency against chimeric replicons containing genotype 1b but not genotype 1a patient sequences, manuscript in preparation.

Section II
Biochemical Screening and Structure-based Drug Design to Discover Antiviral Agents

CHAPTER 6

HIV Integrase Inhibitors

BRIAN A. JOHNS,*[a] TAKASHI KAWASUJI[b] AND
EMILE J. VELTHUISEN[a]

[a] GlaxoSmithKline Research & Development, Infectious Diseases
Therapeutic Area Unit, 5 Moore Drive, Research Triangle Park, NC 27709,
USA; [b] Shionogi Pharmaceutical Research Center, Chemistry of Infectious
Diseases, 3-1-1, Futaba-cho, Toyonaka-shi, Osaka 561-0825, Japan
*Email: brian.a.johns@gsk.com

6.1   Introduction

Progress in the discovery and development of human immunodeficiency virus
(HIV) treatment therapy has been among the most significant accomplishments
of medicine in the past quarter century and, arguably, a miracle of science. As a
result, HIV has become a manageable disease under a strict set of conditions
dependent on excellent medicines used in the right combinations along with
obligations for the patient to comply closely with specific dosing guidelines.
However, success does not occur in a vacuum and the situation is still dire for
many who either do not have access to the right drugs for their profile or have
complicated situations where optimal treatment is not possible. The most
recent numbers for 2011 actually show an increasing number of HIV-infected
persons, primarily as a result of a decrease in the number of deaths combined
with a stabilized rate of new infections.[1] The numbers are still daunting: 1.7
million deaths annually, 2.5 million new infections and 34 million people
currently living with the virus. Additionally, what was once primarily a male-
centered disease is now increasingly affecting women, in particular women of
color in the United States are disproportionately affected by HIV infection.
While the therapy regimens in current use are significantly improved over the

RSC Drug Discovery Series No. 32
Successful Strategies for the Discovery of Antiviral Drugs
Edited by Manoj C. Desai and Nicholas A. Meanwell
© The Royal Society of Chemistry 2013
Published by the Royal Society of Chemistry, www.rsc.org

early days of the disease, it has become strikingly evident that the most basic part of the equation – getting patients to take their drugs – is a major reason for a lack of efficacy.[2,3] Although patient behavior seems far removed from the discovery laboratory, the field has advanced sufficiently that building the appropriate properties into new antiretroviral drugs to allow for optimal patient convenience is now recognized as equally important to these drugs' virological properties.

HIV virus is relentless and continually evolving under drug pressure to escape control and continue the infection.[4] The transmission of such a resistant virus is of concern and currently a topic of intense debate.[5–7] To complicate matters further, the long-term tolerability and safety of many of the available therapeutic successes are not optimal.[8] This creates more challenges for patient compliance. Of particular interest is finding regimens that allow delay of the use of nucleoside analogs (nuc sparing) until later lines of therapy.[9,10] Ironically, one of the more interesting rationales for continued introduction of antiretroviral agents comes from the curative approaches that are now being explored. It has been argued that intensification of current regimens will be required for complete suppression during transcriptional activation envisaged during attempts to purge latent reservoirs.[11] As such, there continues to be a need for new drugs either from new classes or from further study of existing targets.

The above arguments support a need for new antiretroviral agents, but ones that are optimized for patient needs and that provide clear benefit and differentiation over existing options. On the surface, the field would appear to be well served, with over 30 treatment options. However, on closer inspection, many of these drugs center on the same mechanisms of action and are sufficiently similar that they do not provide unique characteristics orthogonal to other drugs already available. The key challenge for contemporary HIV antiretroviral drug discovery efforts is to choose the right target and match a fully optimized inhibitor while taking into account patient needs.

6.1.1 HIV Integrase

HIV is a retrovirus from the lentivirus family. The virus encodes 15 proteins, of which only three (reverse transcriptase, integrase and protease) have enzymatic activity.[12] Reverse transcriptase (RT) and protease (PR) are responsible for the conversion of single-stranded viral RNA into double-stranded DNA and the proteolysis of the gag polyprotein, respectively. Both RT and PR have proven to be excellent drug targets, with several antiviral drugs having been available for both for many years. Targeting virally encoded proteins or enzymes has traditionally been viewed as desirable due to a higher likelihood for selectivity, since they are not derived from the host. A third virally encoded enzyme is HIV integrase (IN). Integrase is a 32 kDa protein that is responsible for the integration of reverse-transcribed viral double-stranded DNA (dsDNA) into host chromosomal DNA during the retroviral replication cycle. As a result of this unique retroviral step, integrase has become an attractive target for drug discovery. The IN enzyme consists of a 288 amino acid primary sequence divided into N-terminal, C-terminal and catalytic core domains. The catalytic core domain, consisting of residues 51–212, contains a triad of carboxylate residues at positions Asp64, Asp116 and Glu152, which is known as a D,D,E motif and is highly conserved in the catalytic active sites of the polynucleotidyl transferases superfamily.[13,14] The three acidic residues are essential to enzyme function and coordinate two metal ion cofactors, presumably Mg^{2+}, under physiological conditions.[21,22] These active site metals catalyze phosphodiester bond breaking and formation during the integration of viral and host dsDNA.[15] The biochemistry of the integration catalysis consists of two distinct mechanistic steps involving removal of a terminal dinucleotide from the respective 3′ ends of the viral nucleic acid in the cytoplasm (3′ processing, Stage B to C in Figure 6.1a) and nicking of the host DNA by the newly exposed 3′-hydroxyl groups in the cell nucleus (strand transfer, Stage D to E, Figure 6.1a).[16] Host cell repair mechanisms serve to complete the process of removing the remaining nucleic acid flaps and fill in leftover gaps due to the DNA nicking event. It is the strand transfer step that has been most amenable to inhibition thus far and the target of the molecules discussed below. A short section at the end of this chapter focuses on a newer class of inhibitors that appear, at least in part, to invoke their antiviral effect due to inhibition of the integrase 3′ processing step.

Early screening efforts resulted in the discovery of what have become known as the diketo acid integrase inhibitors (DKAs) and acid isostere-containing molecules by researchers at Merck[17] and Shionogi.[18] The series of chemotypes that were discovered proved seminal discoveries for much research that was to follow; however, their clinical utility was far from optimal. The subsequent structure–activity relationship (SAR) work that was done clearly established some key pharmacophore attributes that include a metal chelating motif and a hydrophobic tail. The so-called two-metal pharmacophore was set forth and has served for more than a decade as the basic model for integrase strand transfer inhibitor design.[19–21] A representation of the model is shown in Figure 6.1b, coordinated to an early DKA-derived drug known as S-1360.[15,22] Further generalization of the model for consideration of additional molecular design is indicated in Figure 6.1c, whereby the basic elements of metal-chelating and hydrophobic motifs are present. Careful biochemical characterization studies demonstrated that these chelating pharmacophore molecules selectively inhibit the strand transfer step and have little effect on the earlier 3′ processing event.[23,24] Again referring to Figure 6.1a, the binding of an inhibitor results in the 3′ processed intermediate, shown in Stage C, proceeding to an active site–drug complex, indicated by Stage X in the diagram. This binding of the chelating drug to the metals in the active site prevents coordination of the host DNA phosphodiester and its eventual cleavage stabilization by the resulting metals.

Recent progress in crystallography studies reported by Cherepanov and co-workers,[25–27] using the prototype foamy virus (PFV) intasome complex with known IN inhibitors, has served to validate the two-metal pharmacophore concept. Moreover, the stacking interaction of the hydrophobic group with a

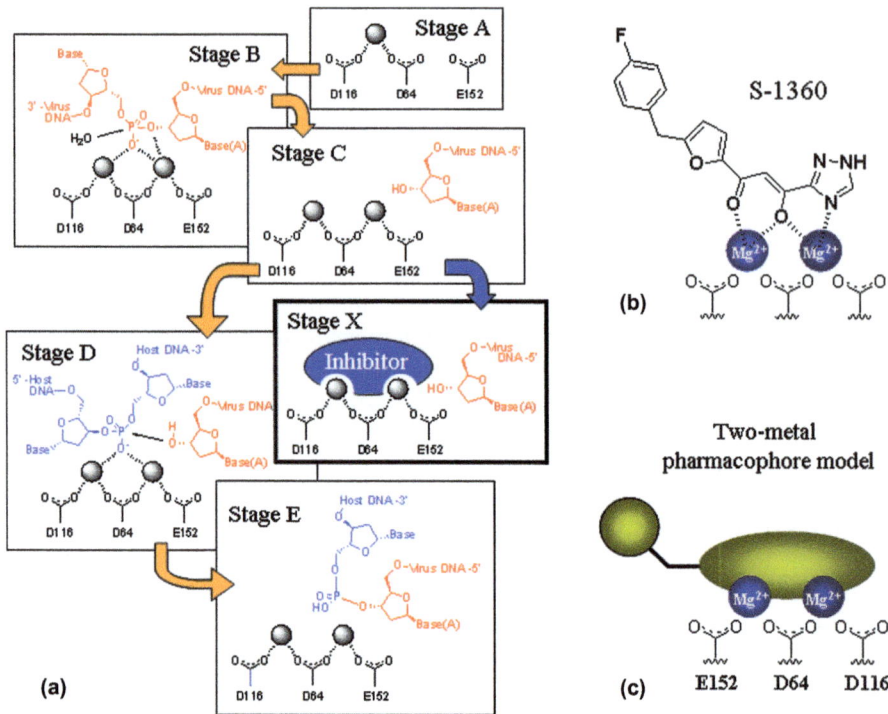

Figure 6.1 (a) A graphical scheme of a two-metal-ion catalysis and inhibition mechanism. Stage A to C indicates the 3′-processing reaction and Stage C to E the strand transfer reaction. Stage A to B: the enzyme recognizes the adenine base conserved in the third position from the 3′ end of viral DNA, then activates the next phosphoric ester with the two metals. Stage B to C: the activated phosphoryl ester is hydrolyzed to excise the terminal dinucleotide and the recognized adenosine is exposed as the new 3′ end, giving a pre-integration complex. Stage C to D: the pre-integration complex non-specifically binds to host DNA to activate a phosphoryl ester by the two metals. Stage D to E: the activated phosphoric ester is attacked by the recessed 3′ end in the manner of an S_N2-like nucleophilic reaction, then the viral DNA and the host DNA are joined with each other. Stage X: an inhibitor chelates to the two metal ions of Stage C to block the host DNA binding. (b) S-1360 binding to the metal cofactors. (c) A graphical depiction of the two-metal pharmacophore model.
Reproduced with permission from reference 20.

viral nucleic acid substrate base observed in the co-crystal structures helped explain some of the empirically determined minimal SAR requirements. Similar concepts of two-metal coordination have been employed to design inhibitors of the RNaseH motif of HIV-1 reverse transcriptase[28,29] and an influenza endonuclease.[30]

The observations described above stimulated extensive research efforts in this area over the past 15 years and have likewise met with considerable success.[31–33] We begin our discussion where most of this work has culminated,

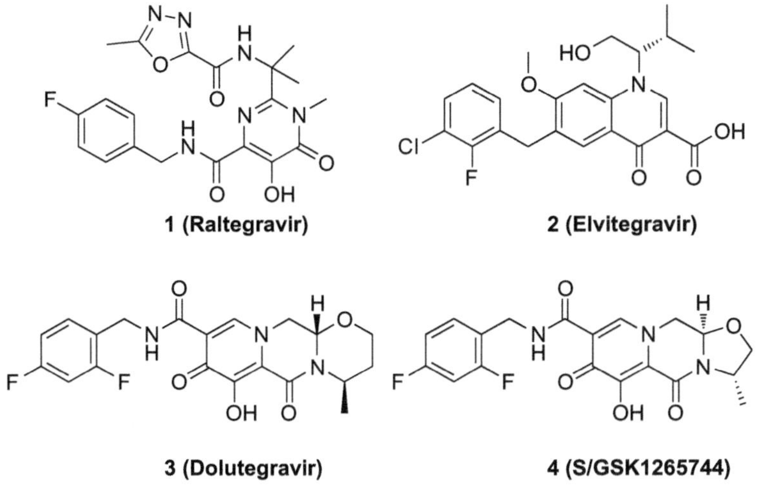

Figure 6.2 Structures of raltegravir, elvitegravir, dolutegravir and S/GSK1265744.

with a brief review of the discovery and development of two first-generation HIV integrase inhibitors (INIs), raltegravir and elvitegravir. The majority of our attention, however, will be centered on the discovery of a series of carbamoylpyridone-based 'next-generation' IN inhibitors, which led to the identification of dolutegravir (S/GSK1349572) and S/GSK1265744, currently under development for the treatment of HIV as oral and long-acting injectable agents, respectively (Figure 6.2).

6.2 First-generation HIV Integrase Drugs Raltegravir and Elvitegravir

As mentioned above, efforts to find a suitable molecule for clinical progression began in earnest after the DKA discoveries in the late 1990s. Several drug candidates progressed into clinical trials after extensive preclinical SAR and safety studies. These include S-1360,[15,22] L-870810,[34] GSK364735,[35] GS-9160[36] and BMS-707035,[37] which were subsequently terminated as a result of insufficient pharmacokinetics (PK) to be either efficacious or differentiated from the competition or because of safety assessment findings that precluded further development. In 2007, however, after a host of failures, raltegravir (RAL) became the first INI to achieve approval and registration successfully. The discovery of raltegravir has been well documented by Rowley[38] and will not extensively be repeated here. To summarize (Figure 6.3), the key lead **5** came from a hepatitis C virus (HCV) program for inhibitors of the NS5B RNA-dependent RNA polymerase. This compound did not inhibit HIV integration, but the metal-binding capacity of the scaffold gave researchers at the Istituto di Ricerche di Biologia Molecolare (IRBM, a Merck subsidary) the impetus to build in the hydrophobic benzylamide (**6**) and they observed remarkable

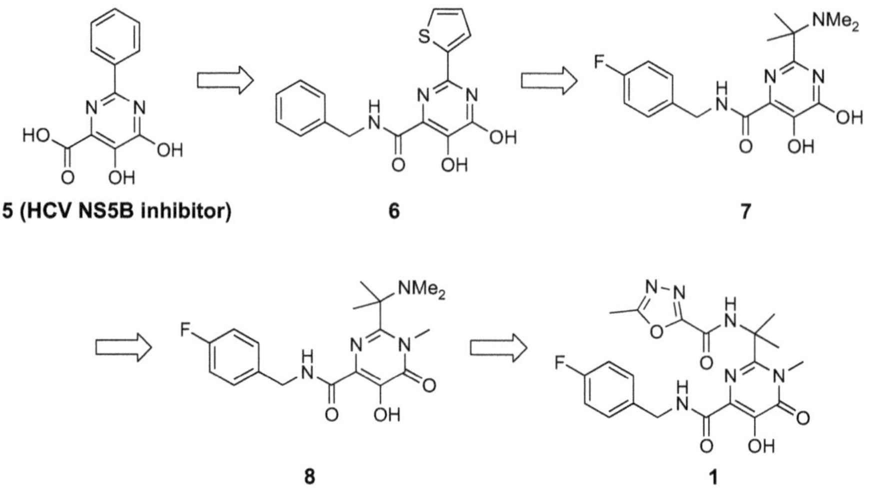

Figure 6.3 Lead progression for the discovery of raltegravir.

potency (85 nM) in an integrase strand transfer assay in return for this very simple modification. Modification of the 2-position of the pyrimidine led to further improvements and substitution of the nitrogen to form the pyrimidinone in moving from **7** to **8** was the last major structural core change of note. Finally, continued optimization around the aminomethyl C2 substituent allowed for the right balance of physicochemical properties while maintaining cell penetration and limiting protein binding effects to deliver exceptional potency. The result was raltegravir (**1**).[39]

Since the approval of RAL, the inclusion of an HIV-1 integrase inhibitor as part of highly active antiretroviral therapy (HAART) regimens has become increasingly attractive.[40,41] Integrase inhibitors have been well tolerated and among the most potent antiretrovirals discovered to date, based on nearly 5 years of post-launch experience with raltegravir, in addition to numerous clinical studies with experimental agents. In addition to exceptional viral load reduction, the class appears to result in a more rapid decrease in RNA copies, although it is still unclear if this results in a meaningful clinical benefit.[42]

Raltegravir had the notable privilege of being a first-in-class agent after several years of relative calm in the development of new antiretroviral drugs. The field was thus primed for an agent with a novel mechanism and the stage was set for a pair of identical pivotal Phase 3 trials, based on geographic regions and with a placebo control in a treatment-experienced population[43,44] that would demonstrate the superiority of raltegravir when added on to optimized background therapy (OBT) with a placebo control in a treatment-experienced population. The combined BENCHMRK 1 and 2 results at 48 and 96 weeks showed clear superiority of a 400 mg twice-daily (bid) dose of raltegravir over OBT alone. In addition to the primary endpoint of the number of patients being below either 400 or 50 copies HIV RNA per mL, a significant increase in $CD4^+$ cell numbers per mm^3 was noted for the raltegravir arms over the placebo. Furthermore, an analysis of resistance provided the first clear information on what has become known as RAL signature mutations: N155, Q148 and Y143, usually in combination with at least one additional mutation, especially in the case of the Q148 pathway, which favors a second G140 mutation to rescue viral fitness.[45–49]

The treatment-naïve patient STARTMRK Phase 3 trial was of non-inferiority design and compared RAL with the non-nucleoside reverse transcriptase inhibitor (NNRTI) efavirenz, both as part of a three-drug regimen that included the two nucleoside RT inhibitors (NRTIs) tenofovir and emtricitabine. This study showed non-inferiority of RAL, with the INI and NNRTI arms achieving 81 and 79% efficacy (<50 copies mL^{-1}) at 96 weeks.[50,51] Similar levels of CD4^{+} cell increase were also observed.

The above studies clearly establish RAL as a bid option for the treatment of naive or experienced patients and investigators at Merck sought to evaluate further the potential of their new agent. The SWITCHMRK study was an attempt to show efficacious non-inferiority to switching from a protease inhibitor (PI) to the INI RAL.[52] Since PIs have a long history of deleterious effects on lipid profiles, albeit in the face of tremendous efficacy and high genetic barrier to resistance, it would be desirable to have a similar virological outcome without the lipid changes. However, RAL was shown to be inferior to lopinavir/ritonavir at 24 weeks. *Post hoc* analysis showed that if previous failure was taken into account, RAL performed similarly to the PI; unfortunately, this was not considered in the trial designs. The effects on cholesterol and triglycerides, however, were significantly more favorable for the RAL arm. Thus, from an efficacy standpoint, there was still interest in further demonstration of the ability to change from a PI to an INI. A slightly redesigned and smaller study called SPIRAL was undertaken in Spain to revisit the switch challenge. Ultimately, this study did show non-inferior efficacy and improved lipids in a 48 week trial, when patients with sustained suppression for more than 6 months were chosen to switch regimens.[53]

These switch studies, along with the levels of resistance observed, particularly in the BENCHMRK studies, began to create concerns about the genetic barrier to resistance for RAL and perhaps, more broadly, for the class in general.[54] The final major clinical study of note involving RAL was an attempt to gain approval for once-daily dosing.[55] The QDMRK study was designed to compare 400 mg bid with 800 mg qd, with a common backbone of tenofovir/emtricitabine. A fully powered Phase 3 study was planned through 96 weeks of dosing, with primary endpoints at both 48 and 96 weeks. The 48 week analysis showed a slightly inferior performance of the qd arm of the study when all participants were included, with an apparent worsening situation for patients with high viral loads of $>100\,000$ copies mL^{-1} or low CD4^{+} cell count.

It is interesting to compare the above results with early clinical and preclinical data around the PK profile of RAL. From the healthy volunteer studies, RAL showed a rapid initial (α) elimination phase ($t_{\frac{1}{2}} = 1$ h) and a longer terminal (β) elimination ($t_{\frac{1}{2}} = 7$–12 h).[56] Similar PK was observed in the Phase 2a study, which established antiviral efficacy after 10 days of monotherapy when trough concentrations ($C_{12\,h}$) were maintained above a 33 nM IC_{95}.[57] Interestingly, all doses tested, 100–600 mg bid, showed similar viral load responses during this study, making a dose–response curve difficult to assess fully. Further review of the preclinical PK data showed a similar biphasic elimination curve, with a rapid initial phase followed by a slower terminal phase. Although specific data were not reported for RAL for the early-phase half-life in rat, dog and monkey studies, it was mentioned that RAL showed similar PK to a closely related compound that nicely mimicked the clinical data from the healthy volunteer and monotherapy trials. Overall, RAL showed a 2, 11 and 4 h terminal elimination phase in rat, dog and rhesus i.v. PK studies, respectively. The corresponding clearance values were 39, 6 and 18 mL min^{-1} kg^{-1}; volume of distribution (V_d) measures were 2.0, 0.9 and 1.2 L kg^{-1}.

Elvitegravir (EVG) is derived from a quinolone antibiotic scaffold originally designed for bacterial DNA gyrase activity. The evolution of the quinolone into a potent INI is shown in Figure 6.4a. Shinkai and co-workers designed the quinolone keto acid motif **10** as a bioisostere of the two-metal binding compound L-870,810 (**9**), which was an early-generation INI and the first to show clinical efficacy.[58] Although the keto acid **10** did not have integrase activity, researchers were surprised to see that the structurally simpler acid **11** had low micromolar inhibition of an enzymatic strand transfer assay. Through significant lead optimization, workers at Japan Tobacco were able to build in substantial potency ($IC_{50} = 0.9$ nM, protein-adjusted $IC_{50} = 9.8$ nM) and develop a suitable PK profile to justify clinical development.[59] Preclinical PK showed moderate oral bioavailability in rats and dogs (34.1 and 29.6%) and also modest i.v. clearance rates (0.5 and 1.0 L h$^-$ kg^{-1}), respectively. Terminal half-life measures were 2.3 and 5.2 h in rats and dogs, respectively.[60]

EVG is predominantly metabolized via CYP3A. This feature became evident during the early clinical investigations and has manifested itself throughout the EVG development program.[61,62] As a result, EVG requires a PK booster co-dose to achieve a clinically meaningful increase in EVG AUC (area under the curve) measures, which is the key to making EVG a once-daily drug.[63,64] Since the very early monotherapy efficacy investigations, all clinical studies have included a co-dose of ritonavir,[65] which itself is an early-generation HIV protease inhibitor, or of the CYP3A inhibitor cobicistat (COBI), which is based on ritonavir but devoid itself of potent HIV antiviral activity.[66] With the pharmacokinetic booster available along with the two nucleoside analogs tenofovir (TDF) and emtricitabine (FTC), the focal point of the development strategy for elvitegravir has been the four-drug combination pill (QUAD) of EVG–COBI–TDF–FTC.[67,68]

Two key Phase 2 studies looked at EVG in either experienced or treatment-naive patients. In the experienced population, a protease inhibitor was compared with EVG both with study arms including ritonavir and OBT.[69] For naives, the QUAD regimen was compared with the three-drug combination containing Atripla (efavirenz–tenofovir–emtricitabine).[70] In both cases, the primary endpoints were robustly met and EVG progressed into pivotal Phase 3 studies. The first Phase 3 trial to report was a study in experienced patients

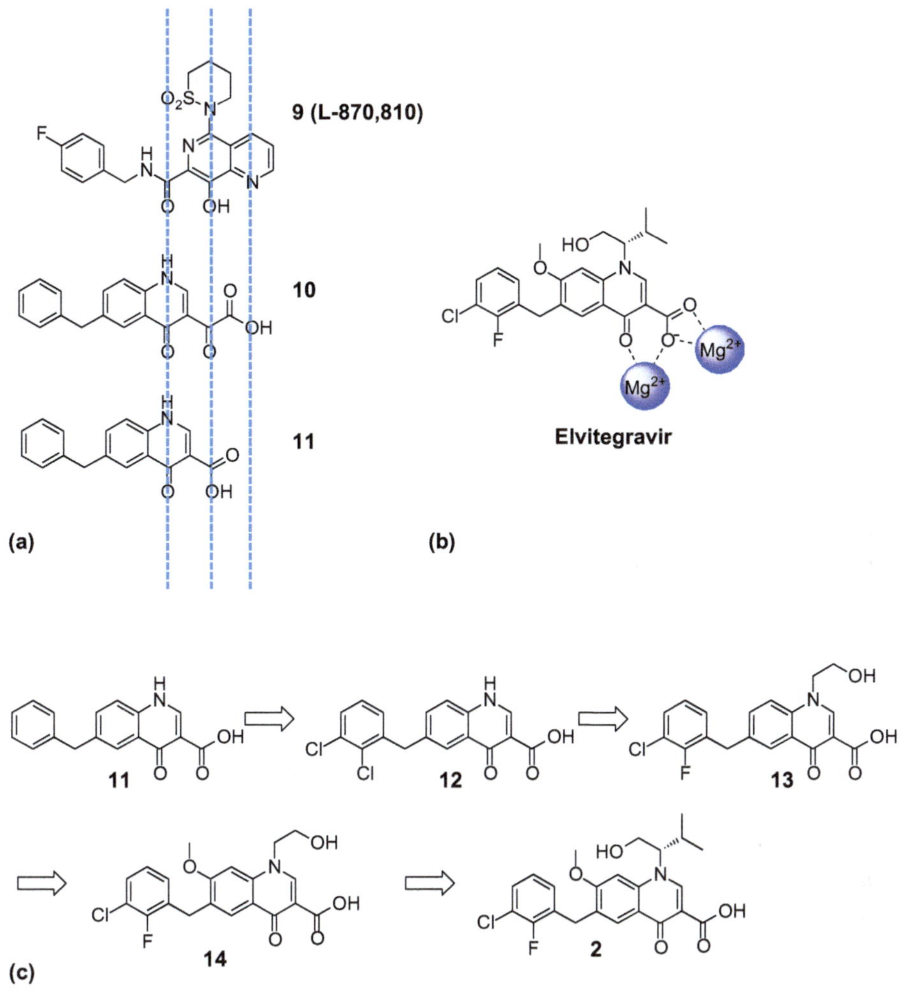

Figure 6.4 (a) Metal-chelating Lewis basic groups in the quinolone series derived from naphthyridone INI L-870,810. (b) Two-metal chelation model for elvitegravir. (c) Quinolone series evolution to discover elvitegravir.

with at least two-class resistance comparing EVG (150 mg qd) with RAL (400 mg bid) with a background regimen of a ritonavir-boosted protease inhibitor. EVG was shown to be non-inferior to RAL in the 48 week analysis.[71] A second study (Study 102) based on the Phase 2 naïve design mentioned above comparing two single-pill regimens has recently reached the 48 week analysis. The results confirm non-inferior efficacy of the QUAD to Atripla, with some advantages reported for CNS, rash and lipid levels favoring the integrase-containing regimen.[72,73] A third study (Study 103), again in naïve patients, reported non-inferior results when the QUAD regimen was compared with the boosted protease inhibitor atazanavir with a Truvada (tenofovir–emtricitabine) backbone, effectively comparing the boosted INI with a boosted PI.[74,75]

With the advent of the QUAD regimen, elvitegravir clearly marks an advance in convenience and, as a result, likely also an improvement in adherence compared with raltegravir. However, EVG is highly cross-resistant to two of the three major RAL signature mutation pathways. The N155H and various residue 148 and 140 combinations were selected for in-cell culture passage experiments[76–79] and have been observed in patients.[80,81] EVG signature mutation pathways also include T66A/I/K and E92Q/V.[77] EVG does retain potency against the Y143 mutants,[82,83] a difference from RAL that is well explained from seminal structural work done on the prototype foamy virus (PFV) surrogate system for HIV integrase by Cherepanov and co-workers.[26,27] Tyrosine 212 in the PFV system corresponds to residue 143 in the HIV integrase protein. Raltegravir makes a π-stacking interaction between its oxadiazole tail with the wild-type tyrosine benzyl ring. EVG lacks an aromatic tail in the same region, hence its potency does not depend on this additional interaction with the protein.

The work mentioned above covering RAL and EVG is just a summary of the early-through late-stage development for the two integrase inhibitors. When looking over the data profiles of these drugs, the opportunities for improvement become apparent, centered around dosing requirements and resistance. These limitations were apparent very early during the development of both compounds and it was our goal to address both of these areas in our attempt to find a truly next-generation INI.

6.3 Discovery and Development of Dolutegravir

6.3.1 Differentiation Objectives

The current financial hurdles to bringing a drug from early discovery to market are enormous: some estimate the cost at over \$1 billion.[84] Although the actual number is hotly debated, the cost is high enough that the decision to develop a promising lead has significant ramifications and investment must be placed where it will have the greatest impact. With RAL and EVG already well in place, the development of an additional integrase strand transfer inhibitor needs to provide something of value to the patient, rather than being just a 'me-too' drug of questionable utility.[85] As our team began to consider the attributes necessary to deliver an impactful drug, we developed four critical primary criteria. These key attributes were (1) unboosted once-daily dosing, (2) a low-milligram dose burden that would be amenable to a variety of fixed-dose combinations (FDCs), (3) a superior resistance profile to retain potency against clinically relevant INI resistance mutations and (4) a high genetic barrier to resistance.

6.3.2 Design of a Next-generation Scaffold

Since the DKA motif was identified, many two-metal binding scaffolds other than a hydroxypyrimidinone (RAL's core unit) or a quinolonecarboxylic acid (EVG's core unit) have been explored in the field of HIV-1 integrase drug discovery. This

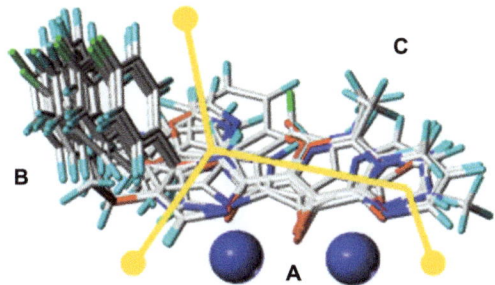

Figure 6.5 Overlay of accumulated two-metal binding scaffolds for refinement of pharmacophore elements. Yellow lines partition the structures into each pharmacophore region, A, B and C.
Reproduced with permission from reference 90. Copyright (2012) American Chemical Society.

long-term accumulation of SAR experience, based on the two-metal pharmacophore concept, opened up a new vista of lead discovery, which would eventually address all the key attributes mentioned above. Figure 6.5 represents a structural overlapping model of many historical two-metal binding scaffolds. As elaborated in the original pharmacophore discussion above, region A is the two-metal binding motif critical to all members of this class of strand-transfer active-site binders. Region B is a hydrophobic pharmacophore space which is consistently a halo-substituted phenyl group for optimal potency based on an empirical review of many molecules in this class from our laboratories and others. Region C is commonly fairly flexible and tolerant to diverse structural modifications that allow for optimization of the PK and other drug-like properties required of an orally dosed small molecule. Additionally, there has been evidence of region C adding significantly to the intrinsic potency of the inhibitor.

Region B was an interesting area in which we began to focus on subtle details during our molecular design of a novel scaffold. Phenyl rings are clearly clustering on region B, but upon closer inspection there appeared to be a range of space occupied by the different rings, which have antiviral potencies ranging from nanomolar to micromolar. It appeared from many examples that a molecule extending the phenyl ring further in the overlap comparison was more likely to show higher potency. Figure 6.6 shows two examples of a comparison with different spacing of the region B hydrophobic benzyl group.

In order to demonstrate the concept, we can use a series of naphthyridines (NAP-A and -B) and naphthyridinones (NTD-A and -B) as examples. Both NAP-A and NAP-B have an identical naphthyridine metal chelation motif, but are effectively rotated on a pseudo C_2 symmetry axis, placing the benzyl group on opposite ends of the heterocycle.[86,87] As such, the NAP-A and NAP-B scaffolds are set up to bind in a reversed orientation to one another, thereby placing the hydrophobic benzyl group in the same pharmacophore space. As seen in an overlapped model in the middle, NAP-A spaces the phenyl ring further from the core region and represents higher potency than NAP-B. The situation is similar for the naphthyridinones NTD-A and -B.[88,89] Again, the

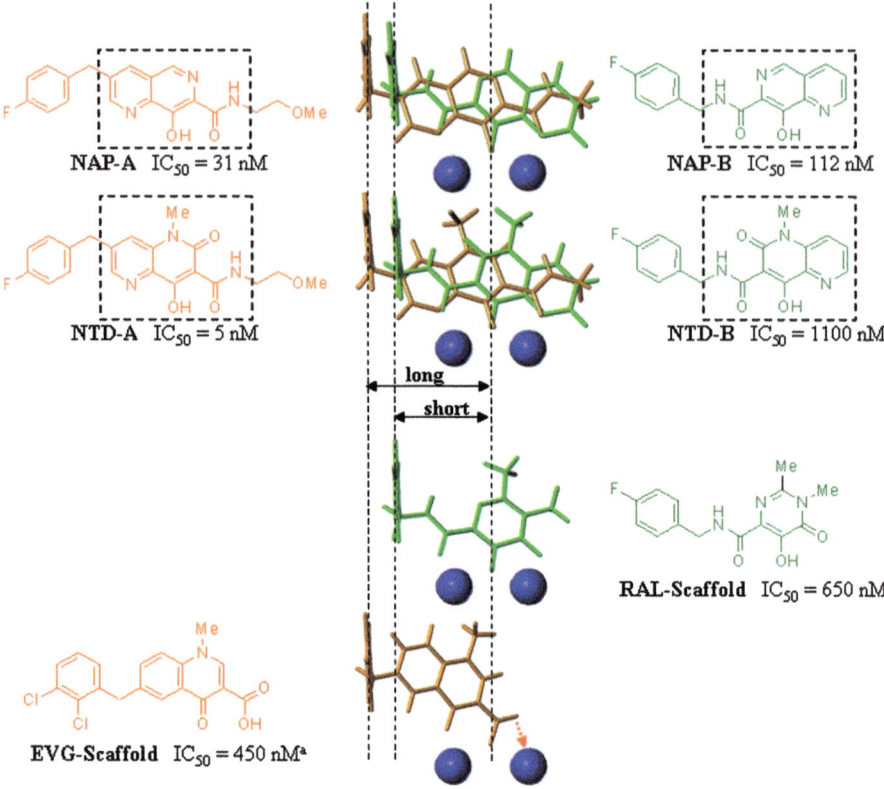

Figure 6.6 Comparison of hydrophobic region B aromatic ring spacing in pharmacophore analysis and design. The basis for increased ring spacing for optimal potency in the design of a next generation inhibitor.
[a]Data from reference 59. Reproduced with permission from reference 90. Copyright (2012) American Chemical Society.

NTD-A further extends the hydrophobic moiety and shows higher potency than NTD-B.

Additional observations with other scaffolds further supported this optimal hydrophobic positioning and, as a result, we adopted this rationale in the course of our molecular design. Interestingly, a simplified RAL scaffold, which has a short benzyl group spacing, has only weak to modest antiviral potency. However, as we saw in the RAL optimization story (referenced in Section 6.2), with additional modifications towards region C, RAL also achieved low nanomolar potency.

In contrast, the EVG scaffold extends a phenyl group similarly to the NAP-A and NTD-A structures. However, metal chelation could not be completed as effectively due to weak bidentate coordination with the carboxylic acid unit. Again, with additional modifications to the quinolone carboxylic acid scaffold, EVG achieved high potency. In recent crystallographic studies of the prototype formy virus (PFV) intasome with some HIV-1 integrase inhibitors reported by

Cherepanov and co-workers, these aromatic rings extended to region B appear to contact with the viral DNA substrate cytosine base.[25] This further supports the hypothesis that RAL could not extend a phenyl group sufficiently deep to stack effectively with the substrate base. These crystallographic views also support the suggestion that the EVG carboxylic acid could not complete an effective bidentate chelation to both divalent magnesium cofactors in the active site. Although these inspections are based on complexes with the PFV intasome, the HIV-1 intasome likely represents a similar scenario.

The next step for molecular design was to consider the metal chelation properties in region A, which is the critical pharmacophore. The DKA original scaffold in this class includes a three-oxygen chelating unit. However, medicinal chemistry optimization led to new two-metal binding scaffolds, including basic heteroaromatic nitrogen atoms involved in metal coordination. The idea is that a more basic chelator is able to provide higher potency, in addition to designing chelating motifs into more 'drug-like' structures, as well as providing easier synthesis. Several of the early candidates mentioned above (S-1360, L-870810, GSK364735, GS-9160, BMS-707035) include at least one nitrogen to co-ordinate a metal cofactor. The early-generation NTD and NAP scaffolds described previously also consist of a nitrogen–oxygen–oxygen chelating unit. As is generally accepted, integrase uses divalent metal cofactors under physiological conditions that are presumed to be Mg^{2+}.[21,22] According to the hard and soft acid and base (HSAB) theory, ionic magnesium is categorized as a hard metal; therefore, a hard base, such as oxygen, is preferred over nitrogen to coordinate a magnesium metal in the active site. hence we proposed replacing the nitrogen in NTD-A with oxygen in the chelation unit. An additional benefit of using a harder base, such as oxygen, is to reduce potential side effects caused by non-specific binding to soft metals utilized in other biological functions in the human body. Our first attempt at replacing the nitrogen with an oxygen is shown in Figure 6.7.[90] Owing to bonding differences in oxygen versus nitrogen, the ring system in the NAP and NTD scaffolds was opened to utilize an 'acyclic' carbonyl as the Lewis basic chelation group. The prior scaffold 'ring' integrity and positioning of the benzyl group were maintained by placing a carboxamide in the 5-position of the pyridone system whereby an intramolecular hydrogen bonding with the pyridone core carbonyl allowed for coplanarity and

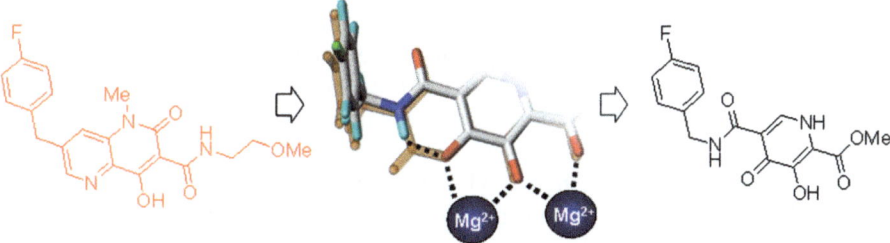

Figure 6.7 Molecular design of N to O chelation motif changes and introduction of the carbamoylpyridone metal chelation region A.

re-establishment of the original scaffold pseudo-ring system. With this design, the benzyl is placed in the desired hydrophobic pharmacophore region, thereby maintaining a favorable region B alignment.

An initial carbamoylpyridone inhibitor 15 provided a high antiviral potency against a wild-type HIV-1 strain from a very simple scaffold. Even in the presence of human serum albumin (HSA), a potency reduction of only 12-fold resulted in 120 nM ($38.4 \, ng \, mL^{-1}$) as the protein adjusted IC_{50} in an MT4 cell-based antiviral assay ($^{MT4}PAIC_{50}$). This agent did not show sufficient efficacy against three clinically relevant RAL-resistant mutants (Table 6.1); nevertheless, it was comparable or slightly improved when compared with RAL or EVG. Furthermore, animal PK studies provided good $t_{\frac{1}{2}}$ (3.79–5.31 h), consistently low i.v. clearance ($Clt = 0.84$–$2.44 \, mL \, min^{-1} \, kg^{-1}$) and moderate bioavailability ($F = 22.2$–53.4%) across species, with the rat data shown in Table 6.2. Most importantly, good coverage of plasma concentration over the $^{MT4}PAIC_{50}$ of $38.4 \, ng \, mL^{-1}$ was observed in every species at 24 h post-dosing ($5 \, mg \, kg^{-1}$ oral dose). Moderate molecular weight and good solubility combined with neither significant CYP inhibition nor metabolic concern suggested that this was an excellent starting point for further optimization.

A first attempt at further modification was the replacement of the C2 ester unit with an amide unit. It was unsuccessful because of an unfavorable intramolecular hydrogen bonding or steric hindrance that breaks the planarity of the chelating core unit, which is required for effective binding to the two active site metals (data not shown). Cyclization of the amide unit was a logical means to address both removal of the ester and alleviation of the amide planarity issue in order to reconstruct the metal chelation motif coplanarity. Hence this bicyclic carbamoylpyridone core unit was derivatized towards cyclized subclasses, represented by 16–18 shown in Tables 6.1 and 6.2. The inhibitors are differentiated by a linkage, by the presence of unsaturation in the new ring or by the addition of a hydroxyl group, in the case of 18. As was expected, every derivative exhibited high antiviral potency against the wild-type strain and acceptable potency shift in the presence of HSA. Although inhibitors 16 and 17 do not have sufficient efficacy against Q148K, one of resistant mutants for RAL, inhibitor 18 showed improvement with only a 2.1-fold change. Conversely, the rat PK profiles for bicyclic analogs 16 and 17 were encouraging, but low oral bioavailability of the hydroxyl derivative 18 raised concerns and left room for improvement. In addition, the hemiaminal functionality is not only a chiral center but quickly racemized after chiral separation, presumably via equilibrium of the open-chain aldehyde form.

As had been done in progressing from the monocyclic amide to the bicyclic derivatives to address the issue around coplanarity and its effects on potency, we again resorted to the formation of a ring to solve a completely different challenge in the case of the hydroxyl analog 18. Formation of the tricyclic ring system 19 again resulted in high potency against the wild-type virus and a low protein-adjusted IC_{50} and, similarly to the hydroxyl derivative, the tricyclic version had only a threefold change against the Q148K site-directed mutant that was being used as a surrogate measure of the ability to address clinically

Table 6.1 Antiviral activity for carbamoylpyridone integrase inhibitors.

Compound	Structure	$^{MT4}IC_{50}$ (μM)	$^{MT4}PAIC_{50}$ (μM)	Q148K (FC)	N155H (FC)	Y143R (FC)
RAL		0.0062	0.029	83	8.4	16
EVG		0.0013	0.029	>1700	25	1.8
15		0.010	0.120	200	21	1.5
16		0.0016	0.027	35	3.7	1.6
17		0.0059	0.018	18	1.5	1.5
18		0.0038	0.014	2.1	1.5	1.5

Table 6.2   Rat pharmacokinetics for carbamoylpyridone integrase inhibitors.

Compound	$Clt\ (ml\,min^{-1}\,kg^{-1})$	$t_{1/2}\ (h)$	$V_{dss}\ (L\,kg^{-1})$	$F\ (\%)$	$C_{24}\,h/^{MT4}PAIC_{50}$
15	1.16	4.67	0.19	53.4	3.4
16	0.02	10.5	0.01	39.5	2600
17	0.51	3.36	0.12	40.6	27
18	0.24	7.6	0.14	13.6	117

Table 6.3   Antiviral activity and pharmacokinetics for **19**.

Compound	Structure	$^{pHIV}IC_{50}$ (μM)	$^{pHIV}PAIC_{50}$ (μM)	$Q148K$ (FC)	Cl $(ml$ min^{-1} $kg^{-1})$	F $(\%)$
19		0.002	0.019	2.8	0.10	51
19S		0.002	0.081	2.4	0.10	79
19R		0.002	0.009	11	0.17	52

relevant resistance mutations. Also encouraging was the fact that the tricyclic compound **19** had a very low i.v. clearance in rats, along with the best half-life observed thus far and over 50% oral bioavailability (Table 6.3).

6.3.3   Execution and Delivery of the Tricyclic Carbamoylpyridone

Once robust activity and scaffold properties had been established, our program moved to a single-round pseudotyped antiviral luciferase reporter assay (pHIV),[91] for improved throughput and biosafety concerns. Tricyclic carba-moylpyridone **19** was also a potent antiviral (IC$_{50}$ 2 nM) in the pHIV assay while demonstrating a similar threefold loss in activity against the Q148K mutant virus, as was observed previously in the MT4 assay system. Additionally, a modest protein binding shift of tenfold resulted in a protein-adjusted IC$_{50}$ (^{PHIV}PAIC$_{50}$) against the resistant virus of 53 nM.

This virological profile was a significant advance toward our goal. However, this data set was from the racemate and the concern over differential potency,

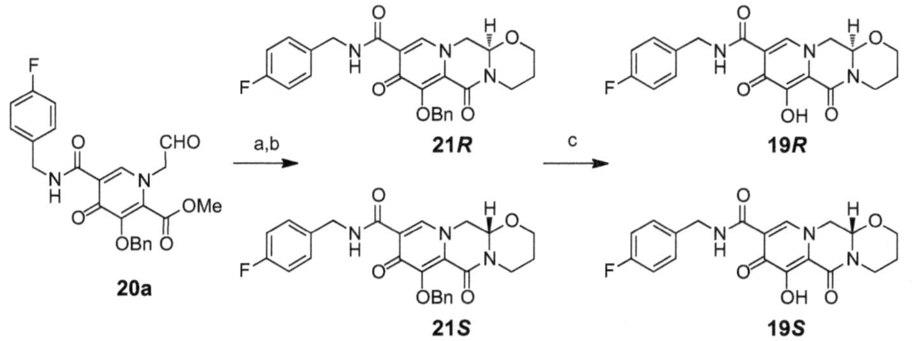

Scheme 6.1 Synthesis and isolation of tricyclic carbamoylpyridone enantiomers **19R** and **19S**. Reagents and conditions: (a) 3-aminopropanol, AcOH, μW; (b) chiral SFC separation; (c) Pd/C, H$_2$, THF.

protein binding and PK properties needed to be addressed immediately. The individual isomers were isolated after chiral chromatographic separation of the *O*-benzyl-protected precursor **21** to give the enantiopure benzyl ethers **21R** and **21S**, which were subsequently deprotected through hydrogenolysis to give the individual pure enantiomers of **19** (Scheme 6.1).

Table 6.3 gives comparison antiviral and key low-dose rat PK parameters for racemic **19** and its individual enantiomers **19S** and **19R**. We were initially concerned that the hemiaminal stability would be an issue, but no inter-conversion or chemical stability issues were observed with the purified isomers of **19** or any other of the tricyclic analogs discussed below. Notably, the isomers had almost identical antiviral activity, although significantly different protein binding shifts, with the *S* isomer showing a 40-fold shift, whereas the *R* isomer had a very modest 5-fold loss of potency with added human serum albumin. The isomers also showed a differential loss of activity against the Q148K mutation, with the *S* derivative having only a 2-fold potency change, whereas the *R* analog had a 11-fold shift in activity. Both purified isomers and the parent racemate showed very low rat i.v. clearance values and good oral bioavailability, further stimulating our interest in the series. However, close inspection of the data set makes it evident that data for the racemate are in fact a convolution of the favorable potency and low protein adjustment from the *R* isomer, while the PK and Q148K fold change are perhaps driven substantially by the *S* isomer.

Although the data may be complicated, the clear fact was that both enantiomers were very promising compounds. The question was how to progress, as separation via chromatographic means was highly undesirable, even for early stages of development. The most attractive way forward would have been an asymmetric synthetic approach but, since the stereocenter of concern was a hemiaminal in a complex ring system, good synthetic methodology was not apparent and would have required extensive experimentation even if it were possible under any conditions. The concept of a chiral auxiliary was among the first to be considered, but this was quickly recognized as an

undesirable approach because there was no precedence for our specific system
and the fact that our core ring system had limited functional handles on which to
append any necessary stereocontrol elements. Consideration of the mechanism
for the condensation and ring formation brought to mind the use of substrate
control as a possible strategy to establish a handle to impact stereochemistry and
also further evaluate SARs. Our hope was that by introducing substitution to the
system we would either further improve the biological activity and/or PK and
allow for robust stereocontrol or minimally reap the synthetic chemistry benefits
and not have any deleterious effects on the above properties. This concept is akin
to an auxiliary approach whereby a diastereomeric situation is established to
discriminate between isomers, with the exception that, in our case, the auxiliary is
retained as a beneficial part of the final drug.

As shown at the top of Scheme 6.2, with the use of an achiral amino alcohol,
as had been done for the synthesis of **21**, there is no stereochemical preference

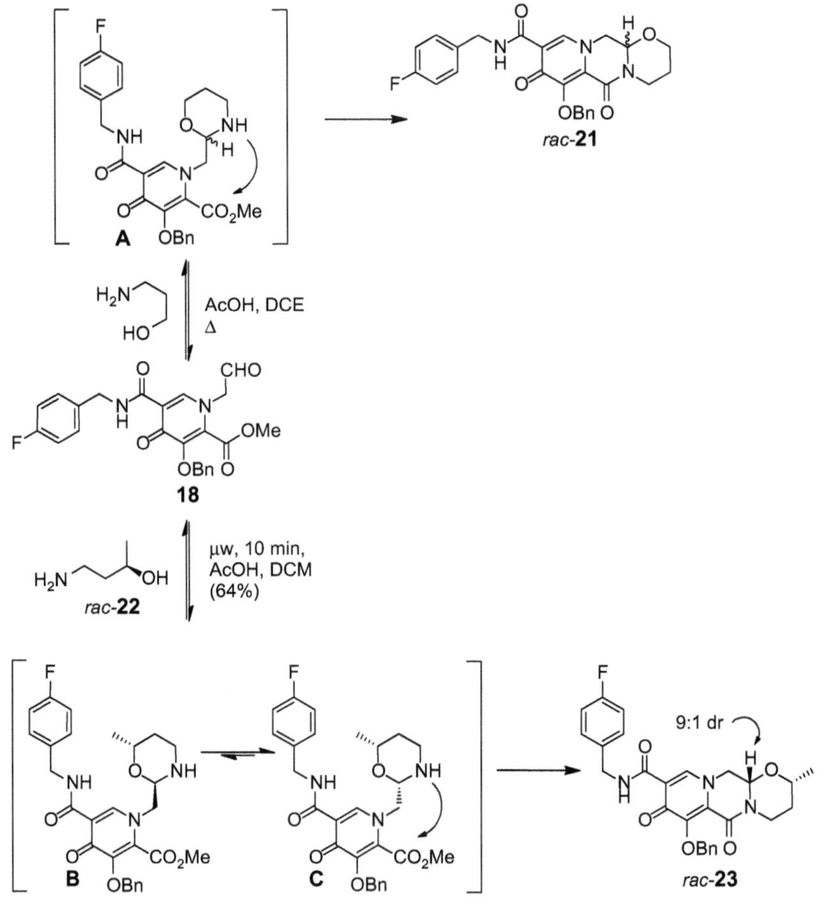

Scheme 6.2   Implementing substrate control of a hemiaminal stereocenter.

in intermediate **A** and, therefore, racemic product is formed. However, if an amino alcohol with a chiral center is employed, diastereomeric intermediates **B** and **C** are formed, resulting in an equilibrium that should favor the most stable chair conformation, as indicated by intermediate **C**. The final step of lactam ring closure should then lead to a biased product distribution related to the intermediates **B** and **C**. In the first attempt at this method using a racemic alcohol, *rac*-**22**, a 9:1 diastereomeric ratio (*dr*) favoring the expected product **23** from having both the methyl and methylene hemiaminal substituent in equatorial orientations was observed. Since our initial experiment was only to prove the relative stereocontrol principle in our system, we immediately repeated the reaction with chiral non-racemic alcohol **22** and obtained enantiomerically pure **23** after removal of the minor diastereomer by achiral silica gel chromatography. Removal of the benzyl ether under standard hydrogenolysis conditions served as a final step to arrive at the desired inhibitor design.

We sought to extend this methodology further to other substituted 1,3-amino alcohols. In the case of the 2-substituted derivative, an erosion of the stereo-control was observed under the conditions employed to give a still useful 4:1 selectivity [eqn (6.1)]. Placement of the stereocenter next to the amino terminus of the 1,3-amino alcohol led to exceptional diastereoselectivity [eqn 6.2)].

Interestingly, regarding the 3-substituted case shown in eqn (6.2), the stereochemical outcome resulted in the methyl and hydrogen of the two stereocenters existing on the same face, which effectively places the methyl group in an axial position. This may not be expected upon casual inspection of the system since, mechanistically, the reaction is believed to proceed through first formation of the hemiaminal with the aldehyde, which then can cyclize to form the tricyclic-fused amido system. This initial hemiaminal would be expected to exist in the most stable chair form indicated under energy local minima B in Figure 6.8. Ring closure of this would then provide the 1,3-diaxial hydrogen-containing system shown as D in Figure 6.8, but that is not observed.

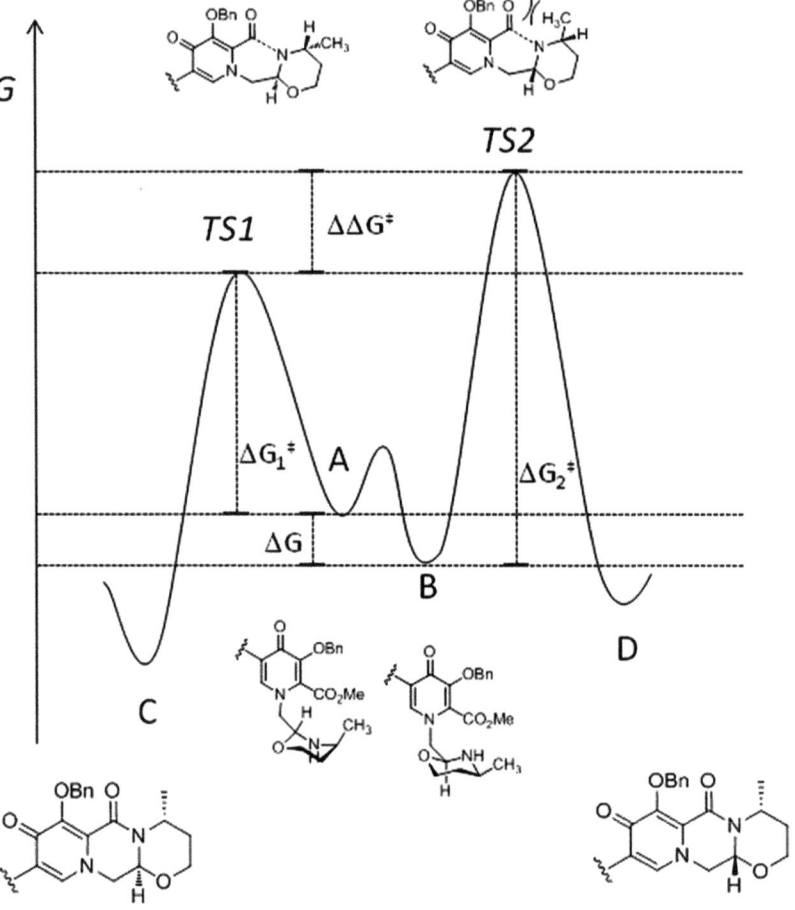

Figure 6.8 Curtin–Hammett explanation of 4-methyl substitution stereochemistry.

In fact, a Curtin–Hammett scenario is established as a result of a significant eclipsing interaction, resulting in a high-energy TS2 compared with the alternative axial methyl-containing TS1. There is a 3.6 kcal mol^{-1} energy difference between the two diastereomeric products favoring C, which is observed experimentally as the predominantly observed product with >20:1 selectivity.

The findings for the 1,3-amino alcohols stimulated our interest to investigate the related 1,2-aminopropanols to explore the diastereoselectivity of this system. Our initial attempt using (*R*)-1-aminopropan-2-ol (**29**) gave disappointing results, with a difficult-to-separate 3:1 mixture of diastereomers (**30**) [eqn (6.3)]. However, switching to the regioisomeric (*R*)-2-aminopropan-1-ol (**31**) gave a >40:1 mixture of diastereomers favoring the *trans* stereochemistry as shown in **32** [eqn (6.4)]. This is again explained as a result of a Curtin–Hammett situation, whereby the energetically favored intermediate is slow to cyclize and the product distribution is driven by the ring-closing step energetics, with the *trans* product having a ~2.24 kcal mol^{-1} lower energy.

6.3.4 Choosing the Optimal Candidate

We were pleased to observe that the methyl isomers (33–37, 3) all had high potency as good as or better than that of the unsubstituted original enantio-merically pure analogs of 19. A detailed examination of the SAR is not included herein, but suffice it to say that the methyl analogs appeared to be the optimal molecules to deliver stereocontrol while preserving or improving the necessary virological and PK properties (Table 6.4). Further complexity, lipophilicity, polarity or size of the substituent did not appear to bring any advantages to justify the added challenges of accessing the necessary amino alcohol and increase in molecular weight.

A key parameter in our triage of compounds was the coverage of the protein-adjusted IC$_{50}$ at 24 h post-dosing from a standard 5 mg kg^{-1} oral dose ($C_{24 h}$/^{pHIV}PAIC$_{50}$). The formulation studies ultimately do become important, but for screening of analogs we typically began with either a solution or suspension dose and eventually moved to a fully crystalline drug in capsules as a measure of our worst-case bioavailability. The combination of the PK profile along with early virology data places compounds 3 and 4 in an area with great potential for more detailed virological assessment.[92]

6.3.5 Tricyclic Carbamoylpyridones Deliver 'Next-generation' Virological Profiles

Compounds 3 and 4 were examined in MT-4 cell and peripheral blood mononuclear cell (PBMC) HIV multiround replication assays (Table 6.5).[93,94] A more robust protein shift value was also determined *via* titrating various concentrations of human serum into the MT4 cell assay system.[35] From these data, a better approximation of the protein-adjusted clinical target was able to be determined. The overall fold shift values were somewhat higher than the

Table 6.4 Antiviral and pharmacokinetic data for diastereomeric carbamoyl-pyridone tricyclic series.

Compound	X	Structure	$^{pHIV}IC_{50}$ (μM)	$^{pHIV}PAIC_{50}$ (μM)	Cl (ml min^{-1} kg^{-1})	$C_{24h}/^{pHIV}PAIC_{50}$
33	H		0.002	0.025	1.4	21
34	H		0.001	0.003	16.5	3.6
35	H		0.0008	0.21	ND	ND
36	H		0.0007	0.056	ND	ND
37	F		0.0005	0.004	0.04	702
3	F		0.002	0.022	0.23	112
4	F		0.0005	0.030	NR[a]	1247
38	F		0.005	0.039	0.20	725

[a] Not reportable due to prolonged exposure and extrapolation in excess of 20% of the total AUC.

pHIV data using purified human serum albumin suggested, but the overall comparison was similar, with **4** showing a significantly higher fold shift than **3**. The final clinical target was determined using the PBMC data and applying a fourfold factor to correct from the measured IC_{50} to approximated IC_{90} and

Table 6.5 Antiviral potency and protein-adjusted IC_{90} of **3** and **4**.

Compound	INST IC_{50} (μM)	$^{MT4}IC_{50}$ (μM)	PBMC IC_{50} (μM)	HuS fold shift	PAIC$_{90}$ (ng mL^{-1})
3	0.003	0.002	0.0005	75	64
4	0.003	0.001	0.0002	408	166

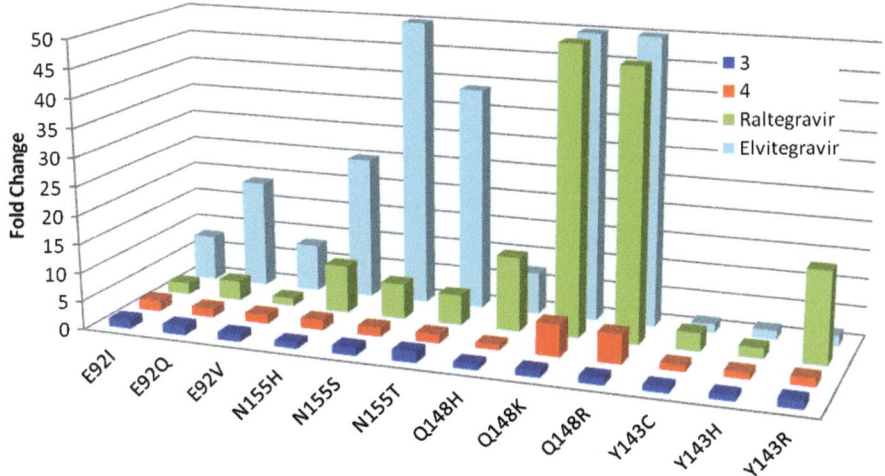

Figure 6.9 Comparison of fold change in HeLa cells IC_{50} of **3** and **4** against key IN resistance mutations.

applying the fold shift factor to that number. For **3** and **4**, the final PAIC$_{90}$ values were determined as 64 and 166 ng mL^{-1}, respectively.

Further evaluation of the resistance profile of **3** and **4** was undertaken in a HeLa cell assay with several site-directed mutant viruses meant to profile the two lead structures against a panel of relevant raltegravir and elvitegravir signature mutations. As can be seen in Figure 6.9, the fold change in activity when challenged with resistant mutations is vastly improved for **3** and **4** compared with RAL and EVG. Further data for **3** have been reported previously in a wider panel of mutants.[93]

6.3.6 Preclinical Pharmacokinetics

The PK profiles of the two leads were determined for Sprague–Dawley rats, beagle dogs and cynomolgus monkeys (Table 6.6). For the purpose of rigorous assessment, oral doses were performed using a standard 5 mg kg^{-1} dose from fully crystalline drug in a capsule to assess solid dosage form performance. For the purpose of this study, **3** was studied as its corresponding sodium salt, whereas **4** was dosed as the neutral molecule. Values for the i.v. data for both compounds appear to correlate well with the free unbound fraction, resulting in higher clearance values relative to hepatic blood flow in species with a larger

Table 6.6 Pharmacokinetic parameters for rats, dogs and cynomolgus monkeys for **3** and **4**.

Compound	Species[a]	Unbound (%)	Cl (mL min^{-1} kg^{-1})	$t_{\frac{1}{2}}$ (h)	V_{dss} (L kg^{-1})	F (%)	C_{24h}/ PAIC$_{90}$[c]
3	Rat	0.1	0.2	6.2	0.1	35	18
	Dog	4.6	2.2	5.2	0.3	35	4.6
	Monkey	0.9	2.1	6.0	0.3	25	0.76
	Human	0.7	–	–	–	–	–
4	Rat	<0.1	NR[b]	>18	NR[b]	NR[b]	57
	Dog	0.7	0.3	5.7	0.1	8	0.9
	Monkey	0.3	0.3	4.0	0.1	6	1.3
	Human	0.4	–	–	–	–	–

[a]Liver blood flow (lbf) = 55/31/44 mL min^{-1} kg^{-1} for rat/dog/monkey.
[b]Not reportable due to prolonged exposure and extrapolation in excess of 20% of the total AUC.
[c]The PAIC$_{90}$ value used was the clinical target of 64 ng mL^{-1} for **3** and 166 ng mL^{-1} for **4**.

unbound drug fraction. This is most pronounced in dogs for **3**, where a 4.6% free fraction results in clearance being ~7% of liver blood flow (lbf) [the lbf values used (mL min^{-1} kg^{-1}) were 55 (rat), 31 (dog) and 44(monkey), according to Davies and Morris[95]] whereas for rats the clearance is extremely low (<0.5% lbf), corresponding to a much lower free drug fraction as determined by equilibrium dialysis measurements. The data for monkeys are between for the rat and dog at 5% of lbf (44 ml min^{-1} kg^{-1} for a monkey). Similar trends are present for **4**, although it has a lower free unbound fraction across all species that results in very low clearance. The oral bioavailability of **3** is good from a suspension dosing formulation. Lower values were measured for **4** that was dosed from an unoptimized solid 'powder in a capsule' formulation, although coverage of the PAIC$_{90}$ was still possible across species. Both compounds were highly permeable (MDCK, P_{app} = 585 nm s^{-1} for **3** and 425 nm s^{-1} for **4**). However, solubility in simulated gastrointestinal (GI) fluid was low (130 and 42 µg mL^{-1} for **3** and **4**, respectively), likely resulting in dissolution-limited absorption from our solid-dose formulation.

No significant inhibition of CYP3A4 or time-dependent inhibition was observed for either **3** or **4**. The compounds were also extremely metabolically stable ($t_{\frac{1}{2}}$ >180 min) in rat, dog, monkey and human S9 fractions and also rat and human hepatocyte incubations ($t_{\frac{1}{2}}$ >220 min). The combination of all the above PK data resulted in low predicted human qd doses (<100 mg d^{-1}) that would achieve the targeted C_{trough} levels.

6.3.7 Choosing a Lead and Back-up

With two advancing leads, it became necessary to make some decisions about a front-runner and a back-up to prioritize resources and aggressively move the best asset forward. However, in our case, both compounds were attractive and there was nothing to suggest clearly that one should be sidelined, although it was not a case of everything being equal. Different PK properties and C_{trough} efficacy targets led us to consider whether perhaps one of these compounds did have an advantage that we had not yet accounted for in preclinical studies. Consequently, both compounds continued to proceed. In the end, the lower clinical target of **3** made it the priority and it quickly gained speed over the remaining compound **4**. There were some concerns about the protein shift and higher C_{trough} target for **4**, but these properties have thus far proven to be significant attributes, albeit with far fewer data in hand than for **3**. A description of how **4** is being developed is described in the next section. Compound **3** ultimately became what is known as dolutegravir (DTG, S/GSK1349572) and compound **4** is known as S/GSK1265744 (S/GSK744).

6.3.8 Clinical Development of Dolutegravir

Early Phase 1 clinical data for DTG demonstrated a 15 h half-life while producing a 25-fold coverage of the PAIC$_{90}$ *versus* C_{trough} from a 50 mg once-daily dose, without the need for a boosting agent.[96] These data were translated into a study design whereby once-daily doses of 2, 10 and 50 mg were given for 10 days as monotherapy to HIV-infected patients who were treatment-naive or off-therapy experienced during a Phase 2a proof-of-concept study. The results from these doses showed log$_{10}$ viral load decreases in HIV-1 RNA copies per milliliter of plasma of 1.51, 2.03 and 2.46, respectively (Figure 6.10).[97] DTG was well tolerated during these studies while delivering an excellent impact on

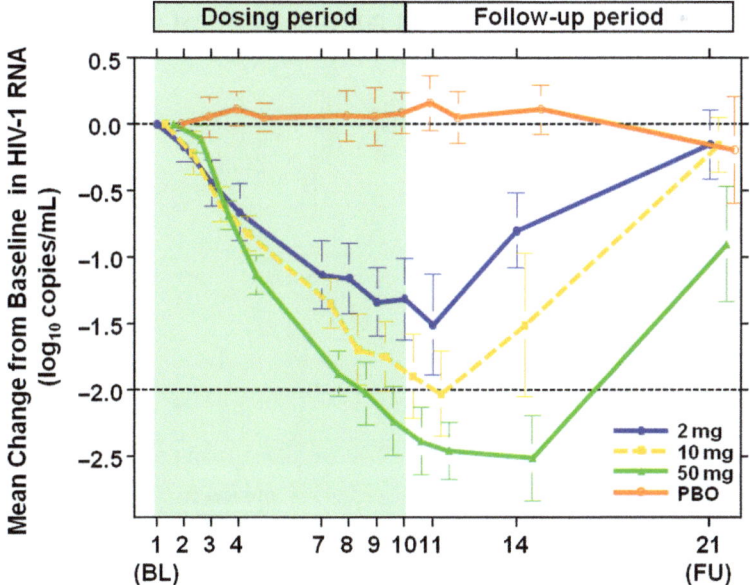

Figure 6.10   Phase 2a 10 day monotherapy viral load data for DTG.

viral load from a low once-daily dose. Clinical pharmacology studies showed no evidence of clinically significant drug–drug interactions with lopinavir/ritonavir (r), atazanavir/r, darunavir/r, tenofovir, the proton pump inhibitor omeprazole or multivitamins.[98–101] Additionally, there appears to be a lack of a clinically significant food effect, further improving patient convenience and dosing flexibility.[102] Antacids have been found to impact the clinical PK sufficiently that either dosing 2 h prior to taking antacids or 6 h after is recommended. This may be due to chelate formation in the stomach/GI tract with metal ions in antacids such as Maalox, which decreased absorption. It could also potentially be a result of pH adjustment in the stomach or upper GI compartment as a result of the basic supplement. It has been observed that co-administration with the NNRTI agent etravirine alone leads to significant decrease in exposure of DTG, likely due to induction of the UGT1A1 enzyme, a primary route of metabolic clearance for DTG.[103] If this combination is co-administered with a ritonavir-boosted protease inhibitor, this effect is greatly reduced.

DTG was further examined in the Phase 2b SPRING-1 dose-ranging study in treatment-naive patients. Doses of 10, 25 or 50 mg qd with either Truvada or Epzicom nucleoside backbone agents were compared with the same background therapy with 600 mg of the NNRTI efavirenz (Figure 6.11).[104] Once-daily, unboosted DTG demonstrated durable antiviral activity for all dosing arms, with 88% responding through 96 weeks[105] and a favorable safety profile at the 50 mg qd dose, the dose chosen for further Phase 3 studies outlined below.

A second smaller but important dose-ranging study, termed VIKING, was designed to determine the antiviral effects of DTG in an integrase-resistant, highly treatment-experienced patient population. A first cohort of 27 patients

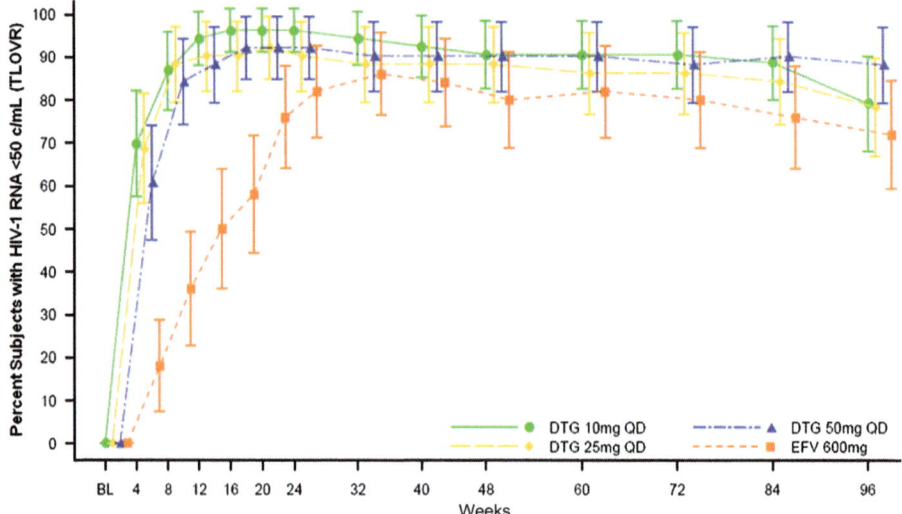

Figure 6.11   SPRING-1 Phase 2b treatment-naive data through 96 weeks.[105]

were given 50 mg qd on top of their existing background regimen for 10 days, followed by OBT on day 11 onwards. A second cohort of 24 patients were dosed under similar conditions, but with 50 mg bid to increase DTG drug levels further and a requirement for at least one fully active background agent. After 10 days with Cohort 1, 21/27 (78%) patients met the primary endpoint of <400 copies mL^{-1} whereas 23/24 (96%) of the Cohort 2 patients achieved the endpoint successfully. After 24 weeks, 41 and 52% of patients in the qd cohort and 75 and 83% of patients in bid cohort achieved plasma HIV RNA levels below either 50 or 400 copies mL^{-1}, respectively.[106,107] These data are critical in addressing the ability of DTG to treat integrase-resistant viruses in an actual clinical environment and serves to validate the preclinical virological data presented above.

Phase 3 evaluation of DTG is well under way at the time of drafting this chapter. Two pivotal trials looking at the treatment-naive population (SPRING-2 and SINGLE) have been reported thus far. SPRING-2 demonstrated that DTG was non-inferior to RAL through 48 weeks of treatment; 88% of study participants on the dolutegravir regimen were virologically suppressed (<50 copies mL^{-1}) *versus* 85% of participants on raltegravir [2.5% adjusted treatment difference, with 95% confidence interval (CI) -2.2% to $+7.1$%].[108] SINGLE is an ongoing double-blind, double dummy study designed to compare the efficacy and safety of two antiretroviral regimens: dolutegravir 50 mg plus abacavir/lamivudine (Kivexa/Epzicom) *versus* Atripla (tenofovir/emtricitabine/efavirenz). The study demonstrated superiority of the dolutegravir-based regimen over Atripla: at 48 weeks, 88% of study participants on the dolutegravir regimen were virologically suppressed (<50 copies mL^{-1}) *versus* 81% of participants on Atripla [difference 7.4% and 95% CI $+2.5$% to $+12.3$%; the difference in the primary endpoint was statistically significant, $p = 0.003$].[109]

There are also two ongoing studies in the treatment-experienced population. The SAILING study compares DTG 50 mg qd OBT with RAL plus OBT in treatment-experienced patients who are failing in their current therapy. A fourth Phase 3 study, VIKING-3, will explore the effect of DTG treatment (plus OBT) in INI-resistant patients. At this point, DTG is in late-stage clinical evaluation, but sufficient data exist to consider the goals of the discovery program to have been robustly achieved.

6.3.9 Long-acting Parenteral INI – S/GSK744

The potency and resistance profile[110] of S/GSK744 also makes it a very intriguing drug. A multipart Phase 1 study was conducted that included a cohort of HIV patients and also healthy volunteers to assess both efficacy and human pharmacokinetics, in addition to safety and tolerability data. In this study, subjects were given doses of 5, 10 and 25 mg once daily for 14 days and showed a near 30 h half life, with robust coverage of the $PAIC_{90}$ value determined as 166 ng mL^{-1}. Even at the 5 mg dose, several multiples of exposure over the clinical target were observed through 72 h. In the monotherapy cohort,

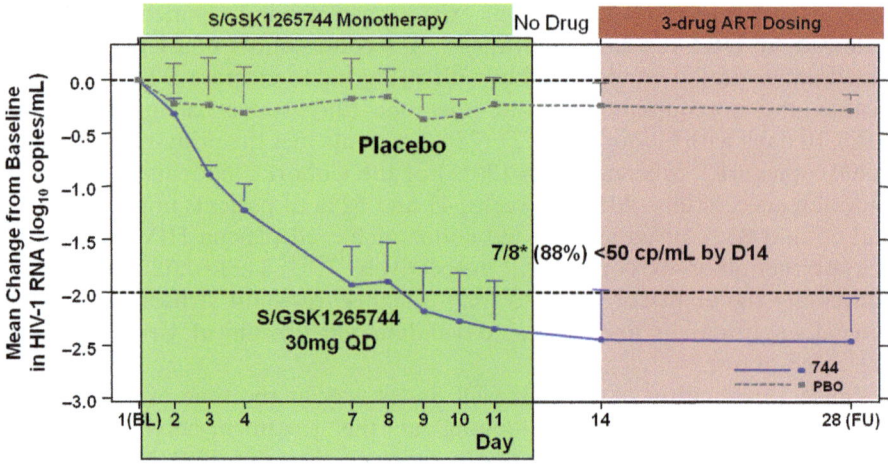

Figure 6.12 S/GSK1265744 Phase 2a monotherapy study.[111]

a 30 mg qd dose provided a 2.6log$_{10}$ decrease in viral load at day 11, thus verifying the pharmacodynamic response predicted from the antiviral and PK data (Figure 6.12).[111] In addition, a study to examine drug–drug interactions between S/GSK744 and etravirine was completed and it was found that ETR does not affect S/GSK744 pharmacokinetics.[112]

The significant viral load impact observed along with the low dose required to achieve such an effect plus long $t_{\frac{1}{2}}$ led to the proposal for development of S/GSK744 as a long-acting parenteral agent. In this scenario, a suspension of drug is introduced *via* subcutaneous or intramuscular injection and a reservoir of drug is established whereby exposure is achieved through slow absorption from the drug depot. The first evidence of clinical PK, safety and tolerability of S/GSK744 in healthy volunteers has recently been disclosed, showing an apparent $t_{\frac{1}{2}}$ of 21–47 and 45–47 days from intramuscular and subcutaneous injections, respectively.[113] Further studies are currently under way to examine this paradigm-shifting approach whereby daily adherence concerns are removed and new options for antiretroviral therapy are introduced in a situation that now requires relentless daily therapy for the remainder of a patient's life.

6.4   Non-catalytic Site Integrase Inhibitors

Despite the success of integrase strand transfer inhibitors, intensive research is ongoing to seek alternative mechanisms to inhibit this enzyme. The primary concern is the emergence of resistance in clinical use as observed with both raltegravir (RAL) and elvitegravir (EVG). To complicate the situation further, multiple studies have demonstrated cross-resistance between first-generation INIs, as alluded to above.[49,77,79] Although the next-generation integrase inhibitor DTG has a superior resistance profile, the potential for resistance remains, as is the case for all antiretroviral agents. Beyond strand transfer inhibitors, momentum has been growing in the field collectively known as non-catalytic site integrase inhibitors (NCINIs). These NCINIs are believed to be allosteric inhibitors of the integrase enzyme that disrupt the interaction of IN with the key cellular co-factor, lens epithelium-derived growth factor (LEDGF/p75). The proposed role for this interaction is to guide the preintegration complex to transcriptionally active regions of endogenous chromatin during integration. The LEDGF/p75 protein consists of 530 amino acids and contains a chromatin-binding N-terminal region that has a Pro–Trp–Trp–Pro (PWWP) motif, two copies of the AT-hook DNA binding motif and a nuclear localization signal (Figure 6.13). A highly conserved region of the C-terminus of LEDGF/p75 binds to IN and is termed the integrase-binding domain (IBD).[114–118] The importance of LEDGF/p75 in HIV-1 replication has been demonstrated by numerous groups and it is well documented that alteration of the cellular co-factor by mutagenesis, knockdown, RNA interference or alteration of the nuclear localization signal sequence has a deleterious effect upon viral replication.[119–122] Furthermore, LEDGF/p75 has been reported both to stimulate HIV-1 IN enzymatic activity (both strand transfer and 3′ processing) and to protect it from ubiquitin–proteasome degradation.[123,124]

In 2005, the first high-resolution crystal structure of the dimeric catalytic core domain (CCD) of HIV-1 IN complexed to the IBD of LEDGF/p75 was disclosed.[125] Of particular interest was a small binding pocket located at the CCD dimer interface where the IBD of LEDGF/p75 contacted two IN subunits (Figure 6.14). Although many of the key contact amino acid residues were validated by site-directed mutagenesis, additional interactions between the full-length proteins were more extensive than observed in the crystal structure.[120] A majority of these additional residues reside at the interface of the IN subunits and may suggest a potential role of LEDGF/p75 with IN multimerization.[126] Overall, these findings established the framework for designing small-molecule inhibitors of the interaction of LEDGF/p75 with HIV-1 integrase.

Debyser and co-workers utilized structure-based design to develop a series of inhibitors based on the aforementioned co-crystal structure of HIV-1 IN CCD with the IBD of LEDGF/p75.[127] Initially they performed a virtual screen

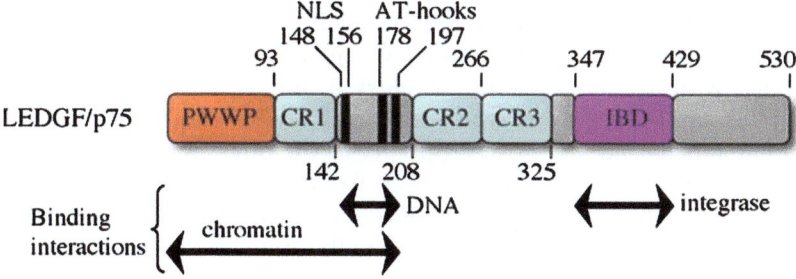

Figure 6.13 Domain organization of the LEDGF/p75 protein.
Reproduced under terms of the CCAL license from reference 118.

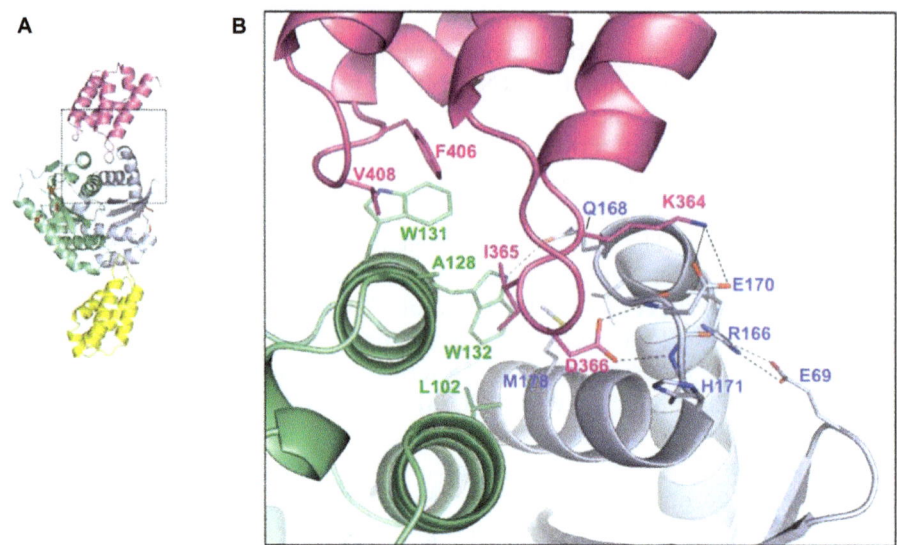

Figure 6.14   Crystal structure of the LEDGF/p75-IN Interaction. (A) Integrase-
binding domain of LEDGF/p75 (in magenta and yellow) binding with
the integrase CCD (in blue and green). (B) Highlighting key residues of
the CCD–IBD interaction: a bidentate hydrogen bond of D366 with IN
residues H171/E170 and I365 interacting with a hydrophobic pocket
formed by IN residues A128, W132, L102 and M178.
Reproduced under terms of the CCAL license from reference 118.

of 200 000 commercially available compounds that afforded 25 hits for further
evaluation. This led to a series of 2-(quinolin-3-yl)acetic acid derivatives that
inhibited the IN-LEDGF/p75 interaction *in vitro* and also demonstrated
micromolar antiviral activity in HIV-1 infected cells with no appreciable
cytotoxicity. Their most potent compound (LEDGIN 6, **39**, Figure 6.15) did
not affect integrase–DNA binding, was a very weak strand-transfer inhibitor
($IC_{50} = 19\,\mu M$) and was inactive in 3′ processing ($IC_{50} > 250\,\mu M$). *In vitro*
resistance passaging of **39** selected for a single point mutation A128T, which is
located at the edge of the LEDGF/p75 binding pocket (Figure 6.14B). The
binding mode was unambiguously assigned, with a co-crystal structure of **39**
revealing that the compound occupies the LEDGF/p75 pocket located at the
CCD dimer interface. Furthermore, **39** demonstrated no significant cross-
resistance with reverse transcriptase, entry or integrase-resistant virus strains.
Conversely, an IBD-resistant strain of virus (A128T/E170G) was fully resistant
to **39** but suffered no loss in susceptibility to the other antiviral agents tested.
Although the lack of antiviral potency for these early inhibitors prevented
further development, these compounds did offer valuable insight into the
potential for this new class of integrase inhibitors.

Independent of the findings reported by Debyser's group, Boehringer
Ingelheim published a patent claiming *tert*-butoxy(4-phenylquinolin-3-yl)acetic
acid (tBPQA) derivatives as inhibitors of HIV replication.[128] This series was

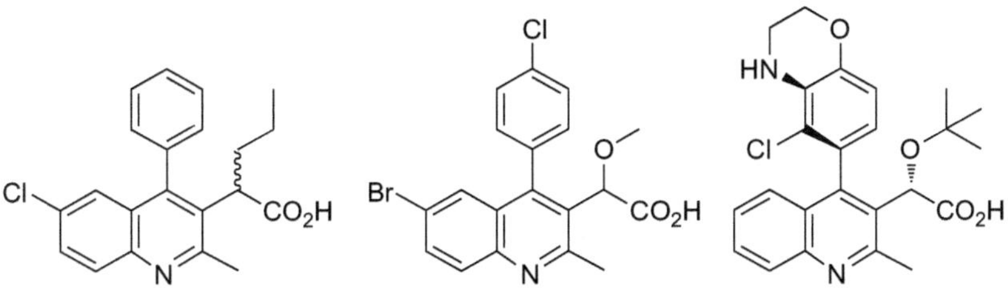

Figure 6.15 Structures of LEDGIN 6, BI-1001 and BI-C.

originally discovered as part of a high-throughput screen for integrase 3′ processing activity (see Figure 6.15, **41** for a representative example).[129] The lead compound, BI224436, has excellent ADME and antiviral properties and was advanced into Phase 1a clinical trials to evaluate safety and dosing in healthy volunteers. Using the PhenoSense assay to measure antiviral activity against 200 clinical isolates, BI224436 had a mean IC_{50} of 13 nM. As observed with the LEDGIN class of inhibitor, there was no decrease in potency against RAL-resistant viruses.[130] In a double-blind, placebo-controlled, single-rising dose escalation, BI224436 proved safe and well tolerated up to the highest dose group of 200 mg. Overall, BI224436 demonstrated a dose-proportional increase in plasma C_{max} and AUC and was rapidly absorbed with a median t_{max} of 0.5 h and a $t_{\frac{1}{2}}$ of 7.1 h. Based on the PK profile, an oral solution dose of 100 mg qd provided the target therapeutic plasma drug concentration of 500 nmol L^{-1} at 24 h.[131] For undisclosed reasons, the development of BI224436 was stopped following the initial phase 1 study. In late 2011, Boehringer Ingelheim entered a licensing agreement with Gilead Sciences to develop this series of NCINI further.

Although the understanding of the mechanism of action for these types of inhibitors will likely evolve over time, it is clear that these molecules inhibit the HIV-1 IN enzyme in a manner unique to the current INIs. Owing to the structural similarity between the LEDGIN and tBPQA class of NCINI, it is difficult to reconcile the reported differences in mechanism of action between these molecules. A recent study refuted the original claim that LEDGIN 6 was devoid of 3′ processing activity and reported it to be a low micromolar inhibitor of 3′ processing ($IC_{50} = 3.9\,\mu M$) similar in activity to BI-1001 ($IC_{50} = 2.3\,\mu M$; **40**, Figure 6.15).[132] Crystal structures for both **39** and **40** bound to the HIV-1 IN CCD reveal that the molecules occupy nearly identical spaces with the major difference being an additional H-bond interaction of the BI-1001 methoxy group with T174 of IN. Additionally, both classes of molecule promote integrase multimerization.[132] This alternative mechanism has been well documented to play an important role in viral replication as precise coordination of the multimeric state of IN is required to bind viral DNA productively.[126,133] The mechanistic details for inhibition are obviously complex, but a consensus appears to be forming around the concept of a

multimodal inhibition in which compounds impair both LEDGF-
IN-dependent binding and LEDGF-independent integrase catalytic activity.

6.5 Conclusion

The emerging field of non-catalytic site integrase inhibitors is starting to gain
significant momentum. The first molecules that inhibit the interaction of
LEDGF/p75 with HIV-1 IN were reported in 2007, and in 2011 we already
witnessed the first clinical trials of this new class of integrase inhibitor. The
rapid progress in this new field is reminiscent of how the development of the
strand transfer inhibitors accelerated over a decade ago. As with the active-site
inhibitors, it would be expected that a number of molecules may be required to
enter advanced preclinical and clinical study before medicinal chemists achieve
success. It is exciting to consider potential regimens using both types of
integrase inhibitors to restrict further the potential for viral escape, possibly by
forcing losses in fitness from having to elude multiple inhibitors. Although
patients are well served at present, it is clear that new ideas and new directions
are still viable and ripe for consideration as worthy investments.

References

1. UNAIDS, United Nations Program on HIV/AIDS (UNAIDS), *UNAIDS
   Report on the Global AIDS Epidemic*, December 2012, www.UNAIDS.org
   (last accessed 23 February 2013).
2. A. Maitland, A. Jackson, J. Osorio, S. Mandalia, B. G. Gazzard and
   G. J. Moyle, *HIV Med.*, 2008, **9**, 667.
3. J. B. Nachega, V. C. Marconi, G. U. van Zyl, E. M. Gardner, W. Preiser,
   S. Y. Hong, E. J. Mills and R. Gross, *Infect. Disord. Drug Targets*, 2011,
   11, 167.
4. A. S. Fauci and D. M. Morens, *N. Engl. J. Med.*, 2012, **366**, 454.
5. H.-H. M. Truong, T. A. Kellogg, W. McFarland, B. Louie,
   J. D. Klausner, S. S. Philip and R. M. Grant, *Plos One*, 2011, **6**, 1.
6. L. Wittkop, H. F. Günthard, D. Dunn, A. Cossi-Lepri, A. de Luca,
   C. Kücherer, N. Obel, V. von Wyl, B. Masquelier, C. Stephan, C. Torti,
   A. Antinori, F. Garcia, A. Judd, K. Porter, R. Thiébaut, H. Castro,
   A. van Sighem, C. Colin, J. Kjaer, J. D. Lundgren, R. Paredes,
   A. Pozniak, B. Clotet, A. Philips, D. Pillay and G. Chêne, *Lancet Infect.
   Dis.*, 2011, **11**, 363.
7. V. Jain, T. Liegler, E. Vittinghoff, W. Hartogensis, P. Bacchetti, L. Poole,
   L. Loeb, C. D. Pilcher, R. M. Grant, S. G. Deeks and F. M. Hecht, *Plos
   One*, 2010, **5**, 1.
8. Panel on Antiretroviral Guidelines for Adults and Adolescents,
   Department of Health and Human Services, *Guidelines for the Use of
   Antiretroviral Agents in HIV-1-Infected Adults and Adolescents*, 2011,
   http://aidsinfo.nih.gov/contentfiles/adultandadolescentGL.pdf.

9. B. Taiwo, L. Zheng, S. Gallien, R. M. Matining, D. R. Kuritzkes, C. C. Wilson, B. I. Berzins, E. P. Acosta, B. Bastow, P. S. Kim and J. J. Eron Jr., *AIDS*, 2011, **25**, 2113.

10. J. Reynes, A. Lawal, F. Pulido, R. Soto-Malave, J. Gathe, M. Tian, L. M. Frerick, T. J. Podsadecki and A. M. Nilius, *HIV Clin. Trials*, 2011, **12**, 255.

11. A. S. Perelson and S. G. Deeks, *Sci. Transl. Med.*, 2011, **3**, 91ps30.

12. A. D. Frankel and J. A. T. Young, *Annu. Rev. Biochem.*, 1998, **67**, 1.

13. P. Rice, R. Craigie and D. R. Davies, *Curr. Opin. Struct. Biol.*, 1996, **6**, 76.

14. F. Dyda, A. B. Hickman, T. M. Jenkins, A. Engelman, R. Craigie and D. R. Davies, *Science*, 1994, **266**, 1981.

15. L. Haren, B. Ton-Hoang and M. Chandler, *Annu. Rev. Microbiol.*, 1999, **53**, 245.

16. J. A. Grobler, K. Stillmock, B. Hu, M. Witmer, P. Felock, A. S. Espeseth, A. Wolfe, M. Egbertson, M. Bourgeois, J. Melamed, J. S. Wai, S. Young, J. Vacca and D. J. Hazuda, *Proc. Natl. Acad. Sci. U. S. A.*, 2002, **99**, 6661.

17. J. S. Wai, M. S. Egbertson, L. S. Payne, T. E. Fisher, M. W. Embrey, L. O. Tran, J. Y. Melamed, H. M. Langford, J. P. Guare Jr., L. Zhuang, V. E. Grey, J. P. Vacca, M. K. Holloway, A. M. Naylor-Olsen, D. J. Hazuda, P. J. Felock, A. L. Wolfe, K. A. Stillmock, W. A. Schleif, L. J. Gabryelski and S. D. Young, *J. Med. Chem.*, 2000, **43**, 4923.

18. T. Fujishita and T. Yoshinaga, *PCT Int. Appl.*, WO1999/50245, 1999; T. Fujishita, T. Yoshinaga and A. Sato, *PCT Int. Appl.*, WO2000/3, 1999, 9086.

19. R. Kiyama and T. Kawasuji, *PCT Int. Appl.*, WO01/95905, 2001.

20. T. Kawasuji, M. Fuji, T. Yoshinaga, A. Sato, T. Fujiwara and R. Kiyama, *Bioorg. Med. Chem.*, 2006, **14**, 8420.

21. T. Kawasuji, T. Yoshinaga, A. Sato, M. Yodo, T. Fujiwara and R. Kiyama, *Bioorg. Med. Chem.*, 2006, **14**, 8430.

22. T. Kawasuji, M. Fuji, T. Yoshinaga, A. Sato, T. Fujiwara and R. Kiyama, *Bioorg. Med. Chem.*, 2007, **15**, 5487; M. J. C. Rosemond, L. John-Williams, T. Yamaguchi, T. Fujishita and J. S. Walsh, *Chem.-Biol. Interact.*, 2004, **147**, 129.

23. D. J. Hazuda, P. Felock, M. Witmer, A. Wolfe, K. Stillmock, J. A. Grobler, A. Espeseth, L. Gabryelski, W. Schleif, C. Blau and M. D. Miller, *Science*, 2000, **287**, 646.

24. A. S. Espeseth, P. Felock, A. Wolfe, M. Witmer, J. A. Grobler, N. Anthony, M. Egbertson, J. Y. Melamed, S. Young, T. Hamill, J. L. Cole and D. J. Hazuda, *Proc. Natl. Acad. Sci. U. S. A.*, 2000, **97**, 11244.

25. L. Krishnan, X. Li, H. L. Naraharisetty, S. Hare, P. Cherepanov and A. Engelman, *Proc. Natl. Acad. Sci. U. S. A.*, 2010, **107**, 15910.

26. S. Hare, S. S. Gupta, E. Valkov, A. Engelman and P. Cherepanov, *Nature*, 2010, **464**, 232.

27. S. Hare, A. M. Vos, R. F. Clayton, J. W. Thuring, M. D. Cummings and P. Cherepanov, *Proc. Natl. Acad. Sci. U. S. A.*, 2010, **107**, 20057.

28. S. Chung, D. M. Himmel, J. -K. Jiang, K. Wojtak, J. D. Bauman, J. W. Rausch, J. A. Wilson, J. A. Beutler, C. J. Thomas, E. Arnold and S. F. J. Le Grice, *J. Med. Chem.*, 2011, **54**, 4462.

29. E. B. Lansdon, Q. Liu, S. A. Leavitt, M. Balakrishnan, J. K. Perry, C. Lancaster-Moyer, N. Kutty, X. Liu, N. H. Squires, W. J. Watkins and T. A. Kirschberg, *Antimicrob. Agents Chemother.*, 2011, **55**, 2905.

30. K. E. B. Parkes, P. Ermert, J. Fassler, J. Ives, J. A. Martin, J. H. Merrett, D. Obrecht, G. Williams and K. Klumpp, *J. Med. Chem.*, 2003, **46**, 1153.

31. A. Pendri, N. A. Meanwell, K. M. Peese and M. A. Walker, *Expert Opin. Ther. Pat.*, 2011, **21**, 1173.

32. B. A. Johns and A. C. Svolto, *Expert Opin. Ther. Pat*, 2008, **18**, 1225.

33. B. A. Johns, *Annu. Rep. Med. Chem.*, 2010, **45**, 263.

34. S. Little, G. Drusano, R. Schooley, D. Haas, P. Kumar, S. Hammer, D. McMahon, K. Squires, R. Asfour, D. Richman, J. Chen, A. Saah, R. Leavitt, D. Hazuda and B. Y. Nguyen, presented at the 12th Conference on Retroviruses and Opportunistic Infections, Boston, MA, February 2005, Abstract 161.

35. E. P. Garvey, B. A. Johns, M. J. Gartland, S. A. Foster, W. H. Miller, R. G. Ferris, R. J. Hazen, M. R. Underwood, E. E. Boros, J. B. Thompson, J. G. Weatherhead, C. K. Koble, S. H. Allen, L. T. Schaller, R. G. Sherrill, T. Yoshinaga, M. Kobayashi, C. Wakasa-Morimoto, S. Miki, K. Nakahara, T. Noshi, A. Sato and T. Fujiwara, *Antimicrob. Agents Chemother.*, 2008, **52**, 901.

36. G. S. Jones, F. Yu, A. Zeynalzadegan, J. Hesselgesser, X. Chen, J. Chen, H. Jin, C. U. Kim, M. Wright, R. Geleziunas and M. Tsiang, *Antimicrob. Agents Chemother.*, 2009, **53**, 1194.

37. Z. Lin, D. Langley, B. Terry, T. Protack, M. A. Walker, B. N. Naidu, M. Patel, N. Meanwell, M. Krystal and I. B. Dicker, presented at the 19th Conference on Retroviruses and Opportunistic Infections, Seattle, WA, 5–8 March 2012, Abstract 690.

38. M. Rowley, *Prog. Med. Chem.*, 2008, **46**, 1.

39. V. Summa, A. Petrocchi, F. Bonelli, B. Crescenzi, M. Donghi, M. Ferrara, F. Fiore, C. Gardelli, O. G. Paz, D. J. Hazuda, P. Jones, O. Kinzel, R. Laufer, E. Monteagudo, E. Muraglia, E. Nizi, F. Orvieto, P. Pace, G. Pescatore, R. Scarpelli, K. Stillmock, M.V. Witmer and M. Rowley, *J. Med. Chem.*, 2008, **51**, 5843.

40. M. A. Thompson, J. A. Adberg, P. Cahn, J. S. G. Montaner, G. Rizzardini, A. Telenti, J. M. Gatell, H. F. Günthard, S. M. Hammer, M. S. Hirsch, D. M. Jacobsen, P. Reiss, D. D. Richman, P. A. Volberding, P. Yeni and R. T. Schooley, *JAMA*, 2010, **304**, 321.

41. A. Zolopa, *Antiviral Res.*, 2010, **85**, 241.

42. J. M. Murray, S. Emery, A. D. Kelleher, M. Law, J. Chen, D. J. Hazuda, B.-Y. T. Nguyen, H. Teppler and D A. Cooper, *AIDS*, 2007, **21**, 2315.

43. R. A. Steigbigel, D. A. Cooper, P. N. Kumar, J. E. Enron, M. Schecter, M. Markowitz, M. R. Loutfy, J. L. Lennox, J. M. Gatell,

J. K. Rockstroh, C. Katlama, P. Yeni, A. Lazzarin, B. Clotet, J. Zhao, J. Chen, D. M. Ryan, R. R. Rhodes, J. A. Killar, L. R. Gilde, K. M. Strohmaier, A. R. Meibohm, M. D. Miller, D. J. Hazuda, M. L. Nessly, M. J. DiNubile, R. D. Isaacs, B.-Y. Nguyen and H. Teppler, *N. Engl. J. Med.*, 2008, **359**, 339.

44. R. A. Steigbigel, D. A. Cooper, H. Teppler, J. E. Enron, P. N. Kumar, J. K. Rockstroh, M. Schecter, C. Katlama, M. Markowitz, P. Yeni, M. R. Loutfy, A. Lazzarin, J. L. Lennox, B. Clotet, J. Zhao, H. Wan, R. R. Rhodes, K. M. Strohmaier, R. J. Barnard, R. D. Isaas and B.-Y. Nguyen, *Clin. Infect. Dis.*, 2010, **50**, 605.

45. D. A. Cooper, R. A. Steigbigel, J. M. Gatell, J. K. Rockstroh, C. Katlama, P. Yeni, A. Lazzarin, B. Clotet, P. N. Kumar, J. E. Enron, M. Schecter, M. Markowitz, M. R. Loutfy, J. L. Lennox, J. Zhao, J. Chen, D. M. Ryan, R. R. Rhodes, J. A. Killar, L. R. Gilde, K. M. Strohmaier, A. R. Meibohm, M. D. Miller, D. J. Hazuda, M. L. Nessly, M. J. DiNubile, R. D. Isaacs and H. Teppler, *N. Engl. J. Med.*, 2008, **359**, 355.

46. D. da Silva, L. Van Wesenbeeck, D. Breilh, S. Reigadas, G. Anies, K. Van Baelen, P. Morlat, D. Neau, M. Dupon, L. Wittkop, H. Fleury and B. Masquelier, *J. Antimicrob. Chemother.*, 2010, **65**, 1262.

47. O. Delelis, S. Thierry, F. Subra, F. Simon, I. Malet, C. Alloui, S. Sayon, V. Calvez, E. Deprez, A.-G. Marcelin, L. Tchertanov and J.-F. Mouscadet, *Antimicrob. Agents Chemother.*, 2010, **54**, 491.

48. M. Métifiot, C. Marchand, K. Maddali and Y. Pommier, *Viruses*, 2010, **2**, 1347.

49. M. Métifiot, K. Maddali, A. Naumova, X. Zhang, C. Marchand and Y. Pommier, *Biochemistry*, 2010, **49**, 3715.

50. J. L. Lennox, E. DeJesus, A. Lazzarin, R. B. Pollard, J. V. R. Madruga, D. S. Berger, J. Zhao, X. Xu, A. Williams-Diaz, A. J. Rodgers, R. J. O. Barnard, R. M. D. Miller, M. J. DiNubile, B.-Y. Nguyen, R. Leavitt and P. Sklar, *Lancet*, 2009, **374**, 796.

51. J. L. Lennox, E. DeJesus, D. S. Berger, A. Lazzarin, R. B. Pollard, J. V. R. Madruga, J. Zhao, H. Wan, C. L. Gilbert, H. Teppler, A. J. Rodgers, R. J. O. Barnard, M. D. Miller, M. J. DiNubile, B.-Y. Nguyen, R. Leavitt and P. Sklar, *J. Acquir. Immune Defic. Syndr.*, 2010, **55**, 39.

52. J. J. Eron, B. Young, D. A. Cooper, M. Youle, E. DeJesus, J. Andrade-Villaneuva, C. Workman, R. Zajdenverg, G. Fätkenheuer, D. S. Berger, P. N. Kumar, A. J. Rodgers, M. A. Shaughnessy, M. L. Walker, R. J. O. Barnard, M. D. Miller, M. J. DiNubile, B.-Y. Nguyen, R. Leavitt, X. Xu and P. Sklar, *Lancet*, 2010, **375**, 396.

53. E. Martinez, M. Larrousse, J. M. Llibre, F. Gutierrez, M. Saumoy, A. Antela, H. Knobel, J. Murillas, J. Berenguer, J. Pich, I. Perez and J. Gatell, *AIDS*, 2010, **24**, 1697.

54. M. A. Wainberg, G. J. Zaharatos and B. G. Brenner, *N. Engl. J. Med.*, 2011, **365**, 637.

55. J. J. Eron, J. K. Rockstroh, J. Reynes, J. Andrade-Villanueva, J. V. Ramalho-Madruga, L.-G. Bekker, B. Young, C. Katlama, J. M. Gatell-Artigas, J. R. Arribas, M. Nelson, H. Campbell, J. Zhao, A. J. Rodgers, M. L. Rizk, L. Wenning, M. D. Miller, D. J. Hazuda, M. J. DiNubile, R. Leavitt, R. Isaacs, M. N. Robertson, P. Sklar and B.-Y. Nguyen, *Lancet Infect. Dis.*, 2011, **11**, 907.

56. M. Iwamoto, L. A. Wenning, A. S. Petry, M. Laethem, M. De Smet, J. T. Kost, S. A. Merschman, K. M. Strohmaier, S. Ramael, K. C. Lasseter, J. A. Stone, K. M. Gottesdiener and J. A. Wagner, *Clin. Pharm. Ther.*, 2007, **83**, 293.

57. M. Markowitz, J. O. Morales-Ramirez, B.-Y. Nguyen, C. M. Kovacs, R. A. Steigbigel, D. A. Cooper, R. Liporace, R. Schwartz, R. Isaacs, L. R. Gilde, L. Wenning, J. Zhao and H. Teppler, *J. Acquir. Immune Defic. Syndr.*, 2006, **43**, 509.

58. M. Sato, T. Motomura, H. Aramaki, T. Matsuda, M. Yamashita, Y. Ito, H. Kawakami, Y. Matsuzaki, W. Watanabe, K. Yamataka, S. Ikeda, E. Kodama, M. Matuoka and H. Shinkai, *J. Med. Chem.*, 2006, **49**, 1506.

59. M. Sato, H. Kawakami, T. Motomura, H. Aramaki, T. Matsuda, M. Yamashita, Y. Ito, Y. Matsuzaki, K. Yamataka, S. Ikeda and H. Shinkai, *J. Med. Chem.*, 2009, **52**, 4869.

60. Y. Matsuzaki, W. Watanabe and K. Yamataka, presented at the 13th Conference on Retroviruses and Opportunistic Infections, Denver, CO, 5–9 February 2006, Abstract 508.

61. T. Willis and V. Vega, *Expert Opin. Invest. Drugs*, 2012, **21**, 395.

62. K. Shimura and E. N. Kodama, *Antiviral Chem. Chemother.*, 2012, **20**, 79.

63. E. DeJesus, D. Berger, M. Markowitz, C. Cohen, T. Hawkins, P. Ruane, R. Elion, C. Farthing, L. Zhong, A. K. Cheng, D. McColl and B. P. Kearney, *J. Acquir. Immune Defic. Syndr.*, 2006, **43**, 1.

64. S. Ramanathan, A. A. Mathias, P. German and B. P. Kearney, *Clin. Pharmacokinet.*, 2011, **50**, 229.

65. S. Ramanathan, G. Shen, A. Cheng and B. P. Kearney, *J. Acquir. Immune Defic. Syndr.*, 2007, **45**, 274.

66. A. A. Mathias, P. German, B. P. Murray, L. Wei, A. Jain, S. West, D. Warren, J. Hui and B. P. Kearney, *Clin. Pharmacol. Ther.*, 2010, **87**, 322.

67. P. German, D. Warren, S. West, J. Hui and B. P. Kearney, *J. Acquir. Immune Defic. Syndr.*, 2010, **55**, 323.

68. C. Marchand, *Expert Opin. Investig. Drugs*, 2012, **21**, 901.

69. A. R. Zolopa, D. S. Berger, H. Lampiris, L. Zhong, S. L. Chuck, J. V. Enejosa, B. P. Kearney and A. K. Cheng, *J. Infect. Dis.*, 2010, **201**, 814.

70. C. Cohen, R. Elion, P. Ruane, D. Shamblaw, E. DeJesus, B. Rashbaum, S. L. Chuck, K. Yale, H. C. Liu, D. R. Warren, S. Ramanathan and B. P. Kearney, *AIDS*, 2011, **25**, F7.

71. J.-M. Molina, A. LaMarca, J. Andrade-Villanueva, B. Clotet, N. Clumeck, Y.-P. Liu, L. Zhong, N. Margot, A. K. Cheng and S. L. Chuck, *Lancet Infect. Dis.*, 2012, **12**, 27.

72. P. Sax, E. DeJesus, A. Mills, A. Zolopa, C. Cohen, D. Wohl, J. Gallant, H. Liu, E. Quirk and B. P. Kearney, presented at the 19th Conference of Retroviruses and Opportunistic Infections, Seattle, WA, 5–8 March 2012, Abstract 101.
73. P. E. Sax, E. DeJesus, A. Mills, A. Zolopa, C. Cohen, D. Wohl, J. E. Gallant, H. C. Liu, L. Zhong, K. Yale, K. White, B. P. Kearney, J. Szwarcberg, E. Quirk and A. K. Cheng, *Lancet*, 2012, **379**, 2439.
74. E. DeJesus, J. Rockstroh, J.-M. Molina, J. Gathe, S. Ramanathan, X. Wei, J. Szwarcberg, A. Jandourek and A. Cheng, presented at the 19th Conference of Retroviruses and Opportunistic Infections, Seattle, WA, 5–8 March 2012, Abstract 627.
75. E. DeJesus, J. K. Rockstroh, J. –M. Molina, J. Gathe, S. Ramanathan, X. Wei, K. Yale, J. Szwarcberg, K. White, A. K. Cheng and B. P. Kearney, *Lancet*, 2012, **379**, 2429.
76. O. Goethals, R. Clayton, M. Van Ginderen, I. Vereycken, E. Wagemans, P. Geluykens, K. Dock, R. Strijbos, V. Smits, A. Vos, G. Meersseman, D. Jochmans, K. Vermeire, D. Schols, A. Hallenberger and K. Hertogs, *J. Virol.*, 2008, **82**, 10366.
77. K. Shimura, E. Kodama, Y. Sakagami, Y. Matsuzaki, W. Watanabe, K. Yamataka, Y. Watanabe, Y. Ohata, S. Doi, M. Sato, M. Kano, S. Ikeda and M. Matuoka, *J. Virol.*, 2008, **82**, 764.
78. A. Hombrouch, A. Voet, B. Van Remoortel, C. Desadeleer, M. De Maeyer, Z. Debyser and M. Witvrouw, *Antimicrob. Agents Chemother.*, 2008, **52**, 2069.
79. J. Marinello, C. Marchand, B. T. Mott, A. Bain, C. J. Thomas and Y. Pommier, *Biochemistry*, 2008, **47**, 9345.
80. J.-L. Blanco, V. Varghese, S.-Y. Rhee, J. M. Gatell and R. W. Shafer, *J. Infect. Dis.*, 2011, **203**, 1204.
81. B. Taiwo, L. Zheng, S. Gallien, R. M. Matining, D. R. Kuritzkes, C. C. Wilson, B. I. Berzins, E. P. Acosta, B. Bastow, P. S. Kim and J. J. Eron, *AIDS*, 2011, **25**, 2113.
82. M. Métifiot, N. Vandegraaff, K. Maddali, A. Naumova, X. Zhang, D. Rhodes, C. Marchand and Y. Pommier, *AIDS*, 2011, **25**, 1175.
83. P. K. Quashie, R. D. Sloan and M. A. Wainberg, *BMC Med.*, 2012, **10**, 34.
84. R. Collier, *CMAJ*, 2009, **180**, 279.
85. E. Serrao, S. Odde, K. Ramkumar and N. Neamati, *Retrovirology*, 2009, **6**, 25.
86. H. Murai, T. Endo, N. Kurose, T. Taishi and H. Yoshida, *PCT Int. Appl.*, WO04/024693, 2004.
87. M. Egbertson, H. M. Moritz, J. Y. Melamed, W. Han, D. S. Perlow, M. S. Kuo, M. Embrey, J. P. Vacca, M. M. Zrada, A. R. Cortes, A. Wallace, Y. Leonard, D. J. Hazuda, M. D. Miller, P. J. Felock, K. A. Stillmock, M. V. Witmer, W. Schlief, L. J. Gabryelski, G. Moyer, J. D. Ellis, L. Jin, W. Xu, M. P. Braun, K. Kassahun, N. N. Tsou and S. D. Young, *Bioorg. Med. Chem. Lett.*, 2007, **17**, 1392.
88. E. E. Boros, C. E. Edwards, S. A. Foster, M. Fuji, T. Fujiwara, E. P. Garvey, P. L. Golden, R. J. Hazen, J. L. Jeffrey, B. A. Johns,

T. Kawasuji, R. Kiyama, C. S. Koble, N. Kurose, W. H. Miller, A. L. Mote, H. Murai, A. Sato, J. B. Thompson, M. C. Woodward and T. Yoshinaga, *J. Med. Chem.*, 2009, **52**, 2754.

89. M. Egbertson, J. Y. Melamed, H. M. Langford and S. D. Young, *PCT Int. Appl.*, WO03/062204, 2003.

90. T. Kawasuji, B. Johns, H. Yoshida, T. Taishi, Y. Taoda, H. Murai, R. Kiyama, M. Fuji, T. Yoshinaga, T. Seki, M. Kobayashi, A. Sato and T. Fujiwara, *J. Med. Chem.*, 2012, **55**, 8735.

91. G. Jarmy, M. Heinelein, B. Weissbrich, C. Jassoy and A. Rethwilm, *J. Med. Virol.*, 2001, **64**, 223.

92. B. Johns, T. Kawasuji, T. Taishi, H. Yoshida, E. Garvey, W. Spreen, M. Underwood, A. Sato, T. Yoshinaga and T. Fujiwara, presented at the 17th Conference of Retroviruses and Opportunistic Infections, San Francisco, CA, 16–19 February 2010, Abstract 55.

93. M. Kobayashi, T. Yoshinaga, T. Seki, C. Wakasa-Morimoto, K. W. Brown, R. Ferris, S. A. Foster, R. J. Hazen, S. Miki, A. Suyama-Kagitani, S. Kawauchi-Miki, T. Taishi, T. Kawasuji, B. A. Johns, M. R. Underwood, E. P. Garvey, A. Sato and T. Fujiwara, *Antimicrob. Agents Chemother.*, 2011, **55**, 813.

94. B. A. Johns, T. Kawasuji, J. G. Weatherhead, T. Taishi, D. P. Temelkoff, H. Yoshida, T. Akiyama, Y. Taoda, H. Murai, R. Kiyama, M. Fuji, N. Tanimoto, J. Jeffrey, S. A. Foster, T. Yoshinaga, T. Seki, M. Kobayashi, A. Sato, M. N. Johnson, E. P. Garvey and T. Fujiwara, *J. Med. Chem.*, submitted for publication.

95. B. Davies and T. Morris, *Pharm. Res.*, 1993, **10**, 1093.

96. S. Min, I. Song, J. Borland, S. Chen, Y. Lou, T. Fujiwara and S. Piscitelli, *Antimicrob. Agents Chemother.*, 2010, **54**, 254.

97. S. Min, L. Sloan, E. DeJesus, T. Hawkins, L. McCurdy, I. Song, R. Stroder, S. Chen, M. Underwood, T. Fujiwara, S. Piscitelli and J. Lalezari, *AIDS*, 2011, **25**, 1737.

98. I. Song, S. Min, J. Borland, Y. Lou, S. Chen, P. Patel, T. Ishibashi and S. Piscitelli, *J. Clin. Pharmacol.*, 2011, **51**, 237.

99. I. Song, J. Borland, S. Chen, Y. Lou, A. Peppercorn, T. Wajima, S. Min and S. Piscitelli, *Br. J. Clin. Pharmacol.*, 2011, **72**, 103.

100. I. Song, S. Min, J. Borland, Y. Lou, S. Chen, T. Ishibashi, T. Wajima and S. Piscitelli, *J. Acquir. Immune Defic. Syndr.*, 2010, **55**, 365.

101. P. Patel, I. Song, J. Borland, A. Patel, Y. Lou, S. Chen, T. Wajima, A. Peppercorn, S. S. Min and S. C. Piscitelli, *J. Antimicrob. Chemother.*, 2011, **66**, 1567.

102. I. Song, J. Borland, S. Chen, P. Patel, T. Wajima, A. Peppercorn and S. C. Piscitelli, *Antimicrob. Agents Chemother.*, 2012, **56**, 1627.

103. I. Song, J. Borland, S. Min, Y. Lou, S. Chen, P. Patel, T. Wajima and S. C. Piscitelli, *Antimicrob. Agents Chemother.*, 2011, **55**, 3517.

104. J. Van Lunzen, F. Maggiolo, J. R. Arribas, A. Rakhmanova, P. Yeni, B. Young, J K. Rockstroh, S. Almond, I. Song, C. Brothers and S. Min, *Lancet Infect. Dis.*, 2012, **12**, 111.

105. H.–J. Stellbrink, J. Reynes, A. Lazzarin, E. Voronin, F. Pulido, F. Felizarta, S. Almond, M. St. Clair, N. Flack and S. Min, presented at the 19th Conference on Retroviruses and Opportunistic Infections, Seattle, WA, 5–8 March 2012, Abstract K-102.
106. V. Soriano, J. Cox, J. Eron, P. Kumar, C. Katlama, A. Lazzarin, I. Poizot-Martin, G. Richmond, M. Ait-Khaled, T. Fujiwara, J. Huang, S. Min, C. Vavro and J. M. Yeo, presented at the 13th European AIDS Conference, Belgrade, Serbia, 12–15 October 2011.
107. J. Eron, P. Kumar, A. Lazzarin, G. Richmond, V. Soriano, J. Huang, C. Vavro, M. Ait-Khaled, S. Min and J. Yeo, presented at the 18th Conference on Retroviruses and Opportunistic Infections, Boston, MA, 27 February–2 March 2011.
108. F. Raffi, A. Rachlis, H.-J. Stellbrink, W. D. Hardy, C. Torti, C. Orkin, M. Bloch, D. Podzamczer, V. Pokrovsky, S. Almond, D. Margolis and S. Min, presented at the XIX International AIDS Conference, Washington, DC, 22–27 July 2012, Abstract THLBB04.
109. S. Walmsley, A. Antela, N. Clumeck, D. Duiculescu, A. Eberhard, F. Gutiérrez, L. Hocqueloux, F. Maggiolo, U. Sandkovsky, C. Granier, B. Wynne and K. Pappa, presented at the 52nd Interscience Conference on Antimicrobials and Chemotherapy (ICAAC), San Francisco, CA, 9–12 September 2012, Abstract H-556b.
110. M. R. Underwood, M. St. Clair, B. A. Johns, A. Sato, T. Fujiwara and W. Spreen, presented at the XVIII International AIDS Conference, Vienna, 18–23 July 2010.
111. S. Min, E. DeJesus, L. McCurdy, G. Richmond, J. Torres, S. Ford, S. Chen, Y. Lou, M. Bomar, T. Cyr, M. St.Clair, T. Fujiwara and S. Piscitelli, presented at the 49th Interscience Conference on Antimicrobial Agents and Chemotherapy (ICAAC), San Francisco, CA, 12–15 September 2009, paper H-1228.
112. S. L. Ford, S. Chen, E. Gould, E. Dumont, Y. Lou, S. Piscitelli, T. Taishi and W. Spreen, presented at the 13th International Workshop on Clinical Pharmacology of HIV Therapy, Barcelona, 2012, paper P-15.
113. W. Spreen, S. L. Ford, S. Chen, E. Gould, D. Wilfret, D. Subich, T. Taishi and Z. Hong, presented at the XIX International AIDS Conference, Washington, DC, 22–27 July 2012, Abstract TUPE040.
114. M. Llano, M. Vanegas, N. Hutchins, D. Thompson, S. Delgado and E. M. Poeschla, *J. Mol Biol.*, 2006, **360**, 760.
115. P. Cherepanov, E. Devroe, P. A. Silver and A. Engelman, *J. Biol. Chem.*, 2004, **279**, 48883.
116. M. Vanegas, M. Llano, S. Delago, D. Thompson, M. Peretz and E. Poeschla, *J. Cell Sci.*, 2005, **118**, 1733.
117. M. Llano, M. Vanegas, O. Fregoso, D. Saenz, S. Chung, M. Peretz and E. M. Poeschla, *J. Virol.*, 2004, **78**, 9524.
118. A. Engelman and P. Cherepanov, *PLoS Pathogen*, 2008, **4**, e1000046.
119. S. Emiliani, A. Mousnier, K. Busschots, M. Maroun, B. Van Maele, D. Tempe, L. Vandekerckhove, F. Moisant, L. Ben-Slama, M. Witvrouw,

F. Christ, J.-C. Rain, C. Dargemont, Z. Debyser and R. Benarous, *J. Biol. Chem.*, 2005, **280**, 25517.

120. L. Vandekerckhove, F. Christ, B. Van Maele, J. De Rijck, R. Gijsbers, C. Van den Haute, M. Witvrouw and Z. Debyser, *J. Virol.*, 2006, **80**, 1886.

121. G. Maertens, P. Cherepanov, W. Pluymers, K. Busschots, E. De Clercq, Z. Debyser and Y. Engelborghs, *J. Biol. Chem.*, 2003, **278**, 33528.

122. G. Maertens, P. Cherepanov, Z. Debyser, Y. Engelborghs and A. Engelman, *J. Biol. Chem.*, 2004, **279**, 33421.

123. G. Yu, G. S. Jones, M. Hung, A. H. Wagner, H. L. MacArthur, X. Liu, S. Leavitt, M. J. McDermott and M. Tsiang, *Biochemistry*, 2007, **46**, 2899.

124. M. Llano, S. Delgado, M. Vanegas and E. M. Poeschla, *J. Biol. Chem.*, 2004, **279**, 55570.

125. P. Cherepanov, A. L. B. Ambrosio, S. Rahman, T. Ellenberger and A. Engelman, *Proc. Natl. Acad. Sci. U. S. A.*, 2005, **102**, 17309.

126. C. J. McKee, J. J. Kessl, N. Shkriabai, M. J. Dar, A. Engelman and M. Kvaratskelia, *J. Biol. Chem.*, 2008, **283**, 31802.

127. F. Christ, A. Voet, A. Marchand, S. Nicolet, B. A. Desimmie, D. Marchand, D. Bardiot, N. J. Van der Veken, B. Van Remoortel, S. V. Strelkov, M. De Maeyer, P. I. Chaltin and Z. Debyser, *Nat. Chem. Biol.*, 2010, **6**, 442.

128. Y. S. Tsantrizos, M. Boes, C. Brochu, C. Fenwick, E. Malenfant, S. Mason and M. Pesant, *PCT Int. Appl.*, 2007, WO2007/131350A1.

129. C. Yoakim, M. Amad, M. D. Bailey, R. Bethell, M. Bös, P. Bonneau, M. Cordingley, R. Coulombe, J. Duan, P. Edwards, L. Fader, A.-M. Faucher, M. Garneau, A. Jakalian, S. Kawai, L. Lamorte, S. LaPlante, L. Luo, S. Mason, M.-A. Poupart, N. Rioux, B. Simoneau, Y. Tsantrizos, M. Witvrouw and C. Fenwick, presented at the 51st Interscience Conference on Antimicrobial Agents and Chemotherapy (ICAAC), *Chicago, IL, 17–20* September 2011, Abstract F1–1369.

130. C. Fenwick, R. Bethell, A. Quinson, P. Robinson, B. Simoneau and C. Yoakim, presented at the 51st Interscience Conference on Antimicrobial Agents and Chemotherapy (ICAAC), Chicago, IL, 17–20 September 2011, Abstract F1-1370.

131. S. Aslanyan, C. Ballow, J. P. Sabo, J. Halbeck, D. Roos, T. R. Macgregor, P. Robinson and J. Kort, presented at the 51st Interscience Conference on Antimicrobial Agents and Chemotherapy (ICAAC), Chicago, IL, 17–20 September 2011, Abstract AI-1725.

132. J. J. Kessl, N. Jena, Y. Koh, S. –H. Taskent, A. Slaughter, L. Feng, S. DeSilva, L. Wu, S. F. Le Grice, A. Engelman, J. R. Fuchs and M. Kvaratskhelia, *J. Biol. Chem.*, 2012, **287**, 16801.

133. Z. Hayouka, J. Rosenbluh, A. Levin, S. Loya, M. Lebendiker, D. Veprintsev, M. Kotler, A. Hizi, A. Loyter and A. Friedler, *Proc. Natl. Acad. Sci., U. S. A.*, 2007, **104**, 8316.

CHAPTER 7

HCV NS3/4a Protease Inhibitors: Simeprevir (TMC-435350), Vaniprevir (MK-7009) and MK-5172

JOHN A. MCCAULEY,* MICHAEL T. RUDD AND NIGEL J. LIVERTON

Department of Medicinal Chemistry, Merck Research Laboratories, West Point, PA 19486, USA
*Email: john_mccauley@merck.com

7.1 Introduction

Hepatitis C virus (HCV) infection continues to represent a major health issue, with estimates of 130–170 million people infected worldwide.[1] Although hepatitis C is slow to progress, the disease ultimately leads to liver damage in up to 50% of individuals and also hepatocellular carcinoma in a significant number of cases.[2] Multiple genotypes of this positive strand RNA Flaviviridae virus exist[3] in addition to virtually unlimited quasi-species.[4,5] Until recently, standard-of-care treatment for HCV-infected patients has been a combination of pegylated interferon alpha (IFN-α) and ribavirin. For the most common genotype 1 infections, modest sustained virologic response (SVR) rates of ~50% are observed after administration for 48 weeks.[6,7] Shorter treatment duration (24 weeks) is possible in genotype 2 and 3 infections and higher SVR rates are observed on the order of 80%.[8]

RSC Drug Discovery Series No. 32
Successful Strategies for the Discovery of Antiviral Drugs
Edited by Manoj C. Desai and Nicholas A. Meanwell
© The Royal Society of Chemistry 2013
Published by the Royal Society of Chemistry, www.rsc.org

The modest genotype 1 SVR rates, coupled with significant adverse events (AEs), have stimulated a large effort to identify inhibitors of several key non-structural proteins essential for viral replication, with the NS3/4A HCV protease inhibitor area being the most well-populated and mature.[9]

Recent developments in the HCV protease area have significantly improved treatment options. Approval of the HCV NS3/4A protease inhibitors, boceprevir (Victrelis),[10,11] **1** (Figure 7.1), and telaprevir (Incivek/Incivo),[12,13] **2**, for treatment of HCV, in combination with pegylated IFN-α and ribavirin, represents a major step forward,[14] with enhanced SVR rates and the potential for shorter duration therapies.[15] In addition, patients who previously failed to achieve SVR with interferon/ribavirin alone can be retreated with these newly approved combinations and achieve significant SVR rates,[16] albeit lower than those observed in treatment-naive patients. These triple combinations now represent a new standard of care in HCV treatment.[17] However, a more dramatic paradigm shift in treatment of HCV infection appears all but certain in the coming years, with a move to all oral combination therapy with direct-acting antivirals (DAAs). HCV protease inhibitors have the potential to play a significant role in these DAA combination therapies, with the development of inhibitors of NS5A and NS5B also advancing. Clinical studies involving multiple permutations of nucleosides, ribavirin, NS3/4A protease and NS5A inhibitors are in progress across a range of patient populations and promising initial results have been obtained, including significant SVR rates.[18] While these studies will provide insight into both the number of drugs needed for combination therapy and the potential for even further shortened treatment duration, the individual profiles of the components in terms of cross genotype activity and resistance profile may make the drawing of general conclusions difficult.

HCV protease inhibitors can be classified by their modes of interaction with the protease, ketoamides that participate in slowly reversible covalent binding with the active site serine and more rapidly reversible inhibitors. Both of the first-generation HCV protease inhibitors, boceprevir and telaprevir, belong to the ketoamide class of molecules. Subsequent research in this area has been relatively limited, with narlaprevir (SCH 900518),[19] **3**, the only other compound disclosed that has moved into clinical development.[20,21]

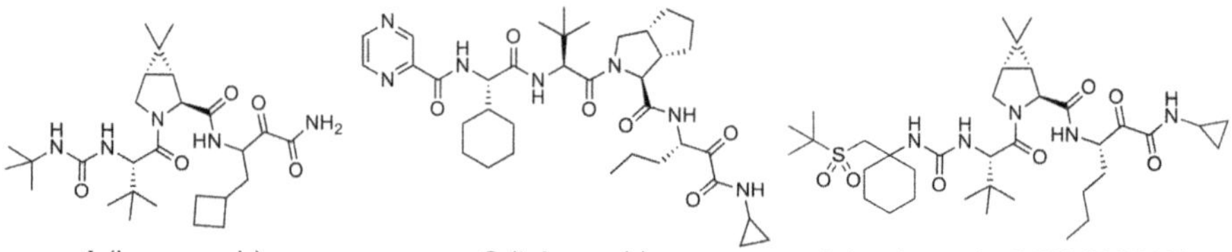

1 (boceprevir)   **2 (telaprevir)**   **3 (narlaprevir, SCH-900518)**

Figure 7.1   Ketoamide HCV NS3 inhibitors.

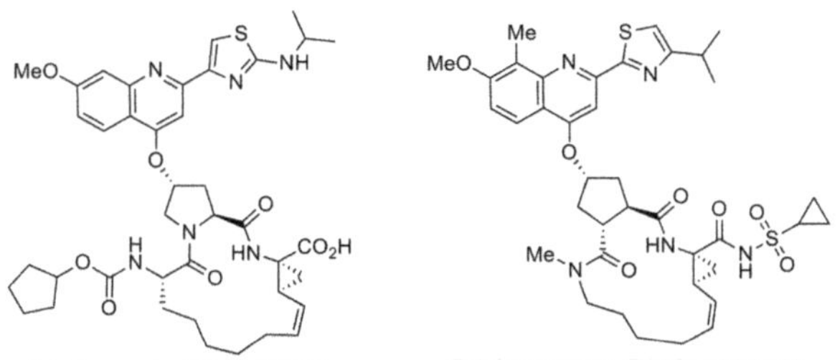

4 (ciluprevir, BILN-2061) **5** (simeprevir, TMC-435350)

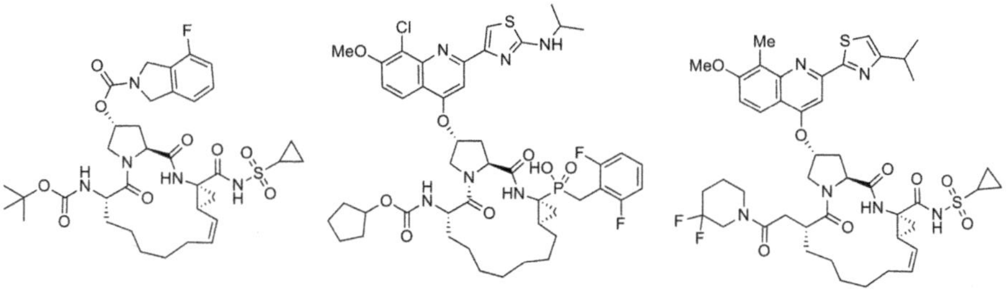

6 (danoprevir, ITMN-191, R7227) **7** (GS-9256) **8** (neceprevir, ACH-2684)

Figure 7.2 P1–P3 macrocyclic HCV NS3 inhibitors.

The reversible inhibitor structural class is significantly larger and can itself be broken down into three sub-classes based on structure: P1–P3 macrocyclic, acyclic and P2–P4 macrocyclic compounds.

The prototypical reversible inhibitor is a P1–P3 macrocyclic inhibitor, ciluprevir (BILN-2061),[22] **4**, (Figure 7.2) which demonstrated multi-log decreases in viral load during a Phase 1b study.[23] Development of BILN-2061 was subsequently discontinued due to cardiac findings in rhesus monkey safety studies.[24] Elaboration and modification of the P1–P3 scaffold core has provided multiple compounds that have advanced into clinical development. In addition to simeprevir (TMC-435350), **5**, which is discussed in more detail below, danoprevir (ITMN-191, R7227),[25–27] **6**, GS-9256,[28,29] **7**, and neceprevir (ACH-2684, **8**)[30] belong to this structural group.

The related acyclic chemical space has also yielded a number of clinical development candidates, including asunaprevir (BMS-650032),[31,32] **9**, faldeprevir (BI-201335),[33–37] **10**, GS-9451,[38,39] **11**, and sovaprevir (ACH-1625),[40] **12** (Figure 7.3).

The alternative P2–P4 macrocyclic constraint, derived from a modeling based approach,[41] led to vaniprevir (MK-7009),[42,43] **13**, and the second-generation inhibitor MK-5172,[44,45] **14**, described in detail below (Figure 7.4).

In addition, there are multiple HCV protease inhibitors in clinical development for which chemical structures have not been disclosed. These include ABT-450,[46] IDX-320[47,48] and PHX-1766.[49,50] In this chapter, we discuss in detail the design and discovery of three HCV NS3/4a protease inhibitors: simeprevir (TMC-435350, **5**), vaniprevir (MK-7009, **13**) and MK-5172 (**14**).

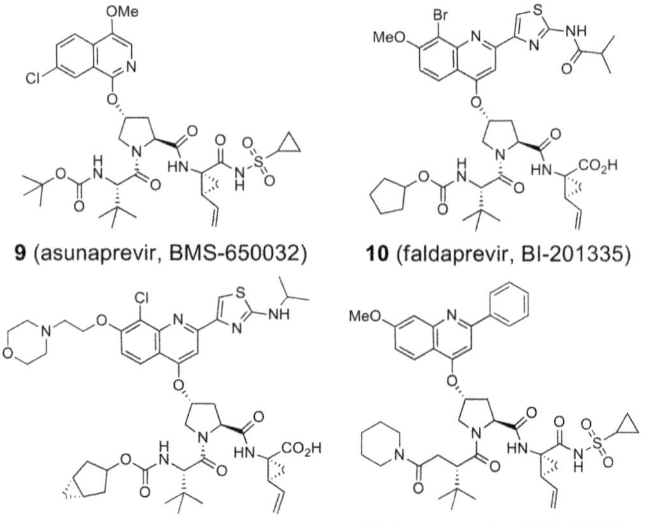

9 (asunaprevir, BMS-650032) **10** (faldaprevir, BI-201335)

11 (GS-9451) **12** (sovaprevir, ACH-1625)

Figure 7.3 Acyclic HCV NS3 inhibitors.

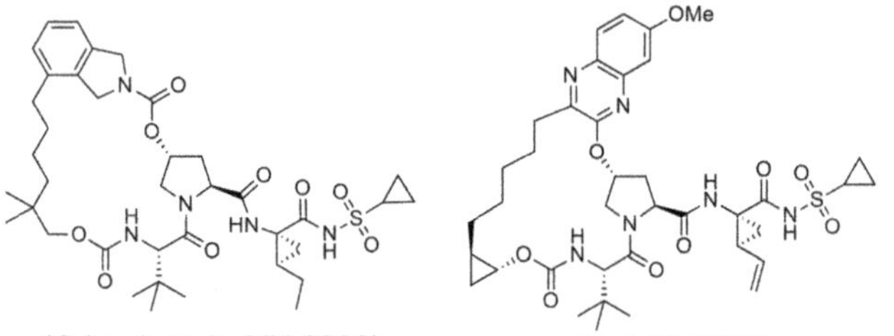

13 (vaniprevir, MK-7009) **14** (MK-5172)

Figure 7.4 P1–P3 macrocyclic HCV NS3 inhibitors.

7.2 Discovery of Simeprevir (TMC-435350)

Following the clinical proof-of-concept established for the inhibition of the HCV NS3/4a protease mechanism of action with BILN-2061[22,51] (**4**, Figure 7.2), a number of research groups utilized this molecule as a starting point for novel compound/series development.

One such avenue of development that Medivir/Tibotec pursued was modification of the central *N*-acyl-(4*R*)-hydroxyproline core present in BILN-2061 and related acyclic analogs (**15**, Figure 7.5).[52] Replacement of proline with isosteric cyclopentane in inhibitors of other proteases has previously been demonstrated to be a successful strategy. Notably, Samuelsson and co-workers utilized dicarbonylcyclopentane **17** and -cyclopentene **18** to replace proline in a class of thrombin inhibitors[53,54] (Figure 7.6).

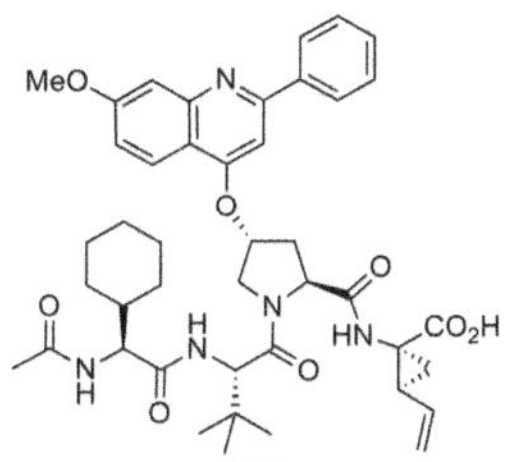

15

Figure 7.5 Example of a potential initial lead for the Medivir/Tibotec NS3 program.

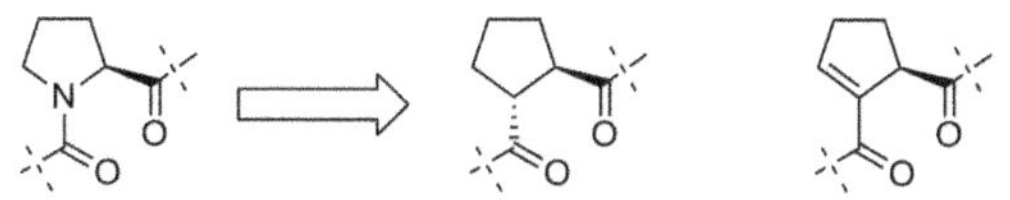

16   **17**   **18**

Figure 7.6 Proline replacement strategy in the thrombin program.

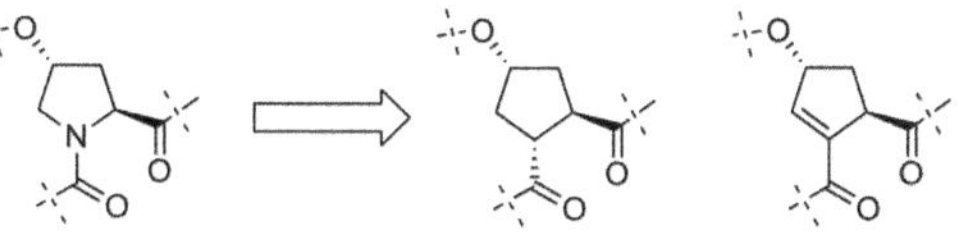

19   **20**   **21**

Figure 7.7 Medivir/Tibotec's hydroxyproline replacement strategy in the HCV NS3 program.

In the case of HCV NS3/4a protease, the more highly substituted (4*R*)-hydroxyproline could potentially be replaced by the corresponding hydroxy-substituted cyclopentanes or cyclopentenes, **20** and **21** (Figure 7.7).

As an initial proof-of-concept using the linear inhibitor **15** as a starting point, **22** (Figure 7.8) was prepared and demonstrated promising activity of 22 nM in the enzymatic inhibitions assay against gt1a NS3.[55] This is especially interesting, because in order to avoid a ketone linker, the P3/P4 side chain is changed to a reverse amide and thus the location of the sterically bulky *tert*-butyl and cyclo-hexane groups, and also the H-bond donor, are shifted by one position relative to the starting core ring structure. Additionally, the hybridization of the connecting atom is changed from proline's sp^2-nitrogen to an sp^3-carbon.

The synthesis of **22** is shown in Scheme 7.1 and begins with the known chiral building-block (−)-*trans*-(3*R*,4*R*)-bis(methoxycarbonyl)cyclopentanone.[56] Reduction of the ketone to give **23**, followed by cyclization and protection of the resultant carboxylic acid as the *tert*-butyl ester, led to bicyclic lactone **24**. Mild opening of the lactam followed by amide coupling to the known vinylcyclopropylamine **25**[57] gave intermediate **26**. Conversion to desired product **22** was accomplished by inverting the hydroxyl chiral center using a

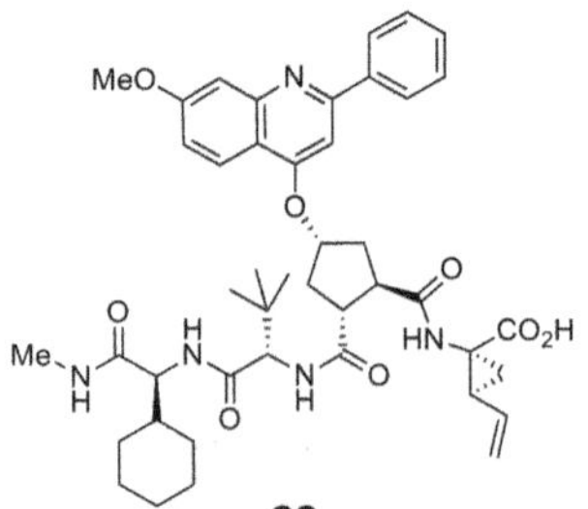

Figure 7.8   Cyclopentyl core-containing compound **22**.

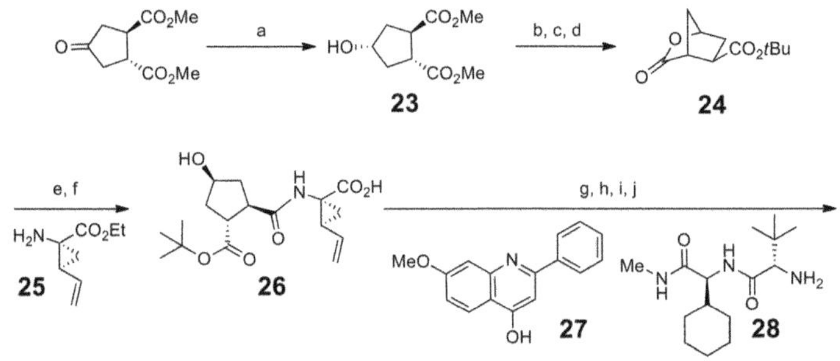

Scheme 7.1   Conditions: (a) NaBH₄, MeOH, 0 °C, 76%; (b) NaOH, MeOH; (c) Ac₂O, pyridine, 88%, two steps; (d) Boc₂O, DMAP, DCM, 52%; (e) LiOH, dioxane–water, 0 °C; (f) **25**, HATU, DIEA, DMF, 89%, two steps; (g) **27**, PPh₃, DIAD, THF, 68%; (h) TFA, Et₃SiH, DCM; (i) **28**, HATU, DIEA, DMF, 74%, two steps; (j) LiOH, THF, MeOH, water, 67%.

Mitsunobu reaction with quinolinol **27**,[52,58] deprotection, amide coupling to chiral amine **28** and hydrolysis. Overall, the original sequence is 10 linear steps and provided **22** in 10% overall yield.

A structure–activity relationship (SAR) study was then undertaken within this series of compounds (Table 7.1). Initially, simplified analogs **29–31** established the stereochemical requirements for active inhibitors as *S*; inhibitors **30** and **31**, containing (*R*)-amino acids at either stereocenter in the P3/P4 region, were significantly less potent. Replacement of the methyl ester with a methylamide also had little effect (**32**). As can been seen in Table 7.1, incorporation of more highly optimized[59,60] P1 groups (**33**, **34**) leads to a large increase in potency, as was previously seen with proline-based HCV NS3/4a protease inhibitors.[59] The importance of the H-bond donors in P3/P4 is also readily seen in **35**, where methylation of the cyclohexyl glycine amine amide results in complete loss of potency. In contrast, methylation of the terminal amide has no significant effect (compare **33** with **34**). Inhibitors with truncated P3/P4 chains have also been demonstrated to have reduced potency.[55]

Table 7.1 Activity of compounds **29–37**.

R =

Compound	Structure	NS3 1a K_i $(\mu M)^a$
29		2.3
30		35% @ 10 µM
31		37% @ 10 µM
32		1.7
33		0.022
34		0.016

Table 7.1 (*Continued*)

Compound	Structure	NS3 1a K_i $(\mu M)^a$
35		>10
36		0.1
37		0.0013

[a]Inhibition constants of full-length NS3/4a protease.

Following the successful implementation of the cyclopentane proline replacement, the corresponding cyclopentene-containing compounds were also prepared by a similar scheme.[61] Compared with the cyclopentane analogs, the cyclopentene analogs are generally more potent (Table 7.1). For example, even simple norvaline P1 analog **36** is a 100 nM inhibitor against gt1a, whereas the corresponding cyclopentane analog **29** is 2.3 µM. As was observed with the cyclopentane core analogs, introduction of vinylcyclopropane-containing P1 groups led to a large increase in potency, with **37** being a 1.3 nM inhibitor of gt1a.

With these promising results in hand, a further evolution was incorporation of the P1–P3 macrocyclization strategy utilized in BILN-2061 into the cyclopentane/cyclopentene core inhibitor series.[62] Given the differences inherent in the switch from proline to cyclopentane that have been discussed, an examination of various ring sizes was undertaken (**38–41,** Table 7.2) and clearly showed that the optimum ring size is 14, which is smaller than is present in BILN-2061, again demonstrating the difference between proline- and cyclopentane-based cores and also the interplay of P3/P4 *N*-substituents and P1–P3 macrocycle. The identity of the nitrogen R_1 substituent is also key to HCV NS3 activity. While a hydrazine carbamate R_1 group (**39**) leads to potent biochemical activity, activity in the cellular replicon activity is lost (>10 µM).[62]

Table 7.2 Activity of compounds **38–47**.

$$R =$$

Compound	Structure	R_1	Ring size	HCV NS3 1a K_i (nM)	HCV NS3 1b EC_{50} (nM)
38		NHBoc	13	130	>10000
39		NHBoc	14	31	>10000
40		NHBoc	15	710	>10000
41		NHBoc	16	>10000	>10000
42		NH_2	14	6.0	7600
43		NH_2	14	15	5400

Table 7.2 (*Continued*)

Compound	Structure	R_1	Ring size	HCV NS3 1a K_i (nM)	HCV NS3 1b EC_{50} (nM)
44		H	14	260	>10000
45		Me	14	44	2200
46		H	14	2.2	4400
47		Me	14	0.41	9.1

In contrast, a free hydrazine (**42**) maintains biochemical potency while also showing some activity in the replicon assay (7.6 µM). Although the cyclopentene-based inhibitor **43** is similarly potent, the difficulty of synthesis and potential of irreversible covalent attachment *via* the core Michael acceptor led to the decision to focus on the cyclopentane analogs.

As has been demonstrated previously by Tibotec/Medivir[63] on proline-based macrocycles, the typically large P3 groups can be truncated significantly, including complete elimination of the R_1 substituent. Within this series of inhibitors, when $R_1 = H$ (**44**, Table 7.2) biochemical potency was maintained, but replicon activity was again lost. Simply capping the free NH with a methyl group (**45**) led to a more than fivefold improvement in replicon potency. Incorporation of the potency-enhancing carboxylic acid bioisostere cyclopropaneacylsulfonamide[64–66] then led to low- to sub-nanomolar inhibitors (**46**, **47**) which also possessed single-digit nanomolar replicon EC_{50} when the P3-capping group was methyl (**47**).

Building on **47**, additional SARs were developed in order to improve the moderate permeability ($P_{app} = 3.8 \times 10^{-6}$ cm s^{-1}), intrinsic clearance

($46 \, \mu L \, min^{-1} \, mg^{-1}$) and poor oral absorption in rats ($F = 2.5\%$), which was driven by a very high rate of excretion of parent drug into bile. Given that the P2 heterocycle has often been reported to be the key modulator of pharmacokinetics (PK), a study of the P2-quinoline was undertaken.[67] The primary focus of this effort was on replacement of the 2-phenyl group and the introduction of additional substituents on the quinoline ring.

Incorporation of the P2 group present in BILN-2061 (isopropylaminothiazol-4-yl substituent) led to a similarly potent inhibitor (**48**, Table 7.3), but with further reduced permeability. A switch to a simpler thiazol-2-yl group increased potency (**49**), but permeability remained low. Introduction of an isopropyl substituent on the 2-thiazole (**50**) and especially along with an 8-methylquinoline (**5**) group, significantly improved permeability while maintaining sub-nanomolar potency. Importantly, the intrinsic clearance of **5** is also fairly low ($<6 \, \mu L \, min^{-1} \, mg^{-1}$). Replacement of the 8-methyl with ethyl (**51**) led to $\sim$10-fold decrease in potency, while a chloro substituent improved potency about threefold, but led to erosion of permeability (**52**). While groups other than methyl did not lead to further improvements, replacing the thiazole group with isopropylpyrazole (**53**) or isopropylpyridine (**54**) led to similarly promising profiles. However, after PK evaluation in rats, **5** (simeprevir, TMC-435350) proved superior in terms of both intravenous and oral parameters [i.v. clearance, liver:plasma ratio, oral area under the curve (AUC), C_{max}] and was therefore chosen for further advancement.[68,69]

TMC-435350 has also been reported[67] to have excellent plasma exposure in dogs following oral dosing; a $6.5 \, mg \, kg^{-1}$ p.o. dose led to 100% F with a C_{max} of $4.72 \, \mu M$, an AUC of $14\,986 \, ng \, h \, mL^{-1}$ and half-life of $5.1 \, h$. The detailed rat PK profile is shown in Table 7.4.[68] As can be seen, simeprevir (**5**) is a moderate clearance compound ($2.3 \, L \, h^{-1} \, kg^{-1}$) in rats and also achieves high plasma exposure upon oral dosing ($40 \, mg \, kg^{-1}$, $C_{max} = 1430 \, ng \, mL^{-1}$; $AUC = 7750 \, ng \, h \, mL^{-1}$) with 44% bioavailability. Given that the liver is the primary site of viral replication, obtaining high liver exposure is likely key to the success of any anti-HCV drug and following the $40 \, mg \, kg^{-1}$ oral dose in rats, the liver:plasma ratio (along with other tissue:plasma ratios) was measured at several time points up to 31 h, where it was 39 : 1 based upon the AUC. The small (128 : 1) and large intestine (29 : 1) also contained significant amounts of compound, but in most other tissues TMC-435350 was found at relatively low levels.

Following the same $40 \, mg \, kg^{-1}$ p.o. dose in rats, TMC-435350 also maintained a liver concentration above the gt1b replicon EC_{99} for over 30 h. The plasma concentration in the same study was higher than the EC_{99} for $\sim$8 h and around the EC_{50} after 24 h, again demonstrating the high liver:plasma ratio, along with significant plasma exposure for an extended period.

In addition to robust pharmacokinetics, TMC-435350 displayed potent inhibition of both HCV NS3/4a protease genotype 1a and 1b replicons[68] (Table 7.5), with EC_{50}s ranging from 8 to 28 nM. The effect of additional protein content in the assay was relatively minimal, with only a 1.4-fold shift in EC_{50} in the presence of human serum.

Table 7.3 Activity of compounds **5** and **47–54**.

Compound	R	R_1	NS3/4a K_i (nM)	Replicon EC_{50} (nM)	Cl (int) ($\mu L\ min^{-1}\ mg^{-1}$)	P_{app} (cm s^{-1})[a]
47		H	0.41	9	46	3.8
48		H	1.4	17	15	1.4
49		H	0.2	11	38	1.6
50		H	0.84	17	16	13
5 (TMC-435350)		Me	0.36	7.8	<6	8.4
51		Et	3.1	66	–	–
52		Cl	0.1	2.9	9.0	5.8
53		Me	0.16	9.7	<6	15
54		Me	0.3	6.8	<6	12

[a]A–B apparent permeability coefficient in Caco-2 cells.

Table 7.4   Rat pharmacokinetic profile of TMC-435350 (**5**).

Dose[a]	Property	TMC-435350
i.v. (4 mg kg^{-1}, $n = 3$)	Cl (L h^{-1} kg^{-1})	2.3
	V_{dss} (L kg^{-1})	5.3
	AUC$_{0-inf}$ (ng h mL^{-1})	1750
p.o. (40 mg kg^{-1}, $n = 21$)	AUC$_{0-inf}$ (ng h mL^{-1})	7750
	C_{max} (ng mL^{-1})	1430
	t_{max} (h)	2.0
	$t_{\frac{1}{2},\ 8-24h}$ (h)	2.6
	F (%)	44

[a]I.v. dosed in 20% 2-hydroxypropyl-β-cyclodextrin, 0.1 M NaOH to pH 8.0, mannitol, pyrogen-free water; p.o. dosed vitamin E acetate–D-α-tocopheryl poly(ethylene glycol) 1000 succinate–poly(ethylene glycol) 400.

Table 7.5   Replicon potency of TMC-435350.

Replicon cell line	Sub-type	EC_{50} (nM)	EC_{90} (nM)
Huh7-Luc (luciferase)	1b	8.1	23.6
Huh7-Luc (qRT-PCR)	1b	13.0	28.7
Huh7-SGcon1b (qRT-PCR)	1b	25.2	102.4
Huh7-SG1a H77 (qRT-PCR)	1a	28.4	120.4

Table 7.6   TMC-435350 genotype potency.[a]

gt	1b				1a		2b	3a	4a	5a	6a
Sample	G1b(con1)	G1b_03	G1b_05	G1b_08	G1a_07	G1a_08	G2b	G3a	G4a	G4a_54	G5a-01 G6l-02
K_i (nM)	3.6	8.7	4.0	7.0	3.6	12.0	1.9	36.5	8.9	4.6	6.2 2.3

[a]NS3 proteases cloned from clinical isolates of HCV, expressed and purified from *Escherichia coli*. The IC$_{50}$ value determined by fluorescence resonance energy transfer induced by the cleavage of the RetS1 peptide when incubated with the NS3 protease domain.

The broad genotype coverage was further demonstrated in a biochemical assay format using protease sequences cloned from patient isolates (Table 7.6).[70] All IC$_{50}$s are less than 10 nM, with the exception of gt3a, which shifts to 36.5 nM.

The selectivity of TMC-435350 was then assessed in a panel of proteases and only against cathepsin S did TMC-435350 show significant inhibition (IC$_{50}$ = 800 nM). Overall, the selectivity index is >1000 and no significant cytotoxicity was seen in a variety of cell lines.[68]

Given the likelihood of combination therapy being part of any anti-HCV treatment, the potential synergy of TMC 435350 was evaluated[68] (Table 7.7) in the gt1b replicon system using the Loewe additivity model. Whereas ribavirin was simply additive, both IFN-α and two different HCV NS5b inhibitors (NM-107 and HCV796) led to synergistic behavior, which could be important as TMC-435350 advances in the clinic.

Table 7.7   Synergy with TMC-435350a.[a]

Combination compound	CI^b value for:			Influence[c]
	ED_{50}	ED_{75}	ED_{90}	
IFN-α	0.77	0.57	0.43	Synergistic
Ribavirin	0.98	0.94	0.92	Additive
NM-107	0.58	0.64	0.64	Synergistic
HCV-796	0.86	0.88	0.92	Additive/synergistic

[a]Huh7-Luc replicon cells treated for 72 h.
[b]Combination index.
[c]CI values of <0.9 (synergy), 0.9–1.1 (additive) and >1.1 (antagonism).

Table 7.8   Mutant enzyme potency profile of TMC435450.

Mutant	EC_{50} (nM)	Fold shift	Mutant	EC_{50} (nM)	Fold shift
WT	11	NA	D168G	42	4.4
F43S	83	12	D168N	79	6.6
F43I	562	89	D168E	302	40
F43V	626	99	D1688T	4089	308
Q80R	49	6.9	D168Y	6238	666
Q80H	41	3.6	D168H	5655	368
Q80G	22	1.8	D168A	6356	594
Q80L	12	2.1	D168V	17917	2591
R155M	3.5	0.4	D168I	23203	1807
R155I	6.6	0.8	F43S + Q80R	1815	286
R155Q	21	1.6	F43S + D168E	3607	694
R155T	314	24	Q80K + R155K	4647	364
R155K	260	30	Q80R + R155K	2853	270
R155G	350	20	Q80R + D168E	5902	412
A156S	3.0	0.3	Q80H + D168E	1362	163
A156G	206	16	Q80R + D168A	16867	2655
A156T	377	44			
A156V	2149	177			

The full *in vitro* resistance profile for TMC-435350 has also been published.[71] Similarly to other macrocyclic HCV NS3/4a protease inhibitors,[72] TMC-435350 gives rise to a number of mutations in replicon-based resistance selection studies. The major location for mutation is Asp168, which primarily mutates to alanine (D168A) or valine (D168V), although a number of others have also been isolated. A156 and Q80 have been observed to mutate less frequently and R155K mutations are produced in gentotype 1a replicons (where a single nucleotide change can lead to R155K, in contrast to gentoype 1b which requires more than one). A variety of double mutants were also generated, with combinations of mutations at residues 43, 80, 155, 156 and 168. The potency of TMC-435350 against a panel of the observed variants was then assessed by engineering the mutations into a con1b replicon construct. Similarly to BILN-2061, ITMN-191, boceprevir and telaprevir, reductions in TMC-435350 potency were observed for mutations at positions 43, 80, 155, 156 and 168, and also for several double mutants (Table 7.8).[71] Overall, the largest decrease in potency was observed with mutations at D168, with losses of up to 1800–2500-fold with D168V and D168I.

Double mutations also led to greater decreases in potency in general compared with the single mutants alone. The relative replicative capacity of each of the mutants varies from A156V, which is only 18% as fit as the wild type (WT), to R109K (which was only observed in combination) which is three times *more* fit than WT. In a colony formation assay,[71] increasing concentrations of TMC-435350 up to 25 times the EC_{50} led to a dose-dependent reduction in colony number, but complete suppression of colony formation was not reached. However, in combination with several HCV NS5B non-nucleoside inhibitors, full inhibition of colony growth was realized.

The medicinal chemistry synthesis[67,73] of TMC-435350 (Scheme 7.2) begins with amide formation on the bicyclic lactone carboxylic acid **55**[55,74] with *N*-methyl-hex-5-enamine,[73,75] which proceeds in 68% yield. Opening of the lactone is then carried out with LiOH to reveal a carboxylic acid, which is subsequently coupled to the vinylcylopropyl amino ester **25**[57] to give cyclopentanol **57** in 53% in two steps. After coupling to the functionalized quinolinol **58**,[67,73] the resulting diene **59** is subjected to ring-closing metathesis using the Hoveyda–Grubbs first-generation catalyst[76] to form the desired 14-membered macrocycle (**60**) in 60% yield. The final cyclopropylacylsulfonamide is then installed using the standard acylsulfonamide formation chemistry to produce TMC-435350.

A 2.4 Å X-ray crystal structure of TMC-435350 bound in the active site of the HCV NS3 protease domain genotype 1b has recently been published,[77] which reveals a previously unobserved extension in the S2 sub-site by the large P2 heterocycle. Additionally, the acylsulfonamide forms a number of hydrogen bonds within the oxyanion hole and nearly all of the macrocycle appears to be involved in binding to the enzyme. The structure also enables understanding of the origin of the potency losses seen with several clinically relevant mutations. Q80, R155 and D168 all form part of the induced pocket in the S2 sub-site formed around the P2 heterocycle and mutations at these residues all lead to variable amounts of potency loss.[71]

Given the overall *in vitro* biological and preclinical PK profile, simeprevir was advanced into clinical development, where it showed attractive human plasma pharmacokinetics[78] and currently is being evaluated as a once-daily anti-HCV agent in a variety of ongoing monotherapy and combination clinical trials.[78,79]

7.3 Discovery of Vaniprevir (MK-7009)

The starting point for subnanomolar P2–P4 macrocyclic NS3/4a inhibitors emerged from examining published views of a close analog of BILN-2061 (**4**) bound to the 1–180 protease domain of NS3 protease.[80] This suggested that the P2 thiazolylquinoline portion of the inhibitor lies on a relatively featureless enzyme surface with binding interactions that provide little apparent basis for the very large contribution to potency derived from that moiety.[81] In an effort

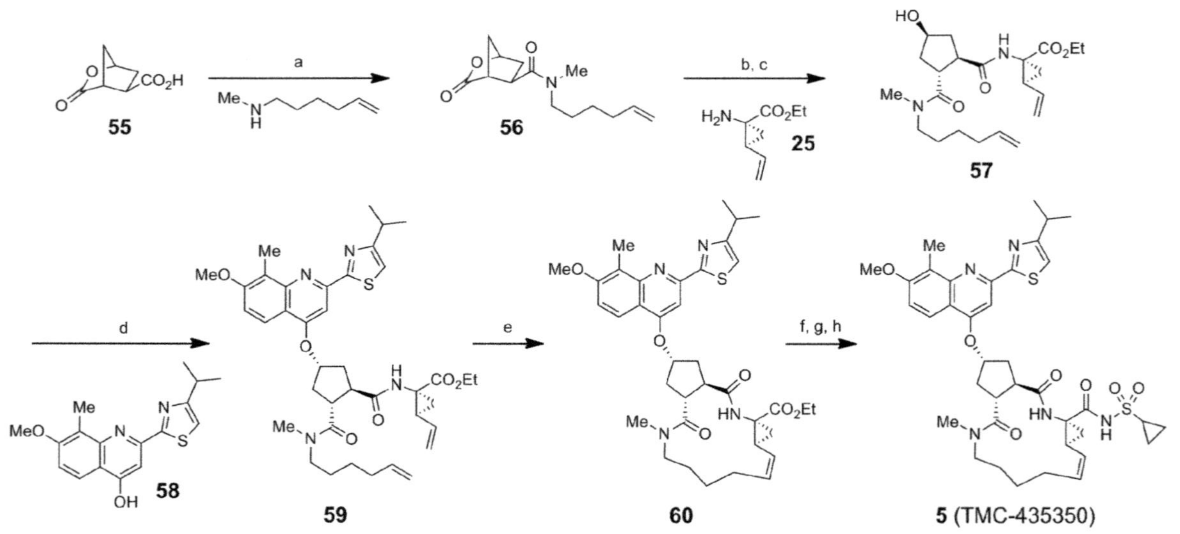

Scheme 7.2 Conditions: (a) HATU, DIEA, DMF, 68%; (b) LiOH, THF–MeOH–water, 88%; (c) HATU, DIEA, **25**, 60%; (d) **58**, PPh₃, DIAD, THF, –15 °C– r.t., 56%; (e) Hoveyda–Grubbs first-generation catalyst, DCE, 70° C, 60%; (f) LiOH, THF–MeOH, water, 88%; (g) CDI, THF, reflux; (h) cyclopropylsulfonamide, DBU, THF, 50 °C, 40%.

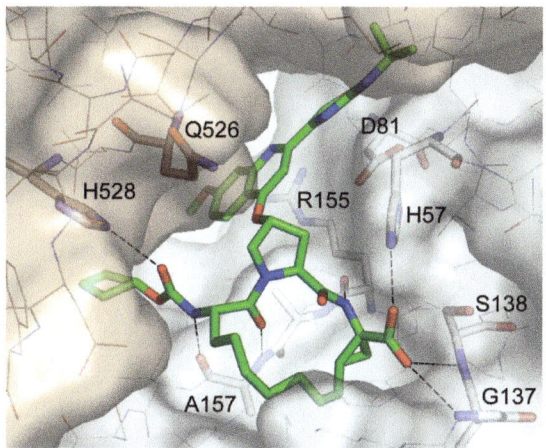

Figure 7.9   Model of **4** (green) bound to full-length NS3/4A (protease, white; helicase, light brown) with key protein–-inhibitor interactions shown.

to rationalize this, Merck chose to model **4** bound to the full-length NS3/4A protein, including the significantly larger helicase domain (Figure 7.9), to determine the extent of any role the helicase could play in inhibitor binding. The specific role of the helicase in binding has been the subject of debate in the literature,[82] although recent X-ray structures have confirmed that the helicase does make contact with NS3 inhibitors.[83] At the time, no full-length structures with inhibitors bound were available; consequently, a published *apo*-enzyme structure[84] was used as the starting point for docking studies. To permit access of the inhibitor to the active site, the six C-terminal residues (DLEVVT) of the helicase domain that lie in the active site of the *apo* X-ray structure were truncated.

Analysis of **4** docked in the latter structure suggested that the helicase domain could provide a surface over the P2 moiety, including a hydrophobic pocket that appeared to accommodate the thiazolyl substituent.

Specific inhibitor–helicase interactions in this model include His528–carbamate oxygen and Gln526–quinoline. The other key finding from this study was that there is space between the carbamate cyclopentane and the quinoline ring that could potentially accommodate a macrocyclic connection. This concept was supported by re-examination of the helicase C-terminus from the *apo* structure, overlaid with BILN-2061 (Figure 7.10) in which the side chain of Glu628 occupies the same space as the proposed linker. Together, these observations suggested that an alternative P2 quinoline–P4 cyclopentyl macrocyclization to form a structurally distinct series of inhibitors was feasible.

In order to test this hypothesis in a readily accessible system, the initial targets were carbamate derivatives **61–64** (**63** docked in Figure 7.11), in which the P1–P3 macrocyclic linker was disconnected, the proposed P2–P4 linker was formed and a simplified 3-phenylquinoline P2 was utilized. Energy scores for the different linker lengths calculated by two methods predicted that the five- and six-carbon linkers would show greatest activity (Table 7.9).

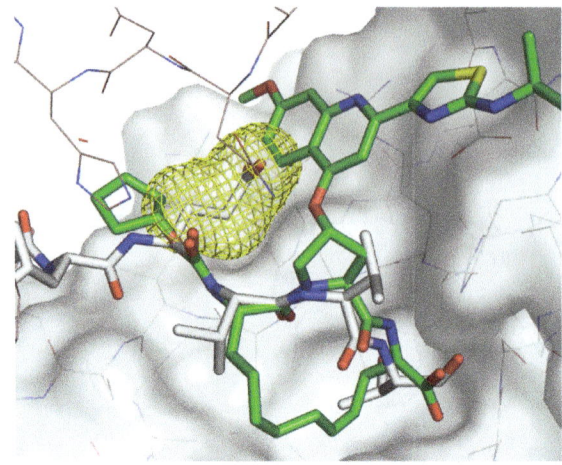

Figure 7.10   Model of **4** bound to full-length NS3/4A with helicase C-terminus (white) restored and Glu628 highlighted with mesh surface.

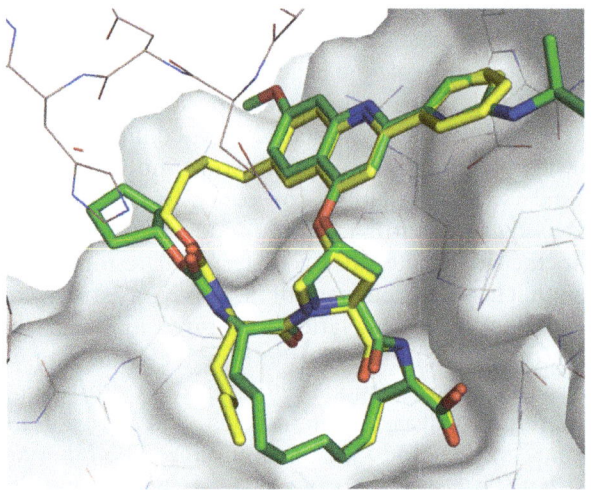

Figure 7.11   Target macrocycle **63** (yellow) overlaid with **4** (green).

The desired compounds were prepared as outlined in Scheme 7.3 *via* a synthetic sequence featuring a ring-closing metathesis (RCM) [85,86] as the key macrocycle-forming event. Displacement of the brosylate of proline derivative **65**[87] with bromohydroxyquinoline **66** yielded ether **67**, which was vinylated *via* a palladium-catalyzed Stille reaction with tributylvinyltin to give **68**. Subsequent removal of the Boc protecting group and coupling with the appropriate norleucine carbamate derivatives **69a–d** afforded key RCM precursors **70a–d**. Ring closure was accomplished in excellent yield (84–93%) with the Zhan 1b catalyst[88] in 1,2-dichloroethane to give macrocyclic olefins **71a–d**. Conversion to

Table 7.9 *In vitro* activity of P2–P4 macrocycles.[a]

Compound	Modeling			1b replicon IC_{50} (nM)	
	E_{inter}	*Xscore*	*1b K_i (nM)*	*10% FBS*	*50% NHS*
4			0.3	3	19
2			93	1100	4800
61	−69.5	8.09	2000	–	–
62	−70.5	8.26	145	6100	>100000
63	−71.1	8.36	8.5	1150	5600
64	−71.4	8.44	25	1200	9100
74			4400	–	–
75			40	4800	>100000
76			<0.016	6.7	26
77			<0.016	13	25

[a]For molecular modeling details, see reference 41.

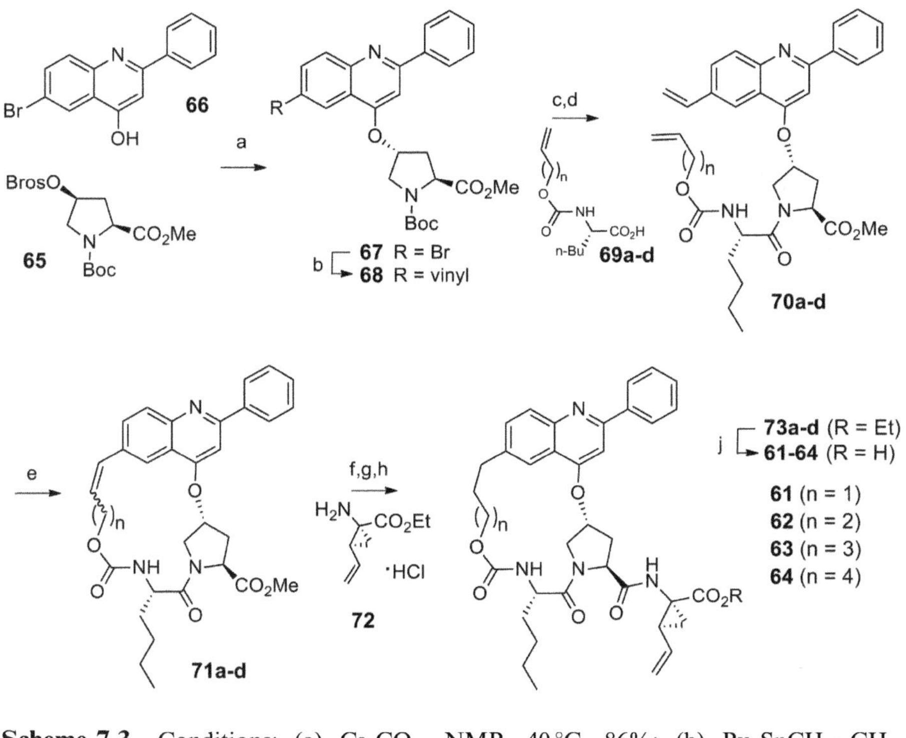

Scheme 7.3 Conditions: (a) Cs$_2$CO$_3$, NMP, 40 °C, 86%; (b) Bu$_3$SnCH=CH$_2$, Pd(PPh$_3$)$_4$, toluene, 100 °C, 79%; (c) 4 M HCl, dioxane; (d) HATU, DIPEA, DMAP, DMF, **69a–d**; (e) Zhan 1b catalyst,[88] DCE, ∼10 mM, 84–93%; (f) H$_2$, 10% Pd/C, EtOAc; (g) LiOH, THF, MeOH, H$_2$O; (h) HATU, DIPEA, DMAP, **72**, DMF; (j) LiOH, THF, MeOH, H$_2$O.

the desired products (**61–64**) was carried out *via* hydrogenation of the styryl olefin, hydrolysis of the proline ester, coupling with cyclopropylamino ester **72**[89] to afford **73a–d** and hydrolysis of the P1 ethyl ester.

Compound **61**, with three carbons between the carbamate oxygen and the P2-quinoline, proved to have very modest activity of 2000 nM in a genotype 1b NS3/4A enzyme inhibition assay[90] (Table 7.9). Incremental lengthening of the linker, however, afforded dramatic improvements with optimized activity of 8.5 nM in the case of the C_5-linked **63**. The improved enzyme inhibition of **63** also translated to a corresponding improvement in genotype 1b cell-based replicon activity.[91] The point of attachment of the macrocyclic linker on the P2-quinoline was critical to activity, as demonstrated by the synthesis of the corresponding 5-substituted derivative **74** (Figure 7.12), which proved much less active (K_i 4400 nM). Synthesis of an acyclic analog, **75**, demonstrated that potency enhancement, particularly in the replicon assay, could be achieved through a P2–P4 macrocyclization strategy.

Replacement of the P1 carboxylic acid functionality in **63** with the cyclopropylacylsulfonamide[65] to afford **76** (Figure 7.12) led to a subnanomolar inhibitor of NS3/4A protease ($K_i < 0.016$ nM). Disappointingly, given the critical need for liver exposure, oral administration of **76** to rat at 5 mg kg^{-1} p.o. provided low (0.2 µM) compound levels in the liver at 4 h with barely detectable plasma exposure (Table 7.10). In a surprising result, when the P3 *n*-butyl residue was replaced with the isomeric *tert*-butyl group, the resultant inhibitor **77**, which had a very similar *in vitro* activity profile, was effectively partitioned into the liver with a tissue concentration at 4 h of 3.9 µM, although plasma levels were unimproved. The dramatic impact of this minor structural change on liver levels strongly suggested that uptake was *via* an active transporter-mediated process.

An additional attraction of this class of P2–P4 macrocycles relative to the P1–P3 structural class is the potential to use a range of macrocyclization

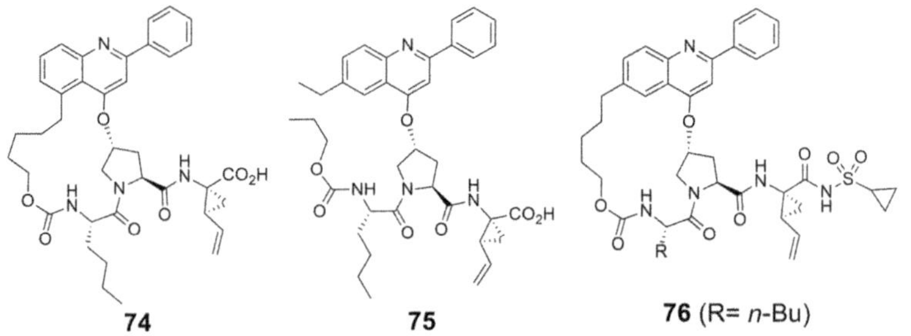

74   **75**   **76** (R= *n*-Bu)
                               77 (R= *t*-Bu)

Figure 7.12   Phenylquinoline analogs.

Table 7.10   Liver exposure of compounds **76** and **77**.[a]

Compound	C_{max} (nM)	Plasma AUC 0–4 h (µM h)	4 h liver concentration (µM)
76	7	0.006	0.2
77	6	0.01	3.9

[a]Compounds dosed at 2 mg kg^{-1} p.o. in PEG400 ($n = 3$).

strategies in advanced process development, including intramolecular Heck or Suzuki reactions, proline amide[92] coupling or RCM. This contrasts with synthetic approaches to P1–P3 compounds, where an RCM macrocyclization step appears to be the only viable strategy.[93] While there are reports of compounds employing a related P2–P4 cyclization strategy,[94,95] the inhibitors had no P2 substituent linked directly to the proline and displayed modest micromolar potencies. Synthesis of the cyclopropylacylsulfonamide analog of one of these P2–P4 compounds had little effect on potency.[41]

With the P2–P4 macrocyclization strategy validated by the attractive potency and liver exposure of **77** (Figure 7.12),[41] further exploration of the P2 heterocyclic portion of this class of compounds was carried out. The literature provided a significant diversity of substituents in the P2 region which could be divided into two main classes. First were the heterocycles directly linked *via* an ether to the proline, such as BILN-2061 and TMC 435350 and **78**;[96,97] second were those linked *via* a carbamate functionality, such as ITMN-191,[25] **79** and **80** (Figure 7.13).[97] The modular synthetic approach to these P2–P4 macrocycles relied on a key RCM reaction,[85,98,99] allowing for the rapid exploration of a diverse array of P2 substituents in the new scaffold.

In one of the initial efforts aimed at exploring the P2 region of the novel P2–P4 macrocycles, Merck chose to explore simple benzylamine-derived carbamates. The 2-linked isomer **81** had modest activity in both the 1b enzyme and cellular replicon assays (Table 7.11). Activity could be significantly improved by moving the linker to the 3-position as in **82**, resulting in a sixfold increase in replicon activity. This increase in activity prompted further constraint of the P2 region of the compounds by cyclizing to the two corresponding isomers of tetrahydroisoquinoline, **83** and **84**, the latter of which demonstrated an even greater difference in 1b enzyme potency of >50-fold. Ring contraction of the tetrahydroisoquinoline in **84** to give the isoindoline in **85** provided a highly potent compound (**85**) with 50 pM activity versus the genotype 1b enzyme and 7 nM activity in the replicon assay.

Compound **85** also showed liver exposure in rat of 910 nM at the 4 h time point following a single 5 mg kg^{-1} oral dose. Based on the need for high liver exposure with any anti-HCV drug,[41] rat liver concentrations were used as the primary pharmacokinetic readout for evaluating compounds. On the basis of these data, **85** became an important lead and the focus turned towards evaluating SARs in other parts of the molecule, while leaving the P2 isoindoline intact.

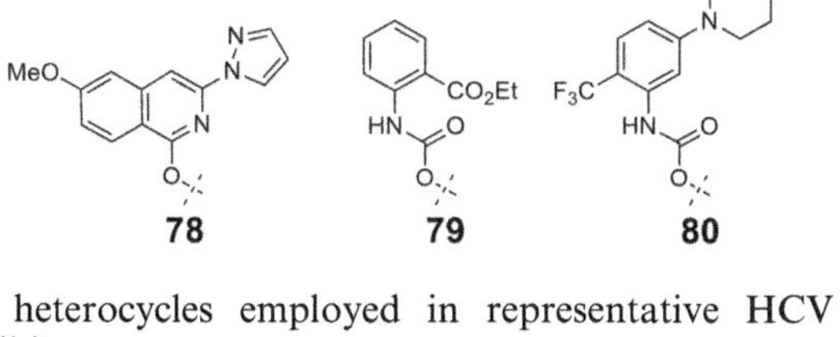

Figure 7.13 P2 heterocycles employed in representative HCV NS3/4a protease inhibitors.

Table 7.11   *In vitro* activity of compounds **81–85**.

Compound	*1b K_i* *(nM)*[a]	*1b replicon IC$_{50}$ (nM)*		*Rat [liver] @* *4 h (μM)*
		10% FBS	*50% NHS*	
81	1.60	98	610	nd
82	0.14	28	42	0.45
83	8.30	730	2900	nd
84	0.15	26	140	3.5
85	0.05	7	32	0.91

[a]NS3/4a protease time-resolved fluorescence assay measurements.

Investigation of the role of the P3 amino acid side chain was carried out *via* the synthesis of the series of compounds **85–90** (Table 7.12). Very little difference in potency was seen with various alkyl and cycloalkyl substituents. Rat liver concentrations after oral dosing showed a slight trend towards increased liver exposure for the most lipophilic substituents. For example, cyclohexyl derivatives **88** and **89** showed 2.8 and 1.6 μM liver concentrations, respectively, in rat liver at 4 h, whereas the isopropyl derivative **87** showed a about a fivefold lower concentration at 0.44 μM. All three compounds, however, showed comparable enzyme and cellular activity.

Contraction of the macrocyclic ring constraint by one carbon gave **91** and led to decreases in both potency and cellular activity. Expansion of the linker by one and two carbons to the six-carbon-linked **92** and seven-carbon-linked **93** gave slightly improved *in vitro* activity, but lower liver exposure relative to the lead **85** (Table 7.13).

A study to examine the effect of saturation of each of the two olefins present in **85** was then undertaken. Reduction of the macrocyclic olefin provided **94**, which had a potency and PK profile virtually identical with those of **85**.

Table 7.12 *In vitro* activity of compounds **85–90**.

85 (R = *t*-Bu)
86 (R = *n*-Bu)
87 (R = *i*-Pr)
88 (R = *cy*-hexyl)
89 (R = 1-Me-*cy*-hexyl)
90 (R = *cy*-pentyl)

Compound	1b K_i (nM)	1b replicon IC_{50} (nM)		Rat [liver] @ 4 h (μM)
		10% FBS	50% NHS	
85	0.05	7	32	0.91
86	0.05	6	26	0.46
87	0.04	12	28	0.44
88	0.02	3	24	2.8
89	0.06	4	38	1.6
90	0.04	4	23	nd

Table 7.13 *In vitro* activity of compounds **85** and **91–93**.

91 (n = 1)
85 (n = 2)
92 (n = 3)
93 (n = 4)

Compound	1b K_i (nM)	1b replicon IC_{50} (nM)		Rat [liver] @ 4 h (μM)
		10% FBS	50% NHS	
91	0.10	16	390	nd
85	0.05	7	32	0.91
92	0.03	5.0	19	0.23
93	<0.02	3.0	23	0.14

Saturation of the P1 vinyl group to the P1 ethyl (**95**) led to about a fourfold increase in rat liver concentration, but this was accompanied by an ∼10-fold decrease in potency. The fully saturated compound (**96**) however, maintained much of the potency of **85** but showed improved liver exposure (10.7 μM) with a 5 mg kg^{-1} oral dose to rats. Encouraged by this result, further profiling of **96** showed that in the replicon activity, when measured against the genotype 2a system, potency was significantly reduced. In fact, only **94** in this series was below 100 nM in the genotype 2a replicon assay (Table 7.14).

Knowing that there was flexibility with regard to macrocyclic ring size, one-carbon expanded ring analogs of **96** were then re-examined (Table 7.15).

Table 7.14 *In vitro* activity of compounds **85** and **94–96**.

Compound	1b K_i (nM)	2a K_i (nM)	Replicon IC$_{50}$ (nM)			Rat [liver] @ 4 h (μM)
			1b 10% FBS	1b 50% NHS	2a 10% FBS	
85	0.05	2.6	7	32	150	0.91
94	0.04	0.9	7	27	66	0.52
95	0.45	19	29	110	720	3.8
96	0.18	4.0	10	35	140	10.7

Table 7.15 *In vitro activity of compounds* **13** *and* **97–101**.

97

13 (MK-7009)

98 (n = 1)
99 (n = 2)
100 (n = 3)
101 (n = 4)

| Compound | 1b K_i (nM) | 2a K_i (nM) | Replicon IC$_{50}$ (nM) | | | | Rat [liver] @ 4 h (µM) |
			1b 10% FBS	1b 50% NHS	2a 10% FBS		
97	0.06	0.9	5.0	27	13		1.9
13 (MK-7009)	0.05	0.9	3.0	19	9.0		9.9
98	0.10	4.3	6	41	35		4.6
99	0.09	3.6	3.0	30	20		45
100	0.09	5.1	7	85	20		6.6
101	1.22	63	43	860	270		25

Simple addition of a methylene group (97) gave a small boost in genotype 1b replicon activity but also led to a >10-fold increase in genotype 2a replicon activity. Somewhat unexpectedly, the liver exposure seen with 97 was about five times lower than with 96, even though the structures differ by only one methylene unit in the linker. Installation of a dimethyl group β to the carbamate oxygen gave 13, which showed a concentration of 9.9 μM in rat liver at 4 h following a 5 mg kg^{-1} oral dose. Furthermore, replicon activity at both genotypes 1b and 2a was under 10 nM in the presence of 10% FBS, and in the presence of 50% NHS the genotype 1b replicon activity showed a very small protein shift to 19 nM. A series of spirocycloalkyl substituents, from cyclopropane to cyclohexane (98–101), showed that increasing the steric bulk and lipophilicity in this region of the molecule led to a decrease in enzyme potency and a marked erosion in replicon activity in the presence of 50% human serum (Table 7.15). Liver exposure, however, was affected in a less predictable manner. The change from a dimethyl substituent (13) to a spirocyclohexyl group (101) increased liver exposure by 2.5-fold but decreased replicon activity in the presence of 50% NHS by 50-fold. Although it has reduced potency, the spirocyclobutyl compound (99) showed remarkable rat liver exposure of >40 μM after the 5 mg kg^{-1} oral dose and illustrates the difficulty in predicting this parameter.

Within the context of 13, the P3 amino acid side chain was revisited. In the case of 85–90 (Table 7.12), varying the P3 group had little effect on activity but modulated liver exposure. With 102–104, however, larger, more lipophilic P3 groups were not tolerated (Table 7.16). For example, the cyclohexyl substituent used in 102 led to good liver exposure, but this change was accompanied by about a fivefold decrease in the serum-shifted replicon IC$_{50}$. This result is in contrast to that with 88, which did not suffer a decrease in activity with the P3 cyclohexyl group. The requirement for liver exposure needs to be carefully balanced with cellular activity. Perhaps the increase in

Table 7.16 *In vitro* activity of compounds 13 and 102–104.

13 (R = *t*-Bu, MK-7009)
102 (R = *cy*-hexyl)
103 (R = 1-Me-*cy*-hexyl)
104 (R = CH$_2$O(t-Bu))

Compound	1b K$_i$ (nM)	Replicon IC$_{50}$ (nM)		Rat [liver] @ 4 h (μM)
		1b 10% FBS	1b 50% NHS	
13 (MK-7009)	0.05	3.0	19	9.9
102	0.09	16	75	7.5
103	0.08	10	90	2.0
104	0.30	16	180	13

lipophilicity of the linker that gives **13** and **98–104** good liver exposure precludes a further increase in overall lipophilicity through contributions from P3 substituents. Based on the exploration of the macrocyclic scaffold in the P2 isoindoline series, **13** was identified as the optimal compound in the series and was subsequently designated MK-7009.[42,43,100]

The synthesis of MK-7009 starts with 3-bromo-*o*-xylene (**105**), which was dibrominated with *N*-bromosuccinimide (NBS) and benzoyl peroxide (Scheme 7.4).[101] Displacement of the bromines with benzylamine with concomitant ring closure gave 2-benzyl-4-bromoisoindoline (**106**). Installation of the vinyl group and removal of the benzyl protecting group with 1-chloroethyl chloroformate[102] and methanol provided **107**. Standard carbamate-forming conditions with **108** and removal of the Boc protecting group gave key intermediate **109**. Compound **109** was deprotected and coupled to linker intermediate **110** to give the bis-olefin **111**. The Zhan 1b metathesis catalyst[88] was used to affect macrocyclization and the newly formed olefin could be hydrogenated, providing **112**. Ester hydrolysis and coupling to **113** gave MK-7009 (**13**). Subsequent optimization of synthetic routes led to improved approaches to MK-7009.[92,103]

In terms of off-target activity, MK-7009 demonstrated a large selectivity window against a broad panel of cathepsins, and also other proteases such as chymase and elastases (Table 7.17). Although MK-7009 does show a modest inhibitory activity against chymotrypsin, the selectivity ratio relative to NS3/4A potency is very large (8700-fold). The > 10 000-fold selectivity observed in an array of 169 pharmacologically relevant ion channel, receptor and enzyme targets conducted at MDS Pharma (Panlabs) serves to demonstrate further the highly selective nature of MK-7009. In an assay to assess potential hERG activity, the ability of MK-7009 had $IC_{50} > 10\,\mu M$ in an MK-499 binding assay. In addition, an assessment of the ability of MK-7009 to inhibit p450 reversibly did not show significant inhibition of isoforms tested (3A4, 2D6 and 2C9), with $IC_{50} > 10\,\mu M$. This lack of significant activity observed against p450 isoforms examined is also clearly an attractive attribute for a drug targeting significant liver exposure.

The pharmacokinetic properties of MK-7009 were evaluated in multiple species (Table 7.18).[42] In rats, the compound showed a plasma clearance of $74\,mL\,min^{-1}\,kg^{-1}$ and a plasma half-life of $\sim 1\,h$. When dosed orally at 5 mg kg^{-1}, the plasma exposure was modest, with an AUC of $0.1\,\mu M\,h$. In contrast to the plasma exposure, the liver exposure of MK-7009 at 24 h after a $5\,mg\,kg^{-1}$ oral dose was significant ($0.4\,\mu M$).

When dosed to dogs, the plasma pharmacokinetics of MK-7009 were greatly improved (Table 7.18), with moderate clearance of $11\,mL\,min^{-1}\,kg^{-1}$ and a 1.2 h half-life after i.v. dosing and good plasma exposure ($AUC = 1.2\,\mu M\,h$) after a $5\,mg\,kg^{-1}$ oral dose. Dog liver biopsy studies showed that the liver concentrations of MK-7009 after the $5\,mg\,kg^{-1}$ oral dose were 34 and $0.5\,\mu M$ at the 2 and 24 h time points, respectively. Similarly to its behavior in rats, MK-7009 demonstrates effective partitioning into liver tissue and maintains high liver concentration, relative to potency, 24 h after oral dosing in dogs.

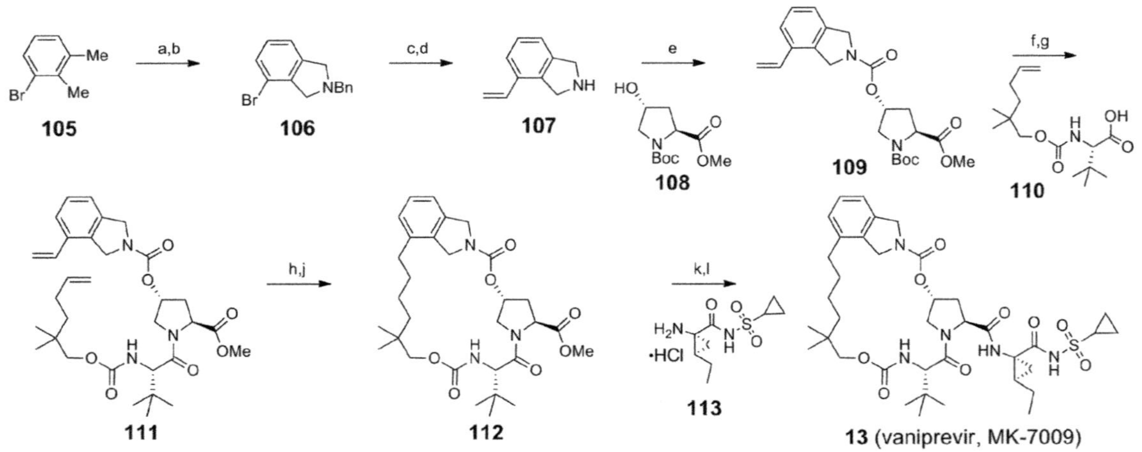

Scheme 7.4 Conditions: (a) NBS, benzoyl peroxide, CCl_4, reflux, 92%; (b) K_2CO_3, benzylamine, MeCN, 77 °C, 50%; (c) $Bu_3SnCH{=}CH_2$, $Pd(PPh_3)_4$, toluene, 100 °C, 83%; (d) 1-chloroethyl chloroformate, DCE, reflux; MeOH, reflux, 86%; (e) **108**, carbonyldiimidazole, DMF; **107**, 60 °C, 81%; (f) HCl, dioxane, 90%; (g) **110**, HATU, DIPEA, DMF, 91%; (h) Zhan 1b catalyst,[88] CH_2Cl_2, 98%; (j) H_2, 10% Pd/C, EtOH, EtOAc; (k) $LiOH \cdot H_2O$, THF, MeOH, H_2O; (l) **113**, HATU, DIPEA, DMAP, CH_2Cl_2, 91%, three steps.

Table 7.17 *In vitro* selectivity of MK-7009.

	MK-7009[a]
Chymotrypsin	520
Trypsin	>10000
Cathepsin B[b]	>10000
Cathepsin F[b]	>10000
Cathepsin K[b]	>10000
Cathepsin L[b]	>10000
Cathepsin S[b]	>10000
Cathepsin V[b]	>10000
Chymase[c]	>10000
Pancreatic elastase 1[c]	>10000
Neutrophil elastase 2[c]	>10000

[a]Data reported as IC_{50} (nM).
[b]$n = 2$.
[c]Assays run at MDS Pharma, quantitative mode, $n = 3$.

Table 7.18 Pharmacokinetic parameters for **13** (MK-7009).[a]

Species	Cl (mL min^{-1} kg^{-1})	V_d (L kg^{-1})	$t_{\frac{1}{2}}$ (h)	p.o. C_{max} (μM)	p.o. AUC (μM h)	p.o. [liver] 2 h (μM)	p.o. [liver] 24 h (μM)
Rat	74	1.9	0.9	0–0.1	0–0.1	9.9 @ 4 h	0.4
Dog	11	0.3	1.2	0.5	1.2	34	0.5
Rhesus	18	0.4	1.3	0.01–0.2	0.05–0.2	3	nd
Chimpanzee	nd	nd	nd	0.9	5.2	nd	31 @ 12 h

[a]Rat, dog and rhesus i.v. (2 mg kg^{-1}, $n = 3$, DMSO), p.o. (5 mg kg^{-1}, $n = 3$, PEG400), chimpanzee (p.o., 10 mg kg^{-1}, $n = 2$, chocolate milk).

Rhesus monkey PK parameters were generally similar to those of rats and characterized by good liver exposure and poor plasma exposure after a 5 mg kg^{-1} oral dose (Table 7.18). As a prelude to efficacy experiments, which are described elsewhere,[104] MK-7009 was dosed to chimpanzees to evaluate plasma and liver pharmacokinetics. Following an oral dose of 10 mg kg^{-1} in chimpanzees, MK-7009 showed excellent plasma exposure, with an AUC of 5.2 μM h and a C_{max} of 0.9 μM. Furthermore, a liver biopsy at this dose level showed a liver concentration of 31 μM at 12 h post-dose for MK-7009. This concentration is >1500-fold greater than the serum-shifted replicon IC_{50} and clearly supported a dosing regimen of, at most, twice per day in chimpanzees.

MK-7009 demonstrates significant liver exposure in multiple preclinical species which persists over a significant timeframe. In particular, the concentrations present represent a significant margin relative to the replicon EC_{50} activity measured in the presence of 50% NHS at 24 h in the case of rats (31-fold), dogs (46-fold) and rhesus monkies (15-fold) and 12 h in chimpanzees (2400-fold), where an additional 24 h sample was not available (Figure 7.14). The excellent exposure seen with MK-7009 in healthy uninfected chimpanzees provided a strong basis for clinical evaluation of an antiviral effect.

In summary, investigation of a series of P2–P4 macrocycles containing a hydroxyproline carbamate led to the identification of 5-linked isoindoline as a promising P2 substituent for a new class of HCV NS3/4a protease inhibitors. Optimization of this series, with the goals of achieving rat liver exposure and balanced activity against both the 1b and 2a genotypes of NS3/4a protease led to the identification of vaniprevir (**13**, MK-7009). Further studies of vaniprevir, including clinical investigations of the PK and efficacy profile, are ongoing.[43]

7.4 Discovery of MK-5172

With the development of vaniprevir progressing, Merck set as a goal for the ongoing discovery program the identification of a second-generation NS3/4a protease inhibitor. While maintaining or improving the PK profile seen with vaniprevir, significant improvements in activity against the gt3a enzyme and key clinically relevant gt1 mutant enzymes were desired. An entry point for enhanced gt3a activity was provided by a consistent trend for improving gt3a activity with the addition of a large substituent to the P2 heterocycle in early compounds. For example, the gt3a activity of **114** ($K_i = 200$ nM, Table 7.19) was improved approximately fourfold with the addition of a phenyl group to give closely related compound **115** ($K_i = 45$ nM). Based on this observation, Merck undertook a systematic investigation of substituents at this position of the P2 quinoline.

In order to explore fully the SAR at the 2-position,[105] a significant number of analogs would be needed and using the classic Conrad–Limpach quinoline synthesis,[106] the substituent in the 2-position was installed in the first step (Scheme 7.5). Utilization of this synthesis would require an additional nine steps to complete each analog. Hoping to avoid this inefficiency, a more versatile route, which introduced the 2-substituent near the end of the synthetic

Table 7.19 *In vitro* activity of compounds **114** and **115**.

Compound	1b K_i (nM)[a]	3a K_i (nM)[a]	1b replicon IC_{50} 50% NHS (nM)	Rat [liver] @ 4 h (µM)
114	0.02	200	10	21
115	0.05	45	19	7.8

[a]NS3/4a protease time-resolved fluorescence assay, mean of $n \geq 3$ measurements.

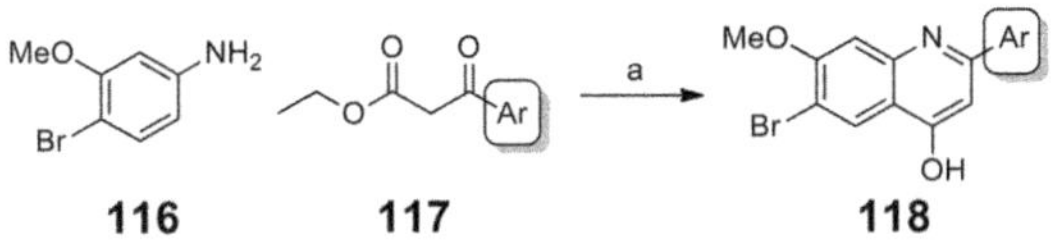

Scheme 7.5 Conditions: (a) polyphosphoric acid, heat.

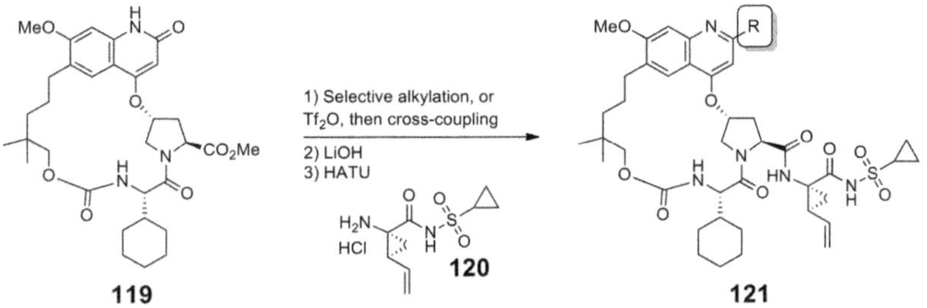

Scheme 7.6

route, was developed. In this vein, **119** could serve as a common intermediate to introduce a variety of groups in the 2-position (Scheme 7.6). These intermediates could then be transformed into the desired compounds in two straightforward steps.

In practice, a regioselective hydroxyquinolone formation utilizing bromomethoxyaniline **122** and malonic acid[107] was followed by chemoselective alkylation[108] at the 4-position with brosylate **124**[87] (Scheme 7.7). Removal of the Boc group, HATU coupling with **126**[109] and vinylation with Molander's vinyl trifluoroborate reagent[110] led to intermediate **127**. RCM using the Zhan 1B catalyst[88] then produced the desired macrocycle in good yield. Hydrogenation of the styrene double bond yielded key intermediate **119** in seven steps with an overall yield of 23%.

In order to access various aryl and heteroaryl groups, **119** was first converted to the corresponding quinoline-2-triflate (**128**) and then Suzuki reactions[111] were carried out with selected boronic acids (Table 7.20). These intermediates were readily converted into the desired target compounds **129**–**134** using standard conditions. As is readily apparent from Table 7.20, substitution at the *ortho*-, *meta*- and *para*-positions (**129**–**133**) led to either no improvement or a reduction in gt3a potency, although similar rat liver concentrations and replicon potency was generally maintained. Changing the simple aryl ring to a pyridine (**134**) did not lead to improved gt3a potency either and also resulted in the loss of rat liver exposure.

Neither phenols nor simple alcohols participated in triflate-displacement reactions or palladium-catalyzed cross-couplings, hence quinolone **119** was utilized in a series of oxygen-selective alkylations to access a range of 2-alkoxyquinolines. Simple alkyl ethers such as methyl and ethyl were readily

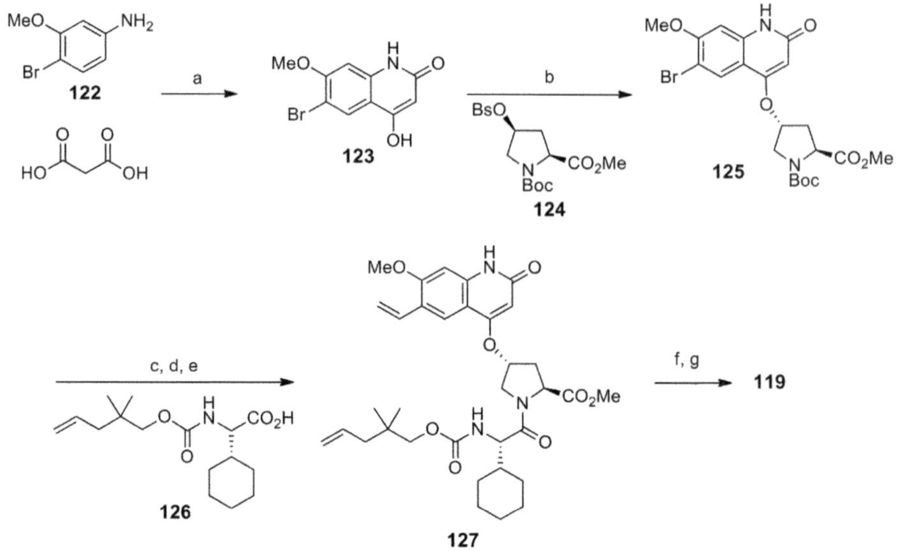

Scheme 7.7 Conditions: (a) 1 equiv. POCl$_3$, 105 °C, 1 h, neat, 74%; (b) **124**,[87] NMP, Cs$_2$CO$_3$, 50%; (c) HCl–dioxane; (d) HATU, DMF, DIEA, 80%, two steps; (e) EtOH, Et$_3$N, 5% Pd(dppf)Cl$_2$, (CH$_2$=CH$_2$BF$_3$)K, reflux, 94%; (f) 10% Zhan 1b,[88] DCE, 45°C, 93%; (g) 5% Pd/C, H$_2$, EtOH, 90%.

available through the use of Meerwein-type reagents in the absence of base, which led to nearly exclusive *O*-alkylation.[112] More functionalized ethers were selectively accessed through the Mitsunobu reaction[113] or alkylation of alkyl halides in the presence of silver salts.[114]

As Table 7.21 illustrates, a variety of 2-alkoxyquinolines showed excellent genotype 3a potency. Methoxy- and ethoxyquinolines (**136**, **137**) were particularly potent and generally exhibited excellent cellular potency and rat liver exposure. Increasing the size or complexity of the alkyl group did not generally lead to any further improvement (**138**). However, methylthiazole analog **139** does have slightly improved gt3a potency (3 nM) and also offers an improved gt1 replicon activity with 50% NHS compared with ethyl analog **137**, and also improved rat liver exposure compared with methyl analog **136**.

In an effort to explore further the space occupied by the alkoxy groups,[105] an effort was made to tie back the substituent into another ring, thus forming a tricyclic P2 group. The versatility of the triflate intermediate **128** was again exploited through simple conversion to imidazo tricycle **140** (Scheme 7.8).[105]

P2 tricyclic analog **140** (Table 7.22) possesses excellent gt3a potency (1.8 nM); unfortunately, it has relatively poor gt1 replicon potency and rat liver exposure. Removal of the C7-methoxy group (**141**) reduces the gt3a potency slightly (5.2 nM) and this compound also suffers from poor replicon potency

Table 7.20 2-Aryl-, heteroaryl-, and aminoquinoline SARs.

Compound	Ar	*1b K_i* (nM)	*3a K_i* (nM)	*1b replicon IC$_{50}$ (nM)* 10% FBS	50% NHS	*Rat [liver] @ 4 h (µM)*
129		0.05	21	3.6	28	9.9
130		0.07	18	3.0	25	6.1
131		0.07	78	6.0	19	6.7
132		0.09	62	7.0	31	6.3
133		0.18	29	11	110	–
134		–	55	27	92	0.2

and rat liver exposure. However, given the promising gt3a enzyme activity of the imidazo tricycles, the related isoquinoline tricycle **142** was synthesized to explore P2 tricycle SARs further. Gratifyingly, **142** displayed excellent gt3a potency (1.7 nM) and, in contrast to **140**, also had good replicon activity (3 nM) and rat liver exposure (27 µM). The effects of substituents on the newly introduced ring of the tricycle were then explored. Substitution at the 2-position was not well tolerated (**143**) in terms of gt3a potency; however, substitution at the 3-position was generally somewhat beneficial. The C3-methyl analog **144** provided similar gt3a potency and rat liver exposure, but was twice as potent in the 50% NHS replicon assay. Cyano (**145**), fluoro (**146**) and methoxy (**147**) groups all boosted gt3a potency by about twofold to

Table 7.21 2-Alkoxyquinoline SARs.

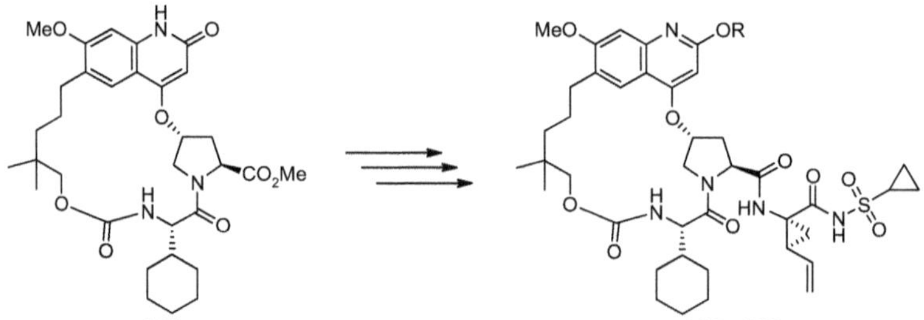

119 → 135–139

Compound	OR	1b K_i (nM)	3a K_i (nM)[a]	1b replicon IC_{50} (nM) 10% FBS	1b replicon IC_{50} (nM) 50% NHS	Rat [liver] @ 4 h (μM)
135	OH	0.07	23	85	460	0.07
136	OMe	0.04	4.1	4	19	5.4
137	OEt	0.09	7.6	6	43	40
138	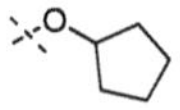	0.05	15	5	34	4.6
139	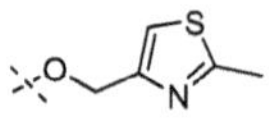	0.1	3.1	3.5	16	19

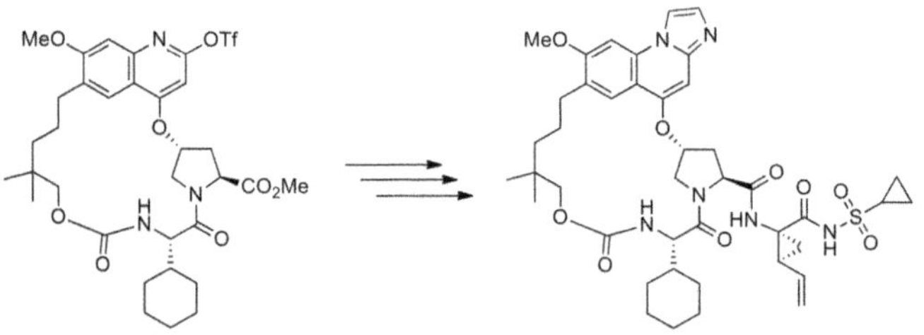

128 → 140

Scheme 7.8

subnanomolar regions, and methoxy analog **147** also provided a slight increase in rat liver exposure compared with unsubstituted **142**.

In addition to the methoxy analogs, unsubstituted tricycle **148** (Table 7.22) exhibited greatly improved gt3a enzyme activity ($K_i = 12\,nM$) compared with P2 bicyclics. This represented a 16-fold increase in gt3a potency for **148** over bicyclic compound **114** and showed that simply a fused ring could give results similar to those seen with 2-quinoline substituents. This increase in potency was

Table 7.22 Tricyclic P2 derivatives.

140–147

Compound	P2	R^a	1b K_i (nM)	3a K_i (nM)	1b replicon IC_{50} (nM) 10% FBS	50% NHS	Rat [liver] @ 4 h (µM)
140		Cyh	0.04	1.8	43	140	0.8
141		Cyh	0.08	5.2	33	180	0.7
142		Cyp	<0.02	1.7	3	29	27
143		Cyp	<0.02	36	7	31	–
144		Cyp	<0.02	1.9	4	15	16
145		Cyp	<0.02	0.8	4	13	7.5
146		Cyp	<0.02	0.9	9	43	22

Table 7.22   (*Continued*)

Compound	P2	R^a	*1b K_i (nM)*	*3a K_i (nM)*	*1b replicon IC$_{50}$ (nM)*		*Rat [liver] @ 4h (µM)*
					10% FBS	*50% NHS*	
147	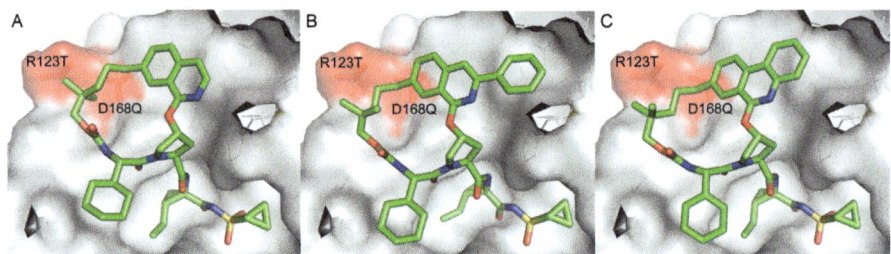	Cyp	<0.02	0.8	4	27	35
148		Cyh	0.05	12	10	–	96

^{a}Cyh = cyclohexyl; Cyp = cyclopentyl.

Figure 7.14   Compounds **114** (A), **115** (B) and **148** (C) docked in the gt1b NS3/4a active site. Pink = areas of diversity between the gt1b and gt3a enzymes. White = conserved areas.

also accompanied by an increase in rat liver exposure to 96 µM at the 4 h time point for **148**, *versus* 21 µM for **114**.

Compounds **114**, **115** and **148** were docked in both the gt1 and gt3a NS3/4a active site *via* molecular modeling in an effort to explain this observed increase in gt3a activity. Figure 7.14 shows the three compounds docked in the gt1b active site and the major differences between the two genotypes are highlighted in pink. All three compounds adopt similar conformations as expected, but it is clear that **115** and **148** extend further into the S2 pocket of the active site than **114**. Furthermore, this pocket is a conserved region between gt1 and gt3a, leading to the hypothesis that increasing contacts with the conserved regions of the active sites of the two genotypes should lead to increases in activity at gt3a without causing a loss in potency versus gt1. It is also clear from the modeling that the P2 to P4 linker falls into a region of the active site that is not conserved between genotypes and is also the region where two important gt3a mutations occur. The docked poses of **115** and **148** suggest that the addition of a larger P2 substituent may also bias the macrocyclic ring to shift towards these mutant residues, possibly improving complementarity to the gt3a site, especially with

Table 7.23 *In vitro* activity of compounds **148–152**.

Compound	1b K_i (nM)	3a K_i (nM)	1b replicon IC_{50} 50% NHS (nM)	Rat [liver] @ 4 h (μM)
148	0.05	12	19	96
149	0.06	130	21	25
150	1.1	1200	nd	nd
151	1.9	520	nd	nd
152	0.08	26	44	0.6

respect to the R123T polymorph, which presumably leads to a significantly different active site surface in this region. In addition to the D168Q difference between gt1b and gt3a, D168 and R155 mutations are observed in the gt1[115] enzyme in clinical studies and these mutations cause potency shifts against both vaniprevir and MK-1220[116] (**153**, see Table 7.24). This second observation led to the hypothesis that removing steric bulk from the P2 region of the molecule in addition to increasing the flexibility in the P2 to P4 macrocyclic linker may allow for further improvements in gt3a activity and mutant profile.

In order to test these hypotheses, **149**, a bicyclic analog of **148**, was targeted. This compound was designed by removing two of the carbons of the tricyclic system, effectively lengthening the P2 to P4 linker by two carbons and maintaining the interaction with the conserved pocket, which seems to enhance gt3a activity. Compound **149** maintained excellent gt1 activity ($K_i = 0.06$ nM) and liver exposure (25 μM) and showed reasonable gt3a activity ($K_i = 130$ nM), considering that a large structural change had been made (Table 7.23). This result with **149** led to the re-evaluation of some compounds made early in the Merck program. Compounds **150** and **151** have simple 2- and 3-pyridyl substituents, respectively, in the P2 position and do not have the gt1 potency seen in advanced NS3/4a inhibitors. However, when just the relative potency of these compounds is considered, compound **151** seems to have an advantage with regard to gt3a potency. Moving the nitrogen from the 2-pyridyl position (**150**) to the 3-pyridyl position (**151**) leads to a twofold *loss* in potency *versus* gt1b ($K_i = 1.1$ *versus* 1.9 nM) but also leads to a twofold *improvement* in potency against the gt3a enzyme ($K_i = 1200$ *versus* 520 nM). Based on modeling, this result is difficult to explain, as there is no apparent hydrogen bonding inter-action associated with the pyridyl nitrogen in either position. However, one hypothesis is a more favorable electrostatic interaction with R155 in the case of **151** or a change in hydrogen bonding to active site water molecules. Based on the results with **150** and **151**, this same nitrogen switch was performed on bicyclic 2-quinoline **149**, giving 3-quinoline **152**, which led to a fivefold increase in gt3a potency ($K_i = 26$ nM). Although liver exposure was markedly reduced (4 h liver exposure: **152** = 0.6 μM *versus* **149** = 26 μM), **152** emerged as a key lead for the second-generation program, with a subsequent focus on improving liver exposure as well as optimizing potency *versus* gt1b mutant enzymes observed in clinical studies with vaniprevir.[117]

As discussed in the context of MK-7009, increasing lipophilicity in the P2–P4 macrocyclic series can lead to increases in liver exposure.[42,116] Adding a dimethyl linker substituent to **152** gave **154** and led to only a slight increase in liver exposure (Table 7.24). This compound, however, had an improved K_i of 2.8 nM against the gt3a enzyme, representing an almost 20-fold improvement over two previous Merck clinical candidates, vaniprevir, **13** (gt3a $K_i = 54$ nM), and MK-1220,[116] **153** (gt3a $K_i = 70$ nM).[116] Further profiling of **154** showed that activity against the key R155K mutant[115] ($K_i = 0.32$ nM) had been improved by almost 50-fold in comparison with both **13** (gt1b R155K, $K_i = 19$ nM) and **153** (gt1b R155K, $K_i = 15$ nM). The activity of **154** *versus* the A156 mutants was improved by 10–30-fold in relation to **153**, but was comparable to that of **13**. Saturation of the linker olefin of **154** gave **155**, which had a similar potency profile, but showed a 40-fold increase in rat liver exposure to 63 μM at 4 h after a 5 mg kg^{-1} oral dose. These dramatic changes in rat liver exposure were observed throughout the program and highlight the need for a rat liver screening strategy that was discussed earlier.[41,42,116] Substituent effects were studied on the quinoline scaffold and a 7-methoxy substituent (**156**) was optimal for activity, giving slight increases in potency against gt3a and 1b mutants R155K, A156T and A156V.

In general, increasing potency against the A156 mutants was difficult, likely because, as seen in Figure 7.16, the P2–P4 macrocycles are in close contact with this residue. Activity against R155 and D168 mutants was more addressable because the P2 substituent in the quinoline series mostly avoids this region of the active site.

While the development of the P2 3-quinoline series was ongoing, work within the related alkoxyquinoline series discussed earlier led to a key observation in the development of MK-5172. Similarly to ethoxyquinoline **137**, the corre-sponding P3 cyclopentyl analog **157** (Figure 7.15) was identified as a potent gt1

Table 7.24 *In vitro activity of compounds* **13** *and* **153–156**.

153 (MK-1220)   154   155   156

R =

Compound	1b K_i (nM)	3a K_i (nM)	1b R155K K_i (nM)	1b A156T K_i (nM)	1b A156V K_i (nM)	1b D168V K_i (nM)	1b replicon IC_{50} 50% NHS (nM)	Rat [liver] @ 4 h (µM)
13	0.05	54	19	22	88	77	19	9.9
153	0.09	70	15	540	790	27	11	23
154	0.02	2.8	0.32	18	79	nd	37	1.5
155	0.02	4.0	0.20	27	80	nd	18	63
156	0.02	1.9	0.16	12	64	nd	28	9.8

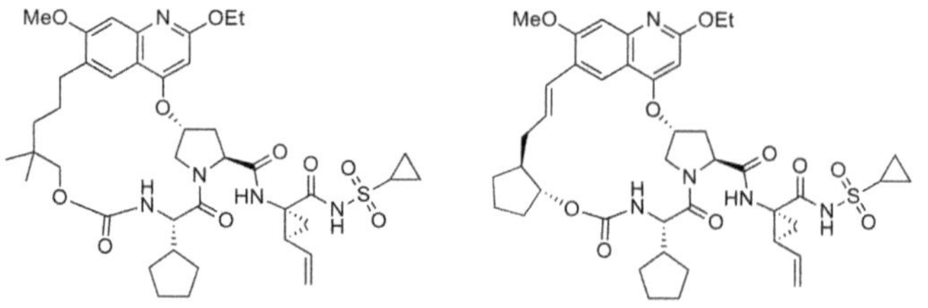

Figure 7.15 2-Ethoxyquinolinecarbamate constrained linkers.

and gt3 inhibitor, which lost a good deal of potency against the gt1 A156T mutant. During the SAR development, which led to the identification of **157**, the effect of alternative carbamates on the P2–P4 scaffold was also examined,[118] given the P4 structural diversity seen in the literature for P1–P3 macrocyclic and non-macrocyclic cores.[119] Introduction of a cyclopentylcarbamate related to that present in BILN-2061 led to **158** (Figure 7.15). While **158** lost about half of the gt3a potency, the activity against gt1b A156T improved >40-fold relative to **157** while still maintaining reasonable rat liver exposure (Table 7.25). The tight SAR around gt1b A156T potency is demonstrated *via* introduction of a methyl group beta to the carbamate oxygen in **158** leading to **159**. Whereas **159** maintains similar gt3a activity, gt1b A156T potency is decreased >40-fold.

Further contraction of cyclic constraints to 1,2- and 1,3-cyclobutane (**160**, **161**) leads to different profiles (Table 7.25). 1,2-Cyclobutane **160** loses potency versus gt3a and gt1b A156T, whereas the 1,3-linked isomer **161** has moderate gt3a activity, but good potency *versus* A156 mutants, including only about a threefold shift on going from A156T to A156V. Moving the fused ring constraint adjacent to the P2 heterocycle led to a compound with the most balanced overall profile within this sub-series. Cyclopropyl analog **162** possesses excellent gt3a ($K_i = 1$ nM) potency, gt1b replicon activity ($IC_{50} = 11$ nM) and rat liver exposure (27 000 nM at 4 h, 5 mg kg^{-1}). In terms of gt1b mutant activity, **162** displays <15 nM K_i against both A156T (2.8 nM) and A156V (14 nM), while maintaining subnanomolar activity against gt1b R155K.

Within the context of the P2 2-quinoline series related to **149**, the introduction of cyclic constraints on the linker was also very effective at improving activity *versus* key gt1 mutant enzymes (Table 7.26). Given the altered orientation of the P2 group, the number of atoms connecting the linker ring to the P2 group was generally increased by two atoms. While introduction of two of the better carbamate linkers from the P2 ethoxyquinoline series (**163**, **164**) led to only moderate gt3a and gt1b A156T/V potency, introduction of an alkoxypyrrolidineurea (**165**, **166**) gave compounds with good overall profiles in terms of potency and rat liver exposure. Compound **165** ($K_i = 3$ nM) is twice as potent as **166** against gt3a ($K_i = 5.9$ nM), but is less potent against gt1b A156 mutants.

A comparison across gt1–3 enzymes and gt1 mutants is shown in Table 7.27 for two different linker-constrained compounds compared with a compound

Table 7.25 2-Ethoxyquinoline carbamate constrained linkers.

158–162

Compound	Linker	3a K_i (nM)	1b replicon (nM)	1b mutants K_i (nM)			Rat [liver] @ 4 h (μM)
				R155K	A156T	A156V	
157		2.2	11	1.3	41	–	35
158		4.6	17	0.5	0.8	15	10.2
159		4.1	38	0.9	49	–	–
160[a]		18	–	2.2	52	–	–
161[b]		7.7	16	2.4	4.8	16	–
162[b]		1.0	11	0.2	2.8	14	27.0

[a]Mixture of stereoisomers.
[b]Single isomer; absolute stereochemistry not established.

with a linear linker. Both **162** and **165** show sub-nanomolar to low-nanomolar potency across genotypes 1, 2 and 3, and also three different key mutation positions (R155, A156, D168).

In the 3-quinoline case, fusion of a cyclopentyl ring gave **168** and resulted in a broad activity profile against genotypes and key mutant enzymes (Table 7.28).

Table 7.26 P2 2-alkoxy quinolines with constrained linkers.

163–166

Compound	Linker	3a K_i (nM)	1b replicon (nM)	1b mutants K_i (nM) R155K	A156T	A156V	Rat [liver] @ 4 h (μM)
163		11.6	58	0.2	17	–	30.8
164		5.7	5.0	0.1	7.8	64	–
165		3.0	32	0.1	2.9	37	8.6
166		5.9	31	0.1	1.5	22	12.0

Table 7.27 Key compound comparison.

Compound	1b K_i (nM)a	2a K_i (nM)a	3a K_i (nM)a	1b replicon IC_{50} (nM) 10% FBS	50% NHS	gt 1b K_i (nM)[a] R155K	A156T	A156V	D168V
157	0.016	0.14	2.1	2.0	11	1.2	41	–	2.2
162	0.016	0.14	1.3	2.3	11	0.2	2.8	14	1.8
165	0.016	0.17	3.1	3.1	32	0.1	2.9	37	1.2

[a]NS3/4a protease time-resolved fluorescence assay.

Table 7.28 *In vitro* activity of compounds **167–169** and **14**.

167

168

169

14 (MK-5172)

R =

Compound	1b K_i (nM)	3a K_i (nM)	1b R155K K_i (nM)	1b A156T K_i (nM)	1b A156V K_i (nM)	1b D168V K_i (nM)	1b replicon IC_{50} 50% NHS (nM)	Rat [liver] @ 4h (μM)
167	0.02	1.9	0.16	12	64	nd	28	9.8
168	0.02	0.39	0.07	2.4	11	0.06	16	6.3
169	0.02	0.31	0.02	2.9	5.7	0.05	13	20
14 (MK-5172)	0.02	0.70	0.07	5.3	12	0.14	7.4	23

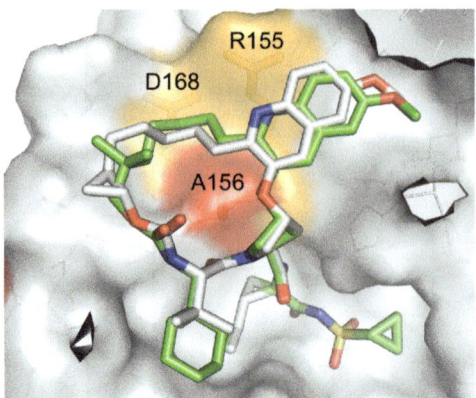

Figure 7.16   Comparison of the energy-minimized conformations of **167** (magenta) and **169** (green) docked in the gt1b NS3/4a active site.

Ring contraction to a fused cyclopropyl further improved the activity and gave a compound (**169**) with excellent liver exposure. Modeling results with **167** and **169** give some insight into the reasons for the 4–10-fold improvement against A156 mutant enzymes (Figure 7.16). The cyclopropyl constraint of **169** seems to shift the macrocycle away from the A156 residue compared with **167**. Since the A156 residue is positioned close to the middle of the macrocyclic ring, this slight shift may allow more space for the macrocycle of **169** to accommodate the A156T or A156V mutation. Compound **169** met second-generation HCV protease potency and PK criteria and was evaluated in a battery of studies as a prelude to its selection as a development candidate.

During this candidate work-up, an important pharmacokinetics-related issue regarding the behavior of pharmaceutical salts of **169** emerged, which represented a major development hurdle. Although the crystalline potassium salt of **169** has good aqueous solubility (9.7 mg mL^{-1}), it readily disproportionates to a crystalline zwitterionic form that has a greatly reduced aqueous solubility of <0.009 mg mL^{-1}. This behavior is driven by the moderate basicity of the quinoline P2 group of **169** (calculated $pK_a = 4.47$),[120] along with moderate acidity of the acylsulfonamide (calculated $pK_a = 3.67$). *In vivo*, this process is promoted in the stomach where the native pH of the dosing solution is attenuated, giving rise to non-reproducible pharmacokinetics.[121] Poor absorption of the crystalline zwitterionic form of **169** that is formed *in vivo* significantly capped the achievable plasma and liver exposures of **169** in preclinical species. Given the excellent overall profile of **169**, Merck examined alternative P2 heterocycles with less basic functionality that potentially would maintain a similar potency and PK profile. From this work, P2 quinoxalines emerged as the most promising subclass, with lower pK_a reducing the risk of zwitterion formation (calculated $pK_a \sim 1.2$). Incorporating a quinoxaline into the optimized framework gave **14** (Table 7.28). Overall, **14** maintained the excellent potency against the gt3a enzyme and also a broad panel of mutant enzymes and showed excellent rat liver exposure. The potassium salt of **14**

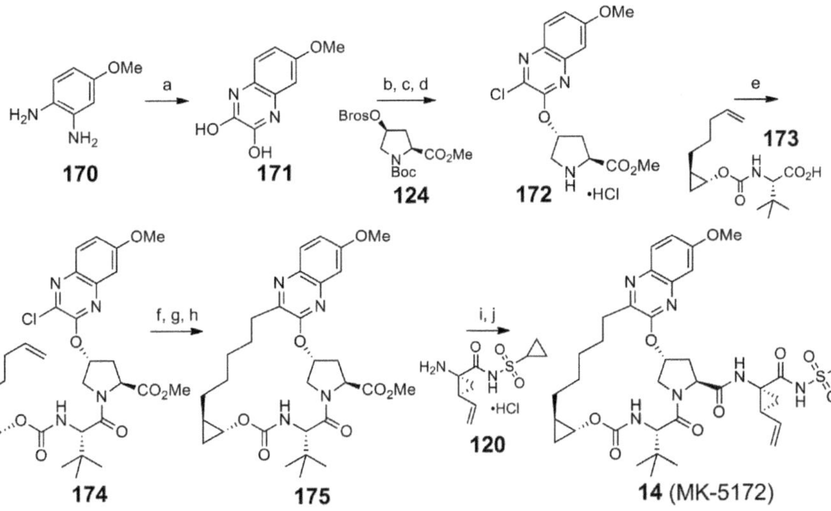

Scheme 7.9 Conditions: (a) diethyl oxalate, TEA, 150°C, 69%; (b) thionyl chloride, DMF, 110 °C, 69%; (c) **124**,[87] Cs$_2$CO$_3$, NMP; (d) HCl, dioxane; (e) **173**, HATU, DIPEA, DMF, 78%; (f) potassium vinyltrifluoroborate, TEA, dichloro[1,1-bis(diphenylphosphino)ferrocene]palladium(II) chloride, EtOH; (g) Zhan 1B catalyst,[88] DCE, 25%, two steps; (h) H$_2$, 10% Pd/C, MeOH, dioxane, 99%; (i) LiOH, THF, water; (j) **120**,[66] TBTU, DIPEA, DMAP, DMF, 89%, two steps.

showed an aqueous solubility of 1.8 mg mL^{-1}, somewhat lower than that of **169**; however, no disproportionation was observed, making this compound (**14**, MK-5172) a viable development candidate.[44]

MK-5172 (**14**) was prepared by a route broadly analogous to the synthetic sequence used for previous compounds.[42,116] Specifically, the synthesis of **14** starts with 4-methoxyphenylenediamine hydrochloride (**170**), which was heated with diethyl oxalate to give **171** in 69% yield (Scheme 7.9). Chemoselective chlorination with thionyl chloride followed by addition to activated *cis*-hydroxyproline derivative **124**[87] and deprotection then led to intermediate **172** in 69% yield. Straightforward amide coupling with **173**,[44] prepared in enantiopure form in six steps, led to chloroquinoxaline **174** in 78% yield. Next, while vinylation proceeded smoothly, the key RCM using the Zhan 1b catalyst[88] was problematic and the desired macrocycle was isolated in 25% yield. Reduction of the resulting olefin gave macrocycle **175**, which was readily converted to **14** (MK-5172) in 89% yield by hydrolysis and coupling to the known cyclopropylaminoacylsulfonamide **120**.[66]

The pharmacokinetic properties of the potassium salt of MK-5172 were evaluated in multiple species (Table 7.29). In the rat, MK-5172 showed a plasma clearance of 28 mL min^{-1} kg^{-1} and a plasma half-life of 1.4 h. When dosed orally at 5 mg kg^{-1}, the plasma exposure of MK-5172 was good with an AUC of 0.7 μM h. The liver exposure of the compound was also significant

Table 7.29 Pharmacokinetic parameters for the potassium salt of **14** (MK-5172)[a]

Species	Cl (mL min^{-1} kg^{-1})	V_d (L kg^{-1})	$t_{\frac{1}{2}}$ (h)	p.o. AUC (μM h)	p.o. [liver] 4 h (μM)	p.o. [liver] 24 h (μM)
Rat	28	3.1	1.4	0.7	23	0.2
Dog	5	0.7	3.0	0.4	nd	1.4

[a]Rat i.v. (2 mg kg^{-1}, $n = 3$, DMSO), dog i.v. (0.5 mg kg^{-1}, $n = 3$, DMSO), rat p.o. (5 mg kg^{-1}, $n = 3$, PEG400), dog p.o. (1 mg kg^{-1}, $n = 3$, PEG400).

(23 μM at 4 h), and at 24 h the liver concentration of MK-5172 was 0.2 μM, which is >25-fold higher than the IC$_{50}$ in the replicon assay with 50% NHS.

When dosed intravenously to dogs, MK-5172 shows a low clearance of 5 mL min^{-1} kg^{-1} and a half-life of 3 h; it also has good plasma exposure (AUC = 0.4 μM h) after a 1 mg kg^{-1} oral dose (Table 7.29). Dog liver biopsy studies showed that the liver concentration of MK-5172 after the 1 mg kg^{-1} oral dose is 1.4 μM at the 24 h time point. Similarly to its behavior in rats, MK-5172 demonstrates effective partitioning into liver tissue and maintains a high liver concentration, relative to potency, 24 h after oral dosing in dogs.

In summary, initial screening for gt3a activity along with molecular modeling led to the discovery of a series of P2 quinoline macrocycles with excellent broad activity against NS3/4a genotypes and clinically observed gt1b mutant enzymes. This series was optimized for enzyme activity and liver exposure in preclinical species and led to the second-generation NS3/4a protease inhibitor MK-5172. Further studies of MK-5172, including clinical investigations of the PK and efficacy profile, are ongoing.[45]

7.5 Conclusion

The path to the development of HCV NS3/4A protease inhibitors has been a long one, but has now successfully yielded two marketed drugs, Victrelis and Incivek/Incivo, with multiple other compounds entering into late-stage clinical development, including the three compounds simeprevir (TMC-435350), vaniprevir (MK-7009) and MK-5172 that are the focus of this discussion. Ongoing and future clinical studies with these compounds, particularly in combination treatment regimes, will provide additional data, but HCV protease inhibitors appear well-positioned to play an important role in the rapidly evolving standard of care for treatment of HCV infections.

From a medicinal chemistry perspective, the structural properties of the inhibitors described in detail here do not conform well to what has been considered druggable space.[122–125] With high molecular weights in the 750 g mol^{-1} range and large polar surface areas, identification of compounds that show good plasma pharmacokinetics represents a significant accomplishment. Although structural chemistry has provided some key guidance in the design of new inhibitor structures, a significant portion of the optimization, particularly against mutant enzymes, has been empirical in nature. In each case, the design of modular synthetic approaches has led to an efficient and rapid exploration of SARs in these relatively complex molecules.

References

1. WHO, WHO hepatitis C – global prevalence (update), *Wkly. Epidemiol. Rec.*, 1999, 425–427.
2. T. J. Liang and T. Heller, Pathogenesis of hepatitis C-associated hepato-cellular carcinoma, *Gastroenterology*, 2004, **127**, S62–S71.
3. P. Simmonds, Genetic diversity and evolution of hepatitis C virus – 15 years on, *J. Gen. Virol.*, 2004, **85**, 3173–3188.
4. F. Maggi, C. Fornai, M. L. Vatteroni, M. Giorgi, A. Morrica, M. Pistello, G. Cammarota, S. Marchi, P. Ciccorossi, A. Bionda and M. Bendinelli, Differences in hepatitis C virus quasispecies composition between liver, peripheral blood mononuclear cells and plasma, *J. Gen. Virol.*, 1997, **78**, 1521–1525.
5. J. Bukh, R. H. Miller and R. H. Purcell, Biology and genetic heterogeneity of hepatitis C virus, *Clin. Exp. Rheumatol.*, 1995, **13**(Suppl. 13), S3–7.
6. S. J. Hadziyannis, H. Sette Jr., T. R. Morgan, V. Balan, M. Diago, P. Marcellin, G. Ramadori, H. Bodenheimer Jr., D. Bernstein, M. Rizzetto, S. Zeuzem, P. J. Pockros, A. Lin and A. M. Ackrill, Peginterferon-α2a and ribavirin combination therapy in chronic hepatitis C: a randomized study of treatment duration and ribavirin dose, *Ann. Intern. Med.*, 2004, **140**, 346–355.
7. S. Zeuzem, T. Berg, B. Moeller, H. Hinrichsen, S. Mauss, H. Wedemeyer, C. Sarrazin, D. Hueppe, E. Zehnter and M. P. Manns, Expert opinion on the treatment of patients with chronic hepatitis C, *J. Viral Hepat.*, 2009, **16**, 75–90.
8. A. Andriulli, A. Mangia, A. Iacobellis, A. Ippolito, G. Leandro and S. Zeuzem, Meta-analysis: the outcome of anti-viral therapy in HCV genotype 2 and genotype 3 infected patients with chronic hepatitis, *Aliment. Pharmacol. Ther.*, 2008, **28**, 397–404.
9. C. P. Gordon and P. A. Keller, Control of hepatitis C: a medicinal chemistry perspective, *J. Med. Chem.*, 2005, **48**, 1–20.
10. B. A. Malcolm, R. Liu, F. Lahser, S. Agrawal, B. Belanger, N. Butkiewicz, R. Chase, F. Gheyas, A. Hart, D. Hesk, P. Ingravallo, C. Jiang, R. Kong, J. Lu, J. Pichardo, A. Prongay, A. Skelton, X. Tong, S. Venkatraman, E. Xia, V. Girijavallabhan and F. G. Njoroge, SCH 503034, a mechanism-based inhibitor of hepatitis C virus NS3 protease, suppresses polyprotein maturation and enhances the antiviral activity of alpha interferon in replicon cells, *Antimicrob. Agents Chemother.*, 2006, **50**, 1013–1020.
11. F. Poordad, J. McCone Jr., B. Bacon, S. Bruno, M. P. Manns, M. S. Sulkowski, I. M. Jacobson, R. K. Rajender, Z. D. Goodman, N. Boparai, M. J. DiNubile, V. Sniukiene, C. A. Brass, J. K. Albrecht and

J.-P. Bronowicki, Boceprevir for untreated chronic HCV genotype 1 infection, *N. Engl. J. Med.*, 2011, **364**, 1195–1206.

12. R. B. Perni, S. J. Almquist, R. A. Byrn, G. Chandorkar, P. R. Chaturvedi, L. F. Courtney, C. J. Decker, K. Dinehart, C. A. Gates, S. L. Harbeson, A. Heiser, G. Kalkeri, E. Kolaczkowski, K. Lin, Y.-P. Luong, B. G Rao., W. P. Taylor, J. A. Thomson, R. D. Tung, Y. Wei, A. D. Kwong and C. Lin, Preclinical profile of VX-950, a potent, selective and orally bio-available inhibitor of hepatitis C virus NS3-4A serine protease, *Antimicrob. Agents Chemother.*, 2006, **50**, 899–909.

13. I. M. Jacobson, J. G. McHutchison, G. Dusheiko, B. Di, M. Adrian, K. R. Reddy, N. H. Bzowej, P. Marcellin, A. J. Muir, P. Ferenci, R. Flisiak, J. George, M. Rizzetto, D. Shouval, R. Sola, R. A. Terg, E. M. Yoshida, N. Adda, L. Bengtsson, A. J. Sankoh, T. L. Kieffer, S. George, R. S. Kauffman and S. Zeuzem, Telaprevir for previously untreated chronic hepatitis C virus infection, *N. Engl. J. Med.*, 2011, **364**, 2405–2416.

14. I. M. Jacobson, J.-M. Pawlotsky, N. H. Afdhal, G. M. Dusheiko, X. Forns, D. M. Jensen, F. Poordad and J. Schulz, A practical guide for the use of boceprevir and telaprevir for the treatment of hepatitis C, *J. Viral Hepat.*, 2012, **19**(Suppl. 2), 1–26.

15. K. E. Sherman, S. L. Flamm, N. H. Afdhal, D. R. Nelson, M. S. Sulkowski, G. T. Everson, M. W. Fried, M. Adler, H. W. Reesink, M. Martin, A. J. Sankoh, N. Adda, R. S. Kauffman, S. George, C. I. Wright and F. Poordad, Response-guided telaprevir combination treatment for hepatitis C virus infection, *N. Engl. J. Med.*, 2011, **365**, 1551.

16. B. R. Bacon, S. C. Gordon, E. Lawitz, P. Marcellin, J. M. Vierling, S. Zeuzem, F. Poordad, Z. D. Goodman, H. L. Sings, N. Boparai, M. Burroughs, C. A. Brass, J. K. Albrecht and R. Esteban, Boceprevir for previously treated chronic HCV genotype 1 infection, *N. Engl. J. Med.*, 2011, **364**, 1207–1217.

17. W. P. Hofmann and S. Zeuzem, A new standard of care for the treatment of chronic HCV infection, *Nat. Rev. Gastroenterol. Hepatol.*, 2011, **8**, 257–264.

18. L. Y. Lee, C. Y. W. Tong, T. Wong and M. Wilkinson, New therapies for chronic hepatitis C infection: a systematic review of evidence from clinical trials, *Int. J. Clin. Pract.*, 2012, **66**, 342–355.

19. X. Tong, A. Arasappan, F. Bennett, R. Chase, B. Feld, Z. Guo, A. Hart, V. Madison, B. Malcolm, J. Pichardo, A. Prongay, R. Ralston, A. Skelton, E. Xia, R. Zhang and F. G. Njoroge, Preclinical char-acterization of the antiviral activity of SCH 900518 (narlaprevir), a novel mechanism-based inhibitor of hepatitis C virus NS3 protease, *Antimicrob. Agents Chemother.*, 2010, **54**, 2365–2370.

20. H. Reesink, J. Bergmann, J. de Bruijne, C. Weegink, J. van Lier, A. van Vliet, A. Keung, J. Li, E. O'Mara, M. Treitel, E. Hughes, H. Janssen and R. de Knegt, Safety and antiviral activity of SCH 900518 administered as monotherapy and in combination with PEGinterferon alpha-2b to naive and treatment-experienced HCV-1 infected pateients, *J. Hepatol.*, 2009, **50**(Suppl. 1), S35.

21. J. de Bruijne, J. F. Bergmann, H. W. Reesink, C. J. Weegink, R. Molenkamp, J. Schinkel, X. Tong, J. Li, M. A. Treitel, E. A. Hughes, J. J. van Lier, A. A. van Vliet, H. L. Janssen and R. J. de Knegt, Antiviral activity of narlaprevir combined with ritonavir and pegylated interferon in chronic hepatitis C patients, *Hepatology*, 2010, **52**, 1590–1599.

22. M. Llinas-Brunet, M. D. Bailey, G. Bolger, C. Brochu, A.-M. Faucher, J. M. Ferland, M. Garneau, E. Ghiro, V. Gorys, C. Grand-Maitre, T. Halmos, N. Lapeyre-Paquette, F. Liard, M. Poirier, M. Rheaume, Y. S. Tsantrizos and D. Lamarre, Structure–activity study on a novel series of macrocyclic inhibitors of the hepatitis C virus NS3 protease leading to the discovery of BILN 2061, *J. Med. Chem.*, 2004, **47**, 1605–1608.

23. D. Lamarre, P. C. Anderson, M. Bailey, P. Beaulieu, G. Bolger, P. Bonneau, M. Bos, D. R. Cameron, M. Cartier, M. G. Cordingley, A.-M. Faucher, N. Goudreau, S. H. Kawai, G. Kukolj, L. Lagace, S. R. LaPlante, H. Narjes, M.-A. Poupart, J. Rancourt, R. E. Sentjens, R. St. George, B. Simoneau, G. Steinmann, D. Thibeault, Y. S. Tsantrizos, S. M. Weldon, C.-L. Yong and M. Llinas-Brunet, An NS3 protease inhibitor with antiviral effects in humans infected with hepatitis C virus, *Nature*, 2003, **426**, 186–189.

24. J. H. Stoltz, J. O. Stern, Q. Huang, R. W. Seidler, F. D. Pack and B. L. Knight, A twenty-eight-day mechanistic time course study in the rhesus monkey with hepatitis C virus protease inhibitor BILN 2061, *Toxicol. Pathol.*, 2011, **39**, 496–501.

25. S. D. Seiwert, S. W. Andrews, Y. Jiang, V. Serebryany, H. Tan, K. Kossen, P. T. R. Rajagopalan, S. Misialek, S. K. Stevens, A. Stoycheva, J. Hong, S. R. Lim, X. Qin, R. Rieger, K. R. Condroski, H. Zhang, M. G. Do, C. Lemieux, G. P. Hingorani, D. P. Hartley, J. A. Josey, L. Pan, L. Beigelman and L. M. Blatt, Preclinical characteristics of the hepatitis C virus NS3/4A protease inhibitor ITMN-191 (R7227), *Antimicrob. Agents Chemother.*, 2008, **52**, 4432–4441.

26. N. Forestier, D. Larrey, D. Guyader, P. Marcellin, R. Rouzier, A. Patat, P. Smith, W. Bradford, S. Porter, L. Blatt, S. D. Seiwert and S. Zeuzem, Treatment of chronic hepatitis C patients with the NS3/4A protease inhibitor danoprevir (ITMN-191/RG7227) leads to robust reductions in viral RNA: a phase 1b multiple ascending dose study, *J. Hepatol.*, 2011, **54**, 1130–1136.

27. N. Forestier, D. Larrey, P. Marcellin, D. Guyader, A. Patat, R. Rouzier, P. F. Smith, X. Qin, S. Lim, W. Bradford, S. Porter, S. D. Seiwert and S. Zeuzem, Antiviral activity of danoprevir (ITMN-191/RG7227) in combination with pegylated interferon α-2a and ribavirin in patients with hepatitis C, *J. Infect. Dis.*, 2011, **204**, 601–608.

28. X. C. Sheng, A. Casarez, R. Cai, M. O. Clarke, X. Chen, A. Cho, W. E. Delaney, E. Doerffler, M. Ji, M. Mertzman, R. Pakdaman, H.-J. Pyun, T. Rowe, Q. Wu, J. Xu and C. U. Kim, Discovery of GS-9256: a novel phosphinic acid derived inhibitor of the hepatitis C virus

NS3/4A protease with potent clinical activity, *Bioorg. Med. Chem. Lett.*, 2012, **22**, 1394–1396.

29. E. J. Lawitz, T. C. Marbury, B. D. Vince, N. Grunenberg, M. Rodriguez-Torres, M. P. D. Micco, J. N. Tarro, M. J. Shelton, S. West, J. Zong, A. Bae, K. Wong, H.-M. Mo, D. Oldach, W. Delaney and F. Rousseau, Dose-ranging, three-day monotherapy study of the HCV NS3 protease inhibitor GS-9256, *J. Hepatol.*, 2010, **52**, S466–S467.

30. M. Huang, S. Podos, D. Patel, G. Yang, J. L. Fabrycki, Y. Zhao, C. Marlor, P. Kapoor, X. Wang, A. Hashimoto, V. Gadhachanda, G. Pais, D. Chen, A. Agarwal, M. Deshpande, K. L. Stauber and A. Phadke, ACH-2684: HCV NS3 protease inhibitor with potent activity against multiple genotypes and known resistant variants, *Hepatology*, 2010, **52**, 1204A.

31. C. Pasquinelli, F. McPhee, T. Eley, C. Villegas, K. Sandy, P. Sheridan, A. Persson, S.-P. Huang, D. Hernandez, A. K. Sheaffer, P. Scola, T. Marbury, E. Lawitz, R. Goldwater, M. Rodriguez-Torres, M. DeMicco, D. Wright, M. Charlton, W. K. Kraft, J.-C. Lopez-Talavera and D. M. Grasela, Single- and multiple-ascending-dose studies of the NS3 protease inhibitor asunaprevir in subjects with or without chronic hepatitis C, *Antimicrob. Agents Chemother.*, 2012, **56**, 1838–1844.

32. J. P. Bronowicki, S. Pol, P. J. Thuluvath, D. Larrey, C. T. Martorell, V. K. Rustgi, D. W. Morris, Z. Younes, M. W. Fried, M. Bourliere, C. Hezode, O. Massoud, G. A. Abrams, V. Ratziu and A. Thiry, BMS-650032, an NS3 inhibitor, in combination with PEGinterferon alpha-2A and ribavirin in treatment-naive subjects with genotype 1 cronic hepatitis C infection, *J. Hepatol.*, 2011, **54**, S472.

33. M. Llinas-Brunet, M. D. Bailey, N. Goudreau, P. K. Bhardwaj, J. Bordeleau, M. Bos, Y. Bousquet, M. G. Cordingley, J. Duan, P. Forgione, M. Garneau, E. Ghiro, V. Gorys, S. Goulet, T. Halmos, S. H. Kawai, J. Naud, M.-A. Poupart and P. W. White, Discovery of a potent and selective noncovalent linear inhibitor of the hepatitis C virus NS3 protease (BI 201335), *J. Med. Chem.*, 2010, **53**, 6466–6476.

34. P. W. White, M. Llinas-Brunet, M. Amad, R. C. Bethell, G. Bolger, M. G. Cordingley, J. Duan, M. Garneau, L. Lagace, D. Thibeault and G. Kukolj, Preclinical characterization of BI 201335, a C-terminal carboxylic acid inhibitor of the hepatitis C virus NS3–NS4A protease, *Antimicrob. Agents Chemother.*, 2010, **54**, 4611–4618.

35. S. Zeuzem, T. Asselah, P. Angus, J.-P. Zarski, D. Larrey, B. Muellhaupt, E. Gane, M. Schuchmann, A. Lohse, S. Pol, J.-P. Bronowicki, S. Roberts, K. Arasteh, F. Zoulim, M. Heim, J. O. Stern, G. Kukolj, G. Nehmiz, C. Haefner and W. O. Boecher, Efficacy of the protease inhibitor BI 201335, polymerase inhibitor BI 207127 and ribavirin in patients with chronic HCV infection, *Gastroenterology*, 2011, **141**, 2047–2055.

36. M. P. Manns, M. Bourliere, Y. Benhamou, S. Pol, M. Bonacini, C. Trepo, D. Wright, T. Berg, J. L. Calleja, P. W. White, J. O. Stern, G. Steinmann, C.-L. Yong, G. Kukolj, J. Scherer and W. O. Boecher, Potency, safety and

pharmacokinetics of the NS3/4A protease inhibitor BI201335 in patients with chronic HCV genotype-1 infection, *J. Hepatol.*, 2011, **54**, 1114–1122.

37. M. S. Sulkowski, E. Ceasu, T. Asselah, F. A. Caruntu, J. Lalezari, P. Ferenci, A. Streinu-Cercel, H. Fainboim, H. Tanno, L. Preotescu, B. Leggett, F. Bessone, S. Mauss, J. O. Stern, C. Hafner, Y. Datsenko, G. Nehmiz, W. Böcher and G. Steinmann, Sustained virologic response (SVR) and safety of BI201335 combined with pegintereron alfa-2a and ribavirin (P/R) in treatment-naive patients with chronic genotype 1 HCV infection, *J. Hepatol.*, 2011, **54**(Suppl. 1), S27.

38. X. C. Sheng, T. Appleby, T. Butler, R. Cai, X. Chen, A. Cho, M. O. Clarke, J. Cottell, W. E. Delaney IV, E. Doerffler, J. Link, M. Ji, R. Pakdaman, H.-J. Pyun, Q. Wu, J. Xu and C. U. Kim, Discovery of GS-9451: an acid inhibitor of the hepatitis C virus NS3/4A protease, *Bioorg. Med. Chem. Lett.*, 2012, **22**, 2629–2634.

39. E. Lawitz, J. M. Hill, T. C. Marbury, M. Rodriguez-Torres, M. DeMicco, J. Quesada, P. Shaw, S. C. Gordon, M. J. Shelton, D. H. Coombs, J. A. Zong, A. Bae, K. A. Wong, H. Mo, E. Mondou, K. Hirsch and W. Delaney, Three-day, dose-ranging study of the HCV NS3 protease inhibitor GS-9451, *Hepatology*, 2010, **52**(Suppl), 714A–715A.

40. V. Detishin, W. Haazen, H. Robison, L. Robarge and E. Olek, Virological response, safety and pharmacokinetic profile following single- and multiple-dose administration of ACH-0141625 protease inhibitor to healthy volunteers and HCV genotype-1 patients, *J. Hepatol.*, 2010, **52**(Suppl. 1), S468.

41. N. J. Liverton, M. K. Holloway, J. A. McCauley, M. T. Rudd, J. W. Butcher, S. S. Carroll, J. DiMuzio, C. Fandozzi, K. F. Gilbert, S.-S. Mao, C. J. McIntyre, K. T. Nguyen, J. J. Romano, M. Stahlhut, B.-L. Wan, D. B. Olsen and J. P. Vacca, Molecular modeling based approach to potent P2–P4 macrocyclic inhibitors of hepatitis C NS3/4A protease, *J. Am. Chem. Soc.*, 2008, **130**, 4607–4609.

42. J. A. McCauley, C. J. McIntyre, M. T. Rudd, K. T. Nguyen, J. J. Romano, J. W. Butcher, K. F. Gilbert, K. J. Bush, M. K. Holloway, J. Swestock, B.-L. Wan, S. S. Carroll, J. M. DiMuzio, D. J. Graham, S. W. Ludmerer, S.-S. Mao, M. W. Stahlhut, C. M. Fandozzi, N. Trainor, D. B. Olsen, J. P. Vacca and N. J. Liverton, Discovery of vaniprevir (MK-7009), a macrocyclic hepatitis C virus NS3/4a protease inhibitor, *J. Med. Chem.*, 2010, **53**, 2443–2463.

43. N. J. Liverton, S. S. Carroll, J. Dimuzio, C. Fandozzi, D. J. Graham, D. Hazuda, M. K. Holloway, S. W. Ludmerer, J. A. McCauley, C. J. McIntyre, D. B. Olsen, M. T. Rudd, M. Stahlhut and J. P. Vacca, MK-7009, a potent and selective inhibitor of hepatitis C virus NS3/4A protease, *Antimicrob. Agents Chemother.*, 2010, **54**, 305–311.

44. S. Harper, J. A. McCauley, M. T. Rudd, M. Ferrara, M. DiFilippo, B. Crescenzi, U. Koch, A. Petrocchi, M. K. Holloway, J. W. Butcher, J. J. Romano, K. J. Bush, K. F. Gilbert, C. J. McIntyre, K. T. Nguyen, E. Nizi, S. S. Carroll, S. W. Ludmerer, C. Burlein, J. M. DiMuzio,

D. J. Graham, C. M. McHale, M. W. Stahlhut, D. B. Olsen, E. Monteagudo, S. Cianetti, C. Giuliano, V. Pucci, N. Trainor, C. M. Fandozzi, M. Rowley, P. J. Coleman, J. P. Vacca, V. Summa and N. J. Liverton, Discovery of MK-5172, a macrocyclic hepatitis C virus NS3/4a protease inhibitor, *ACS Med. Chem. Lett.*, 2012, **3**, 332–336.

45. V. Summa, S. W. Ludmerer, J. A. McCauley, C. Fandozzi, C. Burlein, G. Claudio, P. J. Coleman, J. M. Dimuzio, M. Ferrara, M. Di Filippo, A. T. Gates, D. J. Graham, S. Harper, D. J. Hazuda, C. McHale, E. Monteagudo, V. Pucci, M. Rowley, M. T. Rudd, A. Soriano, M. W. Stahlhut, J. P. Vacca, D. B. Olsen, N. J. Liverton and S. S. Carroll, MK-5172, a selective inhibitor of hepatitis C virus NS3/4a protease with broad activity across genotypes and resistant variants, *Antimicrob. Agents Chemother.*, 2012, **56**, 4161–4167.

46. E. Lawitz, I. Gaultier, F. Poordad, E. DeJesus, K. Kowdley, G. Sepulveda, D. Cohen, R. Menon, L. M. Larsen, T. J. Podsadecki and B. Bernstein, 4-week virologic response and safety of ABT-450 given with low-dose ritonavir (ABT-450r) in combination with pegylated intereron alfa-2a and ribavirin after 3-day monotherapy in genotype 1 HCV-infected treatment-naive subjects, *Hepatology*, 2010, **52**(Suppl), 878A–879A.

47. S. S. Good, X. R. Pan-Zhou, M. Larsson, S. Luo, J. R. Selden, H. Rashidzadeh, S. Bhadresa, M. Camire, C. Parsy and D. Surleraux, Preclinical pharmacokinetic profile of IDX320, a novel and potent HCV protease inhibitor, *J. Hepatol.*, 2010, **52**(Suppl. 1), S292.

48. J. de Bruijne, A. van Vliet, C. J. Weegink, W. Mazur, A. Wiercinska-Drapalo, K. Simon, G. Cholewinska-Szymanska, J. Kapocsi, I. Varkonyi, X.-J. Zhou, M.-F. Temam, J. Molles, J. Chen, K. Pietropaolo, J. F. McCarville, J. Z. Sullivan-Bolyai, D. Mayers and H. Reesink, Rapid decline of viral RNA in chronic hepatitis C patients treated once daily with IDX320: a novel macrocyclic HCV protease inhibitor, *Antiviral Ther.*, 2012, **17**, 633–642.

49. D. M. Hotho, J. de Bruijne, A. M. O'Farrell, T. Boyea, J. Li, M. Bracken, X. Li, D. Campbell, H.-P. Guler, C. J. Weegink, J. Schinkel, R. Molenkamp, J. van de Wetering de Ruij, A. van Vliet, H. L. Janssen, R. J. de Knegt and H. W. Reesink, Pharmacokinetics and antiviral activity of PHX1766, a novel HCV protease inhibitor, using an accelerated Phase I study design, *Antiviral Ther.*, 2012, **17**, 365–375.

50. D. Hotho, J. de Bruijne, A. O'Farrell, T. Boyea, J. Li, C. J. Weegink, J. Schinkel, R. Molenkamp, J. van de Wetering, A. A. van Vliet, H. L. Janssen, R. J. de Knegt and H. W. Reesink, Accelerated clinical trial design to assess the safety, tolerability and anti-viral activity of PHX1766, a novel HCV NS4/4A protease inhibitor, in healthy volunteers and chronic hepatitis C patients, *Hepatology*, 2009, **50**(Suppl), 1031A–1032A.

51. D. Lamarre, P. C. Anderson, M. Bailey, P. Beaulieu, G. Bolger, P. Bonneau, M. Boes, D. R. Cameron, M. Cartier, M. G. Cordingley, A.-M. Faucher, N. Goudreau, S. H. Kawai, G. Kukolj, L. Lagace,

S. R. LaPlante, H. Narjes, M.-A. Poupart, J. Rancourt, R. E. Sentjens, R. St. George, B. Simoneau, G. Steinmann, D. Thibeault, Y. S. Tsantrizos, S. M. Weldon, C.-L. Yong and M. Llinas-Brunet, An NS3 protease inhibitor with antiviral effects in humans infected with hepatitis C virus, *Nature*, 2003, **426**, 186–189.

52. M. Llinas-Brunet, M. D. Bailey, E. Ghiro, V. Gorys, T. Halmos, M. Poirier, J. Rancourt and N. Goudreau, A systematic approach to the optimization of substrate-based inhibitors of the hepatitis C virus NS3 protease: discovery of potent and specific tripeptide inhibitors, *J. Med. Chem.*, 2004, **47**, 6584–6594.

53. D. Noeteberg, J. Brnalt, I. Kvarnstroem, M. Linschoten, D. Musil, J.-E. Nystroem, G. Zuccarello and B. Samuelsson, New proline mimetics: synthesis of thrombin inhibitors incorporating cyclopentane- and cyclopentenedicarboxylic acid templates in the P(2) position. Binding conformation investigated by X-ray crystallography, *J. Med. Chem.*, 2000, **43**, 1705–1713.

54. F. Thorstensson, I. Kvarnstroem, D. Musil, I. Nilsson and B. Samuelsson, Synthesis of novel thrombin inhibitors. Use of ring-closing metathesis reactions for synthesis of P2 cyclopentene- and cyclohexenedicarboxylic acid derivatives, *J. Med. Chem.*, 2003, **46**, 1165–1179.

55. P.-O. Johansson, M. Baeck, I. Kvarnstroem, K. Jansson, L. Vrang, E. Hamelink, A. Hallberg, A. Rosenquist and B. Samuelsson, Potent inhibitors of the hepatitis C virus NS3 protease: use of a novel P2 cyclopentane-derived template, *Bioorg. Med. Chem.*, 2006, **14**, 5136–5151.

56. A. Rosenquist, I. Kvarnstroem, S. C. T. Svensson, B. Classon and B. Samuelsson, Synthesis of enantiomerically pure *trans*-3,4-substituted cyclopentanols by enzymic resolution, *Acta Chem. Scand.*, 1992, **46**, 1127–1129.

57. M. Llinas-Brunet, M. D. Bailey, D. Cameron, A.-M. Faucher, E. Ghiro, N. Goudreau, T. Halmos, M.-A. Poupart, J. Rancourt, Y. S. Tsantrizos, D. M. Wernic and B. Simoneau, Preparation of hepatitis C inhibitory tripeptides, *Patent Application*, WO2000-009543A2, 2000.

58. G. A. M. Giardina, H. M. Sarau, C. Farina, A. D. Medhurst, M. Grugni, L. F. Raveglia, D. B. Schmidt, R. Rigolio, M. Luttmann, V. Vecchietti and D. W. P. Hay, Discovery of a novel class of selective non-peptide antagonists for the human neurokinin-3 receptor. 1. Identification of the 4-quinolinecarboxamide framework, *J. Med. Chem.*, 1997, **40**, 1794–1807.

59. J. Rancourt, D. R. Cameron, V. Gorys, D. Lamarre, M. Poirier, D. Thibeault and M. Llinas-Brunet, Peptide-based inhibitors of the hepatitis C virus NS3 protease: structure–activity relationship at the C-terminal position, *J. Med. Chem.*, 2004, **47**, 2511–2522.

60. M. Llinas-Brunet, M. Bailey, G. Fazal, E. Ghiro, V. Gorys, S. Goulet, T. Halmos, R. Maurice, M. Poirier, M.-A. Poupart, J. Rancourt, D. Thibeault, D. Wernic and D. Lamarre, Highly potent and selective peptide-based inhibitors of the hepatitis C virus serine protease: towards smaller inhibitors, *Bioorg. Med. Chem. Lett.*, 2000, **10**, 2267–2270.

61. F. Thorstensson, F. Wangsell, I. Kvarnstroem, L. Vrang, E. Hamelink, K. Jansson, A. Hallberg, A. Rosenquist and B. Samuelsson, Synthesis of novel potent hepatitis C virus NS3 protease inhibitors: discovery of 4-hydroxycyclopent-2-ene-1,2-dicarboxylic acid as an *N*-acyl-ʟ-hydroxyproline bioisostere, *Bioorg. Med. Chem.*, 2007, **15**, 827–838.

62. M. Baeck, P.-O. Johansson, F. Waangsell, F. Thorstensson, I. Kvarnstroem, S. Ayesa, H. Waehling, M. Pelcman, K. Jansson, S. Lindstroem, H. Wallberg, B. Classon, C. Rydergaard, L. Vrang, E. Hamelink, A. Hallberg, A. Rosenquist and B. Samuelsson, Novel potent macrocyclic inhibitors of the hepatitis C virus NS3 protease: use of cyclopentane and cyclopentene P2-motifs, *Bioorg. Med. Chem.*, 2007, **15**, 7184–7202.

63. P. Raboisson, T.-I. Lin, H. de Kock, S. Vendeville, W. Van de Vreken, D. McGowan, A. Tahri, L. Hu, O. Lenz, F. Delouvroy, D. Surleraux, P. Wigerinck, M. Nilsson, A. Rosenquist, B. Samuelsson and K. Simmen, Discovery of novel potent and selective dipeptide hepatitis C virus NS3/4A serine protease inhibitors, *Bioorg. Med. Chem. Lett.*, 2008, **18**, 5095–5100.

64. A. Johansson, A. Poliakov, E. Akerblom, K. Wiklund, G. Lindeberg, S. Winiwarter, U. H. Danielson, B. Samuelsson and A. Hallberg, Acyl sulfonamides as potent protease inhibitors of the hepatitis C virus full-length NS3 (protease–helicase/NTPase): a comparative study of different C-terminals, *Bioorg. Med. Chem.*, 2003, **11**, 2551–2568.

65. Y. Tu, P. M. Scola, A. C. Good and J. A. Campbell, Preparation of prolinamide peptides as hepatitis C virus inhibitors, *Patent Application*, WO2005-054430A2, 2005.

66. J. Li, D. Smith, H. S. Wong, J. A. Campbell, N. A. Meanwell and P. M. Scola, A facile synthesis of 1-substituted cyclopropylsulfonamides, *Synlett*, 2006, 725–728.

67. P. Raboisson, H. de Kock, A. Rosenquist, M. Nilsson, L. Salvador-Oden, T.-I. Lin, N. Roue, V. Ivanov, H. Wahling, K. Wickstrom, E. Hamelink, M. Edlund, L. Vrang, S. Vendeville, W. Van de Vreken, D. McGowan, A. Tahri, L. Hu, C. Boutton, O. Lenz, F. Delouvroy, G. Pille, D. Surleraux, P. Wigerinck, B. Samuelsson and K. Simmen, Structure–activity relationship study on a novel series of cyclopentane-containing macrocyclic inhibitors of the hepatitis C virus NS3/4A protease leading to the discovery of TMC435350, *Bioorg. Med. Chem. Lett.*, 2008, **18**, 4853–4858.

68. T.-I. Lin, O. Lenz, G. Fanning, T. Verbinnen, F. Delouvroy, A. Scholliers, K. Vermeiren, A. Rosenquist, M. Edlund, B. Samuelsson, L. Vrang, H. de Kock, P. Wigerinck, P. Raboisson and K. Simmen, *In vitro* activity and preclinical profile of TMC435350, a potent hepatitis C virus protease inhibitor, *Antimicrob. Agents Chemother.*, 2009, **53**, 1377–1385.

69. S. Davies, TMC–435350. HCV NS3/4A protease inhibitor anti-hepatitis C virus Drug, *Drugs Future*, 2009, **34**, 545–554.

70. T.-I. Lin, B. Devogelaere, O. Lenz, O. Nyanguile, E. van der Helm, K. Vermeiren, G. Vandercruyssen, E. Cleiren, J. Lindberg, M. Edlund,

P. Raboisson, H. de Kock, M. Cummings, G. Fanning and K. Simmen, Inhibitory activity of TMC435350 on HCV NS3/4A proteases from genotypes 1 to 6, *Hepatology*, 2008, **48**, 1166A.

71. O. Lenz, T. Verbinnen, T.-I. Lin, L. Vijgen, M. D. Cummings, J. Lindberg, J. M. Berke, P. Dehertogh, E. Fransen, A. Scholliers, K. Vermeiren, T. Ivens, P. Raboisson, M. Edlund, S. Storm, L. Vrang, H. de Kock, G. C. Fanning and K. A. Simmen, *In vitro* resistance profile of the hepatitis C virus NS3/4A protease inhibitor TMC435, *Antimicrob. Agents Chemother.*, 2010, **54**, 1878–1887.

72. G. Koev and W. Kati, The emerging field of HCV drug resistance, *Expert Opin. Invest. Drugs*, 2008, **17**, 303–319.

73. P. J.-M. B. Raboisson, H. A. de Kock, L. Hu, S. M. H. Vendeville, A. Tahri, D. L. N. G. Surleraux, K. A. Simmen, K. M. Nilsson, B. B. Samuelsson, A. A. K. Rosenquist, V. Ivanov, M. Pelcman, A. K. G. L. Belfrage and P.-O. M. Johansson, Preparation of macrocyclic inhibitors of hepatitis C virus, *Patent Application*, WO2007-014926A1, 2007.

74. P. A. Bartlett and F. R. Green III, Total synthesis of brefeldin A, *J. Am. Chem. Soc.*, 1978, **100**, 4858–4865.

75. C. Meyer, O. Piva and J.-P. Pete, [2 + 2] photocycloadditions and photorearrangements of 2-alkenylcarboxamido-2-cycloalken-1-ones, *Tetrahedron*, 2000, **56**, 4479–4489.

76. J. S. Kingsbury, J. P. A. Harrity, P. J. J. Bonitatebus and A. H. Hoveyda, A recyclable Ru-based metathesis catalyst, *J. Am. Chem. Soc.*, 1999, **121**, 791–799.

77. M. D. Cummings, J. Lindberg, T.-I. Lin, H. de Kock, O. Lenz, E. Lilja, S. Fellander, V. Baraznenok, S. Nystrom, M. Nilsson, L. Vrang, M. Edlund, A. Rosenquist, B. Samuelsson, P. Raboisson and K. Simmen, Induced-fit binding of the macrocyclic noncovalent inhibitor TMC435 to its HCV NS3/NS4A protease target, *Angew. Chem. Int. Ed.*, 2010, **49**, 1652–1655.

78. H. W. Reesink, G. C. Fanning, K. A. Fartha, C. Weegink, A. Van Vliet, G. Van't Klooster, O. Lenz, F. Aharchi, K. Marien, P. Van Remoortere, H. de Kock, F. Broeckaert, P. Meyvisch, E. Van Beirendonck, K. Simmen and R. Verloes, Rapid HCV–RNA decline with once daily TMC435: a phase I study in healthy volunteers and hepatitis C patients, *Gastroenterology*, 2010, **138**, 913–921.

79. M. Manns, H. Reesink, T. Berg, G. Dusheiko, R. Flisiak, P. Marcellin, C. Moreno, O. Lenz, P. Meyvisch, M. Peeters, V. Sekar, K. Simmen and R. Verloes, Rapid viral response of once-daily TMC435 plus pegylated interferon/ribavirin in hepatitis C genotype-1 patients: a randomized trial, *Antiviral Ther.*, 2011, **16**, 1021–1033.

80. Y. S. Tsantrizos, G. Bolger, P. Bonneau, D. R. Cameron, N. Goudreau, G. Kukolj, S. R. LaPlante, M. Llinas-Brunet, H. Nar and D. Lamarre, Macrocyclic inhibitors of the NS3 protease as potential therapeutic agents of hepatitis C virus infection, *Angew. Chem. Int. Ed.*, 2003, **42**, 1356–1360.

81. S. R. LaPlante and M. Llinas-Brunet, Dynamics and structure-based design of drugs targeting the critical serine protease of the hepatitis C virus – from a peptidic substrate to BILN 2061, *Curr. Med. Chem. Anti-Infect. Agents*, 2005, **4**, 111–132.

82. D. Thibeault, M.-J. Massariol, S. Zhao, E. Welchner, N. Goudreau, R. Gingras, M. Llinas-Brunet and P. W. White, Use of the fused NS4A peptide–NS3 protease domain to study the importance of the helicase domain for protease inhibitor binding to hepatitis C virus NS3-NS4A, *Biochemistry*, 2009, **48**, 744–753.

83. N. Schiering, A. D'Arcy, F. Villard, O. Simic, M. Kamke, G. Monnet, U. Hassiepen, D. I. Svergun, R. Pulfer, J. Eder, P. Raman and U. Bodendorf, A macrocyclic HCV NS3/4A protease inhibitor interacts with protease and helicase residues in the complex with its full-length target, *Proc. Natl. Acad. Sci. U. S. A.*, 2011, **108**, 21052–21056.

84. N. Yao, P. Reichert, S. S. Taremi, W. W. Prosise and P. C. Weber, Molecular views of viral polyprotein processing revealed by the crystal structure of the hepatitis C virus bifunctional protease–helicase, *Structure (London)*, 1999, **7**, 1353–1363.

85. R. H. Grubbs, Olefin metathesis, *Tetrahedron*, 2004, **60**, 7117–7140.

86. R. H. Grubbs, Olefin-metathesis catalysts for the preparation of molecules and materials (Nobel Lecture), *Angew. Chem. Int. Ed.*, 2006, **45**, 3760–3765.

87. A. Arasappan, K. X. Chen, F. G. Njoroge, T. N. Parekh and V. Girijavallabhan, Novel dipeptide macrocycles from 4-oxo, -thio- and -amino-substituted proline derivatives, *J. Org. Chem.*, 2002, **67**, 3923–3926.

88. Strem Chemicals, 1,3-Bis(2,4,6-trimethylphenyl)-4,5-dihydroimidazol-2-ylidene[2-(isopropoxy)-5-(*N*,*N*-imethylaminosulfonyl)phenyl]methylene-ruthenium(II) dichloride, Zhan Catalyst-1B, CAS No. 918870-76-5, Catalog No. 44-0082, Strem Chemicals, Newburyport, MA.

89. P. L. Beaulieu, J. Gillard, M. D. Bailey, C. Boucher, J.-S. Duceppe, B. Simoneau, X.-J. Wang, L. Zhang, K. Grozinger, I. Houpis, V. Farina, H. Heimroth, T. Krueger and J. Schnaubelt, Synthesis of (1*R*,2*S*)-1-amino-2-vinylcyclopropanecarboxylic acid vinyl-ACCA) derivatives: key intermediates for the preparation of inhibitors of the hepatitis C virus NS3 protease, *J. Org. Chem.*, 2005, **70**, 5869–5879.

90. S.-S. Mao, J. DiMuzio, C. McHale, C. Burlein, D. Olsen and S. S. Carroll, A time-resolved, internally quenched fluorescence assay to characterize inhibition of hepatitis C virus nonstructural protein 3–4 Å protease at low enzyme concentrations, *Anal. Biochem.*, 2008, **373**, 1–8.

91. G. Migliaccio, J. E. Tomassini, S. S. Carroll, L. Tomei, S. Altamura, B. Bhat, L. Bartholomew, M. R. Bosserman, A. Ceccacci, L. F. Colwell, R. Cortese, R. De Francesco, A. B Eldrup., K. L. Getty, X. S. Hou, R. L. LaFemina, S. W. Ludmerer, M. MacCoss, D. R. McMasters, M. W. Stahlhut, D. B. Olsen, D. J. Hazuda and O. A. Flores, Characterization of resistance to non-obligate chain-terminating ribonucleoside analogs that inhibit hepatitis C virus replication *in vitro*, *J. Biol. Chem.*, 2003, **278**, 49164–49170.

92. Z. J. Song, D. M. Tellers, M. Journet, J. T. Kuethe, D. Lieberman, G. Humphrey, F. Zhang, Z. Peng, M. S. Waters, D. Zewge, A. Nolting, D. Zhao, R. A. Reamer, P. G. Dormer, K. M. Belyk, I. W. Davies, P. N. Devine and D. M. Tschaen, Synthesis of vaniprevir (MK-7009): lactamization to prepare a 20-membered macrocycle, *J. Org. Chem.*, 2011, **76**, 7804–7815.

93. T. Nicola, M. Brenner, K. Donsbach and P. Kreye, First scale-up to production scale of a ring closing metathesis reaction forming a 15-membered macrocycle as a precursor of an active pharmaceutical ingredient, *Org. Process Res. Dev.*, 2005, **9**, 513–515.

94. A. Marchetti, J. M. Ontoria and V. G. Matassa, Synthesis of two novel cyclic diphenyl ether analogs of an inhibitor of HCV NS3 protease, *Synlett*, 1999, 1000–1002.

95. K. X. Chen, F. G. Njoroge, J. Pichardo, A. Prongay, N. Butkiewicz, N. Yao, V. Madison and V. Girijavallabhan, Potent 7-hydroxy-1,2,3,4-tetrahydroisoquinoline-3-carboxylic acid-based macrocyclic inhibitors of hepatitis C virus NS3 protease, *J. Med. Chem.*, 2006, **49**, 567–574.

96. F. McPhee, J. A. Campbell, W. Li, S. D'Andrea, Z. B. Zheng, A. C. Good, D. J. Carini, B. L. Johnson and P. M. Scola, Preparation of macrocyclic isoquinoline peptide inhibitors of hepatitis C virus, *Patent Application*, WO2004-094452A2, 2004.

97. A. Ripka, J. A. Campbell, A. C. Good, P. M. Scola, N. Sin and B. Venables, Preparation of hydroxyprolinamide peptides as hepatitis C virus inhibitors, *US Patent Application*, 20040048802A1, 2004.

98. P. Van de Weghe and J. Eustache, The application of olefin metathesis to the synthesis of biologically active macrocyclic agents, *Curr. Top. Med. Chem.*, 2005, **5**, 1495–1519.

99. A. Gradillas and J. Perez-Castells, Macrocyclization by ring-closing metathesis in the total synthesis of natural products: reaction conditions and limitations, *Angew. Chem. Int. Ed.*, 2006, **45**, 6086–6101.

100. J. P. Vacca, M. T. Rudd, D. B. Olsen, C. J. McIntyre, J. A. McCauley, S. W. Ludmerer, N. J. Liverton and M. K. Holloway, to Merck and Co., Inc., HCV NS3 protease inhibitors, *US Patent*, 7 470 664, 2008.

101. H. Tsue, S. Nakashima, Y. Goto, H. Tatemitsu, S. Misumi, R. J. Abraham, T. Asahi, Y. Tanaka, T. Okada, N. Mataga and Y. Sakata, Synthesis of rigid porphyrin–quinone compounds for studying mutual orientation effects on electron transfer and their photophysical properties, *Bull. Chem. Soc. Jpn.*, 1994, **67**, 3067–3075.

102. R. A. Olofson, J. T. Martz, J. P. Senet, M. Piteau and T. Malfroot, A new reagent for the selective, high-yield *N*-dealkylation of tertiary amines: improved syntheses of naltrexone and nalbuphine, *J. Org. Chem.*, 1984, **49**, 2081–2082.

103. J. Kong, C.-y. Chen, J. Balsells-Padros, Y. Cao, R. F. Dunn, S. J. Dolman, J. Janey, H. Li and M. J. Zacuto, Synthesis of the HCV protease inhibitor vaniprevir (MK-7009) using ring-closing metathesis strategy, *J. Org. Chem.*, 2012, **77**, 3820–3828.

104. D. B. Olsen, M.-E. Davies, L. Handt, K. Koeplinger, N. R. Zhang, S. W. Ludmerer, D. Graham, N. Liverton, M. MacCoss, D. Hazuda and S. S. Carroll, Sustained viral response in a hepatitis C virus-infected chimpanzee via a combination of direct-acting antiviral agents, *Antimicrob. Agents Chemother.*, 2011, **55**, 937–939.

105. M. T. Rudd, J. A. McCauley, J. J. Romano, J. W. Butcher, K. Bush, C. J. McIntyre, K. T. Nguyen, K. F. Gilbert, T. A. Lyle, M. K. Holloway, B.-L. Wan, J. P. Vacca, V. Summa, S. Harper, M. Rowley, S. S. Carroll, C. Burlein, J. M. DiMuzio, A. Gates, D. J. Graham, Q. Huang, S. W. Ludmerer, S. McClain, C. McHale, M. Stahlhut, C. Fandozzi, A. Taylor, N. Trainor, D. B. Olsen and N. J. Liverton, Development of potent macrocyclic inhibitors of genotype 3a HCV NS3/4A protease, *Bioorg. Med. Chem. Lett.*, 2012, **22**, 7201–7206.

106. R. H. Reitsema, The chemistry of 4-hydroxyquinolines, *Chem. Rev.*, 1948, **43**, 47.

107. A. Knierzinger and O. S. Wolfbeis, Syntheses of fluorescent dyes. IX. New 4-hydroxycoumarins, 4-hydroxy-2-quinolones, 2*H*,5*H*-pyrano[3,2-*c*]benzopyran-2,5-diones and 2*H*,5*H*-pyrano[3,2-*c*]quinoline-2,5-diones, *J. Heterocycl. Chem.*, 1980, **17**, 225–229.

108. Y. Tagawa, T. Kawaoka and Y. Goto, A convenient preparation of 4,8-dimethoxy-3-substituted-2(1*H*)-quinolones by an electrophilic reaction through base-induced deprotonation and its synthetic application for the synthesis of new alkaloids, 3,4,8-trimethoxy-2(1*H*)-quinolone and 3-formyl-4,7,8-trimethoxy-2(1*H*)-quinolone (glycocitridine), *J. Heterocycl. Chem.*, 1997, **34**, 1677–1683.

109. M. K. Holloway, N. J. Liverton, S. W. Ludmerer, J. A. McCauley, D. B. Olsen, M. T. Rudd and J. P. Vacca, Preparation of macrocyclic compounds as HCV NS3 protease inhibitors, *Patent Application*, WO2006-119061A2, 2006.

110. G. A. Molander and N. Ellis, Organotrifluoroborates: protected boronic acids that expand the versatility of the Suzuki coupling reaction, *Acc. Chem. Res.*, 2007, **40**, 275–286.

111. N. Miyaura and A. Suzuki, Palladium-catalyzed cross-coupling reactions of organoboron compounds, *Chem. Rev.*, 1995, **95**, 2457–2483.

112. P. Beak, J. B. Covington, S. G. Smith, J. M. White and J. M. Zeigler, Displacement of protomeric equilibriums by self-association: hydroxy-pyridine–pyridone and mercaptopyridine–thiopyridone isomer pairs, *J. Org. Chem.*, 1980, **45**, 1354–1362.

113. D. L. Hughes, The Mitsunobu reaction, *Org. React.*, 1992, **42**, 335–656.

114. R. Gonzalez, M. T. Ramos, E. de la Cuesta and C. Avendano, Base-catalyzed electrophilic substitution in 2(1*H*)-quinolinones, *Heterocycles*, 1993, **36**, 315–322.

115. The R155K mutant arises clinically in gt1a-infected patients; however, Merck chose to screen against the gt1b mutant such that the same background was utilized in relation to the main screening genotype 1

subtype. Where comparison data exist for both gt1a and 1b R155K, the K_i values are similar. C. Sarrazin, T. L. Kieffer, D. Bartells, B. Hanzelka, U. Muh, M. Welker, D. Wincheringer, Y. Zhou, H.-M. Chu, C. Lin, C. Weegink, H. Reesink, S. Zeuzem and A. D. Kwong, Dynamic Hepatitis C Virus Genotypic and Phenotypic Changes in Patients Treated With the Protease Inhibitor Telaprevir, *Gastroenterology*, 2007, **132**, 1767–1777.

116. M. T. Rudd, J. A. McCauley, J. W. Butcher, J. J. Romano, C. J. McIntyre, K. T. Nguyen, K. F. Gilbert, K. J. Bush, M. K. Holloway, J. Swestock, B.-L. Wan, S. S. Carroll, J. M. DiMuzio, D. J. Graham, S. W. Ludmerer, M. W. Stahlhut, C. M. Fandozzi, N. Trainor, D. B. Olsen, J. P. Vacca and N. J. Liverton, Discovery of MK-1220: a macrocyclic inhibitor of hepatitis C virus NS3/4A protease with improved preclinical plasma exposure, *ACS Med. Chem. Lett.*, 2011, **2**, 207–212.

117. M. P. Manns, E. J. Gane, M. Rodriguez-Torres, A. D. Stoehr, C. T. Yeh, P. Marcellin, R. T. Wiedmann, P. Hwang, R. J. Barnard and A. W. Lee, Sustained viral response (SVR) rates in genotype 1 treatment-naive patients with chronic hepatitis C (CHC) infection treated with vaniprevir (MK-7009), an NS3/4a protease inhibitor, in combination with pegylated interferon alfa-2a and ribavirin for 28 days, *Hepatology*, 2010, **52**(Suppl), 361a.

118. M. T. Rudd, C. J. McIntyre, J. J. Romano, J. W. Butcher, M. K. Holloway, K. Bush, K. T. Nguyen, K. F. Gilbert, T. A. Lyle, N. J. Liverton, B.-L. Wan, V. Summa, S. Harper, M. Rowley, J. P. Vacca, S. S. Carroll, C. Burlein, J. M. DiMuzio, A. Gates, D. J. Graham, Q. Huang, S. W. Ludmerer, S. McClain, C. McHale, M. Stahlhut, C. Fandozzi, A. Taylor, N. Trainor, D. B. Olsen and J. A. McCauley, Development of macrocyclic inhibitors of HCV NS3/4A protease with cyclic constrained P2–P4 linkers, *Bioorg. Med. Chem. Lett.*, 2012, **22**, 7207–7213.

119. K. X. Chen and F. G. Njoroge, A review of HCV protease inhibitors, *Curr. Opin. Invest. Drugs*, 2009, **10**, 821–837.

120. ChemAxon, *MarvinSketch*, Version 3.5.2, ChemAxon Ltd., Budapest, 2006.

121. E. Monteagudo, M. Fonsi, X. Chu, K. Bleasby, R. Evers, V. Pucci, M. V. Orsale, S. Cianetti, M. Ferrara, S. Harper, R. Laufer, M. Rowley and V. Summa, The metabolism and disposition of a potent inhibitor of hepatitis C virus NS3/4A protease, *Xenobiotica*, 2010, **40**, 826–839.

122. C. A. Lipinski, Lead- and drug-like compounds: the rule-of-five revolution, *Drug Discov. Today: Technol.*, 2004, **1**, 337–341.

123. M. P. Gleeson, Generation of a set of simple, interpretable ADMET rules of thumb, *J. Med. Chem.*, 2008, **51**, 817–834.

124. P. D. Leeson and B. Springthorpe, The influence of drug-like concepts on decision-making in medicinal chemistry, *Nat. Rev. Drug Discov.*, 2007, **6**, 881–890.

125. M. J. Waring, Lipophilicity in drug discovery, *Expert Opin. Drug Discov.*, 2010, **5**, 235–248.

CHAPTER 8

Design and Development of NS5B Polymerase Non-nucleoside Inhibitors for the Treatment of Hepatitis C Virus Infection

PIERRE L. BEAULIEU

Boehringer Ingelheim (Canada) Ltd., 2100 Cunard Street, Laval, Québec, Canada, H7S 2G5
Email: resgeneral.lav@boehringer-ingelheim.com

8.1 Introduction

Hepatitis C virus (HCV) infects an estimated 170–200 million people worldwide and is associated with life-threatening liver diseases such as cirrhosis and hepatocellular carcinomas. Although transmission of the infection at present is mostly limited to unsafe medical practices and the use of illicit injectable drugs, the number of individuals diagnosed with HCV infection is expected to increase in the future. It is a growing concern for healthcare authorities; in the USA alone, more than 15 000 people are expected to succumb yearly to conditions associated with HCV infection and 3–5 million new cases are documented worldwide. The viral infection has become the leading cause of liver transplantation in industrialized nations.[1]

RSC Drug Discovery Series No. 32
Successful Strategies for the Discovery of Antiviral Drugs
Edited by Manoj C. Desai and Nicholas A. Meanwell
© The Royal Society of Chemistry 2013
Published by the Royal Society of Chemistry, www.rsc.org

HCV, initially uncovered more than 20 years ago[2] as the etiological agent responsible for non-A/non-B hepatitis, is a member of the (+)-RNA virus Flaviviridae (*Hepacivirus* genus) and occurs as seven major genotypes, which are comprised, in turn, of several subtypes and a multitude of quasi-species.[3] Genotypes 1a/1b are predominant, particularly in America, Europe and Japan. Until recently, the HCV standard of care (SOC) was limited to a combination of subcutaneously administered pegylated interferon-α (PegIFN) and an oral broad-spectrum antiviral, ribavirin (RBV). This treatment has suboptimal efficacy, particularly in the genotype 1 patient population (45–50%), and suffers from severe side effects and contraindications.[4] As scientists have gained understanding of the HCV life cycle, several attractive targets have emerged for alternative therapeutic intervention.[5] Protease inhibitors (PIs) rapidly took the lead as direct-acting antivirals (DAAs) and in 2011 two agents, Vitrelis from Merck and Incivek from Vertex, provided patients with improved treatment options, raising efficacy up to 75% (as measured by sustained virologic response, SVR) when administered in combination with PegIFN and RBV to treatment-naive patients.[6] Unfortunately, owing to its error-prone polymerase and the lack of a proof-reading mechanism, the use of single DAAs, even in combination with immunotherapy, results in the rapid emergence of resistant virus. Furthermore, these two drugs have not addressed tolerance issues, as specific drug-related side effects have merely reinforced those associated with the use of PegIFN/RBV. In the near future, more potent second-generation PIs (*e.g.*, faldaprevir and simeprevir) currently in Phase 3 clinical trials, in combination with PegIFN/RBV, may be expected to provide simplified dosing regimens and further improvement over current options.[7]

The virally encoded NS5B RNA-dependent RNA polymerase (RdRp) is a vital component of the replicase complex that orchestrates the replication process leading to the production of progeny virus.[8] The RdRp also attracted early interest from the research community and drug developers, since it does not have a mammalian counterpart and inhibition of this target is not expected to cause host toxicity. Furthermore, much precedent was available from HIV research to suggest that inhibition of polymerase activity offered an attractive opportunity for therapeutic intervention. Indeed, many classes of NS5B inhibitors (*e.g.*, nucleoside or nucleotide analogs, allosteric inhibitors) have progressed into clinical trials in recent years and have shown promising efficacy in chronically infected HCV patients.[9] However, in the case of non-nucleoside inhibitors (NNIs), the rapid emergence of resistant virus has hampered progress of these agents in monotherapy or in combination with SOC. Highly successful HIV HAART strategies, if applied to HCV, suggest, however, that a combination of DAAs targeting complementary and essential viral functions should provide a powerful approach to address the resistance issue. NS5B inhibitors are therefore expected to become an important component in the future of more effective and better tolerated anti-HCV regimens. Finally, a recent paradigm shift towards the development of interferon-free therapies using combinations of DAAs with complementary modes of action (*e.g.*, PI + NS5B or NS5A and also NS5B + NS5A combinations) is envisaged to provide efficacy with improved safety and tolerability, while minimizing emergence of resistant virus.[10] NS5B polymerase inhibitors are therefore expected to play a key role in the future treatment of chronic HCV infection.

8.2 The NS5B RNA-dependent RNA Polymerase

NS5B's primary function is to synthesize a virion's complementary negative RNA strand, which it then uses as a template to produce (+)-RNA for further translation into viral proteins necessary for replication and virion morphogenesis.[8] As previously mentioned, NS5B exhibits low fidelity and is responsible for the broad heterogeneity observed in circulating virus and, hence, occurrence of drug-resistant viral strains. The 65 kDa protein shares the common folds of other nucleic acid polymerases with characteristic thumb, finger and palm domains, with two-finger loop residues enclosing the active site through interactions with the thumb domain (Figure 8.1).[11] Recently, the crystal structure of an NS5B construct bound to an RNA primer–template has shed some insight into structural requirements for substrate recognition and RNA

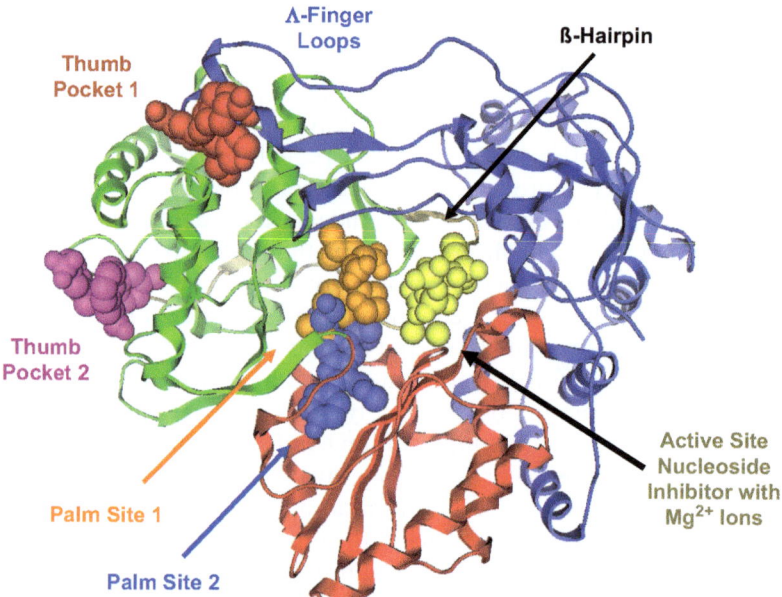

Figure 8.1 Three-dimensional structure of NS5B with allosteric inhibitor binding sites. The three-dimensional structure of NS5B is shown in ribbon representation with the palm domain colored red, the thumb domain green and the finger domain with the Λ finger loops blue. Inhibitor binding sites are depicted using CPK models of representative inhibitors: thumb pocket 1 (indoleacetamidecarboxylic acid; red), thumb pocket 2 (phenyl-alanine derivative; magenta), palm site 1 (acylpyrrolidine; orange), palm site 2 (benzofuran derivative; blue), active site (nucleoside analog with two Mg^{2+} ions; yellow). Copyright 2009, Informa Healthcare.[9a]

synthesis.[11b] Like HIV-RT, HCV NS5B can be inhibited by nucleoside and nucleotide analogs (particularly those carrying 2'-ribose substitution) that are anabolized to their triphosphate derivatives and are incorporated by the enzyme to act as non-obligate chain terminators.[9] In addition, however, as drug companies began probing their corporate compound collections for potential NS5B inhibitors, a surprising array of structurally diverse chemotypes began to emerge that were shown through reversed genetics and resistance signatures or X-ray crystallography to bind to one of four distinct allosteric sites.[9] These non-nucleoside inhibitors (NNI), which bind in allosteric sites referred to as thumb pockets 1 and 2 and palm sites 1 and 2 depending on their location (Figure 8.1), interfere with protein conformational movements, which are thought to be necessary for RNA synthesis.

NNIs from all four allosteric sites have progressed into the clinic and demonstrated antiviral potency in HCV-infected patients, thus validating the mechanism by which they interfere with NS5B-mediated RNA synthesis. To date, more than 430 patent applications have been filed on structurally diverse molecules that claim to inhibit NS5B polymerase, qualifying this protein as the most 'druggable' HCV target. Recently, the topics of both nucleoside and non-nucleoside HCV NS5B inhibitors was comprehensively reviewed.[9]

This chapter does not attempt to provide exhaustive coverage of all published classes of NS5B inhibitors; rather, it focuses on successful drug design strategies that have led to the identification and the development of drug candidates from all four classes of allosteric NNIs (nucleoside and nucleotide inhibitors are discussed in chapter 12). The discussion will be structured according to inhibitor binding site.

8.3   Non-nucleoside NS5B Polymerase Inhibitors

8.3.1   Thumb Pocket 1 Inhibitors

8.3.1.1   Discovery of BILB 1941 and BI 207127

Approximately 12 years ago, Japan Tobacco and Boehringer Ingelheim (Canada) published consecutive patent applications disclosing closely related benzimidazole-5-carboxamide derivatives as specific inhibitors of HCV NS5B polymerase.[12] The initial hits (**1** and **2**, Figure 8.2) displayed similar potencies and were discovered through screening of proprietary corporate sample collections.[13]

Interestingly, optimization of these hits followed divergent paths. The observation that carboxylic acid derivatives were slightly more potent than amide derivatives (*e.g.*, **3**), led Japan Tobacco to focus their structure–activity relationship (SAR) work on optimization of the left-hand side C2 substitution of the benzimidazole-5-carboxylic acids.[13a] On the other hand, NMR-based techniques, such as differential line broadening (DLB), were used at Boehringer Ingelheim to identify portions of the initial hit that appeared to be interacting with the target protein. The data suggested that only portions of the hit's

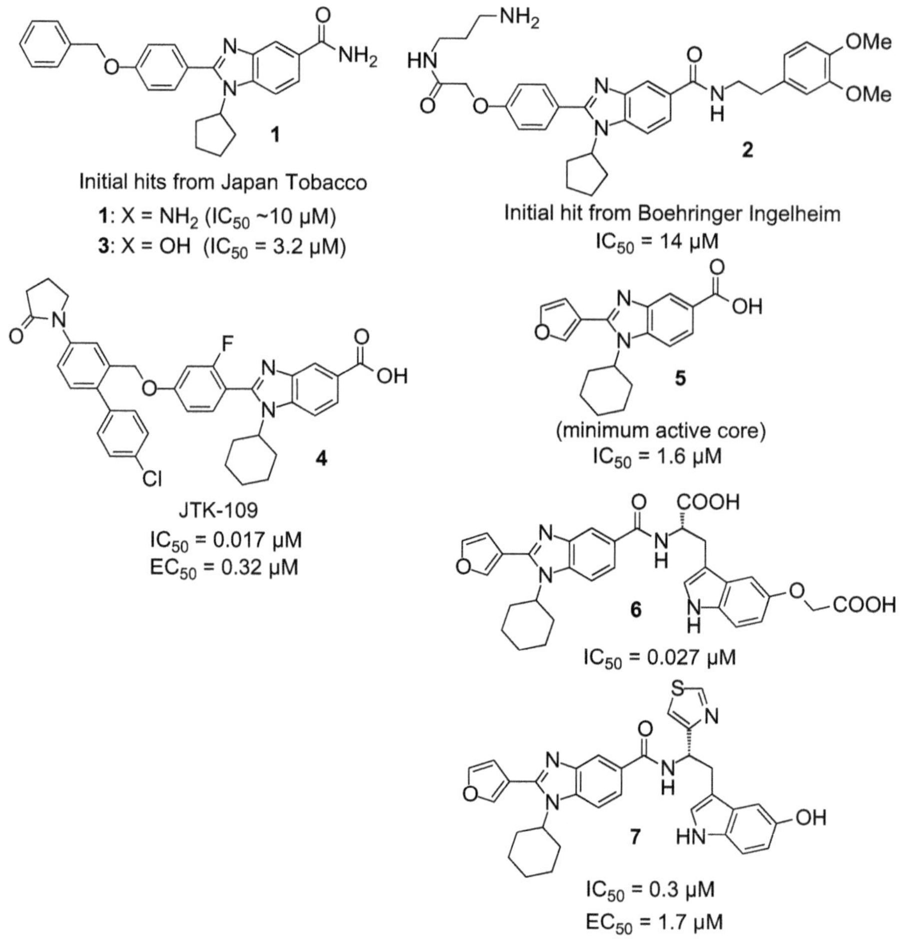

Figure 8.2 Non-nucleoside benzimidazole inhibitors from Japan Tobacco and Boehringer Ingelheim.

left-hand side aryloxy substituent interacted with protein. Guided by this information and using high-throughput-directed libraries, small benzimidazole-5-carboxylic acid analog **5** was initially established as the minimum core for biochemical activity.[13b] At the time these discoveries were made, cell culture systems were not yet available to establish whether the mechanism by which these benzimidazole derivatives were inhibiting NS5B function *in vitro* was relevant to HCV replication. Characterization of the biochemical profile of **2** indicated that inhibition by this compound was non-competitive with the process of nucleotide triphosphate incorporation on to the nascent RNA chain. Order of addition experiments suggested that the compounds lacked the ability to inhibit processive chain elongation, instead inhibiting the initiation phase of HCV polymerase activity.[14]

Optimization proceeded at Boehringer Ingelheim with analogs such as **6**, in which the minimum core **5** was coupled to various amino acids to improve potency and address solubility issues associated with neutral amide substituents (*e.g.*, **2**).[15] The potency of analogs such as **6** allowed the design of a photo-affinity probe, which was used in cross-linking experiments with NS5B to establish that the binding site of benzimidazole inhibitors was located in the thumb region of the polymerase in proximity to Λ1 finger loop residues and a postulated regulatory GTP-binding site.[11a,16] This was the first piece of evidence that suggested the presence of an allosteric site at the top of the thumb domain. Intriguingly, however, the presumed bioactive conformation of inhibitors such as **6** could not be docked in this region of the enzyme, as it lacked suitable binding pockets in the *apo*-enzyme structure.[17]

In early, 2000, Japan Tobacco selected a compound for clinical evaluation (JTK-003, structure undisclosed) that presumably belonged to the benz-imidazole class. The outcome of Phase 2 trials with this drug were never published, but around the same time, a cell-based replicon system was developed that reconstituted replication of subgenomic HCV RNA in cell culture.[18] The best inhibitors reported at the time by Japan Tobacco exhibited only weak antiviral potency in the replicon ($\sim 1 \, \mu M$), which could explain in part why the development of JTK-003 was discontinued. SAR studies were extended to analogs bearing large lipophilic biphenyl substituents on the left-hand side, one of which (JTK-109, **4**, Figure 8.2) inhibited the 1b replicon at submicromolar concentration ($EC_{50} = 0.32 \, \mu M$) but with a high EC_{50}/IC_{50} ratio of 19. Nevertheless, the compound displayed favorable pharmacokinetic (PK) and safety profiles and also good plasma-to-liver distribution in rats.[19] In 2003, JTK-109 was reported to have initiated a Phase 1 clinical trial, but the outcome was never reported. Development was likely stopped due to still modest replicon potency and unsatisfactory antiviral clinical efficacy, safety and/or unsatisfactory pharmacokinetics.

In Boehringer Ingelheim's series, the presence of two ionizable carboxylic acid functions in compounds such as **6** prevented cell permeability and activity in the replicon system. Replacement of one of the carboxylic acid functions, which was shown to act as an orienting rather than binding group[17] by a small heterocycle, provided derivatives such as **7**, which were active in cell culture at micromolar concentrations and provided the possibility for further optimization.[20]

Inhibitors such as **7** exhibited low solubility and high lipophilicity, which compromised their ability to progress as drug candidates. As a result, an effort was initiated to identify alternative right-hand sides with potential to confer superior drug-like properties. Figure 8.3 summarizes the knowledge that had been acquired so far on benzimidazole inhibitors (structure **8**), pointing out the role of some of the structural features embedded in the molecules.

Retaining the cyclohexyl ring as a potency anchor and the right-hand side carboxylic acid as a solubilizing group, linkers between the two were explored using high-throughput synthesis techniques. Ultimately, an (*R*)-alanine–cinnamic acid conjugate (*e.g.*, **9**) provided inhibitors with promising

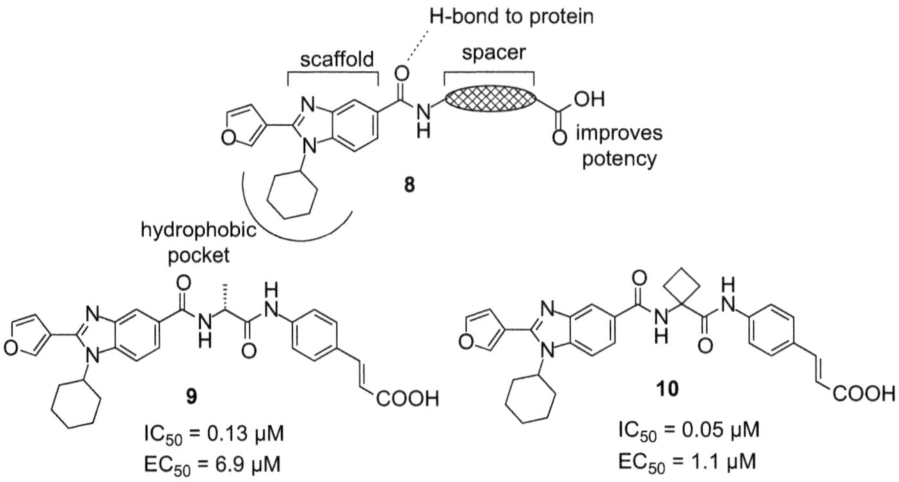

Figure 8.3   Discovery of benzimidazole inhibitors with novel diamide right-hand sides.

intrinsic and replicon potency, which upon further optimization led to achiral derivatives such as **10** with excellent intrinsic activity (IC$_{50}$ = 50 nM), comparable antiviral potency to inhibitor **7** and improved physicochemical properties.[21] Diamide derivatives similar to **10** were used to select and characterize resistant variants using the replicon system. Gratifyingly, resistant mutations that emerged with this class of benzimidazole diamides were found to localize in the NS5B thumb domain and encoded amino acid substitutions at positions P495, P496 and V499, proximal to the region previously described in cross-linking experiments. Inhibitors were most sensitive to P495 mutants, which conferred the highest shifts in single-mutant replicon sensitivity (>80-fold shift).[16]

A key finding for the program occurred when efforts to improve permeability and cell culture potency led to the replacement of the benzimidazole scaffold with a more lipophilic indole isostere. Unexpectedly, this modification improved both intrinsic ands cell-based potency by two orders of magnitude, as shown in Figure 8.4 for the *N*-methylindole-6-carboxylic acid **11**, suggesting a more optimal interaction of the lipophilic indole nucleus compared with the more polar benzimidazole version. At this time, benzimidazole-based NS5B inhibitors had aroused interest from other companies and the IRBM-Merck group began reporting on similar observations.[22]

In addition to providing leads with greatly improved potency, the availability of an easily accessible position on the scaffold to explore new interactions with the protein quickly led to the simultaneous discovery by the Boehringer Ingelheim and IRBM-Merck groups of indole-*N*-acetamide derivatives such as **12** and **13**. An unexpected benefit from this class of inhibitors came about when both groups succeeded in generating X-ray data on NS5B–indoleacetamide complexes, revealing for the first time the exact location of the

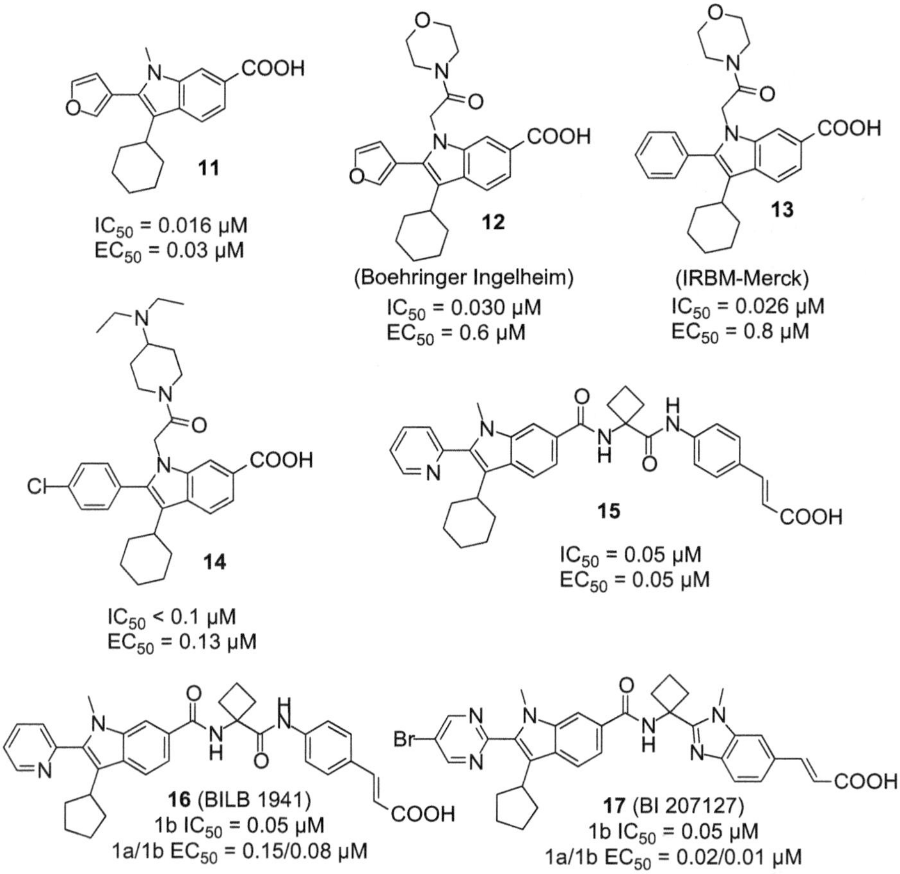

11

IC$_{50}$ = 0.016 μM
EC$_{50}$ = 0.03 μM

12

(Boehringer Ingelheim)
IC$_{50}$ = 0.030 μM
EC$_{50}$ = 0.6 μM

13

(IRBM-Merck)
IC$_{50}$ = 0.026 μM
EC$_{50}$ = 0.8 μM

14

IC$_{50}$ < 0.1 μM
EC$_{50}$ = 0.13 μM

15

IC$_{50}$ = 0.05 μM
EC$_{50}$ = 0.05 μM

16 (BILB 1941)
1b IC$_{50}$ = 0.05 μM
1a/1b EC$_{50}$ = 0.15/0.08 μM

17 (BI 207127)
1b IC$_{50}$ = 0.05 μM
1a/1b EC$_{50}$ = 0.02/0.01 μM

Figure 8.4 From benzimidazole to indole-based thumb pocket inhibitors.

benzimidazole/indole allosteric binding site and shedding light on a plausible mechanism of inhibition.[23] All previous efforts to co-crystallize or soak NS5B constructs with benzimidazole inhibitors had met with frustrating failure due to crystal cracking and poor diffraction data. These observations suggested possible protein conformational changes upon inhibitor binding, consistent with an allosteric inhibition mechanism. The structure of an indole-*N*-acetamide inhibitor bound to NS5B polymerase is shown in Figure 8.5.

Most striking is the loss of electron density for the Λ1 finger loop upon inhibitor binding, the displacement of which reveals a well-defined binding site, previously occupied by lipophilic side chains of loop residues. The cyclohexyl ring of the inhibitor occupies most of that lipophilic pocket, while the scaffold is stacked against the well-conserved proline residues (P495 and P496), providing a rationale for the preference of the indole scaffold over the more polar benzimidazole system. In addition, the carboxylic acid of indole-based inhibitors interacts through a salt bridge with the basic guanidinium side chain

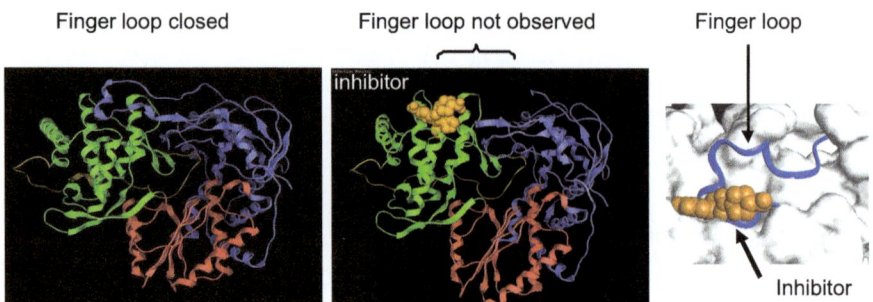

Figure 8.5   Apo NS5B enzyme (left) is depicted in ribbon form with the palm, thumb and finger domains shown in red, green and blue, respectively. The *apo* form shows an enclosed active site tunnel with the Λ finger loops protruding from the finger domain and interacting with the thumb domain. Upon binding of an inhibitor from the indoleacetamide class (center), the inhibitor (yellow) displaces the Λ finger loop and occupies the liberated binding pocket (thumb pocket 1). Finger loops are no longer visible in the structures as a result of increased mobility. An overlap between the inhibitor and a portion of the Λ finger loop (right) shows how they compete for the same binding space on the thumb domain. Copyright 2006, Thomson Reuters.[101]

of R503. The structural data were consistent with previously described cross-linking data and resistant-replicon selection experiments. Furthermore, hypotheses were beginning to emerge on a plausible mechanism of action for this class of allosteric NS5B inhibitors. The overlap between the inhibitor and the Λ1 loop binding sites suggested that inhibitors prevent closure of the finger loop on to the thumb domain and the formation of an enclosed active site tunnel, which may be necessary for RNA synthesis. Compounds from this class are now commonly referred to as 'finger loop inhibitors' and the allosteric binding site as 'thumb pocket 1.'

Optimization of indole-*N*-acetamide derivatives proceeded with matrices of compounds exploring various indole C2 substituents and acetamide combinations, eventually leading to very potent zwitterionic derivatives.[24] Following optimization of plasma protein binding/serum shift, off-target activity (*e.g.*, PXR activation associated with the nature of the acetamide side chain) and *in vivo* PK parameters, the IRBM-Merck group identified analog **14** (Figure 8.4), which, despite low bioavailability in rats, displayed improved properties in monkeys and dogs and was predicted to be a low-clearance drug in humans. This class of compound had only weak potential to form reactive acylglucuronide metabolites that could lead to toxicity and idiosyncratic reactions. The potential of **14** for further development was highlighted, but there have not been reports of this compound progressing into clinical development.[25]

At Boehringer Ingelheim, most of the focus remained on diamide derivatives (Figure 8.3). The excellent intrinsic potency of **10** was rationalized by NMR studies which established the ability of α,α-disubstituted amino acid linkers to induce an L-shaped bioactive conformation of ligands in the free state, which complemented the protein bioactive binding pocket as revealed in X-ray structures.[26] While diamide derivatives such as **10** were excellent inhibitors of the polymerase *in vitro*, their modest potency in the cell-based replicon compromised further progression of the benzimidazole class. Replacement of the benzimidazole left-hand side of compounds, such as **10,** by the isosteric *N*-methylindole **11** depicted in Figure 8.4, increased cell activity by 20–30-fold. A large array of indole diamides bearing variations at C2, amino acid linker and the right-hand side capping group were subsequently synthesized, eventually leading to the discovery of **15** (Figure 8.5).[27] Compound **15** displayed comparable enzymatic and cell-based potency (IC_{50} and $EC_{50} \approx 50$ nM), reflecting improved cellular permeability compared with benzimidazole analogs. The C2 furyl substituent was replaced with a more drug-like 2-pyridyl moiety, which also provided a significant decrease in lipophilicity and improved solubility.

Whereas Caco-2 permeability for this compound was excellent and predictive of good oral absorption (Caco-$2_{A \rightarrow B} = 39 \times 10^{-6}$ cm s^{-1}), metabolic stability upon incubation with human and rat liver microsomes was only modest (HLM/RLM $t_{\frac{1}{2}} = 50$ and 17 min, respectively) and compounds in this series showed disappointing oral pharmacokinetics in rats.[27] As poor plasma exposures following oral dosing were in part due to rapid clearance of compound, metabolite ID studies were performed and hydroxylation of the cyclohexyl ring was identified as a central metabolic pathway for 3-cyclohexylindole analogs.[28] Fortunately, replacement of the cyclohexyl moiety with the less lipophilic cyclopentyl group provided compounds with significantly improved *in vitro* metabolic stability and only modest decrease in potency.

High-throughput parallel synthesis was used once more to provide a sparse matrix of 3-cyclopentylindole inhibitors that harbored modifications at C2, the central α,α-disubstituted amino acid and the right-hand side acidic capping group. Out of 110 combinations, 84 compounds were identified with $EC_{50} < 150$ nM in the cell-based replicon. Following Caco-2 permeability profiling and metabolic stability assessment in human and rat liver microsomes, 71 compounds were selected for oral absorption studies in rats. This was accomplished through screening of compounds in cassette mode, where mixtures of 3–4 compounds were orally administered to animals and plasma exposure was measured at 1 and 2 h time points. This rapid screening provided a short list of compounds that were then evaluated as single compounds in rats, dogs and monkeys, culminating with the discovery of BILB 1941 (**16**, Figure 8.4) as a development candidate.[28] BILB 1941 is a specific and reversible inhibitor of HCV NS5B polymerase with good cell-based potency in gt1a/1b replicons ($EC_{50} = 153$ and 84 nM, respectively). BILB 1941 displays an attractive pan-genotype profile in chimeric replicon assays with moderate (up to threefold) shifts in genotypes 3a, 4a, 5a and 6a, while a more significant shift (14–32-fold) was observed for gt2a/2b.[29] BILB 1941 is highly protein bound (98–99%), but a modest threefold EC_{50} shift was measured in replicon assays performed in the presence of serum. The PK profile in rats, monkeys and dogs was favorable, with clearances of 12, 21 and 6% Q_h (% hepatic blood flow) and good bioavailability (54–70 %). The compound was favorably distributed to the liver target organ in rats with a liver/plasma ratio of ~9.

Based on its preclinical and safety profile, BILB 1941 was evaluated in 5 day multiple rising dose monotherapy in gt1-HCV-infected patients, where it produced significant (up to 2.5$\log_{10}$ reduction in viremia) and HCV subtype-dependent antiviral activity when dosed at 450 mg q8h.[30] Unfortunately, increased virological response was limited by gastrointestinal intolerance, which precluded testing of the compound at higher doses. This proof-of-concept trial suggested that more potent analogs could result in a more consistent antiviral response, and efforts at Boehringer Ingelheim were subsequently focused on identifying analogs with improved potency against both gt1a and gt1b replicons, while maintaining a favorable PK profile. These efforts culminated in the discovery of a follow-up compound, BI 207127 (**17**, Figure 8.4; 1b/1a $EC_{50} = 11$ and 23 nM), which maintained replicon potency <100 nM across genotypes 1–6.[31] BI 207127 given as monotherapy at doses up to 1200 mg q8h for 5 days exhibited strong and dose-dependent antiviral activity (>3$\log_{10}$) and a low frequency for emergence of resistance mutations in gt1-HCV-infected patients.[32] In combination with PegIFN-α2a and ribavirin, BI 207127 decreased HCV RNA in gt1-HCV patients in a dose-dependent manner [5.6$\log_{10}$ in treatment-naïve (TN) and 4.2$\log_{10}$ in treatment-experienced (TE) patients] at a dose of 600 mg administered tid for a period of 28 days. Breakthroughs were not observed in the TN group and 3/30 TE patients experienced rebound due to a P495 mutation.[33] BI 207127 has also been investigated in a Phase 2 trial in an interferon-free combination with protease inhibitor faldaprevir and ribavirin. In a 4 week study with BI 207127 (600 mg tid) + faldaprevir (120 mg qd) + RBV (1000–1200 mg d^{-1}), 100% of patients achieved a lower limit of quantification (25 IU mL^{-1}) independent of gt1-HCV subtype, and the combination was generally well tolerated. In a larger Phase 2b study of faldaprevir (120 mg qd) + BI 207127 (600 mg bid) + RBV, 68% SVR_{12} was achieved after 28 weeks of treatment in all gt1 TN patients; this included an 82% SVR_{12} among all gt1b (IL28B: CC and non-CC) and gt1a (IL28B: CC only).[34b] The data support further evaluation of these two drugs in interferon-free regimens for the treatment of HCV infection, including patients with cirrhosis and phase 3 trials are currently ongoing.[34]

In 2006, Japan Tobacco reported the discovery of a new series of conformationally constrained tetracyclic indolecarboxylic acid derivatives with improved potency. The novel thumb pocket 1 inhibitors exhibited low serum shifts in replicon assays performed in the presence of human serum albumin.[35] Installation of a bridge between the indole nitrogen and the *ortho*-position of an aromatic ring at C2 provided rigidification of the dihedral angle between the indole scaffold and the C2 substituent. The optimal three-atom tether and dihedral angles approximating 50° (similar to the bioactive conformation revealed in the crystal structure of a tetracyclic analog in complex with NS5B) provided up to 10-fold gains in biochemical potency (see, for example, **18**, Figure 8.6).[35a]

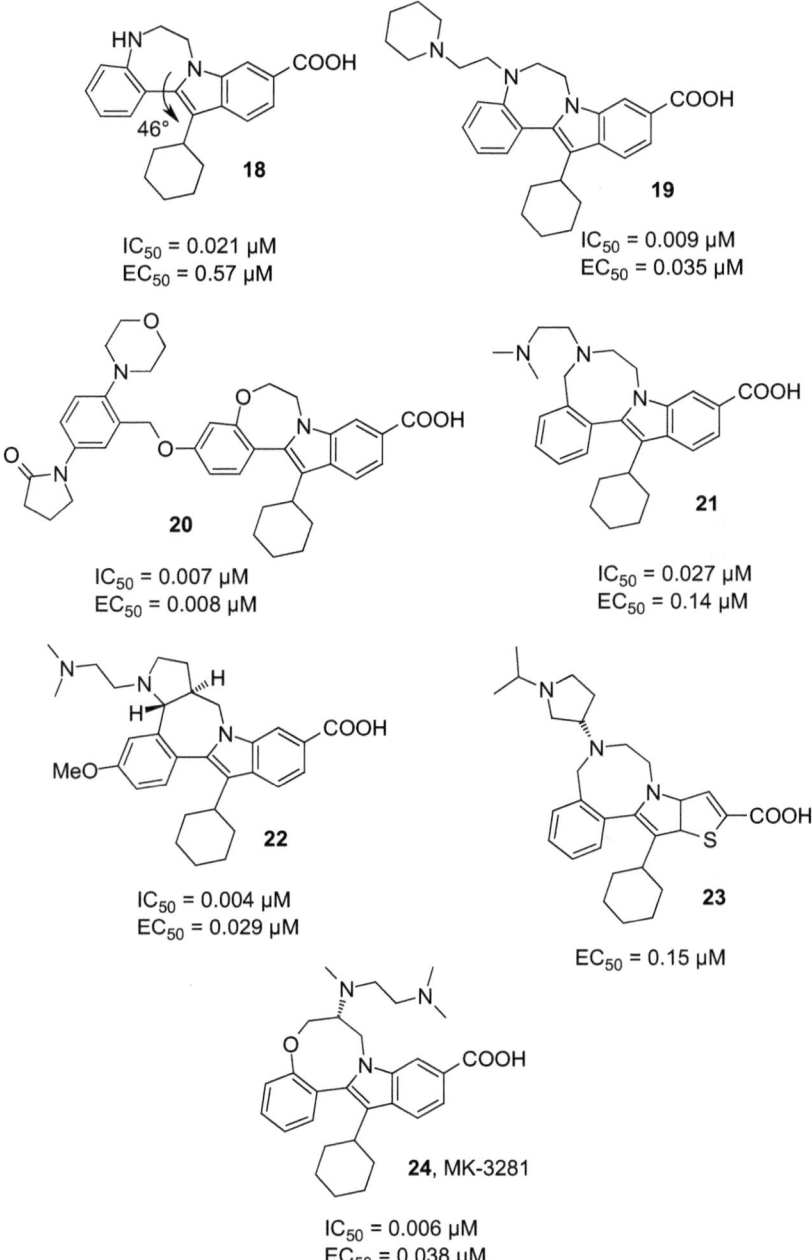

Figure 8.6 Constrained tetracyclic and pentacyclic indole inhibitors from Japan Tobacco and Merck.

Further substitution of the bridge with solvent-exposed basic side chains produced zwitterionic derivatives with excellent potency in cell culture (*e.g.*, **19**, $EC_{50} = 35$ nM). Introduction of this scaffold-rigidifying element into JTK-109 analogs provided compounds with impressive potencies: IC_{50} and $EC_{50} < 10$ nM (*e.g.*, **20**).[35b] No further reports from Japan Tobacco have appeared since these initial findings.

8.3.1.2   Discovery of MK-3281

Other companies also reported similarly rigidified analogs of thumb pocket 1 NS5B inhibitors. The IRBM-Merck group in Italy reported tetracyclic indole derivatives containing a four-atom bridge and basic side chains (*e.g.*, **21**) in which the anilinic function of analogs such as **19** (considered as a potential liability from a drug safety point of view) was eliminated.[36a] Such tetracyclic indole derivatives featuring an eight-membered ring displayed potency comparable to that of previously published indole-*N*-acetamides such as **14,** but with improved PK properties in preclinical species relative to open-chain analogs. Indeed, analog **21** had a cross-species PK profile consistent with low clearance in humans and had good distribution to the liver target organ in rats (liver/plasma ratio = 16). In human hepatocytes, elimination was mainly through CYP3A4-mediated phase 1 metabolism (hydroxylation of the cyclohexyl ring). Despite the identification of glucuronide conjugates, the compound had low potential for covalent binding to human microsomal protein. The overall profile of **21** justified advancement as a preclinical candidate, but no report is available on the outcome of such activities for this compound.[36a]

Subsequent reports from the IRBM-Merck group focused on identifying analogs with improved potency. These included pentacyclic indole derivatives in which the number of rotatable bonds of tetracyclic analogs such as **21** was reduced by inclusion of the solvent-exposed basic side chain into a five-membered ring that was fused on to a benzazepine rather than a benzo-diazepine framework (*e.g.*, **22**, Figure 8.6).[36b] In general, *trans*-annelated systems retained comparable or improved potency relative to tetracyclic systems and did not exhibit notable differences between low- and high-serum conditions in the cell-based replicon system. Attempts were made to optimize the inhibitor core itself and thienopyrrole-based inhibitors (*e.g.*, **23**) were also described, but no improvement was achieved over previously described analogs.[37] Ultimately, a tetracyclic benzoxazocine derivative (**24**, MK-3281) was selected for clinical development as it combined improved potency with good oral bioavailability in preclinical animal models. Furthermore, MK-3281 showed efficacy in a chimeric mouse model of HCV infection where a bid dose of 50 mg kg^{-1} produced a 3.1log$_{10}$ drop in viremia. On the basis of this promising profile, the compound was evaluated in the HCV-infected chimpanzee model and was progressed into Phase 1 clinical trials.[38] For undisclosed reasons, however, the development of MK-3281 has been discontinued.

8.3.1.3 Discovery of BMS-791325

More recently, the virology group at Bristol Myers Squibb (BMS) has begun reporting on the discovery of their own version of conformationally rigidified polycyclic indole thumb pocket 1 inhibitors.[39] Early examples include lactam-bridged analogs such as **25** (Figure 8.7). Hybridization with Boehringer Ingelheim's cinnamic acid right-hand sides (*e.g.*, **26**) resulted in compounds with 10–20-fold increases in replicon potency ($EC_{50} = 10$ nM for **26**). The results were rationalized with X-ray crystal structure data maintaining the typical dihedral angle between the indole and C2 aromatic substitution ($\sim 46°$), a hydrogen bond between the indole C6 amide and R503 and an interaction of the cinnamic acid carboxylate with the side chain of R498.[39]

Early inhibitors displayed poor PK profiles in rats, due in part to high clearance. In a subsequent report, replacement of the phenyl ring at the C2

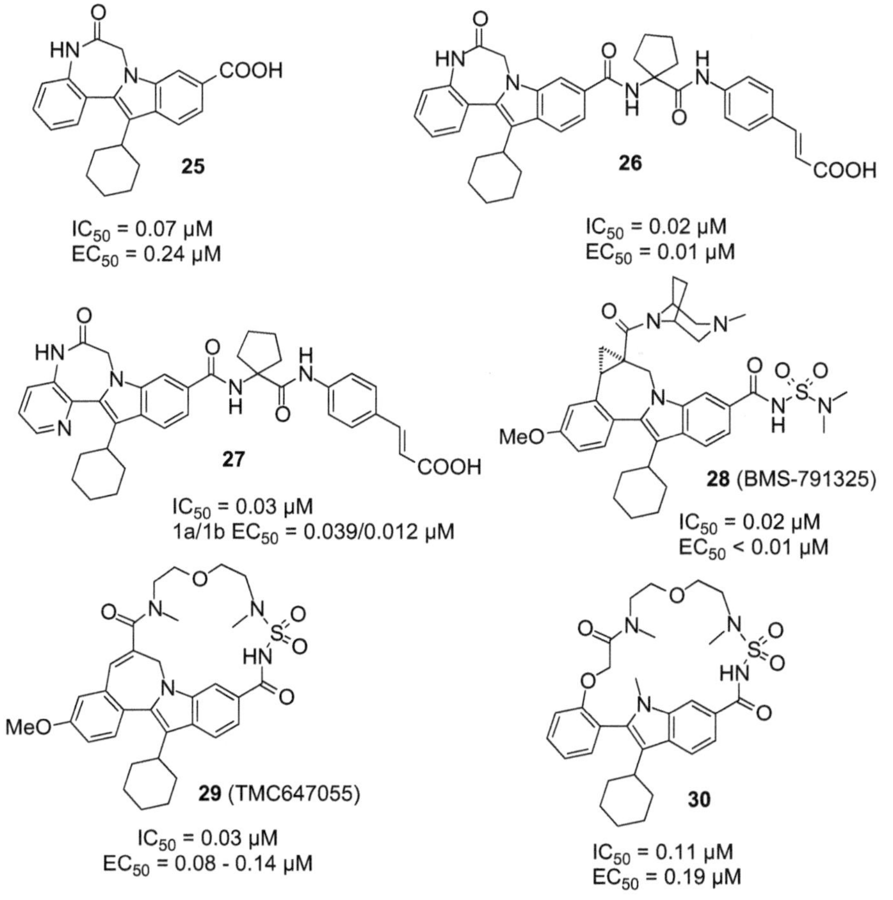

25

$IC_{50} = 0.07$ μM
$EC_{50} = 0.24$ μM

26

$IC_{50} = 0.02$ μM
$EC_{50} = 0.01$ μM

27

$IC_{50} = 0.03$ μM
1a/1b $EC_{50} = 0.039/0.012$ μM

28 (BMS-791325)

$IC_{50} = 0.02$ μM
$EC_{50} < 0.01$ μM

29 (TMC647055)

$IC_{50} = 0.03$ μM
$EC_{50} = 0.08 - 0.14$ μM

30

$IC_{50} = 0.11$ μM
$EC_{50} = 0.19$ μM

Figure 8.7   Constrained polycyclic indole inhibitors from Bristol Myers Squibb and Tibotec.

bridge with heterocyclic moieties was explored in an attempt to improve physicochemical properties. Although several analogs exhibited good cell-based potency in both 1a and 1b replicon assays (*e.g.*, pyridine analog **27**), PK profiles were not improved relative to previously described analogs, presumably due to low solubility and poor permeability.[40] While the findings that guided further optimization of this class of thumb pocket 1 inhibitors await future publications, BMS recently disclosed the chemical structure of a clinical candidate currently in Phase 2 clinical trials in combination with SOC. BMS-791325 (**28**) is an indole-based pentacyclic thumb pocket 1 inhibitor in which the indolecarboxylic acid moiety was replaced with an *N,N*-dimethylsulfamide moiety that presumably reduces the potential to form reactive acylglucuronide conjugates. The molecule features a basic side chain substituent that, in combination with the ionizable acylsulfamide moiety, confers zwitterionic character to the molecule. Compound **28** has reported $EC_{50} = 3–14$ nM for genotypes 1a/1b, 3a and 5 while genotypes 2a/2b and 6 are in the 112–270 nM range. It is highly protein bound (97.8–98.8%) and the liver distribution for this compound is species dependent, ranging from 10–15-fold in rats to twofold in dogs. In an ongoing Phase 2a study, BMS-791325 (75 or 150 mg bid) in combination with PegIFN and RBV was well tolerated. During the first 12 weeks of therapy, responses were highest in the 75 mg bid group, where 92% of patients had undetectable HCV RNA at week 4 (RVR). At week 12, 77% of patients (75 mg bid group) had undetectable HCV RNA (complete early virological response, cEVR) *versus* 54% in the 150 mg dose group, which also suffered from higher discontinuation rates (15% *versus* 8%). Three of 26 patients in the 150 mg dose group experienced viral breakthrough with A421V and P495L/S resistant variants emerging in the gt1a and gt1b groups, respectively. No virologic escape was detected in the 75 mg-dose group.[41]

8.3.1.4   Discovery of TMC647055

Several groups have noted the potential of 6-indolecarboxylic acid derivatives (*e.g.*, indole-*N*-acetamide **14**) to form reactive circulating acylglucuronide conjugates and have therefore searched for suitable carboxylic acid isosteres.[42] Certain heterocycles (*e.g.*, oxadiazolones), and especially acylsulfonamides and acylsulfamides (*e.g.*, **28**), have been reported with excellent intrinsic potency and have provided opportunities to modulate physicochemical properties (*e.g.*, pK_a) and explore additional interactions with the enzyme.[43]

Recent reports from Tibotec describe the rational design of NS5B thumb pocket 1 allosteric inhibitors that combine both concepts of carboxylic acid replacement and conformational rigidification with the opportunity to exploit new biochemical space and modulate the properties of the molecules. These studies ultimately led to the discovery of TMC647055 (**29**, Figure 8.7), a macrocyclic molecule with the necessary potency and preclinical profile for selection as a clinical candidate.[44] Using knowledge derived from the overlap of publicly available X-ray structures of thumb pocket 1 inhibitors in complex with NS5B, Tibotec researchers elaborated several macrocyclization strategies to design novel molecules with improved properties. Macrocyclization is a well-established approach to improving potency (both intrinsic and cell-based) of inhibitors and enhancing the drug-like properties of molecules.[45] This approach has already borne fruit in the structure-based design of successful HCV protease drug candidates, as described in chapter 10 of this book. In the particular case of the Tibotec NS5B inhibitors, focus was on exploring carboxylic acid bio-isosteres, introducing PK-enhancing substituents in solvent-exposed areas while avoiding zwitterionic species and exploiting conformational rigidification to improve potency. These objectives were partially fulfilled with inhibitors in which a 13-atom bridge was installed that linked the C6 position of the indole scaffold to the *ortho*-position of an aromatic ring at the C2 position (**30**). This 20-membered ring macrocycle, containing an acylsulfamide link to the indole core, exhibited potency comparable to that of previously described indole-*N*-acetamides (*e.g.*, **14**) and tetracyclic derivatives (*e.g.*, **21**) with $IC_{50} = 0.11$ µM and $EC_{50} = 0.19$ µM. This class of derivatives had good distribution to the liver in rats; however, early analogs suffered from suboptimal PK properties in rodents, such as short half-life ($t_{\frac{1}{2}} = 1$ h) and low bioavailability (14%).[44a]

These deficiencies were addressed in a subsequent round of design which led to the discovery of **29**. In this phase of the optimization process, pre-orientation of the C2 indole substituent in the bioactive conformation using previously described tetracyclic analogs was exploited to provide analogs with increased overall rigidity. The introduction of a bridge in the tetracyclic indole series proved to be beneficial, as a twofold improvement in cell-based potency was generally observed compared with previous macrocycles. The structural features present in the bridge had a strong impact on the *in vivo* preclinical profiles of these molecules. Compound **29** (TMC647055), with an unsaturated amide bridge, provided the best overall profile. The compound inhibited the 1a/1b replicons with $EC_{50} = 74$–166 nM, depending on assay readout (luciferase or qRT-PCR),[46] and had improved clearance, liver distribution (liver/plasma = 46 in rats) and bioavailability (66% in rats). The X-ray structure of a macrocyclic analog bound to NS5B revealed the presence of additional protein residues interacting with the molecule compared with previous structures, providing a rationale for the improved intrinsic potency.[44c] Despite the deviation of these molecules from what are generally considered drug-like features (*e.g.*, high molecular weight and clog*P*), the macrocyclic structures designed by the Tibotec group feature improved PK behavior relative to some previously described series of inhibitors. The antiviral activity of TMC647055 in 5 day monotherapy was studied in gt1a/1b HCV patients. Doses of 500 or 1000 mg bid produced median maximum decrease in HCV RNA of $3.3\log_{10}$ and $3.4\log_{10}$ in 1b patients. In 1a genotypes, the effect was dose dependent ($1.4\log_{10}$ and $2.4\log_{10}$ reductions in HCV RNA depending on the dose). These results support further development of TMC647055 for the treatment of hepatitis C.[46]

In summary, thumb pocket 1 allosteric NS5B inhibitors, discovered jointly by Japan Tobacco and Boehringer Ingelheim, have evolved over the years from benzimidazole-5-carboxylic acids, which exhibited modest potency in cell culture, to more potent indole-6-carboxylic acid derivatives and, more recently, conformationally restricted polycyclic analogs. This class of allosteric inhibitors was shown through biochemical studies to inhibit viral replication at the initiation stage of RNA synthesis by preventing the formation of a productive polymerase:RNA primer–template complex. Structural studies suggest that this effect results from inhibitors binding to the thumb domain, competing with an interaction with the protein loop that extends from the finger domain, thus interfering with the formation of an enclosed active site, which is thought to be important for RNA synthesis. BILB 1941 is the first compound in this class to provide proof of concept for this mechanism of action in genotype 1 HCV-infected patients. Among NS5B allosteric inhibitors, those that bind to thumb pocket 1 (finger-loop inhibitors) display an attractive cross-genotype profile, with comparable potency observed against genotypes 1, 3, 4, 5 and 6 (within threefold) but somewhat larger shifts against genotype 2 (15–30-fold). Consistent with clinical observations, the *in vitro* resistance signature of this class of allosteric inhibitors is characterized by mutations at residues P495 and P496. Today, three thumb pocket 1 inhibitors (BI 207127, BMS-791325 and TMC647055) are progressing through clinical evaluation and may become potential combination partners with complementary DAAs in all-oral interferon-free regimens.

8.3.2 Thumb Pocket 2 Inhibitors

The existence of a second allosteric pocket in the thumb domain of NS5B (thumb pocket 2) was initially revealed in 2002 in a patent from Agouron Pharmaceuticals describing the X-ray structure of the dihydropyrone derivative **31** (Figure 8.8) in complex with the enzyme.[47a,b] In 2003, Shire Biochem described the crystal structure of a different chemotype (phenylalanine derivative **32**) bound to the same hydrophobic pocket, 30–35 Å away from the polymerase active site and, subsequently, a thiophenecarboxylic acid derivative, **33**, bound to a gt2a NS5B.[47c,d] These two findings served as starting points for two successful programs that each led to the discovery of development candidates (filibuvir and lomibuvir, respectively).

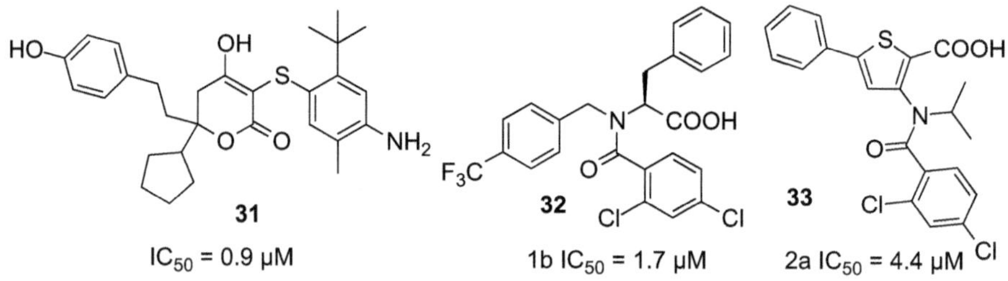

Figure 8.8   Thumb pocket 2 starting points from Agouron and Shire Biochem.

8.3.2.1 Discovery of Filibuvir (PF-00868554)

The dihydropyrone **31** was discovered as a weak and reversible NS5B inhibitor (IC$_{50}$ = 0.9 μM) in a screening campaign at Agouron Pharmaceuticals (now Pfizer).[47b] The core of this molecule is reminiscent of tipranavir, an orally active HIV protease inhibitor.[48] Biochemical studies revealed that dihydropyrones behaved similarly to thumb pocket 1 inhibitors in that they did not compete with nucleotide substrates. A crystal structure of **31** bound to NS5B revealed a new, well-conserved binding site in the thumb domain, ∼30 Å away from the active site. In addition to hydrophobic contacts, key protein–inhibitor interactions comprised direct hydrogen bonds between the enol–ketone oxygen of the inhibitor with the backbone amide NH of L476 in addition to a water-mediated H-bond to Y477 NH. Preliminary data suggested that dihydropyrones interfered with NS5B function through perturbation of protein dynamics, interference with RNA binding or disruption of enzyme oligomerization.[47b] Initial hit-to-lead activities guided by the available structural data allowed a 30-fold improvement of intrinsic potency, providing analogs such as **34** (Figure 8.9) with sub-micromolar potency in biochemical assays but that lacked significant antiviral potency in cell-based replicon assays.[49]

The large discrepancy between the nanomolar potencies achieved in enzymatic assays and the moderate activity in cell culture, and the poor rat PK properties (bioavailability) associated with sulfur-linked dihydropyrone

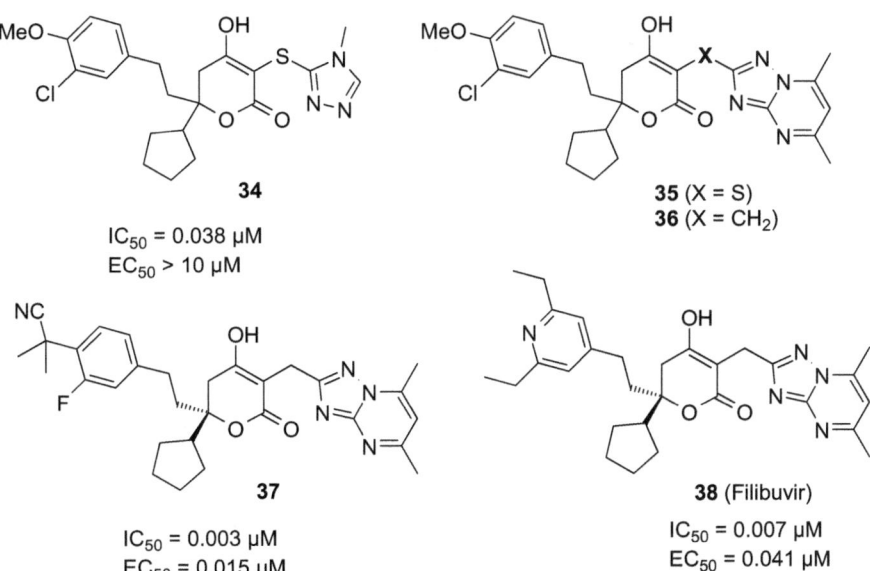

34

IC$_{50}$ = 0.038 μM
EC$_{50}$ > 10 μM

35 (X = S)
36 (X = CH$_2$)

37

IC$_{50}$ = 0.003 μM
EC$_{50}$ = 0.015 μM

38 (Filibuvir)

IC$_{50}$ = 0.007 μM
EC$_{50}$ = 0.041 μM

Figure 8.9   Evolution of dihydropyranone-based thumb pocket 2 inhibitors – discovery of filibuvir.

Table 8.1 Comparative profiles of S- and C-linked dihydropyranone thumb pocket 2 inhibitors from Pfizer.

Property	35 (X = S)	36 (X = CH₂)
pK_a	4–5	5–7
IC_{50} (mm)	0.036	0.020
EC_{50} (mm)	3.25	0.33
Cl $(mL\,min^{-1}\,kg^{-1})$	14.8 ± 2.6	8.3 ± 1.1
V_{SS} $(L\,kg^{-1})$	2.0	0.3
Caco-2 AB/BA $(\times 10^6\,cm\,s^{-1})$	3/2.3	10/6.8
F (%)	2	42

derivatives, were ascribed to a lack of permeability. These issues were resolved through modulation of the physicochemical properties and, in particular, through adjustment of the acidity of the enol moiety present in the molecules. Indeed, S-linked derivatives were found to be 10–100-fold more acidic than the corresponding C-linked analogs ($pK_a \sim 4$–5 *versus* ~ 5–7, respectively) resulting in reduced permeability and compromising *in vivo* profiles.[50] The comparative profiles of an S-linked (**35**) and corresponding C-linked (**36**) analog are shown in Table 8.1.

Simple S→C atom replacement increased cell permeability and provided a 10-fold improvement in cell-based potency while maintaining intrinsic potency in the nanomolar range, thus narrowing the IC_{50}/EC_{50} ratio. Furthermore, oral absorption of C-linked analogs was also improved to set the stage for further optimization. Introduction of *gem*-dimethylcyanomethyl and fluorine substituents on the aromatic ring, followed by evaluation of the two separate enantiomers, led to the identification of analog **37**, which was selected as the first compound from the class for preclinical evaluation.[50] Development of **37** was soon compromised, as a low-solubility crystal form that significantly reduced exposure in animal studies was identified during scale-up. Furthermore, **37** was identified as a strong CYP2D6 inhibitor ($IC_{50} = 0.3\,\mu M$), which generated concerns for potential drug–drug interaction issues during clinical development and led to a decision to resume optimization to resolve these liabilities.[51] Through a systematic SAR exploration, the culprit for CYP2D6 inhibition was soon identified as the aromatic ring, the substitution of which had a profound impact on the undesired off-target activity. In particular, it became apparent that the cyano group in **37** was likely involved in an interaction with the CYP isozyme and removal of this group or hindrance through steric interaction had a beneficial impact. Fortunately, modifications in this part of the molecule were well tolerated and compound potency was not affected. Leveraging structural information, key protein–inhibitor hydrophobic interactions were optimized. Replacement of the phenyl ring by ethyl-substituted pyridines was well tolerated and provided a basic center that decreased lipophilicity and improved solubility.

Compound **38** (filibuvir) is a potent inhibitor of HCV polymerase ($IC_{50} = 7\,nM$) and the replicon ($EC_{50} = 41\,nM$). It did not inhibit any CYP450 isoforms ($IC_{50} > 30 \, \mu M$) and had good solubility at pH 6.2 (2.55 mg mL^{-1}). The cross-species PK profile of **38** suggested that this compound was amenable to a twice-daily dosing regimen.[52] Filibuvir was tested in a panel of genotype 1 isolates in which it displayed an average $EC_{50} = 75$ nM and modest shifts were observed in the presence of human serum (3.5-fold). Amino acid positions associated with resistance to filibuvir include M423, M426 and I482. M423 substitutions confer a 715- to >2200-fold increase in EC_{50}, confirming the importance of this residue for interaction with dihydropyrones. Filibuvir was weakly active against non-genotype 1 HCV polymerases ($IC_{50} \geq 1 \, \mu M$) and not active against human DNA and RNA polymerases. Despite its initial structural similarity to tipranavir, it was not significantly active against HIV protease and a panel of human aspartyl and serine proteases.[52] In the light of its broad-spectrum antiviral activity against genotype-1 HCV and favorable PK profile in preclinical animal species, in 2007 Pfizer initiated clinical trials with filibuvir in genotype-1 HCV patients. At a dose of 450 mg administered bid for 8 days, filibuvir was well tolerated and after 48 h produced a 1.74log$_{10}$ mean maximal reduction in viral load in TN gt1 HCV patients. A 450 mg bid dose provided similar results in TE patients. In agreement with *in vitro* studies, resistance to filibuvir in HCV-infected patients was observed and primary mutations identified at amino acid M423.[53] Filibuvir is a weak time-dependent inhibitor and inducer of CYP3A4 and potential for drug–drug interactions has been evaluated in clinical trials where ketoconazole and midazolam both had measurable effects on the PK of filibuvir.[54]

In a Phase 2a study, filibuvir was tested for 4 weeks in combination with the SOC (PegIFN-α2a and RBV), followed by SOC for an additional 44 weeks. Filibuvir significantly increased the proportion of patients achieving undetectable HCV RNA levels at week 4 compared with SOC alone (up to 75% RVR in the 300 mg bid group). However, a high rate of virological breakthrough was observed (20–50% of patients with undetectable levels at week 48 relapsed by week 60) such that SVR$_{12}$ was similar for the filibuvir- and placebo-treated groups.[55] These observation suggest that a longer treatment duration is required. A 24 week Phase 2b trial assessing the safety and efficacy of filibuvir in combination with SOC was completed, but the outcome has not yet been reported.

In summary, filibuvir is a non-nucleoside inhibitor of HCV polymerase that binds in the thumb domain of NS5B (thumb pocket 2) and likely interferes with protein conformational changes that are required for initiating RNA synthesis. The resistance signature for this compound is characterized by mutations at M423. The main advantage of the drug is its potency against major 1a/1b HCV genotypes. Weaknesses of this compound include its drastically reduced potency against non-1 virus, potential for drug–drug interactions and low barrier to resistance within 4 week therapy in combination with SOC. NS5B inhibitors are currently positioning themselves to be used in combination with complementary DAAs in IFN-free regimens rather than SOC, and no information is available suggesting that this compound is being considered for any such strategy.

8.3.2.2   Discovery of Lomibuvir (VX-222)

Very little information is available on the progression of *N,N*-disubstituted phenylalanine derivatives (*e.g.*, **32**) identified by Shire BioChem several years ago through screening of compound libraries, suggesting that this series was abandoned. Although compounds from this class permitted the identification of a second chemotype with affinity for thumb pocket 2, data reported so far do not suggest that intrinsic potency could be improved significantly and cell-based activity for this class has not been reported.[56]

Shire BioChem was more successful with a second screening hit based on a thiophene-2-carboxylic acid structure. The initial carboxamide hit, **39** (Figure 8.10), showed only modest potency in the polymerase assay ($IC_{50} = 14\,\mu M$), but preliminary SAR studies rapidly revealed that the corresponding carboxylic acid was three times more potent and provided weak cell-based activity in the replicon ($EC_{50} = 14\,\mu M$) with modest selectivity (selectivity index, $SI = 5.7$). Further exploration of this series provided analogs with submicromolar enzymatic activity (*e.g.*, **40**) and confirmed efficacy in cell-based assays where analogs with $SI \approx 12$ were identified.[57] With these encouraging preliminary results in hand, the team's focus was directed towards further improving potency in both biochemical and cellular assays. Removal of H-bond donors by methylation often has a positive impact on cell permeation and cell-based potency. Unfortunately, in this case, *N*-methylation of the sulfonamide NH had a negative impact on intrinsic potency and did not improve replicon potency. A key discovery for the program was made when the sulfonamide linkage was replaced by a tertiary amide moiety, producing the first analogs with

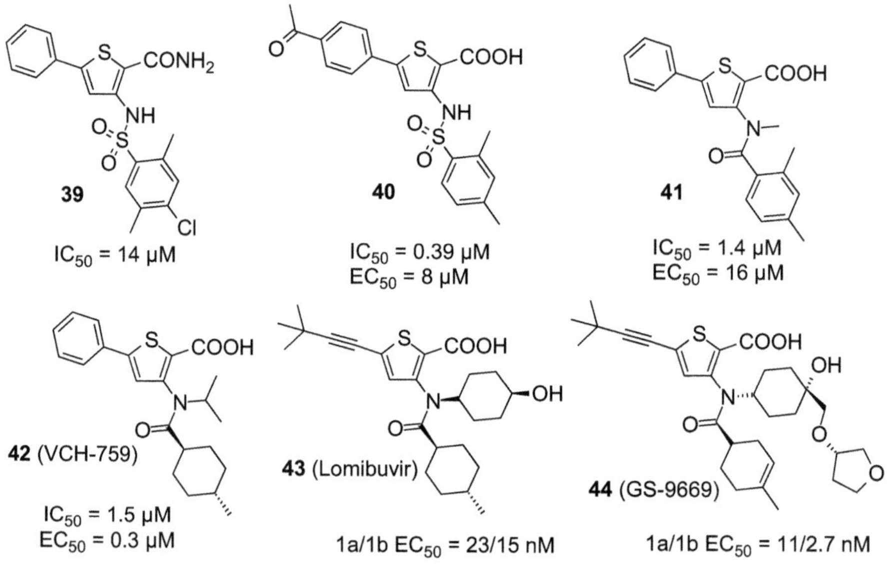

Figure 8.10   Evolution of thiophene-2-carboxylic acid thumb pocket 2 inhibitors and discovery of lomibuvir (VX-222) and GS-9669.

potency comparable to that of sulfonamide derivatives (*e.g.*, **41**); in this case, potency could be improved further by bulking up the amide linkage with more lipophilic entities (*e.g.*, **42**) to provide the first non-cytotoxic thiophene-2-carboxylic acid analogs with submicromolar replicon potency (1b $EC_{50} = 0.3$ µM; SI > 200).[58] Trends suggested that the carboxamide substituent was positioned into a well-defined hydrophobic pocket in the enzyme, as the SAR at this position was very sensitive to the nature and orientation of substituents (*e.g.*, the *cis*-4-methylcyclohexyl analog of **42** is 16-fold less potent).

Whereas NMR data were used to provide clues on the bioactive conformation of these tertiary amides,[59] crystal structures of **33** and several other analogs bound to genotype 2a or 1b NS5B provided information on the inhibitor binding site and hypotheses on plausible modes of action for this class of inhibitors.[47d,60] The binding site of thiophene-2-carboxylic acid derivatives is located in a hydrophobic depression at the base of the thumb domain, 35 Å away from the active site. This allosteric site (thumb pocket 2) coincides with that of filibuvir and exploits similar interactions with the protein, including hydrophobic interactions with L419, M423 and two direct H-bonds between the carboxylate group of the inhibitor and main-chain NHs of S476 and Y477. Both classes of inhibitors are non-competitive with NTP substrates and inhibit an early step in primer-dependent RNA synthesis. While binding of phenyl-alanine analogs did not induce local conformational changes in the 1b enzyme, the tighter binding of thiophene derivatives causes significant changes in the side-chain conformations of several binding site residues relative to *apo*-enzyme (*e.g.*, M423). The conformational changes induced in thumb pocket 2 have significant structural repercussions throughout the thumb domain, in particular residues P496–R505, which are part of a positively charged helix (helix T) in proximity of a presumed GTP binding site that may be an important regulator of HCV RNA replication *in vivo*.[11a] Conformational disturbances induced by inhibitor binding in thumb pocket 2 may reduce GTP affinity and the stability of a polymerase state capable of undergoing RNA synthesis. The shift in conformation of helix T may also impact open/closed dynamics of the polymerase which are required for initiation of RNA synthesis. It is conceivable that thumb pocket 2 inhibitors lock the polymerase into an inactive 'open' form, interfering with the interaction of the thumb domain with the Λ1 finger loop and inhibit polymerase function through a mechanism related to that of thumb pocket 1 inhibitors. Alternatively, the observed conformational changes could have a profound effect on the ability of the polymerase to oligomerize to *in vivo*-relevant states.

In 2006, ViroChem Pharma took over Shire Biochem's NS5B polymerase inhibitor portfolio and a 10 day Phase 1 proof-of-concept clinical study was initiated with **42** (VCH-759; 1a/1b replicon $EC_{50} = 0.34/0.27$ µM).[61] Dose-dependent maximum viral load decreases of 1.9–2.5$\log_{10}$ were achieved for both subtypes and genotypic analysis of NS5B variants from patients treated with VCH-759 revealed mutations at the expected residues M423 and L419, as previously observed in replicon experiments. In 2010, Virochem Pharma was acquired by Vertex Pharmaceuticals, which was nearing market introduction of their HCV protease inhibitor telaprevir and became interested in ViroChem's thiophene-2-carboxylic acid thumb pocket 2 NS5B inhibitors as potential partners for combination with telepravir and the development of IFN-free regimens. By that time, ViroChem's portfolio of thumb pocket 2 inhibitors had grown to include more potent inhibitors, such as VCH-916 (structure not disclosed; 1a/1b $EC_{50} = 79/110$ nM) and VCH-222/VX-222 (1a/1b $EC_{50} = 23/15$ nM). VCH-916 had produced a maximum $1.5\log_{10}$ viral load decline when administered at a dose of 300 or 400 mg bid in a 3 day monotherapy trial.[62] VX-222 (**43**, lomibuvir) is a *tert*-butylacetylene-substituted 2-thiophenecarboxylic acid thumb pocket 2 inhibitor that is significantly more potent *in vitro* than its predecessors and exhibited favorable pharmacokinetics in preclinical animal species.[63] In HCV genotype-1 patients, VX-222 monotherapy was well tolerated and produced a mean $3.7\log_{10}$ decrease in plasma HCV RNA at day 4, with a dose of 750 mg bid administered for 3 days.[64] These encouraging results led Vertex Pharmaceuticals to initiate the evaluation of lomibuvir (100 or 400 mg bid) in combination with PI inhibitor telaprevir (1125 mg bid), with and without PegIFN and RBV.[65] While the evaluation of the dual all-oral NS5B + PI combination was discontinued due to viral breakthrough in some patients within 4 weeks of treatment, ongoing Phase 2b studies of lomibuvir in combination with telaprevir and SOC have achieved high SVR_{12} rates in gt1 patients (64–100% and 83–94% for the 100 and 400 mg bid dose groups, respectively), independent of genetic predisposition to IFN therapy.[66]

Filibuvir and lomibuvir are two structurally distinct non-nucleoside NS5B inhibitors that bind in the same allosteric site at the base of the thumb domain (thumb pocket 2) ~ 35 Å away from the enzyme binding site. Both compounds inhibit primer-dependent RNA synthesis at nanomolar concentrations and show similar efficacy against 1a and 1b HCV genotypes.[67] On the other hand, thumb pocket 2 inhibitors show drastically reduced activity against non-1 HCV genotypes (lomibuvir gt2a $EC_{50} = 4.6$ μM). Mutations encoding substitutions at M423 are characteristic of the resistance profile of thumb pocket 2 inhibitors. Both compounds have demonstrated efficacy in HCV-infected patients and showed susceptibility to select for resistant virus in the clinic. Lomibuvir is currently undergoing clinical evaluation in an all-oral triple combination with protease inhibitors telaprevir and ribavirin, towards the development of an IFN-free HCV therapy.

Recently, Gilead Sciences has disclosed the structure of GS-9669 (**44**, Figure 8.10), a thiophene-2-carboxylic acid analog with an improved preclinical *in vitro* profile against single, double and triple class mutants that confer resistance to other major classes of DAAs. GS-9669 is a potent inhibitor of HCV gt 1a, 1b and 5a replicons (1a/1b/5a $EC_{50} = 11/2.7/8.0$ nM) and is significantly less active against gt 2, 3 and 4 ($EC_{50} = 0.33$ to >50 μM). In gt1a/1b replicons, **44** selected for thumb pocket 2 resistance mutations that included I482L and L419M (also R422K and F429L in gt1b), conferring 20–40-fold shifts in potency, whereas only modest shifts were observed for the more common M423 variants (3–12-fold shifts). The safety, tolerability and antiviral activity of GS-9669 were recently investigated in gt1 HCV patients. A median viral load decrease of 3.25–4.09$\log_{10}$ was observed for all tested doses (50, 100 and 500 mg bid and 500 mg qd) following 3 days of treatment in both gt 1a and 1b subjects.[68] The potency, pharmacokinetic and safety profile of GS-9669 support once-per-day dosing and use in combination with other DAAs.

8.3.3 Palm Site 1 Inhibitors

8.3.3.1 Benzothiadiazines from GlaxoSmithKline

One of the first literature reports describing the discovery of non-nucleoside NS5B polymerase inhibitors was published in 2002 by the GlaxoSmithKline group. Benzothiadiazine derivative **45** (Figure 8.11), identified as a screening hit, was disclosed as a potent and specific inhibitor of HCV NS5B with submicromolar potency in cell-based replicon assays ($IC_{50} = 0.10 \,\mu M$; $EC_{50} = 0.5 \,\mu M$).[69]

Resistance selection experiments were performed in the replicon and mutations encoding amino acid changes at a palm site residue close to the interface with the thumb domain (M414) were found to be sufficient to confer resistance to the benzothiadiazine chemical class.[69b] As for other allosteric inhibitors, benzothiadiazines were found to interfere with initiation of RNA synthesis rather than elongation, regardless of whether replication was *de novo* or primer–template dependent. Compounds were non-competitive with respect to NTP substrates and could bind independently to *apo*-NS5B or RNA–NS5B complex and thus did not require disruption of NS5B–RNA interactions to mediate inhibition.[69c] SAR studies performed on the quinoline portion of **45** led to the discovery of analogs with significantly improved antiviral potency.[70] The nature of the N1 substituent (length, sterics and polarity) was found to be critical for NS5B inhibition, with a cyclopropylethyl chain providing optimum potency over polar, aromatic or heterocyclic substituents. X-ray crystal structures of several benzothiadiazine inhibitors in complex with NS5B were obtained and confirmed binding in proximity to the interface between palm and thumb domains, in agreement with the previously selected M414-resistant

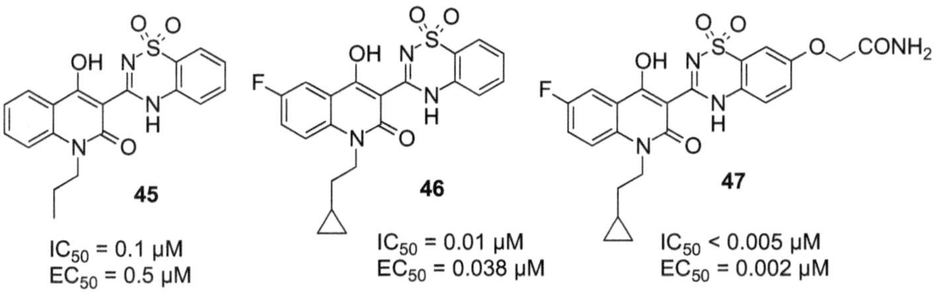

IC$_{50}$ = 0.1 μM
EC$_{50}$ = 0.5 μM

45

IC$_{50}$ = 0.01 μM
EC$_{50}$ = 0.038 μM

46

IC$_{50}$ < 0.005 μM
EC$_{50}$ = 0.002 μM

47

Figure 8.11 Benzothiadiazine palm site 1 NS5B inhibitors from GlaxoSmithKline.

mutants (referred to as palm site 1). The quinolone portion of the inhibitor fits in a tight pocket which is lined by residues from the thumb, palm and finger domains and restricts substitution at the C5 and C6 positions. While small- to medium-sized substitution was tolerated at C5, the effect on NS5B inhibition was negligible. On the other hand, C6 tolerated greater variety and allowed optimization of replicon potency.

Hydrogen-bonding interactions between the benzothiadiazine core and the protein were probed through modification of the core itself. Removal of acidic H-bond donors led in all cases to inactive compounds, suggesting that critical direct or water-mediated interactions with the protein were affected or a presumed bioactive planar conformation of the ring systems was no longer favorable. Indeed, analysis of structural data revealed the presence of several conserved water molecules, mediating important H-bonding interactions between the inhibitor and protein. In addition, an edge-to-face π-interaction between the benzothiadiazine ring system and a phenylalanine residue (F193) was apparent. Whereas the free state of benzothiadiazines locks the two ring systems into coplanarity through internal hydrogen bonds, this conformation appears to be somewhat distorted when bound to protein.

Overall, analog **46** provided the best potency profile, inhibiting gt1a, 1b and 2a NS5B with comparable efficiency ($IC_{50} = 10$–49 nM), while the compound was significantly less potent towards the 3a genotype (60% inhibition at $10 \mu M$).[70] Analog **46** exhibited a favorable combination of high oral bioavailability and low intrinsic clearance in preclinical animal studies with favorable distribution to the liver in rats (liver/plasma ratio ≥ 2–5-fold). Benzothiadiazines had relatively low volumes of distribution and were extensively bound to human plasma proteins ($>99\%$). In the presence of human serum albumin, the EC_{50} of **46** increased ~ 350-fold. Analog **46** was evaluated in a 4 day toxicology study in rats at doses up to 300 mg kg^{-1} d^{-1} with no adverse findings and was advanced into preclinical development. The outcome has not been disclosed and there are no reports of this analog progressing into clinical trials.[70] Subsequent SAR studied on the benzothiadiazine portion of inhibitors was particularly productive at the C7 position. For example, addition of an acetamide chain through an ether linkage provided further potency enhancements that could in part compensate for some of the observed potency shifts in high-protein replicon assays and also improve overall developability of the compounds (*e.g.*, **47**). Unfortunately, no further information is currently available on the GSK benzothiadiazine program.[71]

8.3.3.2 *Benzothiadiazines from Abbott*

Several companies have followed suit on benzothiadiazine NS5B leads; among them, Abbott Laboratories have published on the discovery, development and clinical evaluation of its own versions of this chemotype. Initial reports by Abbott described the effect of replacing the C8 carbon of the quinoline ring system by a nitrogen atom and linking the aliphatic chain at N1 through an oxygen or nitrogen atom. Quinolone and 1,8-naphthyridone cores were shown

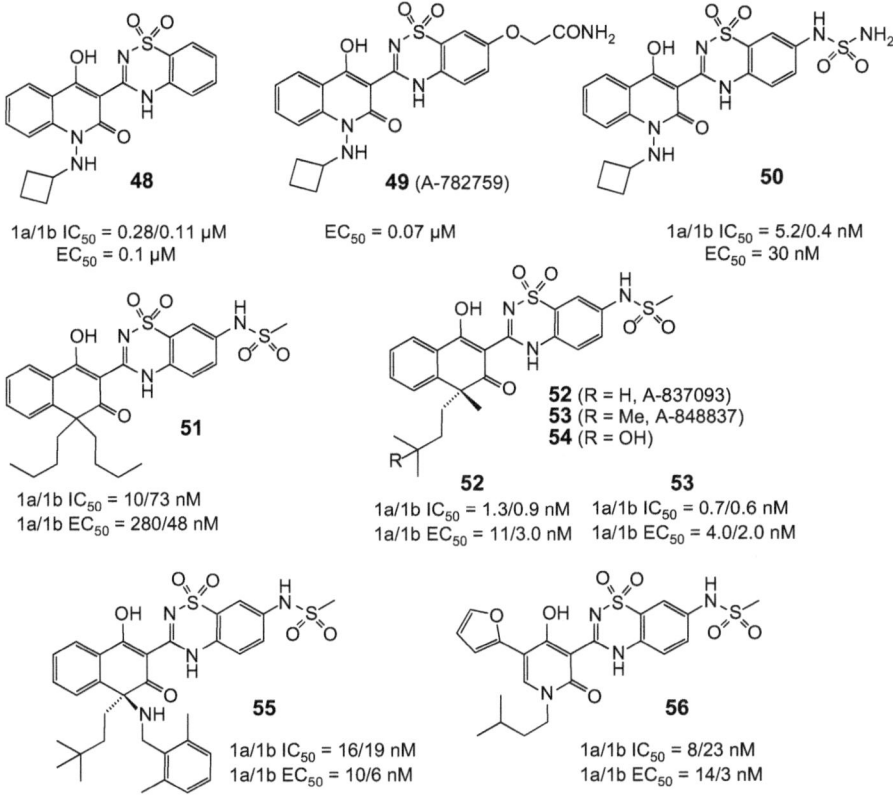

Figure 8.12   Benzothiadiazine palm site 1 inhibitors from Abbott Laboratories.

to be roughly equivalent in potency and the N1 methylene side-chain linker was shown to be replaceable by a nitrogen atom (*e.g.*, **48**, Figure 8.12). Interestingly, as biochemical potencies were being tracked using both gt1a and 1b enzymes, the 1a polymerase was often found to be more discriminating than the 1b enzyme and more sensitive to N1 structural modifications.[72a] Abbott also described C7-substituted analogs with improved profiles.[72b,c] For example, 7-alkoxyacetamide analog **49** (A-782759) displayed improved replicon potency and pharmacokinetic profiles compared with previously described analogs,[73a] while the corresponding sulfonamide or sulfamide (**50**) provided dramatically improved potency in this series ($EC_{50} = 3$ nM), remaining <100 nM when the replicon assay was performed in presence of 40% serum.[73b]

Whereas in replicon experiments A-782759 (**49**) and a protease inhibitor (BILN, 2061, celuprevir) alone were not able to reduce HCV RNA to undetectable levels due to the emergence of resistant mutants (M414T in the case of **49**), a combination of NS5B inhibitor **49** and celuprevir resulted in undetectable replicon RNA after 16 days, suggesting that such combinations could be effective at suppressing emergence of pre-existing resistant mutations in the patient population, thereby improving the efficacy of HCV therapy by reducing the likelihood of developing resistance and rapid viral rebound.[73]

Even though this generation of benzothiadiazine derivatives was successfully optimized to provide analogs with exquisite cell-based potency in gt1a and 1b replicons and some analogs (*e.g.*, **48**) exhibited favorable pharmacokinetic profiles in preclinical species, others suffered from extremely poor aqueous solubility, complicating formulation development and compromising further *in vivo* evaluation. A strategy was therefore developed to address this issue, which entailed replacing the ring nitrogen of analogs such as **48** with a quaternary carbon center to reduce planarity. This core modification was expected to minimize π-stacking of molecules in the solid state, reducing crystallinity and thus improving solubility. The relationship between topology and increased sp^3 character of molecules and improved solubility and ADME properties is nowadays a well-established paradigm within the medicinal chemistry community.[74] Encouraging initial results were obtained with more synthetically accessible achiral vinylogous acid derivatives such as **51** (Figure 8.12) that demonstrated the viability of this approach in terms of gt1a and 1b replicon potency, pharmacokinetic profile in rats (low clearance and high bioavailability) and plasma-to-liver distribution (maximum liver/plasma ratio = 48 at 6 h).[75a]

Chemistry was subsequently developed to access synthetically more challenging unsymmetric derivatives and investigate substitution patterns that would provide optimal interactions within the benzothiadiazine binding pocket and also antiviral potency in the replicon assays. The effort produced racemic and subsequently enantiomerically pure *R*-isomers of analogs **52** (A-837093) and **53** (A-848837) that exhibited excellent single-digit nanomolar potency.[75b] The corresponding *S*-distomers were 500–2500-fold less potent in biochemical assays. In the presence of human serum, EC_{50} values shifted 14–17-fold, compared with >50-fold for aza analogs such as **50**. While the effect on solubility of introducing a quaternary carbon center in the scaffold was somewhat ambiguous depending on pH and solvent, a significant decrease in melting point was observed for the free form of **53**, suggesting weaker association of molecules within the crystalline lattice.

Both analogs **52** and **53** were profiled *in vivo*. In the rat, isoamyl analog **52** suffered from a short elimination half-life (1.2 h). A cannulated bile duct experiment indicated that biliary excretion was the major route of excretion and hydroxylated derivative **54** was identified as the main metabolite in this species.[75b] The sodium salt of *tert*-butyl analog **53**, on the other hand, displayed an improved i.v. profile with a 3–4-fold improvement in clearance and elimination half-life in rats (4.5 h). Prolonged exposure was observed following oral administration, presumably as a consequence of the low solubility of the precipitated form of the neutralized species in the acidic environment of the stomach. The liver distribution was similar for both compounds, with liver/plasma ratio $\sim$20 at 12 h in rats. In higher species, *in vivo* parameters correlated well for both compounds with *in vitro* clearance data obtained in microsomes and hepatocytes. Overall, analog **53** is more potent than **52** and both compounds exhibited attractive pharmacokinetic profiles (**53** featuring improved metabolic stability and slightly superior oral *in vivo* profiles), justifying further investigation in the clinic.

The chimpanzee remains the only animal model susceptible to long-term HCV infection and has been used on several occasions to investigate the pharmacokinetic–pharmacodynamic effect of nucleoside and non-nucleoside HCV polymerase inhibitors. The effect of **52** (A-837093) in HCV-infected chimpanzees has been reported.[76] In this proof-of-concept study, two animals were treated orally with A-837093 at a dose of 30 mg kg^{-1} administered bid for 14 days. Maximum viral load reductions of $1.4\log_{10}$ and $2.5\log_{10}$ were observed in gt1a and gt1b HCV-infected chimpanzees, respectively, within 2 days of administration. Viral rebound was observed in both cases and clonal analysis of NS5B gene sequences revealed several mutations associated with resistance to A-837093, including, among others, changes that encode C316Y and G554D in the gt1b-infected animal, in agreement with *in vitro* selection studies. This study validated the antiviral efficacy and resistance selection of benzothiadiazine palm site 1 NS5B inhibitors *in vivo*.

A computer-generated model of **52** bound to palm site 1 suggested that the methyl group of this class of inhibitors projected towards a polar region of the enzyme (R386 and S367), providing a new opportunity to explore this region using slightly modified analogs. To this end, the methyl group was replaced by substituted amine analogs with potentially improved properties, such as *in vitro* potency, PK or physicochemical properties. Benzylic amine **55** is an example of an analog that met these objectives, providing an attractive balance among low nanomolar potency in cell-based replicons (1a/1b $EC_{50} = 10/6$ nM with a 17-fold shift in presence of human serum), improved elimination half-life in rats (3.1 h) and bioavailability (56%) compared with **53**.[77] No information is currently available on further progression of these compounds.

Several companies in addition to GSK and Abbott Laboratories (*e.g.*, Anadys, InterMune/WuXi AppTec, Roche/Array BioPharma, Idenix) have investigated functional equivalents to benzothiadiazine palm site 1 NS5B inhibitors. These endeavors have been reviewed recently and some are discussed later.[78] Efforts by the Abbott group to simplify the scaffold through removal of the A-ring have produced analogs of potency comparable to some of the best tetracyclic systems.[79] These compounds were designed using structural information by initial truncation of the A-ring, which resulted in a significant decrease in potency, although addition of substituents on the remaining pyridone ring provided an avenue for compensation by refilling the space that was left unoccupied. Thus, inhibitors such as **56** (Figure 8.12) provided structurally diverse alternatives to the original quinolone systems, while maintaining a comparable level of potency against both gt1a and gt1b enzymes and replicons with somewhat reduced serum shifts (~6–24-fold for des-A-ring analogs *versus* 60–360-fold for the four-ring systems). The PK profile of this class of compounds has not yet been reported. Finally, modification to the benzothiadiazine ring system of palm site 1 inhibitors described by Abbott and others suggests that the sulfonyl moiety implanted within the benzothiadiazine ring system may be less critical than originally thought, opening the door to still new structurally diverse analogs.[78,80]

8.3.3.3 Benzothiadiazines from Anadys Pharmaceuticals

In 2008, Anadys Pharmaceuticals began to disclose their work on benzothiadiazine palm site 1 NS5B inhibitors, which ultimately led to the discovery of setrobuvir, currently in Phase 2 clinical trials. Based on a computational analysis of NS5B allosteric sites, which suggested that the palm domain was well conserved across gt1 HCV, efforts were focused on molecules that bind to this site and, more particularly, on the benzothiadiazine class for which potent starting points and structural information were available from the literature. Using a structure-guided approach and building on previously published knowledge, initial efforts led to a series of pyridazin-2-one des-A-ring analogs (*e.g.*, **57**, Figure 8.13) reminiscent of the pyridone analogs (*e.g.*, **56**) described by Abbott Laboratories.[79,81] While very potent inhibitors of the gt1b-HCV replicon were rapidly identified, most of the challenges lay in the optimization of the pharmaceutical and DMPK profiles of this class of molecules (*e.g.*, aqueous solubility, lipophilicity and bioavailability). Indeed, advancement of molecules such as **57** was compromised by their high polar surface area (PSA $= 203\,\text{Å}^2$), resulting in low permeability, high efflux ratio and very low bioavailability in preclinical animal species.[81] Although some understanding of the parameters that negatively impacted the DMPK profile of these molecules led to some modest improvements (10–15% bioavailability), these could not be achieved without compromising the antiviral potency profile (EC$_{50} = 110$–320 nM) and focus was switched to an alternative strategy.

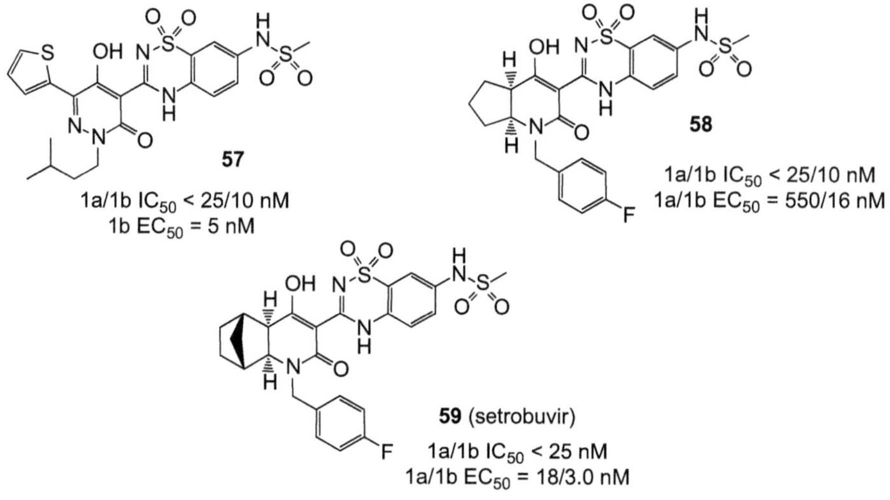

57

1a/1b IC$_{50}$ < 25/10 nM
1b EC$_{50}$ = 5 nM

58

1a/1b IC$_{50}$ < 25/10 nM
1a/1b EC$_{50}$ = 550/16 nM

59 (setrobuvir)

1a/1b IC$_{50}$ < 25 nM
1a/1b EC$_{50}$ = 18/3.0 nM

Figure 8.13   Palm site 1 inhibitors from Anadys Pharmaceuticals and discovery of setrobuvir.

Several structure-based modifications to the AB-ring system (*e.g.*, pyrrolo[1,2-*b*]pyridazin-2-ones) combined with modifications of the benzo-thiadiazine moiety (*e.g.*, dioxoisothiazoles, benzothiophene dioxides, benzo-thiazines) aimed at reducing the PSA and improving permeability were explored, again leading to the discovery of very potent molecules, but low bioavailability remained a persistent issue for these derivatives also.[78] The subsequent strategy investigated saturation of the A-ring system as a means to disrupt planarity in this part of the molecule, decrease PSA and improve bioavailability. Dihydropyridin-2-ones offered the first hints of success for this approach as the modification was well tolerated and excellent replicon potency was maintained for gt1b. Furthermore, analogs such as **58** (Figure 8.13) exhibited good metabolic stability, low clearance and some improvement in permeability ($P_{\text{app}} = 1.6 \times 10^{-6}\,\text{cm s}^{-1}$), resulting in moderate bioavailability (24%) in cynomolgous monkeys.[82]

In order to address the lower potency seen in the gt1a replicon compared with 1b ($\text{EC}_{50} = 550$ and $16\,\text{nM}$, respectively), a novel AB-ring modification containing a saturated tricyclic dihydropyridinone motif was designed, which led to the selection of the clinical candidate setrobuvir (**59**). The rationale behind the design of this structural motif was derived from examining X-ray crystal structures of analogs such as **58** bound to palm site 1 of NS5B.[83] Analysis of the shallow surface pocket interacting with the cyclopentane moiety of **58** suggested that more space-filling groups could be accommodated in this area and possibly lead to potency improvements. Modeling of compounds such as **59** into palm site 1 suggested that the orientation and all key interactions would be maintained between the ligand and the protein. Moreover, the bicyclic moiety of **59** appeared to fulfill the objective for which it was designed, *i.e.*, increasing occupancy of the sub-pocket. Several saturated, unsaturated and heterocyclic versions of fused pyridinones were evaluated in addition to different bridge sizes. In the end, compound **59** (setrobuvir) emerged as an analog with excellent antiviral potency against gt1a/1b-HCV ($\text{EC}_{50} = 18/3\,\text{nM}$), good solubility ($>100\,\mu\text{g mL}^{-1}$) and good plasma exposure levels in cyno-molgous monkeys, consistent with its *in vitro* metabolic profile and low *in vivo* clearance. Oral bioavailability (52% in monkeys) was also improved over previous analogs, even though apparent permeability in Caco-2 cells remained low for this compound ($1.3 \times 10^{-6}\,\text{cm s}^{-1}$). Unfortunately, further information on the compound's profile must await future publication. Setrobuvir (formally ANA598) is currently in Phase 2 clinical trials and is being developed by Roche following their acquisition of Anadys in October 2011 (see below).

8.3.3.4 Benzothiadiazines in Clinical Development

Four palm site 1 NS5B inhibitors are currently in clinical development. Two benzothiadiazine derivatives, ABT-072 and ABT-333 (the structures of these compounds have not yet been disclosed), were simultaneously progressed into the clinic by Abbott Laboratories. Both compounds are potent inhibitors of gt1a/1b HCV with $\text{EC}_{50} < 10\,\text{nM}$ and are considerably less potent against non-1 genotypes ($>2\,\mu M$). They exhibit favorable PK profiles and good distribution to the liver (liver/plasma ratio ranging from 10 in dogs to 70 in rats and monkeys at 12 h for ABT-072).[84]

ABT-072 and ABT-333 were evaluated in parallel in Phase 2a studies where treatment-naive patients were treated for 12 weeks with either 100, 300 or 600 mg qd of ABT-072 or 400 mg or 800 mg bid of ABT-333, both in combination with PegIFN/RBV following 3 days of monotherapy with one of the two drugs alone. SVR_{12} was achieved in up to 86% of subjects in the ABT-072 high-dose group, while similar 63% SVR_{12} was obtained with both ABT-333 doses, although these studies had fewer than 10 patients per treatment arm and larger studies are required to elucidate response rates in a broader population.[85]

Having demonstrated *in vitro* that combinations of ABT-333 with either NS3/4A protease or NS5A inhibitors produced additive to synergistic effects and were effective at increasing the genetic barrier to resistance, Abbott initiated Phase 2 studies on IFN-free combinations of ABT-072 (400 mg qd) or ABT-333 (400 mg bid) with a ritonavir-boosted PI (ABT-450) and RBV. Treatment-naive gt1-HCV patients who were administered ABT-072 for 12 weeks achieved 91% SVR_{24} and 82% SVR_{36}. Two ABT-333 (400 mg bid) arms in combination with either 250/100 mg ABT450/r qd or 150/100 mg ABT-450/r qd achieved 95% and 93% SVR_{12}, respectively. In the case of treatment-experienced patients, treatment with ABT-450/r (150/100 mg qd) + ABT-333 (400 mg bid) + RBV resulted in a lower SVR_{12} rate of 47%, with 6/17 patients experiencing viral breakthrough.[86]

The third benzothiadiazine palm site 1 compound currently in Phase 2 clinical development is setrobuvir (ANA598). Anadys initiated a Phase 1 study of setrobuvir in 2008 and the drug was reported to be well tolerated, achieving 2.4, 2.3 and $2.9\log_{10}$ reductions in HCV RNA at doses of 200, 400 and 800 mg administered tid for 4 days to gt1 HCV patients.[87] An ongoing Phase 2b study is investigating the safety, tolerability and efficacy of setrobuvir (200 and 400 mg bid) in combination with PegIFN and RBV in TN gt1-HCV subjects. Twelve week data have been reported for the setrobuvir 200 mg bid dose group, for which 78% of treatment-naive patients had undetectable HCV RNA (cEVR). Partial responders and relapsers experienced similar outcomes (cEVR = 76%). Viral breakthrough rates were very low (~ 3%) in both patient categories.[88]

The structure of the fourth clinical candidate, IDX375 (**60**, Figure 8.14; Idenix Pharmaceuticals) was recently disclosed.[89] This palm site 1 NS5B inhibitor features some significant structural departures from the original benzothiadiazine scaffold. The left-hand side of the molecule is a monocyclic tetramic acid derivative whereas the benzothiadiazine moiety was modified to a phosphorus isostere. The compound exhibits high antiviral potency in the gt1b replicon ($EC_{50} = 2.3\,nM$) and a 2.7-fold shift ($EC_{50} = 6.2\,nM$) against gt1a. IDX375 showed adequate exposure in preclinical PK, favorable liver distribution and moderate bioavailability ranging from 16 to 42%. In a first-in-man study, **60** was well tolerated and well absorbed with long elimination

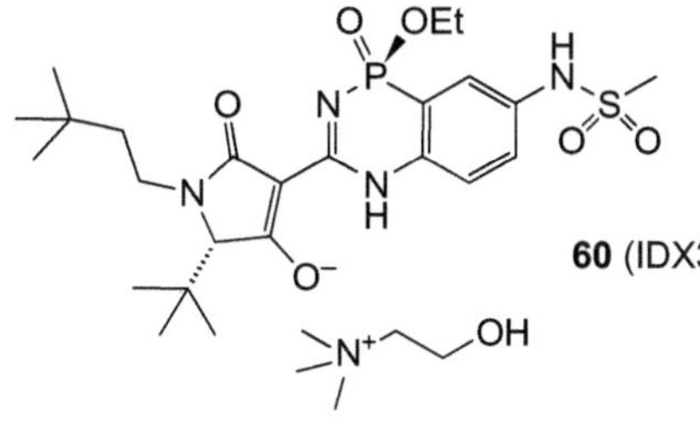

60 (IDX375)

1a/1b EC$_{50}$ = 6/2 nM

Figure 8.14 Structure of palm site 1 NS5B inhibitor IDX375 from Idenix Pharmaceuticals.

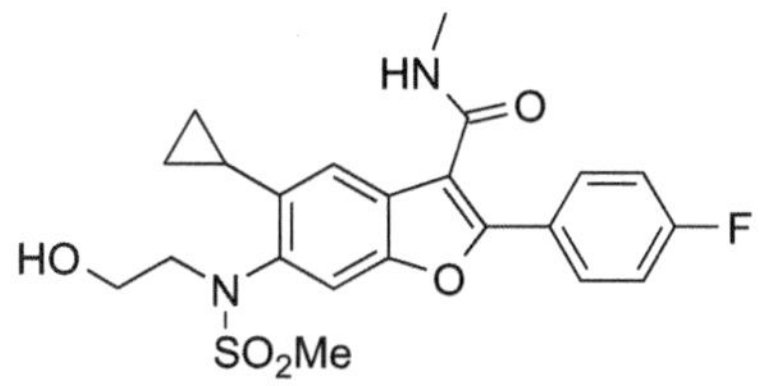

61 (HCV-796)
IC$_{50}$ (NS5B) = 0.04 μM
1a/1b EC$_{50}$ = 9/5 nM

Figure 8.15 Benzofuran palm site 2 NS5B inhibitor HCV-796 from Wyeth/ ViroPharma.

half-lives ($t_{\frac{1}{2}}$ = 30–40 h) at doses of 200 mg administered once or twice daily. A 200 mg bid dose for 1 day resulted in a maximum 0.5–1.1log$_{10}$ reduction in plasma HCV RNA.

8.3.4 Palm Site 2 Inhibitors

As part of a research collaboration that took place in the early 2000s between ViroPharma and Wyeth Research, a fourth series of allosteric NS5B inhibitors was discovered through screening of sample collections and produced clinical candidates. Early compounds (*e.g.*, HCV-086, EC$_{50}$ = 200 nM) exhibited modest activity, poor metabolic stability and low solubility and had low efficacy in the clinic. Subsequent optimization to improve potency, ADME-PK and genotype profiles led to the discovery of HCV-796 (**61**, Figure 8.15) a benzofurancarboxamide derivative that displayed a very attractive preclinical profile and was the first NS5B allosteric inhibitor for which promising clinical activity was demonstrated in HCV-infected patients.

Unfortunately, neither the structure of the original screening hit nor the lead optimization process which led to the discovery of HCV-086 and HCV-796 has been reported in the literature. HCV-796 is a specific and reversible inhibitor of HCV NS5B polymerases from genotypes 1a/1b, 3 and 4 (IC$_{50}$ = 0.01–0.57 μM) and has reduced potency against genotype 2 (IC$_{50}$ = 1.7 μM). It is non-competitive with respect to substrates NTPs and RNA template and acts at the initiation step of RNA synthesis. In the replicon cell-based assay, **61** inhibited gt1a/1b at single-digit nanomolar concentrations ($EC_{50} = 9/5$ nM). Using the replicon system, HCV variants that harbor mutations which confer resistance to HCV-796 were isolated at several amino acid residues, mapping a new binding pocket adjacent to the enzyme active site. Sensitivity to major C316Y mutant in close proximity to the catalytic triad of the polymerase (GDD motif) in particular was rationalized using X-ray structural data.[90] Binding of HCV-796 analogs to NS5B was found to induce a conformational change in the R200 hinge site, revealing a new pocket that only partially overlaps with that of benzothiadiazines (palm site 1 binders), as depicted in Figure 8.1.[9a] HCV-796 was evaluated in a chimeric mouse model of HCV infection and, unlike its predecessors, produced a $2\log_{10}$ reduction in HCV titers.[91]

In 2006, HCV-796 was advanced into a Phase 1 proof-of-concept clinical trial and a modest $1.4\log_{10}$ reduction in viral load was obtained for a 1000 mg bid dose in gt1 HCV patients. Furthermore, on-treatment rebound was observed in this monotherapy trial, with the C316Y mutant emerging in half of the patients. In combination with PegIFN/RBV, a more robust antiviral effect was obtained, with 2.6–$3.2\log_{10}$ reductions in virus for gt1 patients. Unfortunately, soon after Phase 2 trials were initiated with HCV-796 in combination with SOC, clinically significant liver enzyme elevations were noted in 8% of patients following 8 weeks of therapy. Development of this compound was subsequently terminated owing to safety concerns.[90]

In the last few years, there seems to have been renewed interest in this class of palm site 2 NS5B inhibitors, as several patent applications have been filed by companies (*e.g.*, Abbott Laboratories, Biota, Bristol-Myers Squibb, GlaxoSmithKline, Presidio Pharmaceuticals) claiming analogs of the benzofurancarboxamide chemotype. One such compound, from Presidio Pharmaceuticals (PPI-383, structure undisclosed), was recently presented as a next-generation NS5B non-nucleoside inhibitor with potent replicon activity against major HCV genotypes 1a/b, 2a, 3 and 4a ($EC_{50} = 3.0$–13 nM).[92] Resistant variants with decreased susceptibility to PPI-383 were isolated and encoded S365T/A and C316Y mutants, consistent with inhibitors targeting the palm site 2 binding site. Unlike many NS5B allosteric inhibitors, PPI-383 is only moderately protein bound (66%) and exhibited good oral bioavailability in preclinical animal species and plasma half-lives consistent with once- or twice-daily dosing in humans. Time will tell if these new derivatives are able to provide good efficacy and, at the same time, address the safety issues encountered during the development of HCV-796.

8.3.5 Covalent NS5B Inhibitors

Imidazopyridines, shown several years ago to possess potent cell-culture activity against pestivirus replication (*e.g.*, BVDV, **62**, Figure 8.16), were optimized to a series of specific HCV replication inhibitors through a collaboration between Gilead Sciences and the Rega Institute in Leuven. Key structural features necessary for selectivity and high anti-HCV potency in the

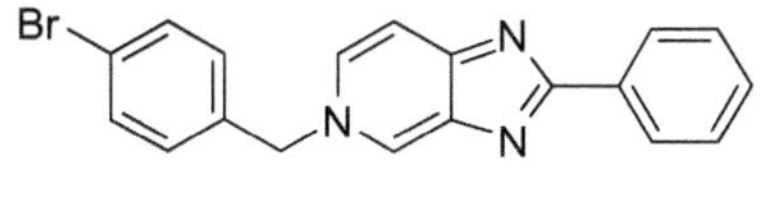

62

BVDV EC_{50} = 0.07 µM
HCV EC_{50} > 50 µM

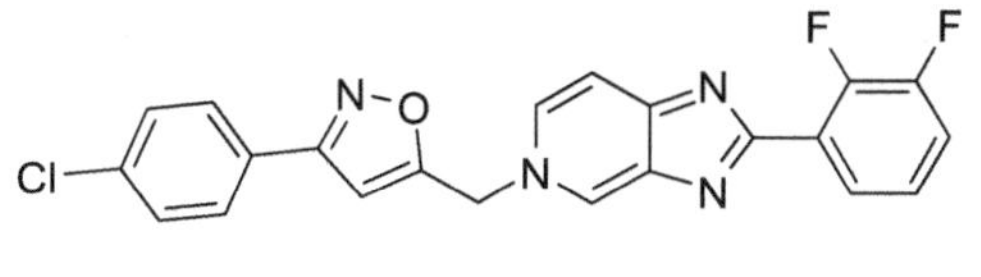

63

BVDV EC_{50} = 1.4 µM
HCV EC_{50} = 0.004 µM

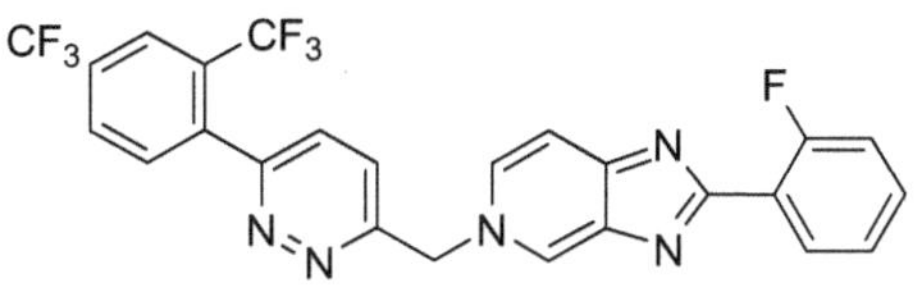

64 (Tegobuvir/GS-9190)
1a/1b EC_{50} = 2.5/0.7 nM

Figure 8.16 Imidazopyridine HCV replication inhibitors that interact with NS5B – discovery of tegobuvir (GS-9180).

low nanomolar range included the presence of a halogen atom at the *ortho*-position of the 2-phenyl ring and extension from the N5 position with aryl-substituted heterocycles (*e.g.*, **63**). Tegobuvir (**64**, formerly GS-9190) is a member of this class that was selected for clinical evaluation.[93]

Interestingly, although this class of highly potent HCV replication inhibitors did not inhibit any of the usual viral targets in biochemical assays, virus specificity and significantly reduced sensitivity to non-1 HCV genotypes (*e.g.*, gt2–6) strongly suggested that a host target was unlikely to be involved in its mechanism of action. Furthermore, selection experiments performed with the replicon led to the identification of NS5B mutants (C316Y, C445F, Y448H and Y452H) which conferred resistance to imidazopyridines and were not cross-resistant to other classes of HCV DAAs.[94] Although biophysical methods such as NMR, X-ray crystallography and photoaffinity labeling did not provide definitive evidence that inhibitors could bind to NS5B, several clues about the mechanism of action were gathered from mutational studies which suggested that the β-hairpin region of the thumb domain played an important role in the mechanism of inhibition. Based on computer docking experiments, SAR studies and the mutation signature of imidazopyridines, a putative binding site was proposed for the class that encompasses both palm sites 1 and 2.[94a] Interestingly, further understanding of the mechanism of action of tegobuvir was derived from metabolic studies, which suggested that the antiviral effect is mediated through a unique oxidative chemical activation pathway and subsequent interaction with NS5B to form a covalent complex. The formation of NS5B–tegobuvir covalent adducts was dependent on cellular glutathione levels and CYP1A activity. A mechanistic pathway was proposed that begins with CYP-mediated oxidation of the 2-fluorophenyl ring to produce a reactive epoxide, followed by glutathione adduct formation and conjugation to NS5B to yield a replication-deficient enzyme.[95]

In Phase 1 clinical trials, tegobuvir provided modest ($1.6\log_{10}$ and $1.95\log_{10}$ at 40 and 120 mg bid) viral load reductions following 8 days of monotherapy, but concerns over QT prolongation observed at the high dose limited future studies to the lower 40 mg dose.[96] NS5B resistance analysis revealed that 58% of patients in the monotherapy trial harbored a Y448H alone or combined with Y452H that seriously compromised the efficacy of the drug. In a Phase 2b clinical trial, supplementing tegobuvir (40 mg bid for 24 or 48 weeks) with PegIFN/RBV enhanced early virologic response but overall did not improve the SVR rate over SOC alone. Y448H in NS5B was the primary resistant mutant in gt1a/1b patients, reverting to wild-type in half the patients following the 12 week end of treatment.[97] *In vitro* experiments performed with the replicon showed that **64** was not cross-resistant to other classes of DAAs and combinations of tegobuvir with protease inhibitors or complementary NS5B inhibitors were either additive or slightly synergistic in curing cells from HCV replicons.[98] For that reason, the safety, tolerability and efficacy of tegobuvir when combined with a protease inhibitor (GS-9256) with and without PegIFN/RBV were investigated in subsequent trials. After 4 weeks of treatment, the two DAA combinations (with and without SOC followed by SOC alone) provided high early virological response rates (EVRs).[99a] High SVR_{12} rates (92%) were reported with 16 weeks of tegobuvir (20 mg bid) + GS-9256 (150 mg bid) with PegIFN/RBV. However, two severe pancytopenia adverse events were reported in two separate trials in patients administered tegobuvir/DAA/PegIFN/RBV, and all ongoing trials involving such quadruple combinations have been stopped.[99b] The most recent report on tegobuvir trials presented 12 week interim results for an interferon-free, all-oral combination of tegobuvir (30 mg bid) + NS5A inhibitor (GS-5885, 30 or 90 mg per day) + a new once-per-day NS3 PI (GS-9451, 200 mg qd) and RBV in gt1 HCV patients.[100] High SVR_4 rates (96%) were observed for patients who stopped therapy at week 12 and breakthrough occurred in only one subject (gt1a).

8.4 Conclusion

HCV NS5B has provided researchers with fertile ground for drug discovery. An unusual diversity of structurally disparate chemotypes has been uncovered over the years that were successfully optimized to provide development candidates. In addition to the nucleoside active-site inhibitors discussed elsewhere in this book (chapter 12), non-nucleoside inhibitors that bind to four distinct allosteric sites on the enzyme (thumb pockets 1 and 2, palm sites 1 and 2) have been identified. Representative compounds from these families are thought to interfere with important NS5B conformational changes at various stages of the initiation steps that are necessary for productive RNA replication. The antiviral efficacy of inhibitors from each of the four allosteric classes has been validated in the clinic, but the emergence of resistant virus has not provided a path forward for the use of these agents in monotherapy. While some of the drugs could be pursued in combination with PegIFN/RBV SOC, this approach does not provide any clear advantage over the more advanced second-generation PI + SOC combinations currently in Phase 3. As a result, non-nucleoside NS5B inhibitors are positioning themselves as combination partners in all-oral interferon-free regimens. So far, this paradigm shift has provided encouraging results in Phase 2, offering hope for the development of more efficacious and better-tolerated HCV therapies.

Acknowledgements

Drs George Kukolj and Timothy Stammers are thanked for their critical reading of the manuscript and suggestions.

References

1. (a) D. Lavanchy, *Clin. Microbiol. Infect.*, 2011, **17**, 107; (b) M. Wise, S. Bialek, L. Finelli, B. Bell and F. Sorvillo, *Hepatology*, 2008, **47**, 1128; (c) G. L. Davis, J. E. Albright, S. F. Cook and D. M. Rosenberg, *Liver Transpl.*, 2003, **9**, 331.
2. Q.-L. Choo, G. Kuo, A. J. Weiner, L. R. Overby, D. W. Bradley and M. Houghton, *Science*, 1989, **244**, 359.
3. P. Simmonds, *J. Gen. Virol.*, 2004, **85**, 3173.
4. (a) S. Zeuzem, T. Berg, B. Moeller, H. Hinrichsen, S. Mauss, H. Wedemeyer, C. Sarrazin, D. Hueppe, E. Zehnter and M. P. Manns, *J. Virol. Hepat.*, 2009, **16**, 75; (b) E. Björnsson, H. Verbaan, A. Oksanen, A. Frydén, J. Johansson, S. Friberg, O. Dalgård and E. Kalaitzakis, *Scand. J. Gastroenterol.*, 2009, **44**, 878.
5. S.-L. Tan and Y. He, *Hepatitis C antiviral Drug Discovery and Development*, Caister Academic Press, Norwich, 2011.
6. (a) I. Gentile, M. A. Carleo, F. Borgia, G. Castaldo and G. Borgia, *Expert Opin. Investig. Drugs*, 2010, **19**, 151; (b) D. A. Kwong, R. S. Kauffman and P. Mueller, *Nat. Biotechnol.*, 2011, **29**, 993; (c) K. Berman and P. Y. Kwo, *Clin. Liver Dis.*, 2009, **13**, 429; (d) T. Asselah and P. Marcellin, *Liver Int.*, 2011, **31**, 68.
7. (a) P. W. White, M. Llinàs-Brunet, M. Amad, R. C. Bethell, G. Bolger, M. G. Cordingley, J. Duan, M. Garneau, L. Lagacé, D. Thibeault and G. Kukolj, *Antimicrob. Agents Chemother.*, 2010, **54**, 4611; (b) T.-L. Lin, O. Lenz, G. Fanning, T. Verbinnen, F. Delouvroy, A. Scholliers, K. Vermeiren, Å. Rosenquist, M. Edlund, B. Samuelsson, L. Vrang, H. de Kock, P. Wigerinck, P. Raboisson and K. Simmen, *Antimicrob. Agents Chemother.*, 2009, **53**, 1377.
8. (a) A. A. Kolykhalov, K. Mihalik, S. M. Feinstone and C. M. Rice, *J. Virol.*, 2000, **74**, 2046; (b) S. Chinnaswamy, I. Yarbrough, S. Palaninathan, C. T. R. Kumar, V. Vijayaraghavan, B. Demeler, S. M. Lemon, J. C. Sacchettini and C. C. Kao, *J. Biol. Chem.*, 2008,

283, 20535; (c) D. Harrus, N. Ahmed-El-Sayed, P. C. Simister, S. Miller, M. Triconnet, C. H. Hagedorn, K. Mahias, F. A. Rey, T. Astier-Gin and S. Bressanelli, *J. Biol. Chem.*, 2010, **285**, 32906.

9. For recent reviews on NS5B inhibitors, see for example:(a) P. L. Beaulieu, *Expert Opin.Ther. Pat.*, 2009, **19**, 145; (b) W. J. Watkins, A. S. Ray and L. S. Chong, *Curr. Opin. Drug Discov. Dev.*, 2010, **13**, 441; (c) M. J. Sofia, W. Chang, P. A. Furman, R. T. Mosley and B. S. Ross, *J. Med. Chem.*, 2012, **55**, 2532.

10. (a) V. A. Soriano, *ACS Med. Chem. Lett.*, 2012, **3**, 440; (b) D. Gane, *Liver Int.*, 2011, **31**, 62; (c) D. G. Cordek, J. T. Bechtel, A. T. Maynard, W. M. Kazmierski and C. E. Cameron, *Drugs Future*, 2011, **36**, 691; (d) M. Sulkowski, D. Gardiner, E. Lawitz, F. Hinestrosa, D. Nelson, P. Thuluvath, M. Rodriguez-Torres, A. Lok, H. Schwartz, K. R. Reddy, T. Eley, M. Wind-Rotolo, J.-P Huang, M. Gao, F. McPhee, R. Hindes, B. Symonds, C. Pasquinelli and D. Grasela, *J. Hepatol.*, 2012, **56**(Suppl. 2), S560.

11. (a) S. Bressanelli, L. Tomei, A. Roussel, I. Incitti, R. L. Vitale, M. Mathieu, R. De Francesco and F. A. Rey, *Proc. Natl. Acad. Sci. U. S. A.*, 1999, **96**, 13034; (b) R. T. Mosley, T. E. Edwards, E. Murakami, A. M. Lam, R. L. Grice, J. Du, M. J. Sofia, P. A. Furman and M. J. Otto, *J. Virol.*, 2012, **86**, 6503.

12. (a) H. Hashimoto, K. Mizutani and A. Yoshida, *Patent Application*, WO01/47883, 2001; (b) P. L. Beaulieu, G. Fazal, J. Gillard, G. Kukolj and A. Austel, *US Patent*, 06448281, 2002.

13. (a) T. Ishida, T. Suzuki, S. Hirashima, K. Mizutani, A. Yoshida, I. Ando, S. Ikeda, T. Adachi and H. Hashimito, *Bioorg. Med. Chem. Lett.*, 2006, **16**, 1859; (b) P. L. Beaulieu, M. Bös, Y. Bousquet, G. Fazal, J. Gauthier, J. Gillard, S. Goulet, S. LaPlante, M.-A. Poupart, S. Lefebvre, G. McKercher, C. Pellerin, V. Austel and G. Kukolj, *Bioorg. Med. Chem. Lett.*, 2004, **14**, 119.

14. G. McKercher, P. L. Beaulieu, D. Lamarre, S. LaPlante, S. Lefebvre, C. Pellerin, L. Thauvette and G. Kukolj, *Nucleic Acids Res.*, 2004, **32**, 422.

15. P. L. Beaulieu, M. Bös, Y. Bousquet, P. DeRoy, G. Fazal, J. Gauthier, J. Gillard, S. Goulet, G. McKercher, M.-A. Poupart, S. Valois and G. Kukolj, *Bioorg. Med. Chem. Lett.*, 2004, **14**, 967.

16. G. Kukolj, G. A. McGibbon, G. McKercher, M. Marquis, S. Lefèbvre, L. Thauvette, J. Gauthier, S. Goulet, M.-A. Poupart and P. L. Beaulieu, *J. Biol. Chem.*, 2005, **280**, 39260.

17. S. LaPlante, A. Jakalian, N. Aubry, Y. Bousquet, J.-M. Ferland, J. Gillard, S. Lefebvre, M. Poirier, Y. Tsantrizos, G. Kukolj and P. L. Beaulieu, *Angew. Chem. Int. Ed.*, 2004, **43**, 4406.

18. V. Lohmann, F. Körner, J. Koch, U. Herian, L. Theilmann and R. Bartenschlager, *Science*, 1999, **285**, 110.

19. S. Hirashima, T. Suzuki, T. Ishida, S. Noji, S. Yata, I. Ando, M. Komatsu, S. Ikeda and H. Hashimoto, *J. Med. Chem.*, 2006, **49**, 4721.

20. P. L. Beaulieu, Y. Bousquet, J. Gauthier, J. Gillard, M. Marquis, G. McKercher, C. Pellerin, S. Valois and G. Kukolj, *J. Med. Chem.*, 2004, **47**, 6884.
21. (a) S. Goulet, M.-A. Poupart, J. Gillard, M. Poirier, G. Kukolj and P. L. Beaulieu, *Bioorg. Med. Chem. Lett.*, 2010, **20**, 196; (b) P. L. Beaulieu, N. Dansereau, J. Duan, M. Garneau, J. Gillard, G. McKercher, S. LaPlante, L. Lagacé, L. Thauvette and G. Kukolj, *Bioorg. Med. Chem. Lett.*, 2010, **20**, 1825.
22. (a) P. L. Beaulieu, J. Gillard, D. Bykowski, C. Brochu, N. Dansereau, J.-S. Duceppe, B. Haché, A. Jakalian, L. Lagacé, S. LaPlante, G. McKercher, E. Moreau, S. Perreault, T. Stammers, L. Thauvette, J. Warrington and G. Kukolj, *Bioorg. Med. Chem. Lett.*, 2006, **16**, 4987; (b) S. Harper, B. Pacini, S. Avolio, M. Di Filippo, G. Migliaccio, R. Laufer, R. De Francesco, M. Rowley and F. Narjes, *J. Med. Chem.*, 2005, **48**, 1314.
23. (a) R. Coulombe, P. L. Beaulieu, E. Jolicoeur, G. Kukolj, S. LaPlante and M.-A. Poupart, *Patent Application*, WO-04/099241, 2004; (b) S. Di Marco, C. Volpari, L. Tomei, S. Altamura, S. Harper, F. Narjes, U. Koch, M. Rowley, R. De Francesco, G. Migliaccio and A. Carfì, *J. Biol. Chem.*, 2005, **280**, 29765.
24. (a) S. Harper, S. Avolio, B. Pacini, M. Di Filippo, S. Altamura, L. Tomei, G. Paonessa, S. Di Marco, A. Carfì, C. Giuliano, J. Padron, F. Bonelli, G. Migliaccio, R. De Francesco, R. Laufer, M. Rowley and F. Narjes, *J. Med. Chem.*, 2005, **48**, 4547; (b) P. L. Beaulieu, E. Jolicoeur, J. Gillard, C. Brochu, R. Coulombe, N. Dansereau, J. Duan, M. Garneau, A. Jakalian, P. Kühn, L. Lagacé, S. LaPlante, G. McKercher, S. Perreault, M. Poirier, M.-A. Poupart, T. Stammers, L. Thauvette, B. Thavonekham and G. Kukolj, *Bioorg. Med. Chem. Lett.*, 2010, **20**, 857.
25. C. Giuliano, F. Fiore, A. Di Marco, J. Padron Velazquez, A. Bishop, F. Bonelli, O. Gonzalez-Paz, I. Marcucci, S. Harper, F. Narjes, B. Pacini, E. Monteagudo, G. Migliaccio, M. Rowley and R. Laufer, *Xenobiotica*, 2005, **35**, 1035.
26. S. R. LaPlante, J. R. Gillard, A. Jakalian, N. Aubry, R. Coulombe, C. Brochu, Y. S. Tsantrizos, M. Poirier, G. Kukolj and P. L. Beaulieu, *J. Am. Chem. Soc.*, 2010, **132**, 15204.
27. (a) P. L. Beaulieu, J. Gillard, E. Jolicoeur, J. Duan, M. Garneau and G. Kukolj, *Bioorg. Med. Chem. Lett.*, 2011, **21**, 3658; (b) P. L. Beaulieu, C. Chabot, J. Duan, M. Garneau, J. Gillard, E. Jolicoeur, M. Poirier, M.-A. Poupart, T. A. Stammers, G. Kukolj and Y. S. Tsantrizos, *Bioorg. Med. Chem. Lett.*, 2011, **21**, 3664.
28. P. L. Beaulieu, M. Bös, M. G. Cordingley, C. Chabot, G. Fazal, M. Garneau, J. R. Gillard, E. Jolicoeur, S. LaPlante, G. McKercher, M. Poirier, M.-A. Poupart, Y. S. Tsantrizos, J. Duan and G. Kukolj, *J. Med. Chem.*, 2012, **55**, 7650.
29. G. Kukolj, P. C. Anderson, M. Bös, M. G. Cordingley, R. Coulombe, J. Duan, M. Garneau, J. Gillard, S. Goulet, E. Jolicoeur, L. Lagacé,

D. Lamarre, S. LaPlante, M. Marquis, G. McKercher, C. Pellerin, M.-A. Poupart and P. L. Beaulieu, *J. Hepatol.*, 2011, **54**(Suppl. 1), S480.

30. A. Erhardt, K. Deterding, Y. Benhamou, M. Reiser, X. Forns, S. Pol, J.-L. Calleja, S. Ross, H.-C. Spangenberg, J. Garcia-Samaniego, M. Fuchs, J. Enríquez, J. Wiegand, J. Stern, K. Wu, G. Kukolj, M. Marquis, P. Beaulieu, G. Nehmiz and J. Steffgen, *Antiviral Ther.*, 2009, **14**, 23.

31. P. L. Beaulieu, P. C. Anderson, C. Brochu, M. Bös, M. G. Cordingley, J. Duan, M. Garneau, L. Lagacé, M. Marquis, G. McKercher, M.-A. Poupart, J. Rancourt, T. Stammers, Y. S. Tsantrizos and G. Kukolj, *J. Hepatol.*, 2012, **56**(Suppl. 2), S321.

32. D. G. Larrey, Y. Benhamou, A. W. Lohse, C. Trepo, C. Moelleken, J.-P. Bronowicki, M. H. Heim, K. Arastéh, J.-P. Zarski, M. Boulière, R. Wiest, J. L. Calleja, J. Enríquez, A. Erhardt, H. Wedemeyer, T. lah, G. Nehmiz, W. O. Boecher, F. M. Berger and J. Steffgen, *Hepatology*, 2009, **50**, 1044A.

33. D. Larrey, A. W. Lohse, V. de Ledinghen, C. Trepo, T. Gerlach, J.-P. Zarski, A. Tran, P. Mthurin, R. Thimme, K. Arastéh, C. Trautwein, A Cerny, N. Dikopoulos, M. Schuchmann, M. H. Heim, G. Gerken, J. O. Stern, K. Wu, N. Abdallah, B. Girlich, J. Scherer, F. Berger, M. Marquis, G. Kukolj, W. Böcher and J. Steffgen, *J. Hepatol.*, 2012, **57**, 39.

34. (a) S. Zeuzem, T. Asselah, P. Angus, J.-P. Zarski, D. Larrey, B. Müllhaupt, E. Gane, M. Schuchmann, A. Lohse, S. Pol, J.-P. Bronowicki, S. Roberts, K. Arastéh, F. Zoulim, M. Heim, J. O. Stern, G. Kukolj, G. Nehmiz, C. Haefner and W. O. Böcher, *Gastroenterology*, 2011, **141**, 2047; (b) S. Zeuzem, V. Soriano, T. Asselah, J.-P. Bronowicki, A. Lohse, B. Müllhaupt, M. Schuchmann, M. Bourlière, M. Buti, S. Roberts, E. Gane, J. O. Stern, G. Kukolj, L. Dai, W. O. Böcker and F. J. Mensa, *J. Hepatol.*, 2012, **56**(Suppl. 2), S45; (c) V. Soriano, E. Gane, P. Angus, F. Stickel, J.-P. Bronowicki, S. Roberts, M. Manns, S. Zeuzem, L. Dai, W. Boecher, J. Stern and F. Mensa, *J. Hepatol.*, 2012, **56**(Suppl. 2), S559.

35. (a) K. Ikegashira, T. Oka, S. Hirashima, S. Noji, H. Yamanaka, Y. Hara, T. Adachi, J.-I. Tsuruha, S. Doi, Y. Hase, T. Noguchi, I. Ando, N. Ogura, S. Ikeda and H. Hashimoto, *J. Med. Chem.*, 2006, **49**, 6950; (b) S. Hirashima, T. Oka, K. Ikegashira, S. Noji, H. Yamanaka, Y. Hara, H. Goto, R. Mizojiri, Y. Niwa, T. Noguchi, I. Ando, S. Ikeda and H. Hashimoto, *Bioorg. Med. Chem. Lett.*, 2007, **17**, 3181.

36. (a) I. Stanfield, C. Ercolani, A. Mackay, I. Conte, M. Pompei, U. Koch, N. Gennari, C. Giuliano, M. Rowley and F. Narjes, *Bioorg. Med. Chem. Lett.*, 2009, **19**, 627; (b) J. Habermann, E. Capitò, M. del Rosario Rico, U. Ferreira, Koch and F. Narjes, *Bioorg. Med. Chem. Lett.*, 2009, **19**, 633.

37. J. I. M. Hernando, J. M. Ontoria, S. Malancona, B. Attenni, F. Fiore, F. Bonelli, U. Koch, S. Di Marco, S. Colarusso, S. Ponzi, N. Gennari, S. E. Vignetti, M. del Rosario Rico, J. Ferreira, M. Habermann, Rowley and F. Narjes, *ChemMedChem*, 2009, **4**, 1695.

38. F. Narjes, B. Crescenzi, M. Ferrara, J. Habermann, S. Colarusso, M. del Rosario Rico Ferreira, I. Stanfield, A. C. Mackay, I. Conte, C. Ercolani, S. Zaramella, M.-C. Palumbi, P. Meuleman, G. R. Leroux-Roels, C. Giuliano, F. Fiore, S. Di Marco, P. Baiocco, U. Koch, G. Migliaccio, S. Altamura, R. Laufer, R. De Francesco and M. Rowley, *J. Med. Chem.*, 2011, **54**, 289.

39. X. Zheng, T. W. Hudyma, S. W. Martin, C. Bergstrom, M. Ding, F. He, J. Romine, M. A. Poss, J. F. Kadow, C.-H. Chang, J. Wan, M. R. Witmer, P. Morin, D. M. Camac, S. Sheriff, B. R. Beno, K. L. Rigat, Y.-K. Wang, R. Fridell, J. Lemm, D. Qiu, M. Liu, S. Voss, L. Pelosi, S. B. Roberts, M. Gao, J. Knipe and R. G. Gentles, *Bioorg. Med. Chem. Lett.*, 2011, **21**, 2925.

40. M. Ding, F. He, T. W. Hudyma, X. Zheng, M. A. Poss, J. F. Kadow, B. R. Beno, K. L. Rigat, Y.-K. Wang, R. A. Fridell, J. A. Lemm, D. Qiu, M. Liu, S. Voss, L. A. Pelosi, S. B. Roberts, M. Gao, J. Knipe and R. Gentles, *Bioorg. Med. Chem. Lett.*, 2012, **22**, 2866.

41. (a) J. F. Kadow, R. Gentles, M. Ding, J. Bender, C. Bergstrom, K. Grant-Young, P. Hewawasam, T. Hudyma, S. Martin, A. Nickel, A. Regueiro-Ren, Y. Tu, Z. Yang, K.-S. Yeung, X. Zheng, B.-C. Chen, S. Chao, J.-H. Sun, J. Li, A. Mathur, D. Smith, D.-R. Wu, B. Beno, U. Hanumegowda, J. Knipe, D. D. Parker, X. Zhuo, J. Lemm, M. Liu, L. Pelosi, K. Rigat, S. Voss, Y. Wang, Y.-K. Wang, R. Colonno, M. Gao, S. B. Roberts and N. A. Meanwell, presented at the 243rd ACS National Meeting, San Diego, CA, 25-29 March 2012, Abstract MEDI-23; (b) H. Tatum, P. Thuluvath, E. Lawitz, C. T. Martorell, M. Demicco, S. Cohen, V. Rustgi, N. Ravendhran, R. Ghalib, J. Hanson, J. Zamparo, R. Yang, E. Hughes and E. Cooney, *J. Hepatol.*, 2012, **56**(Suppl. 2), S460; (c) F. McPhee, P. Falk, P. Fracasso, J. Lemm, M. Liu, M. Kirk, D. Hernandez, E. Cooney, E. Hughes and M. Gao, *J. Hepatol.*, 2012, **56**(Suppl. 2), S473.

42. N. A. Meanwell, *J. Med. Chem.*, 2011, **54**, 2529.

43. (a) I. Stanfield, M. Pompei, I. Conte, C. Ercolani, G. Migliaccio, M. Jairaj, C. Giuliano, M. Rowley and F. Narjes, *Bioorg. Med. Chem. Lett.*, 2007, **17**, 5143; (b) P. L. Beaulieu, R. Coulombe, J. Gillard, C. Brochu, J. Duan, M. Garneau, E. Jolicoeur, P. Kuhn, M.-A. Poupart, J. Rancourt, T. A. Stammers, B. Thavonekham and G. Kukolj, *Can. J. Chem.*, 2013, **91**, 66.

44. (a) D. McGowan, S. Vendeville, T.-I. Lin, A. Tahri, L. Hu, M. D. Cummings, K. Amssoms, J. M. Berke, M. Canard, E. Cleiren, P. Dehertogh, S. Last, E. Fransen, E. van der Helm, I. van den Steen, L. Vijgen, M.-C. Rouan, G. Fanning, O. Nyanguile, K. van Emelen, K. Simmen and P. Raboisson, *Bioorg. Med. Chem. Lett.*, 2012, **22**, 4431; (b) S. Vendeville, T.-I. Lin, L. Hu, A. Tahri, D. McGowan, M. D. Cummings, K. Amssoms, M. Canard, S. Last, I. van den Steen, B. Devogelaere, M.-C. Rouan, L. Vijgen, J. M. Berke, P. Dehertogh, E. Fransen, E. Cleiren, L. van der Helm, G. Fanning, K. van Emelen,

O. Nyanguile, K. Simmen and P. Raboisson, *Bioorg. Med. Chem. Lett.*, 2012, **22**, 4437; (c) M. D. Cummings, T.-I. Lin, L. Hu, A. Tahri, D. McGowan, K. Amssoms, S. Last, B. Devogelaere, M.-C. Rouan, L. Vijgen, J. M. Berke, P. Dehertogh, E. Fransen, E. Cleiren, L. van der Helm, G. Fanning, K. van Emelen, O. Nyanguile, K. Simmen, P. Raboisson and S. Vendeville, *Angew. Chem. Int. Ed.*, 2012, **51**, 4637.

45. E. Marsault and M. L. Peterson, *J. Med. Chem.*, 2011, **54**, 1961.

46. (a) B. Devogelaere, J. M. Berke, L. Vijgen, P. Dehertogh, E. Fransen, E. Cleiren, L. van der Helm, O. Nyanguile, A. Tahri, K. Amssoms, O. Lenz, M. D. Cummings, R. F. Clayton, S. Vendeville, P. Raboisson, K. A. Simmen, G. C. Fanning and T.-I. Lin, *Antimicrob. Agents Chemother.*, 2012, **56**, 4676; (b) J. Leempoels, H. Reesink, S. Bourgeois, L. Vijgen, M.-C. Rouan, A. Vandebosch, G. Ispas, K. Marien, P. van Remoortere, G. Fanning, K. Simmen and R. Verloes, *Hepatology*, 2011, **54**(Suppl. 1), 136A.

47. (a) R. A. Love, X. Yu, W. Diehl, M. J. Hickey, H. E. Parge, J. Gao and S. Fuhrman, *European Patent*, EP 12566628, 2002; (b) R. A. Love, H. E. Parge, X. Yu, M. J. Hickey, W. Diehl, J. Gao, H. Wriggers, A. Ekker, L. Wang, J. A. Thomson, P. S. Dragovich and S. A. Fuhrman, *J. Virol.*, 2003, **77**, 7575; (c) M. Wang, K. K.-S. Ng, M. M. Cherney, L. Chan, C. G. Yannopoulos, J. Bedard, N. Morin, N. Nguyen-Ba, M. H. Alaoui-Ismaili, R. C. Bethell and M. N. G. James, *J. Biol. Chem.*, 2003, **278**, 9489; (d) B. K. Biswal, M. M. Cherney, M. Wang, L. Chan, C. G. Yannopoulos, D. Bilimoria, O. Nicolas, J. Bedard and M. N. G. James, *J. Biol. Chem.*, 2005, **280**, 18202.

48. S. R. Turner, J. W. Strohbach, R. A. Tommasi, P. A. Aristoff, P. D. Johnson, H. I. Skulnick, L. A. Dolak, E. P. Seest, P. K. Tomich, M. J. Bohanon, M.-M. Horng, J. C. Lynn, K.-T. Chong, R. R. Hinshaw, K. D. Watenpaugh, M. N. Janakiraman and S. Thaisrivongs, *J. Med. Chem.*, 1998, **41**, 3467.

49. H. Li, J. Tatlock, A. Linton, J. Gonzalez, A. Borchardt, P. Dragovich, T. Jewell, T. Prins, R. Zhou, J. Blazel, H. Parge, R. Love, M. Hickey, C. Doan, S. Shi, R. Duggal, C. Lewis and S. Fuhrman, *Bioorg. Med. Chem. Lett.*, 2006, **16**, 4834.

50. H. Li, A. Linton, J. Tatlock, J. Gonzalez, A. Borchardt, M. Abreo, T. Jewell, L. Patel, M. Drowns, S. Ludlum, M. Goble, M. Yang, J. Blazel, R. Rahavendran, H. Slor, S. Shi, C. Lewis and S. Fuhrman, *J. Med. Chem.*, 2007, **50**, 3969.

51. H. Li, J. Tatlock, A. Linton, J. Gonzalez, T. Jewell, L. Patel, S. Ludlum, M. Drowns, S. V. Rahavendran, H. Slor, R. Hunter, S. T. Shi, K. J. Herlihy, H. Parge, M. Hickey, X. Yu, F. Chau, J. Nonomiya and C. Lewis, *J. Med. Chem.*, 2009, **52**, 1255.

52. (a) S. T. Shi, K. J. Herlihy, J. P. Graham, J. Nonomiya, S. V. Rahavendran, H. Skor, R. Irvine, S. Binford, J. Tatlock, H. Li, J. Gonzalez, A. Linton, A. K. Patick and C. Lewis, *Antimicrob. Agents Chemother.*, 2009, **53**, 2544; (b) S. T. Shi, K. J. Herlihy, J. P. Graham,

S. A. Fuhrman, C. Doan, H. Parge, M. Hickey, J. Gao, X. Yu, F. Chau, J. Gonzalez, H. Li, C. Lewis, A. K. Patick and R. Duggal, *Antimicrob. Agents Chemother.*, 2008, **52**, 675.

53. (a) F. Wagner, R. Thompson, C. Kantaridis, P. Simpson, P. J. F. Troke, S. Jagannatha, S. Neelakantan, V. S. Purohit and J. L. Hammond, *Hepatology*, 2011, **54**, 50; (b) P. J. F. Troke, M. Lewis, P. Simpson, K. Gore, J. Hammond, C. Craig and M. Westby, *Antimicrob. Agents Chemoth.*, 2012, **56**, 1331.

54. V. S. Purohit, D. Fairman, J. Fang, M. Dickins, M. Rosario and J. Hammond, *Clin. Pharmacol. Ther.*, 2010, **87**(Suppl. 1), Abstract PI-73.

55. I. Jacobson, P. J. Pockros, J. Lalezari, E. Lawitz, M. Rodriguez-Torres, E. DeJesus, F. Haas, C. Martorell, R. Pruitt, V. S. Purohit, S. Srinivasan, S. Jagannatha, K. Rana and J. Hammond, *J. Hepatol.*, 2010, **52**, S465.

56. (a) L. Chan, T. J. Reddy, M. Proulx, S. K. Das, O. Pereira, W. Wang, A. Siddiqui, C. G. Yannopoulos, C. Poisson, N. Turcotte, A. Drouin, M. H. Alaoui-Ismaili, R. Bethell, M. Hamel, L. L'Heureux, D. Bilimoria and N. Nguyen-Ba, *J. Med. Chem.*, 2003, **46**, 1283; (b) T. J. Reddy, L. Chan, N. Turcotte, M. Proulx, O. Z. Pereira, S. K. Das, A. Siddiqui, W. Wang, C. Poisson, C. G. Yannopoulos, D. Bilimoria, L. L'Heureux, H. M. Alaoui and N. Nguyen-Ba, *Bioorg. Med. Chem. Lett.*, 2003, **13**, 3341.

57. L. Chan, S. K. Das, T. J. Reddy, C. Poisson, M. Proulx, O. Pereira, M. Courchesne, C. Roy, W. Wang, A. Siddiqui, C. G. Yannopoulos, N. Nguyen-Ba, D. Labrecque, R. Bethell, M. Hamel, P. Courtemanche-Asselin, L. L'Heureux, M. David, O. Nicolas, S. Brunette, D. Bilimoria and J. Bédard, *Bioorg. Med. Chem. Lett.*, 2004, **14**, 793.

58. L. Chan, O. Pereira, T. J. Reddy, S. K. Das, C. Poisson, M. Courchesne, M. Proulx, A. Siddiqui, C. G. Yonnapoulos, N. Nguyen-Ba, C. Roy, D. Nasturica, C. Moinet, R. Bethell, M. Hamel, L. L'Heureux, M. David, O. Nicolas, P. Courtemanche-Asselin, S. Brunette, D. Bilimoria and J. Bédard, *Bioorg. Med. Chem. Lett.*, 2004, **14**, 797.

59. C. G. Yonnapoulos, P. Xu, F. Ni, L. Chan, O. Z. Pereira, T. J. Reddy, S. K. Das, C. Poisson, N. Nguyen-Ba, N. Turcotte, M. Proulx, L. Halab, W. Wang, J. Bédard, N. Morin, M. Hamel, O. Nicolas, D. Bilimoria, L. L'Heureux, R. Bethell and G. Dionne, *Bioorg. Med. Chem. Lett.*, 2004, **14**, 5333.

60. B. K. Biswal, M. Wang, M. M. Cherney, L. Chan, C. G. Yonnapoulos, D. Bilimoria, J. Bedard and M. N. G. James, *J. Mol. Biol.*, 2006, **361**, 33.

61. C. Cooper, E. J. Lawitz, P. Ghali, M. Rodriguez-Torres, F. H. Anderson, S.S. Lee, J. Bédard, N. Chauret, R. Thibert, I. Boivin, O. Nicolas and L. Proulx, *J. Hepatol.*, 2009, **51**, 39.

62. (a) N. Chauret, C. Chagnon-Labelle, M. Diallo, J. Laquerre and S. Plante, *J. Hepatol.*, 2008, **48**(Suppl. 2), S317; (b) E. Lawitz, C. Cooper, M. Rodriguez-Torres, R. Ghalib, R. Lalonde, A. Sheikh, B. Bourgault,

N. Chauret, L. Proulx and J. McHutchison, *J. Heptol.*, 2009, **50**(Suppl. 1), S37.

63. (a) J. Bedard, O. Nicolas, D. Bilimoria, L. L'Heureux, P. Fex, M. David and L. Chan, *J. Hepatol.*, 2009, **50**(Suppl. 1), S340; (b) N. Chauret, C. Chagnon-Labelle, M. Diallo, J Laquerre, S. Laterreur, S. May and L. Ste-Marie, *J. Hepatol.*, 2009, **50**(Suppl. 1), S341.

64. (a) C. Cooper, R. Larouche, B. Bourgault, N. Chauret and L. Proulx, *J. Hepatol.*, 2009, **50**(Suppl. 1), S342; (b) M. Rodriguez-Torres, E. Lawitz, B. Conway, K. Kaita, A. M. Sheikh, R. Ghalib, R. Adrover, C. Cooper, M. Silva, M. Rosario, B. Bourgault, L. Proulx and J. G. McHutchison, *J. Hepatol.*, 2010, **52**(Suppl. 1), S14.

65. A. M. Di Bisceglie, D. R. Nelson, E. Gane, K. Alves, M. J. Koziel, C. De Souza, T. L. Kieffer, S. George, R. S. Kauffman, I. M. Jacobson and M. S. Sulkowski, *J. Hepatol.*, 2011, **54**(Suppl. 1), S540.

66. M. S. Penney, C. De Souza, S. Seepersaud, K. Alves, M. J. Koziel, R. S. Kauffman, A. M. Di Bisceglie and M. C. Botfield, *J. Hepatol.*, 2012, **56**(Suppl. 2), S476.

67. G. Yi, J. Deval, B. Fan, H. Cai, C. Soulard, C. T. Ranjith-Kumar, D. B. Smith, L. Blatt, L. Beigelman and C. C. Kao, *Antimicrob. Agents Chemother.*, 2012, **56**, 830.

68. (a) M. Fenaux, Y. Tian, M. Matles, E. Mabery, J. Zhang, S. Eng, B. Murray, J. Mwangi, S. Lazerwith, W. Lew, E. Canales, Q. Liu, D. Byun, E. Doerffler, H. Ye, M. Clarke, M. Mertzman, P. Morganelli, J. Zhang, S. Leavitt, T. Appleby, A. Hashash, A. Bidgood, S. Krawczyk and W. Watkins, *J. Hepatol.*, 2011, **54**(Suppl. 1), S476; (b) W. J. Watkins, T. C. Appleby, A. M. Bidgood, D. Byun, E. Canales, M. O. Clarke, E. Doerffler, S. Eng, M. Fenaux, P. German, A. Hashas, S. H. Krawczyk, S. E. Lazerwith, S. A. Leavitt, W. Lew, Q. Liu, E. M. Mabery, M. Matles, M. Mertzman, P. Morganelli, B. P. Murray, J. Mwangi, S. Rossi, Y. Tian, H. Ye and J. Zhang, presented at the 244th ACS National Meeting, Philadelphia, PA, 1923, August 2012, Abstract MEDI-18; (c) E. Lawitz, L. Hazan, D. Gruener, H. Hack, M. Backonja, J. Hill, P. German, H. Dvory-Sobol, A. Jain, S. Arterburn, W. J. Watkins, S. Rossi, J. McHutchison and M. Rodriguez-Torres, *J. Hepatol.*, 2012, **56**(Suppl. 2), S471.

69. (a) D. Dhanak, K. J. Duffy, V. K. Johnston, J. Lin-Goerke, M. Darcy, A. N. Shaw, B. Gu, C. Silverman, A. T. Gates, M. R. Nonnemacher, D. L. Earnshaw, D. J. Casper, A. Kaura, A. Baker, C. Greenwood, L. L. Gutshall, D. Maley, A. DelVecchio, R. Macarron, G. A. Hofmann, Z. Alnoah, H.-Y. Cheng, G. Chan, S. Khanfekar, R. M. Keenan and R. T. Sarisky, *J. Biol. Chem.*, 2002, **277**, 38322; (b) T. T. Nguyen, A. T. Gates, L. L. Gutshall, V. K. Jojnston, B. Gu, K. J. Duffy and R. T. Sarisky, *Antimicrob. Agents Chemother.*, 2003, **47**, 3525; (c) B. Gu, V. K. Johnston, L. L. Gutshall, T. T. Nguyen, R. R. Gontarek, M. G. Darcy, R. Tedesco, D. Dhanak, K. J. Duffy, C. C. Kao and R. T. Sarisky, *J. Biol. Chem.*, 2003, **278**, 16602.

70. R. Tedesco, A. N. Shaw, R. Bambal, D. Chai, N. O. Concha, M. D. Darcy, D. Dhanak, D. M. Fitch, A. Gates, W. G. Gerhardt, D. L. Halegoua, C. Han, G. A. Hofmann, V. K. Johnston, A. C. Kaura, N. Liu, R. M. Keenan, J. Lin-Goerke, R. T. Sarisky, K. J. Wiggall, M. N. Zimmerman and K. J. Duffy, *J. Med. Chem.*, 2006, **49**, 971.

71. A. N. Shaw, R. Tedesco, R. Bambal, D. Chai, N. O. Concha, M. G. Darcy, D. Dhanak, K. J. Duffy, D. M. Fitch, A. Gates, V. K. Johnston, R. M. Keenan, J. Lin-Goerke, N. Liu, R. T. Sarisky, K. J. Wiggall and M. N. Zimmerman, *Bioorg. Med. Chem. Lett.*, 2009, **19**, 4350.

72. (a) J. K. Pratt, P. Donner, K. F. McDaniel, C. J. Maring, W. M. Kati, H. Mo, T. Middleton, Y. Liu, T. Ng, Q. Xie, R. Zhang, D. Montgomery, A. Molla, D. J. Kempf and W. Kohlbrenner, *Bioorg. Med. Chem. Lett.*, 2005, **15**, 1577; (b) A. C. Krueger, D. L. Madigan, W. W. Jiang, W. M. Kati, D. Liu, Y. Liu, C. J. Maring, S. Masse, K. F. McDaniel, T. Middleton, H. Mo, A. Molla, D. Montgomery, J. K. Pratt, T. W. Rockway, R. Zhang and D. J. Kempf, *Bioorg. Med. Chem. Lett.*, 2006, **16**, 3367; (c) T. W. Rockway, R. Zhang, D. Liu, D. A. Betebenner, K. F. McDaniel, J. K. Pratt, D. Beno, D. Montgomery, W. W. Jiang, S. Masse, W. M. Kati, T. Middleton, A. Mola, C. J. Maring and D. J. Kempf, *Bioorg. Med. Chem. Lett.*, 2006, **16**, 3833.

73. (a) H. Mo, L. Lu, T. Pilot-Matias, R. Pithawalla, R. Mondal, S. Masse, T. Dekhtyar, T. Ng, G. Koev, V. Stoll, K. D. Stewart, J. Pratt, P. Donner, T. Rockway, C. Maring and A. Molla, *Antimicrob. Agents Chemother.*, 2005, **49**, 4305; (b) G. Koev, T. Dekhtyar, L. Han, P. Yan, T. I. Ng, C. T. Lin, H. Mo and A. Molla, *Antiviral Res.*, 2007, **73**, 78; (c) L. Lu, H. Mo, T. J. Pilot-Matias and A. Molla, *Antimicrob. Agents Chemother.*, 2007, **51**, 1889; (d) R. L. Tripathi, P. Krishnan, Y. He, T. Middleton, T. Pilot-Matias, C.-M. Chen, D. T. Y. Lau, S. M. Lemon, H. Mo, W. Kati and A. Molla, *Antiviral Res.*, 2007, **73**, 40.

74. (a) F. Lovering, J. Bikker and C. Humblet, *J. Med. Chem.*, 2009, **52**, 6752; (b) M. Ishikawa and Y. Hashimoto, *J. Med. Chem.*, 2011, **54**, 1539; (c) Y. Yang, O. Engkvist, A. Llinàs and H. Chen, *J. Med. Chem.*, 2012, **55**, 3667.

75. (a) D. K. Hutchinson, T. Rosenberg, L. L. Klein, T. D. Bosse, D. P. Larson, W. He, W. W. Jiang, W. M. Kati, W. E. Kohlbrenner, Y. Liu, S. V. Masse, T. Middleton, A. Molla, D. A. Montgomery, D. W. A. Beno, K. D. Stewart, V. S. Stoll and D. J. Kempf, *Bioorg. Med. Chem. Lett.*, 2008, **18**, 3887; (b) R. Wagner, D. P. Larson, D. W. A. Beno, T. D. Bosse, J. F. Darbyshire, Y. Gao, B. D. Gates, W. He, R. F. Henry, L. E. Hernandez, D. K. Hutchinson, W. W. Jiang, W. M. Kati, L. L. Klein, G. Koev, W. Kohlbrenner, A. C. Krueger, J. Liu, Y. Liu, M. A. Long, C. J. Maring, S. V. Masse, T. Middleton, D. A. Montgomery, J. K. Pratt, P. Stuart, A. Molla and D. J. Kempf, *J. Med. Chem.*, 2009, **52**, 1659.

76. C.-M. Chen, Y. He, L. Lu, H. B. Lim, R. L. Tripathi, T. Middleton, L. E. Hernandez, D. W. A. Beno, M. A. Long, W. M. Kati, T. D. Bosse,

D. P. Larson, R. Wagner, R. E. Landford, W. E. Kohlbrenner, D. J. Kempf, T. J. Pilot-Matias and A. Molla, *Antimicrob. Agents Chemother.*, 2007, **51**, 4290.

77. J. T. Randolph, C. A. Flentge, P. P. Huang, D. K. Hutchinson, L. L. Klein, H. B. Lim, R. Mondal, T. Reisch, D. A. Montgomery, W. W. Jiang, S. V. Masse, L. E. Hernandez, R. F. Henry, Y. Liu, G. Koev, W. M. Kati, K. D. Stewart, D. W. A. Beno, A. Molla and D. J. Kempf, *J. Med. Chem.*, 2009, **52**, 3174.

78. D. Das, J. Hong, S.-H. Chen, G. Wang, L. Beigelman, S. D. Seiwert and B. O. Buckman, *Bioorg. Med. Chem.*, 2011, **19**, 4690.

79. P. L. Donner, Q. Xie, J. K. Pratt, C. J. Maring, W. Kati, W. Jiang, Y. Liu, G. Koev, S. Masse, D. Montgomery, A. Molla and D. J. Kempf, *Bioorg. Med. Chem. Lett.*, 2008, **18**, 2735.

80. D. K. Hutchinson, C. A. Flentge, P. L. Donner, R. Wagner, C. J. Maring, W. M. Kati, Y. Liu, S. V. Masse, T. Middleton, H. Mo, D. Montgomery, W. W. Jiang, G. Koev, D. W. A. Beno, K. D. Stewart, V. S. Stoll, A. Molla and D. J. Kempf, *Bioorg. Med. Chem. Lett.*, 2011, **21**, 1876.

81. (a) L.-S. Li, Y. Zhou, D. E. Murphy, N. Stankovic, J. Zhao, P. S. Dragovich, T. Bertolini, Z. Sun, B. Ayida, C. V. Tran, F. Ruebsam, S. E. Webber, A. M. Shah, M. Tsan, R. E. Showalter, R. Patel, L. A. LeBrun, D. M. Bartkowski, T. G. Nolan, D. A. Norris, R. Kamran, J. Brooks, M. V. Sergeeva, L. Kirkovsky, Q. Zhao and C. R. Kissinger, *Bioorg. Med. Chem. Lett.*, 2008, **18**, 3446; (b) M. V. Sergeeva, Y. Zhou, D. M. Bartkowski, T. G. Nolan, D. A. Norris, E. Okamoto, L. Kirkovsky, R. Kamran, L. A. LeBrun, M. Tsan, R. Patel, A. M. Shah, M. Lardy, A. Gobbi, L.-S. Li, J. Zhao, T. Bertolini, N. Stankovic, Z. Sun, D. E. Murphy, S. E. Weber and P. S. Dragovich, *Bioorg. Med. Chem. Lett.*, 2008, **18**, 3421.

82. F. Ruebsam, C. V. Tran, L.-S. Li, S. H. Kim, A. X. Xiang, Y. Zhou, J. K. Blazel, Z. Sun, P. S. Dragovich, J. Zhao, H. M. McGuire, D. E. Murphy, M. T. Tran, N. Stankovic, D. A. Ellis, A. Gobbi, R.E. Showalter, S. E. Webber, A. M. Shaw, M. Tsan, R. A. Patel, L. A. LeBrun, H. J. Hou, R. Kamran, M. V. Sergeeva, D. M. Bartkowski, T. G. Nolan, D. A. Norris and L. Kirkovsky, *Bioorg. Med. Chem. Lett.*, 2009, **19**, 451.

83. (a) F. Ruebsam, D. E. Murphy, C. V. Tran, L.-S. Li, J. Zhao, P. S. Dragovich, H. M. McGuire, A. X. Xiang, Z. Sun, B. K. Ayida, J. K. Blazel, S. H. Kim, Y. Zhou, Q. Han, C. R. Kissinger, S. E. Webber, R. E. Showalter, A. M. Shah, M. Tsan, R. A. Patel, P. A. Thompson, L. A. LeBrun, H. J. Hou, R. Kamran, M. V. Sergeeva, D. M. Bartkowski, T. G. Nolan, D. A. Norris, J. Khandurina, J. Brooks, E. Okamoto and L. Kirkovsky, *Bioorg. Med. Chem. Lett.*, 2009, **19**, 6404; (b) P. S. Dragovich, P. A. Thompson and F. Ruebsam, *Patent Application*, WO-2010/042834, 2010.

84. (a) R. Wagner, C. Maring, P. L. Donner, J. T. Randolph, A. C. Krueger, T. W. Rockway, J. K. Pratt, D. Liu, M. Tufano, W. M. Kati, Y. Liu,

H. B. Lim, G. Koev, J. M. Beyer, R. Mondal, T. J. Reisch, P. Krishnan, D. W. A. Beno, Y. Y. Lau, J. Shen, J. S. Fisher, Y. Gao and A. Molla, *J. Hepatol.*, 2009, **50**(Suppl. 1), S352; (b) C. W. R. Maring, D. Hutchinson, C. Flentge, W. Kati, G. Koev, Y. Liu, D. Beno, J. Shen, Y. Lau, Y. Gao, J. Fisher, S. Vaidyanathan, H. Lim, J. Beyer, R. Mondal and A. Molla, *J. Hepatol.*, 2009, **50**(Suppl. 1), S347.

85. E. Poordad, E. Lawitz, E. DeJesus, K. V. Kowdley, I. Gaultier, D. E. Cohen, W. Xie, L. Larsen, T. Pilot-Matias and G. Koev, *J. Hepatol.*, 2012, **56**(Suppl. 2), S478.

86. (a) E. Lawitz, F. Poordad, K. V. Kowdley, D. Jensen, D. E. Cohen, S. Siggelkow, K. Wikstrom, L. Larsen, R. M. Menon, T. Podsadecki and B. Bernstein, *J. Hepatol.*, 2012, **56**(Suppl. 2), S7; (b) F. Poordad, E. Lawitz, K. V. Kowdley, G. T. Everson, B. Freilich, D. Cohen, S. Siggelkow, M. Heckaman, R. Menon, T. Pilot-Matias, T. Podsadecki and B. Bernstein, *J. Hepatol.*, 2012, **56**(Suppl. 2), S549.

87. E. Lawitz, M. Rodriguez-Torres, M. DeMicco, T. Nguyen, E. Godofsky, J. Appleman, M. Rahimy, C. Crowley and J. Freddo, *J. Hepatol.*, 2009, **50**(Suppl. 1), S384.

88. E. Lawitz, M. Rodriguez-Torres, V. K. Rustgi, T. Hassanein, M. H. Rahimy, C. A. Crowley, J. L. Freddo, A. Muir and J. McHutchison, *J. Hepatol.*, 2010, **52**(Suppl. 1), S467.

89. J. de Bruijne, J. van de Wetering de Rooij, A. A. van Vliet, X. J. Zhou, M. F. Temam, J. Molles, J. Chen, K. Pietropaolo, J. Z. Sullivan-Bólyai, D. Mayers and H. W. Reesink, *Antimicrob. Agents Chemother.*, 2012, **56**, 4525.

90. A. Y. M. Howe, H. Cheng, S. Johann, S. Mullen, S. K. Chunduru, D. C. Young, J. Bard, R. Chopra, C. Krishnamurthy, T. Mansour and J. O'Connell, *Antimicrob. Agents Chemother.*, 2008, **52**, 3327.

91. N. M. Kneteman, A. Y. M. Howe, T. Gao, J. Lewis, D. Pavear, G. Lund, D. Douglas, D. F. Mercer, D. L. J. Tyrell, F. Immermann, I. Chaudhary, J. Speth, S. A. Villano, J. O'Connell and M. Collett, *Hepatology*, 2009, **49**, 745.

92. R. Colonno, N. Huang, M. Lau, Q. Huang, A. Huq, E. Peng, M. Bencsik, M. Zhong and L. Li, *J. Hepatol.*, 2012, **56**(Suppl. 2), S464.

93. J. Vliegen, J. Paeshuyse, T. De Burghgraeve, L. S. Lehman, M. Paulson, I. H. Shih, E Mabery, N. Boddeker, E. De Clercq, H. Reiser, D. Oare, W. A. Lee, W. Zhong, S. Bondy, G. Purstinger and J. Neyts, *J. Hepatol.*, 2009, **50**, 999.

94. (a) I.-H. Shih, I. Vliegen, B. Peng, H. Yang, C. Hebner, J. Paeshuyse, G. Pürstinger, M. Fenaux, Y. Tian, E. Mabery, X. Qi, G. Bahador, M. Paulson, L. S. Lehman, S. Bondy, W. Tse, H. Reiser, W. A. Lee, U. Schmitz, J. Neyts and W. Zhong, *Antimicrob. Agents Chemother.*, 2011, **55**, 4196; (b) K. A. Wong, S. Xu, R. Martin, M. D. Miller and H. Mo, *Virology*, 2012, **429**, 57.

95. C. M. Hebner, B. Han, K. M. Brendza, M. Nash, M. Sulfab, Y. Tian, M. Hung, W. Fung, R. W. Vivian, J. Trenkle, J. Taylor, K. Bjornson,

S. Bondy, X. Liu, J. Link, J. Neyts, R. Sakowicz, W. Zhong, H. Tang and U. Schmitz, *PLos One*, 2012, **7**, e39163.

96. (a) J. Harris, A. Bae, S. Sun, E. Svarovskaia, M. Miller and H. Mo, presented at the 61st Annual Meeting of the American Association for the Study of the Liver Diseases, Boston, MA, 29 October–3 November 2010, Abstract 833; (b) L. Bavisotto, C. Wang, I. Jacobson, P. Marcellin, S. Zeuzem, S. Lawitz, N. M. Lunde, P. Sereni, C. O'Brien, D. Oldach, G. Rhodes and the GS-9190 Study Team, *Hepatology*, 2007, **46**(Suppl. 1), 255A; (c) Gilead Sciences, *Gilead Announces Phase I Data for GS 9190, an Investigational Compound for the Treatment of Chronic Hepatitis C*, Press Release, 4 November 2007, http://www.gilead.com/pr_1072088 (last accessed 23 February 2012).

97. (a) E. Lawitz, I. Jacobson, E. Godofsky, G. R. Foster, R. Flisiak, M. Bennett, M. Ryan, J. Hinkle, J. Simpson, J. McHutchison and D. Oldach, *J. Hepatol.*, 2011, **54**(Suppl. 1), S181; (b) C. Hebner, J. Harris, D. Oldach, M. D. Miller and H. Mo, *J. Hepatol.*, 2011, **54**(Suppl. 1), S478.

98. I. Vliegen, L. Delang, J. Paeshuyse, I.-H. Shih, G. Pürstinger, S. Bondy, W. A. Lee, W. Zhong and J. Neyts, *J. Hepatol.*, 2011, **54**(Suppl. 1), S197.

99. (a) G. R. Foster, P. Buggisch, P. Marcellin, S. Zeuzem, K. Agarwal, M. Manns, D. Sereni, H. Klinker, C. Moreno, J. P. Zarski, Y. Horsmans, M. Shelton, S. Arterburn, W. Lee, J. McHutchison, W. Delaney and D. Oldach, *J. Hepatol.*, 2011, **54**(Suppl. 1), S172; (b) D. R. Nelson, E. Lawitz, V. Bain, N. Gitlin, T. Hawkins, P. Marotta, K. Workowski, S. Flamm, P. S. Pang, J. G. McHutchison, J. McNally, P. Urbanek and E. M. Yoshida, *J. Hepatol.*, 2012, **56**(Suppl. 2), S6.

100. M. Sulkowski, M. Rodriguez-Torres, E. Lawitz, M. Shiffman, S. Pol, R. Herring, J. McHutchison, P. Pang, D. Brainard, D. Wyles and F. Habersetzer, *J. Hepatol.*, 2012, **56**(Suppl. 2), S560.

101. P. L. Beaulieu, *IDrugs*, 2006, **9**, 39.

CHAPTER 9

Virus-coded Ion Channels as Antiviral Targets

STEPHEN GRIFFIN

Leeds Institute of Molecular Medicine, Faculty of Medicine and Health, University of Leeds, Wellcome Trust Brenner Building, St. James' University Hospital, Beckett Street, Leeds LS9 7TF, UK
Email: s.d.c.griffin@leeds.ac.uk

9.1 Introduction

Since the 1960s, it has been known that the adamantane drugs amantadine (1-adamantylamine) and rimantadine [1-(1-adamantyl)ethanamine] (see Table 9.2) exerted an antiviral effect against the entry of influenza A viruses, which led to them being one of the first antiviral prophylactic treatments licensed for use in humans. However, it was not until the 1980s that the molecular target of adamantanes was revealed by resistance mapping to be the M2 protein, a minor component of the infectious virion. M2 was subsequently shown to form a tetrameric proton channel, acting to mediate acidification of the virion interior and so promote efficient uncoating of the viral nucleoprotein. Today, despite a precise understanding of the M2 channel atomic structure, its key functional determinants and its role during the influenza life cycle, amantadine and rimantadine remain the only licensed antivirals targeting the M2 ion channel, and their clinical use is heavily restricted by resistant viral variants which now populate the majority of circulating isolates. It is hard to recollect another example of a class of antivirals where such initial progress has not been further capitalised upon up to the present day, leaving the therapeutic potential of an essential drug target untapped.

RSC Drug Discovery Series No. 32
Successful Strategies for the Discovery of Antiviral Drugs
Edited by Manoj C. Desai and Nicholas A. Meanwell
© The Royal Society of Chemistry 2013
Published by the Royal Society of Chemistry, www.rsc.org

M2 now serves as the prototypic member of an ever-growing family of virus-encoded ion channels, termed 'viroporins.' Examples of viroporins herald from a wide range of diverse viruses infecting humans and animals alike, including some of the most clinically and economically important viral pathogens on the planet. Although their sequences, structures and roles during viral life cycles vary, viroporins are unified by almost unanimously being essential proteins, supporting their potential as drug targets. However, despite the progress with ion channel inhibitors in other aspects of medicine, the development of novel viroporin inhibitors has rarely progressed beyond the early stages of target/hit discovery and the current repertoire of inhibitory small molecules is narrow indeed. Nevertheless, recent examples of targeted discovery are beginning to yield clinically viable compounds that should serve as pathfinders in the expansion of viroporins as drug targets and unlock their potential for the treatment of myriad viral diseases.

9.1.1   Discovery and Expansion of the Viroporin Family

Although the activity of amantadine against influenza virus entry, and at early stages in replication for some strains, was known for some time, its relationship with a virus-coded ion channel was unknown.[1] The first indications that virus infection altered host cell membrane permeability came about in the 1970s, when small-molecule antibiotics and other compounds were shown to enter virus-infected mammalian cells and inhibit translation.[2] This was initially presumed to result from the budding of nascent viral particles from the cell surface and was proposed as a means of selectively targeting virus infection using commonly available drugs. Such membrane effects have since formed the basis of many indirect methods of assessing viroporin function (see Section 9.1.3), although it is not always clear whether molecules pass directly through viral channels or enter cells indirectly as a result of viroporin-induced perturbations of cellular membrane homeostasis.

The 1990s saw a rapid expansion of viroporin research following the demonstration that M2 displayed proton channel activity in the membranes of *Xenopus* oocytes,[3] recapitulated by peptides corresponding to its *trans*-membrane region in artificial bilayers.[4] Importantly, this activity was directly linked to the amantadine-mediated blockade of influenza virus infection at both early[5,6] (entry) and late (assembly) stages[7,8] in various strains[9,10] by the requirement for virion acidification and a monensin-like dissipation of Golgi pH, respectively (see Section 9.2.1). This provided a functional context to the observed channel activity of M2 in isolation. A number of other small hydrophobic proteins were also proposed to display viroporin activity due to their effects on membrane trafficking and permeability in various systems (see Sections 9.2–9.5), including picornavirus 2B, 2BC, 3A and VP4 proteins,[11–15] alphavirus 6K,[16,17] paramyxovirus small hydrophobic (SH) protein,[18–20] rotavirus NSP4[21–23] and human immunodeficiency virus type 1 (HIV-1) Vpu.[24–26] These proteins, with M2 as the prototype, formed the core of viroporin literature and this subsequently expanded in the new millennium to include many more proteins such as those from coronaviruses (E, 3A),[27-34] flaviviruses (M),[35,36] hepaciviruses (p7, p13/7),[37-40] pestiviruses (p7)[41] and orthomyxoviruses (BM2, CM2).[42,43] More recently, proteins from DNA viruses, including polyomaviruses (VP4/Agnoprotein)[44] and papillomaviruses (E5),[45] have been identified. In addition, a number of other proteins have been proposed to display viroporin activity because of their behaviour in one or more experimental systems where membrane integrity/permeability was altered; however, often this cannot be reproduced in other assays, has no obvious context in the virus life cycle, or was later found to result from membrane destabilisation rather than the formation of a discrete, selective channel complex. Table 9.1 summarises those proteins either confirmed as, or highly likely to comprise, members of the viroporin family at the time of writing. Whilst intended to be as comprehensive as possible, the wide structural and functional diversity of viroporins and the assays used to identify them mean that omissions and future discoveries will no doubt add to, or modify, this list in the near future.

9.1.2 Characteristics Inherent to Viroporins

The unifying characteristic of viroporins is that they are small, usually around 100 amino acids or less, and contain at least one hydrophobic *trans*-membrane helical region that displays some degree of amphipathic character, separating residues into hydrophobic and polar/hydrophilic faces. Their small size necessitates that they form oligomers in order to form a continuous channel through a membrane, a process primarily driven by hydrophobic interactions, although disulfide linkages can also occur. Examples of tetrameric (*e.g.* M2)[46,47] through to heptameric [*e.g.* hepatitis C virus (HCV) p7][37,48-50] have been identified, generating membrane helical bundles of a minimum of four and up to 18 or more *trans*-membrane domains (TMDs), depending on the number present within each protomer.

A recent review proposed a classification system based on the number of TMDs within viroporins, focused primarily on those with single or double membrane-spanning topologies in either orientation.[51] Class 1 comprised those with single TMDs and was subdivided into 1a/b dependent on the lumenal (*e.g.* M2) or cytosolic (*e.g.* SH) topology of the amino-terminus. A similar stratification applied to class 2, where double TMD proteins such as HCV p7 (class 2a, lumenal termini) and picornavirus 2B (class 2b, cytosolic termini) served as exemplars. Class 1 viroporins are usually translated in isolation, whereas class 2 proteins are generally derived following the cleavage of a polyprotein product by host or viral proteases. Although this is a useful classification in many respects, it should be noted that there is often little functional similarity between the proteins within each grouping and that the membrane topologies of some class 2 viroporins can potentially flip between one or two TMDs.[52] Furthermore, growing evidence now exists that supports the inclusion of a third class of viroporins with three TMDs, comprising coronavirus 3a,[27]

Table 9.1 Summary of known viroporin characteristics. Viroporins are encoded by numerous viruses from many different families, including both RNA and DNA viruses. This table summarises current consensus from the literature regarding viroporin function, size (AA, amino acids), ion specificity (Ion?) and the number of *trans*-membrane domains (TM).

Class	Family	Virus	Name	AA	TM	Ion?	Role of Channel Activity
ssRNA (+)	Picornaviridae	Poliovirus	2B	97	2	Ca^{2+}	Particle Production, cell lysis
			VP4	68	1	-	Entry
		Coxsackievirus B3	2B	99	2	Ca^{2+}	Particle Production, cell lysis
		EV71	2B	99	2	Cl^-	Virus Spread
		Human Rhinovirus	VP4	68	1	-	Entry
	Flaviviridae	Hepatitis C virus	p7	63	2	H^+	Particle Production, Entry?
		BVDV	p7	63	2	$?H^+$	Particle Production
		CSFV	p7	63	2	-	Particle Production
	Togaviridae	Dengue Virus	M	75	1	K^+/Na^+	Particle Production
		Semliki Forest Virus	6K	60	2	K^+/Na^+	Particle Production
		Sindbis Virus	6K	55	1*	K^+/Na^+	Particle Production
		Ross River Virus	6K	62	1*	K^+/Na^+	Particle Production
	Coronaviridae	SARS CoV	E	76	1	K^+/Na^+	Particle Production
			3a	274	3	K^+	Virus Spread
			8a	39	1	K^+	-
		MHV	E	83	1	K^+/Na^+	Particle Production
ssRNA(-)	Paramyxoviridae	hRSV	SH	64	1	K^+/Na^+	TNF antagonist, Pathogenesis
	Orthomyxoviridae	Influenza A virus	M2	97	1	H^+	Entry, Particle Production (some)
		Influenza B virus	BM2	115	1	H^+	Entry, Particle Production (some)
			NB	100	1	H^+	-
		Influenza C virus	CM2	115	1	H^+	Entry, Particle Production (some)
dsRNA	Reoviridae	Rotavirus	NSP4	175	1/3	Ca^{2+}	Particle Production, Endotoxin
RT (RNA)	Retroviridae	HIV-1	Vpu	81	1	K^+/Na^+	Particle Production
		HTLV-1	P13ii	87	2	$?K^+$	Mitochondrial Permeability
dsDNA	Polyomaviridae	SV40	VP4	125	1	Ca^{2+}	Particle Production
		JC	Agno	71	1	Ca^{2+}	Particle Production
	Papillomaviridae	HPV-16	E5	83	3	$? H^+$	Oncogene, Signalling/Trafficking

Symbols: *, indicates computer prediction data only; ?, Indicates inferred from indirect assays.

Abbreviations: AA, number of amino acid residues; TM, number of *trans*-membrane domains; Ion, consensus ion species specificity based on available literature; EV71, enterovirus 71; BVDV, bovine viral diarrhoea virus; CSFV, classical swine fever virus; SARS CoV, severe acute respiratory distress syndrome-associated coronavirus; MHV, murine hepatitis virus; hRSV, human respiratory syncitial virus; HIV-1, human immunodeficiency virus type 1; HTLV-1, human T-lymphotropic virus type 1; SV40, simian vacuolating virus 40; JC, John Cunningham polyomavirus; HPV-16, human papillomavirus type 16.

papillomavirus E5[45] and potentially rotavirus NSP4, although whether this possesses one or three TMDs is not confirmed.[21]

Viroporins were formerly thought of solely as mediators of virus assembly and/or entry, based primarily on M2 as a prototype, and this is indeed the case for many members of the family. However, with the addition of new members from multiple virus families, it is clear that viroporins have evolved to perform multiple roles which, even if effecting the same outcome, generally do so by distinct mechanisms appropriate to the life cycle of the virus in question. For example, whilst M2 (in certain strains) and 2B proteins both promote virion release, M2 achieves this by dissipating Golgi/endosome proton gradients to protect acid-sensitive envelope haem-agglutinin proteins on the virus particle,[8] whereas 2B alters Ca^{2+} gradients to disrupt membranes and allow the release of non-enveloped particles from the cell.[13] It is also the case for several viroporins that it is not always clear how ion channel function relates to phenotypes observed following deletion or mutation of the protein in question or the application of a small-molecule inhibitor. Recently, the diversity of viroporin function was well illustrated by the finding that small-molecule inhibitors of the human papillomavirus 16 (HPV16) E5 protein prevented effects on growth factor signalling,[45] providing the first potential link between viroporin activity and oncogenesis.

One final shared characteristic of viroporins that often confounds both their identification and subsequent characterisation is their often weak ion selectivity and/or indeterminate gating behaviour compared with cellular ion channels. Viroporins rarely behave as classical voltage- or ligand-gated channels and lack the highly exclusive ion specificity displayed by cellular proteins. This is likely due to their inherent simplicity and the limited coding capacity of viruses, but has led to scepticism concerning whether they form true channels or merely pores across membranes. Nevertheless, the majority of viroporins do retain a degree of selectivity for ionic species, with M2 proton conductance being by far the best characterised.[53,54] Others show clear preferences for cations over anions in artificial bilayers (*e.g.* HCV p7, HIV-1 Vpu),[38,39,49,55] yet the limitations of such systems generally require further experimental confirmation of their ion selectivity within cells. Even so, the simplicity of viroporins allows them to serve as model systems for the study of channel gating, despite the apparent channel–pore dualism thought to apply to many. As such, viroporins can generally be thought of as membrane pores possessing selectivity filters exerting a greater or lesser influence on the passage of ions through their lumen along an electrochemical gradient. This definition is in many ways applicable to all ion channels as defined by conventional electrophysiology, but also encompasses the wide spectrum of viroporins ranging from those more ion channel-like to pore-like in nature. Furthermore, the notion that viroporins merely form non-selective openings across membranes is countered by the action of inhibitory small molecules, which belie the existence of defined gating and conductance mechanisms, however simplistic they may be.

9.1.3   Experimental Approaches to Identifying and Understanding Viroporin Function

The simplistic and variable nature of viroporin channel function generally requires the corroboration and correlation between several experimental systems in order to assign such activity confidently to a particular protein. The gold standard for putative viroporins entails the direct link between a requirement during the virus life cycle for alterations in ion homeostasis and the functionality of the protein in question, attested by well-characterised null mutations and/or specific inhibitors with accompanying resistance polymorphisms. This is naturally often a difficult outcome to achieve and is lacking for many proteins which are otherwise generally well accepted as viroporins. Unsurprisingly, the most extensively studied protein, M2, fulfils these criteria due primarily to the relative simplicity of class 1 viroporins, where activity can usually be recapitulated *in vitro* using short peptides corresponding to the TMD.[4] This circumvents issues regarding bacterial expression and is also greatly enabling for structural studies. Nevertheless, despite the considerable amount of work undertaken on M2, additional features of this protein are continually identified, not least of which is the recent discovery of its ability to mediate membrane scission independent of the cellular ESCRT pathway[56] (see Section 9.2.1), plus the way in which inhibitors prevent the flow of protons through M2 channels is still a matter of some debate (see Section 9.2.1.3).

The most immediate issue for the study of viroporins *in vitro* is their hydrophobic nature. Although class 1 proteins are often amenable to the use of peptides representing minimal TMDs, they often remain difficult to synthesise and may behave differently to longer peptides or the full-length protein. However, for the most part, peptides are sufficient to study most aspects of ion channel function and can be very useful in structural studies. For class 2 or 3 viroporins, it is usually the case that the complete protein is necessary for channel activity, necessitating recombinant expression, although some groups have successfully generated examples of full-length, biologically active peptides for HCV p7.[38,39] Circumventing the inherent toxicity associated with bacterial viroporin expression, which itself can form the basis of functional assays (see below), often requires intensive protocols or fusion of the viroporin in question to a partner that prevents its cytopathic effect. This has been successfully pursued in our own laboratory using glutathione-*S*-transferase (GST) as a fusion partner to p7,[37,49] SH[18] and E5,[45] amongst others. Such fusions result in targeting of the protein to bacterial inclusion bodies rather than the inner membrane, significantly reducing toxicity and expediting purification.

Obtaining recombinant protein or peptide permits the demonstration of channel-forming activity using artificial bilayer systems, primarily so-called 'black' lipid membranes.[57,58] Here, highly insulating bilayers separate two buffer chambers containing electrodes that allow manipulation of potential difference. This set-up has been used to analyse many viroporins, including M2,[4] Vpu[26] and p7,[37–39,49] incorporating both small-molecule inhibitors and null mutations. However, these systems can be limited by sensitivity to non-specific membrane disruption or perturbation, which can manifest as current traces and has led to some proteins being controversially nominated as viroporins.[59,60] In addition, although general ionic preferences can be determined, *e.g.* for cations compared with anions, some viroporins have been shown to conduct ions that are later shown not to be physiologically relevant in cell-based assays – many ion channels will transport ionic species that are present if their favoured ligand is absent, albeit with reduced efficiency/specificity. Finally, it is not generally possible to determine the amount of *functional* peptide/protein present within bilayers at a given time, making it difficult to calculate the kinetics of small-molecule interactions in most cases.

An alternative to artificial bilayers is the use of liposome dye release assays as indirect measurements of channel activity.[1] Although these do not monitor single channel events and are subject to many of the limitations of bilayer systems, they have the advantage of being technically straightforward, reliable and readily expanded into multi-well formats, as illustrated by a recent high-throughput screen conducted by Boehringer Ingelheim for HCV p7.[62] Although indirect with regard to channel activity *per se*, such assays have been used to assess reliably the effects of null mutants and small molecules against a variety of viroporins. The use of liposomes also permits biochemical analysis of proteins as bulk populations, which can often provide additional insight into functional determinants and membrane association/insertion. Dye release assays, like bilayers, illustrate the flexibility of viroporin conductance; p7,[61,62] SH,[18] E5[45] and also M2 and NSP4 (unpublished observations) readily permit the release of anionic carboxyfluorescein dye. However, the usefulness of such assays to drug discovery is supported by the identification of effective small molecules in liposomes that also display efficacy in virus culture.[63,64]

Assessment of viroporin activity in cells has been undertaken in prokaryotes, yeast and mammalian cells using a variety of strategies. Early work identifying membrane permeability for picornavirus 2B, alphavirus 6K, HIV-1 Vpu and others was based on bacterial systems.[24,65–67] These included bacterial lysis assays and also viroporin-induced permeability of bacterial membranes to otherwise impermeant antibiotics, with resultant diminished protein translation. Such techniques have also been used recently to map functional domains in rotavirus NSP4.[23] In addition, Vpu-induced disruption of bacterial membrane Na^+ gradients was demonstrated using a system where this led to leakage of proline from *Escherichia coli* cells, thereby promoting the growth of an adjacent proline-auxotroph indicator strain.[26] These systems are readily amenable to the testing of multiple viroporin variants, making them a convenient means of screening viral proteins for this activity. However, one drawback is the potential for indirect induction of membrane permeability, through either membrane disruption or perturbation of host membrane proteins.

Experiments in non-prokaryotic systems have primarily involved classical electrophysiological set-ups employing *Xenopus laevis* oocyte surface expression, including seminal studies on M2.[3] This powerful, adaptable technique permits the analysis of both single channels and/or populations and also the manipulation of membrane potentials, ionic conditions and reversible changes in other solute and/or drug molecule concentrations. However, this technique is not appropriate for certain viroporins that either are not tolerated by oocytes or are not surface targeted with high efficiency. For this situation, technically challenging procedures can be followed, such as the patching of enlarged mammalian endosomes or isolated nuclear membranes. Other mammalian cell techniques involve the use of ion-sensitive fluorophores to detect the change or loss of ionic gradients within cellular compartments. This technique was originally used to demonstrate the monensin-like activity of M2 in the Golgi,[8,68] and has been used extensively in the study of picornavirus 2B[13,69] and rotavirus NSP4 calcium channel activities.[21] Recently, it was adapted to demonstrate p7 proton channel activity within HCV-infected hepatoma cells.[40] Lastly, viroporin activity can also be measured indirectly through effects on cell vesicular trafficking, often using fluorescent-tagged vesicular stomatitis virus (VSV) glycoprotein (G), a commonly employed marker for the secretory pathway.[70] Viroporin proton channel effects have also been measured indirectly using the surface trafficking of 'multi-basic' HA proteins. These are cleaved intracellularly by furin and their transport to the cell surface in a non-fusogenic, receptor-binding conformation is reliant on M2 activity. This not only has been used to investigate M2 from varied influenza strains,[9,10] but also provided the first indication of p7 proton channel activity.[71]

From a drug discovery perspective, the availability of high-resolution structural data is greatly enabling to the understanding of both viroporin function and protein–drug interactions. However, the difficulties involved in membrane protein structure determination are considerable and often rely on membrane-mimetic environments, notwithstanding the limitations discussed above concerning protein availability. However, recent M2 solid-state NMR studies provide the first viroporin structural data in true membrane bilayers, setting a new standard and potentially enabling rational drug design.[72] However, studies of class 2 or 3 viroporins are in their infancy by comparison and only two structures of a complete viroporin molecule currently exist: electron microscopy 3D reconstructions of HCV p7 at 16 Å resolution[50] and the RSV SH protein solved by solution NMR in detergent micelles,[20] although the latter study does not appear to have an associated entry in the protein data bank.

Ultimately, the validation of drug discovery efforts targeting viroporins requires infectious culture assays where the specific effects of inhibitors can be unambiguously assigned to the functional requirements of the protein in question. However, to date only two viroporins exist where the activity of small molecules has been validated by specific resistance mutations, namely M2[5] and HCV p7.[63]

9.1.4   The Current Viroporin Inhibitor Chemical Toolbox

To date, the majority of viroporin inhibitors have been identified by simply testing compounds shown to exert effects in other systems, and little targeted discovery has taken place. Even the adamantane M2 inhibitors were selected in whole virus assays and it was many years before their target was known. As a result, virtually all viroporin inhibitors described to date originate from three chemical classes, namely adamantanes, amilorides and alkylated imino sugars. Adamantanes were, of course, identified for influenza A viruses,[5] imino sugars were explored as HIV-1 therapeutics targeting glycosylation[73] and amiloride is a well-known inhibitor of the epithelial Na^+ channel, commonly used as a diuretic in hypertensive patients.[74] Table 9.2 shows a list of prototype viroporin inhibitors, their targets and resistance polymorphisms where known. The majority of prototypic viroporin inhibitor studies merely applied one or more variants of such molecules with little or no idea regarding modes of action. Surprisingly, this small repertoire of molecules appears promiscuous in its ability to block channels from multiple unrelated viruses, including examples from all three classes of viroporins to a greater or lesser extent. However, with the exception of some adamantane variants which are effective at low micromolar concentrations against certain influenza strains,[75–77] the majority of prototype viroporin inhibitors are only effective at concentrations of tens to hundreds of micromoles. Further studies have generally involved the derivatisation and modification of prototype scaffolds rather than rational design or screening, even where atomic structures are available.

At first glance, the wide-ranging effects of individual prototype inhibitors may seem appealing for drug discovery, indicative that 'magic bullet' compounds might be possible to generate targeting multiple viroporins. However, derivatives rarely improve on the relatively low potency of prototypes to any great extent, indicating that the molecular fit and binding of such molecules within prospective binding pockets may be inefficient. Even for M2, where atomic structures exist, the structure–activity relationships (SARs) for the multiple adamantane derivatives generated to date are not clear. Indeed, there is even controversy surrounding the binding modes of amantadine and rimantadine themselves as a result of conflicting structural studies where either lumenal[78,79] or peripheral binding sites[80] were proposed. Such ambiguity is likely a result of adamantane structures comprising small hydrophobic cages capable of fitting within multiple protein cavities, where their amine extensions H-bond with nearby residues. This, in turn, is probably also responsible for the promiscuity of these molecules and their relatively low potency; although capable of binding to multiple sites, they probably do not do so with the same affinity as would a theoretical molecule designed to fill the entire cavity and make multiple stabilising contacts with the protein. This view is supported by the enhanced potency of certain adamantane derivatives with bulky R groups, which can also inhibit some amantadine-resistant M2[81,82] and p7 channels.[63] Given the relatively similar potency and promiscuity of prototypic imino sugars and amilorides, the same scenario is likely to apply.

The non-targeted nature of the current prototype viroporin inhibitor repertoire is also reflected by their clinical efficacy. On the one hand, adamantanes serve as a clinical precedent for viroporins as drug targets, yet their limited clinical benefits and the rapid selection of resistant polymorphisms with seemingly little or no fitness cost to the virus, whether it be influenza A[83] or HCV,[84,85] has generated significant scepticism regarding the potential of

Table 9.2 The viroporin chemical toolbox. Summary of the major prototypic and derivative viroporin inhibitors reported in the literature to date. Effective/inhibitory concentrations have been omitted due to difficulty comparing between various systems.

Class	Compound	Structure	Target	System	Resistance
Adamantane	"Amantadine" (1-adamantylamine) (5, 36, 37)		Influenza M2	M, B, C, V, O, P, S	L26F, L28F, V27A, A30T, S31N, G34E
			HCV p7	M, D, C, V, P, S	L20F, genotype 1a (H77), 2a (H77), (JFH-1)
			Dengue M (C-terminus)	M	
	"Rimantadine" 1-(1-adamantyl)ethanamine (5, 64, 255)		Influenza A M2	M, B, C, V, O, P, S	L26F, L28F, V27A, A30T, S31N, G34E
			HCV p7	M, D, C, V, P	L20F, genotype 1a (H77)
	"H" 5-(1-adamantyl)-2-methyl-1H-imidazole (63)		HCV p7 (L20F)	V	
	Spiro[piperidine-2,2'-adamantane] (158)		Influenza A M2	V, S	S31N

	Structure			
"Spiroadamantane" (155)		Influenza A M2 (V27A, L26F)	V, S	S31N
Spirane-amine	"BL-1743"(2-[3-azaspiro (5,5)undecanol]-2-imidazoline), (156)	Influenza A M2	Y, V	I35T
Alkyl Imino-Sugar	"*N*N-NDNJ":N-nonyl deoxynojirimycin (38)	HCV p7	M, D, V	F25A, Genotype 3a (452)
	"*N*N-DGJ": N-Nonyl deoxygalactonojirimycin (38)	HCV p7	M, V	F25A, Genotype 3a (452)
	UT-231b	HCV p7	P	

Table 9.2 (*Continued*)

Class	Compound	Structure	Target	System	Resistance
Amiloride	"HMA": 5-(N,N-hexamethylene)amiloride (29, 36, 39,181)		HCV p7 SARS CoV E Dengue M (C-terminus) HIV-1 Vpu	M M M M, C	
	"BIT-225": (N-[5-(1-methyl-1H-pyrazol-4-yl)-napthalene-2-carbonyl]-guanidine (86, 87)	BIT225	HCV p7 BVDV p7 HIV-1 Vpu	M V V	
Other	"CD": 1,3dibenzyl 5(2H1,2,3,4tetraazol5yl) hexahydropyrimidine (63)		HCV p7	S, D, V	L20F

Compound	Structure	Virus	System
"Emodin": 6-Methyl-1,3,8-trihydroxyanthraquinone (34)		SARS CoV 3a	M, V
Verapamil (41)		CSFV p7	M, V
"DIDS": 4,4'-diisothiocyano-2,2'-stilbenedisulfonic acid (273)		EV71 2B	M, V
MV006(45)	?	HPV-16 E5	D, C

Virus abbreviations as in Table 9.1. 'System' indicates methods used for viroporin analysis, namely: B, bacterial toxicity/permeability; C, cell culture; D, liposome dye release; M, membrane bilayers; O, *Xenopus* oocytes; P, patient trials; S, structural and/or biophysical techniques; V, virus culture; Y, yeast.

viroporin-targeted drugs. However, the fact that resistant variants are selected at all suggests that even prototype inhibitors exert a selective pressure on the virus, meaning that druggable sites almost certainly exist on viroporin molecules. As such, prototype inhibitors serve to highlight regions on viroporin molecules suitable for further exploration through targeted drug discovery or which may select high-affinity ligands through library screens. This view is supported by the reported nanomolar potency of the BIT225 molecule {*N*-[5-(1-methyl-1*H*-pyrazol-4-yl)-naphthalene-2-carbonyl]]guanidine, Biotron, Australia; see Table 9.2} in certain systems, which was selected through a limited screening programme in bacteria. Although this molecule shares some traits of prototypes with reported activity against both p7 and Vpu, it appears more potent in cell culture models for HIV-1[86] and a pestivirus model for HCV,[87] although its binding mode and ability to select resistance are not yet apparent. Results of both HIV-1 and HCV clinical trials are currently pending. As more information is obtained on viroporin biochemistry and atomic structures, there is every chance that novel molecules will be generated that show significant improvements over existing prototypes, suitable for drug development and incorporation into modern antiviral regimens.

9.1.5 Non-ion Channel Functions of Viroporins: Confounding Factors in the Study of Virus-coded Ion Channels

The variability and small coding capacity of many viruses require that their limited number of gene products be capable of functional redundancy. As such, it is not surprising that many viroporins display additional roles during the virus life cycle that are separate to ion channel activity. In addition, their small size often means that overlapping regions and residues can be involved in both aspects, leading to great difficulty in the interpretation of mutagenesis studies and in ascribing phenotypes to either the lack of channel activity or other interactions. In addition, viroporins produced following the cleavage of polyprotein precursors (*e.g.* picornavirus 2B, HCV p7) add a further layer of complexity as mutations can affect the production or topology of the protein. As such, although this chapter does not discuss these additional aspects of viroporin function in great detail, it is pertinent always to consider that phenotypes reported for particular mutations may well result from an indirect effect on viroporin activity or indeed not apply to channel function in any regard.

9.2 Viroporins Encoded by Pathogenic Human RNA Viruses with Known Small-molecule Inhibitors

9.2.1 Influenza A Virus M2: Clinical Precedent and Prototype Viroporin

Influenza A virus represents one of the most prevalent human viral infections. Although 'seasonal' influenza epidemics cause around 500 000 deaths each year worldwide, these are controlled by vaccination and herd immunity. Of greater concern is the emergence of 'pandemic' influenza strains to which humans carry little or no immunity that can spread rapidly around the world. Such pandemics vary in severity, ranging from the 'Spanish flu' of 1918 (50 million deaths), to the 1957 pandemic (2 million deaths). Pathogenicity is impossible to predict prior to the emergence of pandemic strains. The recent H1N1 'swine flu' pandemic, although less pathogenic than the Spanish flu, spread rapidly owing to a lack of herd immunity and had both significant health and socio-economic impacts. H5N1 'avian flu' shows considerable pandemic potential and has infected hundreds of individuals with a >50% mortality rate, yet human-to-human transmission is currently rare. The possibility that avian flu could acquire the ability to spread efficiently from human to human is a global cause for concern, and recent controversial experiments demonstrated that only a limited number of genetic changes may be necessary for this to occur.[88,89]

As pandemics occur too rapidly for vaccination programmes to prevent worldwide spread of influenza, antivirals represent the best way to limit both the initial spread and the severity of virus infection for individual patients. Classically, influenza was treated using amantadine and rimantadine,[90] specific inhibitors of the viral M2 proton channel which plays vital roles during both virus assembly and cell entry. These drugs are now clinically redundant owing to viral resistance, occurring via defined point mutations in the M2 protein (e.g. S31N).[83] Current anti-influenza therapy focuses on a single target, neuraminidase, using the drugs oseltamivir[91–93] and zanamivir.[94–96] Recent detection of drug resistance for the more commonly used oseltamivir has raised concerns over the longevity of both compounds.[97] As M2 and neuraminidase inhibitors act synergistically in culture and suppress viral resistance,[98] new M2-targeted antivirals would be valuable additions to future combinatorial regimens.

9.2.1.1 Identifying M2 as an Ion Channel Target of Amantadine

Amantadine was approved by the US Food and Drug Administration (FDA) in 1966 for clinical use in the treatment of influenza virus infection under the trade-name Symmetrel. Amantadine was known to block an early stage of the influenza infectious cycle, but was also shown to affect later stages in a number of influenza strains.[1] Its methylated derivative, rimantadine (trade-name Flumadine), was later licensed (1994, FDA) in many countries as a more effective alternative with less pronounced side effects. It was not until 1985 that the molecular target of amantadine was identified through the selection of drug resistance in several amantadine-sensitive strains.[5] Mutations clustered within the trans-membrane region of the M2 protein, which is encoded on segment 7 of the influenza RNA genome and generated by alternative splicing.[99–101] In addition, the origin of the HA protein also impacted on amantadine sensitivity in strains where it affected the later stages of the life cycle, leading to the proposal that the drug disrupted an M2–HA protein–protein interaction.[5]

The demonstration that M2 formed stable, disulfide-linked tetramers[102] combined with M2-mediated raised Golgi/endosomal pH[7,8,103–105] provided the first clue to M2 channel formation. This was confirmed in seminal studies of M2 expressed in *Xenopus* oocytes, demonstrating amantadine-sensitive monovalent cation conductance that was activated upon reduced external pH.[3] Mutations within M2 known to confer resistance to amantadine also rendered isolated channel activity insensitive to the drug. Amantadine-sensitive channel activity was also shown for M2 peptides corresponding to the minimal TMD (amino acids 21–46) using artificial bilayers.[4]

Thus, M2 became the first example of a virus-encoded ion channel protein where activity of the protein in isolation was directly related to a functional requirement at specific stages in the virus life cycle, which in turn was known to be targeted by small-molecule inhibitors. Several studies showed that M2 displayed selectivity for protons and that gating was both activated by acidic pH and dependent on a highly conserved His37 residue, as demonstrated by rational mutagenesis with subsequent effects on channel gating.[106] The convergent conclusions of multiple studies led to the proposed role of M2 during virus entry as mediating acidification of the virion interior following endosome acidification,[6] which also triggers membrane fusion mediated by HA.[107] Reduced intra-virion pH is thought to uncouple interactions between RNPs and the matrix protein, thereby promoting efficient uncoating following release of the virion contents into the cytoplasm. The role of M2 during particle release applied to influenza strains where the HA0 precursor contained a multibasic region that enabled intracellular cleavage by furin. This results in HA becoming prematurely primed to undergo fusogenic change in reduced pH environments, such as in the TGN and endosomes where vATPase is active. M2, present in the membrane of such compartments, negates vATPase activity by promoting the flow of protons out of the acidifying secretory compartment, thereby allowing HA to traffic to particle assembly sites on the cell surface in a functional form, able to bind to sialic acid receptors.[7–10,103–105]

9.2.1.2 Structure and Gating of Influenza A Virus M2

M2 is a 97 amino acid (aa) protein with a single TMD which forms disulfide-linked tetramers in membranes.[46,102,108] The 25 N-terminal residues are located on the surface of the plasma/virion membrane and are highly conserved. This has recently been exploited in vaccine development due to the similarity between most influenza A isolates.[109,110] The TMD extends from aa 25 to 46 and is followed by an amphipathic helix (aa 47–62) and the remaining cytosolic domain. Ion channel activity can be recapitulated by a minimal domain including the TMD (aa 22–46),[4] although a longer 'conductance domain' (CD), which includes the amphipathic helices (aa 18–60 or 22–62, depending on the study), displays enhanced channel properties in oocytes.[111] The amphipathic helices are also critical to a recently described function for M2 in mediating membrane curvature and scission in a cholesterol-dependent mechanism that renders virus budding independent of the cellular ESCRT system.[56] Finally, the C-terminus of the protein interacts with the M1 matrix protein during the formation of the virus particle.[112–115]

Reversal potentials for the M2 TMD in oocyte membranes indicated a 10-fold preference for protons over sodium, potassium or chloride ions.[116] The observation that channel conductance is reversible was consistent with pore-like behaviour,[3,116] yet a slow rate of conductance ($\sim 200\,H^+\,s^{-1}$) in both oocytes and liposomes[117,118] combined with the lack of transport of alkali metal ions were indicative that the water column through the channel complex was interrupted by a selectivity filter. Furthermore, proton conductance only occurs when the external pH is acidic,[116] implicating protonation of ionisable residues as an important step. In this regard, mutagenesis of the highly conserved His37 residue within the TMD resulted in enhanced conductance[106] as well as a loss of selectivity[116] while retaining amantadine sensitivity, indicating that this residue played a fundamental role during M2 gating. Functionally defined models for the M2 TMD channel complex predicted that His37 did indeed line the channel lumen.[119] consistent with Cu^{2+}-mediated inhibition of channel activity.[120] Furthermore, the major gating residue was predicted to comprise the similarly conserved Trp41 residues,[120] resulting in a now well-accepted HxxxW tetrad comprising the gating mechanism for all M2 proteins, although Val27 may also form a secondary gate towards the N-terminal neck of the TMD.[53,72,121] This functional unit has been borne out by extensive structural studies and a range of biophysical techniques employed to study the gating behaviour of M2, which has become a model for prototypic proton channels.

M2 structural studies were expedited by the fact that peptides representing minimal domains were able to recapitulate channel function, allowing the study of tetrameric complexes in membrane-mimetic environments. In all cases, M2 forms a left-handed four-helix bundle[122] with a defined hydrophilic lumen containing both His37 and Trp41 tetrads. Figure 9.1 shows the major M2 structures solved to date in chronological order and separated by technique/conditions.

The first reported structures were solved by solid-state NMR (ssNMR) in Cross's laboratory for the TM domain (22–46), in both the absence (PDB: 1NYJ)[123,124] and the presence of amantadine (PDB: 2H95).[125] This TM structure was subsequently refined by Cady and co-workers in the presence of amantadine (PDB: 2KQT).[79,126] The TM construct was also the subject of crystallographic studies, producing an amantadine-bound structure (PDB: 3BKD/3C9J),[78] which again has been recently refined (PDB: 3LBW).[53] What was apparent from early ssNMR and X-ray studies was that, in the absence of drug or near-neutral pH required to stabilise the closed form of the channel, the C-terminal region comprising the Trp41 tetrad was splayed, such that that the 'gate' could not be formed. However, this was not the case either in refined ssNMR/X-ray structures or for structures comprising the conductance domain where the additional presence of the amphipathic helix resulted in a more compact channel structure where the Trp41 tetrad restricted the C-terminal end

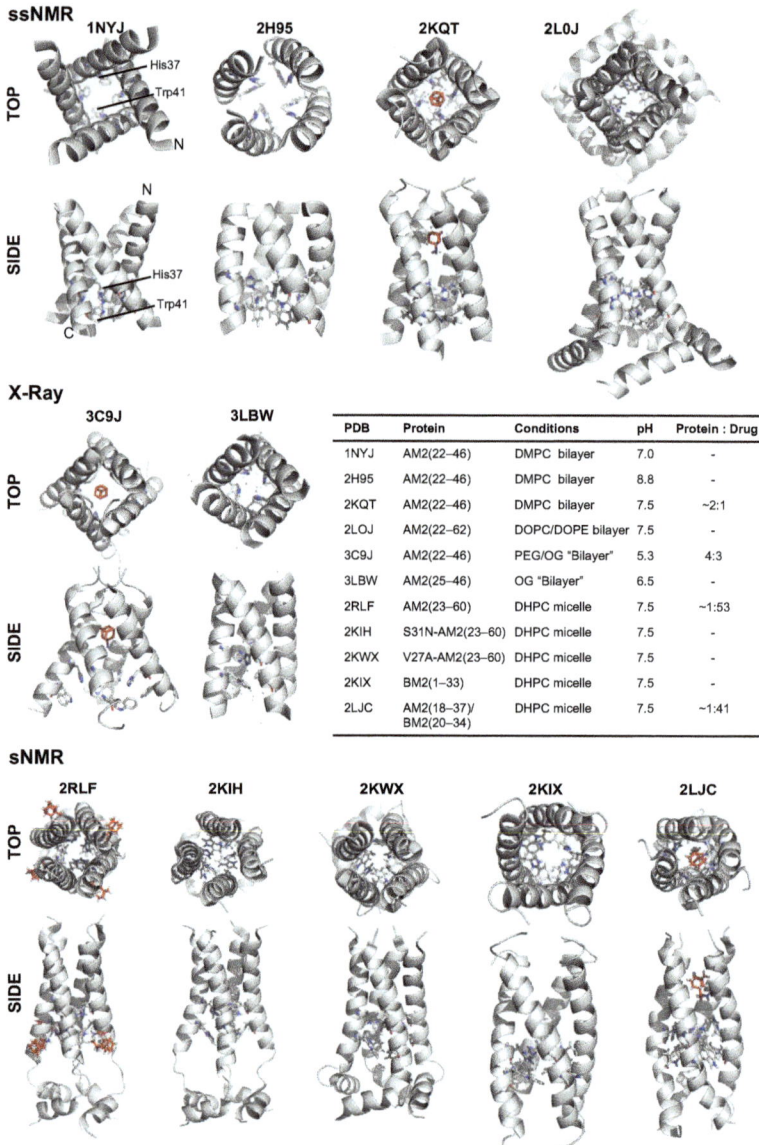

Figure 9.1   Structural studies of influenza M2 proton channels. Summary of key
AM2, BM2 and chimeric channel structures reported to date in the
literature, separated by technique and shown in chronological order right
to left. The inset summary table describes experimental conditions for each
study, listed by PDB identification. Structures are shown in both a side
and a top view, representing the N-terminus of the protein facing the
cell/virion exterior. The key proton sensor and gating residues are shown
as stick representations (His17, Trp41 for AM2, His19, Trp23 for BM2).
Adamantane-based inhibitors amantadine and rimantadine are highlighted in red for drug-bound structures.[†] Images generated in PyMol
using available PDB structures.

of the channel. This was seen in a structure for a rimantadine-bound channel complex from Chou's laboratory solved by solution NMR (PDB: 2RLF).[80] This was followed by two solution structures for amantadine-resistant variants, S31N (PDB: 2KIH)[127] and V27A (2KWX).[128] Finally, an ssNMR structure of the conductance domain was recently obtained in the presence of lipid bilayers (PDB: 2L0J).[72] In this structure, the position of the amphipathic helices was significantly altered compared with the solution structures, bringing them into closer proximity to the TM region. This is likely due to the influence of various membrane-mimetic environments on protein structure compared with an authentic membrane bilayer. Differences in drug binding are discussed in Section 9.2.1.3; however, these structures combined with related biophysical techniques have provided insight into the mechanism underpinning M2 gating, although this also varies with the construct and experimental system employed.[129] Interestingly, protonation of the His37 tetrad appears to greatly enhance the stability of the assembled tetramer,[47] this being three orders of magnitude higher at pH 6 than pH 9.[111] His37 protonation first occurs at a much higher pH (8.2) compared with His in free solution.[130] One explanation for this comes from molecular mechanics studies around the 2L0J structure, which support a 'dimer of dimers' model for the His37 tetrad, allowing sharing of a single proton between two His residues within each pair.[72] This both results in a strong hydrogen bond between the two residues and also permits one of each pair to undergo cation–π interactions with the indole of the adjacent Trp41. The third protonation event, which occurs at pH $\sim$6, then induces an unstable activated state where indole–imidazole cation–π interactions are disrupted by alterations in the helical bundle, allowing opening of the Tryp41 gate and the flow of protons to occur.[116] Kinetic analyses combined with solution NMR support synchronicity between His37 dimer protonation and Trp41 opening, with conformational changes detected in solution for both residues concomitantly upon reduced pH.[131] A similar mechanism is also supported by molecular dynamics and mathematical models, which also indicate a rate-limiting role for the secondary gate formed by Val27.[132] However, other studies support a 'shuttle' mechanism of proton conductance inconsistent with the formation of His37 dimers, whereby exchange of protons between His37 and water residues are facilitated by imidazole ring reorientations.[133–136] As such, the precise molecular mechanisms of even this 'simple' model for proton channel conductance remain a matter of some debate, potentially due to the different structural orientations, membrane environments and peptides employed affecting the His37 and Trp41 tetrads in distinct structural studies.

---

†Abbreviations: DHPC, 1,2-diheptanoyl-*sn*-glycero-3-phosphatidylcholine; DMPC, 1,2-dimyristyl-*sn*-glycero-3-phosphatidylcholine; DOPC, 1,2-dioleoyl-*sn*-glycero-3-phosphatidylcholine; DOPE, 1,2-dioleoyl-*sn*-glycero-3-phosphatidylethanolamine; PEG, poly(ethylene glycol); OG, octyl β-D-glucopyranoside; ssNMR, solidstate nuclear magnetic resonance; sNMR, solution nuclear magnetic resonance; X-ray, X-ray crystallography; AM2, influenza A virus M2; BM2, influenza B virus M2.

9.2.1.3 *A Tale of Two Sites: the Controversy of Allosteric Versus Lumenal Binding for M2 Inhibitors and Implications for Drug Development*

The clustering of amantadine resistance mutations within the M2 TMD,[5,7,103,137,138] combined with early biophysical studies[139] and the first structures for the TM domain,[125] led to the commonly held view that both amantadine and rimantadine bound within the lumen in close proximity to His37, thereby preventing proton transport by occlusion or *via* interfering with associated steric changes in the His37/Trp41 tetrads. However, the first solution structure for the conductance domain sparked considerable controversy and renewed interest surrounding the mechanism by which these classical antivirals bound to the M2 channel complex.[80] In place of a single adamantane molecule in the lumen, four were located on the channel periphery bound to a pocket formed by the C-terminal region of the TMD and the beginnings of the amphipathic helices. This binding site was positioned so as to be readily accessible to rimantadine molecules partitioned within membranes, a phenomenon for which these compounds are well known.[140] In this scenario, the hydrophobic adamantyl cage resides in the hydrophobic region of the bilayer alongside phospholipid side chains, with the polar amine partitioned with head groups at the membrane–aqueous interface. However, a study was published in the same issue of *Nature* describing a TM domain crystal structure at low pH where density corresponding to amantadine was located within the channel lumen.[78] These parallel reports have created an upsurge in M2-related publications attempting to resolve this controversy in addition to renewed interest in M2 as an antiviral target.

The mechanism underpinning inhibition of M2 *via* binding at the allosteric site was proposed to involve stabilisation of the closed state of the channel, with drug binding preventing the conformational changes in the TM helix necessary for opening. In support of this, several well-known resistance mutations were seen to destabilise the tetramer, making drug binding less likely.[127] In addition, magic angle spinning NMR studies on conductance domain peptides in bilayers supported that rimantadine bound the channel at a stoichiometry greater than one and interacted with residues corresponding to the allosteric site.[141]

However, a large number of studies have argued against binding of adamantanes to the M2 periphery. These include functional studies where chimeric proteins were generated between the M2 proteins of influenza A (AM2) and that of influenza B (BM2), which is not sensitive to amantadine or rimantadine.[142] Transfer of part of the AM2 N-terminus and TM domains (aa 19–36) into BM2 conferred amantadine susceptibility in oocyte membranes, arguing that drug binding within the pore was the primary interaction; transferring aa 37–45 did not confer sensitivity. In addition, a series of biophysical and structural investigations using the TM domain peptide favoured amantadine binding within the lumen, including the refined ssNMR structure of amantadine-bound M2.[79,126,143,144] Furthermore, the solution structure of a chimeric A/BM2 conductance domain construct with the TM domain from AM2 showed rimantadine binding within the lumen,[145] and the allosteric binding site present on the conductance domain structure solved in bilayers is significantly altered by the positioning of the amphipathic helices.[72]

Hence a seemingly overwhelming body of evidence has accumulated in recent years favouring the lumenal M2–adamantane interaction. However, a number of studies on both conductance domain and TM peptides have identified that these compounds are in fact capable of binding to both sites, although the affinity for the allosteric, membrane-exposed site is lower as judged by surface plasmon resonance (SPR)[146] and allosteric binding is only seen in solid-state studies of TM peptides upon the partitioning of the drug into the lipidic compartment.[79] *In silico* docking to the conductance domain solution NMR structure also exclusively favours binding to the allosteric site,[147] whereas lumenal binding is predicted when performed on the solid-state TM structure.[82] Interestingly, molecular dynamics simulations on the conductance domain again predict binding to both sites, although the nature of binding to each was different.[148] Binding within the lumen resulted in a much more stable interaction, yet had to overcome a significant energy barrier, while binding to the allosteric site was less stable yet binding was significantly kinetically more favourable than binding to the lumen. One unifying possibility could involve a cooperative mechanism of inhibition, whereby rapid kinetic drug binding to the allosteric site serves to stabilise channel complexes such that more stable interactions within the lumen might take place.

In addition, the mechanisms by which the most commonly described adamantane resistance mutants, such as L26F, L28F, V27A, A30T, S31N and G34E, prevent drug binding are not clear. Some, such as V27A, A30T and G34E, affect residues predicted to interact with drug molecules following lumenal binding, making the likelihood of disrupting a stabilising interaction the most obvious mechanism. However, others, such as L26F and L28F, affect residues involved in helical stacking in most of the described structures, making their role less clear. Interestingly, the most significant resistance polymorphism, S31N, found in both pandemic H1N1 and highly pathogenic H5N1 avian strains, has been proposed potentially to affect binding at both sites.[126,127] For lumenal binding, this mutation induces broadening of the channel aperture, which has been proposed to allow drugs to bind only loosely within the lumen, thereby not affecting the flow of protons. However, this mutation also disrupts the peripheral binding site in solution NMR structures (PDB: 2KIH) and induces the formation of less stable tetramers.

Which, then, is the most relevant adamantane binding site in Nature? One confounding issue resulting from the many structural and biophysical studies is the different peptides and conditions used when studying M2–drug interactions. The most obvious difference is that conductance domain peptides contain additional M2 sequence compared with TM peptides and the inclusion of the amphipathic helices appears to stabilise the structure, making it more compact. Although TM peptides contain some key elements of the peripheral binding site such as Asp45, they lack distal residues such as Phe47 and Phe48, which would likely make stabilising hydrophobic contacts with adamantyl cages if drugs bound to the periphery. Conversely, conductance domain solution structures in detergent micelles display narrowed lumenal apertures, which would therefore disfavour drug binding at this site.[149] The conductance domain structure from bilayers (2LOJ) does not suffer greatly from either of these issues and likely represents the most physiologically relevant system in which to study drug interactions,[72] although, unfortunately, a drug-bound version of this structure has not yet been determined.

Another consideration is the format in which adamantane–drug interactions are studied. Several divergent studies point to the weaker allosteric binding occurring only following drug partitioning within the bilayer,[79] yet systems such as SPR[146] and patch clamping of oocyte membranes[142] transiently introduce compounds at relatively (with respect to the number of channels present in the membrane) high concentrations in the aqueous phase, likely favouring rapidly measurable lumenal interactions. By contrast, notwithstanding differences between peptides, structural studies are usually performed at roughly equimolar ratios of drug and protein, possibly favouring lumenal interactions due to less membrane partitioning. The exception to this is the allosteric rimantadine-bound solution structure (2RLF), where an $\sim 50:1$ drug:protein molar ratio was used, potentially favouring saturation within the micelle and resultant peripheral binding.[80] In patients, adamantane concentrations within the plasma rarely exceed 2–3 µM,[150,151] yet it is not clear either how these drugs partition within aqueous and membranous compartments in the body or how they arrive at the respiratory epithelium to exert their antiviral effects.

Perhaps the greatest obstacle to resolving these issues is the characteristic properties of amantadine and rimantadine themselves. Neither molecule was specifically selected to target M2, as reflected by their potency in cell culture. Their small size and amphiphilic properties lend themselves to promiscuous, yet relatively inefficient, binding within both solvent and membrane-exposed cavities. They therefore act as poor structure probes as their selectivity to a particular binding site is difficult to determine; it should be remembered that *in silico* docking and/or molecular dynamic simulations of drug–protein interactions in a membrane environment are technically challenging and are often not comparable to those in solution. However, while the controversy over the potential binding of prototype adamantanes to M2 may continue, it does serve to illustrate that two regions on the M2 channel complex are potentially amenable to targeted drug design.

9.2.1.4 Improving on Amantadine: the Search for Modern Influenza Drugs and Their Potential as New Therapies

Efforts to improve upon the amantadine and rimantadine M2 inhibitors have primarily concerned the derivatisation of these prototype molecules, adding one or more R groups of varying size and molecular composition. Although this has defined several potent series of compounds that principally act against amantadine-sensitive strains,[75–77,81,152–155] some examples of inhibitors that act against resistant M2 variants also exist.[82] However, the search for potent compounds targeting the most clinically-relevant S31N channels continues.

One example of an M2 compound screen that successfully generated a hit against M2 has been published. The screen was conducted in yeast and was based on the growth-inhibitory effects of M2 expression, which was restored by amantadine.[156] The compound, BL-1743 {2-(3-azaspiro[5,5]undecano)-2-imidazoline} (Figure 9.2), caused reversible inhibition of A/Udorn 72 M2 in oocytes over a similar micromolar range to amantadine, compared with the irreversible action of the adamantane. In addition, whereas several amantadine-resistant mutations (V27S, A30T, S31N, G34E) also prevented the action of BL-1743, another, L38F, did not. Furthermore, BL-1743 selected a unique resistant variant, I35T, which, in turn, did not affect amantadine sensitivity.[157]

Hence BL-1743 potentially shares a distinct, but overlapping, M2 binding mode with adamantanes and this has prompted research into other spiranamines and/or spiropiperidines as M2-selective molecules. This has generated numerous compounds with improved antiviral potency relative to rimantadine or amantadine. A similar story holds true for derivatised adamantanes with improved antiviral potency, although both classes of compound have yet to yield variants with sub-micromolar IC_{50} values.[75,76,81,82,154,155,158] Examples of spiranamines and/or spiropiperidines have shown some activity ($IC_{50} \approx 30$–$80\,\mu M$) against some adamantane-resistant channels (V27A, L26F mutants, H3N2 background).[82] Other adamantane derivatives also exist with extremely limited activity against S31N channels, reducing channel activity by 22% at a high drug concentration of $100\,\mu M$, $IC_{50} \approx 250\,\mu M$;[81] however, this was less effective than high concentrations of amantadine, which reduced S31N activity by 35.6% at $100\,\mu M$ with an IC_{50} of $\sim 200\,\mu M$. Recently, these two families of M2 inhibitors have been combined following molecular dynamics simulations of drug binding to the refined ssNMR TM domain structure (2KQT), to identify spiroadamantane as a compound capable of inhibiting not only Ser31-containing H3N2 M2, but also V27A- and L26F-resistant variants at low micromolar concentrations.[82] However, no activity was seen for this molecule when tested against H3N2 M2 with the S31N mutation.

This last example illustrates a shift in the field from a focus on compound-centred derivatisation towards the employment of the available atomic structures in rational programmes for the identification of novel M2-targeted compounds. This will begin to identify more potent molecules with clear site preferences and the establishment of SARs should enable clarification over the binding modes for each compound. However, as discussed above, the choice of the atomic structures employed as a template may influence outcomes with regard to which region of the channel is targeted. Ideally, templates should incorporate resistance mutations as starting points in drug discovery programmes. In parallel, screening of M2 peptide channel activity could be a means by which to identify novel inhibitors with improved characteristics, potentially with chemotypes unrelated to prototypes. In this respect, liposome assays based on fluorescent dye release,[61] or indeed those based on proton flux, as described for conductance domain peptides,[127] may be highly productive.

9.2.1.5   Influenza B and C Proton Channels

Due to their lesser clinical importance, the proton channels of influenza B and C are less well characterised compared with AM2. The 100 amino acid NB protein of influenza B was first shown to display cation channel activity in bilayer systems, which could be inhibited by high concentrations of amantadine.[159] Like AM2, NB is a minor virion component[160] and was proposed to play the equivalent role during the life cycle. However, it is now generally accepted that the BM2 protein of influenza B is in fact the functional homologue of AM2.[43,161] BM2 has been shown to critical for virus replication and to display proton channel activity in membranes which is insensitive to amantadine. A composite solution NMR structure of this channel was reported combining the cytosolic (PDB: 2KJ1) and the TM (PDB: 2KIX) regions.[162] BM2 gating is regulated by the His19 and Trp23, equivalents of AM2 His37 and Trp41.[163–165] As described above, the amantadine-insensitive activity of BM2 has been exploited in the creation of A/BM2 chimeras during investigations of drug binding sites.[142,145] For influenza C viruses, the CM2 protein[166] is produced by an internal cleavage event from an M1–2 precursor[167,168] and has also been directly demonstrated to display channel activity,[169] to alter vesicular pH[42,170] and to promote uncoating during the entry process.[171] To date, no small-molecule inhibitors of B/CM2 or NB activity have been described, although the availability of BM2 structural information could expedite future rational development.

9.2.2   Human Immunodeficiency Virus Type 1 (HIV-1) Vpu

The HIV-1 pandemic has been the major driver of antiviral drug discovery in past decades and around half of all currently licensed antivirals are targeted towards this virus. As a result, highly active antiretroviral therapy (HAART) successfully prolongs the life of the majority of HIV-1 patients in the developed world and is slowly being extended to developing countries, although availability is still greatly short of demand. However, despite extensive permutations available for drug combinations, virus resistance continues to arise in response to such intense selective pressure. Even when on effective treatment, many patients exhibit low level viraemia as a result of virus shedding from infected macrophages and other long-lived infected cells such as memory T lymphocytes. Life-long HAART will therefore require many more drug combinations to continue to be available, unless a strategy can be found to target viraemia derived from virus reservoirs.

HIV-1 and related simian viruses (chimpanzee lineage) encode the Vpu accessory protein.[172,173] This small, multifunctional protein is not a virion

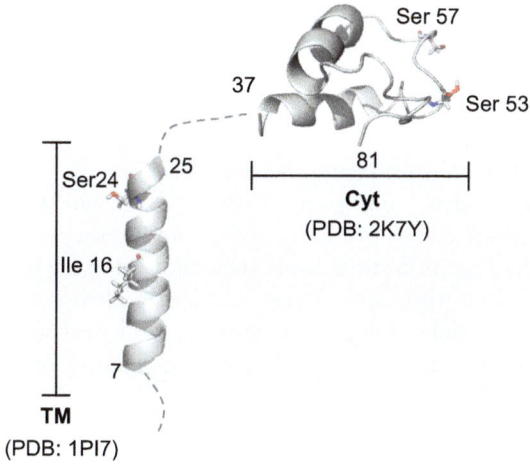

Figure 9.2 Structure of the HIV-1 Vpu channel. No structure for the complete Vpu protein is present on the PDB, yet multiple structures exist for peptides representing the Vpu *trans*-membrane (TM) domain, along with two structures for the cytosolic domain. The structure of the cytosolic domain is regulated by phosphorylation of two conserved serines, Ser53 and Ser57, while Ser23 has been shown to be essential for channel activity *in vitro* (Section 9.2.2.2). Ile16 is also believed to reside within the channel lumen (REF). Images generated in PyMol using available PDB structures.

component, yet plays a pivotal role in the release of infectious virions.[174–178] This function of the protein has been attributed in part to its ability to mediate membrane permeability and the observed channel activity of an N-terminal hydrophobic TM region in bilayers, which is sensitive to amilorides such as HMA [5-(*N,N*-hexamethylene)amiloride].[24–26,55,179–187] This viroporin activity of Vpu has been the subject of much research and has also been targeted by the Australian Biotech company Biotron, *via* a discovery programme which has culminated in one of the first clinical trial investigations of a viroporin-targeted small-molecule since rimantadine. The lead molecule, BIT225, displays activity in culture against HIV-1 infection of monocyte-derived macrophages, leading to the company to propose that it may further enhance HAART by targeting the long-lived macrophage virus reservoir.[86,181]

9.2.2.1 Vpu Forms an Amiloride-sensitive Channel Involved in HIV-1 Particle Production

Vpu was initially shown to mediate HIV-1 release from infected cells *via* mediating proteasomal CD4 degradation, thereby promoting trafficking of the gp120 envelope protein to the cell surface.[176,177] More recently, it was shown to ameliorate antiviral restriction conferred by the interferon-inducible factor

BST-2/tetherin.[188] However, a hydrophobic region present in the N-terminal region of the protein, reminiscent of M2, led investigators to examine possible viroporin activity.

In 1996, recombinant and oocyte-expressed Vpu was shown to display channel activity in bilayers and cell membranes, respectively, with N-terminal TMD peptides also recapitulating activity *in vitro*.[25,26] Channels showed selectivity for Na^+ and K^+ compared with Cl^-, yet whilst a bacterial cross-feeding assay supported a preference for Na^+, oocyte experiments did not confirm such discrimination and also showed partial permeability to divalent cations. Mutating the TMD sequence abrogated Vpu-mediated enhancement of particle release, leading to viroporin function being implicated during this process.[189,190] This was supported by the finding that the amiloride derivatives 5-(N,N-hexamethylene)amiloride and 5-(N,N-dimethyl)amiloride, but not amiloride itself or amantadine, blocked Vpu channel activity in bilayers and also the release of HIV virus-like particles from HeLa cells.[181]

Interestingly, further studies on Vpu indicated that it disrupted, rather than enhanced, the conductance of K^+ ions across oocyte membranes.[187] This may be related to an interaction between the Vpu TMD and mammalian TASK channels, whereby sequence homology between the two proteins led to Vpu abolishing TASK-1 current and, conversely, over-expression of TASK led to a marked impairment of the ability of Vpu to enhance viral particle release.[191] However, multiple studies have demonstrated that the Vpu TMD retained its own channel activity in multiple *in vitro* scenarios[24–26,55,179,181,183,184,192] and scrambling this sequence in simian human immunodeficiency viruses (SHIVs) reduced pathogenicity in macaques.[189] It should be noted, however, that disrupting the TMD may also interfere with Vpu actions on CD4 or tetherin degradation. Interestingly, the same group found that replacing the Vpu TMD with that of M2 generated a pathogenic virus *in vivo* that was sensitive to rimantadine, and this could also be recapitulated by introducing a single histidine residue in place of alanine at position 19.[193,194] This was put forward as proof of principle for the potential of viroporin inhibitors in HIV-1 therapy, yet these mutant/chimeric viruses have not been followed up subsequently. However, in 2010 a novel Vpu inhibitor selected from a bacterial toxicity screen was reported to display an EC_{50} of $\sim 2\,\mu M$ in subtype B HIV-1-infected macrophages.[86] BIT225 is an amiloride derivative and has been reported to block the activity of other viroporins in addition to Vpu, including pestivirus and HCV p7 channels.[87] The targeting of Vpu channel activity by BIT225 is supported by its action on Vpu peptides *in vitro* and also by the compound being ineffective against HIV-2 in culture, which lacks a Vpu protein,[86] although no specific resistance mutations have been described either through rational methods or following long-term virus culture. Nevertheless, Biotron have progressed the compound into clinical Phase 1/2 trials in treatment-naive HIV-1-infected patients to assess safety and tolerability (see www.biotron.com.au). If successful, this will provide a landmark for other viroporin drug development programmes.

9.2.2.2 Structure and Function of Vpu Membrane Channels

Vpu comprises a class 1 viroporin made up of 81 amino acids with a mass of ~9 kDa. The protein is separated into a short *ca.* nine-residue N-terminal ectodomain, a single TMD and a cytosolic domain containing two (or more) α-helices (33–49 and 57–70).[195] Channel activity can be recapitulated by peptides corresponding to the first 30 or so residues *in vitro*. A region located between the two cytosolic helices contains two highly conserved serine residues, which can be phosphorylated by casein kinase II, resulting in structural changes in the cytosolic domain.[196,197] Although it was thought previously that the cytosolic domain primarily mediated interactions with both CD4 and tetherin, it is now clear that the TM domain is also involved in protein–protein interactions,[198] although whether this occurs in the context of monomeric or oligomeric protein is not clear.

High-resolution information exists for both the cytosolic (PDB: 1VPU, 2K7Y) and TM domains (PDB: 2JPX, 2GOF, 2GOH, 1PJE), both in solution and in the presence of detergent micelles, although a structure for the complete protein is not yet available.[180,182,185,199–205] However, molecular dynamics and modelling combining available structures have provided detailed insight into its overall structure and functional determinants.[195] Both phosphorylation and the presence of membranes lead to the cytosolic domain adopting a compact conformation with both helices juxtaposed, which is probably important for protein–protein interactions. In terms of the oligomeric Vpu complex, ssNMR of the TM domain favoured a tetrameric assembly (PDB: 1PI7);[182] however, pentamers could also form and other modelling and biochemical studies support the formation of a pentameric helical bundle comprising the channel complex[206,207] and both species may form and play functional roles.[205,208] The predicted lumen is lined by ionisable, and also hydrophobic and/or bulky aromatic side chains, potentially involved in channel gating/opening.[179,180] *In vitro*, Vpu TM peptides display relatively weak channel-like properties, adopting more of a pore-like character with Michaelis–Menten characteristics in the presence of increasing salt concentration.[55] However, cation preferences, and a critical requirement for Ser24 in the TM domain for channel activity, imply that a level of selectivity and a defined gating mechanism exists for these channels.[26,55,179,183,184] Interestingly, neither mutation of a lumenal Arg31 nor Trp23 affected gating behaviour, confirming the importance of Ser24 and implying that alternate aromatic/hydrophobic residues may serve as gates in the channel complex.[55]

9.2.2.3 Targeting Vpu Channel Activity as an Antiviral Strategy

Although it has been demonstrated that amilorides such as HMA specifically block efficient production of HIV-1 particles from infected macrophages,[86,181] what is not clear is how Vpu channel activity specifically relates to this process. This is especially difficult given the likely involvement of the TMD in protein–protein interactions involving CD4 and tetherin, both of which are also essential events required during efficient particle production. However, recent studies have shown that BIT225, which specifically blocks Vpu activity *in vitro*, does not interfere with Vpu-tetherin interactions[209] and others have shown that mutations blocking tetherin interactions have no effect on channel activity.[186] Therefore, Vpu, like other viroporins, demonstrates significant functional redundancy for such a small protein, possessing three seemingly distinct mechanisms of modulating HIV-1 particle production, although the role of channel activity *per se* is not clear. Interestingly, a recent study revisited the destructive interaction between the Vpu TM domain and cellular TASK channels, as reported previously by the same group. TASK channels help set the resting cell membrane potential and their disruption by both Vpu and/or a dominant negative TASK fragment serves to enhance HIV-1 particle release *via* enhancing depolarization-stimulated exocytosis.[210] Accordingly, HIV-1 release was accelerated by externally imposed depolarization alone, which reduced the barrier to membrane fusion/fission during budding. However, the relationship between this phenomenon and Vpu channel activity itself is not clear as the effects of Vpu inhibitors were not characterised.

Ambiguity surrounding the role of Vpu channel function should be resolvable *via* the use of inhibitors, yet specific resistance mutations to control for drug off-target effects have not been selected to date. Interestingly, *in silico* docking and molecular modelling predicted that HMA bound to a lumenal site in proximity to the functionally significant Ser24 residue,[211] yet unfortunately no biophysical or genetic evidence exists to support this. Similarly, no information as to how BIT225 interacts with Vpu channels has been published and no resistance mechanisms have been defined that may shed light on its mode of action. Nevertheless, BIT225 does not affect HIV-2, which lacks a Vpu protein, and it is reported to display activity against a wide range of both CCR5- ('R5') and CXCR4-tropic ('R4') strains.[86] However, whereas the compound appears potent against virus-infecting macrophages, its potency against T-cell-tropic cultured HIV-1 appears diminished, resulting in a far narrower selectivity index. This may reflect cell type-dependent effects whereby differential requirements for Vpu channel activity occur, consistent with the notion that BIT225 selectively targets macrophage resident virus reservoirs. Alternatively, it is possible that amino acid variation in the Vpu protein between R5 and X4 tropic strains might account for observed differences in potency, although how such strain variance may affect drug interactions has not been addressed *in vitro* or *in silico*. Such polymorphisms may point to sites of drug–protein interactions in lieu of cell culture-generated resistance mutants. Alternatively, output data from forthcoming BIT225 clinical trials may also provide resistance data for the drug in patients displaying viral rebound during treatment. Such trials represent a benchmark in the development of viroporin inhibitors as drug targets and their outcomes will be monitored with great anticipation. This may represent a stepping stone towards future Vpu inhibitors, selected from large-scale high-throughput screens or through the characterisation of protein–drug interactions and use of available TM domain structures in rational programmes.

9.2.3 Hepatitis C Virus p7

9.2.3.1 Hepatitis C Virus Infection: a Uniquely 'Drugable' Target

Hepatitis C virus (HCV) chronically infects over 170 million individuals, more than five times the number infected by HIV-1. Geographical variation in prevalence ranges from around 1% in many developed nations to much higher values in developing countries. The extreme case is that of Egypt, where poor needle hygiene during a mass Bilharzia vaccination programme in the 1960s resulted in nearly one-quarter of the population being infected at present.[212] An additional level of geographical separation exists with the distribution of the seven known HCV genotypes around the planet, which is further divided into many more subtypes.[213] Furthermore, HCV variability at the level of individuals manifests as the virus existing as quasi-species, much like HIV-1, although HCV is many times more variable and replicates at a far higher rate (10–100×). Chronic HCV infection leads in many cases to severe liver fibrosis, cirrhosis and extra-hepatic complications, with cirrhotics showing high levels of progression to hepatocellular carcinoma. Usually, asymptomatic acute infection means that affected patients remain undetected until chronic infection manifests clinically, resulting in uncertainty as to the true extent of the HCV epidemic.

Antiviral therapy exists for HCV infection and is the only known example of such an approach being able to clear a chronic virus infection.[214] Current standard of care (SOC) comprises pegylated interferon alpha combined with the guanosine analogue ribavirin, administered for up to 48 weeks depending on virus genotype. In some countries (*e.g.* USA), SOC has recently been extended to include newly licensed protease inhibitor drugs for patients infected with genotype 1 viruses, as discussed below. Virus genotype also largely influences treatment outcomes, with only 20–25% of cases infected with insensitive genotypes such as 1 and 4 being cleared, ranging up to nearly 80% for 2 and 3. Overall response rates are therefore usually quoted as around 50%. Success of this primarily immune-stimulatory regimen is also highly dependent on host genetics, as illustrated by a strong link between treatment outcome and polymorphisms in the IL-28β gene promoter region.[215,216] Therefore, this unsatisfactory success rate combined with the high cost, invasiveness and toxicity associated with the treatment prompted a huge drive towards the development of virus-targeted, direct-acting antivirals (DAAs), the first of which targeting the virus protease were licensed for human use in 2011.[217]

Recent estimates place the market for new HCV DAAs at $6 billion, and this is expected to double by 2015. Retreatment of previous non-responders to interferon/ribavirin will also serve to increase the drug-treated population and generate greater market demand. Currently, two protease inhibitors, telaprevir and boceprevir, administered with SOC improve responses in genotype 1 patients to around 70%, generating great optimism. In addition, many more compounds targeting protease, polymerase and the NS5A protein are at advanced stages of development and several are expected to be licensed in the next 5 years. Of these, nucleoside polymerase inhibitors show strong pan-genotype activity with good potency and very high genetic barriers to resistance. As such, great optimism exists amongst medical professionals and drug companies alike that the burden of HCV disease will be dramatically reduced over coming years and that the gold standard of an effective, interferon-free, DAA combination regimen will be achieved.

However, a note of caution is pertinent, despite tremendous recent progress, and HCV may still take many years to be controlled effectively. Currently, single DAA efficacy, necessarily given with SOC, remains dependent on response to interferon and it will be several years before sufficient DAAs are available clinically to utilise combinations that lessen this dependence, although several interferon-free trials are ongoing. DAAs are also very expensive and will not be available in many developing countries where they may be needed most, plus patient compliance and identification remain a critical determinant of treating the considerable disease population. In addition, side effects and drug interactions[218] may be significant, with some severe reactions already documented for protease inhibitors.[219] Additionally, drug resistance remains a significant concern owing to the highly variable nature of HCV and its endemicity within populations likely to both rapidly mix and disseminate new strains, while failing to comply adequately with antiviral treatment, such as injecting drug users. Even highly potent nucleosides capable of initially clearing virus in combination with ribavirin for 8 out of 10 patients, in fact result in rebound several weeks later; however, additional combinations and/or interferon may prevent this in the long term. As such, even in the best-case scenario it is likely that HCV drug development may need to continue for many years and accommodate additional targets, much like the case of HIV-1. This will be especially important for those unable to tolerate existing treatments or infected with multi-drug-resistant virus strains that could emerge under such intense selective pressure. Finally, recent high-profile Phase 3 drug trials have been halted owing to severe adverse events, including patient deaths. Notably, the HCV nucleoside polymerase inhibitor BMS-986094 (formerly known as INX-189, developed by Inhibitex) has been discontinued following the identification of both cardiac and kidney toxicity which resulted in patient deaths. Trials of another nucleoside from Idenix Pharmaceuticals with similar chemistry are currently on hold, but studies are continuing for chemically distinct Gilead and Vertex lead drugs in this class. In addition, trials of Alisporivir (Novartis, formerly DEBIO-025 developed by the Swiss company Debiopharm[220,221]), which targets HCV–cyclophilin interactions, have been halted owing to a possible exacerbation of interferon-associated pancreatitis, again involving fatalities. The most up-to-date information on HCV drug trials can be found at www.natap.org.

9.2.3.2   Discovery of the HCV p7 Viroporin

HCV is the prototype for the *Hepacivirus* genus of the Flaviviridae, with a 9.6 kb positive sense ssRNA genome and forming an enveloped particle

~60 nm in diameter. A single open reading frame (ORF) is cap-independently translated from an internal ribosome entry site (IRES) yielding an ~3000 amino acid polyprotein, which is processed into the structural [core, envelope (E)1/2] and non-structural (NS) proteins by host and viral proteases. NS3/4A/4B/5A/5B forms the minimal viral replicase, as defined by sub-genomic replicon RNAs, whereas the NS2 protein acts as an autoprotease and also during virion production.

In 1994, a tenth HCV protein was discovered by groups analysing HCV polyprotein processing in the region between E2 and NS2.[222,223] A 63 amino acid protein, termed p7 according to its molecular mass, was produced following inefficient cleavage by signal peptidase at both the E2 C-terminus and, to a lesser extent, at the NS2 N-terminus; E2–p7 and E2–p7–NS2 precursors are therefore present within HCV-infected cells and over-expression studies have shown that cleavage at these sites is influenced by the preceding helical regions in E2 and NS2, respectively.[52,224] p7 was found to be highly hydrophobic and was predicted to contain two TMDs, separated by a short cytosolic (in accordance with polyprotein membrane topology) loop region containing two highly conserved basic residues.[225] Double membrane spanning topology was supported by cellular expression studies using epitope tags located at either termini and within the loop region, confirming p7 as a class 2 viroporin with its termini being lumenally oriented.[225] However, evidence that the C-terminal region of the protein may flip topology has also been demonstrated where p7 exists as an E2–p7 precursor,[52] and this has more recently been proposed to occur *in vitro* when p7 peptides are exposed to altered membrane compositions.[226]

In 2003, our laboratory demonstrated that recombinant p7 from HCV genotype 1b displayed channel activity in artificial bilayers with activity sensitive to low micromolar concentrations of amantadine.[37] Experiments were performed in a K^+ electrolyte and displayed both single-channel and burst-like activity. Activity was further enhanced by replacing K^+ with Ca^{2+}; however, it could not be ruled out that this was due to effects on membrane bilayers. p7 channel activity was subsequently confirmed later in the same year for synthetic peptides from genotype 1a HCV in K^+ electrolyte containing a low concentration of Ca^{2+}. Both single-channel and burst-like activity were again observed, with high conductance readings compared with recombinant protein.[38] In addition, nonylated but not butyl imino-sugars also effectively blocked channel activity when present at concentrations >100 μM. A year later, another synthetic genotype 1a peptide study also demonstrated p7 channel activity and highlighted 10-fold ion selectivity for K^+ and Na^+ over Cl^-, although this was lessened in the presence of even low levels of Ca^{2+}.[39] Conductance also adopted both single-channel and burst-like activity, with conductances similar to those in recombinant studies. This study also revealed that HMA at 100 μM also specifically inhibited p7 activity. Combined with the finding that p7 was essential for HCV to replicate in chimpanzees,[227] these studies sparked great interest in the development of p7 ion channel inhibitors as novel HCV therapies.

Assigning the role of p7 channel activity within the context of the virus life cycle only became possible following the development of infectious HCV culture in 2005[228] as the protein was not required for genomic RNA replication in replicon systems. Prior to this, p7 had been shown to localise to predominantly ER-associated membrane compartments in over-expression studies,[71,225,229,230] with a small fraction potentially present at the cell surface. Furthermore, following expression in 293T cells, a proportion was found to associate with mitochondrion-associated 'heavy' ER membranes with differential exposure of the protein termini in this or rough ER compartments; however, a functional consequence of this distribution has not been demonstrated in virus-infected cells.[71,229] More recently, tagged p7 expressed in the context of infectious HCV genomes was seen to localise to the ER, although such tags were not genetically stable.[230] The first clue to a possible role for p7 during the HCV life cycle as a proton channel came from a study demonstrating its ability to replace the monensin-like activity of M2 within transfected cells, thereby promoting the surface transport of influenza HA protein.[71] This also, by implication, meant that p7 likely resides within the TGN and also endosomal compartments in addition to the ER, as these are the compartments that become acidified by vATPase. In this study, genotype 1b p7 activity was again sensitive to amantadine at low micromolar concentrations, consistent with *in vitro* effects, and mutation of the conserved basic residues in the cytosolic loop also abrogated activity.

With the advent of HCV infectious culture, p7 was shown to play a critical role during the production of infectious virions.[231,232] Basic loop mutations were originally reported to cause either late or early defects in particle production, although more recent *trans*-complementation studies favour a late-acting role,[233] which is consistent with the effects of small-molecule inhibitors blocking secreted but not intracellular HCV infectivity.[63] However, dissecting ion channel-specific roles for p7 during the HCV life cycle has been complicated by its processing from the polyprotein precursor,[232] relatively poor definition of many point/other mutant phenotypes and the recent finding that p7 also engages in protein–protein interactions with the NS2 protein.[234–240] Therefore, observed defects in intracellular or secreted infectivity may result from mutant effects on multiple processes. As an example, the commonly employed basic loop mutation, where Lys33 and Arg/Lys35 (p7 sequence) are substituted by alanine or glutamine, causes defective proteolytic processing of the polyprotein[232] and has not been ascribed a channel gating defect *in vitro*. Indeed, the only phenotype attributed to this mutation to date is an inability of recombinant p7 to insert into liposomal membranes,[241] consistent with a proposed membrane insertion motif,[242] likely linked to protein processing defects observed when this mutation is introduced into infectious HCV. Nevertheless, p7 channel activity appears separate to its interactions with NS2, as they are not affected by p7-specific small molecules.[237] Recently, p7 proton channel activity was directly demonstrated within virus-infected cells using fluorescent proton ionophores, whereby p7-mediated raising of vesicular pH correlated with the survival of acid-labile intracellular virus particles.[40] Accordingly, reacidification occurred in the presence of p7-specific small-molecule inhibitors, correlating with a reduction in secreted infectivity. Interestingly, p7 basic loop mutations were observed to be rescued *in trans* by influenza M2 in this study,[40] whereas others showed that a p7 deletion mutant lacking the N-terminal portion of the protein could not be rescued in this fashion.[233] This same phenomenon was also observed when bafilomycin A was employed to block vATPase activity, infectivity being restored to loop mutants but not deletions.[40] It therefore appears likely that p7 deletants incur additional defects, probably involving NS2 interactions, which cannot be resolved by restoring vesicular alkalinisation in isolation. Thus, p7 proton channels appear to act at a similar stage to the monensin-like activity of M2 during particle secretion, although the molecular basis underpinning the sensitivity of intracellular particles to reduced pH has yet to be determined.

One final area where p7 has been proposed to function is during virus entry, again mimicking M2, although its presence within an infectious virus particle has not been directly demonstrated.[64,243] If absent from the particle, p7 would be the only HCV protein processed by signal peptidase that was not incorporated. Whilst retroviruses pseudotyped with HCV glycoproteins are not dependent on p7 for efficient cellular entry, the uncoating and subsequent replication of retroviral nucleoproteins are unlikely to resemble those of HCV. Interestingly, p7 enhances the cell entry of VLPs produced from insect cells[244] and C-terminally Myc-tagged protein has been detected within these structures by immunogold electron microscopy (EM), albeit where E2–p7 processing was absent.[52] Nevertheless, these studies show that intact HCV-like particles can therefore form when p7 is contained therein. Evidence against p7 functioning during virus entry comes from specific infectivity measurements of HCV virions secreted by a high-efficiency particle-producing chimeric HCV where a basic loop mutation (K33Q, R35S) reduces virion release by $\sim$3log$_{10}$, rather than abrogating it entirely.[232] In this case, the specific infectivity of particles, normalised for core protein content, was equivalent to the wild-type chimera. However, these experiments made the assumption that loop mutations directly affect p7 channel gating, rather than reducing the efficiency of membrane insertion. In favour of a role for p7 during entry, p7-specific small molecules have been shown to inhibit the entry of HCV in culture in a strain-specific fashion, whereas they did not affect the entry of retroviruses pseudotyped with HCV glycoproteins.[64] Definitive assignment of an entry role for p7 will require both improved immunological reagents and potent inhibitors that, unlike adamantanes, are not able to potentially raise endosomal pH.

9.2.3.3 Structure and Gating of p7 Proton Channels

p7 comprises 63 amino acids with a predominantly helical secondary structure, based on both *in silico* predictions[48,49,225,226,241,245] and recent NMR studies of the monomeric protein.[246–248] Helical wheel plots predict significant amphipathic character in both helices, but most notably in the N-terminal TMD, where a number of hydrophilic residues are thought to align and form the channel lumen upon oligomerisation. Amongst these is a His17 residue which is highly conserved across the majority of HCV isolates, yet is absent from several sequences, as discussed below. Accordingly, biochemical inhibition of genotype 1a p7 channel activity using Cu^{2+} ions strongly indicated the presence of His17 within the channel lumen,[249] agreeing with previous models of the p7 channel complex where the lumen was formed by the N-terminal TMD.

The stoichiometry of the p7 oligomer was initially estimated to be hexameric following low-resolution EM studies of genotype 1b GST–His–p7 fusion proteins and near-native His–p7 in liposomes.[37] Improved expression and purification of 1b GST–FLAG–p7 later permitted rotational averaging of fusion protein complexes in liposomes, which revealed predominantly heptameric channel complexes, although a minority of hexamers were also apparent.[49] Heptameric genotype 1b p7 complexes have also been observed by both chemical cross-linking and native polyacrylamide gel electrophoresis (PAGE) studies conducted in both detergent (DHPC) and liposomes.[63,241] However, recent elegant EM single-particle reconstructions undertaken by Zitzmann's group of genotype 2a p7 peptides in DHPC micelles revealed a clear hexameric stoichiometry of resultant p7 channel complexes.[50] The derived 16 Å resolution channel reconstruction represented the first structure for a complete viroporin and revealed a flower petal-like structure where the p7 termini formed the wider part of the molecule, confirmed by specific immuno-gold labelling. Interestingly, work by the same group revealed a clear influence of the lipid environment on both the helical composition and activity of p7 channels *in vitro*, which was proposed to relate to changes in stoichiometry and/or the membrane topology of the p7 C-terminus.[226] Such topological changes have also been proposed for dermaseptins, whose activity is influenced by the bilayer lipid composition[250] and the oligomeric state of the mechanosensitive channel (MsCL) from *Saccharomyces aureus* also depends on the membrane-mimetic environment.[251] It is therefore possible that p7 channel populations are highly plastic and change depending on membrane environment, protein abundance and genotype-dependent sequence variations, all of which could significantly alter channel behaviour. In support of this notion, recent molecular dynamics simulations revealed a structural plasticity to p7 channel complexes.[48]

Although no atomic structure for full-length p7 exists, solution and solid-state NMR studies on monomeric protein have been reported. One solution study analysed full-length and truncated peptides in 50% TFE derived from genotype 1b p7 incorporating a C27A mutation.[247] A solution structure for one of the truncated peptides corresponding to the complete C-terminal region was generated under these conditions (PDB: 2K8J). Unfortunately, long-range interactions between the two TMDs were not detectable, indicating that p7 probably did not form a hairpin in 50% TFE. However, NMR data were incorporated into a molecular dynamics-derived model of the hairpin in a membrane bilayer, allowing the prediction of secondary structure domains and the fitting of NMR data into a hairpin context. This model predicted two N-terminal helices (aa 2–14 and 19–31) separated by a turn containing the conserved His17 residue. This resided in the virtual membrane such that the second helix comprised the TMD. The loop region (aa 33–39) resided at the membrane interface and the more dynamic and flexible C terminal TM region comprised residues 40–56, followed by another short helix (aa 57–63). Opella's group has undertaken extensive solid-state studies of genotype 1b p7 peptides (C27S change incorporated) in detergent micelles, isotropic bicelles and lipid bilayers, which provided further insight into the likely secondary and tertiary structure of p7 monomers.[246,248,252] p7 was shown to possess two TM regions between residues 6–27 and 41–57, each of which consisted of two distinct helical segments. The second and fourth helical segments adopted a tilt relative to the membrane normal of 25° and 10°, respectively. Consequentially, the cytosolic loop in this scenario is extended relative to the solution NMR/molecular dynamics model and a greater proportion of the N- and C-termini extend out with the membrane interface. The search therefore continues for a complete p7 structure with multiple laboratories worldwide focussing on the challenge of both the monomer and the multimeric complex. Given the effects of environment and genotype discussed above, this endeavour may well yield multiple structures along the lines of previous M2 studies.

The gating mechanism of p7 channels is not well defined. Although shown to conduct protons in cellular membranes,[40] *in vitro* experiments have demonstrated that p7 is capable of conducting alkali metal ions, calcium ions and carboxyfluorescein (CF) dye.[37–39,49,61–64,241,247,253] This may be due to the more 'pore-like' qualities of the protein, represent an artefact of individual experimental systems where these are the major ionic species present, or, especially in the case of anionic CF dye, represent an inefficient secondary conductance serving as an indirect measure of channel opening. Nevertheless, proton channel activity was supported by the finding that His17 was essential for genotype 1b p7 activity in liposomes and that increasing pH gradients such that external buffer pH became more acidic led to markedly increased channel activity.[241] However, mutation of His17 in genotype 1a p7 did not abrogate activity *in vitro*[249] and also did not affect virion production by genotype 2a JFH-1 HCV in culture.[233,254] Furthermore, some HCV isolates such as the genotype 2a J6 strain do not contain His residues in the N-terminal helix. Hence genotype-specific differences clearly appear to influence p7 channel activity, making its behaviour seemingly more complex than a standard M2-like model.

Based on available homology and structural information, key positions lining the p7 lumen are likely to comprise ionisable residues such as His17 (Asn in some isolates), Ser21 (Tyr in some) and Tyr31 (His31 in many genotype 2a isolates), which, although located outside the TMD according to solid-state studies,[248] is likely positioned near the mouth of the channel. In addition, several conserved bulky/hydrophobic residues such as Phe25/26 (located at the end of the TMD in solid-state models) and Trp30 could represent gating candidates (see Figure 9.3), supported by a hyperactive channel phenotype for Phe25Ala mutant p7 channels.[48,63] Ionisable residues in particular vary according to HCV genotype, likely affecting channel gating. Interestingly, genotype-dependent differences in acid-mediated p7 channel activation were

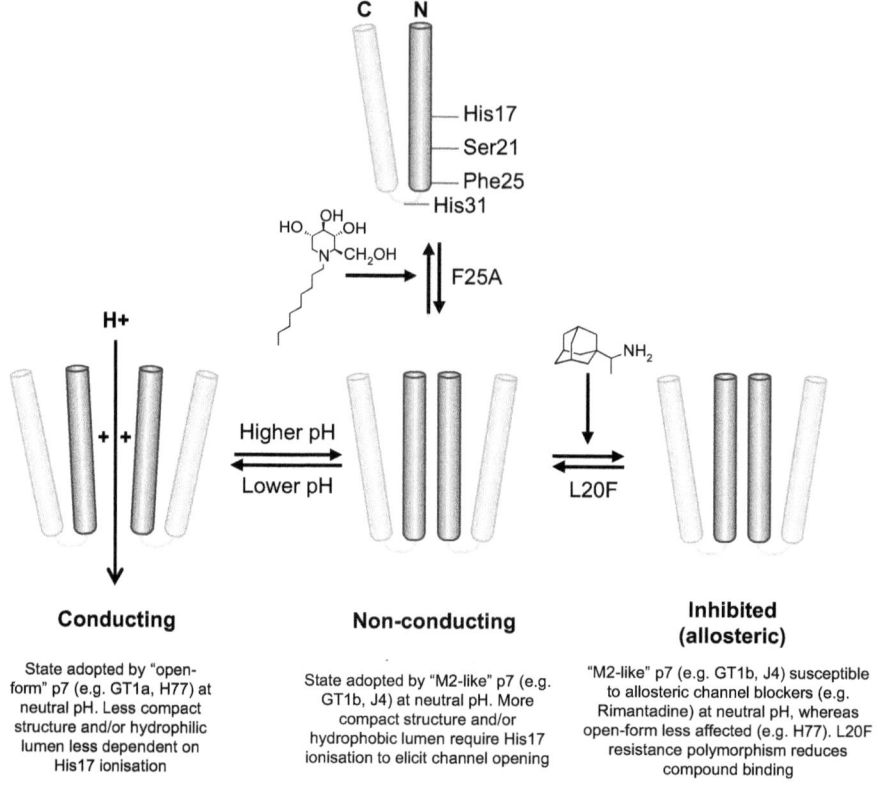

Figure 9.3   Model for p7 channel activity and drug inhibition. Based on *in vitro* studies of p7 channel gating and drug susceptibility, channels originating from certain HCV genotypes (*e.g.* 1b) behave similarly to M2, where reduced external pH enhances channel opening in a unidirectional fashion, whereas others (*e.g.* 1a) behave in more of an 'open-form' fashion, allowing protons to pass through according to electrochemical gradients in either direction (see Section 9.2.3.3). One explanation is that M2-like channels form more compact structures with hydrophobic constrictions within their lumens, which therefore require energy from side-chain ionisation to open efficiently. This is not so for open-form channels, in which the channel equilibrium is shifted towards the open state independent of pH. M2-like channels are therefore more likely to be inhibited by compounds targeting the allosteric site on the channel periphery, such as rimantadine, whereas both channel types are equally likely to be susceptible to alkylated imino sugars. However, amino acid changes within the respective binding sites also dictate the resistance/susceptibility of individual channels, including L20F for rimantadine and F25A for imino sugars (Section 9.2.3.4).

recently demonstrated, whereby genotype 1a channels, which possess a non-essential His17,[249] displayed substantially altered activity in liposomes upon alteration of external buffer pH.[253] Unlike genotype 1b channels, where lowering the external pH, thereby increasing the electrochemical gradient unidirectionally relative to the liposome interior, led to channel activation,[61,253] 1a channels increased activity when gradients were altered in either direction. Thus, genotype 1a channels displayed a more open-form, pore-like behaviour compared with genotype 1b, explaining why His17 may be dispensable for channel opening. Intriguingly, variation amongst patient-derived 1a sequences at key ionisable positions switched p7 gating behaviour back to a genotype 1b pattern, whereas others enhanced/reduced channel activity. Specifically, Ser21Pro and Tyr31His induced 1b-like behaviour, whereas His17Asn dramatically enhanced channel activity, but in a pore-like pattern.[253] Interestingly, Asn17 is also present in the genotype 2a J6 isolate and transferring this p7 into other genotype backgrounds dramatically enhances virion production.[232]

One hypothesis explaining the observed behaviour of p7 from different genotypes and mutant backgrounds is that channel opening is dependent on the hydrophobic/hydrophilic balance of the channel interior, which in turn is influenced by both overall channel structure and variation of the lumenal amino acid sequence. Hydrophobic/bulky residues brought into close proximity would represent a barrier to the formation of a water column through the channel, and this must be overcome for ions to flow. The energy required to achieve this may be minimal in a pore-like fashion or may require events such as ionisation to effect necessary structural changes. In the case of genotype 1a channels, this barrier appears low, resulting in a large proportion of the channel population tending towards an open state, even at neutral pH where His17 remains uncharged. This barrier may be further lowered by the presence of Asn17. Changes where polar residues are altered such as Ser21Pro and Tyr31His would increase the hydrophobicity of the lumen at neutral pH (His pK_a in solution is 6.04) and so make the energy obtained from His17 protonation at lower pH now necessary to effect opening. This scenario seems to be the baseline for genotype 1b channels, so explaining why His17Ala mutations abrogate activity.[241] Lastly, some isolates such as genotype 2a JFH-1 possess both His17 and His31, generating a hydrophobic lumen at neutral pH with a potentially redundant means of effecting channel opening *via* protonation of one or other His residue; JFH-1 p7 is potently activated by reduced external pH in liposomes (E. Atkins and S. Griffin, unpublished observations). Therefore, mutation of one or other JFH-1 His residue in isolation would not prevent channel opening and explains the lack of virion production phenotypes when individual changes are introduced into the JFH-1 genome.[233,254] Extensive mutagenesis programmes and characterisation of pH-dependent gating will be necessary to formalise this hypothesis, and future investigations must allow for the clear genotype- and subtype-dependent differences in p7 function, which appear to affect virtually all facets of this channel examined to date.

9.2.3.4 How Do Adamantanes, Amilorides and Alkyl Imino-Sugars Inhibit p7 Channels?

Work done using prototype p7 inhibitors initially generated considerable controversy, particularly in the light of a perceived lack of clinical efficacy.[84,85]

In particular, amantadine, which was initially shown to inhibit genotype 1b p7 channels at low micromolar concentrations *in vitro*,[37] was required at much higher concentrations to inhibit genotype 1a p7 peptides *in vitro*[243] and appeared ineffective in cell culture at standard concentrations, even against genotype 1b chimeric viruses (up to ~ 50 μM).[64,243] By contrast, 50 μM nonyl-imino sugars were able to cause up to 80% reductions in secreted infectivity in one-step assays and could eliminate HCV from sequential supernatant passage.[243] However, in assessing the susceptibility of multiple HCV sub/genotypes both in culture and *in vitro*, it became clear that p7 inhibitors exerted their effects in a p7 sequence-dependent fashion, with p7 displaying subtle resistance behaviour even to very similar compounds, generating specific patterns for each isolate tested.[64] For example, whilst rimantadine has been shown to inhibit most strains tested,[64,255] both genotype 2a JFH-1 and 1a H77 p7 appeared highly resistant to amantadine, despite the obvious similarity between the two compounds.[64] In addition, genotype 3a p7 appeared resistant to the nonylated imino sugar *N*-nonyldeoxynojirimycin (*N*N-DNJ; see Table 9.2).[64] Although effective *in vitro* against various isolates, HMA toxicity prevented its effects in culture from being characterised. Such genotype-dependent patterns of susceptibility were reproduced in studies assessing proton conductance of channels present within isolated cellular membranes;[40] it also became apparent that cellular uptake/turnover of certain compounds likely affected their cell culture efficacy as activity against channels in purified vesicles was greater compared with intact cells; this may be related to hepatic metabolic turnover *in vivo*.

Genotype-dependent differences in compound susceptibility have subsequently been levied by some as a criticism of p7 inhibitors as a whole. However, from a medicinal chemistry perspective, they point to specific drug–protein interactions that may be defined through resistance and so permit mapping of inhibitor binding sites and the potential definition of compound modes of action. Early clues to the action of adamantanes came from the partial amantadine resistance phenotype for a genotype 1b p7 mutant where a series of C-terminal leucines was changed to alanine.[241] However, like the amantadine-resistant JFH-1 HCV, this mutant retained sensitivity to rimantadine, as is the case for all HCV genotypes to date to a greater or lesser extent (*e.g.* genotype 1a H77 is less sensitive).[64,255] Interaction of these C-terminal leucine residues with amantadine was also shown in ssNMR studies,[246] and together this suggested that adamantanes might bind to a peripheral, membrane-exposed site that potentially stabilises the closed form of channel complexes, as proposed for M2.[80,127] This notion gathered considerable momentum when, in the absence of a channel structure, *de novo* molecular models of the p7 channel complex were employed as templates for *in silico* compound docking studies.[63] These predicted adamantane interactions at a peripheral cavity, which comprised several of the C-terminal leucines and was formed by residues from both adjacent protomers and the N-terminal TMD. Intriguingly, the predicted site contained Leu20, which was previously observed to undergo a non-synonymous Leu20Phe change in non-responsive genotype

1b infected patients receiving amantadine alongside SOC (pegylated interferon and ribavirin for 48 weeks).[256] Convincingly, engineering this mutation into Leu20-containing, rimantadine susceptible strains (genotype 1b, 2a) conferred resistance both *in vitro* and in culture, providing the first point mutation definition of a p7–drug interaction and supporting the binding of adamantanes to a peripheral site.[63]

In the same study, docking of nonylated imino sugars was predicted to differ markedly from that of adamantanes, whereby these molecules were predicted to intercalate between p7 protomers and so potentially prevent the formation of the channel complex.[63] One key interaction was predicted to occur between the imino sugar head group and Phe25. This residue is highly conserved across most HCV isolates with the exception of some genotype 3a (also some 4) sequences, where Ala is present. Genotype 3a p7 was shown to be less sensitive to imino sugars than other HCV strains,[64] and this was shown to correlate with the disruption/preservation of stable oligomers in the presence of drug by native PAGE, compared with sensitive p7 proteins.[63] Introduction of Phe25Ala into susceptible strains again conferred resistance to this inhibitor class, with no cross-resistance to adamantanes, confirming an entirely distinct mode of action. As discussed above, this mutation also caused a hyper-active channel phenotype *in vitro* when transferred into genotype 1b/2a proteins, consistent with a potential gating role for Phe25 in these p7 proteins,[48] but also showing that genotype 3a p7 likely utilises an alternative bulky gating residue. Neither Leu20Phe nor Phe25Ala imbued any significant fitness cost to the virus in culture when introduced into either a 1b or a 2a background, although infectious particle production was reduced by between 30 and 60%. Of course, the consequences of this *in vivo* are unknown, yet this indicates that the genetic barrier to escape from prototype inhibitors may be low.

Critically, the distinct modes of action and lack of cross-resistance between adamantane and imino sugar resistant-variants indicate that compounds targeting these sites could potentially be used in drug combinations.[63] It is important to point out that the mutations identified in this study are likely to represent only one of several sequence changes that may occur to effect resistance within the context of varying p7 sequences. Accordingly, other p7 sequence changes have been observed in patient amantadine studies[257] and even minor variations between 1b p7 sequences around the predicted binding site appear to reduce markedly amantadine sensitivity *in vitro*.[247] Hence the p7 sequence context must be borne in mind when interpreting inhibitor studies, and also for the future development of improved molecules. However, as discussed in previous sections for other viroporins, prototype p7 inhibitors serve to illustrate the presence of druggable sites that could be exploited as drug targets. Rimantadine inhibits all p7 sequences tested to date, albeit with moderate potency,[64,255] and the only exception to imino sugar susceptibility has been genotype 3a.[63] Hence improved molecules targeting these sites may be selected from screens or could be selected by rational design, whereby more efficient binding modes and stabilising contacts lead to improved potency. However, rational design will necessitate atomic models of the channel complex, as attempts to generate bespoke compounds targeting the peripheral site on *de novo* models, although successful, could not improve on the potency of rimantadine.[63]

Screening-based alternatives to rational development have been employed for p7 with some success. Genotype 1b channels were tested in a large-scale liposome dye release screen of 3520 compounds by investigators at Boehringer Ingelheim.[62] This identified multiple hits and also validated this assay system as a robust means of identifying novel inhibitors with low sensitivity to fluorescence artefacts. However, these compounds have not been pursued to date. In addition, BIT225 was selected from a bacterial toxicity screen in cells expressing genotype 1a p7.[87] BIT225 inhibits 1a p7 at 100 μM *in vitro* and was shown to display a cell culture IC_{50} of ~300 nM against the related pestivirus bovine viral diarrhoea virus (BVDV). Given the varied susceptibility of HCV isolates to p7 inhibitors, it is perhaps surprising that BIT225 displays efficacy against the highly sequence-divergent BVDV p7, although, like rimantadine, this may indicate that a binding site is present across multiple HCV genotypes to which the drug binds with reasonable efficacy. In addition, whether BIT225 adopts the same mode of action against p7 proteins as it does for Vpu is unknown. However, recent studies on p7 from the related pestivirus classical swine fever virus (CSFV) revealed marked differences between this and the HCV protein, making it all the more surprising that BIT225 displays efficacy in this system.[41] Data for BIT225 in HCV cell cultures are yet to be presented and no resistance information for this or other amilorides such as HMA, is available to assign a potential mode of action. Based on Vpu studies,[211,258] binding may occur at a lumenal site involving Ser21, yet this has not been analysed *in silico* or otherwise. However, results from early clinical trials appear encouraging (see below), making BIT225 a first step towards targeting p7 clinically for HCV treatment.

9.2.3.5   Clinical Use of p7 Inhibitors: Past, Present and Future

The majority of prototype p7 inhibitor trials have concerned amantadine alongside SOC and were conducted prior to the identification of p7 as a viroporin. Many patient studies of amantadine (usually 200 mg d^{-1}) for 12 weeks in combination with SOC have revealed little improvement compared with dual therapy, as assessed by sustained virological response (SVR) at week 60, 12 weeks after completing a 48 week course of interferon/ribavirin. However, meta-analyses stratifying results for patients who had previously failed treatment – usually genotype 1 HCV – did reveal a measurable benefit of the triple therapy regimen.[84,85] Amantadine monotherapy studies measuring the kinetics of patient responses revealed rapid viral rebound in a matter of days[259] and, when given with SOC, a loss of clinical benefit was observed upon withdrawal of amantadine at 12 weeks, where improved patient responses reverted to the SOC controls.[260] Similarly disappointing results were obtained using a nonylated imino sugar p7 inhibitor based on *N*N-DNJ, UT-231b, developed between United Therapeutics and researchers in Oxford (http://www.clinicaltrials.gov/ct2/show/NCT00069511), although many of the patients involved were at advanced disease stages, making effective responses unlikely. Such studies understandably led to considerable scepticism concerning the worth of p7 inhibitors for HCV treatment; however, in retrospect in the light of the new era of HCV DAA, these trials were seemingly destined to fail according to the criteria by which they were measured. We now know, and probably should have guessed in the past, that the response to even potent HCV DAA (nucleosides perhaps being the exception), given singly with SOC, remains dependent on the interferon response.[261] Interferon-insensitive HCV therefore rapidly selects resistance to what essentially becomes DAA monotherapy. This would be expected to occur all the more rapidly for a DAA not targeting genome replication, such as an inhibitor of particle release. Furthermore, amantadine is known to be the least potent of the prototype p7 inhibitors identified to date,[64,243] which, combined with relatively low serum concentrations being achievable at standard doses (2–3 µM at best), likely results in a very low genetic barrier to resistance occurring, should the patient not already be infected with a naturally resistant strain. The potency of UT-231b was likely far higher than that of amantadine, yet again would face the rapid selection of resistance in interferon-insensitive HCV. In addition, the majority of patients in this trial had already failed SOC and were also clinically less likely to respond to therapy due to their disease state. Nevertheless, the selection of the Leu20Phe resistance polymorphism,[63,256] and possibly others,[257] in genotype 1b patients by an inhibitor as poor as amantadine is strong evidence that a selective pressure was exerted by the drug due to a specific antiviral effect, however weak it may have been. Therefore, potent compounds targeting p7 should be able to dramatically improve clinical outcome. Initially, effects will necessarily need to be measured in combination with SOC, yet ideally p7 inhibitors should be given alongside DAA targeting replication in order to minimise the potential for HCV to select resistant variants.

BIT225 represents the first bespoke p7 inhibitor to be tested in clinical trials and early reports presented at HepDART 2011 (http://www.natap. org/2011/hepDART/hepDART_22.htm) of its efficacy are encouraging: 87% of patients receiving 400 mg of BIT225 alongside SOC achieved an early virological response (12 weeks, >3$\log_{10}$ drop in virus load) compared with 61% for SOC alone. Furthermore, patients on 400 mg of BIT225 also showed an additional median $\log_{10}$ reduction in virus load after the 28 day dosing period compared with SOC alone. However, a greater proportion of the treatment group were infected with genotype 1a HCV, against which BIT225 was selected, compared with the control group (6/8 in the 400 mg group, only 2/8 in the controls), which may skew these data towards a favourable response in such a small cohort. Similarly, the small cohort size means that one or two patients may greatly affect the percentage median response rates; 5/8 SOC controls achieved an EVR, one less than the 6/7 in the 400 mg arm. Nevertheless, linear regression supported a statistically significant benefit ($p = 0.05$) and the treatment was well tolerated. It will be of great interest to determine whether virus carrying BIT225 resistance mutations may be present in non-responsive patients or in those where virus load may rebound at later stages of therapy. This will provide clinical clues as to its mode of action and also whether it may be useful against other virus genotypes. BIT225 and future p7 inhibitors may therefore become a useful addition to the HCV inhibitor repertoire, with applications both for the direct treatment of chronically infected patients and in a post-transplant and/or perinatal setting, where curtailing the amount of secreted infectious virus may be critical. Compounds with improved potency compared with prototypes, or even BIT225, may be derived from extended screening programmes or through solving p7 structures, plus the potential for combining separate classes of p7 inhibitor targeting peripheral/protomer/lumenal binding sites may yield significant clinical benefit by raising the genetic barrier to resistant HCV.

9.3 Other RNA Virus Viroporins: Prospective Targets for Emerging and Clinically Important Viruses

Many examples of (candidate) viroporins have been identified in RNA viruses where details pertaining to inhibitory molecules are lacking. Nevertheless, these essential virus proteins represent potential candidates for viroporin-focused therapy, especially in cases such as arboviruses where disease is self-limiting in humans, thereby minimising resistance concerns and where curtailment of symptoms is paramount for patient survival.

9.3.1 Viroporin Activities in Picornaviruses

Interest in developing picornavirus-targeted antivirals is considerable, particularly in the light of the programme intended for the eradication of poliovirus infection worldwide. The use of live vaccines has already been stopped in most countries, yet killed vaccines do not prevent gastrointestinal infection and virus shedding by chronic carriers, including some derived from live vaccines. Antiviral prophylaxis would represent an excellent means of curtailing outbreaks as the withdrawal of vaccines continues. Other clinically important picornaviruses such as enterovirus 71 (EV71), which causes widespread and severe disease epidemics in Southeast Asia,[262] generate an urgent need for vaccines and antiviral interventions. Lastly, the direct treatment of the common cold, primarily caused by rhinoviruses, would have huge socio-economic benefit.

Membrane permeability is markedly increased during the mid to late phase of picornavirus infection, culminating in cell lysis and release of infectious non-enveloped virions from the cell.[13] In particular, proteins encoded by enteroviruses (*e.g.* poliovirus, Coxsackie B virus, human rhinovirus) were shown to alter both bacterial and mammalian cell membrane permeability and also cellular membrane trafficking within infected cells.[12,13,66,69,70,263–268] Initially, the non-structural 2B, 2C and their precursor 2BC proteins were demonstrated to induce membrane permeability, although 2B is now thought of as the mature viroporin effector molecule.[12,13,264] Enterovirus 2B is a class 2 viroporin with two helical TMDs separated by a stretch of highly polar residues. Enterovirus 2B fused to maltose binding protein forms tetramers with a radius of ~ 6 Å capable of permeabilising vesicles[264] and multimerisation has also been observed in mammalian cells.[269,270] In agreement, molecular modelling of channel complexes predicts tetrameric pores of 5–7 Å radius with a lumen lined by a stretch of three lysines followed by a serine.[271] Expression of enterovirus 2B within cells leads to elevated cytosolic Ca^{2+}, which alters vesicle trafficking, induces apoptosis and leads to eventual cell lysis, reminiscent of a membrane-active toxin.[13] Localisation to the Golgi is necessary for this to occur as 2B from distantly related picornaviruses, such as hepatitis A virus, are ER localised and do not alter Ca^{2+} levels. Recently, enterovirus 2B was shown to activate the NLRP3 inflammasome,[272] revealing an as-yet uncharacterised role for viroporins influencing immune activation. Interestingly, the 2B protein of EV71 was recently proposed to act as a chloride channel rather than a Ca^{2+} channel.[273] Accordingly, a chloride channel inhibitor, DIDS (4,4′-diiso-thiocyano-2,2′-stilbenedisulfonic acid), both inhibited chloride conductance of EV71 2B expressed in *Xenopus* oocyte membranes and retarded the spread of EV71 in cell culture. Direct channel conductance of other 2B proteins has not been characterised, making it possible that they all act to raise Ca^{2+} levels *via* an indirect effect on chloride ion homeostasis and providing an opportunity to apply chloride channel inhibitors as therapeutics. No structural information exists for 2B and no other examples of 2B-specific compounds have been described at the time of writing.

Another picornavirus protein capable of inducing membrane permeability is the VP4 capsid component.[15] This small protein is retained on the inside of the virion particle until uncoating during virus entry, where it becomes externalised and inserted into the endosomal membrane. Here, in cooperation with VP1, it is thought to enable the passage of viral RNA into the cytosol.[274] Such pore-like activity is at the very extreme of the pore–channel dualism spectrum for viroporin function; however, VP4 forms defined channels that do not disrupt membranes and induce discrete channel events in artificial bilayers.[14] Furthermore, VP4 channels can be reconstituted with recombinant protein *in vitro* and their activity is amenable to assessment *via* liposome dye release assays.[15] Hence expanded screens may yield small molecules capable of interfering in this process.

9.3.2 Coronavirus (CoV) E, 3a and ORF8a Proteins

Coronavirus (CoV) infection was brought to the fore in the early part of the century following the outbreak of severe acute respiratory syndrome (SARS), which was subsequently directly linked to the zoonotic transfer of the SARS CoV.[275,276] Although this infection has not re-emerged to date, much research has focused on its molecular biology and the coronavirus field has been reinvigorated as a result. However, CoV also represent significant agricultural pathogens in addition to causing less severe human disease. Three proteins encoded by CoV, primarily from SARS CoV but often with homologues in other viruses, have been proposed to act as viroporins, namely E, 3a and 8a.

Peptides corresponding to the SARS CoV E protein were first shown to display cation channel activity in planar bilayers,[32] which was subsequently shown to be sensitive to HMA.[29] In turn, HMA blocked the replication of the animal coronavirus mouse hepatitis virus (MHV), but not where the entire E ORF had been deleted; such viruses were also attenuated in culture. E has been proposed to form pentameric bundles and comprise a type 1 viroporin, although its topology is a matter of some debate.[277] E channels have also been shown to be inhibited by very high concentrations of amantadine, although the relevance of this interaction in the millimolar range is not clear.[278] Solution NMR studies have yielded models of the pentameric helical bundle derived from peptides corresponding to the TM domain.[31] These were similarly inhibited by HMA in whole 293T cell patch-clamp and the drug was shown to interact with both the N-terminal and C-terminal neck of the TM domain by NMR. E is a structural component of the virion and is highly involved in both their assembly and secretion, although how channel function relates to these activities is unclear as the protein undergoes several protein–protein interactions and may adopt various topologies.[277] Interestingly, a recent report found that the lipid environment directly influenced the ionic conductance of E channels with no cation preference evident in non-charged lipid membranes, whereas negatively charged lipids induced mild cation selectivity. In addition, Asn15Ala and Val25Phe mutations located in the TM domain abrogated channel activity.[28]

In addition to E, the 3a CoV protein was shown to display channel activity.[27] The 3a formed potassium-selective channels in oocytes and recombinant protein tetramerised in membranes, stabilised through disulfide linkages. Emodin, a constituent of various plant extracts including Polygonaceae (Japanese knotweed), inhibited 3a channels with an EC_{50} of $\sim 20\,\mu M$ and this molecule also prevented the spread of a model human coronavirus at similar concentrations.[34] However, the action of this drug against multiple cellular kinases (*e.g.* p56[lck])[279–281] makes it difficult to assign specificity to 3a in such experiments. Finally, a recent report showed that ORF8a single TM peptides form pentameric cation-selective channels *in vitro*, with modelling predicting a lumen lined by cysteines, serines and threonines.[282]

It remains unclear whether one, some or all of the proteins reported to act as viroporins in CoV perform this function during human infection; however, the large genomes of CoV certainly provide sufficient coding capacity to retain multiple channel proteins. It would be of great interest to determine precisely which aspects of the CoV life cycle are sensitive to HMA and/or emodin and whether resistant viruses can be selected in culture for either compound.

9.3.3   The Small Hydrophobic (SH) Proteins of Paramyxoviridae

Three genera of the Paramyxoviridae encode small hydrophobic (SH) proteins, namely the pneumoviruses, metapneumoviruses and rubulaviruses. Of these, respiratory syncitial virus (RSV) from the pneumoviruses and mumps virus (MuV) from the rubulaviruses represent significant human pathogens. Whereas vaccination programmes for mumps are widely available, uptake of the triple measles, mumps and rubella (MMR) vaccine has declined owing to misrepresentation in the media concerning autism risk, leading to increased seroprevalence and multiple regional outbreaks among children. RSV remains a major issue for both juvenile health as a leading cause of severe respiratory tract infections with strong links to asthma, and also in the elderly where it is often misdiagnosed as influenza. No vaccine exists for RSV, but passive immunisation with palivizumab (Synagis) is used for high-risk infants.

Deletion of MuV SH has been shown to induce an attenuated phenotype with respect to neurovirulence; however, recent studies suggest that this is likely due to disrupted virus transcription rather than SH deficiency *per se*.[283] SH appears to be dispensable for the growth of MuV[284] or RSV[285] in many cell culture systems, yet has also been shown to antagonise TNFα-mediated apoptosis.[286,287] SH-deleted RSV replicated to 10-fold reduced titres in small animal models[288] and 40-fold lower titres than wild-type in chimpanzees with considerably reduced rhinorrhoea.[289] SH therefore appears to act as a virulence factor, playing a host-specific role in virus growth and pathology. Within infected cells, SH is present both as an unmodified 7.5 kDa species and in carbohydrate-modified forms. All forms of SH oligomerise in cells and both *in vitro* and *in silico* studies support pentameric stoichiometry.[19,20,290] although recent EM studies of recombinant SH in DHPC detergent showed the formation of both pentamers and hexamers, with pore diameters of 1.9 and 2.6 nm, respectively.[18]

SH is predicted to contain a single TM domain and so comprise a class 1 viroporin of 64 amino acids. Although no information regarding viroporin activity has been reported for MuV SH, RSV SH was shown to induce bacterial membrane permeability to small molecules such as hygromycin,[291] and peptides corresponding to both the TM domain and full-length protein form cation-selective channels in bilayers.[19] Interestingly, activity for TM peptides is reduced by low pH,[19] whereas full-length channels are activated when the pK_a is lowered below that of histidine.[20] NMR-derived models of pentameric SH protein reveal the presence of His22 within the predicted channel lumen and also His51 located towards the C-terminus of the protein; the latter is absent from TM peptides.[20] Of these, His51 is the more highly conserved of the two residues, although mutation of either in isolation to alanine retains pH-activated channel activity. Only double His mutants are defective for channel activity, suggesting that the two residues may cooperate in a redundant fashion to effect channel opening. Although inactivation of TM peptides in response to lowered pH would seemingly argue against this hypothesis, it seems that the functionality of these residues as sensors/gates is likely to be context dependent. Computer models of SH hexamers in a DOPA–DOPC bilayer predict a potential Trp15 gate residue to be present in close proximity to His22, although this would lie outside the predicted TM region (His22 to Cys43) defined by hydrogen–deuterium exchange in DMPC bilayers.[18] To date, small-molecule inhibitors have not been defined for SH channels, although the poor efficacy of available vaccines makes antivirals an attractive strategy to curtail acute respiratory illness, particularly in children.

9.3.4 Alphavirus 6K Proteins

The alphavirus genus of the Togaviridae are insect-borne arboviruses, usually transmitted by mosquitoes. They include significant human pathogens such as Chikungunya virus, Ross River virus and Venezuelan equine encephalitis virus, in addition to the genetic workhorses Sindbis and Semliki Forest viruses. Alphavirus structural proteins are expressed as a single polyprotein which is post-translationally cleaved by signal peptidase to generate the capsid, E1, E2 and 6K proteins; 6K is also acylated,[292,293] involved in both the trafficking and processing of the two major glycoproteins, and plays a major role in virion production as a minor virion component as well as trafficking functions; mutation and/or deletion of 6K results in the formation of aberrant virions with reduced thermal stability.[292,294–298]

Expression of 6K was first shown to induce bacterial hygromycin permeability, indicative of channel formation.[67] Topological predictions of its 61 amino acid sequence support its inclusion within class 2 viroporins, although *in vitro* studies supported a single TM topology.[16,17] These studies also demonstrated cation channel activity with preference for Na^+ and K^+ over Ca^{2+} and a 15-fold preference for Na^+ over Cl^-. The precise role of 6K channel activity has again been difficult to determine as mutations generally affect polyprotein processing, targeting, accumulation within virions or their budding.[294–296] Moreover, no small-molecule inhibitors have been described. Nevertheless, as limiting the disease course in Chikungunya and other alphavirus infections of humans can ameliorate long-term neurological damage occurring after encephalitis has passed, antiviral prophylaxis targeting 6K would be highly desirable.

9.3.5 Flavivirus M proteins

Flaviviruses comprise many significant human pathogens, including yellow fever virus, dengue virus, West Nile virus, Japanese encephalitis virus and tick-borne encephalitis virus. These arboviruses are primarily spread by mosquitoes or ticks and cause severe human disease, including encephalitis and haemorrhagic fevers. Much like alphaviruses, no antiviral therapies exist and patients are treated by symptom management. Ribavirin is sometimes administered in severe cases, although efficacy is limited.

Flavivirus M proteins are cleaved from the other envelope (E) protein by signal peptidase as a prM precursor,[299–301] then processed by resident proteases in the Golgi to generate mature M, a 75 amino acid protein which resides within the virion.[299,302] This cleavage event is essential to the production of mature flavivirus particles. Peptides corresponding to a proposed TM region in the C-terminus of dengue virus M protein displayed cation-selective channel activity in bilayers, which was sensitive to both HMA (100 μM) and amantadine (10 μM).[36] Similar peptides introduced artificially into cells also induce mitochondrial membrane permeability,[303] although prM and/or M do not localise to these organelles during natural infection.[304] However, a recent report found no evidence of acid-stimulated channel activity for the complete M protein expressed in oocyte membranes.[35] Here, M was proposed to adopt a dual-membrane spanning topology when expressed in cells, based on the detection of N- and C-terminally located Myc tags and this topology is also observed in the structure of the dengue virus particle.[305,306] However, this topology requires a very tight turn of only three residues to occur between the TM helices, which would not be expected to be energetically favourable. Some TM domain prediction programmes also favour the formation of a single pass protein, with a TM domain approximating to the C-terminal peptide with channel activity *in vitro*. It is therefore possible that M might adopt dual topologies or that its expression in oocytes does not favour the formation of channels for other reasons such as membrane composition. Interestingly, mutation of His39 in the M protein sequence, which lies in the first TM of the protein, was shown to disrupt the spread of dengue virus in culture without affecting polyprotein processing or the formation of prM-E heterodimers.[302] However, mutations did prevent the secretion of glycoproteins, which could be related to channel-forming activity. It will be of interest to determine whether full-length M displays channel activity within bilayers or alters membrane permeability in other systems before categorically assigning or withdrawing viroporin function from flavivirus M proteins. As for alphaviruses, the opportunity for channel targeting compounds as antivirals for these viruses would be a tantalising prospect.

9.3.6 Human T-cell Lymphotrophic Virus Type 1 (HTLV-1) p13ii Protein

HTLV-1 encodes an 87 amino acid integral membrane protein, p13ii, which localises to the mitochondrial inner membrane.[307] p13ii contains a stretch of basic residues reminiscent of picornaviral 2B proteins and these are present within a predicted amphipathic α-helix, which is also directly responsible for mitochondrial localisation. The presence of the protein within mitochondria leads to a valinomycin-like increase in K^+ permeability, resulting in increased respiratory chain activity and the production of increased reactive oxygen species (ROS).[308] This, in turn, sensitises cells to various apoptotic stimuli. It is not currently clear whether p13ii retains inherent viroporin activity or in fact influences the function of mitochondrion-resident K^+ channels.[309]

9.3.7 The Rotavirus NSP4 Enterotoxin

Rotaviruses are non-enveloped segmented dsRNA viruses from the Reoviridae family. They are the leading cause of life-threatening viral gastroenteritis among children worldwide, causing ∼500 000 deaths per year, and currently available vaccines show poor efficacy in developing countries where the need is greatest. Rotaviruses spread faeco-orally, infecting enterocytes in the small intestine and causing cell death, villus blunting and profuse diarrhoea leading to life-threatening dehydration. A significant hallmark of rotavirus replication is the elevation of cytosolic calcium levels, resulting from increased ER and plasma membrane permeability, which is essential for virus replication and underpins many facets of intestinal disruption. Expression of a single viral non-structural protein, NSP4, is sufficient to recapitulate these effects on calcium homeostasis.[22,310–313]

NSP4 is a 175 amino acid protein that is subdivided into a helical domain at the N-terminus, a coiled-coil region (aa 95–146) for which pentameric crystal structures have been solved,[314–318] and a C-terminal double-layered particle receptor domain, which is essential for its role during the assembly and egress of rotavirus capsids. NSP4 has been found to elevate calcium levels *via* a number of mechanisms, including perturbation of cell signalling,[311] and a secreted C-terminal cleavage product (aa 112–175) induces calcium elevation *via* an endotoxin-like mechanism.[319] However, recent work has attributed direct viroporin activity to the N-terminal region of the protein.[21,23] This region contains multiple helical domains. A short N-terminal helix precedes a TM region that is inserted through an uncleaved signal sequence (H2). This is followed by an extended region (aa 47–92) containing two predicted helical domains, one containing five conserved lysines and the other forming an amphipathic *a*-helix. This region in four sequence-divergent rotavirus subtypes has been shown to mediate membrane permeability in both mammalian and bacterial cells, which is dependent on both the pentalysine motif and the amphipathic character of the second helix, defined by a disruptive mutation.[23] The pentalysine domain also drives insertion of this minimal viroporin domain into membranes, leading to a three TM domain topology being proposed recently for the complete protein.[21] The pentalysine and amphipathic domains are also necessary for cytosolic calcium elevation by full-length protein in mammalian cells and also for induction of membrane permeability, so directly linking direct viroporin activity to this function. Thus, while NSP4 retains multiple activities involving several regions of the protein, it should be feasible to develop small-molecule inhibitors of NSP4 viroporin function that would significantly impair its major role of inducing elevated cytosolic calcium, thereby acting both to inhibit virus replication and to ameliorate diarrhoea symptoms. This could readily be achieved *via* incorporation of minimal viroporin domain peptides into liposome-based screens, potentially incorporating known inhibitors of calcium channels that are already utilised clinically.

9.4   Viroporins Encoded by DNA Viruses

9.4.1   Viroporins of Polyomaviruses

Polyomaviruses are small, non-enveloped dsDNA viruses linked to a number of human diseases such as Merkel cell carcinoma (Merkel cell polyomavirus) and progressive multifocal leuko-encephalopathy (JC virus), but also contain simian vacuolating virus 40 (SV40), which was used as the first viral tool to demonstrate nuclear localisation signals. Recently, two polyomavirus proteins were shown to demonstrate viroporin activity. The first, VP4, is a late-acting protein containing a central hydrophobic core and a nuclear localisation domain. These target the protein to the nuclear envelope at late times of infection in order to mediate membrane disruption and so enable release of assembled virions from the nucleus. VP4 forms defined integral membrane pores of ~ 3 nm diameter and its activity is abrogated by mutation of the hydrophobic domain.[320]

The second polyomavirus protein to display viroporin activity is the agno-protein of JC virus.[44] This small accessory protein of 71 amino acids also possesses a central hydrophobic TM domain which, along with an N-terminal region, is required for ER/plasma membrane localisation and membrane integration. Agnoprotein expression increases plasma membrane permeability and elevates cytosolic calcium, resulting in enhanced virion release. As such, agnoprotein appears functionally reminiscent of picornaviral 2B and rotavirus NSP4. No structural information exists for either VP4 or agnoprotein, although the lack of treatments for polyomavirus associated disease may drive further investigation into these potential drug targets.

9.4.2 The E5 Protein of Human Papillomavirus 16 (HPV16) is an Oncogenic Viroporin

Human papillomavirus (HPV) is the most prevalent sexually transmitted infection worldwide and high-risk viruses, such as HPV16, are the major cause of cervical cancer, which kills over 275 000 women per year. HPV also causes a growing number of anal, penile and head and neck carcinomas in both sexes. Recently introduced vaccines will take decades to affect cancer incidences, are not available to men and cannot aid many millions of currently infected women. HPV-associated carcinomas are virus-driven tumours and persist as a result of viral oncogene expression. In addition to the well-characterised E6 and E7 proteins, HPV encodes a third oncogenic protein, E5.[321,322] The role of E5 during the virus life cycle is not clear, although its oncogenic properties appear strongly linked to its ability to up-regulate growth factor signalling *via* the perturbation of endosomal pH.[323–327]

HPV16 E5 is a highly hydrophobic 9 kDa protein with three predicted TM domains. This was recently demonstrated to form hexameric complexes in membrane-mimetic detergents by native PAGE and EM studies of E5 fusion proteins.[45] E5 complexes were active in liposome dye release assays with sensitivity to relatively high concentrations of rimantadine. As for HCV p7, *de novo* modelling of E5 complexes was successfully employed in the development of a novel, bespoke E5 inhibitor, MV6, with a markedly improved *in vitro* IC$_{50}$ compared with rimantadine. Importantly, both MV6 and rimantadine were able to repress E5-mediated enhancement of EGFR signalling in culture, providing a direct link between E5 viroporin activity and the processes involved during oncogenic change. Development of potent E5 inhibitors could therefore provide a means of targeting both HPV infection and also potentially the proliferation tumours.

9.5 Viroporins of Animal Viruses

Antiviral prophylaxis is not generally employed in agriculture owing to prohibitive costs and the effectiveness of culling programmes. However, the ambiguity and uncertainty surrounding the question of vaccination against many agricultural pathogens continue in many areas, with foot and mouth disease virus (FMDV) being the prime exemplar. Unfortunately, FMDV 2B proteins are dissimilar to those of enteroviruses and are not likely to possess viroporin activity, yet a number of animal viruses with significant agricultural impact are likely to encode viroporins by analogy with their mammalian counterparts. Antiviral therapy, if rendered cost-effective, could provide an alternative to ring vaccination strategies in the event of disease outbreaks.

Concerns surrounding the zoonotic transfer of resistant viruses may be a relevant concern if antiviral strategies were adopted, yet effectively suppressing virus replication in some animal reservoirs may represent a means to limit the jumping of viruses from animals into humans. Of note, reports of the use of amantadine in commercial poultry farms in Southeast Asia were linked by the WHO to acquisition of drug resistance by highly pathogenic H5N1 strains in the early part of the century (http://www.who.int/foodsafety/micro/avian_antiviral/en/index.html). Clearly, sub-optimal dosing of a weak-potency monotherapy is highly likely to drive the selection of resistance in such variable viruses. Nevertheless, more effective therapies may be productive in this regard.

Pestiviruses are an example of an agricultural pathogen where viroporin function has been documented. These express a p7 protein[328,329] with very limited sequence homology (<10%) to that of HCV, yet BVDV p7 has been shown to form oligomers and to promote surface transport of influenza HA, implying proton conductance.[71] BVDV was also the surrogate system used to demonstrate cell culture efficacy of BIT225,[87] providing proof of principle that viroporin inhibitors are able to reduce the spread of infection. A recent investigation of the related classical swine fever virus (CSFV) p7 protein demonstrated its importance for virus replication in swine,[41] and also confirming channel-forming activity *in vitro*. Interestingly, unlike HCV p7, peptides corresponding to the CSFV p7 C-terminus also displayed channel-forming activity, which, combined with the presence of a highly conserved His47 residue, supported the view that the C-terminal helix forms the channel lumen in these proteins. The activity of full-length peptides was sensitive to high concentrations of amantadine ($EC_{50} \approx 2\,mM$), and also to lower concentrations ($EC_{50} \approx 200\,\mu M$) of the phenylalkylamine calcium channel inhibitor, verapamil. In addition, both drugs significantly reduced ($\sim 2\log_{10}$ reduction for $500\,\mu M$ amantadine and $17\,\mu M$ verapamil) in virus titre when applied during the initial infection process. This may suggest a role for CSFV p7 during entry, although others have failed to demonstrate the presence of BVDV p7 in particles.[328] Other economically important animal viruses, such as the avian CoV infectious bronchitis virus, are also likely to encode viroporins, yet these are not yet characterised. However, although showing potential as drug targets, understanding of animal viroporins lags behind those of mammalian viruses with the exception of M2, making it unlikely that these will be exploited in the near future.

9.6 Conclusion and Future Perspectives: How Can Viroporin Inhibitors Fit into Modern Clinical Drug Discovery Scenarios?

The spectre of amantadine and rimantadine in the treatment of influenza represents a double-edged sword when proposing viroporins as targets for modern-day drug development. Although providing a precedent that such a strategy can be deployed in a clinical setting, their relatively poor potency combined with resistance issues (likely linked to application as monotherapies) has led to a perception that viroporin-targeted programmes will not generate effective antivirals. Moreover, this has been compounded by inappropriate application of both adamantanes and imino sugars during HCV trials. However, this flies in the face of successful drugs developed targeting cellular ion channels, which have been critical for cardiac medicine, *etc*. Few other scenarios exist where application of an extremely limited chemical toolbox to poorly characterised proteins encoded by diverse and highly variable viruses has led to such dismissive attitudes. Clearly, characterisation of membrane channels is technically challenging and investigators must start somewhere, yet preliminary, often disappointing, results of studies using prototype viroporin inhibitors should be interpreted with appropriate caveats when assessing their viability as targets.

The commonly employed tools of inhibitor discovery are often not forthcoming in the case of viroporins, namely robust protein production and screening assays, atomic structures and defined SARs based on medicinal chemistry and drug resistance mutations. In addition, the primitive nature of some viroporins often excludes approaches used to characterise cellular ion channels, where ion specificity and gating are usually far more selective. However, recent progress in some fields is beginning to bring these proteins into the drug discovery spotlight. Most notable, of course, is M2, where decades of research have culminated in atomic structures of the protein within membranes that represent ideal templates for drug design. M2 also represents a unique scenario where resistance is already commonplace amongst clinically relevant viruses, necessitating the search for novel compounds to overcome such mutations which will likely be chemically distinct from adamantane prototypes. The ongoing controversy surrounding adamantane binding to these channels merely serves to illustrate the need for improved inhibitors. Structural studies are also ongoing for many other viroporins, including HCV p7, HIV-1 Vpu,

RSV SH and SARS CoV E, potentially expediting rational inhibitor design in the near future. In lieu of high-resolution structures, molecular modelling and dynamics have also proven to be useful tools providing insight into viroporin–compound interactions and, in some cases, permitting the design of bespoke inhibitory molecules, albeit with limited potency.

In parallel with rational approaches, incorporation of viroporins into medium- and high-throughput compound screens will be highly desirable if drug discovery programmes are to be undertaken. Clearly, this requires robust protein production and a simple assay format appropriate for multiplexing. To date, a medium-scale screen undertaken by Boehringer Ingelheim targeting HCV p7 represents the only such attempt to select viroporin inhibitors from a compound library, yet this yielded encouraging results that point to such systems being amenable to larger scales. Importantly, the liposome dye release assay adapted for this screen displayed specificity and a low level of sensitivity to artefacts. Given that such assays have been applied at a standard laboratory scale to a number of viroporins, potential for conducting screens for multiple targets therefore exists, despite the assay read-out not corresponding directly to the preferred ionic species of each channel. The other example of viroporin screening has been provided by the pharmaceutical company Biotron, where a bacterial assay led to the selection of BIT225, for which the results of both HIV-1 and HCV clinical trials are forthcoming.

To date, only M2 and HCV p7 have been described to obtain the specific resistance mutations usually required to validate the specific antiviral effects for any given compound, and for the latter case this has not yet been selected in chronic virus culture. Again, this reflects both the stage of research for many viroporins and the lack of potency/specificity for the three main classes of prototype inhibitors. Combined with a lack of structural information in many cases, this has generally precluded the development of robust SARs through medicinal chemistry. However, as time advances, the tools required to undertake such investigations become closer to being within reach. Of course, the identification of potent novel hit molecules and their optimisation represent the very early stages involved in drug production, yet robust ADME and toxicological screens are widely available for compounds targeting cellular ion channels, and should be readily amenable for the testing and development of viroporin inhibitors. Should these become a reality, it will be important to administer viroporin inhibitors appropriately such that issues concerning resistance are less likely to arise. This may be less of a concern for DNA viruses or perhaps for emerging viruses where humans are a dead-end host and limiting disease severity is paramount. However, for highly variable RNA viruses such as HIV-1 and HCV, viroporin inhibitors should ideally be deployed in combination with other DAAs owing to their lack of effect on virus genome replication in most cases.

Taken together, it is the author's viewpoint – albeit biased –that viroporins in general represent excellent prospective antiviral drug targets, not least owing to their essential roles in the life cycles of many viruses. It is now down to researchers to provide the setting in which the considerable firepower of modern-day drug discovery can be brought to bear upon these targets, subjecting them to full and thorough scrutiny and discovering new, exciting virology in the process. Recent progress should buoy enthusiasm for many viroporin targets and the results of BIT225 trials will ideally rejuvenate this class of antivirals, which has not realistically changed since the discovery of amantadine in the 1960s.

References

1. W. L. Davies, R. R. Grunert, R. F. Haff, J. W. McGahen, E. M. Neumayer, M. Paulshock, J. C. Watts, T. R. Wood, E. C. Hermann and C. E. Hoffmann, *Science*, 1964, **144**, 862–863.
2. L. Carrasco, *Nature*, 1978, **272**, 694–699.
3. L. H. Pinto, L. J. Holsinger and R. A. Lamb, *Cell*, 1992, **69**, 517–528.
4. K. C. Duff and R. H. Ashley, *Virology*, 1992, **190**, 485–489.
5. A. J. Hay, A. J. Wolstenholme, J. J. Skehel and M. H. Smith, *EMBO J.*, 1985, **4**, 3021–3024.
6. S. A. Wharton, R. B. Belshe, J. J. Skehel and A. J. Hay, *J. Gen. Virol.*, 1994, **75**(Pt 4), 945–948.
7. D. A. Steinhauer, S. A. Wharton, J. J. Skehel, D. C. Wiley and A. J. Hay, *Proc. Natl. Acad. Sci. U. S. A.*, 1991, **88**, 11525–11529.
8. F. Ciampor, P. M. Bayley, M. V. Nermut, E. M. Hirst, R. J. Sugrue and A. J. Hay, *Virology*, 1992, **188**, 14–24.
9. K. Takeuchi and R. A. Lamb, *J. Virol.*, 1994, **68**, 911–919.
10. K. Takeuchi, M. A. Shaughnessy and R. A. Lamb, *Virology*, 1994, **202**, 1007–1011.
11. C. Hohenadl, K. Klingel, P. Rieger, P. H. Hofschneider and R. Kandolf, *J. Virol. Methods*, 1994, **47**, 279–295.
12. R. Aldabe, A. Barco and L. Carrasco, *J. Biol. Chem.*, 1996, **271**, 23134–23137.
13. F. J. van Kuppeveld, J. G. Hoenderop, R. L. Smeets, P. H. Willems, H. B. Dijkman, J. M. Galama and W. J. Melchers, *EMBO J.*, 1997, **16**, 3519–3532.
14. P. Danthi, M. Tosteson, Q. H. Li and M. Chow, *J. Virol.*, 2003, **77**, 5266–5274.
15. M. P. Davis, G. Bottley, L. P. Beales, R. A. Killington, D. J. Rowlands and T. J. Tuthill, *J. Virol.*, 2008, **82**, 4169–4174.
16. A. F. Antoine, C. Montpellier, K. Cailliau, E. Browaeys-Poly, J. P. Vilain and J. Dubuisson, *J. Membr. Biol.*, 2007, **215**, 37–48.
17. J. V. Melton, G. D. Ewart, R. C. Weir, P. G. Board, E. Lee and P. W. Gage, *J. Biol. Chem.*, 2002, **277**, 46923–46931.
18. S. D. Carter, K. C. Dent, E. Atkins, T. L. Foster, M. Verow, P. Gorny, M. Harris, J. A. Hiscox, N. A. Ranson, S. Griffin and J. N. Barr, *FEBS Lett.*, 2010, **584**, 2786–2790.
19. S. W. Gan, L. Ng, X. Lin, X. Gong and J. Torres, *Protein Sci.*, 2008, **17**, 813–820.

20. S. W. Gan, E. Tan, X. Lin, D. Yu, J. Wang, G. M. Tan, A. Vararattanavech, C. Y. Yeo, C. H. Soon, T. W. Soong, K. Pervushin and J. Torres, *J. Biol. Chem.*, 2012, **287**, 24671–24689.

21. J. M. Hyser, M. R. Collinson-Pautz, B. Utama and M. K. Estes, *MBio*, 2010, **1**(5), DOI: 10.1128/mBio.00265-10.

22. P. Tian, J. M. Ball, C. Q. Zeng and M. K. Estes, *J. Virol.*, 1996, **70**, 6973–6981.

23. J. M. Hyser, B. Utama, S. E. Crawford and M. K. Estes, *J. Virol.*, 2012, **86**, 4921–4934.

24. M. E. Gonzalez and L. Carrasco, *Biochemistry*, 1998, **37**, 13710–13719.

25. U. Schubert, A. V. Ferrer-Montiel, M. Oblatt-Montal, P. Henklein, K. Strebel and M. Montal, *FEBS Lett.*, 1996, **398**, 12–18.

26. G. D. Ewart, T. Sutherland, P. W. Gage and G. B. Cox, *J. Virol.*, 1996, **70**, 7108–7115.

27. W. Lu, B. J. Zheng, K. Xu, W. Schwarz, L. Du, C. K. Wong, J. Chen, S. Duan, V. Deubel and B. Sun, *Proc. Natl. Acad. Sci. U. S. A.*, 2006, **103**, 12540–12545.

28. C. Verdia-Baguena, J. L. Nieto-Torres, A. Alcaraz, M. L. DeDiego, J. Torres, V. M. Aguilella and L. Enjuanes, *Virology*, 2012, **432**, 485–494.

29. L. Wilson, P. Gage and G. Ewart, *Virology*, 2006, **353**, 294–306.

30. L. Wilson, P. Gage and G. Ewart, *Adv. Exp. Med. Biol.*, 2006, **581**, 573–578.

31. J. Torres, K. Parthasarathy, X. Lin, R. Saravanan, A. Kukol and D. X. Liu, *Biophys. J.*, 2006, **91**, 938–947.

32. L. Wilson, C. McKinlay, P. Gage and G. Ewart, *Virology*, 2004, **330**, 322–331.

33. C. M. Chan, H. Tsoi, W. M. Chan, S. Zhai, C. O. Wong, X. Yao, W. Y. Chan, S. K. Tsui and H. Y. Chan, *Int. J. Biochem. Cell Biol.*, 2009, **41**, 2232–2239.

34. S. Schwarz, K. Wang, W. Yu, B. Sun and W. Schwarz, *Antiviral Res.*, 2011, **90**, 64–69.

35. S. S. Wong, M. Chebib, G. Haqshenas, B. Loveland and E. J. Gowans, *Virology*, 2011, **412**, 83–90.

36. A. Premkumar, C. R. Horan and P. W. Gage, *J. Membr. Biol.*, 2005, **204**, 33–38.

37. S. D. Griffin, L. P. Beales, D. S. Clarke, O. Worsfold, S. D. Evans, J. Jaeger, M. P. Harris and D. J. Rowlands, *FEBS Lett.*, 2003, **535**, 34–38.

38. D. Pavlovic, D. C. Neville, O. Argaud, B. Blumberg, R. A. Dwek, W. B. Fischer and N. Zitzmann, *Proc. Natl. Acad. Sci. U. S. A.*, 2003, **100**, 6104–6108.

39. A. Premkumar, L. Wilson, G. D. Ewart and P. W. Gage, *FEBS Lett.*, 2004, **557**, 99–103.

40. A. L. Wozniak, S. Griffin, D. Rowlands, M. Harris, M. Yi, S. M. Lemon and S. A. Weinman, *PLoS Pathogens*, 2010, **6**, e1001087.

41. D. P. Gladue, L. G. Holinka, E. Largo, I. Fernandez Sainz, C. Carrillo, V. O'Donnell, R. Baker-Branstetter, Z. Lu, X. Ambroggio, G. R. Risatti, J. L. Nieva and M. V. Borca, *J. Virol.*, 2012, **86**, 6778–6791.
42. T. Betakova and A. J. Hay, *J. Gen. Virol.*, 2007, **88**, 2291–2296.
43. J. A. Mould, R. G. Paterson, M. Takeda, Y. Ohigashi, P. Venkataraman, R. A. Lamb and L. H. Pinto, *Dev. Cell*, 2003, **5**, 175–184.
44. T. Suzuki, Y. Orba, Y. Okada, Y. Sunden, T. Kimura, S. Tanaka, K. Nagashima, W. W. Hall and H. Sawa, *PLoS Pathogens*, 2010, **6**, e1000801.
45. L. F. Wetherill, K. K. Holmes, M. Verow, M. Muller, G. Howell, M. Harris, C. Fishwick, N. Stonehouse, R. Foster, G. E. Blair, S. Griffin and A. Macdonald, *J. Virol.*, 2012, **86**, 5341–5351.
46. T. Sakaguchi, Q. Tu, L. H. Pinto and R. A. Lamb, *Proc. Natl. Acad. Sci. U. S. A.*, 1997, **94**, 5000–5005.
47. D. Salom, B. R. Hill, J. D. Lear and W. F. DeGrado, *Biochemistry*, 2000, **39**, 14160–14170.
48. D. E. Chandler, F. Penin, K. Schulten and C. Chipot, *PLoS Comput. Biol.*, 2012, **8**, e1002702.
49. D. Clarke, S. Griffin, L. Beales, C. S. Gelais, S. Burgess, M. Harris and D. Rowlands, *J. Biol. Chem.*, 2006, **281**, 37057–37068.
50. P. Luik, C. Chew, J. Aittoniemi, J. Chang, P. Wentworth Jr., R. A. Dwek, P. C. Biggin, C. Venien-Bryan and N. Zitzmann, *Proc. Natl. Acad. Sci. U. S. A.*, 2009, **106**, 12712–12716.
51. J. L. Nieva, V. Madan and L. Carrasco, *Nat. Rev. Microbiol.*, 2012, **10**, 563–574.
52. B. J. Isherwood and A. H. Patel, *J. Gen. Virol.*, 2005, **86**, 667–676.
53. R. Acharya, V. Carnevale, G. Fiorin, B. G. Levine, A. L. Polishchuk, V. Balannik, I. Samish, R. A. Lamb, L. H. Pinto, W. F. DeGrado and M. L. Klein, *Proc. Natl. Acad. Sci. U. S. A.*, 2010, **107**, 15075–15080.
54. T. Leiding, J. Wang, J. Martinsson, W. F. DeGrado and S. P. Arskold, *Proc. Natl. Acad. Sci. U. S. A.*, 2010, **107**, 15409–15414.
55. T. Mehnert, A. Routh, P. J. Judge, Y. H. Lam, D. Fischer, A. Watts and W. B. Fischer, *Proteins*, 2008, **70**, 1488–1497.
56. J. S. Rossman, X. Jing, G. P. Leser and R. A. Lamb, *Cell*, 2010, **142**, 902–913.
57. T. E. Andreoli, J. A. Bangham and D. C. Tosteson, *J. Gen. Physiol.*, 1967, **50**, 1729–1749.
58. T. E. Andreoli, M. Tieffenberg and D. C. Tosteson, *J. Gen. Physiol.*, 1967, **50**, 2527–2545.
59. S. C. Piller, G. D. Ewart, D. A. Jans, P. W. Gage and G. B. Cox, *J. Virol.*, 1999, **73**, 4230–4238.
60. S. C. Piller, G. D. Ewart, A. Premkumar, G. B. Cox and P. W. Gage, *Proc. Natl. Acad. Sci. U. S. A.*, 1996, **93**, 111–115.
61. C. StGelais, T. J. Tuthill, D. S. Clarke, D. J. Rowlands, M. Harris and S. Griffin, *Antiviral Res.*, 2007, **76**, 48–58.

62. C. Gervais, F. Do, A. Cantin, G. Kukolj, P. W. White, A. Gauthier and F. H. Vaillancourt, *J. Biol. Screen.*, 2011, **16**, 363–369.

63. T. L. Foster, M. Verow, A. L. Wozniak, M. J. Bentham, J. Thompson, E. Atkins, S. A. Weinman, C. Fishwick, R. Foster, M. Harris and S. Griffin, *Hepatology*, 2011, **54**, 79–90.

64. S. Griffin, C. Stgelais, A. M. Owsianka, A. H. Patel, D. Rowlands and M. Harris, *Hepatology*, 2008, **48**, 1779–1790.

65. R. Guinea and L. Carrasco, *FEBS Lett.*, 1994, **343**, 242–246.

66. J. Lama and L. Carrasco, *J. Biol. Chem.*, 1992, **267**, 15932–15937.

67. M. A. Sanz, L. Perez and L. Carrasco, *J. Biol. Chem.*, 1994, **269**, 12106–12110.

68. F. Ciampor, C. A. Thompson, S. Grambas and A. J. Hay, *Virus Res.*, 1992, **22**, 247–258.

69. A. S. de Jong, H. J. Visch, F. de Mattia, M. M. van Dommelen, H. G. Swarts, T. Luyten, G. Callewaert, W. J. Melchers, P. H. Willems and F. J. van Kuppeveld, *J. Biol. Chem.*, 2006, **281**, 14144–14150.

70. J. R. Doedens and K. Kirkegaard, *EMBO J.*, 1995, **14**, 894–907.

71. S. D. Griffin, R. Harvey, D. S. Clarke, W. S. Barclay, M. Harris and D. J. Rowlands, *J. Gen. Virol.*, 2004, **85**, 451–461.

72. M. Sharma, M. Yi, H. Dong, H. Qin, E. Peterson, D. D. Busath, H. X. Zhou and T. A. Cross, *Science*, 2010, **330**, 509–512.

73. G. B. Karlsson, T. D. Butters, R. A. Dwek and F. M. Platt, *J. Biol. Chem.*, 1993, **268**, 570–576.

74. K. Pyorala and I. Rantanen, *Ann. Med. Intern. Fenn.*, 1968, **57**, 91–97.

75. G. Stamatiou, A. Kolocouris, N. Kolocouris, G. Fytas, G. B. Foscolos, J. Neyts and E. De Clercq, *Bioorg. Med. Chem. Lett.*, 2001, **11**, 2137–2142.

76. I. Stylianakis, A. Kolocouris, N. Kolocouris, G. Fytas, G. B. Foscolos, E. Padalko, J. Neyts and E. De Clercq, *Bioorg. Med. Chem. Lett.*, 2003, **13**, 1699–1703.

77. D. Tataridis, G. Fytas, A. Kolocouris, C. Fytas, N. Kolocouris, G. B. Foscolos, E. Padalko, J. Neyts and E. De Clercq, *Bioorg. Med. Chem. Lett.*, 2007, **17**, 692–696.

78. A. L. Stouffer, R. Acharya, D. Salom, A. S. Levine, L. Di Costanzo, C. S. Soto, V. Tereshko, V. Nanda, S. Stayrook and W. F. DeGrado, *Nature*, 2008, **451**, 596–599.

79. S. D. Cady, K. Schmidt-Rohr, J. Wang, C. S. Soto, W. F. Degrado and M. Hong, *Nature*, 2010, **463**, 689–692.

80. J. R. Schnell and J. J. Chou, *Nature*, 2008, **451**, 591–595.

81. M. D. Duque, C. Ma, E. Torres, J. Wang, L. Naesens, J. Juarez-Jimenez, P. Camps, F. J. Luque, W. F. DeGrado, R. A. Lamb, L. H. Pinto and S. Vazquez, *J. Med. Chem.*, 2011, **54**, 2646–2657.

82. J. Wang, C. Ma, G. Fiorin, V. Carnevale, T. Wang, F. Hu, R. A. Lamb, L. H. Pinto, M. Hong, M. L. Klein and W. F. DeGrado, *J. Am. Chem. Soc.*, 2011, **133**, 12834–12841.

83. R. B. Belshe, B. Burk, F. Newman, R. L. Cerruti and I. S. Sim, *J. Infect. Dis.*, 1989, **159**, 430–435.
84. P. Deltenre, J. Henrion, V. Canva, S. Dharancy, F. Texier, A. Louvet, S. De Maeght, J. C. Paris and P. Mathurin, *J. Hepatol.*, 2004, **41**, 462–473.
85. A. Mangia, G. Leandro, B. Helbling, E. L. Renner, M. Tabone, L. Sidoli, S. Caronia, G. R. Foster, S. Zeuzem, T. Berg, V. Di Marco, N. Cino and A. Andriulli, *J. Hepatol.*, 2004, **40**, 478–483.
86. G. Khoury, G. Ewart, C. Luscombe, M. Miller and J. Wilkinson, *Antimicrob. Agents Chemother.*, 2010, **54**, 835–845.
87. C. A. Luscombe, Z. Huang, M. G. Murray, M. Miller, J. Wilkinson and G. D. Ewart, *Antiviral Res.*, 2010, **86**, 144–153.
88. M. Imai, T. Watanabe, M. Hatta, S. C. Das, M. Ozawa, K. Shinya, G. Zhong, A. Hanson, H. Katsura, S. Watanabe, C. Li, E. Kawakami, S. Yamada, M. Kiso, Y. Suzuki, E. A. Maher, G. Neumann and Y. Kawaoka, *Nature*, 2012, **486**, 420–428.
89. S. Herfst, E. J. Schrauwen, M. Linster, S. Chutinimitkul, E. de Wit, V. J. Munster, E. M. Sorrell, T. M. Bestebroer, D. F. Burke, D. J. Smith, G. F. Rimmelzwaan, A. D. Osterhaus and R. A. Fouchier, *Science*, 2012, **336**, 1534–1541.
90. W. L. Wingfield, D. Pollack and R. R. Grunert, *N. Engl. J. Med.*, 1969, **281**, 579–584.
91. F. G. Hayden, R. L. Atmar, M. Schilling, C. Johnson, D. Poretz, D. Paar, L. Huson, P. Ward and R. G. Mills, *N. Engl. J. Med.*, 1999, **341**, 1336–1343.
92. K. G. Nicholson, F. Y. Aoki, A. D. Osterhaus, S. Trottier, O. Carewicz, C. H. Mercier, A. Rode, N. Kinnersley and P. Ward, *Lancet*, 2000, **355**, 1845–1850.
93. J. J. Treanor, F. G. Hayden, P. S. Vrooman, R. Barbarash, R. Bettis, D. Riff, S. Singh, N. Kinnersley, P. Ward and R. G. Mills, *JAMA*, 2000, **283**, 1016–1024.
94. R. C. Read, Treating influenza with zanamavir, *Lancet*, 1998, **352**(9144), 1872–1873.
95. F. G. Hayden, A. D. Osterhaus, J. J. Treanor, D. M. Fleming, F. Y. Aoki, K. G. Nicholson, A. M. Bohnen, H. M. Hirst, O. Keene and K. Wightman, *N. Engl. J. Med.*, 1997, **337**, 874–880.
96. A. S. Monto, D. P. Robinson, M. L. Herlocher, J. M. Hinson Jr., M. J. Elliott and A. Crisp, *JAMA*, 1999, **282**, 31–35.
97. T. G. Sheu, A. M. Fry, R. J. Garten, V. M. Deyde, T. Shwe, L. Bullion, P. J. Peebles, Y. Li, A. I. Klimov and L. V. Gubareva, *J. Infect. Dis.*, 2011, **203**, 13–17.
98. N. A. Ilyushina, E. Hoffmann, R. Salomon, R. G. Webster and E. A. Govorkova, *Antiviral Ther.*, 2007, **12**, 363–370.
99. R. A. Lamb and P. W. Choppin, *Virology*, 1981, **112**, 729–737.
100. R. A. Lamb, C. J. Lai and P. W. Choppin, *Proc. Natl. Acad. Sci. U. S. A.*, 1981, **78**, 4170–4174.
101. R. A. Lamb, S. L. Zebedee and C. D. Richardson, *Cell*, 1985, **40**, 627–633.

102. L. J. Holsinger and R. A. Lamb, *Virology*, 1991, **183**, 32–43.
103. S. Grambas, M. S. Bennett and A. J. Hay, *Virology*, 1992, **191**, 541–549.
104. S. Grambas and A. J. Hay, *Virology*, 1992, **190**, 11–18.
105. R. J. Sugrue, G. Bahadur, M. C. Zambon, M. Hall-Smith, A. R. Douglas and A. J. Hay, *EMBO J.*, 1990, **9**, 3469–3476.
106. C. Wang, R. A. Lamb and L. H. Pinto, *Biophys. J.*, 1995, **69**, 1363–1371.
107. J. J. Skehel, R. S. Daniels, A. J. Hay, R. Ruigrok, S. A. Wharton, N. G. Wrigley, W. Weiss and D. C. Willey, *Biochem. Soc. Trans.*, 1986, **14**, 252–253.
108. R. J. Sugrue and A. J. Hay, *Virology*, 1991, **180**, 617–624.
109. B. S. Shim, Y. K. Choi, C. H. Yun, E. G. Lee, Y. S. Jeon, S. M. Park, I. S. Cheon, D. H. Joo, C. H. Cho, M. S. Song, S. U. Seo, Y. H. Byun, H. J. Park, H. Poo, B. L. Seong, J. O. Kim, H. H. Nguyen, K. Stadler, D. W. Kim, K. J. Hong, C. Czerkinsky and M. K. Song, *PLoS One*, 2011, **6**, e27953.
110. S. Neirynck, T. Deroo, X. Saelens, P. Vanlandschoot, W. M. Jou and W. Fiers, *Nat. Med.*, 1999, **5**, 1157–1163.
111. C. Ma, A. L. Polishchuk, Y. Ohigashi, A. L. Stouffer, A. Schon, E. Magavern, X. Jing, J. D. Lear, E. Freire, R. A. Lamb, W. F. DeGrado and L. H. Pinto, *Proc. Natl. Acad. Sci. U. S. A.*, 2009, **106**, 12283–12288.
112. S. L. Zebedee and R. A. Lamb, *Proc. Natl. Acad. Sci. U. S. A.*, 1989, **86**, 1061–1065.
113. A. I. Klimov, N. I. Sokolov, N. G. Orlova and V. P. Ginzburg, *Virus Res.*, 1991, **19**, 105–114.
114. P. C. Roberts, R. A. Lamb and R. W. Compans, *Virology*, 1998, **240**, 127–137.
115. B. J. Chen, G. P. Leser, D. Jackson and R. A. Lamb, *J. Virol.*, 2008, **82**, 10059–10070.
116. I. V. Chizhmakov, F. M. Geraghty, D. C. Ogden, A. Hayhurst, M. Antoniou and A. J. Hay, *J Physiol*, 1996, **494**(Pt 2), 329–336.
117. J. A. Mould, J. E. Drury, S. M. Frings, U. B. Kaupp, A. Pekosz, R. A. Lamb and L. H. Pinto, *J. Biol. Chem.*, 2000, **275**, 31038–31050.
118. T. I. Lin and C. Schroeder, *J. Virol.*, 2001, **75**, 3647–3656.
119. L. H. Pinto, G. R. Dieckmann, C. S. Gandhi, C. G. Papworth, J. Braman, M. A. Shaughnessy, J. D. Lear, R. A. Lamb and W. F. DeGrado, *Proc. Natl. Acad. Sci. U. S. A.*, 1997, **94**, 11301–11306.
120. Y. Tang, F. Zaitseva, R. A. Lamb and L. H. Pinto, *J. Biol. Chem.*, 2002, **277**, 39880–39886.
121. M. Yi, T. A. Cross and H. X. Zhou, *J. Phys. Chem. B*, 2008, **112**, 7977–7979.
122. F. A. Kovacs and T. A. Cross, *Biophys. J.*, 1997, **73**, 2511–2517.
123. J. Wang, S. Kim, F. Kovacs and T. A. Cross, *Protein Sci.*, 2001, **10**, 2241–2250.
124. K. Nishimura, S. Kim, L. Zhang and T. A. Cross, *Biochemistry*, 2002, **41**, 13170–13177.

125. J. Hu, T. Asbury, S. Achuthan, C. Li, R. Bertram, J. R. Quine, R. Fu and T. A. Cross, *Biophys. J.*, 2007, **92**, 4335–4343.
126. S. D. Cady, T. V. Mishanina and M. Hong, *J. Mol. Biol.*, 2009, **385**, 1127–1141.
127. R. M. Pielak, J. R. Schnell and J. J. Chou, *Proc. Natl. Acad. Sci. U. S. A.*, 2009, **106**, 7379–7384.
128. R. M. Pielak and J. J. Chou, *Biochem. Biophys. Res. Commun.*, 2010, **401**, 58–63.
129. T. A. Cross, H. Dong, M. Sharma, D. D. Busath and H. X. Zhou, *Curr. Opin. Virol.*, 2012, **2**, 128–133.
130. J. Hu, R. Fu, K. Nishimura, L. Zhang, H. X. Zhou, D. D. Busath, V. Vijayvergiya and T. A. Cross, *Proc. Natl. Acad. Sci. U. S. A.*, 2006, **103**, 6865–6870.
131. R. M. Pielak and J. J. Chou, *J. Am. Chem. Soc.*, 2010, **132**, 17695–17697.
132. H. X. Zhou, *Biophys. J.*, 2011, **100**, 912–921.
133. E. Khurana, M. Dal Peraro, R. DeVane, S. Vemparala, W. F. DeGrado and M. L. Klein, *Proc. Natl. Acad. Sci. U. S. A.*, 2009, **106**, 1069–1074.
134. F. Hu, W. Luo and M. Hong, *Science*, 2010, **330**, 505–508.
135. M. Hong and W. F. Degrado, *Protein Sci.*, 2012, **21**, 1620–1633.
136. S. Phongphanphanee, T. Rungrotmongkol, N. Yoshida, S. Hannongbua and F. Hirata, *J. Am. Chem. Soc.*, 2010, **132**, 9782–9788.
137. R. B. Belshe, M. H. Smith, C. B. Hall, R. Betts and A. J. Hay, *J. Virol.*, 1988, **62**, 1508–1512.
138. F. G. Hayden and A. J. Hay, *Curr. Top. Microbiol. Immunol.*, 1992, **176**, 119–130.
139. K. C. Duff, P. J. Gilchrist, A. M. Saxena and J. P. Bradshaw, *Virology*, 1994, **202**, 287–293.
140. J. Wang, J. R. Schnell and J. J. Chou, *Biochem. Biophys. Res. Commun.*, 2004, **324**, 212–217.
141. L. B.Andreas, M. T.Eddy, R. M.Pielak, J.Chou and R. G.Griffin, *J. Am. Chem. Soc.*, 2010, **132** , 10958–10960.
142. Y. Ohigashi, C. Ma, X. Jing, V. Balannick, L. H. Pinto and R. A. Lamb, *Proc. Natl. Acad. Sci. U. S. A.*, 2009, **106**, 18775–18779.
143. S. D. Cady and M. Hong, *Proc. Natl. Acad. Sci. U. S. A.*, 2008, **105**, 1483–1488.
144. S. D. Cady, J. Wang, Y. Wu, W. F. DeGrado and M. Hong, *J. Am. Chem. Soc.*, 2011, **133**, 4274–4284.
145. R. M. Pielak, K. Oxenoid and J. J. Chou, *Structure*, 2011, **19**, 1655–1663.
146. M. R. Rosenberg and M. G. Casarotto, *Proc. Natl. Acad. Sci. U. S. A.*, 2010, **107**, 13866–13871.
147. Q. S. Du, R. B. Huang, S. Q. Wang and K. C. Chou, *PLoS One*, 2010, **5**, e9388.
148. R. X. Gu, L. A. Liu, D. Q. Wei, J. G. Du, L. Liu and H. Liu, *J. Am. Chem. Soc.*, 2011, **133**, 10817–10825.

149. G. Y. Chuang, D. Kozakov, R. Brenke, D. Beglov, F. Guarnieri and S. Vajda, *Biophys. J.*, 2009, **97**, 2846–2853.
150. F. Y. Aoki and D. S. Sitar, *Clin. Pharmacol. Ther.*, 1985, **37**, 137–144.
151. F. Y. Aoki, D. S. Sitar and R. I. Ogilvie, *Clin. Pharmacol. Ther.*, 1979, **26**, 729–736.
152. N. Kolocouris, A. Kolocouris, G. B. Foscolos, G. Fytas, J. Neyts, E. Padalko, J. Balzarini, R. Snoeck, G. Andrei and E. De Clercq, *J. Med. Chem.*, 1996, **39**, 3307–3318.
153. G. Zoidis, N. Kolocouris, L. Naesens and E. De Clercq, *Bioorg. Med. Chem.*, 2009, **17**, 1534–1541.
154. J. Wang, S. D. Cady, V. Balannik, L. H. Pinto, W. F. DeGrado and M. Hong, *J. Am. Chem. Soc.*, 2009, **131**, 8066–8076.
155. J. Wang, C. Ma, Y. Wu, R. A. Lamb, L. H. Pinto and W. F. DeGrado, *J. Am. Chem. Soc.*, 2011, **133**, 13844–13847.
156. S. Kurtz, G. Luo, K. M. Hahnenberger, C. Brooks, O. Gecha, K. Ingalls, K. Numata and M. Krystal, *Antimicrob. Agents Chemother.*, 1995, **39**, 2204–2209.
157. Q. Tu, L. H. Pinto, G. Luo, M. A. Shaughnessy, D. Mullaney, S. Kurtz, M. Krystal and R. A. Lamb, *J. Virol.*, 1996, **70**, 4246–4252.
158. A. Kolocouris, D. Tataridis, G. Fytas, T. Mavromoustakos, G. B. Foscolos, N. Kolocouris and E. De Clercq, *Bioorg. Med. Chem. Lett.*, 1999, **9**, 3465–3470.
159. W. B. Fischer, M. Pitkeathly and M. S. Sansom, *Eur. Biophys. J.*, 2001, **30**, 416–420.
160. T. Betakova, M. V. Nermut and A. J. Hay, *J. Gen. Virol.*, 1996, **77**(Pt 11), 2689–2694.
161. V. Balannik, R. A. Lamb and L. H. Pinto, *J. Biol. Chem.*, 2008, **283**, 4895–4904.
162. J. Wang, R. M. Pielak, M. A. McClintock and J. J. Chou, *Nat. Struct. Mol. Biol.*, 2009, **16**, 1267–1271.
163. T. Betakova and A. J. Hay, *Arch. Virol.*, 2009, **154**, 1619–1624.
164. K. Otomo, A. Toyama, T. Miura and H. Takeuchi, *J. Biochem.*, 2009, **145**, 543–554.
165. S. L. Rouse, T. Carpenter, P. J. Stansfeld and M. S. Sansom, *Biochemistry*, 2009, **48**, 9949–9951.
166. S. Hongo, K. Sugawara, Y. Muraki, F. Kitame and K. Nakamura, *J. Virol.*, 1997, **71**, 2786–2792.
167. A. Pekosz and R. A. Lamb, *Proc. Natl. Acad. Sci. U. S. A.*, 1998, **95**, 13233–13238.
168. S. Hongo, K. Sugawara, Y. Muraki, Y. Matsuzaki, E. Takashita, F. Kitame and K. Nakamura, *J. Virol.*, 1999, **73**, 46–50.
169. S. Hongo, K. Ishii, K. Mori, E. Takashita, Y. Muraki, Y. Matsuzaki and K. Sugawara, *Arch. Virol.*, 2004, **149**, 35–50.
170. S. M. Stewart and A. Pekosz, *J. Virol.*, 2012, **86**, 1277–1281.
171. T. Furukawa, Y. Muraki, T. Noda, E. Takashita, R. Sho, K. Sugawara, Y. Matsuzaki, Y. Shimotai and S. Hongo, *J. Virol.*, 2011, **85**, 1322–1329.

172. E. A. Cohen, E. F. Terwilliger, J. G. Sodroski and W. A. Haseltine, *Nature*, 1988, **334**, 532–534.
173. K. Strebel, T. Klimkait and M. A. Martin, *Science*, 1988, **241**, 1221–1223.
174. E. F. Terwilliger, E. A. Cohen, Y. C. Lu, J. G. Sodroski and W. A. Haseltine, *Proc. Natl. Acad. Sci. U. S. A.*, 1989, **86**, 5163–5167.
175. T. Klimkait, K. Strebel, M. D. Hoggan, M. A. Martin and J. M. Orenstein, *J. Virol.*, 1990, **64**, 621–629.
176. R. L. Willey, F. Maldarelli, M. A. Martin and K. Strebel, *J. Virol.*, 1992, **66**, 7193–7200.
177. R. L. Willey, F. Maldarelli, M. A. Martin and K. Strebel, *J. Virol.*, 1992, **66**, 226–234.
178. X. J. Yao, H. Gottlinger, W. A. Haseltine and E. A. Cohen, *J. Virol.*, 1992, **66**, 5119–5126.
179. A. L. Grice, I. D. Kerr and M. S. Sansom, *FEBS Lett.*, 1997, **405**, 299–304.
180. F. S. Cordes, A. Kukol, L. R. Forrest, I. T. Arkin, M. S. Sansom and W. B. Fischer, *Biochim. Biophys. Acta*, 2001, **1512**, 291–298.
181. G. D. Ewart, K. Mills, G. B. Cox and P. W. Gage, *Eur. Biophys. J.*, 2002, **31**, 26–35.
182. S. H. Park, A. A. Mrse, A. A. Nevzorov, M. F. Mesleh, M. Oblatt-Montal, M. Montal and S. J. Opella, *J. Mol. Biol.*, 2003, **333**, 409–424.
183. W. Romer, Y. H. Lam, D. Fischer, A. Watts, W. B. Fischer, P. Goring, R. B. Wehrspohn, U. Gosele and C. Steinem, *J. Am. Chem. Soc.*, 2004, **126**, 16267–16274.
184. T. Mehnert, Y. H. Lam, P. J. Judge, A. Routh, D. Fischer, A. Watts and W. B. Fischer, *J. Biomol. Struct. Dyn.*, 2007, **24**, 589–596.
185. S. H. Park and S. J. Opella, *Protein Sci.*, 2007, **16**, 2205–2215.
186. S. Bolduan, J. Votteler, V. Lodermeyer, T. Greiner, H. Koppensteiner, M. Schindler, G. Thiel and U. Schubert, *Virology*, 2011, **416**, 75–85.
187. M. J. Coady, N. G. Daniel, E. Tiganos, B. Allain, J. Friborg, J. Y. Lapointe and E. A. Cohen, *Virology*, 1998, **244**, 39–49.
188. S. J. Neil, T. Zang and P. D. Bieniasz, *Nature*, 2008, **451**, 425–430.
189. D. R. Hout, M. L. Gomez, E. Pacyniak, L. M. Gomez, S. H. Inbody, E. R. Mulcahy, N. Culley, D. M. Pinson, M. F. Powers, S. W. Wong and E. B. Stephens, *Virology*, 2005, **339**, 56–69.
190. M. Paul, S. Mazumder, N. Raja and M. A. Jabbar, *J. Virol.*, 1998, **72**, 1270–1279.
191. K. Hsu, J. Seharaseyon, P. Dong, S. Bour and E. Marban, *Mol. Cell*, 2004, **14**, 259–267.
192. C. Ma, F. M. Marassi, D. H. Jones, S. K. Straus, S. Bour, K. Strebel, U. Schubert, M. Oblatt-Montal, M. Montal and S. J. Opella, *Protein Sci.*, 2002, **11**, 546–557.
193. D. R. Hout, L. M. Gomez, E. Pacyniak, J. M. Miller, M. S. Hill and E. B. Stephens, *Virology*, 2006, **348**, 449–461.

194. D. R. Hout, M. L. Gomez, E. Pacyniak, L. M. Gomez, B. Fegley, E. R. Mulcahy, M. S. Hill, N. Culley, D. M. Pinson, W. Nothnick, M. F. Powers, S. W. Wong and E. B. Stephens, *Virology*, 2006, **344**, 541–559.

195. V. Lemaitre, D. Willbold, A. Watts and W. B. Fischer, *J. Biomol. Struct. Dyn.*, 2006, **23**, 485–496.

196. P. Henklein, R. Kinder, U. Schubert and B. Bechinger, *FEBS Lett.*, 2000, **482**, 220–224.

197. M. Wittlich, B. W. Koenig and D. Willbold, *J. Pept. Sci.*, 2008, **14**, 804–810.

198. M. Skasko, Y. Wang, Y. Tian, A. Tokarev, J. Munguia, A. Ruiz, E. B. Stephens, S. J. Opella and J. Guatelli, *J. Biol. Chem.*, 2012, **287**, 58–67.

199. V. Wray, T. Federau, P. Henklein, S. Klabunde, O. Kunert, D. Schomburg and U. Schubert, *Int. J. Pept. Protein Res.*, 1995, **45**, 35–43.

200. T. Federau, U. Schubert, J. Flossdorf, P. Henklein, D. Schomburg and V. Wray, *Int. J. Pept. Protein Res.*, 1996, **47**, 297–310.

201. D. Willbold, S. Hoffmann and P. Rosch, *Eur. J. Biochem.*, 1997, **245**, 581–588.

202. A. Kukol and I. T. Arkin, *Biophys. J.*, 1999, **77**, 1594–1601.

203. S. H. Park, A. A. De Angelis, A. A. Nevzorov, C. H. Wu and S. J. Opella, *Biophys. J.*, 2006, **91**, 3032–3042.

204. S. Sharpe, W. M. Yau and R. Tycko, *Biochemistry*, 2006, **45**, 918–933.

205. J. X. Lu, S. Sharpe, R. Ghirlando, W. M. Yau and R. Tycko, *Protein Sci.*, 2010, **19**, 1877–1896.

206. P. B. Moore, Q. Zhong, T. Husslein and M. L. Klein, *FEBS Lett.*, 1998, **431**, 143–148.

207. A. Hussain, S. R. Das, C. Tanwar and S. Jameel, *Virology J.*, 2007, **4**, 81.

208. C. F. Becker, M. Oblatt-Montal, G. G. Kochendoerfer and M. Montal, *J. Biol. Chem.*, 2004, **279**, 17483–17489.

209. B. D. Kuhl, V. Cheng, D. A. Donahue, R. D. Sloan, C. Liang, J. Wilkinson and M. A. Wainberg, *PLoS One*, 2011, **6**, e27660.

210. K. Hsu, J. Han, K. Shinlapawittayatorn, I. Deschenes and E. Marban, *Biophys. J.*, 2010, **99**, 1718–1725.

211. C. G. Kim, V. Lemaitre, A. Watts and W. B. Fischer, *Anal. Bioanal. Chem.*, 2006, **386**, 2213–2217.

212. C. Frank, M. K. Mohamed, G. T. Strickland, D. Lavanchy, R. R. Arthur, L. S. Magder, T. El Khoby, Y. Abdel-Wahab, E. S. Aly Ohn, W. Anwar and I. Sallam, *Lancet*, 2000, **355**, 887–891.

213. P. Simmonds, *J. Gen. Virol.*, 2004, **85**, 3173–3188.

214. J. M. Pawlotsky, *Hepatology*, 2006, **43**, S207–S220.

215. D. Ge, J. Fellay, A. J. Thompson, J. S. Simon, K. V. Shianna, T. J. Urban, E. L. Heinzen, P. Qiu, A. H. Bertelsen, A. J. Muir, M. Sulkowski, J. G. McHutchison and D. B. Goldstein, *Nature*, 2009, **461**, 399–401.

216. D. L. Thomas, C. L. Thio, M. P. Martin, Y. Qi, D. Ge, C. O'Huigin, J. Kidd, K. Kidd, S. I. Khakoo, G. Alexander, J. J. Goedert, G. D. Kirk, S. M. Donfield, H. R. Rosen, L. H. Tobler, M. P. Busch, J. G. McHutchison, D. B. Goldstein and M. Carrington, *Nature*, 2009, **461**, 798–801.

217. J. M. Pawlotsky, *Gastroenterology*, 2011, **140**, 746–754.

218. V. Garg, R. van Heeswijk, J. Eun Lee, K. Alves, P. Nadkarni and X. Luo, *Hepatology*, 2011, **54**, 20–27.

219. H. Montaudie, T. Passeron, N. Cardot-Leccia, N. Sebbag and J. P. Lacour, *Dermatology*, 2010, **221**, 303–305.

220. R. Flisiak, S. V. Feinman, M. Jablkowski, A. Horban, W. Kryczka, M. Pawlowska, J. E. Heathcote, G. Mazzella, C. Vandelli, V. Nicolas-Metral, P. Grosgurin, J. S. Liz, P. Scalfaro, H. Porchet and R. Crabbe, *Hepatology*, 2009, **49**, 1460–1468.

221. J. Paeshuyse, A. Kaul, E. De Clercq, B. Rosenwirth, J. M. Dumont, P. Scalfaro, R. Bartenschlager and J. Neyts, *Hepatology*, 2006, **43**, 761–770.

222. C. Lin, B. D. Lindenbach, B. M. Pragai, D. W. McCourt and C. M. Rice, *J. Virol.*, 1994, **68**, 5063–5073.

223. H. Mizushima, M. Hijikata, S. Asabe, M. Hirota, K. Kimura and K. Shimotohno, *J. Virol.*, 1994, **68**, 6215–6222.

224. S. Carrere-Kremer, C. Montpellier, L. Lorenzo, B. Brulin, L. Cocquerel, S. Belouzard, F. Penin and J. Dubuisson, *J. Biol. Chem.*, 2004, **279**, 41384–41392.

225. S. Carrere-Kremer, C. Montpellier-Pala, L. Cocquerel, C. Wychowski, F. Penin and J. Dubuisson, *J. Virol.*, 2002, **76**, 3720–3730.

226. T. Whitfield, A. J. Miles, J. C. Scheinost, J. Offer, P. Wentworth Jr, R. A. Dwek, B. A. Wallace, P. C. Biggin and N. Zitzmann, *Mol. Membr. Biol.*, 2011.

227. A. Sakai, M. S. Claire, K. Faulk, S. Govindarajan, S. U. Emerson, R. H. Purcell and J. Bukh, *Proc. Natl. Acad. Sci. U. S. A.*, 2003, **100**, 11646–11651.

228. T. Wakita, T. Pietschmann, T. Kato, T. Date, M. Miyamoto, Z. Zhao, K. Murthy, A. Habermann, H. G. Krausslich, M. Mizokami, R. Bartenschlager and T. J. Liang, *Nat. Med.*, 2005, **11**, 791–796.

229. S. Griffin, D. Clarke, C. McCormick, D. Rowlands and M. Harris, *J. Virol.*, 2005, **79**, 15525–15536.

230. G. Haqshenas, J. M. Mackenzie, X. Dong and E. J. Gowans, *J. Gen. Virol.*, 2007, **88**, 134–142.

231. C. T. Jones, C. L. Murray, D. K. Eastman, J. Tassello and C. M. Rice, *J. Virol.*, 2007, **81**, 8374–8383.

232. E. Steinmann, F. Penin, S. Kallis, A. H. Patel, R. Bartenschlager and T. Pietschmann, *PLoS Pathogens*, 2007, **3**, e103.

233. C. Brohm, E. Steinmann, M. Friesland, I. C. Lorenz, A. Patel, F. Penin, R. Bartenschlager and T. Pietschmann, *J. Virol.*, 2009, **83**, 11682–11693.

234. Y. Ma, M. Anantpadma, J. M. Timpe, S. Shanmugam, S. M. Singh, S. M. Lemon and M. Yi, *J. Virol.*, 2011, **85**, 86–97.
235. C. I. Popescu, N. Callens, D. Trinel, P. Roingeard, D. Moradpour, V. Descamps, G. Duverlie, F. Penin, L. Heliot, Y. Rouille and J. Dubuisson, *PLoS Pathogens*, 2011, **7**, e1001278.
236. K. A. Stapleford and B. D. Lindenbach, *J. Virol.*, 2011, **85**, 1706–1717.
237. P. Tedbury, S. Welbourn, A. Pause, B. King, S. Griffin and M. Harris, *J. Gen. Virol.*, 2011, **92**, 819–830.
238. V. Jirasko, R. Montserret, N. Appel, A. Janvier, L. Eustachi, C. Brohm, E. Steinmann, T. Pietschmann, F. Penin and R. Bartenschlager, *J. Biol. Chem.*, 2008, **283**, 28546–28562.
239. V. Jirasko, R. Montserret, J. Y. Lee, J. Gouttenoire, D. Moradpour, F. Penin and R. Bartenschlager, *PLoS Pathogens*, 2010, **6**, e1001233.
240. B. Boson, O. Granio, R. Bartenschlager and F. L. Cosset, *PLoS Pathogens*, 2011, **7**, e1002144.
241. C. StGelais, T. L. Foster, M. Verow, E. Atkins, C. W. Fishwick, D. Rowlands, M. Harris and S. Griffin, *J. Virol.*, 2009, **83**, 7970–7981.
242. A. J. Perez-Berna, J. Guillen, M. R. Moreno, A. Bernabeu, G. Pabst, P. Laggner and J. Villalain, *J. Biol. Chem.*, 2008, **283**, 8089–8101.
243. E. Steinmann, T. Whitfield, S. Kallis, R. A. Dwek, N. Zitzmann, T. Pietschmann and R. Bartenschlager, *Hepatology*, 2007, **46**, 330–338.
244. B. Saunier, M. Triyatni, L. Ulianich, P. Maruvada, P. Yen and L. D. Kohn, *J. Virol.*, 2003, **77**, 546–559.
245. G. Patargias, N. Zitzmann, R. Dwek and W. B. Fischer, *J. Med. Chem.*, 2006, **49**, 648–655.
246. G. A. Cook and S. J. Opella, *Eur. Biophys. J.: EBJ*, 2010, **39**, 1097–1104.
247. R. Montserret, N. Saint, C. Vanbelle, A. G. Salvay, J. P. Simorre, C. Ebel, N. Sapay, J. G. Renisio, A. Bockmann, E. Steinmann, T. Pietschmann, J. Dubuisson, C. Chipot and F. Penin, *J. Biol. Chem.*, 2010, **285**, 31446–31461.
248. G. A. Cook and S. J. Opella, *Biochim. Biophys. Acta*, 2011, **1808**, 1448–1453.
249. C. F. Chew, R. Vijayan, J. Chang, N. Zitzmann and P. C. Biggin, *Biophys. J.*, 2009, **96**, L10–12.
250. H. Duclohier, *Eur. Biophys. J.: EBJ*, 2006, **35**, 401–409.
251. J. Carney, J. M. East and A. G. Lee, *Biophys. J.*, 2007, **92**, 3556–3563.
252. G. A. Cook, H. Zhang, S. H. Park, Y. Wang and S. J. Opella, *Biochim. Biophys. Acta*, 2011, **1808**, 554–560.
253. H. Li, E. Atkins, J. Bruckner, S. McArdle, W. C. Qiu, L. V. Thomassen, J. Scott, M. C. Shuhart, S. Livingston, L. Townshend-Bulson, B. J. McMahon, M. Harris, S. Griffin and D. R. Gretch, *Virology*, 2012, **423**, 30–37.
254. Z. Meshkat, M. Audsley, C. Beyer, E. J. Gowans and G. Haqshenas, *J. Viral Hepat.*, 2008.

255. J. M. Gottwein, T. B. Jensen, C. K. Mathiesen, P. Meuleman, S. B. Serre, J. B. Lademann, L. Ghanem, T. K. Scheel, G. Leroux-Roels and J. Bukh, *J. Virol.*, 2011, **85**, 8913–8928.

256. U. Mihm, N. Grigorian, C. Welsch, E. Herrmann, B. Kronenberger, G. Teuber, M. von Wagner, W. P. Hofmann, M. Albrecht, T. Lengauer, S. Zeuzem and C. Sarrazin, *Antivir. Ther.*, 2006, **11**, 507–519.

257. S. Castelain, D. Bonte, F. Penin, C. Francois, D. Capron, S. Dedeurwaerder, P. Zawadzki, V. Morel, C. Wychowski and G. Duverlie, *J. Med. Virol.*, 2007, **79**, 144–154.

258. V. Lemaitre, R. Ali, C. G. Kim, A. Watts and W. B. Fischer, *FEBS Lett.*, 2004, **563**, 75–81.

259. J. Chan, K. O'Riordan and T. E. Wiley, *Dig. Dis. Sci.*, 2002, **47**, 438–442.

260. M. Maynard, P. Pradat, F. Bailly, F. Rozier, C. Nemoz, S. N. Si Ahmed, P. Adeleine and C. Trepo, *J. Hepatol.*, 2006, **44**, 484–490.

261. N. Akuta, F. Suzuki, M. Hirakawa, Y. Kawamura, H. Yatsuji, H. Sezaki, Y. Suzuki, T. Hosaka, M. Kobayashi, S. Saitoh, Y. Arase, K. Ikeda, K. Chayama, Y. Nakamura and H. Kumada, *Hepatology*, 2010, **52**, 421–429.

262. C. C. Yip, S. K. Lau, B. Zhou, M. X. Zhang, H. W. Tsoi, K. H. Chan, X. C. Chen, P. C. Woo and K. Y. Yuen, *Arch. Virol.*, 2010, **155**, 1413–1424.

263. I. V. Sandoval and L. Carrasco, *J. Virol.*, 1997, **71**, 4679–4693.

264. A. Agirre, A. Barco, L. Carrasco and J. L. Nieva, *J. Biol. Chem.*, 2002, **277**, 40434–40441.

265. A. S. de Jong, E. Wessels, H. B. Dijkman, J. M. Galama, W. J. Melchers, P. H. Willems and F. J. van Kuppeveld, *J. Biol. Chem.*, 2003, **278**(1012–1021).

266. M. Campanella, A. S. de Jong, K. W. Lanke, W. J. Melchers, P. H. Willems, P. Pinton, R. Rizzuto and F. J. van Kuppeveld, *J. Biol. Chem.*, 2004, **279**, 18440–18450.

267. A. S. de Jong, W. J. Melchers, D. H. Glaudemans, P. H. Willems and F. J. van Kuppeveld, *J. Biol. Chem.*, 2004, **279**, 19924–19935.

268. S. Sanchez-Martinez, N. Huarte, R. Maeso, V. Madan, L. Carrasco and J. L. Nieva, *Biochemistry*, 2008, **47**, 10731–10739.

269. A. S. de Jong, I. W. Schrama, P. H. Willems, J. M. Galama, W. J. Melchers and F. J. van Kuppeveld, *J. Gen. Virol.*, 2002, **83**, 783–793.

270. F. J. van Kuppeveld, W. J. Melchers, P. H. Willems and T. W. Gadella Jr., *J. Virol.*, 2002, **76**, 9446–9456.

271. G. Patargias, T. Barke, A. Watts and W. B. Fischer, *Mol. Membr. Biol.*, 2009, **26**, 309–320.

272. M. Ito, Y. Yanagi and T. Ichinohe, *PLoS Pathogens*, 2012, **8**, e1002857.

273. S. Xie, K. Wang, W. Yu, W. Lu, K. Xu, J. Wang, B. Ye, W. Schwarz, Q. Jin and B. Sun, *Cell Res.*, 2011, **21**, 1271–1275.

274. T. J. Tuthill, E. Groppelli, J. M. Hogle and D. J. Rowlands, *Curr. Top. Microbiol. Immunol.*, 2010, **343**, 43–89.

275. P. A. Rota, M. S. Oberste, S. S. Monroe, W. A. Nix, R. Campagnoli, J. P. Icenogle, S. Penaranda, B. Bankamp, K. Maher, M. H. Chen, S. Tong, A. Tamin, L. Lowe, M. Frace, J. L. DeRisi, Q. Chen, D. Wang, D. D. Erdman, T. C. Peret, C. Burns, T. G. Ksiazek, P. E. Rollin, A. Sanchez, S. Liffick, B. Holloway, J. Limor, K. McCaustland, M. Olsen-Rasmussen, R. Fouchier, S. Gunther, A. D. Osterhaus, C. Drosten, M. A. Pallansch, L. J. Anderson and W. J. Bellini, *Science*, 2003, **300**, 1394–1399.

276. M. A. Marra, S. J. Jones, C. R. Astell, R. A. Holt, A. Brooks-Wilson, Y. S. Butterfield, J. Khattra, J. K. Asano, S. A. Barber, S. Y. Chan, A. Cloutier, S. M. Coughlin, D. Freeman, N. Girn, O. L. Griffith, S. R. Leach, M. Mayo, H. McDonald, S. B. Montgomery, P. K. Pandoh, A. S. Petrescu, A. G. Robertson, J. E. Schein, A. Siddiqui, D. E. Smailus, J. M. Stott, G. S. Yang, F. Plummer, A. Andonov, H. Artsob, N. Bastien, K. Bernard, T. F. Booth, D. Bowness, M. Czub, M. Drebot, L. Fernando, R. Flick, M. Garbutt, M. Gray, A. Grolla, S. Jones, H. Feldmann, A. Meyers, A. Kabani, Y. Li, S. Normand, U. Stroher, G. A. Tipples, S. Tyler, R. Vogrig, D. Ward, B. Watson, R. C. Brunham, M. Krajden, M. Petric, D. M. Skowronski, C. Upton and R. L. Roper, *Science*, 2003, **300**, 1399–1404.

277. T. R. Ruch and C. E. Machamer, *Viruses*, 2012, **4**, 363–382.

278. J. Torres, U. Maheswari, K. Parthasarathy, L. Ng, D. X. Liu and X. Gong, *Protein Sci.*, 2007, **16**, 2065–2071.

279. L. Zhang and M. C. Hung, *Oncogene*, 1996, **12**, 571–576.

280. L. Zhang, Y. K. Lau, W. Xia, G. N. Hortobagyi and M. C. Hung, *Clin. Cancer Res.*, 1999, **5**, 343–353.

281. L. Zhang, Y. K. Lau, L. Xi, R. L. Hong, D. S. Kim, C. F. Chen, G. N. Hortobagyi, C. Chang and M. C. Hung, *Oncogene*, 1998, **16**, 2855–2863.

282. C. C. Chen, J. Kruger, I. Sramala, H. J. Hsu, P. Henklein, Y. M. Chen and W. B. Fischer, *Biochim. Biophys. Acta*, 2011, **1808**, 572–579.

283. T. Malik, C. W. Shegogue, K. Werner, L. Ngo, C. Sauder, C. Zhang, W. P. Duprex and S. Rubin, *J. Virol.*, 2011, **85**, 6082–6085.

284. K. Takeuchi, K. Tanabayashi, M. Hishiyama and A. Yamada, *Virology*, 1996, **225**, 156–162.

285. B. He, G. P. Leser, R. G. Paterson and R. A. Lamb, *Virology*, 1998, **250**, 30–40.

286. Y. Lin, A. C. Bright, T. A. Rothermel and B. He, *J. Virol.*, 2003, **77**, 3371–3383.

287. S. Fuentes, K. C. Tran, P. Luthra, M. N. Teng and B. He, *J. Virol.*, 2007, **81**, 8361–8366.

288. A. Bukreyev, S. S. Whitehead, B. R. Murphy and P. L. Collins, *J. Virol.*, 1997, **71**, 8973–8982.

289. S. S. Whitehead, A. Bukreyev, M. N. Teng, C. Y. Firestone, M. St. Claire, W. R. Elkins, P. L. Collins and B. R. Murphy, *J. Virol.*, 1999, **73**, 3438–3442.

290. P. L. Collins and G. Mottet, *J. Gen. Virol.*, 1993, **74**(Pt 7), 1445–1450.
291. M. Perez, B. Garcia-Barreno, J. A. Melero, L. Carrasco and R. Guinea, *Virology*, 1997, **235**, 342–351.
292. K. Gaedigk-Nitschko and M. J. Schlesinger, *Virology*, 1990, **175**, 274–281.
293. K. Gaedigk-Nitschko, M. X. Ding, M. A. Levy and M. J. Schlesinger, *Virology*, 1990, **175**, 282–291.
294. K. Gaedigk-Nitschko and M. J. Schlesinger, *Virology*, 1991, **183**, 206–214.
295. S. Lusa, H. Garoff and P. Liljestrom, *Virology*, 1991, **185**, 843–846.
296. M. J. Schlesinger, S. D. London and C. Ryan, *Virology*, 1993, **193**, 424–432.
297. M. A. Sanz and L. Carrasco, *J. Virol.*, 2001, **75**, 7778–7784.
298. G. M. McInerney, J. M. Smit, P. Liljestrom and J. Wilschut, *Virology*, 2004, **325**, 200–206.
299. P. Keelapang, R. Sriburi, S. Supasa, N. Panyadee, A. Songjaeng, A. Jairungsri, C. Puttikhunt, W. Kasinrerk, P. Malasit and N. Sittisombut, *J. Virol.*, 2004, **78**, 2367–2381.
300. J. Junjhon, M. Lausumpao, S. Supasa, S. Noisakran, A. Songjaeng, P. Saraithong, K. Chaichoun, U. Utaipat, P. Keelapang, A. Kanjanahaluethai, C. Puttikhunt, W. Kasinrerk, P. Malasit and N. Sittisombut, *J. Virol.*, 2008, **82**, 10776–10791.
301. J. Junjhon, T. J. Edwards, U. Utaipat, V. D. Bowman, H. A. Holdaway, W. Zhang, P. Keelapang, C. Puttikhunt, R. Perera, P. R. Chipman, W. Kasinrerk, P. Malasit, R. J. Kuhn and N. Sittisombut, *J. Virol.*, 2010, **84**, 8353–8358.
302. M. J. Pryor, L. Azzola, P. J. Wright and A. D. Davidson, *J. Gen. Virol.*, 2004, **85**, 3627–3636.
303. A. Catteau, O. Kalinina, M. C. Wagner, V. Deubel, M. P. Courageot and P. Despres, *J. Gen. Virol.*, 2003, **84**, 2781–2793.
304. S. S. Wong, G. Haqshenas, E. J. Gowans and J. Mackenzie, *FEBS Lett.*, 2012, **586**, 1032–1037.
305. R. J. Kuhn, W. Zhang, M. G. Rossmann, S. V. Pletnev, J. Corver, E. Lenches, C. T. Jones, S. Mukhopadhyay, P. R. Chipman, E. G. Strauss, T. S. Baker and J. H. Strauss, *Cell*, 2002, **108**, 717–725.
306. W. Zhang, P. R. Chipman, J. Corver, P. R. Johnson, Y. Zhang, S. Mukhopadhyay, T. S. Baker, J. H. Strauss, M. G. Rossmann and R. J. Kuhn, *Nat. Struct. Biol.*, 2003, **10**, 907–912.
307. R. Biasiotto, P. Aguiari, R. Rizzuto, P. Pinton, D. M. D'Agostino and V. Ciminale, *Biochim. Biophys. Acta*, 2010, **1797**, 945–951.
308. M. Silic-Benussi, O. Marin, R. Biasiotto, D. M. D'Agostino and V. Ciminale, *FEBS Lett.*, 2010, **584**, 2070–2075.
309. E. Tibaldi, A. Venerando, F. Zonta, C. Bidoia, E. Magrin, O. Marin, A. Toninello, L. Bordin, V. Martini, M. A. Pagano and A. M. Brunati, *Biochem. J.*, 2011, **439**, 505–516.

310. P. Tian, M. K. Estes, Y. Hu, J. M. Ball, C. Q. Zeng and W. P. Schilling, *J. Virol.*, 1995, **69**, 5763–5772.

311. Y. Dong, C. Q. Zeng, J. M. Ball, M. K. Estes and A. P. Morris, *Proc. Natl. Acad. Sci. U. S. A.*, 1997, **94**, 3960–3965.

312. N. Halaihel, V. Lievin, J. M. Ball, M. K. Estes, F. Alvarado and M. Vasseur, *J. Virol.*, 2000, **74**, 9464–9470.

313. F. Tafazoli, C. Q. Zeng, M. K. Estes, K. E. Magnusson and L. Svensson, *J. Virol.*, 2001, **75**, 1540–1546.

314. A. R. Chacko, P. H. Zwart, R. J. Read, E. J. Dodson, C. D. Rao and K. Suguna, *Acta Crystallogr. D Biol. Crystallogr.*, 2012, **68**, 1541–1548.

315. A. R. Chacko, J. Jeyakanthan, G. Ueno, K. Sekar, C. D. Rao, E. J. Dodson, K. Suguna and R. J. Read, *Acta Crystallogr. D Biol. Crystallogr.*, 2012, **68**, 57–61.

316. A. R. Chacko, M. Arifullah, N. P. Sastri, J. Jeyakanthan, G. Ueno, K. Sekar, R. J. Read, E. J. Dodson, D. C. Rao and K. Suguna, *J. Virol.*, 2011, **85**, 12721–12732.

317. G. D. Bowman, I. M. Nodelman, O. Levy, S. L. Lin, P. Tian, T. J. Zamb, S. A. Udem, B. Venkataraghavan and C. E. Schutt, *J. Mol. Biol.*, 2000, **304**, 861–871.

318. J. A. O'Brien, J. A. Taylor and A. R. Bellamy, *J. Virol.*, 2000, **74**, 5388–5394.

319. A. W. Einerhand, *J. Pediatr. Gastroenterol. Nutr.*, 1998, **27**, 123–124.

320. S. Raghava, K. M. Giorda, F. B. Romano, A. P. Heuck and D. N. Hebert, *PLoS Pathogens*, 2011, **7**, e1002116.

321. P. Leechanachai, L. Banks, F. Moreau and G. Matlashewski, *Oncogene*, 1992, **7**, 19–25.

322. C. L. Halbert and D. A. Galloway, *J. Virol.*, 1988, **62**, 1071–1075.

323. N. Ganguly, *Cell. Oncol. (Dordr.)*, 2012, **35**, 67–76.

324. A. Venuti, F. Paolini, L. Nasir, A. Corteggio, S. Roperto, M. S. Campo and G. Borzacchiello, *Mol. Cancer*, 2011, **10**, 140.

325. F. A. Suprynowicz, E. Krawczyk, J. D. Hebert, S. R. Sudarshan, V. Simic, C. M. Kamonjoh and R. Schlegel, *J. Virol.*, 2010, **84**, 10619–10629.

326. S. H. Kim, Y. S. Juhnn, S. Kang, S. W. Park, M. W. Sung, Y. J. Bang and Y. S. Song, *Cell. Mol. Life Sci.: CMLS*, 2006, **63**, 930–938.

327. G. L. Disbrow, J. A. Hanover and R. Schlegel, *J. Virol.*, 2005, **79**, 5839–5846.

328. T. Harada, N. Tautz and H. J. Thiel, *J. Virol.*, 2000, **74**, 9498–9506.

329. K. Elbers, N. Tautz, P. Becher, D. Stoll, T. Rumenapf and H. J. Thiel, *J. Virol.*, 1996, **70**, 4131–4135.

Section III
Host Targets

TLR-7 Agonists for the Treatment of Viral Hepatitis

RANDALL L. HALCOMB

Gilead Sciences Inc., Department of Medicinal Chemistry, 333 Lakeside
Drive, Foster City, CA 94404, USA
Email: randall.halcomb@gilead.com

10.1 Introduction: TLR-7 and the Antiviral Effects of Interferon-α Induction

Toll-like receptor-7 (TLR-7) has attracted interest as a target for the treatment of viral infections and also cancer and respiratory diseases.[1,2] Particular attention has been focused on chronic hepatitis C virus infection, but also chronic hepatitis B, where agonists that activate TLR-7 signaling could have utility to induce the production of endogenous interferons, enhance antiviral immune responses and facilitate control or clearance of replicating virus.

Toll-like receptors are components of the innate immune system and serve as the first line of defense against invading pathogens.[3–8] They also serve to link early-stage innate responses to adaptive T and B cell responses. To date, 10 members of the TLR family have been identified in humans, although other species can have greater or fewer numbers. Different TLRs have evolved to recognize different unique molecular patterns that are fragments of bacteria, viruses or other pathogens, including lipopolysaccharides, nucleic acids, lipids and lipopeptides.

As would be expected, given their role, TLRs are predominantly expressed in cells of the immune system[9] and barrier organs and tissues.[10,11] Some TLRs are

RSC Drug Discovery Series No. 32
Successful Strategies for the Discovery of Antiviral Drugs
Edited by Manoj C. Desai and Nicholas A. Meanwell
© The Royal Society of Chemistry 2013
Published by the Royal Society of Chemistry, www.rsc.org

expressed at detectable levels more broadly. TLR-7 is found predominantly in plasmacytoid dendritic cells, which are the primary source of interferon-α (IFN-α).[12] TLR-7 is also expressed at significant levels in B cells and to a lower extent in various other immune cells. TLR-7 is localized intracellularly in endosomes and lysosomes.[13] The protein contains a ligand-binding ectodomain within the lumen of the organelle, a short transmembrane sequence and a cytoplasmic signaling domain referred to as the Toll-interleukin 1 receptor (TIR) domain. The native ligand for TLR-7 is single-stranded RNA, particularly that of viral origin.[14] The definitive pathway by which viral RNA becomes localized with the TLR binding domain in the lumen of endosomes is not fully understood, but it could involve endocytosis, autophagy or some other mechanism. Following binding to ssRNA, adaptor proteins, including MyD88, are recruited to the cytoplasmic TIR domain and a signaling cascade is initiated.[15] This results in the activation and translocation of transcription factors, including IRF-7 and NF-κB, to the nucleus and the transcription of IFN-α and other antiviral cytokine genes is initiated. Subsequently, these dendritic cells are activated to become antigen-producing cells and an adaptive immune response is initiated.

IFN-α is well known to have antiviral effects and pegylated versions, Pegasys and Pegintron, are currently used as components of the standard of care to treat HCV infection, in combination with ribavirin and one of the two protease inhibitors telaprevir or boceprevir.[16–18] HCV patients can achieve sustained viral response (SVR), or an effective cure, after a 24–48 week treatment with the combination. Pegylated interferon is also used to treat chronic HBV infection and is an alternative to antiviral nucleotide analogs.[19–21] Pegasys can induce an immune response in HBV patients, culminating in HBs antigen seroconversion or generation of antibodies to the HBV surface antigen with concomitant loss of circulating surface antigen.[22] This generally results in a finite duration of therapy for HBV, without the need for life-long treatment with direct-acting antiviral agents. However, the percentage of patients who achieve sero-conversion is low and Pegasys is poorly tolerated owing to flu-like symptoms as a side effect, among many others.

TLR-7 activation can trigger additional immunostimulatory antiviral effects, such as T-cell responses and adaptive immunity. Therefore, small-molecule TLR-7 agonists have attracted attention as potential antiviral agents that could replace or complement pegylated IFN-α.[23] Hypotheses have been presented that TLR-7 agonists could have better side effect profiles or could have greater efficacy than pegylated IFN-α. In general, validation of this hypothesis awaits further clinical investigation.

This chapter summarizes selected small-molecule TLR-7 agonists that have been developed for viral hepatitis infections. A summary of the characteristics and structure–activity relationships (SARs) in each of the major classes of agonists are presented, together with findings in clinical trials for hepatitis C infection where available. A significant body of literature exists regarding oligonucleotide agonists of TLR-7 and other endosomal TLRs, which are not discussed here.[24] Finally, an outlook for this class of agents is presented.

10.2 Nucleoside Analogs and Prodrugs

Most of the published agonists of TLR-7 have chemical structures that are reminiscent of nucleoside bases. This is not unexpected, given that the native ligand is a nucleic acid (ssRNA). Some of the first TLR-7 agonists discovered were in fact nucleosides, although their mechanism of action was not realized until much later. Loxoribine[25] (**1**) and isatoribine[26] (**2**) are two examples of nucleoside analogs that originally attracted interest because they were found to have *in vivo* antiviral effects in animal infection models (Figure 10.1). Interestingly, they had little to no direct antiviral activity and the effects seemed to be indirect. Ultimately, the antiviral activity was determined to be due, for the most part, to IFN-α that was induced following *in vivo* administration of the compounds. It was subsequently discovered that the mechanism by which these compounds elicited IFN-α production was by agonist activity on TLR-7.[27]

Isatoribine was originally discovered at ICN,[26] and later became the basis of a discovery program at Anadys Pharmaceuticals. In a preliminary proof of concept clinical trial, intravenous administration of isatoribine to patients infected with HCV resulted in a dose-related reduction of viral levels.[28] The compound was administered at doses of 200, 400, 600 and 800 mg and a dose-related reduction in plasma HCV levels was observed, with maximum reduction at 800 mg. At the highest dose, 8/12 patients had a decrease of $>0.5\log_{10}$ in HCV RNA level. The decline in HCV level was correlated with an increase in plasma IFN-α and blood interferon-stimulated gene (ISG) levels. This clinical trial was informative; however, when dosed orally, isatoribine had poor bioavailability and was associated with gastrointestinal side effects.

Anadys subsequently undertook an effort to optimize isatoribine for oral delivery. The successful approach was the identification of an appropriate prodrug to enhance bioavailability, which also addressed the issue of gut side effects (Figure 10.2). The prodrug strategy was to block multiple sites on the structure.[29] The 2'- and 3'-hydroxyl groups of the ribose were converted to their acetate esters to increase the lipophilicity and therefore presumably increase the degree of absorption in the gastrointestinal tract. Additionally, the guanine oxo group at C4 was removed, further increasing lipophilicity and rendering the

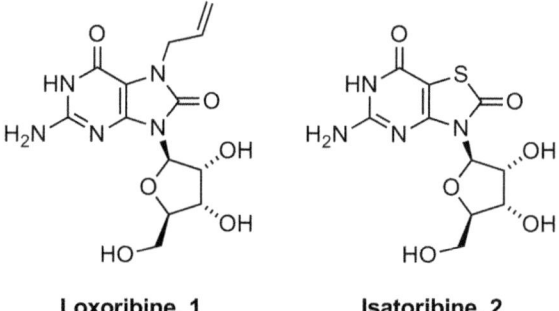

Loxoribine, 1   **Isatoribine, 2**

Figure 10.1 Examples of nucleoside TLR-7 agonists.

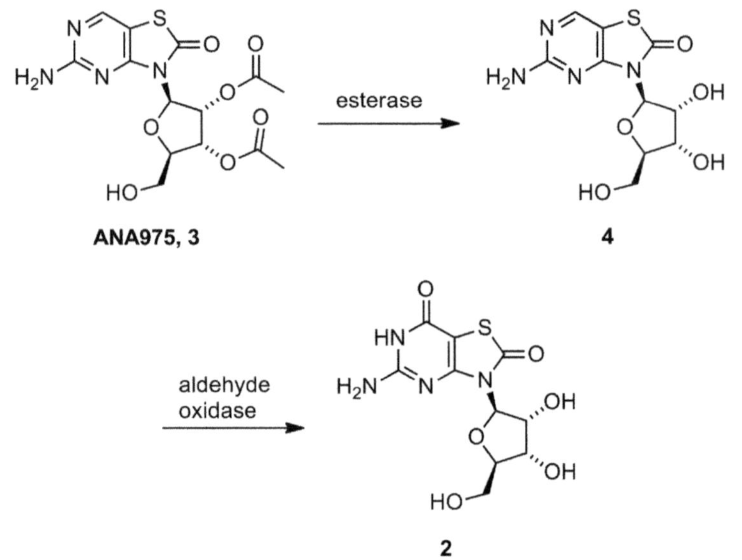

Figure 10.2   Conversion of ANA975, a prodrug of isatoribine, to the active species.

prodrug inactive as a TLR-7 agonist. This resulted in ANA975, **3**, which was further investigated clinically.[30]

In vitro experiments suggested that ANA975 would be metabolized to isatoribine *in vivo*. The mechanism by which this occurred involved removal of the acetate esters by an esterase to produce compound **4** and oxidation of C4 of the guanine-like base by aldehyde oxidase to ultimately yield isatoribine (**2**). This metabolic pathway was supported by the pharmacokinetics of ANA975 in HCV patients. Following oral administration, deacetylation and hepatic oxidative metabolism produced isatoribine, which is the predominant species detected in plasma, thus delivering the active TLR-7 agonist. Oral doses of 975 mg of ANA975 as a solution or powder-in-capsule produced plasma levels of isatoribine that were comparable to those from intravenous administration of isatoribine at 245 mg. Furthermore, the fact that ANA975 and its deacetylated analog **4** are both significantly less active as TLR-7 agonists than isatoribine results in minimization of any gastrointestinal side effects that may result from stimulating TLR-7 in the gut before absorption into systemic circulation and conversion to the active agonist.

ANA975 was evaluated clinically with once-per-day dosing, but the clinical trial was suspended owing to toxicology findings in animal species upon chronic daily dosing. A prodrug of an analog of isatoribine, ANA773, which is reported to have a similar structure to ANA975 and that utilized the same prodrug strategy, was subsequently taken into clinical development with an every-other-day oral dosing frequency.[31] In HCV patients, at the highest dose of 2000 mg, a mean reduction in serum HCV RNA of $1.26\log_{10}$ units was achieved, with a maximum reduction of $3.1\log_{10}$. The decrease in HCV RNA levels was correlated with dose-related increases in IFN markers. This study provided support to the viability of this prodrug strategy.

10.3 Imidazoquinoline Agonists

The imidazoquinoline class of TLR-7 agonists was discovered somewhat fortuitously at 3M Pharmaceuticals.[32] The original program was directed toward finding inhibitors of herpes virus replication that were based on the known anti-herpes adenine derivative **5** (Figure 10.3). A series of adenine analogs based on the imidazoquinoline structures **6** and **7** were synthesized and found to inhibit replication of herpes virus in a mouse model of infection, but had no direct antiviral activity *in vitro*. The antiviral activity of these compounds *in vivo* was determined to result from induction of IFN-α by a specific mechanism that was unknown at the time. The *in vitro* assay used for subsequent compound optimization was based on IFN-α induction in human PBMCs. Ultimately this class of compounds was validated to act as agonists of TLR-7, which is consistent with their IFN-induction property.[33] Some analogs were found to be selective for TLR-7 and others were dual agonists of both TLR-7 and TLR-8.

A general overview of the SAR of this series is summarized in Figure 10.4. The amine group at C4 is required for activity. Comparing the series in Figure 10.5, the parent amine **9** has a minimum effective concentration (MEC) for IFN-α induction of $0.5\,\mu g\,mL^{-1}$. Replacement of the amine with other groups, such as OH or H, in compounds **10** and **11**, respectively, among others, results in inactive compounds.

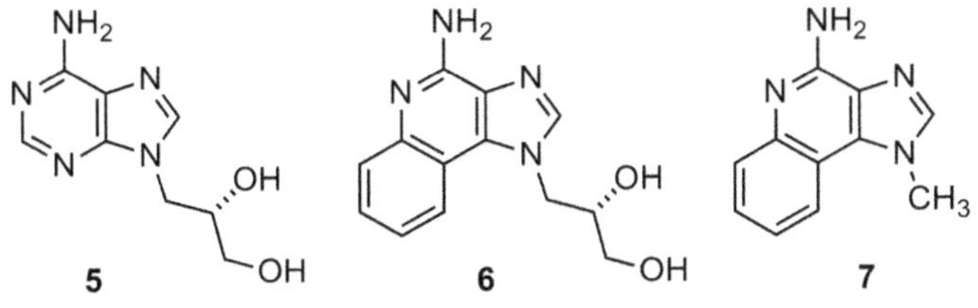

Figure 10.3 Basis for the discovery of imidazoquinolines.

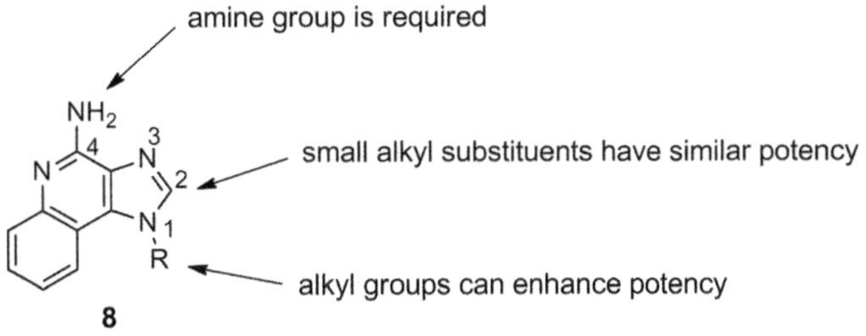

Figure 10.4 Overview of imidazoquinoline SARs.

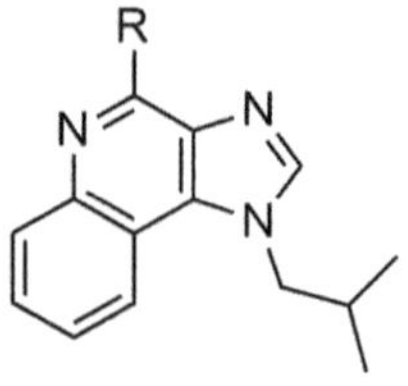

	R	MEC (μg/mL)
9	NH$_2$	0.5
10	OH	no induction
11	H	no induction

Figure 10.5 Effects of replacing the amino group.

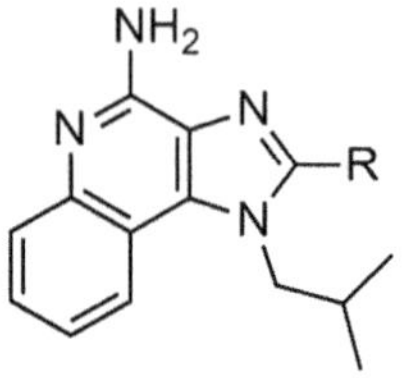

	R	MEC (μg/mL)
11	H	0.5
12	CH$_2$CH$_3$	0.05
13	n-Bu	0.01

Figure 10.6 Substitution of imidazoquinolines at C2.

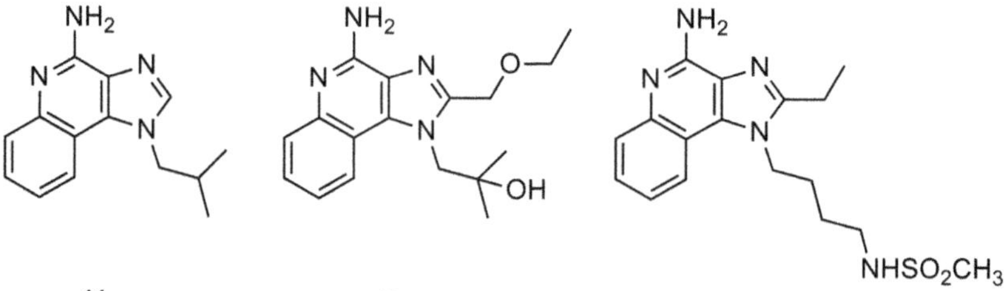

11	15	16
Imiquimod	R-848	R-852
	Resiquimod	PF-4878691

Figure 10.7 Imidazoquinolines evaluated clinically. Resiquimod and PF-4878691 were evaluated in HCV.

Compounds with small alkyl and substituted alkyl groups at C2 trended towards having greater potency relative to their unsubstituted parent compounds. Some representative examples are shown in Figure 10.6. Replacing the H at C2 of **11** with an ethyl group resulted in a 10-fold improvement in potency. A butyl group at C2 gave a 50-fold improvement in potency relative to H. Finally, a significant number of analogs with varying substituents at N1 had similar or improved potency.

This effort resulted in the discovery of imiquimod, **11** (Figure 10.7) now marketed under the trade-name Aldara for the treatment of genital warts associated with human papillomavirus infection.[34] Imiquimod is administered topically and, consistent with its mechanism of action, induces local production of IFN-α. Because it is topical, the antiviral effects also occur locally and the systemic exposure of the drug and systemic pharmacodynamic effects are minimal. Based on these IFN-α induction properties, 3M subsequently pursued this class of compounds further to identify interferon inducers for the treatment of HCV.

An early compound from this series that was investigated clinically for HCV is resiquimod, also known as R-848.[35] Resiquimod is more potent than imiquimod and is suited to oral administration based on pharmacokinetic parameters. Whereas imiquimod is selective for TLR-7, resiquimod is a dual agonist of TLR-7 and TLR-8.[36] Like TLR-7, TLR-8 also is known to sense ssRNA but the receptor has some differences in the nature of the response that it initiates. The distribution of TLR-8 is different: it is predominantly expressed in myeloid dendritic cells and monocytes, which do not produce high levels of interferon.[9] Additionally, TLR-8 signals primarily through NF-κB, resulting in the production of cytokines such as TNF-α that have a more pro-inflammatory signature than the antiviral interferon-type of signature of TLR-7. Hypotheses have been presented that suggest that TLR-8 agonist activity may result in more side effects than TLR-7 activity, although definitive data to support this notion are lacking.

Resiquimod was evaluated in a placebo-controlled Phase 2 clinical trial in HCV patients to establish proof of concept for antiviral activity.[35] It was administered orally twice per week for 4 weeks at 0.01 and 0.02 mg kg^{-1}. In this key clinical study, resiquimod was found to have antiviral efficacy, with mean viral RNA reductions of $\sim 0.4 \log_{10}$ and $1.3 \log_{10}$ units for the 0.01 and 0.02 mg kg^{-1} doses, respectively, with maximum reductions occurring at about 24 h. At the higher dose, 1/11 subjects achieved a viral load reduction of $>3 \log_{10}$ units. The antiviral effects were transient and the virus levels rebounded after 24 h. The 0.01 mg kg^{-1} dose was well tolerated, but some subjects discontinued treatment at the higher dose. Adverse events were dose related and correlated with serum resiquimod levels and serum levels of IFN-α. Based on this study, resiquimod appeared to have a steep dose response both for antiviral activity and for side effects.

In order to address some of the issues with resiquimod, a second-generation compound, R-852, which later became PF-4878691 (**12**, Figure 10.7), was advanced to clinical trials by Pfizer.[37] PF-4878691 is selective for TLR-7, which would address any detrimental issues that might be associated with TLR-8 activity. In order to model the therapeutic window of PF-4878691, *in vitro* experiments were conducted in which human PBMCs were stimulated with PF-4878691 and the supernatants were evaluated for cytokine levels and antiviral activity in the HCV replicon. These experiments suggested that there would be a separation between antiviral effects and the induction of proinflammatory cytokines and served to guide the dose levels in clinical trials.

In a Phase 1 clinical study, PF-4878691 was administered orally to healthy volunteers at doses of 3, 6 and 9 mg twice per week for 2 weeks. PF-4878691 induced biomarkers of immune and IFN responses, such as mRNA level for the interferon-stimulated gene OAS-1. A surrogate biomarker of antiviral activity was the *in vitro* inhibitory activity of serum from these subjects against the HCV replicon. Serum from subjects administered 3 mg did not have activity greater than the placebo controls. There was a dose-dependent increase in replicon activity in serum from subjects in the 6 and 9 mg groups. The conclusion was that the subjects in the 6 and 9 mg doses had significant pharmacologic activation of TLR-7. Subjects in all groups experienced adverse events and one subject in the 9 mg group experienced serious adverse events. The side effects were generally consistent with flu-like symptoms that result from administration of interferon and PEG-IFN. Given that the serum antiviral activity was only observed at doses that also resulted in adverse events, the investigators concluded that the therapeutic window would provide a substantial challenge to PF-4878691 for use in HCV therapy. An unanswered question from this study is whether doses of PF-4878691 that are below the threshold for systemic pharmacologic activity and therefore below levels that cause adverse events would have beneficial antiviral effects in HCV patients.

10.4   8-Oxopurine and 8-Oxodeazapurine Agonists

With the knowledge that the imidazoquinoline interferon inducers identified by 3M bear a resemblance to purine nucleoside bases, Dainippon Sumitomo Pharmaceuticals together with Gifu University sought to identify compounds that induce interferon by screening purine and pyrimidine derivatives in their compound library. From this exercise, a class of 8-oxopurine derivatives (**17a**), also referred to as their alternative tautomer 8-hydroxyadenines (**17b**), was identified (Figure 10.8).[38,39] The initial compounds were weak inducers of IFN-α *in vitro* in mouse splenocytes. This series became the basis of an optimization program to reach potent and selective TLR-7 agonists.

Initially, a systematic exploration of the SARs of substituents around the 8-oxopurine core was undertaken using an assay that measured production of IFN-α in mouse splenocytes as a function of compound concentration.[38] The carbonyl group at C8 was found to be important for activity and was a key feature of future analogs. Small alkyl groups at N9 were found to improve potency, as illustrated in Figure 10.9. The unsubstituted parent **18** was inactive at or below 10 μM, whereas the *n*-butyl (**19**) and benzyl substituents (**20**) had measurable interferon induction activity with an MEC of 10 μM. For comparison, the MEC for imiquimod in this assay was 1 μM. A benzyl group at N9 was generally found to provide good activity, so **20** was used as the basis for the next round of optimization.

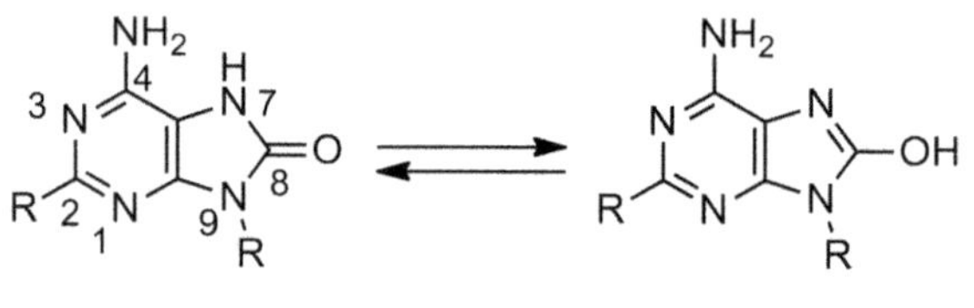

Figure 10.8   The 8-oxopurine/8-hydroxyadenine series identified by Dainippon Sumitomo Pharmaceuticals.

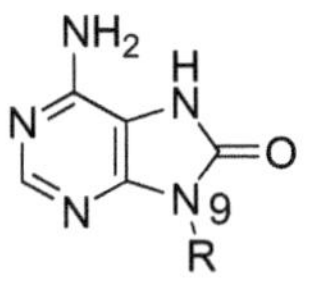

	R	MEC (μM)
18	H	>10
19	n-Bu	10
20	CH₂Ph	10

Figure 10.9 SAR of the N9 substituent.

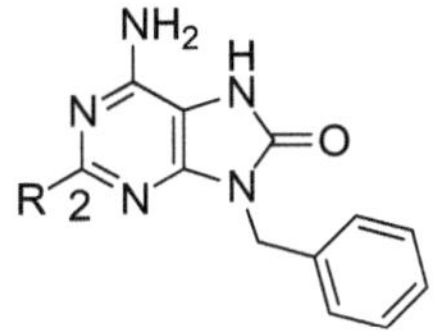

	R	MEC (μM)
20	H	10
21	CH₃CH₂	1
22	n-Bu	0.03
23	Ph	10
24	n-PrO	0.01
25	n-BuO	0.001
26	n-C₅H₁₁O	0.01
27	n-BuNH	0.1
28	n-BuS	0.01

Figure 10.10 SAR of the C2 substituent.

Some selected examples to illustrate the SAR at C2 are shown in Figure 10.10. Small alkyl-substituted compounds were more potent than the C2-unsubstituted parent **20**, with an apparent optimum size of the group. To illustrate, the ethyl-substituted analog **21** had MEC = 1 μM, which was a 10-fold improvement over **20**. The potency further improved with the larger *n*-butyl group (**22**), resulting in MEC = 0.03 μM. However, potency was substantially lost when the substituent was phenyl (**23**, MEC = 10 μM). Alkoxy groups at C2 were more potent than alkyl groups, with the optimum group being *n*-butoxy (**25**, MEC = 0.001 μM). Smaller or larger groups such as *n*-propoxy (**24**) and *n*-pentoxy (**26**) were less potent, with MEC values of 0.01 μM each. Apparently, the *n*-butoxy group is near optimum in chain length and size. Other heteroatoms to link to the *n*-butyl group were also tolerated, albeit with slightly less potency. The *n*-butylamino analog **27** had MEC = 0.1 μM and the *n*-butylthio analog **28** had MEC = 0.01 μM.

In vivo pharmacologic activity was evaluated by administering compounds **25** and **28** orally to mice, followed by measurement of serum levels of IFN-α. Both compounds dose-dependently induced IFN-α at dose levels of 0.1–10 mg kg⁻¹, with a plateau at 3 and 10 mg kg⁻¹. The levels of IFN-α in mice were >250 U mL⁻¹ at doses of 0.3 and 1.0 mg kg⁻¹, respectively, for **28** and **25**, which was stated to be higher than levels required for HCV antiviral activity in humans. Correlations of mouse and human IFN activity used in this analysis were not presented.

The Dainippon Sumitomo group postulated that some of the side effects seen from oral dosing of the 3M imiquimod analogs in clinical trials might be due to off-target activity.[40] For example, some of the imidazoquinoline analogs led to emesis, which was postulated to be due to off-target activity in the CNS. Therefore, a focus of their medicinal chemistry optimization effort was to address this issue and led to an *in vivo* comparison of **22** with imiquimod.[41] It was suggested that an advantage of the oxopurine chemotype over the imidazoquinoline series is its polarity, which might limit penetration into the CNS. *In vivo* pharmacodynamic experiments in mice found that **22**, when dosed orally at 3 mg kg^{-1}, induced significantly more IFN-α at the 2 h peak post-dose (1476 U mL^{-1}) than imiquimod at the same dose (67 U mL^{-1}). This is consistent with **22** being ∼30-fold more potent than imiquimod *in vitro*, but could also be potentially due to better oral absorption of **22**. Pharmacokinetic comparisons were not disclosed. To investigate selectivity and side effects, these two compounds were further compared in a ferret model of emesis. Oral administration of **22** at 10 mg kg^{-1} did not result in emesis in any of the five animals in this group, but oral imiquimod at the same dose resulted in emesis in four of five animals. The conclusion was that the emesis was not due to interferon and that the oxopurine series might have a more favorable side effect profile than imidazoquinolines.

Additional *in vivo* optimization of this series led to the identification of some key compounds. In particular, SM-360320 (**29**, Figure 10.11) demonstrated potent *in vivo* interferon induction in mice at doses as low as 0.03 mg kg^{-1}.[42–44] In rat pharmacokinetic studies, this compound had good oral absorption and bioavailability (40%) and was suitable for further *in vivo* studies. Combining favorable pharmacokinetics and pharmacodynamics with low emetic properties led to SM-276001 (**30**, Figure 10.11). *In vitro* in mouse splenoytes, SM-276001 had an MEC for interferon induction of 0.003 µM. *In vivo* in cynomolgus monkeys, SM-276001 induced interferon in a dose-related manner at doses as low as 0.1 mg kg^{-1}, whereas resiquimod showed no appreciable interferon induction below a dose of 1 mg kg^{-1}. Finally, no emesis was observed in the ferret model up to doses of 30 mg kg^{-1}, whereas resiquimod caused emesis in all animals at 3 mg kg^{-1}.

A program at Pfizer sought to improve on the oxopurine-based TLR-7 agonists developed at Dainippon Sumitomo with a series that was less purine-like and therefore had less potential risk of polypharmacology. Towards this end, a systematic investigation of the oxopurine core scaffold was undertaken

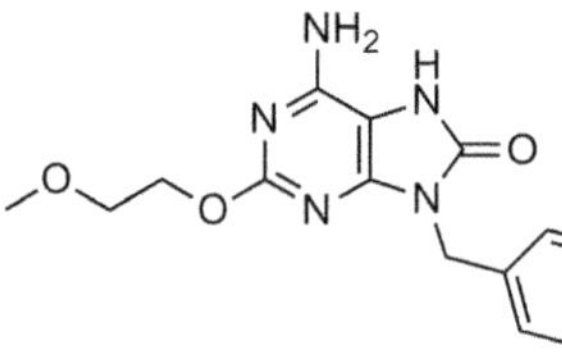

29
SM-360320

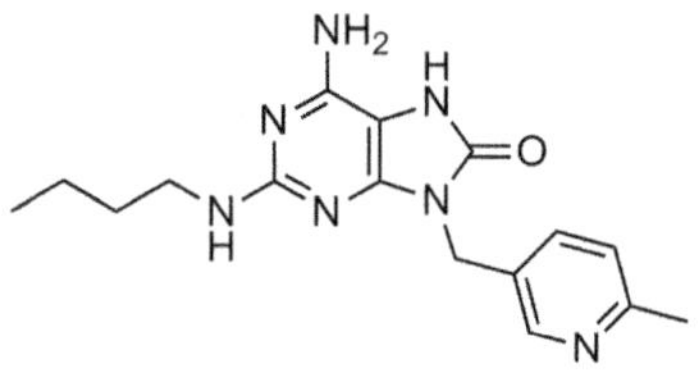

30
SM-276001

Figure 10.11   Optimized compounds in the 8-oxopurine series.

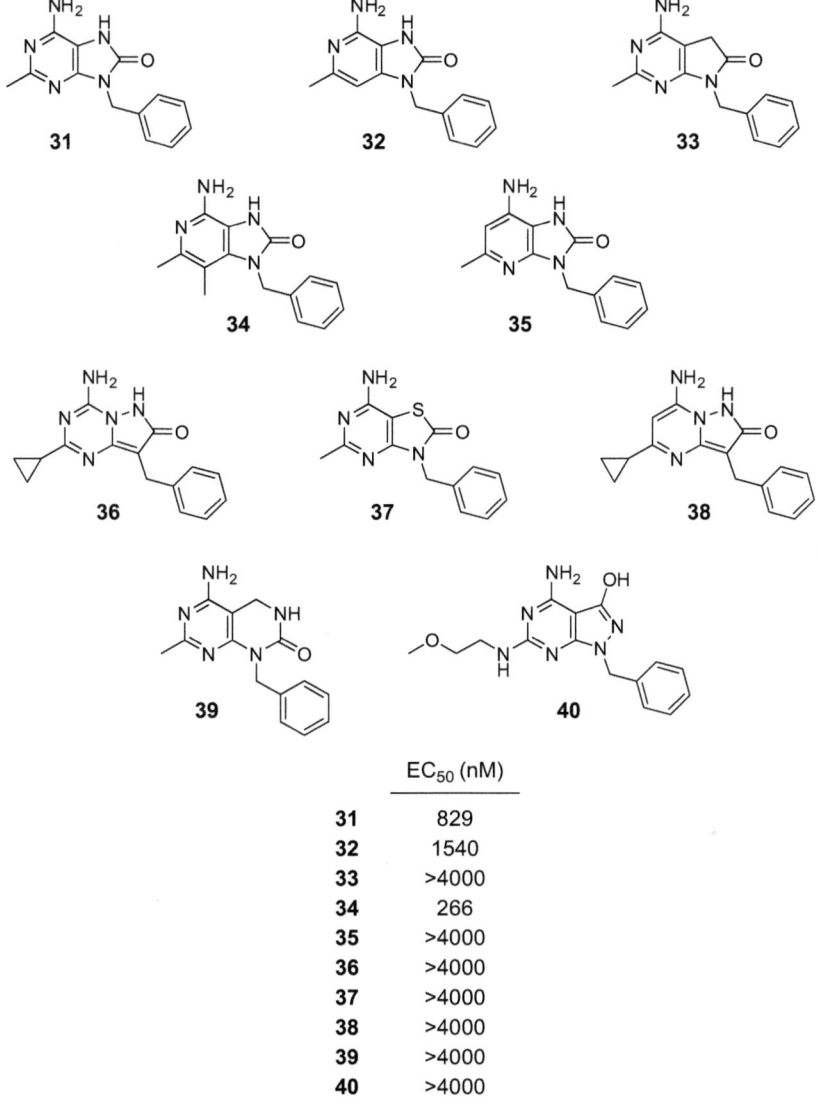

	EC$_{50}$ (nM)
31	829
32	1540
33	>4000
34	266
35	>4000
36	>4000
37	>4000
38	>4000
39	>4000
40	>4000

Figure 10.12   Evaluation of changes to the core scaffold of 8-oxopurines.

with targeted modifications.[45] Figure 10.12 summarizes some representative changes to the core that were made. To compare more easily, the substituent at the purine C2 position was maintained as a methyl or other small substituent. The assay involved stimulation of human PBMCs and measuring the activity of the supernatants in an HCV replicon system to determine EC$_{50}$ values of the tested compounds. Presumably a significant amount, if not all, of the antiviral activity in the PBMC supernatants was due to IFN induced by activation of TLR-7. Of the core modifications evaluated, only the deazapurine cores **32** and

34 were active as TLR-7 agonists. The potency was generally similar to or slightly better than the oxopurine parent **31**. Although **34**, which has an additional methyl group, was more active, for synthetic ease **32** was further optimized.

The next phase of optimization addressed the issue of metabolic stability.[46] As shown in Figure 10.13, the parent core was reasonably stable based on intrinsic *in vitro* clearance by human liver microsomes (HLMs). Compound **41** is illustrative; however, this compound has an EC_{50} >4000 nM, indicating that a substituent at C2 is required for TLR-7 agonist activity. Methyl and small linear alkyl groups at C2 did not introduce metabolic liabilities relative to H, as illustrated by **32** and **42**. However branched alkyl substituents were susceptible to metabolism, such as the isopropyl in **43**. In this series, a trifluoromethyl group (**44**) gave the best combination of potency ($EC_{50} = 102$ nM) and metabolic stability, despite a somewhat high logD (3.3).

Compound **44** became PF-4171455 and was the subject of further investigation. In rat pharmacokinetic studies, **44** had a clearance of $33 \, \mu L \, min^{-1} \, kg^{-1}$ and an oral bioavailability as high as 71% when administered as a nano-suspension. When administered intravenously to dogs, the clearance was $6 \, \mu L \, min^{-1} \, kg^{-1}$. A prediction of the human half-life based on allometric scaling was 3–7 h. Fitting to a pharmakokinetics–pharmacodynamics model[47] indicated that **44** would generate efficacious levels of IFN-α in HCV patients at daily doses less than 50 mg.

PF-4171455 (**44**) met complications during development due to formulation problems associated with poor solubility. A series of analogs was explored to improve this property.[48] The initial strategy was to lower the lipophilicity by adding polar groups in the N9 substituent. Towards this end, a number of heterocyclic replacements for the phenyl ring were evaluated (Figure 10.14). Of several isomeric methylpyridines reported, **46** had the best balance of potency (50 nM) and solubility ($22 \, \mu g \, mL^{-1}$). Isomers **45** and **47** were inferior in both of

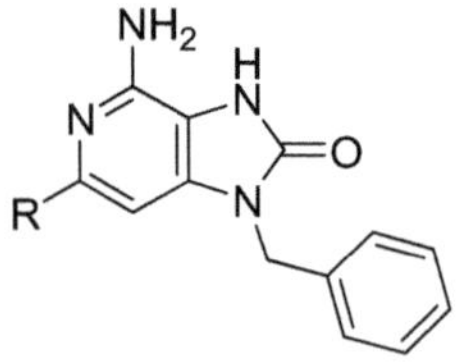

	R	EC_{50} (nM)	HLM (mL/min/mg)
41	H	> 4000	< 7
32	CH_3	1540	< 7
42	*i*-Pr	304	< 7
43	*n*-Pr	480	14
44	CF_3	102	< 7

Figure 10.13   SARs of the C2 substituent of 8-oxodeazapurines. Compound **44** is PF-4171455.

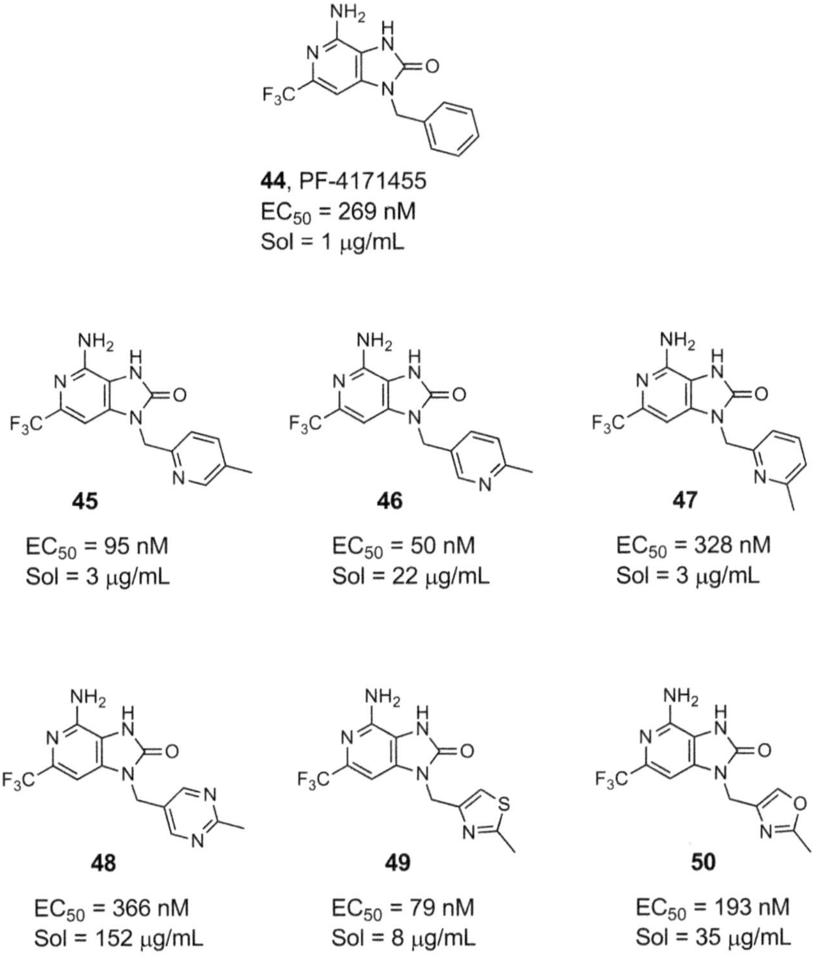

44, PF-4171455
EC$_{50}$ = 269 nM
Sol = 1 µg/mL

45

EC$_{50}$ = 95 nM
Sol = 3 µg/mL

46

EC$_{50}$ = 50 nM
Sol = 22 µg/mL

47

EC$_{50}$ = 328 nM
Sol = 3 µg/mL

48

EC$_{50}$ = 366 nM
Sol = 152 µg/mL

49

EC$_{50}$ = 79 nM
Sol = 8 µg/mL

50

EC$_{50}$ = 193 nM
Sol = 35 µg/mL

Figure 10.14 Replacement of the 8-oxodeazapurine benzyl group with more polar groups.

these properties. Introducing additional heteroatoms into the ring, as in pyrimidine **48**, led to increased solubility, but potency was negatively affected. Five-membered heterocycle replacements for the phenyl group, exemplified by **49** and **50**, were generally inferior to pyridines in either solubility, potency or both.

Rat pharmacokinetics studies of **46** demonstrated that it had high oral absorption (100%) and bioavailability (78%), but the *in vivo* half-life of 0.5 h was low relative to predictions due to metabolism of the pyridine ring by aldehyde oxidase.

To improve solubility and potency further and potentially address the aldehyde oxidase liability, the C2 position was investigated (Figure 10.15). The propensity for oxidation by aldehyde oxidase was determined by measuring the compound half-life in rat cytosol as a surrogate assay. A short

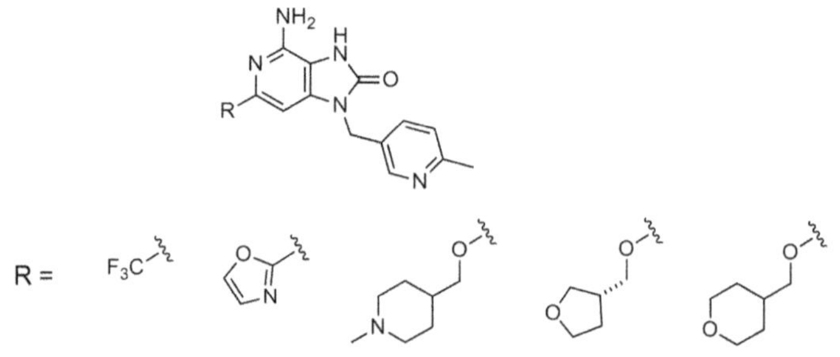

	46	51	52	53	54
EC_{50}(nM)	50	102	>4000	474	99
Sol (μg/mL)	22	66	--	--	17
rat cytosol $t_{1/2}$ (min)	46	46	--	--	>1000

Figure 10.15 SARs of the C2 position of 8-oxodeazapurines.

half-life was suggestive of aldehyde oxidase-mediated metabolism. The strategy to address these issues was to evaluate more polar groups at C2. Replacement of the lipophilic trifluoromethyl group by an oxazole (**51**) led to similar potency and a gain in solubility, but no improvement in aldehyde oxidase substrate liability. Introduction of charged polar groups, as in the tertiary amine of **52**, led to a loss of activity. Cyclic ethers such as **53** were found to retain good potency. The best member of this series was **54**, which contains a cyclic tetrahydropyran. This compound was comparable to the trifluoromethyl starting point **46** regarding potency and solubility, but was significantly more stable in rat cytosol. The half life of >1000 min indicated that **54** was not an appreciable substrate for aldehyde oxidase. Another favorable property of **54** was that it had less risk of polypharmacology as represented by the absence of inhibition activity toward adenosine receptors A1 and A2. Other analogs in this series had measurable, albeit low-level, activity towards these two receptors.

The rat pharmacokinetics of **54** were favorable and comparable to those of **44**. The fraction absorbed (40%) and bioavailability (17%) were somewhat lower and the half-life (0.4 h) was comparable. Based on these parameters and the improved potency, **54** was estimated to have an efficacious clinical dose for induction of IFN-α in HCV patients of <30 mg once per day. Compound **54** therefore satisfied the criteria of having substantially better solubility and thus better pharmaceutics properties than their earlier development compound **44** and accordingly was stated to be a back-up candidate. Additionally, issues of polypharmacology were successfully addressed.

10.5 Conclusion and Outlook

TLR-7 represents an attractive target for the identification of agonists that might possess utility to treat chronic viral hepatitis infection. Several programs have advanced into development and human clinical trials for HCV infection. It has been demonstrated that this class of compounds can induce IFN-α and ISGs at levels appropriate to treat HCV clinically, with demonstrated antiviral effects.

Given the recent dramatic advances in the development of oral direct-acting antiviral agents for HCV, the long-term roles for Pegasys, and by extension interferon-inducing agents such as TLR-7 agonists, in HCV therapy are not clear. A significant hurdle to this class finding an eventual role in chronic HCV therapy could be tolerability at the doses needed for long-term efficacy and SVR. A question still unanswered is whether a TLR-7 agonist at a dose that induces IFN-α at a level lower than that achieved with Pegasys would have acceptable clinical efficacy, especially in combination with current or future direct-acting antiviral agents. If so, and if an immunomodulatory component is needed for some patient populations to achieve SVR, then TLR-7 agonists could have utility.

A much larger and clearer role for TLR-7 agonists could potentially be found in the treatment of chronic HBV infection. Current treatments based on antiviral nucleoside analogs often require life-long therapy to maintain viral suppression and clinical benefit and have very low rates of HBs antigen seroconversion. Treatment of HBV infections with Pegasys is generally for 48 weeks and thus is finite in duration. Nevertheless, this treatment duration is demanding, especially in the light of the poor tolerability of the therapy and the seroconversion rate is suboptimal. A substantial improvement of HBs antigen seroconversion versus that which can be achieved with current therapies would offer significant benefits for HBV patients. If an oral TLR-7 agonist could improve seroconversion rates and lead to a treatment duration that is finite, especially with a better tolerability profile than Pegasys, then a beneficial treatment option to patients may result from this class of compounds.

References

1. A. Thomas, C. Laxton, J. Rodman, N. Myangar, N. Horscroft and T. Parkinson, Investigating Toll-like receptor agonists for potential to treat hepatitis C virus infection, *Antimicrob. Agents Chemother.*, 2007, **51**, 2969.
2. M. Isogawa, M. D. Robek, Y. Furuichi and F. V. Chisari, Toll-like receptor signaling inhibits hepatitis B virus replication *in vivo*, *J. Virol.*, 2005, **79**, 7269–7272.
3. S. Akira, K. Takeda and T. Kaisho, Toll-like receptors: critical proteins linking innate and acquired immunity, *Nat. Immunol.*, 2001, **2**, 675–680.
4. T. Kawai and S. Akira, The role of pattern-recognition receptors in innate immunity: update on Toll-like receptors, *Nat. Immunol.*, 2010, **11**, 373–384.
5. C. A. Janeway and R. Medzhitoov, Innate immune recognition, *Annu. Rev. Immunol.*, 2002, **20**, 197.
6. K. Takeda, T. Kaisho and S. Akira, Toll-like receptors, *Annu. Rev. Immunol.*, 2003, **21**, 335–378.

7. B. Beutler, Inferences, questions and possibilities in Toll-like receptor signalling, *Nature*, 2004, **430**, 257.
8. L. A. O'Neill, After the Toll rush, *Science*, 2004, **303**, 1481.
9. V. Hornung, S. Rothenfusser, S. Britsch, A. Krug, B. Jahrsörfer, T. Giese, S. Endres and G. Hartmann, Quantitative expression of Toll-like receptor 1–10 mRNA in cellular subsets of human peripheral blood mononuclear cells and sensitivity to CpG oligonucleotides, *J. Immunol.*, 2002, **168**, 4531–4537.
10. K. A. Zarember and P. J. Godowski, Tissue expression of human Toll-like receptors and differential regulation of Toll-like receptor mRNAs in leukocytes in response to microbes, their products and cytokines, *J. Immunol*, 2002, **168**, 554–561.
11. M. Nishimura and S. Naito, Tissue-specific mRNA expression profiles of human Toll-like receptors and related genes, *Biol. Pharm. Bull.*, 2005, **28**, 886–892.
12. M. Colonna, G. Trenchieri and Y.-J. Liu, Plasmacytoid dendritic cells: sensing nucleic acids in viral infection and autoimmune disease, *Nat. Immunol.*, 2004, **5**, 1219–1226.
13. T. Nishiya and A. L. DeFranco, Ligand-regulated chimeric receptor approach reveals distinct subcellular localization and signaling properties of the Toll-like receptors, *J. Biol. Chem.*, 2004, **279**, 19008.
14. J. M. Lund, L. Alexopoulou, A. Sato, M. Karow, N. C. Adams, N. W. Gale, A. Iwasaki and R. A. Flavell, Recognition of single-stranded RNA viruses by Toll-like receptor 7, *Proc. Natl. Acad. Sci. U. S. A.*, 2004, **101**, 5598–5603.
15. L. A. J. O'Neill and A. G. Bowie, The family of five: TIR domain-containing adaptors in Toll-like receptor signaling, *Nat. Rev. Immunol.*, 2007, **7**, 353–364.
16. E. Gane, Future perspectives: toward interferon-free regimens for HCV, *Antivir. Ther.*, 2012, **17**, 1201–1210.
17. A. M. Di Bisceglie and J. H. Hoofnagle, Optimal therapy of hepatitis C, *Hepatology*, 2002, **36**, S121–S127.
18. Hepatitis C, *Nature*, 2011, **474**(7350), Suppl., S1–S48.
19. H. Kwon and A. S. Lok, Hepatitis B therapy, *Nat. Rev. Gastroenterol. Hepatol*, 2011, **8**, 275–284.
20. Y.-F. Liaw and C.-M. Chu, Hepatitis B virus infection, *Lancet*, 2009, **373**, 582–592.
21. G. V. Papatheodoridis, S. Manolakopoulos, G. Dusheiko and A. J. Archimandritis, Therapeutic strategies in the management of patients with chronic hepatitis B virus infection, *Lancet Infect. Dis.*, 2008, **8**, 167–178.
22. B. Rehermann and M. Nascimbeni, The immunology of hepatitis B virus and hepatitis C virus infection, *Nat. Rev. Immunol.*, 2005, **5**, 215–229.
23. M. Spyvee, L. D. Hawkins and S. T. Ishizawa, Modulators of Toll-like receptor (TLR) signaling, *Annu. Rep. Med. Chem.*, 2010, **45**, 191–207.
24. W Barchet, V. Wimmenauer, M. Schlee and G. Hartman, *Curr. Opin. Immunol.*, 2008, **20**, 389–395.

25. A. B. Reitz, M. G. Goodman, B. L. Pope, D. C. Argentieri, S. C. Bell, L. E. Burr, E. Chourmouzis, J. C. Come, J. H. Goodman, D. H. Klaubert, B. E. Maryanoff, M. E. McDonnel, M. S. Rampulla, M. R. Schott and R. Chen, Small molecule immunostimulants. synthesis and activity of 7,8-disubstututed guanosines and structurally related compounds, *J. Med. Chem.*, 1994, **37**, 3561–3578.

26. K. Nagahara, J. D. Anderson, G. D. Kini, N. K. Dalley, S. B. Larson, D. F. Smee, A. Jin, B. S. Sharma, W. B. Jolley, R. K. Robins and H. B. Cottam, Thiazolo[4,5-*d*]pyrimidine nucleosides. The synthesis of certain 3-β-D-ribofuranosylthiazolo[4,5-*d*]pyrimidines as potential immuno-therapeutic agents, *J. Med. Chem.*, 1990, **33**, 407–415.

27. J. Lee, T. Chuang, V. Redecke, L. She, P. M. Pitha, D. A. Carson, E. Raz and H. B. Cottam, Molecular basis for the immunostimulatory activity of guanine nucleoside analogs: activation of Toll-like receptor 7, *Proc. Natl. Acad. Sci. U. S. A.*, 2003, **100**, 6646.

28. Y. Horsmans, T. Berg, J.-P. Desager, T. Mueller, E. Schott, S. Fletcher, K. R. Steffy, L. A. Bauman, B. M. Kerr and D. R. Averett, Isatoribine, an agonist of TLR7, reduces plasma virus concentration in chronic hepatitis C infection, *Hepatology*, 2005, **42**, 724–731.

29. S. Fletcher, K. Steffy and D. Averett, Masked oral prodrugs of Toll-like receptor 7 agonists: a new approach for the treatment of infectious disease, *Curr. Opin. Invest. Drugs*, 2006, **7**, 702–708.

30. A. X. Xiang, S. E. Webber, B. M. Kerr, E. J. Ruiden, J. R. Lennox, G. J. Haley, T. Wang, J. S. Ng, M. R. Herbert, D. L. Clark, V. N. Bahn, W. Li, S. P. Fletcher, K. R. Steffy, D. M. Bartkowski, L. I. Kirdovsky, L. A. Bauman and D. R. Averett, Discovery of ANA975: an oral prodrug of the TLR-7 agonist isatoribine, *Nucleosides Nucleotides Nucleic Acids*, 2007, **26**, 625–640.

31. J. F. Bergmann, J. de Bruijne, D. M. Hotho, R. J. de Knegt, A. Boonstra, C. J. Weegink, A. A. van Vliet, J. van de Wetering, S. P. Fletcher, L. A. Bauman, M. Rahimy, J. R. Appleman, J. L. Freddo, H. L. A. Janssen and H. W. Reesink, Randomized clinical trial: anti-viral activity of ANA773, an oral inducer of endogenous interferons acting via TLR7, in chronic HCV, *Aliment. Pharmacol. Ther.*, 2011, **34**, 443–453.

32. J. F. Gerster, K. J. Lindstrom, R. L. Miller, M. A. Tomai, W. Birmachu, S. N. Bomersine, S. J. Gibson, L. M. Imbertson, J. R. Jacobson, R. T. Knafla, P. V. Maye, N. Nikolaides, F. Y. Oneyemi, G. J. Parkhurst, S. E. Pecore, M. J. Reiter, L. S. Scribner, T. L. Testerman, N. J. Thompson, T. L. Wagner, C. E. Weeks, J.-D. Andre, D. Lagain, Y. Bastard and M. Lupu, Synthesis and structure–activity relationships of 1*H*-imidazo[4,5-*c*]quinolines that induce interferon, *J. Med. Chem.*, 2005, **48**, 3481–3491.

33. H. Hemmi, T. Kaisho, O. Takeuchi, S. Sato, H. Sanjo, K. Hoshino, T. Horiuchi, H. Tomizawa, K. Takeda and S. Akira, Small anti-viral compounds activate immune cells via the TLR7 MyD88-dependent signaling pathway, *Nat. Immunol.*, 2002, **3**, 196–200.

34. S. M. Garland, Imiquimod, *Curr. Opin. Infect. Dis.*, 2003, **16**, 85–89.
35. P. J. Pockros, D. Guyader, H. Patton, M. J. Tong, T. Wright, J. McHutchison and T. C. Ment, Oral resiquimod in chronic HCV infections: safety and efficacy in 2 placebo-controlled, double-blind phase IIa studies, *J. Hepatol.*, 2007, **47**, 174–182.
36. M. Jurk, F. Heil, J. Vollmer, C. Schetter, A. M. Kreig, H. Wagner, G. Lipford and S. Bauer, Human TLR7 or TLR8 independently confer responsiveness to the antiviral compound R-848, *Nat. Immunol.*, 2002, **3**, 499.
37. M. D. Fidock, B. E. Souberbielle, C. Laxton, J. Rawal, O. O'Delpuech-Adams, T. P. Corey, P. Colman, V. Kumar, J. B. Cheng, K. Wright, S. Srinivasan, K. Rana, C. Craig, N. Horscroft, M. Perros, M. Westby, R. Webster and E. van der Ryst, The innate immune response, clinical outcomes and *ex vivo* HCV antiviral efficacy of a TLR7 agonist (PF-4878691), *Clin. Pharmacol. Ther.*, 2011, **89**, 821–829.
38. K. Nirota, K. Kazaoka, I. Niimoto, H. Kumihara, H. Sajiki, Y. Isobe, H. Takaku, M. Tobe, H. Ogita, T. Ogino, S. Ichii, A. Kurimoto and H. Kawakami, Discovery of 8-hydroxyadenines as a novel type of interferon inducer, *J. Med. Chem.*, 2002, **45**, 5419.
39. K. Hirota, K. Kazaoka and H. Sajiki, Synthesis and biological evaluation of 2,8-disubstutited 9-benzyladenines: discovery of 8-mercaptoadenines as potent interferon inducers, *Bioorg. Med. Chem.*, 2003, **11**, 2715–2722.
40. N. L. Strominger, R. Brady, G. Gullikson and D. O. Carpenter, Imiquimod-elicited emesis is mediated by the area postrema, but not by direct neuronal activation, *Brain Res. Bull.*, 2001, **55**, 445.
41. Y. Isobe, M. Tobe, H. Ogita, A. Kurimoto, T. Ogino, H. Kawakami, H. Takaku, H. Sajiki, K. Hirota and H. Hayashi, Synthesis and structure–activity relationships of 2-substituted-8-hydroxyadenine derivatives as orally available interferon inducers without emetic side effects, *Bioorg. Med. Chem.*, 2003, **11**, 3641–3647.
42. A. Kurimoto, T Ogino., S. Ichii, Y. Isobe, M. Tobe, H. Ogita, H. Takaku, H. Sajiki, K. Hirota and H. Kawakami, Synthesis and evaluation of 2-substituted 8-hydroxyadenines as potent interferon inducers with improved oral bioavailabilities, *Bioorg. Med. Chem.*, 2004, **12**, 1091–1099.
43. A. Kurimoto, T. Ogino, S. Ichii, Y. Isobe, M. Tobe, H. Ogita, H. Takaku, H. Sajiki, K. Hirota and H. Kawakami, Synthesis and structure–activity relationships of 2-amino-8-hydroxyadenines as orally active interferon inducing agents, *Bioorg. Med. Chem.*, 2003, **11**, 5501–5508.
44. Y. Isobe, A. Kurimoto, M. Tobe, K. Hashimoto, T. Nakamura, K. Norimura, H. Ogita and H. Takaku, Synthesis and biological evaluation of novel 9-substituted-8-hydroxyadenine derivatives as potent interferon inducers, *J. Med. Chem.*, 2006, **49**, 2088–2095.
45. D. C. Pryde, T.-D. Tran, P. Jones, G. C. Parsons, G. Bish, F. M. Adam, M. C. Smith, D. S. Middleton, N. N. Smith, F. Calo, D. Hay, M. Paradowski, K. J. W. Proctor, T. Parkinson, C. Laxton, D. N. A. Fox, N. J. Horscroft, G. Ciaramella, H. M. Jones, J. Duckworth, N. Benson,

A. Harrison and R. Webster, The discovery of a novel prototype small molecule TLR7 agonist for the treatment of hepatitic C virus infection, *Med. Chem. Commun.*, 2011, **2**, 185–189.

46. P. Jones, D. C. Pryde, T.-D. Tran, F. M. Adam, G. Bish, F. Calo, G. Ciaramella, R. Dixon, J. Duckworth, D. N. A. Fox, D. A. Hay, J. Hitchin, N. Horscroft, M. Howard, C. Laxton, T. Parkinson, G. Parsons, K. Proctor, M. C. Smith, N. Smith and A. Thomas, Discovery of a highly potent series of TLR7 agonists, *Bioorg. Med. Chem. Lett.*, 2011, **21**, 5939–5943.

47. N. Benson, J. de Jongh, J. D. Duckworth, H. M. Jones, H. E. Pertinez, J. K. Rawal, T. J. van Steeg and P. H. Van der Graaf, Pharmacokinetic–pharmacodynamic modeling of alpha interferon response induced by a Toll-like receptor 7 agonist in mice, *Antimicrob. Agents Chemother.*, 2010, **54**, 1179.

48. T.-D. Tran, D. C. Pryde, P. Jones, F. M. Adam, N. Benson, G. Bish, F. Calo, G. Ciaramella, R. Dixon, J. Duckworth, D. N. A. Fox, D. A. Hay, J. Hitchin, N. Horscroft, M. Howard, I. Gardner, H. M. Jones, C. Laxton, T. Parkinson, G. Parsons, K. Proctor, M. C. Smith, N. Smith and A. Thomas, Design and optimization of orally active TLR7 agonists for the treatment of hepatitis C virus infection, *Bioorg. Med. Chem. Lett.*, 2011, **21**, 2389–2393.

CHAPTER 11

Optimization of Cyclophilin Inhibitors for Use in Antiviral Therapy

MICHAEL PEEL* AND ANDREW SCRIBNER

SCYNEXIS Inc., Research Triangle Park, NC 27709, USA
*Email: mike.peel@scynexis.com

The discovery that cyclophilin A (CypA) is a requisite binding protein for the immunosuppressive activity of cyclosporine A (CsA) served to place CypA firmly in the sphere of immunology research. The demonstration that the CsA–CypA binary complex inhibits the calcium-dependent phosphatase activity of calcineurin (CaN), leading to inhibition of key activators of T-cells, provides a compelling explanation of the immunosuppressive activity of CsA.[1] However, an expanding range of activity demonstrated by CsA, including mitochondrial function, cell death, chemotaxis and motility, suggested mechanisms that were not easily ascribed to CaN inhibition.[2] The discovery of new CsA analogs that retained potent binding to CypA but prevented ternary complex formation with CaN allowed the biology of cyclophilins (Cyps) to be probed independent of immunosuppressive activities.[3] Finally, the use of knockdown and gene-silencing techniques has revealed a rich, and expanding, biology associated with Cyps. This chapter describes the application of Cyp inhibitors as novel antiviral agents and the factors to be considered when optimization strategies are designed.

RSC Drug Discovery Series No. 32
Successful Strategies for the Discovery of Antiviral Drugs
Edited by Manoj C. Desai and Nicholas A. Meanwell
© The Royal Society of Chemistry 2013
Published by the Royal Society of Chemistry, www.rsc.org

11.1 Cyclophilin Distribution, Function and Inhibition by Cyclosporine A

11.1.1 Cyclophilins

11.1.1.1 Expression

The Cyp family of proteins share a common enzymatic activity, peptidylprolyl isomerase (PPIase), which serves to effect a *cis–trans* isomerization of amide bonds amino-terminal to proline residues, and thus play a regulatory role in protein folding.[4–6] Currently, 19 members of the Cyp family (Figure 11.1) have been identified in humans, with sizes ranging from small, single PPIase-domain Cyps (CypA, CypB, CypC, CypD, CypJ, PPIL1, PPIAL4 and USA-Cyp) to large multi-domain proteins in which a catalytic PPIase activity is linked to various additional activities.[7] CypA is the most abundantly expressed member of the family, representing the majority of the PPIase activity in a cell, with CypB being the only other Cyp being found above trace level. In addition to the Cyps, two other families of proteins with PPIase activity, FKBPs and parvulins (Figure 11.1), have been described.[8,9] Inhibition of the PPIase activity of FKBPs by either FK506 or rapamycin was shown to be required for the immunosuppressive activity of these compounds,[10] and one member of the parvulin family, Pin 1, interacts with phosphoproteins involved in the cell cycle of some cancer cells.[11] However, the lack of potent and selective inhibitors for parvulins has limited research in this area.[12]

The Cyps show widely differing expression profiles, with CypA being found predominantly in the cytoplasm whereas more restricted localization is

Structure	HCV (EC$_{50}$, uM)	IL-2 (EC$_{50}$, uM)	Structure	HCV (EC$_{50}$, uM)	IL-2 (EC$_{50}$, uM)
[O-Acetyl-MeBmt]^{1}CsA^{72}	>10	>2xCsA	[MeIle]^{4}CsA (NIM811)	0.1	>10
[Thr]^2[5'-HOMeLeu]^{9}CsA^{158}	<3	ND	[4'-AcOMeLeu]^{4}CsA (13)	0.25	0.03
[D-Sar-OMe]^{3}CsA^{136}	0.06	<0.1	[MeVal]^{5}CsA	0.77	0.05
[D-Sar-SMe]^3[4'-HOMeLeu]^{4}CsA	0.04	0.06	[BnVal]^{5}CsA	0.5	>10
[D-Sar-SCH$_2$CH$_2$NMe$_2$]^3[4'-HOMeLeu]^{4}CsA	0.1	>2	[Me-D-Ala]^3[EtVal]^{4}CsA (Alisporivir)	0.03	>2
[D-Sar-SCH$_2$(3-pyridyl)]^3[4'-HOMeLeu]^{4}CsA (10)	0.06	>10	[D-Lys]^{8}CsA	2.0	0.1
[D-Sar-SCH$_2$(4-thiazolyl)]^3[4'-HOMeLeu]^{4}CsA (11)	0.1	24	[3-CN-D-Ala]^{8}CsA	0.04	0.005
[D-Sar-CH$_2$SCH$_2$CH$_2$NMe$_2$]^3[4'-HOMeLeu]^{4}CsA (12)^{139}	0.05	ND	[N-ε-Me$_2$-D-Lys]^{8}CsA	0.14	0.4
[MeVal]^{4}CsA	0.15	>2	[MeAla]^{10}CsA^{72}	8.48	>15xCsA

Figure 11.1 Human peptide bond *cis–trans* isomerase enzyme class.

observed for CypB (endoplasmic reticulum), CypD (mitochondrial inner membrane), CypE (nucleus) and RanBP2 (nuclear pore).[13] Studies directed at the functional characterization of CypA, CypB, CypD and Cyp40 have been reported over the last 20 years; however, native roles for the majority of the family and potential roles in disease states have not yet been elucidated.

11.1.1.2 PPIase Activity of Cyclophilins

Several mechanisms by which Cyps catalyze the *cis–trans* isomerization of an Xaa–Pro bond have been proposed; however, the currently accepted mode of catalysis is based on distortion of the planar amide bond (Figure 11.2).[14] Binding of a proline-containing substrate in a manner that twists the amide bond leads to a lower barrier to rotation due to loss of amide resonance.[15] An important protonation of the prolyl amide by an active site residue (side chain of [Arg]55 in CypA) facilitates the formation of a pyramidal nitrogen, which is a key feature of the isomerization.[16] Notably, mutation of [Arg]55 results in a 'catalytically inactive' form of CypA, which has been extensively used as a mechanistic probe.[17] Crystal structure analysis of several proline-containing substrate peptides bound to human CypA revealed that all ligands evaluated bound with the Xaa–Pro bond in the *cis* configuration and with no distortion from planarity.[18] Since most Xaa–Pro amide bonds reside preferably (5–6-fold preference) in the *trans* configuration, Ke and Huai proposed that the principal role of Cyps is to facilitate a *trans* to *cis* isomerization.[18]

While the PPIase activity of Cyps has been well studied, several additional roles for Cyps have been described. Binding of CypA or CypB to CD-147/EMMPRIN (an extracellular receptor for CypA) leads to activation of MAP kinase pathways and induction of matrix metalloproteinase (MMP) production.[19] A role for CypA as an inflammatory chemotactic cytokine, acting through the CD-147/EMMPRIN receptor, was revealed using models of asthma in which blocking the CypA–CD-147 interaction with anti-CD-147 or CsA resulted in a significant decrease in neutrophil and eosinophil migration and improved lung pathology.[20] Further roles for extracellular Cyps in renal fibrosis and ischemia-reperfusion injury have been described with CD-147/EMMPRIN being a key activating receptor.[21]

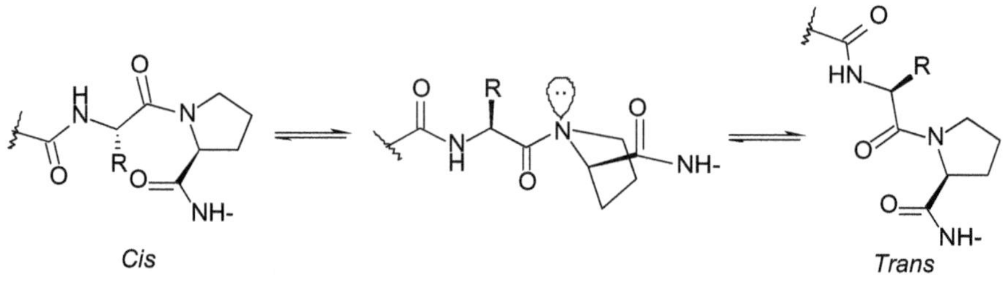

Cis   Trans

Figure 11.2   Proposed mechanism of PPIase action involving a twisted amide bond and a pyramidalized nitrogen.

11.1.2 Cyclosporine A

11.1.2.1 Structure and Conformational Flexibility

Cyclosporine [also known as cyclosporin, cyclosporin(e) A and CsA; **1**, Figure 11.3] is a cyclic undecapeptide first isolated from the fungus *Tolypocladium inflatum* in 1970.[22] Throughout the literature, the structure of CsA, first elucidated in 1976,[23,24] is typically drawn as some variant of that shown in Figure 11.3, with each amino acid numbered 1 through 11 in a clockwise manner starting at 12 o'clock. Ten of the 11 amino acids are known or derivatives of known amino acids, with several having branched lipophilic side chains: $4 \times$ Leu; $2 \times$ Val, $2 \times$ Ala, $1 \times$ Gly (Sar) and $1 \times$ Abu (aminobutyric acid). One of the 11 amino acids is previously unknown: Bmt or (4R)-4-[(E)-2-butenyl]-4-methyl-L-threonine. All 11 amino acids have the natural L-configuration except for [D-Ala]^8 and the achiral [Sar]^3. Seven of the 11 amide nitrogens are capped with a methyl group.

Cyclosporine adopts different conformations depending on the polarity of the solvent in which it is dissolved.[25] In crystalline form and in non-polar solvents such as chloroform or THF, the hydrogens on the four uncapped amide nitrogens form intramolecular hydrogen bonds with the carbonyl oxygens of other amides within the molecule, as shown in Figure 11.3; three of these are transannular ([Abu]^{2}NH–[Val]^{5}C=O; [Val]^{5}NH–[Abu]^{2}C=O and [Ala]^{7}NH–[MeLeu]^{11}C=O), while the fourth is in a γ-turn ([D-Ala]^{8}NH–[MeLeu]^{6}C=O). Residues 1–6 adopt an antiparallel β-pleated sheet conformation, while residues 7–11 form an open loop featuring a *cis* amide bond between [MeLeu]^9 and [MeLeu]^{10}.[26] It is noted that this stable conformation is unique to CsA and close analogs; many synthetic derivatives disrupt this internal hydrogen-bonding network and exist as multiple conformers in non-polar solvent. Examples include derivatives that possess an L-substituent at [Sar]^3,[27,28] and derivatives that are N-capped at the [Val]^5 nitrogen.[29] This has a consequence to the biological activity of CsA analogs that is discussed in Section 11.3.

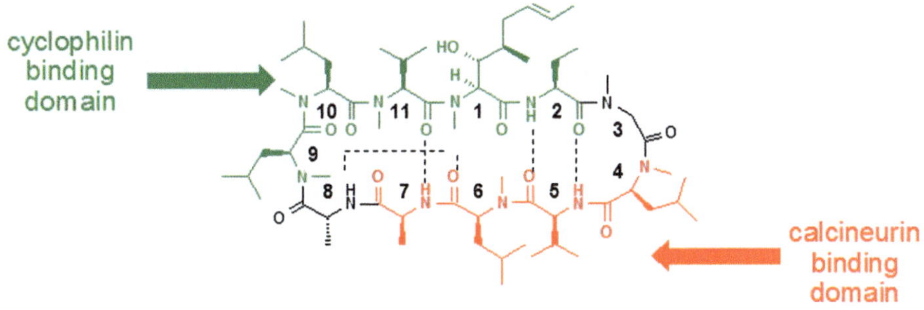

(1)

Figure 11.3 Structure of cyclosporine A (CsA) with the cyclophilin (green) and calcineurin (red) binding regions indicated. Key intramolecular hydrogen bonds that exist in non-polar solvents are indicated by dashed lines.

In polar solvents such as DMSO, water or THF charged with LiCl, this intramolecular hydrogen-bonding network is disrupted and at least in the latter case (THF–LiCl), the amide bond between [MeLeu]^9 and [MeLeu]^{10} is now *trans*.[30] CsA also resides in this conformation when bound to Cyp. In addition, it has been shown that temperature can influence CsA conformational structure in polar solvent.[31]

11.1.2.2 Binding Pharmacophores

CsA exerts its immunosuppressive activity by binding to two proteins sequentially to form a ternary complex (Figure 11.3).[32] The first of these is a Cyp, which is a *cis–trans* proline isomerase, with the most predominant of the human Cyps being CypA.[33,34] The binary CsA–CypA complex binds to and is a potent inhibitor of the phosphatase activity of CaN, a calcium-dependent, serine/threonine phosphatase that promotes the synthesis of T-cell lymphokines such as interleukin-2 (IL-2). Thus, CaN inhibition ultimately suppresses the immune response.

X-ray crystallography has revealed which residues of CsA bind to CypA and CaN.[35–37] Figure 11.3 shows that residues 9, 10, 11, 1 and 2 form the 'Cyp binding domain' that binds to CypA, whereas residues 4, 5, 6 and 7 comprise the 'CaN binding domain' that binds to CaN. Residues 3 and 8 are at interfaces between these two binding domains and can potentially have an impact on both CypA and CaN binding.

In addition to various autoimmune diseases that can be addressed by CaN inhibition, there are a variety of diseases that are treated by Cyp inhibition alone, most notably the infectious diseases caused by human immunodeficiency virus-1 (HIV-1)[38] and HCV,[39] both of which are addressed in Section 11.2. In such cases in which Cyp inhibition alone is sought, it is preferred to use a drug that does not bind CaN and is therefore non-immunosuppressive, particularly when combating a viral disease in which a functional immune response is needed. In such cases, it is necessary to modify the cyclosporine synthetically such that CaN binding is inhibited.

11.2 Cyclophilins Involved in Viral Replication

11.2.1 Human Immunodeficiency Virus

HIV-1, the parasitic virus causing AIDS, depends heavily on host cellular machinery for its survival, replication and infectivity. CypA is the first cellular protein ever found to be incorporated into HIV-1 virions. Subsequently, a significant number of publications have suggested several roles of CypA in the HIV life cycle,[38] all of which are independent of CaN and the NFAT pathway.

Initial reports of CsA inhibiting HIV-1 surfaced in 1988,[40] although the first report specifically implicating CypA involvement in the HIV-1 life cycle was published 5 years later, when it was reported that endogenous CypA colocalizes with the HIV-1 Gag protein in the cytoplasm, binding specifically to a proline-rich sequence in the HIV-1 capsid domain of Gag,[41–43] where residues 85–93 of HIV-1 capsid bind to the active site of CypA, as later revealed by X-ray crystallography.[44] Subsequent studies with CypA in the presence of mutated Gag or competitive Cyp A inhibitors have shown that neither viral budding nor Gag processing are affected; however, virions produced under such conditions were less infectious than those possessing CypA.[44] CypA is then incorporated into nascent HIV-1 virions during the assembly and budding process. As these HIV virions now infect new T cells, CypA from the virion delivered to the new cell facilitates uncoating of the capsid and supports efficient reverse transcription of the viral genome.[45]

More recent studies, however, have suggested that CypA in the infected target cell, and not CypA from HIV-1 virions, is more important to HIV-1 infectivity.[46–48] It was then also found that CypA in the target cell increases HIV-1 infectivity by inhibiting host cell restriction factors mounted as part of an innate immune response that ordinarily combats invading retroviruses.[49] In particular, CypA was found to inhibit tripartite motif 5α (TRIM5α)-mediated restriction of post-entry HIV-1 before reverse transcription *via* binding to HIV-1 capsid, which binds to TRIM5α in the same region that it binds to CypA.[49,50] Reducing the CypA–capsid interaction by either mutations altering binding of CypA to capsid or by the introduction of a CypA inhibitor such as CsA has been shown to increase HIV-1 susceptibility to TRIM5α restriction.[49] Other recent studies have suggested that CypA binds to HIV-1 viral protein R (Vpr), which undergoes *cis–trans* isomerization of proline residues in its N-terminal region. This activation by CypA ultimately governs the functional expression of Vpr, which is needed for translocation of virus to the nucleus and induction of cell cycle arrest and apoptosis in infected T cells.[51,52]

In summary, there are several stages in the HIV-1 life cycle that involve CypA, some understood more clearly than others. Clearly, a CypA inhibitor that bound to CypA alone and not CaN would potentially be an effective anti-HIV therapeutic. Such compounds have been developed and are discussed in Section 11.3.

11.2.2 Hepatitis C Virus

The demonstration of *in vitro* anti-HCV activity, using a sub-genomic replicon cell culture system, for CsA but not for a different CaN inhibitor, FK-506, provided early support for a role of Cyps in HCV replication.[53,54] Clinical studies in HCV-infected patients showing superior virologic response to CsA in combination with interferon-α2b versus interferon-α2b alone[55,56] further implicated Cyp inhibition as an approach to HCV therapy. A definitive role for Cyps in HCV replication was demonstrated by Watashi *et al.* using a set of chemically modified CsA derivatives (Figure 11.4).[56] Both immunosuppressive {CsA (**1**) and [8′-HO-MeBmt]¹CsA (**2**)} and non-immunosuppressive {[MeAla]⁶CsA, [MeVal]⁴CsA (**3**) and [MeIle]⁴CsA (**4**, NIM-811)} derivatives demonstrated potent inhibition of HCV RNA replication while closely related compounds {CsH (**5**) and PSC 833 (**6**)} that do not bind CypA were inactive in

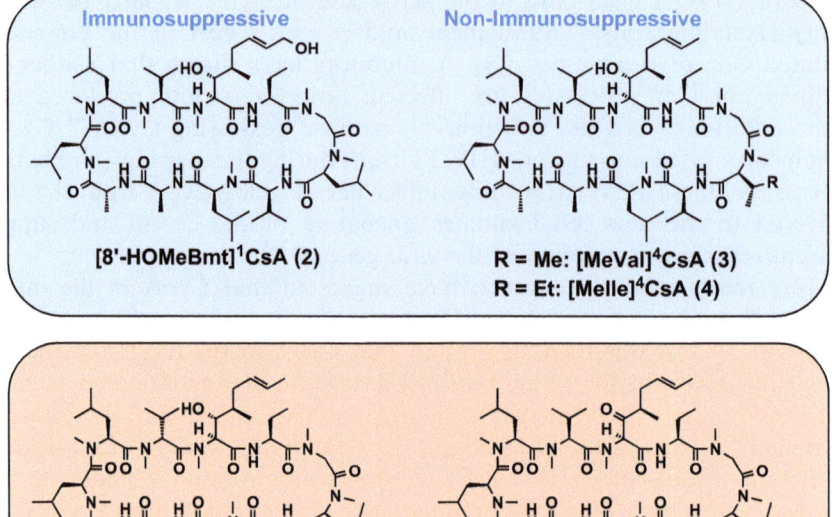

Figure 11.4 Immunosuppressive and non-immunosuppressive CsA derivatives are inhibitors of HCV replication whereas closely related analogs are devoid of CypA binding activity and do not inhibit HCV replication.

the assay. A structurally distinct Cyp inhibitor, sanglifehrin A, was also shown to be an inhibitor of HCV replication.[56]

Initial studies proposed CypB and/or CypC as being crucial to HCV replication;[56–58] however, other studies have suggested that CypA alone is crucial for HCV genotypes 1a, 1b and 2a.[59–63] To complicate matters, a more recent study suggests that several Cyps are important to various stages of HCV replication, including CypA (regulation of transcription and translation), CypB and CypC (protein conformation and transport), CypE and CypH (regulation of mRNA splicing, generation of host proteins necessary for HCV), Cyp40 (regulation of translation, non-vesicular transport of cholesterol, co-chaperone of Hsp90).[64] In any event, a recent consensus has emerged that CypA is likely the principal and perhaps exclusive Cyp crucial for HCV RNA replication and protein expression.[65,66]

There is ample evidence that the HCV life cycle is dependent on host cell Cyp for perhaps multiple processes, as is the case with HIV-1. A Cyp inhibitor, and in particular a CypA inhibitor that was non-immunosuppressive, would be an effective anti-HCV therapy.

Various models have been proposed to explain the role of CypA in the replication of HCV RNA.[66] The finding that *in vitro* resistance to Cyp inhibitors, although slow to emerge, is associated with mutations in both NS5A and NS5B supports the biochemical demonstration of an interaction between these proteins and CypA.[67] However, knockdown of CypA inhibited replication of a full length HCV genome significantly more effectively than the subgenomic replicon system, suggesting that CypA is involved in additional interactions with HCV proteins.[60] Gallay and co-workers provided evidence in support of a model whereby CypA, pre-residing in crude replication complex membrane fractions, serves to associate NS5A–NS5B into an assembling replication complex and that CsA depletes this compartment of CypA, thereby preventing replication.[68] Clues towards an understanding of the role that PPIase activity might play in HCV RNA replication were presented by Bartenschlager and co-workers in a study of the rate of HCV polyprotein processing.[60] The demonstration that a resistance mutation to the Cyp inhibitor DEBIO-025 (alisporivir), occurring near the cleavage site of the NS5A–NS5B polyprotein, results in delayed processing of the protein and impaired replication activity.[60] A role for CypA in the conformational reorganization of multiple regions of the HCV polyprotein can be envisaged that might account for the high barrier to resistance exhibited by Cyp inhibitors due to the need for multiple virus mutations to appear to escape the need for Cyp PPIase activity.

Interestingly, it has been described that treatment of replicon-bearing Huh cell lines with Cyp inhibitors (alisporivir and SCY-635) causes a secretion of CypA and CypB from the cell into the supernatant.[69] This release leads to a depletion of intracellular Cyp levels which may play a role in controlling replication of HCV RNA. Finally, a large-scale siRNA experiment involving Cyp binding compounds provided evidence that NIM-811 (**4**) reduces viral replication *via* inhibition of multiple Cyps and pathways.[64]

The discovery that Cyp inhibition was sufficient to prevent HCV RNA replication quickly led to renewed interest in NIM-811 owing to its prior demonstration of potent Cyp binding activity, oral bioavailability and potential for low nephrotoxicity.[70] While originally discovered as part of an anti-HIV program, NIM-811 was found to be a more potent inhibitor of HCV replication than CsA in a HCV replicon assay, and further, co-treatment with interferon-α led to clearance of the virus.[71] Initial structure–activity relationship (SAR) data presented by Novartis supported a correlation between binding affinity for CypA and activity in HCV replicon assays and highlighted the attractive profile of NIM-811.[72]

NIM-811, at various dose levels, in combination with Peg-IFN/ribavirin was investigated in clinical studies of HCV genotype 1 infected patients over the course of 4 weeks of therapy; however, no subsequent clinical studies have been described.[65]

The most clinically advanced Cyp inhibitor for HCV, alisporivir ([D-MeAla]^3[EtVal]^{4}CsA, **7**, Figure 11.5), was also originally identified as a potent inhibitor of CypA PPIase with good activity against HIV-1 replication.[73] Demonstration that the compound was ~10-fold more potent against HCV replication than CsA led to it being investigated further. Alisporivir binds to CypA in a similar manner as CsA and potently inhibits the PPIase activity of CypA with a K_i of 0.34 nM.[73] *In vitro* studies using an IL-2 reporter assay in Jurkat T-cells or a murine/human mixed lymphocyte proliferation assay demonstrated that alisporivir is up to 7000-fold less immunosuppressive than CsA, consistent with the alisporivir–CypA complex being a weak inhibitor of CaN.[74] Using an HCV genotype 1b subgenomic replicon system, alisporivir was shown to be significantly more effective than CsA at suppressing viral RNA replication (IC_{50} 0.04 µM *versus* 0.3 µM for alisoprivir and CsA, respectively) over the course of 72 h of treatment.[75] Importantly, alisporivir was shown to be equally effective against replicons derived from HCV genotypes 1–4 and to be active against the JFH-1 infectious system.[75] Combinations of alisporivir with interferon, ribavirin, the NS3 protease inhibitor telaprevir, 2'-C-methylcytosine (a nucleoside NS5B inhibitor) or JT-16 (a non-nucleoside NS5B inhibitor) showed additive to slightly synergistic activity against HCV RNA replication with no cytotoxicity toward the host cell.[75] Alisporivir retains wild-type levels of inhibitory activity against replicons resistant to NS5B polymerase (nucleoside and non-nucleoside) and NS3 protease inhibitors and, likewise, these inhibitors retained potency against alisporivir-resistant replicon cell lines, indicating no cross-resistance between Cyp inhibitors and the primary viral target inhibitors.[75] Generation of low-level (about fivefold) *in vitro* resistance to alisporivir was difficult to achieve, requiring a high number of passages, consistent with a very high viral barrier to resistance.[76] Alisporivir was progressed to clinical investigation with initial studies being performed in HIV/HCV co-infected individuals where the compound showed a marginal reduction in HIV viral load; however, a significantly greater reduction in HCV RNA ($3.63\log_{10}$ reduction from baseline) was observed when dosed at 1200 mg bid.[69] Phase 2 clinical studies investigating the effect of alisporivir, at various dose levels, as monotherapy or in combination with Peg-IFN/ribavirin in HCV genotypes 1–4 individuals have been performed.[77] While the details of these clinical studies are beyond the scope of this chapter, alisporivir significantly increased the antiviral effectiveness of Peg-IFN/ribavirin therapy and the compound is currently under evaluation in expanded studies of HCV genotype 1- and 2/3-infected patients.[78]

SCY-635 (**8**, Figure 11.5) is a third Cyp inhibitor to be evaluated in clinical studies of HCV infection and, like NIM-811 and alisporivir, was originally discovered as part of an anti-HIV program.[79] SCY-635 potently inhibits the enzyme activity of CypA ($K_i = 2$ nM), has low immunosuppressive potential (1500-fold lower inhibition of IL-2 production by stimulated Jurkat T cells than CsA) and inhibits replication of a HCV genotype 1b-derived replicon (IC_{50} 0.1 µM) with no evidence of toxicity.[80] When evaluated in a clinical study of activity in HCV genotype 1-infected patients, SCY-635 given as monotherapy resulted in a modest decline of serum HCV RNA, with no evidence of resistant virus being detected.[81] However, the finding that the clinical activity of SCY-635 in HCV-infected patients is closely correlated with an SNP

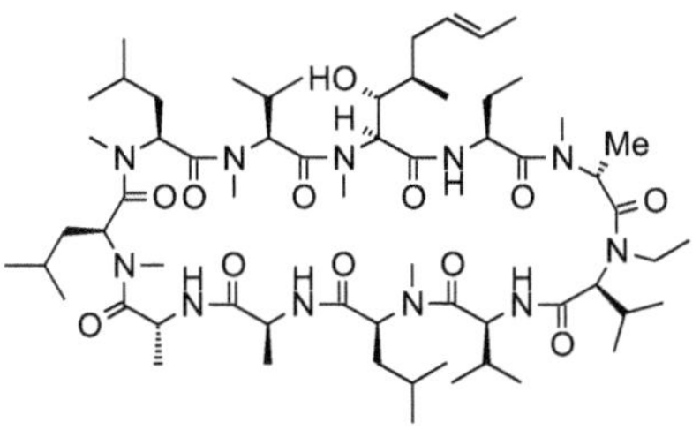

Alisporivir
(Debio-025)
(7)

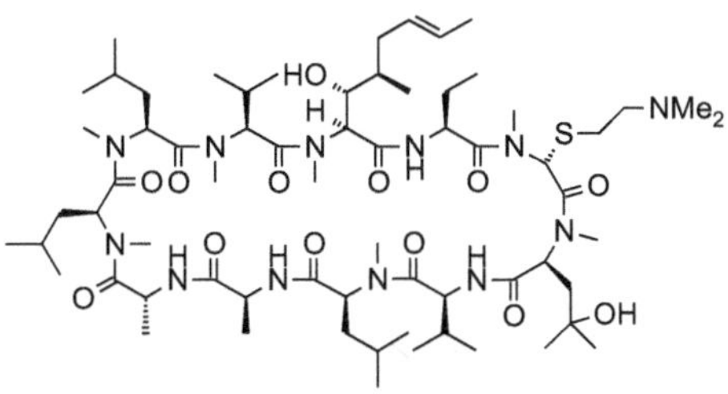

SCY-635
(8)

Figure 11.5 Alisporivir and SCY-635 are currently under clinical investigation as non-immunosuppressive cyclophilin inhibitors for the treatment of hepatitis C virus infection.

(rs12979860) that has been shown to predict responsiveness to pegylated interferon-containing regimens suggests that the mechanism of action of this agent may be more closely aligned with an immunological response than with direct activity on the virus. The additional finding that dosing of HCV-infected patients with SCY-635 results in a transient increase in plasma cytokines (IFNα, IFNl1, IFNl3) points to a potential role for Cyp inhibition in the host innate immune response to HCV infection.[81]

11.2.3   Dengue, West Nile and Other Flaviviruses

Isoform-specific knockdown of CypA, CypB or CypC in Huh-7.5 cells was used by Tang and co-workers[82] to demonstrate a role for Cyps in the replication of West Nile virus (WNV), dengue virus and yellow fever virus. The most significant inhibition of replication of the three flaviviruses was found for a double knockdown of CypA and CypB and the PPIase activity of CypA was shown to be essential for effective replication of dengue virus and WNV. They further showed that CypA directly interacts with the NS5 protein of WNV; however, this interaction did not affect the enzymatic activity of NS5 *in vitro*.[82] Consistent with these findings, CsA was shown to block the interaction of CypA with NS5 of WNV, suppress viral RNA synthesis and inhibit a broad spectrum of RNA viruses at non-toxic concentrations.

CypB was demonstrated to be involved in the replication of Japanese encephalitis virus (JEV) by the rescue of JEV replication by expression of wild-type CypB, but not PPIase-deficient CypB, in various mammalian cells in which endogenous CypB was suppressed.[83] This finding was again supported by the finding that CsA potently inhibits the replication of JEV *in vitro*.[83]

11.2.4   Non-flaviviruses

While a role for Cyp inhibition in restricting flavivirus replication has been demonstrated, the activity of Cyps in other viruses is less clear. A convincing argument for a role for Cyps in the replication of vaccinia virus (VV) was provided by the demonstration that cyclosporine analogs with strong binding affinity for CypA were potent inhibitors of VV replication whereas analogs with weak affinity for CypA lost antiviral activity.[84] Subsequent studies revealed an incorporation of CypA into VV virions as an essential event for successful maturation.[85] A similar incorporation of CypA into viral particles has been reported for vesicular stomatitis virus[86] and SARS coronavirus;[87] however, in each case additional functions of CypA have been proposed.

A novel activity of a Cyp in viral entry was described for human papillomavirus type 16 involving a cell-surface fraction of CypB which, following attachment, serves to effect a conformational change in capsid proteins L1 and L2 leading to efficient infection.[88]

Incorporation of CypB into virions of measles virus (MV) has been shown to be an important requirement for expanding the tropism of the virus to include epithelial and neuronal cells.[89] Neutralizing antibodies directed at CD147/EMMPRIN, a member of the Ig superfamily and a functional receptor for CypA and CypB, served to limit infection of epithelial cells by MV.[89] Inhibition of MV infection was also demonstrated by treatment with CsA and has led to the proposal that MV incorporates CypB, but not CypA, into mature virions and that this CypB serves to allow effective infection of CD147/EMMPRIN expressing epithelial cells.[89]

A role for Cyps in the replication of herpes simplex virus has been suggested by the finding that CsA inhibits virus production *in vitro*;[90] however, the immunosuppressive properties of CsA need to be taken into consideration when assessing such results.

While Cyps play a supportive function in the replication of numerous viruses, some evidence of Cyps being restrictive towards viral replication has been reported. CypA was shown to bind to the M1 protein of influenza A and to restrict viral replication by inducing ubiquitin/proteosome-dependent degradation of the M1 protein.[91] Notably, a PPIase-deficient CypA protein could still bind M1 and inhibit viral replication while depletion of CypA was found to enhance the replication of influenza A virus *in vitro*.[91] A similar finding was reported by Wu and co-workers[92] as a result of studies on the effect of CypA on rotavirus replication. Silencing CypA resulted in a decrease in interferon-β production and an increase in viral replication, with both effects being reported to be independent of the PPIase activity of CypA.[92]

A role for Cyps in signaling pathways that are activated following viral infection was first suggested by the demonstration that CypB is involved in regulation of IRF-3.[93] Knockdown of CypB was found to suppress virus (Newcastle disease virus) induced phosphorylation of IRF-3 which blocked IRF-3 dimerization and inhibited the production of interferon-β. A recent report from Hopkins *et al.*[81] described a novel induction of type I and III interferons in HCV-infected patients following treatment with a non-immunosuppressive Cyp inhibitor, SCY-635. These results point to a possible new activity for Cyps in virus-induced blocks on innate immune responses that would otherwise serve to clear the infection and place Cyp inhibition as a potentially important immunotherapy approach to virology.

11.3 Modification of Cyclosporine to Treat Viral Diseases

11.3.1 Factors Affecting Cyclophilin Binding and Selectivity

As discussed in Section 11.1.2, the Cyp binding domain comprises of residues 9, 10, 11, 1 and 2 on the CsA scaffold. Hence modifications to these residues tend to have a significant effect on Cyp binding, whereas modifications to residues in the CaN binding domain (4, 5, 6 and 7) typically have less influence (Figure 11.6).

The MeBmt group at position 1 plays a key role in that it resides in the Cyp binding domain, yet its side chain drapes across the CsA scaffold and ultimately into the CaN binding domain. In general, modifications to MeBmt tend to decrease Cyp binding. Removing or altering the stereochemistry of the 4-methyl group reduces activity, as does introducing a second methyl group at this position.[94,95] Likewise, removing, replacing or capping the 3'-hydroxy group as an ester also reduces potency.[96,97] The butenyl chain appears crucial for binding, as reducing the double bound to a fully saturated butane chain slightly diminishes Cyp binding.[96] More significant changes, such as truncation of the butene chain, cause a significant decrease in activity;[98] however, modification of the 8'-carbon, including simple homologation or oxidation, can

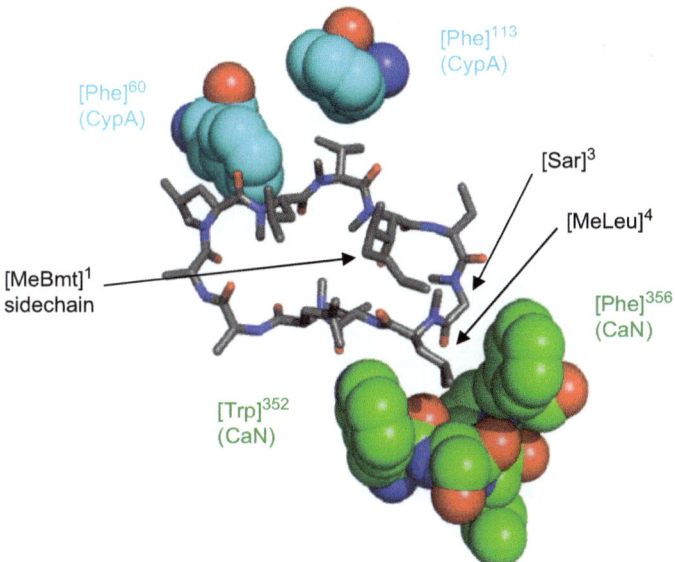

Figure 11.6 Structure of CsA bound to both CypA and CN. The side chain of [Val][11] is bound in a hydrophobic pocket of CypA whereas [Leu][4] binds tightly to CN.

retain Cyp binding potency.[96,99] One case in which Cyp binding is actually improved is when the C=C double bond is replaced with a bioisosteric sulfur atom ('MeThiaBmt'), which was found to bind to Cyp with 178% of the affinity of CsA itself.[95,100]

The [Abu][2] ethyl side chain fits into a large hydrophobic pocket within Cyp.[101] A 3'-OH group on [Abu][2] is tolerated, as is, to a lesser extent, the n-propyl chain of [Nva][2]CsA (CsG); however, larger and/or more branched functionality is not well tolerated,[95,96,101,102] possibly owing to ineffective disruption of hydrogen-bonded water molecules thought to bind in the [Abu][2] pocket.[101,103] Other groups at the 2-position showing modest Cyp binding include Ala, Thr and Val (Figure 11.7).[96] The unsubstituted nitrogen of [Abu][2] is crucial, as capping it with a methyl group obliterates Cyp binding,[104] although some analogs possessing an oxazolidine ring linking the 2-position side chain with this nitrogen have shown Cyp binding affinity as high as 52% of that of CsA.[105]

The [Sar][3] position lies on the periphery on the Cyp binding domain, and consequently structural changes at [Sar][3] can directly or indirectly through conformational bias impact Cyp binding. The N-methyl group of [Sar][3] is crucial, as [Gly][3]CsA has one-sixth the Cyp binding affinity of CsA. Substitution at the 2'- or α-carbon of [Sar][3] is extremely well tolerated, provided that the introduced substituent resides on the β-face to give the Re or D-configuration; substitution on the α-face that gives the Si or L-configuration, results in steric clashing between this substituent and the N-methyl group of [MeLeu][4] and, hence, an altered conformation of the CsA scaffold, which does not bind Cyp as well (Figure 11.7).[106] A significant body of work has demonstrated that a range of polar and non-polar groups can be introduced at this position with minimal impact on Cyp binding.[96,107,108]

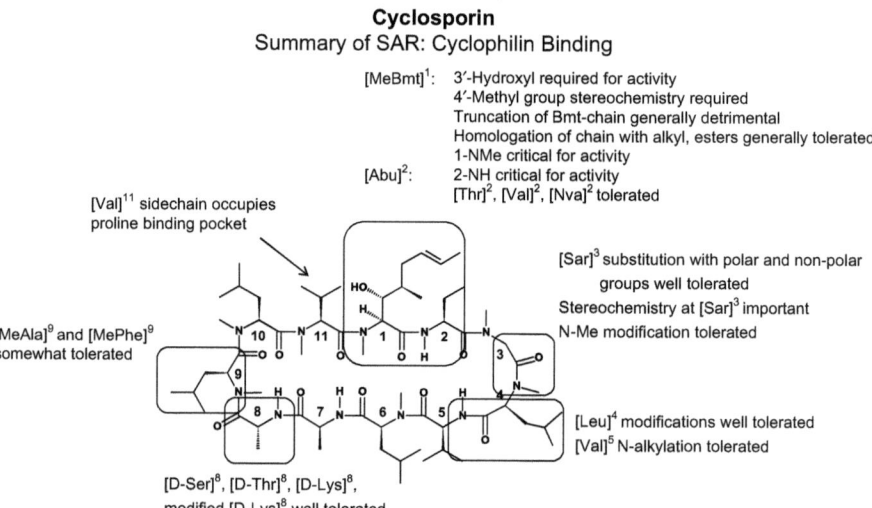

Cyclosporin
Summary of SAR: Cyclophilin Binding

[MeBmt][1]: 3'-Hydroxyl required for activity
4'-Methyl group stereochemistry required
Truncation of Bmt-chain generally detrimental
Homologation of chain with alkyl, esters generally tolerated
1-NMe critical for activity

[Abu][2]: 2-NH critical for activity
[Thr][2], [Val][2], [Nva][2] tolerated

[Val][11] sidechain occupies proline binding pocket

[MeAla][9] and [MePhe][9] somewhat tolerated

[Sar][3] substitution with polar and non-polar groups well tolerated
Stereochemistry at [Sar][3] important
N-Me modification tolerated

[Leu][4] modifications well tolerated
[Val][5] N-alkylation tolerated

[D-Ser][8], [D-Thr][8], [D-Lys][8], modified [D-Lys][8] well tolerated

Figure 11.7 Changes on the CsA scaffold that result in modification of Cyp binding affinity.

As mentioned previously, CsA residues 4, 5, 6 and 7 reside in the CaN binding domain; hence, structural modifications to these residues likely affect Cyp binding indirectly. At the 4-position, [4'-HOMeLeu]⁴CsA has been measured to bind to Cyp comparably to CsA[96] or even twofold better than CsA,[109] while [MeVal]⁴CsA and [MeIle]⁴CsA have been shown to have comparable Cyp binding affinity to CsA itself.[70,72,104] Some modifications cause a drop in Cyp binding, such as the rigid [Pro]⁴CsA, which has no detectable Cyp affinity within 20-fold of that of CsA.[104] Other examples of small alkyl groups, such as methyl, ethyl and CH_2-*c*-Pr, show Cyp affinity within threefold of CsA,[96,110,111] or greater affinity, as shown with *sec*-butyl.[112] Lastly, it is noted that the *N*-methyl group of MeLeu can be replaced with the more metabolically stable *N*-ethyl group without detriment to Cyp binding.[113]

Amino acid replacement at [Val]⁵ generally causes a decrease in Cyp binding affinity, even for relatively modest changes such as [Ala]⁵CsA or [4-F-Val]⁵CsA.[95,104,114,115] Notably, however, alkylation of the nitrogen of [Val]⁵ with a substituted allyl or benzyl group retains Cyp binding while diminishing CaN binding.[29]

Modifications at positions 6 and 7 rely on significant synthetic efforts and have been less well studied. Minor amino acid changes cause a minor decrease in Cyp binding affinity,[17,88,96] while introduction of larger or more rigid residues such as proline cause a greater decrease in binding affinity.[104,116,117] A 'tricyclic CsA' in which a bicyclic dipeptide mimic replaces [Ala]⁷-[D-Ala]⁸, shows threefold greater affinity for CypA and two- to threefold greater immunosuppressive activity than CsA.[118,119]

At the 8-position, we reach the interface between Cyp and CaN binding domains and can now potentially influence Cyp binding by a direct interaction between the modified side chain and Cyp rather than merely indirectly by conformational bias, as seen with modifications to residues 4–7. It has been shown that [D-Ser]⁸CsA, [D-Thr]⁸CsA, [D-Lys]⁸CsA and [Boc-D-Lys]⁸CsA each have a Cyp binding affinity identical with that of CsA.[96] In addition, [D-Gln]⁸CsA and [D-Asn]⁸CsA are also reported to have a Cyp binding activity identical with that of CsA, while the *N*-alkylated analogs [D-MeAla]⁸CsA and [D-Pro]⁸CsA show a 2.5- and 5-fold decrease in Cyp binding, respectively and interestingly [Sar]⁸CsA shows an 8-fold decrease in Cyp binding, underscoring the importance of D-substitution at this residue.[104] Beyond this, sizeable side chains have been tethered to both [D-Ser]⁸CsA[120,121] and [D-Lys]⁸CsA[95,122] that can tune down CaN binding without an appreciable decrease in Cyp binding. As is the case with [Ala]⁷, the nitrogen of [D-Ala]⁸ has also been alkylated to prepare Cyp-binding conjugates.[123,124]

The 9-, 10- and 11-positions that reside in the Cyp binding domain are much more difficult to replace or modify synthetically and there are fewer literature reports of Cyp binding data for such compounds; nevertheless, it is clear that modification in this section of CsA can have a profound affect on Cyp binding. Both [MeAla]⁹CsA and [MePhe]⁹CsA have moderately reduced affinity for Cyp,[95,96] whereas [Pro]⁹CsA has no detectable Cyp binding affinity.[104] At the 10-position, replacing [MeLeu]¹⁰ with MeAla or MePhe also reduces Cyp binding,[95,96] as does MeVal.[2] The 11-position is especially sensitive to Cyp binding, consistent with this residue occupying the 'proline binding region' which is central to the catalytic activity of the protein. Substitution of [MeVal][11] with MeIle, aMeIle or MeLeu completely abolishes Cyp binding,[96] as does MeThr,[104] while [MeAla][11]CsA exhibits only 8.5% of CsA's Cyp binding affinity.[95,96]

11.3.2 Removing Immunosuppressive Potential

In Section 11.2, it was noted that Cyp inhibition alone, and not CaN inhibition, is requisite for inhibiting replication of viruses such as HIV-1 and HCV. Furthermore, one would correctly surmise that inhibiting CaN would be counterproductive towards antiviral therapy since this would compromise the host immune response to an invading virus. Hence, an ideal Cyp-inhibiting antiviral drug should be designed such that it retains high binding affinity for Cyp, but not CaN.

Amino acid 4, where [MeLeu][4] ordinarily resides in CsA, has a profound influence on CaN binding, presumably due to a tight 'aromatic sandwich' that the amino acid side chain must snugly fit into between CaN residues [Trp][352] and [Phe][356]. The first derivative modified at this position found to exhibit a huge disparity between CypA and CaN binding is [MeIle][4]CsA, NIM811.[125] [MeIle][4]CsA was found to have no immunosuppressive activity, while retaining the same affinity for CypA as CsA itself.[125] Subsequent synthesis of analogs revealed that [MeVal][4]CsA (3) exhibited a >2500-fold decrease in immuno-suppressive activity, with a concomitant twofold increase in Cyp binding.[110] Similarly, substituting [MeLeu][4] with [4'-HOMeLeu][4]CsA (7, Figure 11.8) *via*

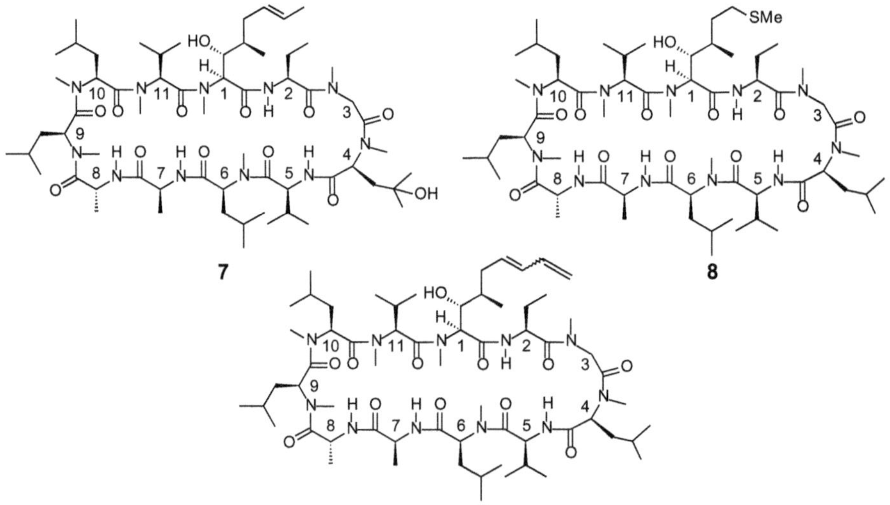

Figure 11.8   Modification of [Leu][4] generally leads to a loss of CN binding, whereas [Bmt][1] modification is less predictable.

microbial oxidation by *Sebekia benihana* significantly diminished the immunosuppressive potential by >100-fold while also increasing CypA binding affinity 1–2-fold.[96,109,110] Several additional CsA analogs bearing alternative groups at position 4 have been prepared,[96,104,109–113] although the three most distinct examples of abolishing immunosuppressive potential without loss of Cyp binding are the three cited above.

Replacing the amide proton at the nitrogen of [Val]⁵ with an allyl or benzyl group was found to attenuate immunosuppressive potential by >100-fold while substantially maintaining CypA binding affinity.[29] Replacing [Val]⁵ altogether with a pseudo-proline residue that cyclically tethered the amino acid side chain to the amide nitrogen introduced a *cis*-amide bond between the 4- and 5-residues and caused a greater reduction in immunosuppressive activity than Cyp binding.[115]

Amino acid 6, [MeLeu]⁶, appears also to be crucial for CaN binding. Replacing [MeLeu]⁶ with [MeVal] causes a 46-fold decrease in immunosuppressivity but only a threefold decrease in Cyp binding.[2] Replacement of [MeLeu]⁶ with Ala, Abu, Ile or Phe were also shown to be more detrimental to immunosuppressive activity than to Cyp binding.[95]

Lastly, amino acid 7, [Ala]⁷, also showed that it could be used to tune down immunosuppressive potential yet retain Cyp binding. Replacing [Ala]⁷ with Val, Ser, Thr or Gly all caused significantly greater reductions in immunosuppressive activity than Cyp binding.[104]

Derivatives bearing modifications on the residues at the interface of Cyp and CaN binding domains, residues 3 and 8, yield more heterogeneous data regarding Cyp and CaN binding. A substituent on the α-face of [Sar]³, resulting in a D-amino acid, typically retains and occasionally improves Cyp binding while also showing some attenuated immunosuppression.[79,95,96] Regarding the 8-position, [D-Ser]⁸CsA derivatives have typically shown Cyp binding that parallels immunosuppressive activity,[120,121] whereas other compounds bearing alkylamines of varying length in place of the Ala methyl group showed a decrease in immunosuppressive potential without a significant reduction in Cyp binding as the chain length increased.[122] The D-stereochemistry of the 8-position substituent appears to be important: [dehydro-Ala]⁸CsA, lacking a stereogenic center at the 2'-α-carbon, exhibits a 3.75-fold reduction in Cyp binding affinity, but is completely non-immunosuppressive.[111]

Beyond this, it is known that the MeBmt side chain drapes over the cavity of cyclosporine when bound to Cyp and consequently modifying this MeBmt side chain at the terminus that approaches CaN can influence CaN binding. [MeThiaBmt]¹CsA (**8**, Figure 11.8), which possesses a thiomethyl bioisosteric replacement of the $CH_3CH=CH-MeBmt$ terminus, exhibits almost a twofold increase in Cyp binding affinity, yet only 9.5% of the immunosuppressive activity of CsA.[95] Other compounds possessing extended alkyl functionality beyond the 8'-position, as found in ISA247 (**9**, Figure 11.8) and related compounds, still possess immunosuppressive activity.[126–128]

Clearly, attenuation of CaN inhibition, and consequent immunosuppressive potential, can also be achieved by various combinations of the modifications described above and the more advanced clinical compounds indeed rely on at least two modifications to the CsA core to remove immunosuppressive activity.

11.3.3   Antiviral Structure–Activity Relationships – HIV, HCV

11.3.3.1   Anti-HIV-1 SARs

The ability to remove immunosuppressive activity associated with the CsA nucleus while retaining potent CypA binding allowed SARs to be developed for inhibition of HIV-1.

As the [MeBmt][1] group is crucial to Cyp binding, there are few reports of MeBmt modification for anti-HIV-1 activity. These include the metabolite [8'-HOMeBmt][1]CsA, which shows comparable anti-HIV-1 activity to CsA.[129–132] Reducing the MeBmt C=C bond, while reducing Cyp binding, does not appear to impede antiviral activity, at least when combined with [4'-HOMeLeu][4], as [dihydro-MeBmt][1][4'-HOMeLeu][4]CsA shows a fourfold greater anti-HIV activity than CsA.[129–132] Interestingly, capping the MeBmt 3'-OH as an acetate does not appear to impede anti-HIV-1 activity, as both [3'-AcO-MeBmt][1][Sar-OAc][3]CsA and [Sar-OAc][3]CsA have identical anti-HIV-1 IC_{50}s of 150 ng mL^{-1}.[133–135]

The 2-position has also not been a point of focus for exploring anti-HIV-1 activity, although two reported examples possessing modification at the 2-position and other positions suggest that anti-HIV-1 activity is retained when [Abu] is replaced with Nva or MeOThr.[129,130]

The 3-position, on the other hand, has been explored extensively, as the [Sar][3] residue is ripe for efficient synthetic modification, without the need for a lengthy total synthesis, such that polar functional groups can be appended to [Sar][3] to augment drug-like properties such as water solubility. Compounds bearing a D-alkylamino thioether side chain at [Sar][3] exhibit anti-HIV-1 often at least as potent as CsA and, in the best case, with IC_{50} of 2.5 nM, which is over 180-fold more potent than CsA,[79,133–136] D-alkylamino ether side chains at [Sar][3] also display anti-HIV-1 activity better than that of CsA.[137]

Combining one of these optimal [Sar][3] D-substituents with functionally at position 4 to reduce immunosuppressive potential has been heavily exploited. Although such modifications at position 4 can be introduced by total synthesis, they can be more rapidly prepared by either microbial oxidation[109] or semi-synthesis starting from CsA itself.[113,138] Of particular interest, [D-substituted-Sar][3][4'-HOMeLeu][4]CsA analogs show excellent anti-HIV activity.[79,136,139] One such example is SCY-635 or [D-SCH$_2$CH$_2$NMe$_2$-Sar][3][4'-HOMeLeu][4]CsA, which inhibits replication of HIV-1 with an IC_{50} of 45 nM.[79] Interestingly, replacing [MeLeu][4] with amino acids known to reduce immunosuppressive activity can improve anti-HIV activity, even without [Sar][3] modification. Examples include [4'-HOMeLeu][4]CsA, [MeVal][4]CsA, [EtVal][4]CsA and [MeIle][4]CsA, each of which is more potent against HIV-1 than CsA itself.[79,129–132,140] Notably, the non-immunosuppressive derivative NIM811 ([MeIle][4]CsA)[3,70,125,141–143] is a potent inhibitor of HIV-1 replication and demonstrates the remarkably subtle effects that translocation of a single methyl group can have on a complex scaffold such as CsA. The [MeVal]⁴CsA derivative served as the starting point for the discovery of the potent non-immunosuppressive anti-HIV-1 compound [D-Ala]³[EtVal]⁴CsA) (alisporivir).[73,144] Addition of a methyl group to the [Sar]³ position may serve to affect conformation distribution patterns and improve CypA binding while the homologation of the [MeVal]⁴ residue to [EtVal]⁴ would be expected to diminish metabolic demethylation of this position, a major metabolism route of CsA itself.

After position 4, little anti-HIV-1 has been published involving modification of positions 5–11. Both [MeVal]⁵CsA and the metabolite [4′-HOMeLeu]⁹CsA exhibit anti-HIV-1 activity roughly equipotent to CsA, while [MeAla]⁶CsA shows about a threefold decrease in anti-HIV-1 activity.[129–132] Beyond this, there is a report of anti-HIV-1 activity with [3′-HOMeLeu]¹[MeAla]⁴[MeAla]⁶CsA, in which costimulation of IL-2 production is observed, lowering HIV-1 infectivity in drug-exposed cells.[145,146]

11.3.3.2 Anti-HCV SARs

The [MeBmt]¹ residue, due to the relative ease of synthetic access, has been investigated by several groups that have included claims to anti-HCV activity,[126–128,147–157] although to date no anti-HCV SAR has been published.

The 2-position, not being as amenable to synthetic modification, has been addressed only sparingly for anti-HCV activity. Astellas Pharma has looked at analogs of CsC ([Thr]²CsA) with additional modifications at either the 1- and/or 9-positions[158] or the 3-, 4-, 5- and 10-positions.[159] In the case of the former, [Thr]²[5′-HOMeLeu]⁹CsA was reported to have an $EC_{50} < 3\,\mu M$, whereas for the latter, nine [Thr]²[Leu]⁵[Leu]¹⁰CsA analogs were found to exhibit $EC_{50} < 0.5\,mg\,mL^{-1}$. These compounds possessed modifications at position 3 (D-SerOMe, D-Ala, D-Abu or D-MeThrO-*t*-Bu) and/or 4 (MeThr, MeThrOMe, MeThrO-*t*-Bu, MeIle, MeVal or MePhe).

As was the case with anti-HIV-1 analogs, [Sar]³ has been explored heavily owing to its relative ease in synthetic modification to analogs well tolerated for CypA binding (Table 11.1). Scynexis has prepared [Sar-*O/S*-alkyl]³CsA analogs with either [MeLeu]⁴ or the preferred [4′-HOMeLeu]⁴, many of which exhibited $EC_{50}s < 400\,nM$, the value observed for CsA.[136] The most potent examples from this set included [Sar-D-OMe]³CsA (60 nM), [Sar-D-SMe]³[4′-HOMeLeu]⁴CsA (40 nM), [Sar-D-OMe]³[4′-HOMeLeu]⁴CsA and [Sar-D-OCH₂CH=CH₂]³[4′-HOMeLeu]⁴CsA. Scynexis subsequently explored [Sar-*O/S*-CH₂aryl]³CsA analogs and found several derivatives, including **10, 11,** [Sar-D-OCH₂-furan-3-yl]³CsA (250 nM) and [Sar-D-OCH₂Ph]³CsA (260 nM), to be more potent against HCV than CsA itself. A recent patent from S & T Global claims [Sar]³ substitution with a methylene (–CH₂–) bridging the 2′-α-carbon to the *O/S*-(CH₂)₂,₃-dialkylamine side chain to give compounds (*e.g.* **12**) with good activity against a subgenomic HCV replicon and improved stability in MeOH at ≥50 °C.[139]

Table 11.1 SARs for anti-HCV activity and immunosuppressive potential for representative derivatives of CsA.

Structure	EC_{50} (μM)		Structure	EC_{50} (μM)	
	HCV	IL-2		HCV	IL-2
[O-Acetyl-MeBmt][1]CsA[72]	>10	>2xCsA	[MeIle][4]CsA (NIM811)	0.1	>2
[Thr][2][5'-HOMeLeu][9]CsA[158]	<3	ND	[4'-AcOMeLeu][4]CsA (13)	0.25	0.03
[D-Sar-OMe][3]CsA[136]	0.06	<0.1	[MeVal][5]CsA	0.77	0.05
[D-Sar-SMe][3][4'-HOMeLeu][4]CsA	0.04	0.06	[BnVal][5]CsA	0.5	>10
[D-Sar-SCH$_2$CH$_2$NMe$_2$][3][4'-HOMeLeu][4]CsA	0.1	>2	[Me-D-Ala][3][EtVal][4]CsA (alisporivir)	0.03	>2
[D-Sar-SCH$_2$(3-pyridyl)][3][4'-HOMeLeu][4]CsA (10)	0.06	>10	[D-Lys][8]CsA	2.0	0.1
[D-Sar-SCH$_2$(4-thiazolyl)][3][4'-HOMeLeu][4]CsA (11)	0.1	24	[3-CN-D-Ala][8]CsA	0.04	0.005
[D-Sar-CH$_2$SCH$_2$CH$_2$NMe$_2$][3][4'-HOMeLeu][4]CsA (12)[139]	0.05	ND	[N-ε-Me$_2$-D-Lys][8]CsA	0.14	0.4
[MeVal][4]CsA[72]	0.15	>10	[MeAla][10]CsA[72]	8.48	>15xCsA

HCV IC$_{50}$ refers to activity in Con-1a sub-genomic replicon systems using Huh-7 hepatoma cells; IL-2 secretion assays utilized Jurkat cells stimulated with phorbol 12-myristate 13-acetate and phytohemagglutinin; IL-2 levels in the medium were determined by ELISA 24 h after stimulation. Calcineurin is required for stimulation of IL-2 production, and inhibition of IL-2 secretion is a surrogate measure of calcineurin inhibition and immunosuppression.

The [MeLeu][4] modification is of importance for controlling immuno-suppression, and several reports have appeared that describe modification of this residue alone to achieve potent anti-HCV activity. Scynexis has explored [4'-ROMeLeu][4]CsA analogs where R = ester[160] (e.g. 13) or ether,[161] from which work [4'-BnOMeLeu][4]CsA was found to have comparable activity to CsA itself. Enanta, in a series of patent applications, described various [MeThrOR][4]CsA derivatives in which the threonine hydroxyl is functionalized with alkyl chains of varying length, often bearing a substituted heteroatom (N, O).[150–155]

Alkylation of the nitrogen of [Val][5] to attenuate immunosuppressive potential without significantly impacting CypA binding was outlined earlier and introduction of simple benzyl or allyl groups was found to retain anti-HCV activity, albeit with a loss of potency.[162] Combination of such [Val][5] N-substitution, including fully saturated heteroatom-substituted alkyl chains, with optimal [Sar][3] functionality such as D-alkyl ethers and D-alkyl thioethers,[163] and also D-alkyl groups,[164] led to several compounds exhibiting potent anti-HCV activity (<200 nM).

Of the remaining positions, 6–11, only for [D-Ala][8] has any detailed description of anti-HCV activity been described. Replacing [D-Ala][8] with D-Lys derivatives of varying length to control immunosuppression yielded compounds with greater anti-HCV potency than CsA itself, including [3'-CN-D-Ala][8]CsA (40 nM) and the less immunosuppressive [3'-CH$_2$CH$_2$CH$_2$NMe$_2$-D-Ala][8]CsA (140 nM).[122,165] Peptolides in which an ester bond containing [Hiv][8] in place of [D-Ala][8] were described by the DebioPharm group.[166] A number of other amino acid changes were incorporated in addition to the changes at position 8 that led to potent anti-HCV compounds such as [Val]^2[D-MeAla]^3[MeVal]^4[Ile]^5[Gly]^7[D-Hiv]^8[Leu]^{10}CsA (22 nM).

11.3.4 ADME Properties of Cyclosporine and Derivatives

Cyclosporine A displays extremely complex pharmacokinetic properties that are still being evaluated over 40 years after its discovery. Cytochrome P450-mediated metabolism and P-gp efflux transport were recognized early as key factors that affect the plasma exposure of CsA.[167] However, additional factors such as a saturable binding to erythrocytes,[168] interaction with a wide range of transporters[169] and a conformation-dependent solubility profile for CsA derivatives make the optimization of new CsA derivatives challenging.

CsA is a relatively hydrophobic compound that is subject to extensive metabolism by Cyp3A4 and Cyp3A5.[167] A comprehensive analysis of the metabolite profile of CsA following dosing to dogs and humans revealed at least nine significant species that were isolated from the bile and feces.[167] Primary routes of metabolism of CsA involve oxidative hydroxylation of the side chains of residues 1, 4, 6 and 9 and a demethylation of the [MeLeu]^4 residue (Figure 11.9).[167] The three predominant metabolites resulted from mono-hydroxylation at the γ-carbon of the [MeLeu]^9 side chain, hydroxylation of the terminal carbon of the Bmt side chain and a monodemethylation at the [MeLeu]^4 residue.[167] Minor metabolic routes included hydroxylation at the γ-carbons of [MeLeu]^4 and [MeLeu]^6, an epoxidation of the MeBmt olefin and products resulting from sequential oxidations.[167] Tetrahydrofuran derivatives resulting from the non-enzymatic ring-closure reaction of Bmt epoxides were also isolated.[167,170]

Limited data have been published regarding the metabolism of derivatives of CsA. Alisporivir appears to be more resistant to metabolism in humans, with up to 66% of the dose being excreted in the feces as unchanged drug,[171] which

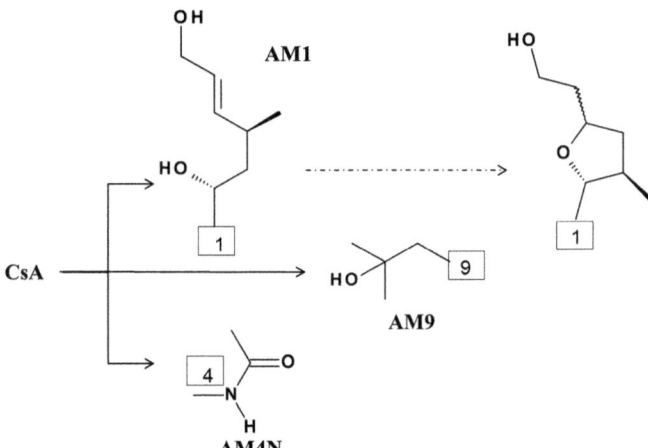

Figure 11.9 Primary human metabolites of CsA.

would be consistent with one of the key metabolism sites in CsA (*N*-demethylation at [MeLeu]^4) being blocked by an ethyl group. However, the remaining portion of the drug undergoes extensive metabolism and no single metabolite has been reported.

CsA displays a selective distribution profile in blood, with the major fraction being found in erythrocytes ($\sim 58\%$), while the portion of drug in the plasma is extensively bound by the lipoprotein component.[172] The distribution of CsA into erythrocytes appears to be a saturable process that can be affected by co-administration of certain excipients. A demonstration that Cremophor EL can significantly increase the plasma concentration in rats of a given dose of CsA was supported by the finding that this excipient also limited the erythrocyte uptake of the drug.[173] Alisporivir demonstrates a concentration-dependent distribution profile when examined *in vitro*. At low blood concentration ($<0.06\,\mu M$), 90% of the compound was found in the erythrocyte fraction, which fell to 30% as the blood concentration was increased to $2\,\mu M$.[174] A commensurate increase in the proportion of the drug in plasma was observed at higher blood concentrations.[174] Pharmacokinetic data published for SCY-635 indicate that the compound has little plasma exposure when delivered at low doses to humans but that a supra-proportional increase in the plasma fraction occurs as the dose level is raised. These data support a model wherein the drug is initially taken into an erythrocyte fraction until such point that this compartment is saturated whereupon the compound 'spills' into the plasma.[175]

CsA was recognized early to be an inhibitor of the P-gp efflux transporter and a modified CsA derivative, PSC-833, was actively studied as a chemo-sensitizer to anti-cancer agents such as vinblastine by preventing the P-gp-mediated efflux of the latter from MDR cancer cells.[176] A study of the SARs of CsA metabolites as inhibitors of P-gp indicated that most of the primary metabolites of CsA are weaker inhibitors of P-gp than the parent molecule; however, for the closely related analog CsG, demethylation at [MeLeu]^4 resulted in a significant increase in P-gp inhibition.[177] CsA is a well-known inhibitor of hepatic transporters, including the organic anion trans-porters (OATPs) and the multi-drug resistance protein transporters, including Mrp-2.[169] The concentration required to inhibit by 50% many of these trans-porters by CsA falls within the range of clinically used blood levels of the drug and notable drug–drug interactions involving CsA and drugs such as statins have been observed.[178] One of the main adverse events noted during clinical studies of alisporivir in HCV-infected patients was an elevation of plasma bilirubin levels that could be ascribed to the inhibitory activity of alisporivir on the key hepatic transporter responsible for clearing bilirubin from blood, Mrp-2.[78] *In vitro* studies confirmed a significant inhibition of Mrp-2 by alis-porivir. However, the finding that SCY-635 is significantly less inhibitory toward several transporters such as Mrp-2 and OATP indicates that SARs can be established to avoid unwanted interactions.

These characteristics of cyclosporine and its derivatives demand that care be taken when initially studying the pharmacokinetic properties of a new compound.

11.4 New Opportunities for Cyclophilin Inhibitors as Antiviral Agents

11.4.1 Non-cyclosporine Cyclophilin Inhibitors

Attempts to discover 'small-molecule' inhibitors of Cyps, using either the protein or ligand structures, have appeared over the last 10 years (Figure 11.10).[179–182] Early attempts to use peptide substrates as starting points identified Succ–Ala–Ala–Pro–Phe–p-NO$_2$–anilide, which is known to bind CypA with a K_m of 870 μM.[183] Replacing the central Ala–Pro core with a (Z)-alkene *cis*-Pro (**14**) mimetic resulted in a compounds with an IC$_{50}$ of 6.5 μM. The demonstration that a CypA–Gag interaction is crucial for effective replication of HIV-1 led to an effort to use Gag sequences as inhibitors of CypA. A pentapeptide Dav–His–Ala–Gly–Pro–Ile–NHBn was found to have higher affinity for CypA ($K_d = 3$ μM) than the entire capsid protein ($K_d = 16$ μM).[184] More recently, a virtual screening exercise performed by Zhang and co-workers identified short peptides (Trp–Gly–Pro, Trp–Ala–Gly–Pro and Tyr–Gly–Pro) that showed inhibition of CypA PPIase activity with potency (33 nM) comparable to that of CsA (10 nM).[185]

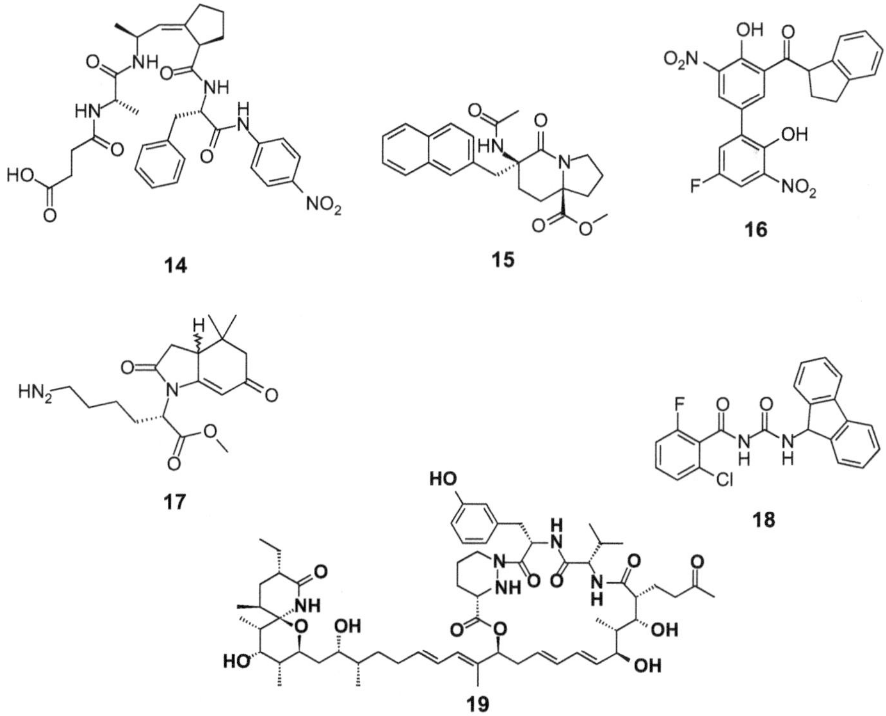

Figure 11.10   Non-CsA inhibitors of CypA.

The design of inhibitors based on the isomerase mechanism of Cyps led to bicyclic lactams that mimic the ground state of a proline-bearing peptide, along with bicyclic amines in which a tertiary amine mimics a 'twisted amide' transition state.[15] The ground-state mimic (15) was found to bind more efficiently to CypA ($K_d = 1.5\,\mu M$ *versus* $K_d = 77\,\mu M$) for this series of derivatives. A series of small-molecule inhibitors based on aryl 1-indanyl ketones was designed by Schiene-Fischer to mimic the 'twisted amide' of a putative transition state and led to compounds (16) that not only bound tightly to CypA (K_i 520 nM) but also demonstrated selectivity for CypA over CypB.[12,13] Very minimal ligands for CypA were developed by Walkinshaw and co-workers by combining a 'twisted amide' concept with a hydrophobic group designed to mimic [MeVal][11] of CsA; however, the affinity of these ligands was poor (>25 mM).[186]

A successful application of fragment-based ligand design identified simple dimedone inhibitors in which the dimethyl group was shown to mimic the isopropyl group present in [MeVal][11] of CsA while one of the carbonyls mimics the [MeLeu][10] carbonyl.[187] Further structure-aided elaboration of these weak ligands led to potent inhibitors of CypA such as 17 (K_D 11 μM), which demonstrated activity against *Caenorhabditis elegans* consistent with inhibition of CypA.[188]

Virtual screening efforts using the key residues in CypA known to be involved in binding CsA ([Trp][121], [Arg][55], [Asn][102], [Gln][63], [Asn][102]) have been employed by a number of groups. A series of thioureas with good potency against CypA (IC_{50} 0.48 μM) were reported by Wu *et al.*[189] as potential neuroprotective/neurotrophic agents, while Chavanieu and co-workers[190] and Jiang and co-workers[191] identified very potent diarylureas ($IC_{50} < 0.05\,\mu M$) (*e.g.* 18) that showed activity in HIV-1 infection assays.

Further studies by Jiang's group using virtual screening in conjunction with surface plasmon resonance to screen for non-CsA-based CypA inhibitors[192–194] identified substituted anilides with modest potency as inhibitors of CypA ($IC_{50} \sim 3\,\mu M$), which showed evidence of immunosuppression. Subsequent elaboration of these leads using X-ray structural data resulted in potent ligands for CypA. Finally, high-throughput screening of diverse chemical libraries has identified new ligands for Cyps[195,196] with modest affinity; however, no further modification of these starting points has been described.

One final compound mentioned in this section is sanglifehrin A (19) (Figure 11.10). Although it is not a small molecule, it is a natural product found to have 60-fold higher affinity for Cyp ($IC_{50} = 6.9$ nM) than CsA itself ($IC_{50} = 420$ nM),[197] and there have been several recent reports of sanglifehrin and synthetic derivatives as potential Cyp-inhibiting drugs, including anti-HIV-1 and/or anti-HCV activity therapeutics.[198–206]

In conclusion, there are several examples of non-CsA-based CypA inhibitors in the recent literature, some of which show CypA binding comparable to that of CsA itself. These synthetically feasible small molecules could foster drug discovery programs towards CypA inhibitors tailored for specific diseases through specific Cyp isoforms, and may have the added benefit of being devoid of CsA-related toxicity.

11.4.2 Immunomodulation by Cyclophilin Inhibition

Although a role for CypA as an essential component of the HCV replication complex has been well established by a number of groups, recent data emerging from clinical studies using the non-immunosuppressive compound SCY-635 suggest additional mechanisms by which this compound may exert antiviral activity. When dosed as monotherapy to HCV-infected subjects, the compound resulted in antiviral responses that were highly dependent on an SNP located near the IL28B gene that has been shown to predict responsiveness to interferon-containing treatment regimens.[81] Further, the compound produced a transient but significant increase of plasma protein concentrations of interferons (IFNα and λ1) and 2′,5′-oligoadenylate synthase (2′,5′-OAS-1) consistent with the restoration of innate immune responses (Figure 11.11).[81] Notably, these effects were observed only in HCV-infected subjects, which suggests that the mechanism involved is distinct from a global immune-stimulation of the type induced by TLR agonists.[207]

Consistent with these findings, it was shown that interferon induction occurs when HCV replicon-bearing Huh7 cells are treated with Cyp inhibitors while no induction is observed in uninfected cells.[81]

Possible new roles for Cyps in the innate immune response to HCV infection are beginning to appear. CypA was shown to be capable of binding to IRF-9, a key component of the interferon-activated JAK/STAT signaling pathway, and that this interaction can be abrogated by expression of the HCV NS5A protein.[175] Gallay and co-workers further showed that inhibition of CypA, using CsA, led to an enhancement of interferon-induced transcriptional activity of the interferon-sensitive response element (ISRE) in uninfected

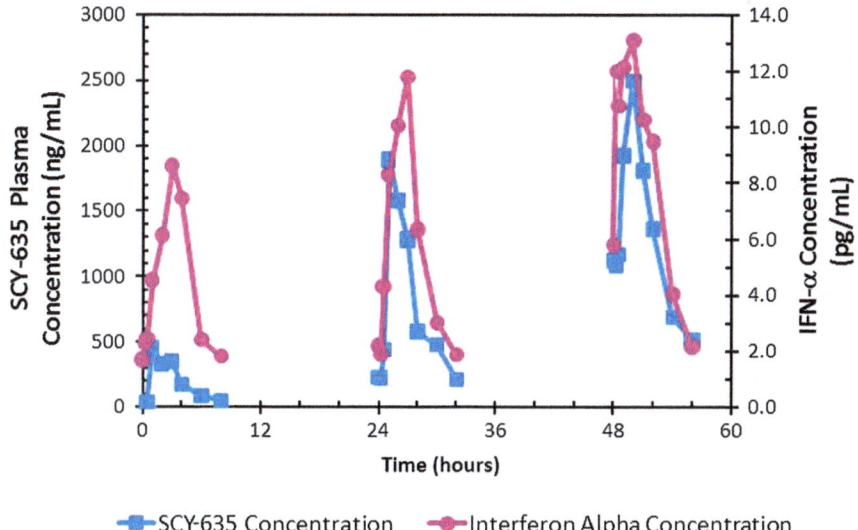

Figure 11.11 Plasma concentrations of SCY-635 (■) and interferon-α (●) following dosing of SCY-635 to an HCV-infected subject.

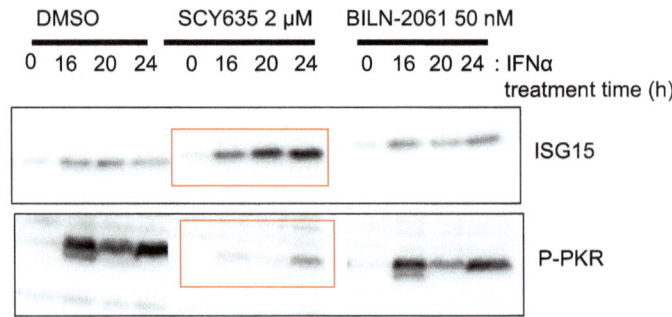

Figure 11.12 Prevention of PKR phosphorylation in response to IFN treatment, and consequent induction of anti-viral ISG protein production, by SCY-635 in HCV (JFH-1)-infected hepatocytes. Inhibition of viral replication using BILN-2061 (HCV NS3/4A protease inhibitor) does not show similar activity.

HepG2 cells.[175] These results clearly point to a role for CypA in the HCV NS5A-mediated interaction with innate immune responses by controlling transcription.[175] A further line of investigation by Watashi *et al.* focused on HCV-mediated blocks on translation and has revealed yet another mechanism by which HCV exploits CypA (Figure 11.12).[208] Interferon-induced activation of PKR resulting in phosphorylation of the translation initiator eIF2α to prevent translation of antiviral proteins, including interferons, is recognized as a mechanism by which HCV subverts innate immune responses.[209] The demonstration that SCY-635, but not an HCV protease inhibitor, significantly reverses the interferon-induced activation leading to expression of antiviral proteins, such as ISG-15, implicates CypA in the HCV-induced block on immune responses.

Clearly, these recent findings of a role for CypA in HCV-mediated avoidance of immune surveillance present opportunities for the use of Cyp inhibitors as immune restoration agents in HCV therapy. The full evaluation of this activity may require human clinical studies since fully immune-competent models of HCV infection are currently lacking.

11.5   Conclusion

The opportunity to separate cyclophilin inhibition from calcineurin activity using modified CsA derivatives has allowed the biology of these abundant proteins to be evaluated more fully over the course of the last 10 years. While the HCV field has seen the use of non-immunosuppressive cyclophilin inhibitors mature into an accepted host-targeted antiviral approach, the recognition that cyclophilins are key players in the replication cycle of many viruses bodes well for future cyclophilin-directed strategies. The ubiquitous use of CsA as the starting point for most of the recent drug discovery efforts is a consequence of the well-established cyclophilin inhibitory activity of this compound; however, the compound carries some significant liabilities. The recent identification of new, small-molecule inhibitors of cyclophilins promises to make the field more acceptable to the medicinal chemist; however, the biology of cyclophilins is clearly far from understood and new therapeutic opportunities for cyclophlin inhibition may emerge in the near future.

References

1. J. Liu, J. D. Farner Jr., W. S. Lane, J. Friedman, I. Weissman and S. L. Schreiber, *Cell*, 1991, **66**, 807.
2. J. Kallen, V. Mikol, V. F. J. Quesniaux, M. D. Walkinshaw, E. Schneider-Scherzer, K. Schorgendorfer, G. Weber and H. G. Fliri, in *Biotechnology, Products of Secondary Metabolism*, H.-J. Rehm, H. Kleinkauf and H. VonDohren, (eds), 1997, vol. 7, p. 535. VCH Weinheim, 1997.
3. A. Billich, F. Hammerschmid, P. Peichl, R. Wenger, G. Zenke, V. Quesniaux and B. Rosenworth, *J. Virol.*, 1995, **69**, 2451.
4. G. Fischer, *Angew. Chem.*, 1994, **106**, 1479.
5. G. Fischer, *Angew. Chem. Int. Ed. Engl.*, 1994, **33**, 1415.
6. T. Kiefhaber, R. Quaas, U. Hahn and F. X. Schmid, *Biochemistry*, 1990, **29**, 3061.
7. A. Galat, *Proteins: Struct. Funct. Bioinf.*, 2004, **56**, 808.
8. C. B. Kang, Y. Hong, S. Dhe-Paganon and H. S. Yoon, *Neuorosignals*, 2008, **16**, 318.
9. C. Schiene and G. Fischer, *Curr. Opin. Struct. Biol.*, 2000, **10**, 40.
10. S. L. Schreiber, *Science*, 1991, **251**, 283.
11. A. Ryo, Y.-C. Liou, K. P. Lu and G. Wulf, *J. Cell Sci.*, 2003, **116**, 773.
12. S. Daum, F. Erdmann, G. Fischer, B. Feaux de Lacroix, A. Hessamian-Alinejad, S. Houben, W. Frank and M. Braun, *Angew. Chem. Int. Ed.*, 2006, **45**, 7454.
13. S. Daum, M. Schumann, S. Mathea, T. Aumueller, M. A. Balsley, S. L. Constant, B. Feaux de Lacroix, F. Kruska, M. Braun and C. Schiene-Fischer, *Biochemistry*, 2009, **48**, 6268.
14. R. K. Harrison and R. L. Stein, *Biochemistry*, 1990, **29**, 1684.
15. H. C. Wang, K. Kim, R. Bakhtiar and J. P. Germanas, *J. Med. Chem.*, 2001, **44**, 2593.
16. Y. Zhao and H. Ke, *Biochemistry*, 1996, **35**, 7356.
17. L. D. Zydowsky, F. A. Etzkom, H. Y. Chang, S. B. Ferguson, L. A. Stolz, S. I. Ho and C. T. Walsh, *Protein Sci.*, 1992, **1**, 1092.
18. H. Ke and Q. Huai, *Front. Biosci., Landmark Ed.*, 2004, **9**, 2285.
19. L. Yan, S. Zucker and B. P. Toole, *Thromb. Haemostasis*, 2005, **93**, 199.
20. E. J. Stemmy, M. A. Balsley, R. A. Jurjus, J. M. Damsker, M. I. Bukrinsky and S. L. Constant, *Am. J. Respir. Cell Mol. Biol.*, 2011, **45**, 991.
21. M. Cour, J. Loufouat, M. Paillard, L. Augeul, J. Goudable, M. Ovize and L. Argaud, *Eur. Heart J.*, 2011, **32**, 226.

22. M. Dreyfuss, E. Haerri, H. Hoffmann, H. Kobel, W. Pache and H. Tscherter, *Eur. J. Appl. Microbiol.*, 1976, **3**, 125.
23. A. Ruegger, M. Kuhn, H. Lichti, H.-R. Loosli, R. Huguenin, C. Quiquerez and A. von Wartburg, *Helv. Chim. Acta*, 1976, **59**, 1075.
24. H. R. Loosli, H. Kessler, H. Oschkinat, H. P. Weber, T. J. Petcher and A. Widmer, *Helv. Chim. Acta*, 1985, **68**, 682.
25. N. El Tayar, A. E. Mark, P. Vallat, R. M. Brunne, B. Testa and W. F. van Gunsteren, *J. Med. Chem.*, 1993, **36**, 3757.
26. R. M. Wenger, *Angew. Chem. Int. Ed. Engl.*, 1985, **24**, 77.
27. D. Seebach, H. G. Bossler, H. Grundler, S.-i. Shoda and R. Wenger, *Helv. Chim. Acta*, 1991, **74**, 197.
28. D. Seebach, A. K. Beck, H. G. Bossler, C. Gerber, S. Y. Ko, C. W. Murtiashaw, R. Naef, S.-i. Shoda and A. Thaler, *Helv. Chim. Acta*, 1993, **76**, 1564.
29. C. Papageorgiou, J. Kallen, J. France and R. French, *Bioorg. Med. Chem.*, 1997, **5**, 187.
30. D. Seebach, H. G. Bossler, R. Flowers and E. M. Arnett, *Helv. Chim. Acta*, 1994, **77**, 291.
31. H. Hasumi, T. Nishikawa and H. Ohtani, *Biochem. Mol. Biol. Int.*, 1994, **34**, 505.
32. S. L. Schreiber and G. R. Crabtree, *Immunol. Today*, 1992, **13**, 136.
33. A. Galat and J. Bua, *Cell. Mol. Life Sci.*, 2010, **67**, 3467.
34. W. Ping and H. Joseph, *Genome Biol.*, 2005, **6**, 226.
35. Q. Huai, H.-Y. Kim, Y. Liu, Y. Zhao, A. Mondragon, J. O. Liu and H. Ke, *Proc. Natl. Acad. Sci. U. S. A.*, 2002, **99**, 12037.
36. H. Ke and Q. Huai, *Biochem. Biophys. Res. Commun.*, 2003, **311**, 1095.
37. L. Jin and S. C. Harrison, *Proc. Natl. Acad. Sci. U. S. A.*, 2002, **99**, 13522.
38. J. Votteler, V. Wray and U. Schubert, *Fut. Virol.*, 2007, **2**, 65.
39. H. Tang, *Viruses*, 2010, **2**, 1621.
40. J. M. Andrieu, P. Even, A. Venet, J. M. Tourani, M. Stern, W. Lowenstein, C. Audroin, D. Eme, D. Masson, H. Sors, D. Israel-Biet and K. Beldjord, *Clin. Immunol. Immunopathol.*, 1988, **47**, 181.
41. J. Luban, K. L. Bossolt, E. K. Franke, G. V. Kalpana and S. P. Goff, *Cell*, 1993, **73**, 1067.
42. E. K. Franke, H. E. H. Yuan and J. Luban, *Nature*, 1994, **372**, 359.
43. M. Thali, A. Butkovsky, E. Kondo, B. Rosenwirth, C. T. Walsh, J. Sodroski and H. G. Goettlinger, *Nature*, 1994, **372**, 363.
44. T. R. Gamble, F. F. Vajdos, S. Yoo, D. K. Worthylake, M. Houseweart, W. I. Sundquist and C. P. Hill, *Cell*, 1996, **87**, 1285.
45. D. Braaten, E. Franke and J. Luban, *J. Virol.*, 1996, **70**, 3551.
46. N. A. Kootstra, C. Munk, N. Tonnu, N. R. Landau and I. M. Verma, *Proc. Natl. Acad. Sci. U. S. A.*, 2003, **100**, 1298.
47. E. Sokolskaja, D. M. Sayah and J. Luban, *J. Virol.*, 2004, **78**, 12800.
48. T. Hatziioannou, D. Perez-Caballero, S. Cowan and P. D. Bieniasz, *J. Virol.*, 2005, **79**, 176.

49. G. J. Towers, T. Hatziioannou, S. Cowan, S. P. Goff, J. Luban and P. D. Bieniasz, *Nat. Med.*, 2003, **9**, 1138.
50. J. Zhang, W. Ge, P. Zhan, E. De Clercq and X. Liu, *Curr. Med. Chem.*, 2011, **18**, 2649.
51. L. Li, H. S. Li, C. D. Pauza, M. Bukrinsky and R. Y. Zhao, *Cell Res.*, 2005, **15**, 923.
52. L. R. Erwann and B. Serge, *Retrovirology*, 2005, **2**, 11.
53. K. Inoue, M. Yoshiba and N. Oomori, presented at the 6th International Symposium on Hepatitis C and Related Viruses, 1999.
54. K. Watashi, M. Hijikata, M. Hosaka, M. Yamaji and K. Shimitohno, *Hepatology*, 2003, **38**, 1282.
55. K. Inoue, K. Sekiyama, M. Yamada, T. Watanabe, H. Yasuda and M. Yoshiba, *J. Gastroenterol.*, 2003, **38**, 567.
56. K. Watashi, N. Ishii, M. Hijikata, D. Inoue, T. Murata, Y. Miyanari and K. Shimitohno, *Mol. Cell.*, 2005, **19**, 111.
57. M. Nakagawa, N. Sakamoto, Y. Tanabe, T. Koyama, Y. Itsui, Y. Takeda, C.-H. Chen, S. Kakinuma, S. Oooka, S. Maekawa, N. Enomoto and M. Watanabe, *Gastroenterology*, 2005, **129**, 1031.
58. J. A. Heck, X. Meng and D. N. Frick, *Biochem. Pharmacol.*, 2009, **77**, 1173.
59. F. Yang, J. M. Robotham, H. B. Nelson, A. Irsigler, R. Kenworthy and H. Tang, *J. Virol.*, 2008, **82**, 5269.
60. A. Kaul, S. Stauffer, C. Berger, T. Pertel, J. Schmitt, S. Kallis, M. Z. Lopez, V. Lohmann, J. Luban and R. Bartenschlager, *PLoS Pathog.*, 2009, **5**, 1.
61. U. Chatterji, M. Bobardt, S. Selvarajah, F. Yang, H. Tang, N. Sakamoto, G. Vuagniaux, T. Parkinson and P. Gallay, *J. Biol. Chem.*, 2009, **284**, 16998.
62. S. Ciesek, E. Steinmann, H. Wedemeyer, M. P. Manns, J. Neyts, N. Tautz, V. Madan, R. Bartenschlager, T. von Hahn and T. Pietschmann, *Hepatology*, 2009, **50**, 1638.
63. K.-I. Abe, M. Ikeda, Y. Ariumi, H. Dansako, T. Wakita and N. Kato, *Arch. Virol.*, 2009, **154**, 1671.
64. L. A. Gaither, J. Borawski, L. J. Anderson, K. A. Balabanis, P. Devay, G. Joberty, C. Rau, M. Schirle, T. Bouwmeester, C. Mickanin, S. Zhao, C. Vickers, L. Lee, G. Deng, J. Baryza, R. A. Fujimoto, K. Lin, T. Compton and B. Wiedmann, *Virology*, 2010, **397**, 43.
65. G. Fischer, P. Gallay and S. Hopkins, *Curr. Opin. Invest. Drugs*, 2010, **11**, 911.
66. P. A. Gallay, *Immunol. Res.*, 2012, **52**, 200.
67. J. M. Robida, H. B. Nelson, Z. Liu and H. Tang, *J. Virol.*, 2007, **81**, 5829.
68. U. Chatterji, M. D. Bobardt, P. Lim and P. A. Gallay, *J. Gen. Virol.*, 2010, **91**, 1189.
69. R. Flisiak, A. Horban, P. Gallay, M. Bobardt, S. Selvarajah, A. Wiercinska-Drapalo, E. Siwak, I. Cielniak, J. Higersberger, J. Kierkus,

C. Aeschlimann, P. Grosgurin, V. Nicolas-Metral, J.-M. Dumont, H. Porchet, R. Crabbe and P. Scalfaro, *Hepatology*, 2008, **47**, 817.

70. B. Rosenwirth, A. Billich, R. Datema, P. Donatsch, F. Hammerschmid, R. Harrison, P. Hiestand, H. Jaksche, P. Mayer, P. Peichl, V. Quesniaux, F. Schatz, H.-J. Schuurman, R. Traber, R. Wenger, B. Wolff, G. Zenke and M. Zurini, *Antimicrob. Agents Chemother.*, 1994, **38**, 1763.

71. K. Goto, K. Watashi, T. Murata, T. Hishiki, M. Hijikata and K. Shimitohno, *Biochem. Biophys. Res. Commun.*, 2006, **343**, 879.

72. S. Ma, J. E. Boerner, C. TiongYip, B. Weidmann, N. S. Ryder, M. P. Cooreman and K. Lin, *Antimicrob. Agents Chemother.*, 2006, **50**, 2976.

73. R. G. Ptak, P. A. Gallay, D. Jochmans, A. P. Halestrap, U. T. Ruegg, L. A. Pallansch, M. D. Bobart, M.-P. de Bethune, J. Neyts, E. De Clercq, J.-M. Dumont, P. Scalfaro, K. Besseghir, R. M. Wenger and B. Rosenwirth, *Antimicrob. Agents Chemother.*, 2008, **52**, 1302.

74. L. M. Plitnick and D. Herzyk, *Methods Mol. Biol.*, 2010, **598**, 159.

75. L. Coelmont, S. Kaptein, J. Paeshuyse, I. Vliegen, J.-M. Dumont, G. Vuagniaux and J. Neyts, *Antimicrob. Agents Chemother.*, 2009, **53**, 967.

76. L. Coelmont, X. Hanoulle, U. Chatterji, C. Berger, J. Snoeck, M. Bobardt, P. Lim, I. Vliegen, J. Paeshuyse, G. Vuagniaux, A.-M. Vandamme, R. Bartenschlager, P. A. Gallay, G. Lippens and J. Neyts, *PLoS One*, 2010, **5**.

77. R. Flisiak and A. Parfieniuk-Kowerda, *Curr. Hepatitis Rep.*, 2012.

78. R. Flisiak, J. Jaroszewicz, I. Flisiak and T. Lapinnski, *Expert Opin. Invest. Drugs*, 2012, **21**, 375.

79. M. Evers, J.-C. Barriere, G. Bashiardes, A. Bousseau, J.-C. Carry, N. Dereu, B. Filoche, Y. Henin, S. Sable, M. Vuilhorgne and S. Mignani, *Bioorg. Med. Chem. Lett.*, 2003, **13**, 4415.

80. S. Hopkins, B. Scorneaux, Z. Huang, M. G. Murray, S. Wring, C. Smitley, R. Harris, F. Erdmann, G. Fischer and Y. Ribeill, *Antimicrob. Agents Chemother.*, 2010, **54**, 660.

81. S. Hopkins, B. DiMassimo, P. Rusnak, D. Heuman, J. Lalezari, A. Sluder, B. Scorneaux, S. Mosier, P. Kowalczyk, Y. Ribeill, J. Baugh and P. Gallay, *J. Hepatol.*, 2012, **57**, 47.

82. M. Qing, F. Yang, B. Zhang, G. Zou, J. M. Robida, Z. Yuan, H. Tang and P.-Y. Shi, *Antimicrob. Agents Chemother.*, 2009, **53**, 3226.

83. H. Kambara, H. Tani, Y. Mori, T. Abe, H. Katoh, T. Fukuhara, S. Taguwa, K. Moriishi and Y. Matsuura, *Virology*, 2011, **412**, 211.

84. C. R. A. Damaso and N. Moussatche, *J. Gen. Virol.*, 1998, **79**, 339.

85. A. P. V. Castro, T. M. U. Carvalho, N. Moussatche and C. R. A. Damaso, *J. Virol.*, 2003, **77**, 9052.

86. S. Bose, M. Mathur, P. Bates, N. Joshi and A. K. Banerjee, *J. Gen. Virol.*, 2003, **84**, 1687.

87. Z. Chen, L. Mi, J. Xu, J. Yu, X. Wang, J. Jiang, J. Xing, P. Shang, A. Qian, Y. Li, P. X. Shaw, J. Wang, S. Duan, J. Ding, C. Fan, Y. Zhang,

Y. Yang, X. Yu, Q. Feng, B. Li, X. Yao, Z. Zhang, L. Li, X. Xue and P. Zhu, *J. Infect. Dis.*, 2005, **191**, 755.

88. B.-H. Malgorzata, P. H. D and S. Martin, *PLoS Pathog.*, 2009, **5**, e1000524.
89. A. Watanabe, M. Yoneda, F. Ikeda, Y. Terao-Muto, H. Sato and C. Kai, *J. Virol.*, 2010, **84**, 4183.
90. A. Vahlne, P. A. Larsson, P. Horal, J. Ahlmen, B. Svennerholm, J. S. Gronowitz and S. Olofsson, *Arch. Virol.*, 1992, **122**, 61.
91. X. Liu, Z. Zhao, C. Xu, L. Sun, J. Chen, L. Zhang and W. Liu, *PLoS One*, 2012, **7**, e31063.
92. H. He, D. Zhou, W. Fan, X. Fu, J. Zhang, Z. Shen, J. Li, J. Li and Y. Wu, *Biochem. Biophys. Res. Commun.*, 2012, **422**, 664.
93. Y. Obata, K. Yamamoto, M. Miyazaki, K. Shimotohno, S. Kohno and T. Matsuyama, *J. Biol. Chem.*, 2005, **280**, 18355.
94. J. D. Aebi, D. T. Deyo, C. Q. Sun, D. Guillaume, B. Dunlap and D. H. Rich, *J. Med. Chem.*, 1990, **33**, 999.
95. N. H. Sigal, F. Dumont, P. Durette, J. J. Siekierka, L. Peterson, D. H. Rich, B. E. Dunlap, M. J. Staruch, M. R. Melino, S. L. Koprak, D. Williams, B. Witzel and J. M. Pisano, *J. Exp. Med.*, 1991, **173**, 619.
96. V. F. J. Quesniaux, M. H. Schreier, R. M. Wenger, P. C. Hiestand, M. W. Harding and M. H. V. Van Regenmortel, *Eur. J. Immunol.*, 1987, **17**, 1359.
97. M. K. Eberle, F. Nuninger and H.-P. Weber, *J. Org. Chem.*, 1995, **60**, 2610.
98. D. H. Rich, M. K. Dhaon, B. Dunlap and S. P. F. Miller, *J. Med. Chem.*, 1986, **29**, 978.
99. J. Bua, A. M. Ruiz, M. Potenza and L. E. Fichera, *Bioorg. Med. Chem. Lett.*, 2004, **14**, 4633.
100. C.-q. Sun, D. Guillaume, B. Dunlap and D. H. Rich, *J. Med. Chem.*, 1990, **33**, 1443.
101. C. Papageorgiou, H.-P. Weber, R. French and X. Borer, *Bioorg. Med. Chem. Lett.*, 1995, **5**, 213.
102. C. Papageorgiou, A. Florineth and R. French, *Bioorg. Med. Chem. Lett.*, 1993, **3**, 2559.
103. V. Mikol, C. Papageorgiou and X. Borer, *J. Med. Chem.*, 1995, **38**, 3361.
104. S. Y. Ko and R. M. Wenger, *Helv. Chim. Acta*, 1997, **80**, 695.
105. M. Keller, T. Wohr, P. Dumy, L. Patiny and M. Mutter, *Chem. Eur. J*, 2000, **6**, 4358.
106. D. Seebach, *Angew. Chem. Int. Ed. Engl.*, 1988, **27**, 1624.
107. M. C. Cruz, M. Del Poeta, P. Wang, R. Wenger, G. Zenke, V. F. J. Quesniaux, N. R. Mowa, J. R. Perfect, M. E. Cardenas and J. Heitman, *Antimicrob. Agents Chemother.*, 2000, **44**, 143.
108. J. Kallen, V. Mikol, P. Taylor and M. D. Walkinshaw, *J. Mol. Biol.*, 1998, **283**, 435.
109. M. Kuhnt, F. Bitsch, J. France, H. Hofmann, J.-J. Sanglier and R. Traber, *J. Antibiot.*, 1996, **49**, 781.

110. C. Papageorgiou, X. Borer and R. R. French, *Bioorg. Med. Chem. Lett.*, 1994, **4**, 267.

111. L. Wei, J. P. Steiner, G. S. Hamilton and Y.-Q. Wu, *Bioorg. Med. Chem. Lett.*, 2004, **14**, 4549.

112. C. Papageorgiou, A. Florineth and V. Mikol, *J. Med. Chem.*, 1994, **37**, 3674.

113. F. Hubler, T. Ruckle, L. Patiny, T. Muamba, J.-F. Guichou, M. Mutter and R. Wenger, *Tetrahedron Lett.*, 2000, **41**, 7193.

114. P. L. Durrette, J. Kollonitsch and A. A. Pessolano, Merck & Co., Inc., *Br. Pat.*, 2 207 678, 1989.

115. L. Patiny, J.-F. Guichou, M. Keller, O. Turpin, T. Ruckle, P. Lhote, T. M. Buetler, U. T. Ruegg, R. M. Wenger and M. Mutter, *Tetrahedron*, 2003, **59**, 5241.

116. M.-K. Hu, A. Badger and D. H. Rich, *J. Med. Chem.*, 1995, **38**, 4164.

117. M. K. Eberle, A.-M. Jutzi-Eme and F. Nuninger, *J. Org. Chem.*, 1994, **59**, 7249.

118. D. G. Alberg and S. L. Schreiber, *Science*, 1993, **262**, 248.

119. P. J. Belshaw, S. D. Meyer, D. D. Johnson, D. Romo, Y. Ikeda, M. Andrus, D. G. Alberg, L. W. Schultz, J. Clardy and S. L. Schreiber, *Synlett*, 1994, 381.

120. Y. Zhang, F. Erdmann, R. Baumgrass, M. Schutkowski and G. Fischer, *J. Biol. Chem.*, 2005, **280**, 4842.

121. M. K. Eberle, P. Hiestand, A.-M. Jutzi-Eme, F. Nuninger and H. Zihlmann, *J. Med. Chem.*, 1995, **38**, 1853.

122. A. Scribner, D. Houck, Z. Huang, S. Mosier, M. Peel and B. Scorneaux, *Bioorg. Med. Chem. Lett.*, 2010, **20**, 6542.

123. Y. F. Zheng, D. K. Vickery and S. C. Swann, Siemens Healthcare Diagnostics, Inc. *Patent Application*, WO2009-088694 A1, 2009.

124. Y. F. Zheng and T. Q. Wei, Siemens Healthcare Diagnostics, Inc., *Patent Application*, WO2009-089262 A1, 2009.

125. R. Traber, H. Kobel, H.-R. Loosli, H. Senn, B. Rosenwirth and A. Lawen, *Antiviral Chem. Chemother.*, 1994, **5**, 331.

126. B. F. Molino, AMR Technology, Inc., *Patent Application*, WO2007-112352 A2, 2007.

127. B. F. Molino, AMR Technology, Inc., *Patent Application*, WO2007-112345 A2, 2007.

128. B. F. Molino, AMR Technology, Inc., *Patent Application*, WO2007-112357 A2, 2007.

129. S. Y. Ko, H. Kobel, B. Besemer-Rosenwirth, D. Seebach, R. P. Traber, R. Wenger and P. Bollinger, Novartis AG, *US Pat.*, 5 767 069, 1998.

130. S. Y. Ko, H. Kobel, B. Besemer-Rosenwirth, D. Seebach, R. P. Traber, R. Wenger and P. Bollinger, Novartis AG, *US Pat.*, 5 981 479, 1999.

131. S. Y. Ko, H. Kobel, B. Besemer-Rosenwirth, D. Seebach, R. P. Traber and P. Bollinger, Novartis AG, *US Pat.*, 6 255 100 B1, 2001.

132. S. Y. Ko, H. Kobel, B. Rosenwirth, D. Seebach, R. P. Traber, R. Wenger and P. Bollinger, Sandoz Ltd., *Eur. Pat.*, 484 281 A2, 1992.

133. E. Ellmerer-Mueller, D. Brossner, N. Maslouh, H.-D. Ambrosi, G. Jas and G. Fischer, C-Chem AG, *Eur. Pat.*, 1 086 124 B1, 1999.

134. E. Ellmerer-Mueller, D. Brossner, N. Maslouh, H. D. Ambrosi, G. Jas and G. Fischer, C-Chem AG, *US Pat.*, 6 583 265 B1, 2003.

135. E. Ellmerer-Mueller, D. Brossner, N. Maslouh, H.-D. Ambrosi and G. Jas, C-Chem AG, *Patent Application*, WO99-65933 A1, 1999.

136. H. G. Fliri and D. R. Houck, Scynexis, Inc., *Patent Application*, WO2006-039668 A2, 2006.

137. D. R. Houck and K. Li, Scynexis, Inc., *Patent Application*, WO2007-041631 A1, 2007.

138. P. Scalfaro, J.-M. Dumont, G. Vuagniaux and R.-Y. Mauvernay, Debiopharm SA, *Patent Application*, WO2006-038088 A1, 2006.

139. Z. Su, Z. Long, Z. Huang and S. Yang, S&T Global, Inc., *Patent Application*, WO2012-075494 A1, 2012.

140. R. M. Wenger, M. Mutter and T. Ruckle, Debiopharm SA, *US Pat.*, 6 927 208 B1, 2005.

141. A. Steinkasserer, R. Harrison, A. Billich, F. Hammerschmid, G. Werner, B. Wolff, P. Peichl, G. Palfi, W. Schnitzel and E. Mlynar, *J. Virol.*, 1995, **69**, 814.

142. T. Dorfman and H. G. Goettlinger, *J. Virol.*, 1996, **70**, 5751.

143. E. Mlynar, D. Bevec, A. Billich, B. Rosenwirth and A. Steinkasserer, *J. Gen. Virol.*, 1997, **78**, 825.

144. D. Daelemans, J.-M. Dumont, B. Rosenwirth, E. De Clercq and C. Pannecouque, *Antiviral Res.*, 2010, **85**, 418.

145. S. R. Bartz, E. Hohenwalter, M.-K. Hu, D. H. Rich and M. Malkovsky, *Proc. Natl. Acad. Sci. U. S. A.*, 1995, **92**, 5381.

146. D. H. Rich, M. Malkovsky and Y. M. Angell, Wisconsin Alumni Research Foundation, *US Pat.*, 5 639 852, 1999.

147. B. F. Molino, AMR Technology, Inc., *US Pat. Appl.*, 2007-0232531 A1, 2007.

148. B. F. Molino, Albany Molecular Research, Inc., *US Pat.*, 7 696 165 B2, 2010.

149. B. F. Molino, Albany Molecular Research, Inc., *US Pat.*, 7 696 166 B2, 2010.

150. Y. S. Or, G. Wang, J. Long and X. Gao, Enanta Pharmaceuticals, Inc., *Patent Application*, WO2010-088573 A1, 2010.

151. Y. S. Or, G. Wang, J. Long and X. Gao, Enanta Pharmaceuticals, Inc., *US Pat. Appl.*, 2010-0209390 A1, 2010.

152. X. Gao, Y. S. Or, G. Wang and J. Long, Enanta Pharmaceuticals, Inc., *US Pat. Appl.*, 2011-0008284 A1, 2011.

153. J. Long, Y. S. Or, X. Gao and G. Wang, Enanta Pharmaceuticals, Inc., *US Pat. Appl.*, 2011-0008285 A1, 2011.

154. G. Wang, Y. S. Or, J. Long and X. Gao, Enanta Pharmaceuticals, Inc., *US Pat. Appl.*, 2011-0008286 A1, 2011.

155. Y. S. Or, I. J. Kim, G. Wang and J. Long, Enanta Pharmaceuticals, Inc., *US Pat. Appl.*, 2011-0206637 A1, 2011.

156. M. D. Abel, D. P. Czajkowski, B. W. Fenske, A. Hegmans, D. McGlade, D. J. Trepanier, D. R. Ure and S. Sugiyama, Isotechnika Labs, Inc., *Patent Application*, WO2010-012073 A1, 2010.

157. A. Hegmans, B. W. Fenske, D. J. Trepanier, M. D. Abel, S. Sugiyama and D. R. Ure, Isotechnika Labs, Inc., *Patent Application*, WO2012-079172 A1, 2012.

158. M. Kobayashi, S. Sasamura, H. Muramatsu, Y. Tsurumi, S. Takase and K. Okaka, Astellas Pharma, Inc., *Patent Application*, WO2006-054801 A1, 2006.

159. M. Neya, S. Yoshimura, K. Kamijyo, T. Makino, M. Yasuda, T. Yamanaka, E. Tsujii and Y. Yamagishi, Astellas Pharma, Inc., *Patent Application*, WO2007-049803 A1, 2007.

160. K. Li, D. R. Houck, C. O. Ogbu, M. R. Peel and A. W. Scribner, Scynexis, Inc., *Patent Application*, WO2010-076329 A1, 2010.

161. C. O. Ogbu and T. E. Richardson, Scynexis, Inc., *Patent Application*, WO2011-082289 A1, 2011.

162. K. Li, A. Mamai and M. R. Peel, Scynexis, Inc., *Patent Application*, WO2009-148615 A1, 2009.

163. K. Li, A. Mamai and M. R. Peel, Scynexis, Inc., *US Pat. Appl.*, 2009-0312300 A1, 2009.

164. K. Li and M. R. Peel, Scynexis, Inc., *Patent Application*, WO2011-070364 A1, 2011.

165. A. W. Scribner and D. R. Houck, Scynexis, Inc., *Patent Application*, WO2008-069917 A2, 2008.

166. R. Wenger, M, Mutter, P. Garrouste, R. Lysek, O. Turpin, G. Vuagniaux, V. Nicolas, L. Novaroli Zanolari and R. Crabbe, Debio Recherche Pharaceutique, SA, *Patent Application*, WO2010-052559 A1, 2010.

167. G. Maurer, H. R. Loosli, E. Schreier and B. Keller, *Drug Metab. Dispos.*, 1984, **12**, 120.

168. N. Shibata, H. Shimakawa, T. Minouchi and A. Yamaji, *Biol. Pharm. Bull.*, 1993, **16**, 702.

169. Y. Shitara, Y. Nagamatsu, S. Wada, Y. Sugiyama and T. Horie, *Drug Metab. Dispos.*, 2009, **37**, 1172.

170. P. Falck, H. Guldseth, A. Aasberg, K. Midtvedt and J. L. E. Reubsaet, *J. Chromatogr. B*, 2007, **852**, 345.

171. R. Crabbe, G. Vuagniaux, J.-M. Dumont, N.-M. Valerie, J. Marfurt and L. Novaroli, *Expert Opin. Invest. Drugs*, 2009, **18**, 211.

172. M. Lemaire and J. P. Tillement, *J. Pharm. Pharmacol.*, 1982, **34**, 715.

173. M. Jin, T. Shimada, K. Yokogawa, M. Nomura, Y. Mizuhara, H. Furukawa, J. Ishizaki and K.-I. Miyamoto, *Int. J. Pharm.*, 2005, **293**, 137.

174. K. Watashi, *Curr. Opin. Invest. Drugs*, 2010, **11**, 213.

175. M. Bobardt, S. Hopkins, J. Baugh, U. Chatterji, F. Hernandez, J. Hiscott, A. Sluder, K. Lin and P. A. Gallay, *J. Hepatol.*, 2013, **58**, 16.

176. S. E. Bates, S. Bakke, M. Kang, R. W. Robey, S. Zhai, P. Thambi, C. C. Chen, S. Patil, T. Smith, S. M. Steinberg, M. Merino, B. Goldspiel,

B. Meadows, W. D. Stein, P. Choyke, F. Balis, W. D. Figg and T. Fojo, *Clin. Cancer Res.*, 2004, **10**, 4724.

177. M. Demeule, A. Laplante, A. Sepehr-Arae, E. Beaulieu, D. Averill-Bates, R. M. Wenger and R. Beliveau, *Biochem. Cell Biol.*, 1999, **77**, 47.
178. P. J. Neuvonen, M. Niemi and J. T. Backman, *Clin. Pharmacol. Ther.*, 2006, **80**, 565.
179. C. Dugave, *Curr. Org. Chem.*, 2002, **6**, 1397.
180. X. J. Wang and F. A. Etzkorn, *Biopolymers*, 2006, **84**, 125.
181. T. Mori and T. Uchida, *Curr. Enzyme Inhib.*, 2010, **6**, 46.
182. S. V. Sambasivarao and O. Acevedo, *J. Chem. Inf. Model.*, 2011, **51**, 475.
183. S. A. Hart and F. A. Etzkorn, *J. Org. Chem.*, 1999, **64**, 2998.
184. Q. Li, M. Moutiez, J.-B. Charbonnier, K. Vaudry, A. Menez, E. Quemeneur and C. Dugave, *J. Med. Chem.*, 2000, **43**, 1770.
185. X. Pang, M. Zhang, L. Zhou, F. Xie, H. Lu, W. He, S. Jiang, L. Yu and X. Zhang, *Eur. J. Med. Chem.*, 2011, **46**, 1701.
186. G. Kontopidis, P. Taylor and M. D. Walkinshaw, *Acta Crystallogr., Sect. D Biol. Crystallogr.*, 2004, **60**, 479.
187. Y. Yang, E. Moir, G. Kontopidis, P. Taylor, M. A. Wear, K. Malone, C. J. Dunsmore, A. P. Page, N. J. Turner and M. D. Walkinshaw, *Biochem. Biophys. Res. Commun.*, 2007, **363**, 1013.
188. C. J. Dunsmore, K. J. Malone, K. R. Bailey, M. A. Wear, H. Florance, S. Shirran, P. E. Barran, A. P. Page, M. D. Walkinshaw and N. J. Turner, *ChemBioChem*, 2011, **12**, 802.
189. Y.-Q. Wu, S. Belyakov, C. Choi, D. Limburg, B. E. Thomas IV, M. Vaal, L. Wei, D. E. Wilkinson, A. Holmes, M. Fuller, J. McCormick, M. Connolly, T. Moeller, J. Steiner and G. S. Hamilton, *J. Med. Chem.*, 2003, **46**, 1112.
190. J.-F. Guichou, J. Viaud, C. Mettling, G. Subra, Y.-L. Lin and A. Chavanieu, *J. Med. Chem.*, 2006, **49**, 900.
191. S. Ni, Y. Yuan, J. Huang, X. Mao, M. Lv, J. Zhu, X. Shen, J. Pei, L. Lai, H. Jiang and J. Li, *J. Med. Chem.*, 2009, **52**, 5295.
192. J. Li, J. Chen, C. Gui, L. Zhang, Y. Qin, Q. Xu, J. Zhang, H. Liu, X. Shen and H. Jiang, *Bioorg. Med. Chem.*, 2006, **14**, 2209.
193. M. Cui, X. Huang, X. Luo, J. M. Briggs, R. Ji, K. Chen, J. Shen and H. Jiang, *J. Med. Chem.*, 2002, **45**, 5249.
194. J. Li, J. Zhang, J. Chen, X. Luo, W. Zhu, J. Shen, H. Liu, X. Shen and H. Jiang, *J. Comb. Chem.*, 2006, **8**, 326.
195. P. D. Dearmond, G. M. West, V. Anbalagan, M. J. Campa, E. F. Patz Jr. and M. C. Fitzgerald, *J. Biomol. Screen.*, 2010, **15**, 1051.
196. T. Mori, M. Hidaka, Y.-C. Lin, I. Yoshizawa, T. Okabe, S. Egashira, H. Kojima, T. Nagano, M. Koketsu, M. Takamiya and T. Uchida, *Biochem. Biophys. Res. Commun.*, 2011, **406**, 439.
197. R. Sedrani, J. Kallen, L. M. M. Cabrejas, C. D. Papageorgiou, F. Senia, S. Rohrbach, D. Wagner, B. Thai, A.-M. J. Eme, J. France, L. Oberer, G. Rihs, G. Zenke and J. Wagner, *J. Am. Chem. Soc.*, 2003, **125**, 3849.

198. M. A. Gregory, M. Bobardt, S. Obeid, U. Chatterji, N. J. Coates, T. Foster, P. Gallay, P. Leyssen, S. J. Moss, J. Neyts, M. Nur-e-Alam, J. Paeshuyse, M. Piraee, D. Suthar, T. Warneck, M.-Q. Zhang and B. Wilkinson, *Antimicrob. Agents Chemother.*, 2011, **55**, 1975.

199. S. J. Moss, M. Bobardt, P. Leyssen, N. Coates, U. Chatterji, 11. Dejian, T. Foster, J. Liu, M. Nur-e-Alam, D. Suthar, C. Yongsheng, T. Warneck, M.-Q. Zhang, J. Neyts, P. Gallay, B. Wilkinson and M. A. Gregory, *MedChemComm*, 2012, **3**, 944.

200. K. Lin, B. Weidmann and K. Zimmermann, Novartis AG, *Patent Application*, WO2006-138507 A1, 2006.

201. S. J. Moss, M. A. Gregory, B. Wilkinson and C. J. Martin, Biotica Technology Limited, *Patent Application*, WO2011-098805 A1, 2011.

202. S. J. Moss, M. A. Gregory and C. J. Martin, Biotica Technology Limited, *Patent Application*, WO2011-098808 A1, 2011.

203. S. J. Moss, M. A. Gregory, B. Wilkinson and C. J. Martin, Biotica Technology Limited, *Patent Application*, WO2011-098809 A1, 2011.

204. S. J. Moss, B. Wilkinson and M.-Q. Zhang, Biotica Technology Limited, *Patent Application*, WO2011-144924 A1, 2011.

205. S. J. Moss, M. A. Gregory, B. Wilkinson, S. G. Kendrew and C. J. Martin, Biotica Technology Limited, *Patent Application*, WO2012-085553 A1, 2012.

206. T. Appleby, H. G. Fliri, A. J. Keats, L. Lazarides, R. L. Mackman, S. N. Pettit, K. G. Poullennec, J. Sanvoisin, V. A. Steadman and G. M. Watt, Gilead Sciences, *Patent Application*, WO2012-078915 A1, 2012.

207. D. Tumas, X. Zheng, B. Lu, G. Rhodes, P. Duatschek, J. Hesselgesser, C. Frey, I. Henne, A. Fosdick, R. Halcomb and G. Wolfgang, presented at the 46th Annual Meeting of the European Association for the Study of the Liver (EASL 2011), Berlin, 30 March–3 April 2011, Abstract 1007.

208. K. Watashi, T. Daito, A. Sluder, K. Borroto-Esoda and T. Wakita, *Presented at the 48th Annual Meeting of the European Association for the Study of the Liver (EASL 2011)*, Amsterdam, 24–28 April 2013, Abstract 10.

209. U. Garaigorta and F. V. Chisari, *Cell Host Microbe*, 2009, **6**, 513.

Section IV
Delivery of Antiviral Agents

CHAPTER 12

Prodrugs in the Treatment of Viral Diseases

MICHAEL J. SOFIA

Oncore Biopharma and Institute for Hepatitis and Virus Research,
The Pennsylvania Commonwealth Institute, 3805 Old Easton Road,
Doylestown, PA 18902, USA
Email: mikes@oncorebiopharma.com

12.1 Introduction

Progression of a molecule with interesting pharmacological properties into clinical development is frequently hampered by undesirable characteristics such as poor pharmacokinetic properties, metabolic or chemical instability or poor pharmaceutical properties. To overcome these limitations, the use of prodrug strategies has become an increasingly important tool in the development of drugs to treat human disease.[1–3] In the realm of drugs used to treat viral diseases, the implementation of prodrug approaches has long been used to improve the drug characteristics of therapeutically interesting molecules that include small molecules, nucleosides and nucleotides.

A prodrug is defined as a compound that undergoes biotransformation or chemical transformation *in vivo* prior to eliciting its pharmacological effect. The administered prodrug is typically pharmacologically inactive and is transformed into the active agent after it is delivered to the biological system. A prodrug approach can be executed either by attaching a promoiety to the active drug or by relying on chemical degradation or biotransformation of a molecular unit integral to the structure of the molecule. A promoiety is generally a temporary structural unit that is ultimately released after its

RSC Drug Discovery Series No. 32
Successful Strategies for the Discovery of Antiviral Drugs
Edited by Manoj C. Desai and Nicholas A. Meanwell
© The Royal Society of Chemistry 2013
Published by the Royal Society of Chemistry, www.rsc.org

purpose is served. The promoiety generally masks some undesired characteristic of the active drug, thus changing the overall properties of the agent to overcome deficiencies that limit the ability of the pharmacologically active agent to function as an effective therapeutic. A molecule can also be classified as a prodrug when a metabolite of the molecules core structure is shown to be the actual pharmacologically active agent. Generally, in these cases the prodrug is not pre-engineered and is classified as such after extensive drug metabolism and pharmacokinetic (PK) studies. Examples of such prodrugs include clopidogrel, mitomycin and omeprazole.[1]

A prodrug strategy is usually implemented to address key issues plaguing the therapeutic utility of a pharmacologically active agent (Figure 12.1).[1–3] In many cases, prodrugs are used to address problems associated with poor oral bioavailability and tissue selectivity. Poor oral bioavailability can be related to poor aqueous solubility, poor passive intestinal absorption or fast metabolism. Therefore, prodrugs have been used to improve aqueous solubility (Figure 12.1B), increase lipophilicity to assist passive diffusion across biological membranes (Figure 12.1A) or to slow metabolism for the purpose of increasing overall exposure of the active agent. Prodrugs have also been used to enhance bioavailability by leveraging transporter-mediated intestinal absorption (Figure 12.1C). Imparting tissue selectivity to a pharmacologically active agent is also an area where prodrug technology can have a significant impact. Prodrug approaches have the potential to enhance tissue selectivity through tissue-specific passive enrichment, transporter-mediated targeting, engagement of tissue- or cell-specific enzymes to release a promoiety or through surface antigen targeting (Figure 12.1D). Each of these uses of prodrug technology has been leveraged in the development of agents to target viral diseases.[4–7]

In initiating a prodrug development effort, the choice of a prodrug strategy needs to be made. The effective implementation of a prodrug strategy requires a comprehensive understanding of the problem to be solved. An in-depth understanding of the absorption, distribution, metabolism, excretion (ADME) and PK profile of the parent drug will provide a clue to the nature of the problem and, therefore, what type of prodrug approach is required. It should be noted that a prodrug can change the overall *in vivo* profile of the parent molecule. Depending on the rate of conversion of the prodrug to the parent, prodrugs may alter the tissue distribution, efficacy and toxicity of the parent drug. For prodrugs containing a promoiety, the fate of the promoiety needs to be determined. It is important that the metabolic products resulting for promoiety release be shown to be safe and rapidly excreted from the body. Furthermore, once a prodrug strategy has been chosen and developed, a full assessment of the ADME and PK characteristics of the prodrug is necessary to ensure that the prodrug is addressing the identified problem and achieving the intended results without introducing additional toxicity concerns.

In developing a prodrug to improve permeability for oral administration, several issues must be taken into consideration. The prodrug must have sufficient chemical stability to be formulated into a solid oral dosage form and be compatible with common excipients for formulation. A prodrug has to have

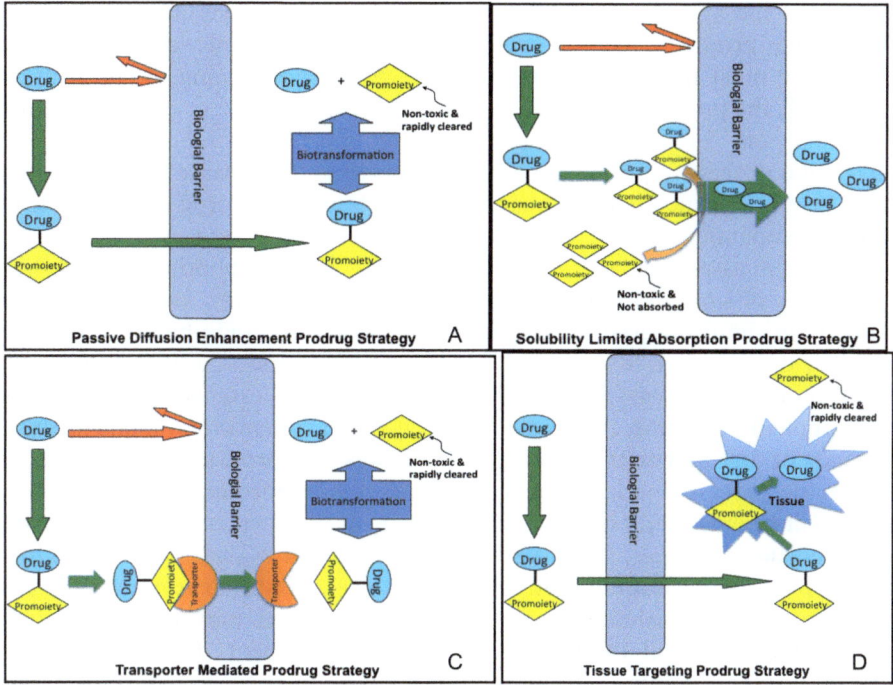

Figure 12.1 (A) Passive diffusion-enhancement prodrug strategy: a prodrug strategy that relies on appending a promoiety to the active drug to enhance the diffusion of the prodrug construct across biological membranes and improve the bioavailability of the drug. The promoiety is released via biotransformation to liberate active drug into the systemic circulation. (B) Solubility-limited absorption prodrug strategy: a prodrug strategy which appends a promoiety to the active drug to improve aqueous solubility in order to increase the local concentration of the drug at the epithelial lining of the gut. The polar promoiety is typically removed by an enzyme found on the gut surface and is not absorbed. (C) Transporter-mediated prodrug strategy: a prodrug of a poorly absorbed drug is developed where the promoiety is recognized by a carrier protein on the surface of a biological membrane and the carrier protein then facilitates the transport of the prodrug across the membrane delivering the prodrug into the systemic circulation. Once in the systemic circulation, the promoiety is removed to reveal the active drug. (D) Tissue-targeting prodrug strategy: a strategy where a promoiety is appended to the active drug. This prodrug construct is absorbed through the gut and, because of the characteristics of the promoiety, the prodrug is selectively recognized by a particular tissue or organ. Once in the tissue, the active drug is released as a result of some characteristic of the tissue that cleaves off the promoiety.

sufficient stability in the gastrointestinal tract to reach the site of absorption intact. The prodrug must facilitate the absorption process through improved physical chemical characteristics, such as solubility or improved passive or active transport through the gut, and deliver the active drug or the prodrug itself into the systemic circulation. In the case of a tissue-targeted prodrug where the prodrug must be delivered into the systemic circulation, the promoiety must not undergo appreciable degradation during the absorption process. If the prodrug is intended to target a specific tissue or organ, it should have sufficient stability in the blood to reach that target organ or tissue. The enzymatic process required to process the prodrug and release the active drug must be efficient such that adequate concentrations of the drug are available to reach therapeutic levels.

To help in assessing the utility of a prodrug strategy and optimize the desired properties, *in vitro* and *in vivo* systems are available. *In vitro* assays evaluating stability in simulated gastric and intestinal fluids, stability in blood or plasma, stability on exposure to intestinal and liver S9 fractions, Caco-2 cell permeability and isolated enzyme stability have been used extensively in prodrug development programs. In using *in vitro* cell-based systems to evaluate intrinsic potency or any potential toxicological liabilities of a prodrug that requires an enzymatic process to release the active drug, it is important to determine that the cell line being used expresses the complement of enzymes required to cleave the prodrug of interest. Otherwise, development and evaluation of a prodrug strategy can be misleading. Ultimately, for the successful development of any prodrug, *in vivo* evaluation needs to be an integral part of the prodrug assessment process at the earliest stages of development. The real test of a prodrug strategy's utility is evaluation in animals where the fate of the prodrug and parent drug can be followed to determine if the prodrug strategy being employed is delivering the desired results when exposed to the complexity of an *in vivo* environment.

The development of antiviral therapies has seen the implementation of a wide variety of prodrug approaches to address a range of drug development problems.[4-7] The field has seen successful implementation of prodrug approaches to address membrane permeability issues, dissolution and solubility-limiting absorption, metabolic deficiencies and tissue targeting. A number of examples that highlight prodrug solutions to problems in antiviral drug discovery and development are presented here.

12.2 Prodrugs of Alcohols and Carboxylic Acids

Polar functionality such as alcohols, amines and carboxylic acids reduce the overall lipophilicity of a molecule and, therefore, can impact diffusion across biological membranes. To mask the polar nature of these functionalities, simple ester, carbonate and carbamate prodrugs are frequently employed (Figure 12.1A).[8] These prodrug constructs are usually cleaved by ubiquitous esterases in the liver, blood or other organs to release the active agent into the systemic circulation or site of action. In developing an ester-type prodrug of a poorly bioavailable molecule, a balance between lipophilicity, solubility and enzymatic cleavage kinetics is required. It should be noted that because of differences in esterase activities in preclinical species, human PK predictions can be difficult.

The use of ester prodrugs has been successfully employed in the development of drugs to treat herpes virus infections. Penciclovir (PCV) (**1**, Figure 12.2) is an acyclic nucleoside analog that was shown to possess potent and selective activity against herpes simplex virus 1 (HSV-1), varicella zoster virus (VZV) and Epstein–Barr virus (EBV) in cell culture and in animal models; however, it demonstrated low oral bioavailability when administered to mice.[9] The poor bioavailability observed for PCV was attributed to both its poor solubility ($3.0 \, \text{mg} \, \text{mL}^{-1}$) and low lipophilicity. Studies showed that PCV was a high-melting, highly crystalline compound which seemed to correlate with its poor bioavailability.

To address these issues, simple ester prodrugs of the primary alcohols were prepared and the guanosine base was replaced with a 2-aminopurine base.[9] These modifications were anticipated to improve the overall lipophilicity of the molecule. It was reasoned based on previous work that the 2-aminopurine base would be oxidized to the guanosine derivative *via* a known oxidation pathway and the esters removed by esterases to provide PCV. Therefore, a double prodrug approach was envisioned to take advantage of both a promoiety strategy and oxidative modification. This rationale led to the development of famciclovir (FCV) (**2**, Figure 12.2).[9] Oral administration of FCV to rats resulted in a bioavailability of 41% compared with 1.5% observed for PCV. Following a single oral dose of FCV in humans, rapid conversion to PCV was observed with peak plasma concentrations detected within 1 h and determined to be 10-fold higher than that observed with PCV.[10] The absolute bioavailability of PCV from FCV was reported to be 77%. The metabolic pathway was determined to proceed *via* initial deacetylation followed by oxidation at the 6-position of the purine base.[11] Famciclovir is used clinically for the treatment of HSV-1 and −2 and for the treatment of shingles caused by herpes zoster.

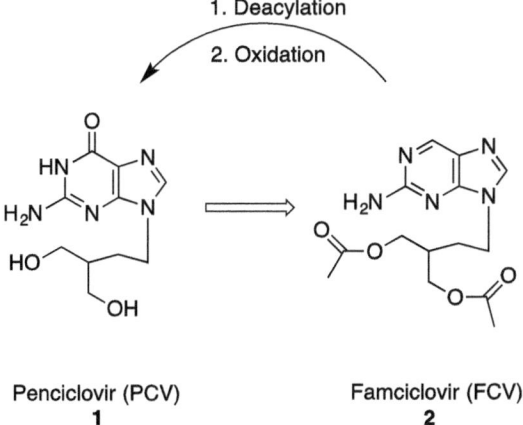

Figure 12.2 Famciclovir (**2**) is the diacyl prodrug of the parent drug penciclovir (**1**). Famciclovir is indicated for the treatment of HSV-1 and -2 and for the treatment of shingles caused by herpes zoster.

In the development of drugs to treat hepatitis C virus (HCV), 2-deoxy-2'-fluoro-2'-*C*-methyl nucleosides have played a prominent role. PSI-6130 (**3**, Figure 12.3), a 2-deoxy-2'-fluoro-2'-*C*-methylcytidine derivative, was the first nucleoside identified in this class and was shown to be a potent and selective inhibitor of hepatitis C virus replication.[12] In animal and human clinical studies, PSI-6130 demonstrated modest oral bioavailability ($\sim$25%) with a major metabolite being the inactive uridine derivative.[13] In an attempt to improve the bioavailability and reduce uridine metabolite formation, prodrugs of the 3'- and 5'-hydroxyl groups were evaluated.[13] It was surmised that increasing lipophilicity would improve absorption and that increasing uptake would reduce uridine metabolite formation in the gut. This effort led to the identification of the 3',5'-diisobutyrate ester prodrug RG7128 (mericitabine) (**4**, Figure 12.3).[13] The diester prodrug mericitabine exhibited clinical proof of concept by achieving a $2.7\log_{10}$ IU mL^{-1} decrease in viral load after 14 days of monotherapy.[13–15] In HCV-infected patients, mericitabine achieved a $5\log_{10}$ IU mL^{-1} reduction in viral load and an 85–90% rapid virological response (undetectable virus after 28 weeks of therapy) when dosed orally at 1500 mg bid in combination with pegylated interferon and ribavirin.[14,15] Mericitabine is currently in Phase 3 clinical development. This ester prodrug approach has been successfully applied to several other nucleosides targeted toward the treatment of HCV (Figure 12.4).[4]

Human influenza virus infections continue to be a major worldwide public health problem. In order to address the need for agents that are effective at combating or controlling influenza infections, several neuraminidase inhibitors have been developed. The first neuraminidase inhibitor, zanamivir (GG167) (**7**, Figure 12.5), showed poor oral bioavailability and rapid renal excretion in clinical trials and, consequently, is administered intranasally.[16] To overcome the limitations associated with zanamivir administration, GS4071 (**8**, Figure 12.5) was developed. GS4071 was shown to be slightly more potent

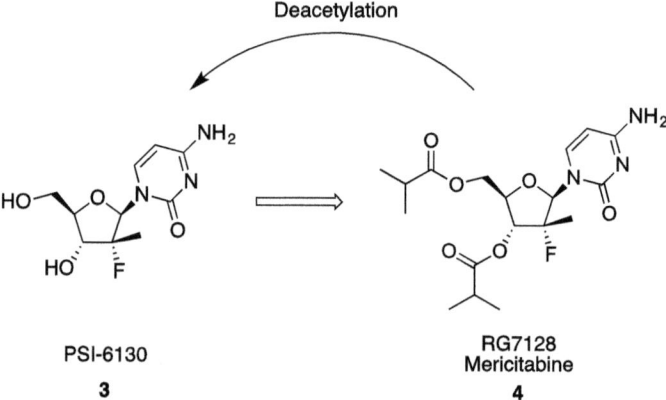

Figure 12.3   Mericitabine (RG7128) (**4**) is a diisobutyrate ester prodrug of the 2'-fluoro-2'-*C*-methylcytidine nucleoside PSI-6130 (**3**). Mericitabine is currently in clinical development for the treatment of HCV infection.

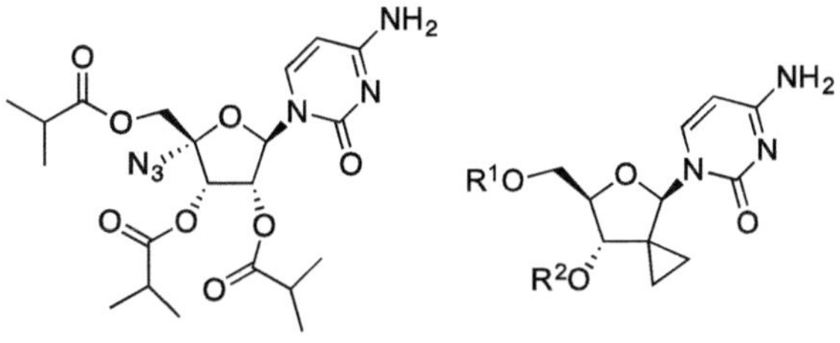

R1626
5

R^1 = H, R^2 = C(O)CH(CH_3)$_2$
R^1, R^2 = C(O)CH(CH_3)$_2$
6

Figure 12.4 R1626 (**5**) and the 2′-cyclopropyl nucleosides (**6**) are ester prodrugs of nucleosides that have been in development for the treatment of HCV infection.

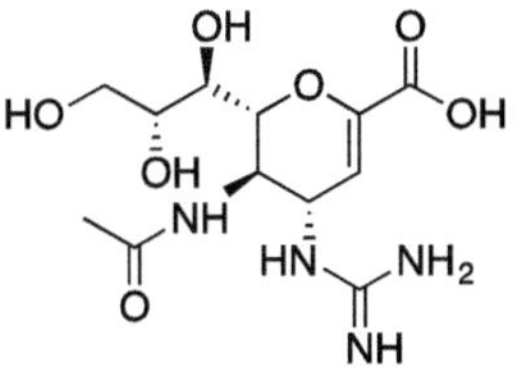

Zanamivir
(GG167)
7

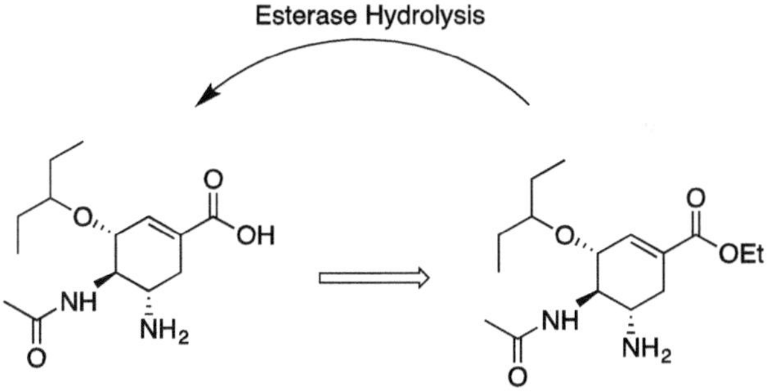

GS4071
8

Oseltamivir
(GS4104)
9

Figure 12.5 Zanamivir (**7** is a neuraminidase inhibitor for the treatment of influenza virus administered as an inhalation or nasally and is not absorbed orally. Oseltamivir (**9**) is an orally active ester prodrug of the neuraminidase inhibitor GS4071 (**8**) currently used for the treatment of influenza virus infection.

than zanamivir *in vitro* and *in vivo*. In addition, GS4071 exhibited significantly longer sustained blood levels when administered to rats.[17] However, even though GS4071 is more lipophilic than zanamivir as a result of replacing the glycerol group with a 3-pentyloxy group and removing a polar guanidino functionality, it still exhibited poor oral bioavailability ($F = 4.3\%$). To improve the oral bioavailability, an ethyl ester prodrug of GS4071's carboxylic acid functionality was prepared.[17] This ester prodrug GS4104 (oseltamivir) (**9**, Figure 12.5) resulted in an oral bioavailability for the parent compound GS4071 of 30–35% in mice and rats, 73% in dog and 11% in ferrets.[17] Rats and mice were shown to convert GS4101 more rapidly to the parent GS4071 than dog and ferret, both of which showed significant circulating prodrug levels.[18] It was shown that this ester prodrug was readily hydrolyzed to the parent by esterases in the blood, but that the prodrug showed poor activity *in vitro* consistent with the need for esterase conversion. Oseltamivir was subsequently shown to be effective in the mouse and ferret models of influenza infection when given orally twice daily.[17,19] Human clinical PK studies showed that after a single dose (20, 50, 100, 200, 500, 1000 mg), oseltamivir levels were observed at ~1 h after administration and declined rapidly with concomitant emergence of the active parent GS4071.[20] Subsequent clinical studies revealed that oseltamivir is effective at inhibiting influenza virus infection in humans and it is now marketed as Tamiflu.

12.3 Prodrugs of Phosphates and Phosphonates

In order for a nucleoside analog to function as an inhibitor of a viral polymerase, it must be converted intracellularly to its 5′-triphosphate *via* the action of three separate kinases. Frequently, a nucleoside analog is a poor substrate for the first and most discriminating kinase in the phosphorylation cascade. In such cases, where the nucleoside triphosphate is a potent inhibitor of the target enzyme, it is desirable to deliver the nucleoside monophosphate (nucleotide) into the cell, thus bypassing the first non-productive phosphorylation step. In addition, methylene phosphonates have been used as stable surrogates for a phosphate moiety, especially in the case of acyclic nucleotide analogs. The presence of a charged phosphate or phosphonate moiety limits a nucleotide's ability to cross biological membranes and, in the case of a nucleotide containing a 5′-phosphate group, introduces a level of chemical and enzymatic instability resulting in poor pharmacokinetic and pharmaceutical properties. Consequently, phosphate and phosphonate prodrugs have been critical to enabling the development of nucleotide analogs to treat human diseases.

In the realm of antiviral therapy, nucleotide analogs have played a prominent role.[5–7,21] They have been shown to be potent inhibitors of viral polymerases in the treatment of viral diseases such as human immunodeficiency virus (HIV), hepatitis B virus (HBV), hepatitis C virus (HCV), herpes viruses (HSV-1, HSV-2), varicella zoster virus (VZV), Epstein–Barr virus (EBV) and cytomegalovirus (CMV). Many prodrug strategies have been developed for delivering nucleotide analogs (Figure 12.6), including the phosphoramidates, acyloxyalkyl esters, *S*-acylthioethyl esters (SATEs), aryl- and lipid phosphate esters and cyclic esters comprising HepDirect, cycloSal and 3′,5′-cyclic phosphates (Figure 12.6).[7,21,22] Each of these has proven effective at delivering

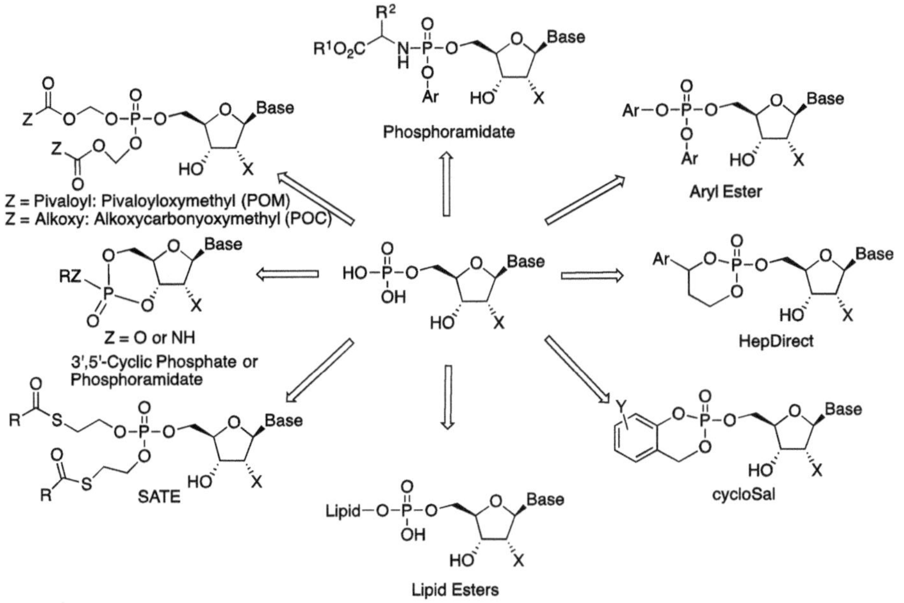

Figure 12.6 Examples of nucleotide prodrug constructs developed to mask the charged nature of phosphate and phosphonate groups to enable diffusion across biological membranes and to stabilize the nature of a phosphate group. Several of these prodrug constructs have been designed to support tissue targeting.

Figure 12.7 Adefovir dipivoxil [adefovir bis(POM)] (**11**) is the phosphonate prodrug of the antiviral adefovir (PMEA) (**10**) developed as an oral agent for the treatment of HBV infection.

nucleotides into cells *in vitro* and *in vivo* in animals, but only a few have translated into human clinical successes.

Adefovir (PMEA) (**10**, Figure 12.7) is an acyclic phosphonate nucleotide analog of adenine. Adefovir is an effective inhibitor of HBV replication *in vitro* following phosphorylation by cellular kinases. Adefovir diphosphate inhibits

HBV DNA polymerase and consequently causes DNA chain termination after incorporation into viral DNA.[23] The clinical use of adefovir is limited by its poor intestinal permeability, presumably due to its charged phosphonate group ($F = 7.8$–11% in rat, 4% in monkey).[24] To circumvent this problem, a bis(pivaloyloxymethyl) ester prodrug [bis(POM)], adefovir dipivoxil (**11**, Figure 12.7), was developed.[25,26] Adefovir dipivoxil demonstrated improved antiviral activity relative to adefovir in cell culture and an improved oral bioavailability ($F = 42\%$ in rat, 27% in cynomolgus monkey).[26] Adefovir dipivoxil was successful in reducing HBV DNA and alanine aminotrans-aminase (ALT) levels in HBV-infected patients at doses of 10 and 30 mg, where 21% of patients at the 10 mg dose and 39% of patients at the 30 mg dose achieved undetectable serum HBV DNA levels. Unfortunately, adefovir dipivoxil was shown to be associated with dose-limiting renal toxicity even at the suboptimal 10 mg dose and was licensed as Hepsera for human use only at the lower dose.[23] In addition, the generation of pivalic acid from decomposition of the pivaloyloxyalkyl ester promoiety has the potential to decrease serum carnitine levels, which is a concern.

In an effort to overcome the toxicity associated with adefovir dipivoxil, a prodrug strategy that attempted to deliver adefovir directly to the target organ, the liver, was developed.[27] The liver targeted prodrug technology, HepDirect, employed a cyclic 1-aryl-1,3-propanyl ester of the phosphonate moiety (Figure 12.6).[28] This prodrug is sensitive to cytochrome P450 (CYP450)-mediated oxidative cleavage. Since CYP450 enzymes are predominantly expressed in the liver, HepDirect technology offered liver targeting potential. By targeting the liver, it was hypothesized that reduced circulating levels of adefovir would result in a lower risk of renal toxicity. Application of HepDirect technology to adefovir led to the identification of the clinical candidate pradefovir (**12**, Figure 12.8).[27,29] Cleavage of pradefovir was shown to occur *via*

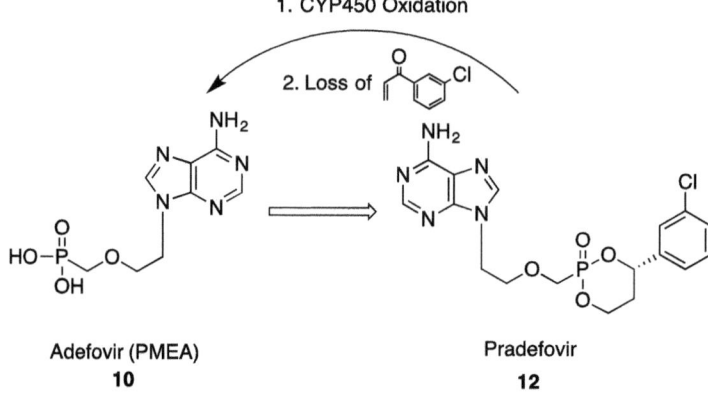

Adefovir (PMEA)
10

Pradefovir
12

Figure 12.8 Pradefovir (**12**) is a prodrug of adefovir (**10**) using the HepDirect prodrug strategy to enable liver targeting for the treatment of HBV infection.

a CYP3A4-mediated oxidation followed by a β-elimination to produce adefovir and an aryl vinyl ketone. Removal of the aryl vinyl ketone is believed to occur by glutathione trapping *in vivo*. Relative to adefovir, pradefovir showed approximately a ninefold increase in adefovir-diphosphate levels in rat hepatocytes *in vitro* and, when given orally to mice and rats, pradefovir demonstrated a 4.5- and 7.5-fold greater exposure of adefovir mono- and diphosphate in the liver relative to kidney and intestine with an oral absorption of 29% and 65% in the rat and monkey, respectively.[30,31] In human clinical studies, pradefovir was rapidly absorbed and converted to adefovir and demonstrated efficacy in reducing serum HBV DNA levels at doses of 10, 20 and 30 mg.[31] Although pradefovir metabolism releases a molecule of aryl vinyl ketone, a nucleophile acceptor, no toxicity concerns were reported in preclinical studies; however, reports of increased cancer risk at high doses in rat and mouse long-term toxicity studies have raised concerns about this drug.[32] Irrespective of these concerns, pradefovir is still in clinical development for the treatment of HBV infection.

Another acyclic nucleotide phosphonate, tenofovir ([9-*R*-(2-phosphonomethoxypropyl)adenine] (PMPA) (**13**, Figure 12.9), was shown to be a potent inhibitor of HIV-1 by inhibiting the viral polymerase.[33] However, despite confirmed limited efficacy both in animal models of HIV-1 and in human clinical trials, PMPA was hampered by poor oral bioavailability (*F* < 2% in mice), presumably because of the presence of a charged phosphonate moiety.[33,34] To overcome this problem, a prodrug effort was initiated which led to the development of bis(isopropyloxymethylcarbonyl)PMPA (tenofovir disoproxil fumarate, TDF (**14**, Figure 12.9).[26,33,34] The isopropyloxymethylcarbonyl (POC) promoiety was chosen instead of the bis(pivaloyloxymethyl) promoiety found in adefovir dipivoxyl (**11**). Release of the POC

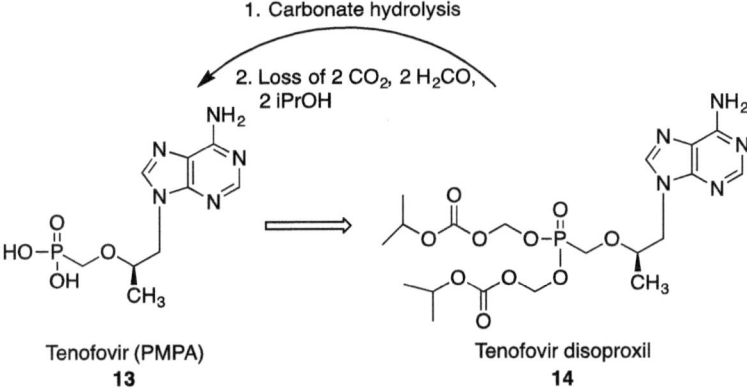

Tenofovir (PMPA)
13

Tenofovir disoproxil
14

Figure 12.9 Tenofovir disoproxil (**14**) is the phosphonate prodrug of acyclic phosphonate nucleotide tenofovir (PMPA) (**12**). It utilizes the isopropyloxymethylcarbonyl (POC) phosphonate prodrug construct to enable more effective oral absorption. Tenofovir disoproxil is currently used for the treatment of HIV and HBV infection.

promoiety produces PMPA and the promoiety byproducts isopropanol, CO_2 and formaldehyde. One factor that contributed to the choice of the POC promoiety was the desire to circumvent the potential issue related to suppression of carnitine levels associated with the release of pivalic acid from the bis(pivaloyloxymethyl) promoiety. In addition, the bis(POC)PMPA derivative demonstrated ~100-fold improvement in potency against HIV-infected cells and at least threefold improvement relative to the bis(POM)PMPA derivative with less cytotoxicity.[33,34] This increased activity could be attributed to the fact that bis(POC)PMPA is rapidly taken up into cells and provides very high intracellular levels of PMPA and its active phosphorylated metabolite.[34] TDF was shown to have an oral bioavailability of 20–36% in mice and dogs and ~20% in monkeys and was shown to be rapidly converted to PMPA.[35,36] In human clinical studies, TDF demonstrated an oral bioavailability of 41% and was highly efficacious in the treatment of HIV-1 infection in humans.[37] Tenofovir disoproxil fumarate was approved for the treatment of HIV-1 infection in humans and is sold under the trade-name Viread. In addition, tenofovir disoproxil fumarate has been approved for the treatment of HBV infection. TDF has also been coformulated with several other antiretroviral drugs. These fixed dose combinations are sold as Truvada, Complera and Atripla.

With the success of highly active antiretroviral therapy (HAART) that combines several drugs, including the prodrug tenofovir disoproxil fumarate (TDF), in fixed-dose combinations for the treatment of HIV-1 infection, HIV-1 has become a long-term manageable disease. However, there is still the risk for the emergence of drug resistance, side effects from long-term use and drug–drug interactions as the patient population ages. The search for new more effective and convenient therapies has continued in the fight against HIV/AIDS. One area of continued effort is in the search for nucleos(t)ide reverse transcriptase inhibitors (N(t)RTIs) that have improved dosing schedules, resistance profiles and reduced long-term toxicity concerns and can be combined with other antiretrovirals in a HAART-type regimen. To this end, the novel nucleotide **16** (Figure 12.10) was identified.[38] This phosphonate nucleotide had an improved resistance profile relative to tenofovir and other N(t)RTIs.[38] However, a prodrug of this novel phosphonate nucleotide analog was desired to deliver the parent ideally to lymphoid cells after successfully traversing the gut and surviving hepatic and plasma enzymes. Early studies showed that the phosphonamidate prodrug of tenofovir GS-7340 (**15**, Figure 12.10) could survive hepatic and plasma enzymes and be preferentially cleaved within lymphoid cells.[39] GS-7340 was shown to be 400-fold more potent than tenofovir in PBMCs and 200-fold more stable in plasma than TDF, resulting in circulating levels of prodrug. Cleavage of the phosphonamidate prodrug was found to be mediated by lysosomal cathepsin A, which is highly expressed in PBMCs.[39–42] Following cathepsin A cleavage of the terminal amino acid ester, cyclization to expel the phenol followed by a series of enzymatic or chemical hydrolyses in the cytosol, the parent drug is released. Studies with tenofovir in dogs and humans confirmed that relative to the POC prodrug TDF, a phosphonamidate

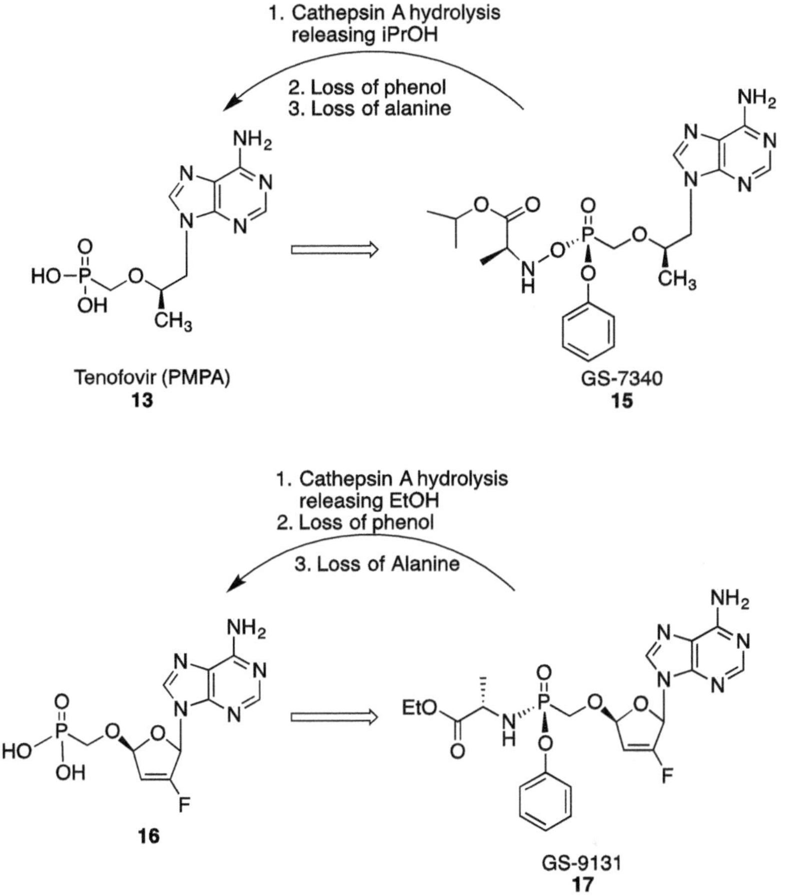

Figure 12.10 GS-7340 (**15**) is the phosphonamidate prodrug of the acyclic phosphonate nucleotide tenofovir (PMPA) (**13**) studied for the treatment of
HIV infection. GS-9131 (**17**) is the phosphonamidate prodrug of the
phosphonate nucleotide (**16**). GS-9131 has demonstrated clinical utility
for the treatment of HIV infection and was shown to target PBMCs.

prodrug (GS-7340, **15**) could dramatically improve the exposure ratio of the
parent nucleotide inside PBMCs relative to the plasma compartment and,
consequently, improve the antiviral response over 14 days of dosing.[38] Patients
receiving GS-7340 at doses of 50 or 150 mg *versus* those receiving TDF at
300 mg for 14 days showed greater decreases in HIV-1 RNA and lower systemic
tenofovir exposure.[43] These results have led to the continued development of
GS-7340 with the potential to improve on both the efficacy and safety of TDF
and possibly to result in the development of new single-tablet regimens not
feasible with TDF. Marriage of this observation regarding the phosphonamidate prodrug GS-7340 with the novel nucleotide **16** ultimately led to
the clinical candidate GS-9131 (**17**, Figure 12.10). GS-9131 was also reported to

be in clinical development and has the potential for providing both lymphocyte targeting and a novel resistance profile.[38]

Prodrugs have played a very important role in the discovery and development of nucleotide inhibitors of HCV replication.[4] The number of reports featuring prodrugs of unique nucleosides and promoiety variations is extensive and has contributed to significant new information in the prodrug field. In particular, the phosphoramidate prodrug strategy for the delivery of nucleoside monophosphate analogs has seen significant application in the area of HCV drug discovery and development.[4] Its prolific use as a prodrug strategy for nucleotide HCV polymerase inhibitors stems from the recognition that phosphoramidate prodrugs could enable liver targeting. Cleavage of a phosphoramidate promoiety requires as its first step the enzymatic cleavage of a carboxylic acid ester by either carboxylesterase or cathepsin A. Leveraging first-pass metabolism and the presence of abundant carboxylesterases in the liver facilitates liver targeting. The use of the phosphoramidate prodrug strategy to enable liver targeting is in contrast to its use to target lyphocytic cells as in the case of GS-7340 (**15**) and GS-9131 (**17**). In the case of liver targeting, prodrug cleavage is optimized against carboxyesterase where, for lymphocytic cells, cathepsin A is the primary enzyme initiating the prodrug degradation cascade. In addition, it has been observed that the parent nucleotide is an important enzyme recognition motif along with the promoiety substituents.

The first example where phosphoramidate technology was implemented in liver targeting of a nucleoside monophosphate analog is exemplified by the discovery and development of sofosbuvir (GS-7977, PSI-7977) (**19**, Figure 12.11).[44] Metabolism studies had shown that the uridine metabolite **18** (Figure 12.11) of the nucleoside PSI-6130 was inactive as an inhibitor of

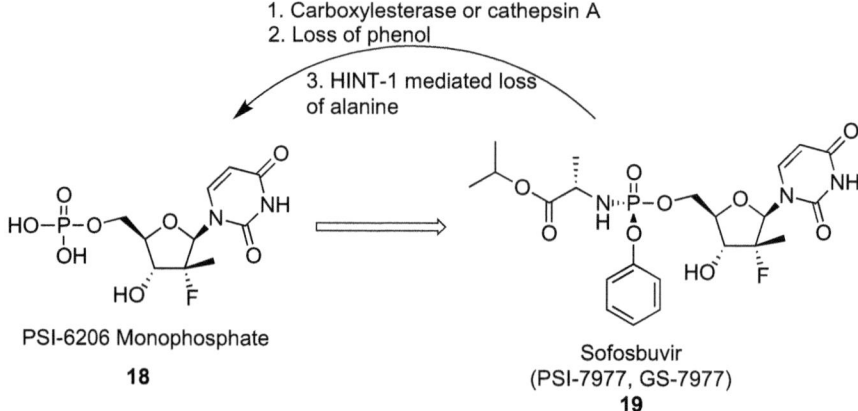

PSI-6206 Monophosphate
18

Sofosbuvir
(PSI-7977, GS-7977)
19

Figure 12.11   Sofosbuvir (PSI-7977, GS-7977) (**19**) is a phosphoramidate prodrug of the 2′-fluoro-2′-*C*-methyluridine 5′-monophosphate (**18**). Sofosbuvir was designed as a liver-targeting prodrug and is currently in Phase 3 clinical trials for the treatment of HCV infection.

HCV replication in the whole cell replicon assay; however, it was also shown that the uridine triphosphate was a potent inhibitor of the HCV polymerase and that it had a long intracellular half-life. The long intracellular half-life was expected potentially to translate into a once-a-day dosing regimen. Further study revealed that the uridine nucleoside was not a substrate for deoxycytidine kinase needed to convert the uridine nucleoside to the 5′-monophosphate, but that the uridine 5′-monophosphate (**18**) was converted to the active triphosphate by mono- and diphosphate kinases.[45,46] To overcome this nonproductive monophosphorylation step and take advantage of first-pass metabolism for liver targeting, a phosphoramidate prodrug strategy was implemented. Execution of this strategy ultimately led to a series of 2′-fluoro-2′-*C*-methyluridine 5′-phosphoramidates that demonstrated potent anti-HCV activity in the whole cell replicon assay, produced high levels of active triphosphate in primary human, rat, dog and monkey hepatocytes and high liver levels of triphosphate when administered orally to rats, dogs and monkeys.[44] Care was also taken in the choice of promoiety substituents to minimize any potential toxicity, as there were concerns regarding the nature of the phenolic substituent and its ultimate release upon prodrug decomposition. The diastereomeric mixture phosphoramidate prodrug having isopropylalanate and phenol substituents, PSI-7851, was chosen as the clinical development candidate and in a Phase 1 human clinical study demonstrated proof of concept by reducing HCV RNA levels in a dose-dependent manner with a maximum mean change of $-1.95\log_{10}$ IU mL^{-1} after qd dosing at 400 mg over 3 days.[13,47] It was also observed that little circulating prodrug was present, indicating rapid conversion of the prodrug and, coupled with the efficacy data, supported the objective of achieving a high liver-to-plasma ratio. Subsequent clinical studies with the more potent pure *S*p diastereomer PSI-7977 (GS-7977, sofosbuvir) (**19**, Figure 12.11) combined with ribavirin produced a sustained virological response (SVR) in HCV genotype 2- and 3-infected patients after 12 weeks of therapy.[4,48] The metabolic pathway for release of the promoiety was elucidated and involved carboxylesterase or cathepsin A hydrolysis of the terminal amino acid ester followed by loss of the phenol and final removal of the amino acid by histidine triad nucleotide-binding protein 1 (HINT-1).[47] Sofosbuvir is currently in Phase 3 clinical studies for the treatment of HCV in all patient populations.

The development of 2′-fluoro-2′-*C*-methyl- and 2′-*C*-methylguanosine nucleotides for the treatment of HCV infection also has benefited from the use of a prodrug strategy employing phosphoramidate technology.[4] In addition, several of these guanosine nucleotide derivatives relied on a double prodrug approach that combines a phosphoramidate promoiety with an enzymatically cleavable C6 substituent on a 2-aminopurine that functions as a masked guanine base. The purpose of the C6-purine substituent is to mask the polar nature of the base unit and increase lipophilicity to improve transport across biological membranes. This strategy of masking the nature of a guanine base has been used previously in the nucleoside field, for example, in the case of abacavir.[49–51] With regard to the C6-substituted purine HCV nucleotide prodrugs, studies demonstrated that the metabolic pathway proceeded first via unmasking of the phosphoramidate prodrug to give the 5′-monophosphate followed by hydrolysis of the C6-purine moiety employing adenosine deaminase-like protein 1 (ADAL-1).[52,53] Examples of guanosine nucleotide analogs where these prodrug strategies have been applied include the clinical candidates BMS-986094 (formerly INX-189)[54–56] (**21**, Figure 12.12), IDX-184[57,58] (**22**, Figure 12.12) and PSI-353661[59] (**24**, Figure 12.13). In each case, their corresponding nucleoside derivatives were weakly active in the HCV replicon whole cell system, yet their triphosphates were good inhibitors of the HCV polymerase *in vitro*. Preparation of these prodrug derivatives led to a 10- to >8000-fold improvement in whole cell replicon potency.[55,57,59] In the case of IDX-184 (**22**), the prodrug moiety incorporates a SATE unit. Decomposition of the IDX-184 promoiety is believed to proceed through both CYP450-dependent and -independent processes, releasing as one of its by-products episulfide. For IDX-184, the formation of intracellular 2′-*C*-methylguanosine triphosphate levels was 100-fold higher in both human and animal hepatocytes than that seen on exposure to 2′-*C*-methylguanosine.[57] Subsequent *in vivo* studies with rats and monkeys showed that oral IDX-184 resulted in a high liver-to-plasma (95% liver extraction) ratio with low plasma levels of the 2′-*C*-methylguanosine metabolite.[57,58] In human clinical studies, oral administration of IDX-184 resulted in low but dose-proportional exposure of the prodrug with a rapid disposition phase and an estimated oral bioavailability

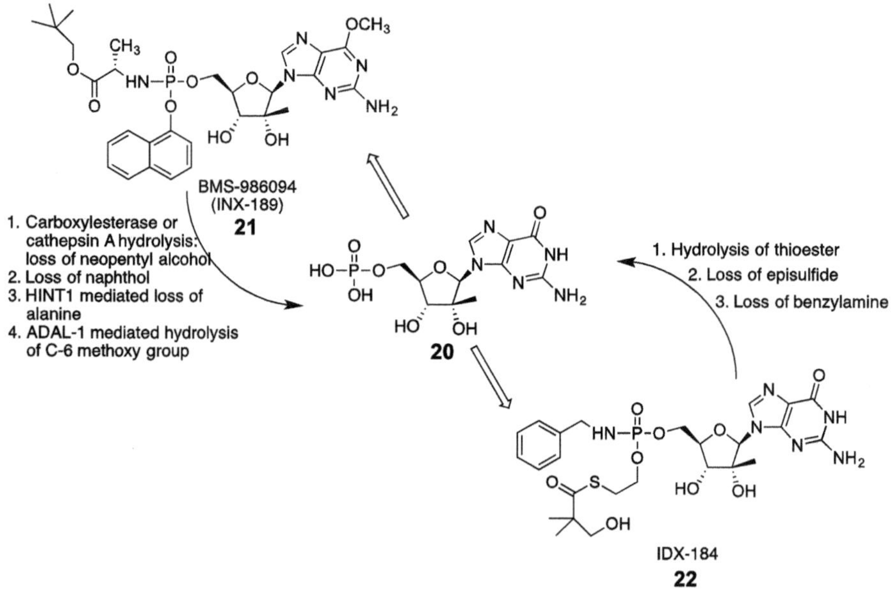

Figure 12.12   BMS-986094 (INX-189) (**21**) and IDX-184 (**22**) are phosphoramidate liver targeting produgs of 2′-*C*-methylguanosine 5′-monophosphate (**20**). BMS-986094 also employs a methoxy group at the C6 position of the purine base as a promoiety to mask the polar nature of the guanine base. Both agents were developed for the treatment of HCV infection.

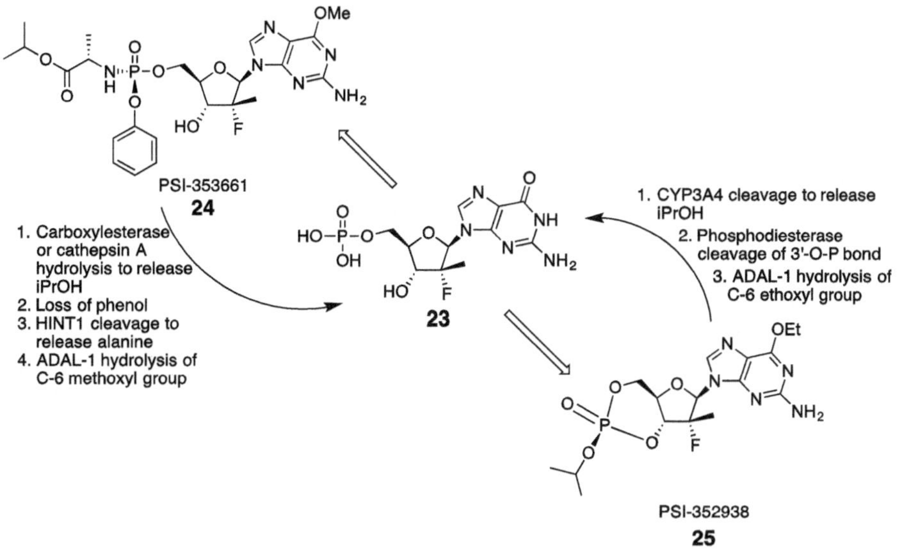

Figure 12.13 PSI-353661 (**24**) and PSI-352938 (**25**) are liver-targeting prodrugs of 2′-fluoro-2′-*C*-methylguanosine 5′-monophosphate (**23**). PSI-353661 uses both a phosphoramidate prodrug strategy to mask the 5′-phosphate group and a C6-methoxy group to mask the polar nature of the guanine base. PSI-352938 utilizes a 3′,5′-cyclic phosphate promoiety and C6-ethoxy promoiety on the guanine base. Both agents were developed for the treatment of HCV infection.

of 20%.[60] In HCV-infected patients, IDX-184 administered at doses from 25 to 100 mg qd over 3 days produced a maximum viral load decline of $0.74\log_{10}$ $IU\,mL^{-1}$ and a subsequent 14 day study in combination with pegylated interferon and ribavirin resulted in a $2.7–4.1\log_{10}$ $IU\,mL^{-1}$ reduction in viral load.[61,62] The PK and efficacy data clearly indicate that IDX-184 preferentially targets the liver.

Like IDX-184, BMS-986094 (**21**, Figure 12.12) demonstrated high dose-proportional liver levels of the same 2′-*C*-methylguanosine triphosphate in rats upon oral administration.[56,63] In portal vein-cannulated cynomolgus monkeys dosed orally with BMS-986094, the prodrug was observed in the portal vein but not in the systemic circulation, where only the 2′-*C*-methylguanosine metabolite was detected, thus supporting efficient liver extraction of the prodrug.[56] A Phase 1 study of BMS-986094 showed that the PK supported qd dosing and in HCV patients demonstrated a mean reduction in HCV RNA levels of 0.71 and $1.3\log_{10}$ $IU\,mL^{-1}$ at doses of 9 and 25 mg, respectively.[64] Unfortunately, BMS-986094 clinical trials were discontinued in August 2012 owing to severe cardiac adverse events. The underlying cause of the adverse events is not known; however, the natures of both the prodrug moiety and parent drug are under scrutiny. These adverse events have also resulted in a clinical hold on IDX-184 because both BMS-986094 and IDX-184 share the same active parent triphosphate although the prodrug substituents are different.

A third member of the guanosine class of phosphoramidate prodrug HCV polymerase inhibitors is represented by PSI-353661 (**24**, Figure 12.13).[59] Like BMS-986094, PSI-353661 also possesses the double prodrug feature where the C6-position of the purine base is substituted with a methoxy group. PSI-353661 is differentiated from BMS-98094 and IDX-184 by the presence of a 2'-fluoro substituent, the phosphoramidate substitution and the fact that it is a single phosphoramidate diastereomer. Like the other guanosine prodrugs, it demonstrated a favorable *in vitro* stability profile and ability to produce high levels of intracellular triphosphate in hepatocytes.[65] A study using radio-labeled prodrug was able to show a favorable 3.5:1 and 4.5:1 liver to plasma ratio of triphosphate and triphosphate precursor metabolites at 1 and 6 h, respectively, after oral administration, thus supporting the liver-targeting potential of the prodrug.[59] PSI-353661 has not yet entered clinical development.

Another unique prodrug construct that was used to leverage the potential of the 2'-fluoro-2'-*C*-methylguanosine triphosphate is the 3',5'-cyclic phosphate. Although 3',5'-cyclic phosphates had been investigated for the delivery of nucleoside 5'-monophosphates, none were shown to be effective in a human clinical setting.[66–68] The first 3',5'-cyclic phosphate prodrug to show clinical proof of concept in the delivery of a nucleotide was PSI-352938 (**25**, Figure 12.13).[69,70] PSI-352938 is a 3',5'-cyclic phosphate of 2'-fluoro-2'-*C*-methyl-6-ethoxy-4-aminopurine and consequently employs the double prodrug approach by masking the C6 position of the purine base. Mechanistic studies showed that the metabolic pathway for release of the cyclic phosphate promoiety employed a CYP3A4-mediated oxidative cleavage of the isopropyl ester followed by phosphodiesterase cleavage of the 3' phosphate–oxygen bond.[71] It was shown that once the 5'-monophosphate was revealed, then cleavage of the C6-ethoxyl group proceeded *via* an ADAL-1 mediated hydrolysis to give the 2'-fluoro-2'-*C*-methylguanosine 5'-monophosphate. Liver targeting of PSI-352938 was enabled by the existence of the CYP3A4 oxidative cleavage of the isopropyl phosphate ester, since CYP enzymes are highly concentrated in the liver. In both rat and dog, high liver levels of the active nucleoside triphosphate were detected after oral dosing. However, unlike with the phosphoramidate prodrug approach, rapid first-pass metabolism was not observed with 3',5'-cyclic phosphate prodrugs.[69] When administered orally to animals or humans, PSI-352938 exhibited high circulating levels of intact prodrug.[72] In a Phase 1 7-day multiple ascending dose human clinical study in HCV-infected patients, doses of PSI-352938 at 100, 200 and 300 mg qd resulted in a mean viral load decline of 4.31, 4.65 and 3.94$\log_{10}$ IU mL–1, respectively and with no adverse side effects.[73] These results demonstrated the first clinical proof of concept study using a 3',5'-cyclic phosphate prodrug strategy to deliver a nucleotide. A 14-day study combining PSI-352938 with PSI-7977 to assess the potential for an interferon-free regimen also resulted in dramatic declines in viral load (4.6–5.5$\log_{10}$ IU mL^{-1}).[74] However, a subsequent extended human clinical study resulted in ALT elevations in patients taking the drug, leading to the suspension of clinical trials. No mechanistic rationale was reported for the ALT elevations and it was not determined if any characteristic of the prodrug may have contributed to the adverse event.

12.4 Prodrugs to Address Solubility-limiting Absorption

Aqueous solubility is an important factor determining the bioavailability of an oral agent. Poorly soluble drugs are not able to present a high concentration of drug at the intestinal lumen and thus lack a concentration gradient driving force for absorption across the epithelial membrane. Charged prodrugs of poorly soluble drugs have been used successfully to improve aqueous solubility and consequently oral absorption (Figure 12.1B).[75] This is accomplished by taking advantage of membrane-associated enzymes that can convert the charged prodrug to the parent drug close to the brush border membrane. Such a strategy has been implemented with several prodrugs developed to treat viral diseases.

Amprenavir (APV, Agenerase) (**26**, Figure 12.14) is an HIV protease inhibitor with several positive attributes that include a favorable resistance profile, lack of food effect and potentially fewer metabolic effects.[76] Amprenavir showed high membrane permeability in a Caco-2 cell model and the clinical formulation demonstrated a relative bioavailability in the dog of 100%.[77,78] However, APV's poor aqueous solubility ($0.04\,\mathrm{mg\,mL^{-1}}$) required a high ratio of excipient to drug to afford gastrointestinal tract solubility and eventually absorption. The clinical formulation required a large number of excipients, some of which posed potential toxicity problems, especially for pediatric administration. Consequently, the resulting high pill burden (16 capsules) in humans had the potential to lead to reduced adherence to therapy. To address the solubility-limiting absorption, a phosphate ester prodrug, fosamprenavir (GW4339808) (**27**, Figure 12.14) with an aqueous solubility of $0.31\,\mathrm{mg\,mL^{-1}}$

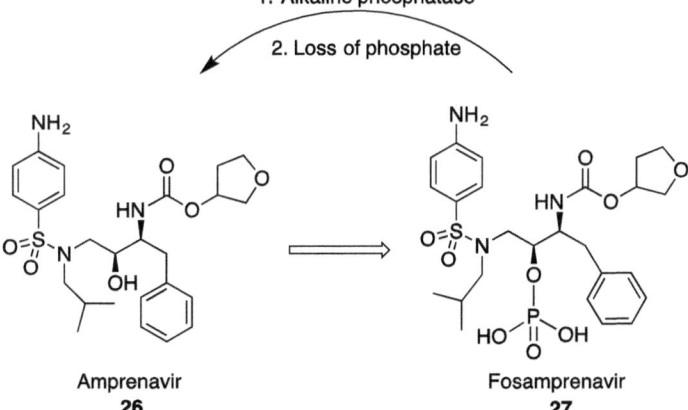

Figure 12.14 Fosamprenavir (**27**) is the phosphate prodrug of the HIV protease inhibitor amprenavir (**26**). Fosamprenavir was developed to address the poor solubility and therefore the poor bioavailability of amprenavir.

was developed.[76,78] Phosphate prodrugs had previously been used to enhance the solubility and, consequently, exposure of small molecules.[2,3] *In vivo*, fosamprenavir was hydrolyzed to APV and inorganic phosphate at the gut epithelium by alkaline phosphatases.[78] In the Caco-2 model, fosamprenavir did not cross the monolayer; however, APV did cross the monolayer >50-fold faster than fosampernavir after Caco-2 exposure to fosamprenavir.[78] Systemic exposure in the rat and dog was shown to be low after oral administration of fosamprenavir, and portal vein-cannulated animals showed minimal levels of fosamprenavir relative to APV, thus supporting rapid hydrolysis in the gut.[78] In a Phase 1 human clinical trial, very low levels of fosamprenavir were detected in plasma (<0.17% of APV concentration) after oral administration, supporting rapid and complete conversion of fosamprenavir to APV during absorption.[79,80] Clinical studies comparing APV and fosamprenavir showed equivalent steady-state AUCs and comparable efficacy and safety profiles, but with a reduced pill burden for fosamprenavir dosing (four tablets).[81] Fosamprenavir is currently marketed as Lexiva.

In the continuing effort to develop inhibitors of HIV-1 with novel mechanisms of action, a series of HIV-1 attachment inhibitors were developed. These agents function by binding to the gp120 protein on the HIV-1 viral envelope, thus interfering with a critical cellular CD4 receptor interaction. The first molecule to demonstrate proof of concept for this approach was BMS-488043 (**28**, Figure 12.15).[82,83] BMS-488043 was shown to be a potent inhibitor of HIV-1 infection *in vitro* and demonstrated good oral bioavailability in rats ($F=90\%$), dogs ($F=57\%$) and monkeys ($F=60\%$) when administered as a solution dose. BMS-488043 given twice daily at a dose of 1800 mg for 8 days produced a viral load decline of $1.0–1.5\log_{10}$ in HIV-1 infected patients.[83] However, to achieve adequate exposure, BMS-488043 required concomitant administration of a high-fat meal. In addition, it was shown that administering a 200 mg dose in solution provided a twofold higher exposure than the solid dosage form.[84] Evidence suggested that poor dissolution and/or poor solubility limited the exposure in humans. To address this dissolution-limited absorption problem, the phosphonooxymethyl prodrug BMS-663749 (**29**, Figure 12.15) was prepared where the promoiety was attached to the indole nitrogen on the parent drug.[84] The attachment of the phosphate promoiety resulted in a dramatic increase in aqueous solubility from 0.04 to $>100\,\mathrm{mg\,mL^{-1}}$. Promoiety cleavage was mediated by alkaline phosphatase (ALP) releasing one molecule of formaldehyde and phosphate.[84] Based on a toxicological assessment, the release of formaldehyde was viewed as acceptable.[85] Animal PK studies showed that BMS-663749 was rapidly converted to the parent BMS-488043 and exhibited absolute bioavailabilities of the parent drug of 62, 93 and 67% in the rat, dog and monkey, respectively.[84] The prodrug also showed a twofold higher AUC of the parent drug in rats compared with oral dosing of the parent. Human clinical studies corroborated the animal PK work and showed rapid conversion of the prodrug to the parent, rapid absorption leading to a shorter $T_{\frac{1}{2}}$ and high bioavailability comparable to solution dosing.[84] Unfortunately, even with the improved PK properties provided by the prodrug, other intrinsic

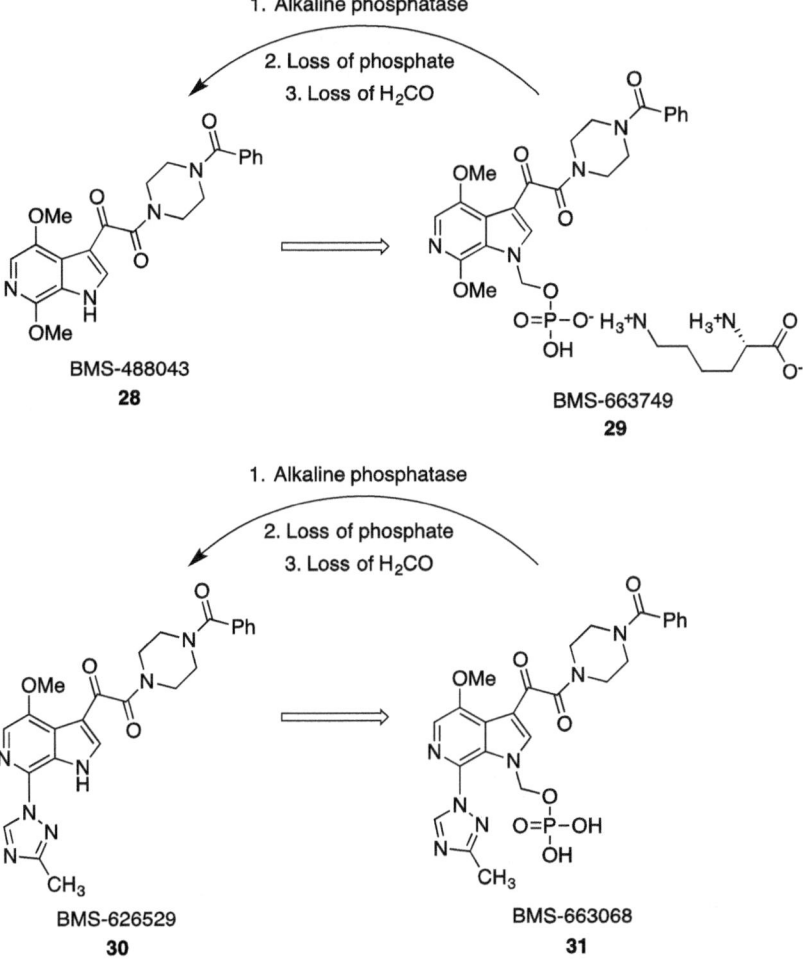

Figure 12.15 BMS-663749 (**29**) and BMS-663068 (**31**) are methylene phosphonate prodrugs of the HIV attachment inhibitors BMS-488043 (**28**) and BMS-626529 (**30**), respectively. They were developed to address the problem of solubility-limiting absorption.

PK characteristics of the parent did not meet a desired twice-daily dosing profile. Consequently, BMS-663749 was not taken further in development. Nevertheless, this prodrug strategy demonstrated its utility in addressing the problem of dissolution and solubility-limited absorption and was subsequently applied to the attachment inhibitor BMS-626529 (**30**, Figure 12.15) that possessed a more favorable intrinsic potency relative to BMS-488043. When administered to healthy volunteers, the phosphonooxymethyl prodrug BMS-663068 (**31**, Figure 12.15)[86] demonstrated rapid conversion to and absorption of the parent drug BMS-626529. An extended-release formulation was subsequently developed to decrease the maximum plasma concentration and increase the trough concentration to enable once- or twice-daily dosing.[87] HIV-1-infected patients treated with BMS-663068 for 8 days with or without ritonavir boosting demonstrated a robust antiviral response with a drop in viral load of 1.21–1.73$\log_{10}$ copies mL^{-1}. BMS-663068 is currently in Phase 2 clinical trials for the treatment of HIV-1 infection.[87]

12.5 Prodrugs Designed to Exploit Carrier-mediated Mechanisms

It is becoming increasingly evident that membrane-associated transporter proteins such as peptide, nucleoside, ion, bile acid or vitamin transporters play an important role in shuttling drugs in and out of cells and consequently they play an important role in drug disposition and toxicity.[88,89] The ability to leverage transport proteins to assist in the delivery of poorly absorbed drugs across biological membranes, either to improve bioavailability or to enable tissue targeting, would be of significant value. Several antiviral prodrugs have taken advantage of removable promoieties that allow prodrug coupling to transport proteins with the objective of improving parent drug bioavailability (Figure 12.1C).

Acyclovir (ACV) (**32**, Figure 12.16) is a selective antiherpetic agent used in the treatment of herpes simplex virus (HSV-1, HSV-2) and varicella zoster virus (VZV) infections.[90,91] Although ACV proved to be very effective at inhibiting HSV in cell culture and was safe when administered systemically, it exhibited low oral bioavailability (15–20%), a short plasma half-life and limited aqueous solubility (~0.2%, 25 °C).[5,92] In an attempt to address these limitations, water-soluble esters of ACV were prepared, leading to the valine ester prodrug valaciclovir (**33**, Figure 16).[92] Valaciclovir demonstrated increased oral bioavailability, attributed to carrier-mediated intestinal absorption. It was

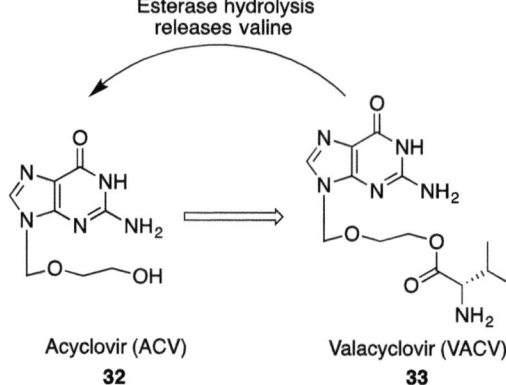

Esterase hydrolysis
releases valine

Acyclovir (ACV)
32

Valacyclovir (VACV)
33

Figure 12.16   Valacyclovir (VACV) (**33**) is the valinate ester prodrug of the antiherpes agent acyclovir (ACV) (**32**). The valine prodrug allows coupling to the intestinal peptide transporter hPEPT1 to improve the oral bioavailability of acyclovir.

determined that the human intestinal peptide transporter hPEPT1 is involved in the facilitated absorption process and that the prodrug valaciclovir is converted rapidly to ACV by ester hydrolysis in the small intestine and also by rapid first-pass metabolism *via* a valacyclovirase.[93–95] Oral valaciclovir has been shown to be safe and efficacious for the treatment of herpes zoster in immunocompetent adults, for episodic treatment of herpes labialis and for suppression of recurrent genital herpes in HIV-infected patients.

A similar facilitated transport approach was applied in the case of the acyclic guanosine nucleoside analog ganciclovir (GCV) (**34**, Figure 12.17), which exhibited poor oral bioavailability ($F = 6$–9%).[96] GVC is administered either by i.v. injection or orally at up to 1000 mg three times daily for the treatment of cytomegalovirus (CMV) infection. Ganciclovir was shown to be a potent inhibitor of CMV infection and more potent than ACV *in vitro*. This enhanced *in vitro* potency is due primarily to the long intracellular half-life of the GCV triphosphate (16.5 h) relative to that for the ACV triphosphate (2.5 h) and the presence of a much higher intracellular concentration of GCV. Significant toxicity is associated with GCV administration, including granulocytopenia, thrombocytopenia and a rise in serum creatinine. To overcome the poor oral bioavailability issue and reduce dose, the valinate ester prodrug of GCV, valganciclovir (**35**, Figure 12.17) was prepared to take advantage of intestinal peptide transporters.[97,98] This prodrug form led to a reduction in dose to 950 mg twice daily. Valganciclovir has been approved to treat CMV retinitis infection in AIDS patients and immunocompromised patients with CMV infection.

In the development of drugs to treat HCV infection in humans, the 2'-*C*-methylcytidine nucleoside NM107 (**36**, Figure 12.18) was discovered as a potent inhibitor of HCV replication *in vitro* and its triphosphate was demonstrated to be a potent inhibitor of the viral polymerase.[99–101] However, animal

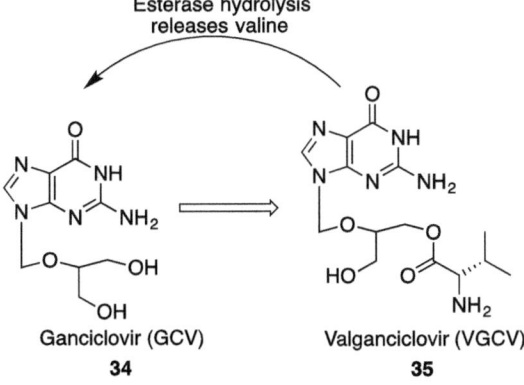

Figure 12.17 Valganciclovir (VGCV) (**35**) is the valinate ester prodrug of the anti-CMV agent ganciclovir (**34**). Valganciclovir takes advantage of the intestinal peptide transporters hPEPT1 and hPEPT2 to improve the oral bioavailability of ganciclovir.

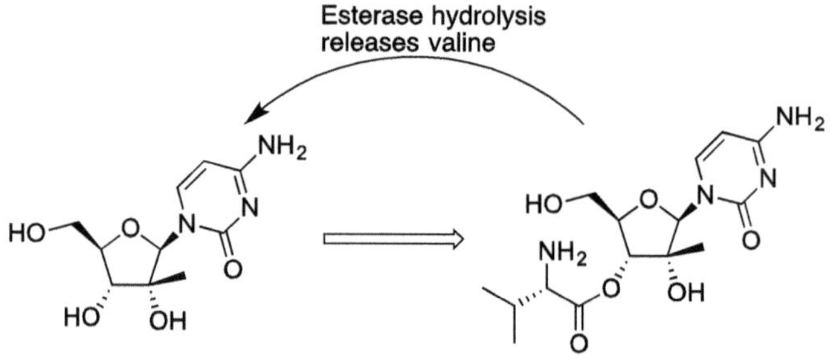

Figure 12.18 Valopicitabine (NM283) (**37**) is the valinate ester prodrug of the 2′-*C*-methylcytidine nucleoside NM107 (**36**). It was developed to improve the bioavailability of the parent drug NM107 by leveraging intestinal peptide transporters.

PK studies showed that NM107 exhibited poor oral bioavailability. Consequently, in the hope of taking advantage of peptide transporters and improve its bioavailability, the 3′-*O*-valinate ester prodrug (NM283, valopicitabine) (**37,** Figure 12.18) was developed.[99] NM283 was shown to have an apparent oral bioavailability of 34% in rats and was subsequently taken into human clinical trials. The proof of concept for the use of this prodrug strategy in humans was obtained when in a monotherapy study, NM283 demonstrated a $1.2\log_{10}$ IU mL^{-1} reduction in viral load after oral administration (800 mg bid) over 14 days and a $>4\log_{10}$ IU mL^{-1} reduction in viral load when coadministered with interferon and ribavirin.[14,15] Unfortunately, this drug was discontinued owing to adverse gastrointestinal side effects.

12.6   Conclusion

The breadth of impact of antiviral prodrugs has been significant in the development of drugs to treat viral diseases. Antiviral prodrugs have shown clinical utility for the treatment of viral diseases that include HIV, HBV, HCV, CMV, VZV, EBV, HSV and influenza. They have shown utility in improving bioavailability of a wide range of molecules that include nucleosides, nucleotides, peptide mimetics and small molecules. Even tissue targeting of nucleotide antivirals has been made possible with the use of designed prodrugs. In fact, there has been no therapeutic area where the implementation of prodrug technology has had more of an impact than it has had in the field of antivirals.

References

1. P. Ettmayer, G. L. Amidon, B. Clement and B. Testa, *J. Med. Chem.*, 2004, **47**, 2393–2404.
2. V. J. Stella, *J. Pharm. Sci.*, 2010, **99**, 4755–4765.

3. J. Rautio, H. Kumpulainen, T. Heimbach, R. Oliyai, D. Oh, T. Jarvinen and J. Savolainen, *Nat. Rev. Drug Discov.*, 2008, **7**, 255–270.
4. M. J. Sofia, W. Chang, P. A. Furman, R. T. Mosley and B. S. Ross, *J. Med. Chem.*, 2012, **55**, 2481–2531.
5. E. De Clercq and H. J. Field, *Br. J. Pharmacol.*, 2006, **147**, 1–11.
6. R. L. Mackman and T. Cihlar, *Annu. Rep. Med. Chem.*, 2004, **39**, 305–321.
7. S. J. Hecker and M. D. Erion, *J. Med. Chem.*, 2008, **51**, 2328–2345.
8. K. Beaumont, R. Webster, I. Gardner and K. Dack, *Curr. Drug Metab.*, 2003, **4**, 461–485.
9. M. R. Harnden, R. L. Jarvest, M. R. Boyd, D. Sutton and R. A. Vere Hodge, *J. Med. Chem.*, 1989, **32**, 1738–1743.
10. M. R. Boyd, R. Boon, S. E. Fowles, K. Pagano, D. Sutton, R. A. Vere Hodge and B. D. Zussman, *Antiviral Res.*, 1988, **9**, 146.
11. R. A. Vere Hodge, D. Sutton, M. R. Boyd and M. Cole, *Antiviral Res.*, 1988, **9**, 146.
12. J. L. Clark, L. Hollecker, J. C. Mason, L. J. Stuyver, P. M. Tharnish, S. Lostia, T. R. McBrayer, R. F. Schinazi, K. A. Watanabe, M. J. Otto, P. A. Furman, W. J. Stec, S. E. Patterson and K. W. Pankiewicz, *J. Med. Chem.*, 2005, **48**, 5504–5508.
13. M. J. Sofia, P. A. Furman and W. T. Symonds, in *RSC Drug Discovery Series*, ed. J. C. Barrish, P. H. Carter, P. T. W. Cheng and R. Zahler, Royal Society of Chemistry, Cambridge, 2010, vol. 4 , pp. 238–266.
14. N. A. Brown, *Expert Opin. Invest. Drugs*, 2009, **18**, 709–725.
15. J. R. Burton, *Clin. Liver Dis.*, 2009, **13**, 12.
16. D. M. Ryan, J. Ticehurst, M. H. Dempsey and C. R. Penn, *Antimicrob. Agents Chemother.*, 1994, **38**, 2270–2275.
17. W. Li, P. A. Escarpe, E. J. Eisenberg, K. C. Cundy, C. Sweet, K. J. Jakeman, J. Merson, W. Lew, M. Williams, L. Zhang, C. U. Kim, N. Bischofberger, M. S. Chen and D. B. Mendel, *Antimicrob. Agents Chemother.*, 1998, **42**, 647–653.
18. D. J. Sweeny, G. Lynch, A. M. Bidgood, W. Lew, K. Y. Wang and K. C. Cundy, *Drug Metab. Dispos.*, 2000, **28**, 737–741.
19. E. J. Eisenberg, A. Bidgood and K. C. Cundy, *Antimicrob. Agents Chemother.*, 1997, **41**, 1949–1952.
20. J. W. Massarella, G. Z. He, A. Dorr, K. Nieforth, P. Ward and A. Brown, *J. Clin. Pharm.*, 2000, **40**, 836–843.
21. M. J. Sofia, *Antiviral Chem. Chemother.*, 2011, **22**, 23–49.
22. K. Y. Hostetler, *Antiviral Res.*, 2009, **82**, A84–A98.
23. R. B. Qaqish, K. A. Mattes and D. J. Ritchie, *Clin. Ther.*, 2003, **25**, 3084–3099.
24. K. C. Cundy, J. P. Shaw and W. A. Lee, *Antimicrob. Agents Chemother.*, 1994, **38**, 365–368.
25. J. E. Starrett Jr., D. R. Tortolani, J. Russell, M. J. Hitchcock, V. Whiterock, J. C. Martin and M. M. Mansuri, *J. Med. Chem.*, 1994, **37**, 1857–1864.

26. M. N. Arimilli, J. Dougherty, K. C. Cundy and N. Bischofberger, in *Advances in Antiviral Drug Design*, ed. E. D. Clercq, JAI Press, Stamford, CT, 1999, vol. 3 , pp. 69–91.
27. H. L. Tillmann, *Curr. Opin. Invest. Drugs*, 2007, **8**, 682–690.
28. M. D. Erion, D. A. Bullough, C.-C. Lin and Z. Hong, *Curr. Opin. Invest. Drugs*, 2006, **7**, 109–117.
29. L. A. Sorbera, N. Serradell and J. Bolos, *Drugs Future*, 2007, **32**, 137–143.
30. C. C. Lin, C. Xu, N. Zhu, D. Lourenco and L. T. Yeh, *Antimicrob. Agents Chemother.*, 2005, **49**, 925–930.
31. C. C. Lin, C. Xu, A. Teng, L. T. Yeh and J. Peterson, *J. Clin. Pharm.*, 2005, **45**, 1250–1258.
32. A. Schreck, Associated Press, San Diego, CA, July 17, 2007.
33. M. N. Arimilli, C. U. Kim, J. Dougherty, A. Mulato, R. Oliyai, J. P. Shaw, K. C. Cundy and N. Bischofberger, *Antiviral Chem. Chemother.*, 1997, **8**, 557–564.
34. B. L. Robbins, R. V. Srinivas, C. Kim, N. Bischofberger and A. Fridland, *Antimicrob. Agents Chemother.*, 1998, **42**, 612–617.
35. L. Naesens, N. Bischofberger, P. Augustijns, P. Annaert, G. Van den Mooter, M. N. Arimilli, C. U. Kim and E. De Clercq, *Antimicrob. Agents Chemother.*, 1998, **42**, 1568–1573.
36. J. P. Shaw, C. M. Sueoka, R. Oliyai, W. A. Lee, M. N. Arimilli, C. U. Kim and K. C. Cundy, *Pharm. Res.*, 1997, **14**, 1824–1829.
37. B. P. Kearney, J. F. Flaherty and J. Shah, *Clin. Pharmacokinet.*, 2004, **43**, 595–612.
38. R. Mackman, in *Accounts in Drug Discovery: Case Studies in Medicinal Chemistry*, ed. J. C. Barrish, P. T.W. Cheng and R. Zahler, Royal Society of Chemistry, Cambridge, 2011, pp. 215–237.
39. W. A. Lee, G. X. He, E. Eisenberg, T. Cihlar, S. Swaminathan, A. Mulato and K. C. Cundy, *Antimicrob. Agents Chemother.*, 2005, **49**, 1898–1906.
40. E. J. Eisenberg, G. X. He and W. Lee, *Nucleosides Nucleotides Nucleic Acids*, 2007, **20**, 1091–1098.
41. G. Birkus, R. Wang, X. Liu, N. Kutty, H. MacArthur, T. Cihlar, C. Gibbs, S. Swaminathan, W. Lee and M. McDermott, *Antimicrob. Agents Chemother.*, 2007, **51**, 543–550.
42. G. Birkus, N. Kutty, G. X. He, A. Mulato, W. Lee, M. McDermott and T. Cihlar, *Mol. Pharmacol.*, 2008, **74**, 92–100.
43. M. Markowitz, A. Zolopa, P. Ruane, K. Squires, L. Zhong, B. P. Kearney and W. Lee, presented at the 18th Conference on Retroviruses and Opportunistic Infections, Boston, 2011.
44. M. J. Sofia, D. Bao, W. Chang, J. Du, D. Nagarathnam, S. Rachakonda, P. G. Reddy, B. S. Ross, P. Wang, H. R. Zhang, S. Bansal, C. Espiritu, M. Keilman, A. M. Lam, H. M. Steuer, C. Niu, M. J. Otto and P. A. Furman, *J. Med. Chem.*, 2010, **53**, 7202–7218.
45. H. Ma, W. R. Jiang, N. Robledo, V. Leveque, S. Ali, T. Lara-Jaime, M. Masjedizadeh, D. B. Smith, N. Cammack, K. Klumpp and J. Symons, *J. Biol. Chem.*, 2007, **282**, 29812–29820.

46. E. Murakami, C. Niu, H. Bao, S. H. M. Micolochick, T. Whitaker, T. Nachman, M. A. Sofia, P. Wang, M. J. Otto and P. A. Furman, *Antimicrob. Agents Chemother.*, 2008, **52**, 458–464.
47. A. M. Lam, E. Murakami, C. Espiritu, H. M. Steuer, C. Niu, M. Keilman, H. Bao, V. Zennou, N. Bourne, J. G. Julander, J. D. Morrey, D. F. Smee, D. N. Frick, J. A. Heck, P. Wang, D. Nagarathnam, B. S. Ross, M. J. Sofia, M. J. Otto and P. A. Furman, *Antimicrob. Agents Chemother.*, 2010, **54**, 3187–3196.
48. E. Lawitz, J. Lalezari, M. Rodriguez-Torres, K. Kowdley, D. Nelson, E. DeJesus, J. McHutchison, M. Cornpropst, M. Mader, E. Albanis, W. Symonds and M. Berrey, presented at the 61st Annual Meeting of the American Association of the Study of Liver Diseases, Boston, 2010.
49. S. M. Daluge, S. S. Good, M. B. Faletto, W. H. Miller, M. H. Clair, L. R. Boone, M. Tisdale, N. R. Parry, J. E. Reardon, R. E. Dornsife, D. R. Averett and T. A. Krenitsky, *Antimicrob. Agents Chemother.*, 1997, **41**, 1082–1093.
50. M. B. Faletto, W. H. Miller, E. P. Garvey, M. H. Clair, S. M. Daluge and S. S. Good, *Antimicrob. Agents Chemother.*, 1997, **41**, 1099–1107.
51. R. Vince and M. Hua, *J. Med. Chem.*, 1990, **33**, 17–21.
52. E. Murakami, H. Bao, R. T. Mosley, J. Du, M. J. Sofia and P. A. Furman, *J. Med. Chem.*, 2011, **54**, 5902–5914.
53. P. A. Furman, E. Murakami, C. Niu, A. M. Lam, C. Espiritu, S. Bansal, H. Bao, T. Tolstykh, H. Micolochick Steuer, M. Keilman, V. Zennou, N. Bourne, R. L. Veselenak, W. Chang, B. S. Ross, J. Du, M. J. Otto and M. J. Sofia, *Antiviral Res.*, 2011, **91**, 120–132.
54. C. McGuigan, K. Madela, M. Aljarah, A. Gilles, S. K. Battina, C. V. Ramamurty, C. Srinivas Rao, J. Vernachio, J. Hutchins, A. Hall, A. Kolykhalov, G. Henson and S. Chamberlain, *Bioorg. Med. Chem. Lett.*, 2011, **21**, 6007–6012.
55. C. McGuigan, K. Madela, M. Aljarah, A. Gilles, A. Brancale, N. Zonta, S. Chamberlain, J. Vernachio, J. Hutchins, A. Hall, B. Ames, E. Gorovits, B. Ganguly, A. Kolykhalov, J. Wang, J. Muhammad, J. M. Patti and G. Henson, *Bioorg. Med. Chem. Lett.*, 2010, **20**, 4850–4854.
56. J. H. Vernachio, B. Bleiman, K. D. Bryant, S. Chamberlain, D. Hunley, J. Hutchins, B. Ames, E. Gorovits, B. Ganguly, A. Hall, A. Kolykhalov, Y. Liu, J. Muhammad, N. Raja, C. R. Walters, J. Wang, K. Williams, J. M. Patti, G. Henson, K. Madela, M. Aljarah, A. Gilles and C. McGuigan, *Antimicrob. Agents Chemother.*, 2011, **55**, 1843–1851.
57. D. Standring, R. Lanford, E. Cretton-Scott, L. Licklinder, M. Larsson, C. Pierra, G. Gosselin, C. Perigaud, D. Surleraux, B. Mayes, A. Moussa and J. Selden, presented at the 43rd Annual Meeting of the European Society for the Study of the Liver, Milan, 2008.
58. D. Standring, R. Lanford, B. Li, R. Panzo, M. Seifer, M. Larsson, S. Good and X. J. Zhou, presented at the 44th Annual Meeting of the European Association for the Study of the Liver, Copenhagen, 2009.

59. W. Chang, D. Bao, B.-K. Chun, D. Naduthambi, D. Nagarathnam, S. Rachakonda, P. G. Reddy, B. S. Ross, H.-R. Zhang, S. Bansal, C. L. Espiritu, M. Keilman, A. M. Lam, C. Niu, H. M. Steuer, P. A. Furman, M. J. Otto and M. J. Sofia, *ACS Med. Chem. Lett.*, 2011, **2**, 130–135.

60. X. J. Zhou, K. Pietropaolo, J. Chen, S. Khan, J. Sullivan-Bolyai and D. Mayers, *Antimicrob. Agents Chemother.*, 2011, **55**, 76–81.

61. J. Lalezari, D. Asmuth, A. Casiro, H. Vargas, G. Dubuc Patrick, W. Liu, K. Pietropaolo, X. J. Zhou, J. Sullivan-Bolyai and D. Mayers, presented at the 60th Annual Meeting of the American Association for the Study of Liver Diseases, Boston, 2009.

62. J. Lalezari, W. O'Riordan, F. Poordad, T. T. Nguyen, G. Dubuc Patrick, J. Chen, X. J. Zhou, J. Sullivan-Bolyai and D. L. Mayers, presented at the 61st Annual Meeting of the American Association for the Study of Liver Diseases, Boston, 2010.

63. C. McGuigan, A. Gilles, K. Madela, M. Aljarah, S. Holl, S. Jones, J. Vernachio, J. Hutchins, B. Ames, K. D. Bryant, E. Gorovits, B. Ganguly, D. Hunley, A. Hall, A. Kolykhalov, Y. Liu, J. Muhammad, N. Raja, R. Walters, J. Wang, S. Chamberlain and G. Henson, *J. Med. Chem.*, 2010, **53**, 4949–4957.

64. J. M. Patti, M. Matson, B. Goehlecke, A. Barry, E. Wensel, H. Pentikis, J. Alam and G. Henson, presented at the 46th Annual Meeting of the European Association for the Study of the Liver, Berlin, 2011.

65. P. A. Furman, E. Murakami, C. Niu, A. M. Lam, C. Espiritu, S. Bansal, H. Bao, T. Tolstykh, H. Micolochick Steuer, M. Keilman, V. Zennou, N. Bourne, R. L. Veselenak, W. Chang, B. S. Ross, J. Du, M. J. Otto and M. J. Sofia, *Antiviral Res.*, 2011, **91**, 120–132.

66. J. Beres, W. G. Bentrude, G. Kruppa, P. A. McKernan and R. K. Robins, *J. Med. Chem.*, 1985, **28**, 418–422.

67. J. Beres, G. Sagi, W. G. Bentrude, J. Balzarini, C. E. De and L. Otvos, *J. Med. Chem.*, 1986, **29**, 1243–1249.

68. J.-L. Girardet, G. Gosselin, C. Perigaud, J. Balzarini, E. De Clercq and J.-L. Imbach, *Nucleosides Nucleotides*, 1995, **14**, 645–647.

69. P. G. Reddy, D. Bao, W. Chang, B. K. Chun, J. Du, D. Nagarathnam, S. Rachakonda, B. S. Ross, H. R. Zhang, S. Bansal, C. L. Espiritu, M. Keilman, A. M. Lam, C. Niu, H. M. Steuer, P. A. Furman, M. J. Otto and M. J. Sofia, *Bioorg. Med. Chem. Lett.*, 2010, **20**, 7376–7380.

70. A. M. Lam, C. Espiritu, S. Bansal, H. M. Micolochick Steuer, V. Zennou, M. J. Otto and P. A. Furman, *J. Virol.*, 2011, **85**, 12334–12342.

71. C. Niu, T. Tolstykh, H. Bao, Y. Park, D. Babusis, A. M. Lam, S. Bansal, J. Du, W. Chang, P. G. Reddy, H. R. Zhang, J. Woolley, L. Q. Wang, P. B. Chao, A. S. Ray, M. J. Otto, M. J. Sofia, P. A. Furman and E. Murakami, *Antimicrob. Agents Chemother.*, 2012, **56**, 3767–3775.

72. W. T. Symonds, J. M. Denning, E. Albanis, M. T. Cornpropst, R. Wright, A. Lai and M. M. Berrey, presented at the 61st Annual Meeting of the American Association for the Study of Liver Diseases, Boston, 2010.

73. M. Rodriguez-Torres, E. Lawitz, J. M. Denning, M. T. Cornpropst, E. Albanis, W. T. Symonds and M. M. Berrey, presented at the 46th Annual Meeting of the European Association for the Study of the Liver, Berlin, 2011.

74. E. Lawitz, M. Rodriguez-Torres, J. M. Denning, M. T. Cornpropst, D. Clemons, L. McNair, M. M. Berrey and W. T. Symonds, presented at the 46th Annual Meeting of the European Association for the Sudy of the Liver, Berlin, 2011.

75. D. Fleisher, R. Bong and B. H. Stewart, *Adv. Drug Deliv. Rev.*, 1996, **19**, 115–130.

76. H. Ouyang, in *Prodrugs: Challenges and Rewards, Part 2*, eds V. J. Stella, R. T. Borchardt, M. J. Hageman, R. Oliyai, H. Magg and J. W. Tilley, Springer, New York, 2007, pp. 541–549.

77. B. M. Sadler and D. S. Stein, *Ann. Pharmacother.*, 2002, **36**, 102–118.

78. E. S. Furfine, C. T. Baker, M. R. Hale, D. J. Reynolds, J. A. Salisbury, A. D. Searle, S. D. Studenberg, D. Todd, R. D. Tung and A. Spaltenstein, *Antimicrob. Agents Chemother.*, 2004, **48**, 791–798.

79. C. Falcoz, J. M. Jenkins, C. Bye, T. C. Hardman, K. B. Kenney, S. Studenberg, H. Fuder and W. T. Prince, *J. Clin. Pharmacol.*, 2002, **42**, 887–898.

80. M. B. Wire, M. J. Shelton and S. Studenberg, *Clin. Pharmacokinet.*, 2006, **45**, 137–168.

81. T. M. Chapman, G. L. Plosker and C. M. Perry, *Drugs*, 2004, **64**, 2101–2124.

82. T. Wang, Z. Zhang, O. B. Wallace, M. Deshpande, H. Fang, Z. Yang, L. M. Zadjura, D. L. Tweedie, S. Huang, F. Zhao, S. Ranadive, B. S. Robinson, Y. F. Gong, K. Ricarrdi, T. P. Spicer, C. Deminie, R. Rose, H. G. Wang, W. S. Blair, P. Y. Shi, P. F. Lin, R. J. Colonno and N. A. Meanwell, *J. Med. Chem.*, 2003, **46**, 4236–4239.

83. T. Wang, Z. Yin, Z. Zhang, J. A. Bender, Z. Yang, G. Johnson, Z. Yang, L. M. Zadjura, C. J. D'Arienzo, D. DiGiugno Parker, C. Gesenberg, G. A. Yamanaka, Y. F. Gong, H. T. Ho, H. Fang, N. Zhou, B. V. McAuliffe, B. J. Eggers, L. Fan, B. Nowicka-Sans, I. B. Dicker, Q. Gao, R. J. Colonno, P. F. Lin, N. A. Meanwell and J. F. Kadow, *J. Med. Chem.*, 2009, **52**, 7778–7787.

84. J. F. Kadow, Y. Ueda, N. A. Meanwell, T. P. Connolly, T. Wang, C. P. Chen, K. S. Yeung, J. Zhu, J. A. Bender, Z. Yang, D. Parker, P. F. Lin, R. J. Colonno, M. Mathew, D. Morgan, M. Zheng, C. Chien and D. Grasela, *J. Med. Chem.*, 2012, **55**, 2048–2056.

85. S. S. Dhareshwar and V. J. Stella, *J. Pharm. Sci.*, 2008, **97**, 4184–4193.

86. B. Nowicka-Sans, Y. F. Gong, B. McAuliffe, I. Dicker, H. T. Ho, N. Zhou, B. Eggers, P. F. Lin, N. Ray, M. Wind-Rotolo, L. Zhu, A. Majumdar, D. Stock, M. Lataillade, G. J. Hanna, J. D. Matiskella, Y. Ueda, T. Wang, J. F. Kadow, N. A. Meanwell and M. Krystal, *Antimicrob. Agents Chemother.*, 2012, **56**, 3498–3507.

87. R. E. Nettles, D. Schurmann, L. Zhu, M. Stonier, S. P. Huang, I. Chang, C. Chien, M. Krystal, M. Wind-Rotolo, N. Ray, G. J. Hanna, R. Bertz and D. Grasela, *J. Infect. Dis.*, 2012, **206**, 1002–1011.
88. M. K. DeGorter, C. Q. Xia, J. J. Yang and R. B. Kim, *Annu. Rev. Pharmacol. Toxicol.*, 2012, **52**, 249–273.
89. K. M. Giacomini, S. M. Huang, D. J. Tweedie, L. Z. Benet, K. L. Brouwer, X. Chu, A. Dahlin, R. Evers, V. Fischer, K. M. Hillgren, K. A. Hoffmaster, T. Ishikawa, D. Keppler, R. B. Kim, C. A. Lee, M. Niemi, J. W. Polli, Y. Sugiyama, P. W. Swaan, J. A. Ware, S. H. Wright, S. W. Yee, M. J. Zamek-Gliszczynski and L. Zhang, *Nat. Rev. Drug Discov.*, 2010, **9**, 215–236.
90. G. B. Elion, P. A. Furman, J. A. Fyfe, P. De Miranda, L. Beauchamp and H. J. Schaeffer, *Proc. Natl. Acad. Sci. U. S. A.*, 1977, **74**, 5716–5720.
91. J. C. Huff, B. Bean, H. H. Balfour Jr, O. L. Laskin, J. D. Connor, L. Corey, Y. J. Bryson and P. McGuirt, *Am. J. Med.*, 1988, **85**, 84–89.
92. K. R. Beutner, D. J. Friedman, C. Forszpaniak, P. L. Andersen and M. J. Wood, *Antimicrob. Agents Chemother.*, 1995, **39**, 1546–1553.
93. A. Guo, P. Hu, P. V. Balimane, F. H. Leibach and P. J. Sinko, *J. Pharmacol. Exp. Ther.*, 1999, **289**, 448–454.
94. I. Kim, X. Y. Chu, S. Kim, C. J. Provoda, K. D. Lee and G. L. Amidon, *J. Biol. Chem.*, 2003, **278**, 25348–25356.
95. C. M. Perry and D. Faulds, *Drugs*, 1996, **52**, 754–772.
96. C. S. Crumpacker, *N. Engl. J. Med.*, 1996, **335**, 721–729.
97. P. Reusser, *Expert Opin. Invest. Drugs*, 2001, **10**, 1745–1753.
98. M. Sugawara, W. Huang, Y. J. Fei, F. H. Leibach, V. Ganapathy and M. E. Ganapathy, *J. Pharm. Sci.*, 2000, **89**, 781–789.
99. C. Pierra, A. Amador, S. Benzaria, E. Cretton-Scott, M. D'Amours, J. Mao, S. Mathieu, A. Moussa, E. G. Bridges, D. N. Standring, J.-P. Sommadossi, R. Storer and G. Gosselin, *J. Med. Chem.*, 2006, **49**, 6614–6620.
100. C. Pierra, S. Benzaria, A. Amador, A. Moussa, S. Mathieu, R. Storer and G. Gosselin, *Nucleosides Nucleotides Nucleic Acids*, 2005, **24**, 767–770.
101. L. A. Sorbera, J. Castaner and P. A. Leeson, *Drugs Future*, 2006, **31**, 320–324.

CHAPTER 13

Cobicistat and Ritonavir as Pharmacoenhancers for Antiviral Drugs

LIANHONG XU* AND MANOJ C. DESAI

Gilead Sciences, Inc., 333 Lakeside Drive, Foster City, CA 94404, USA
*Email: lianhong.xu@gilead.com

13.1 Introduction

During the past half century, there have been significant advances in the development of effective antiviral drugs, and many of these drugs have played key roles in the treatment of viral-infected patients. Great progress has been made in the discovery and development of therapies to treat human immunodeficiency virus (HIV) infection, and also significant improvement in the mortality and morbidity of HIV-infected patients. However, resistance development remains a major obstacle to antiviral therapy and all active antiviral agents have been shown to select for resistance mutations. Resistance evolution is a particular problem under conditions of suboptimal therapy. Studies have shown that an adequate trough concentration at the target site, usually several-fold above its protein-adjusted, effective inhibitory concentration EC_{90} or EC_{95}, corresponds to achieving sustained efficacy and preventing the emergence of resistance. Unfortunately, many drugs need to be dosed frequently and at high dose levels to achieve the targeted trough concentrations, which presents challenges for patients, especially in cases of chronic treatment. High pill burden and dosing frequency result in low adherence to a regimen and,

RSC Drug Discovery Series No. 32
Successful Strategies for the Discovery of Antiviral Drugs
Edited by Manoj C. Desai and Nicholas A. Meanwell
© The Royal Society of Chemistry 2013
Published by the Royal Society of Chemistry, www.rsc.org

consequently, suboptimal drug concentrations at trough, which is a major cause of treatment failure and the emergence of drug resistance. A pharmacoenhancer, which itself is often not active against the therapeutic target but can inhibit the enzymes that metabolize the active drug, can enhance pharmacokinetic (PK) profiles of the therapeutic drug to achieve adequate trough concentrations at lower dosage and with less frequent dosing. This was especially evident in the treatment of HIV infection. The pharmacoenhancers ritonavir (RTV) and cobicistat (COBI, previously known as GS-9350) have contributed to the success of durable viral suppression *via* simplified treatment regimens with favorable tolerability. This chapter reviews the relationship of drug PK, drug resistance and durable viral suppression using HIV therapy as an example to discuss the role of pharmacoenhancers.

13.2 Antiviral Resistances and HIV Protease Inhibitor Ritonavir

13.2.1 Virus and Drug Resistance Mutations

Pathogenic viruses cause a tremendous burden of disease and death worldwide. Viruses can rapidly adapt to selective pressure and exist with multi-quasi-species (polymorphisms) due to their rapid replication cycles, their error-prone replication process and a large amount of viral progeny from a single infected host cell. However, significant progress has been made in the past half century towards the development of effective and specific antivirals. Although some antiviral agents, such as anti-influenza drugs, have been used in acute infections, many of them are generally used in the treatment of persistent and chronic infections. In recent years, major efforts of antiviral research have been focused on viruses that cause chronic infection affecting millions of individuals worldwide, such as HIV, hepatitis C virus (HCV) and hepatitis B virus (HBV). Many drugs have been approved and many more are in advanced stages of clinical studies. Although the ultimate goal in chronic treatment is to eradicate the viral pathogen from the host, in some cases that goal is unattainable. When eradication is not achievable, the focus is redirected towards achieving durable viral suppression to alleviate or prevent the clinical manifestation and long-term consequences of chronic infection and preventing the transmission of the pathogen itself.

Long-term treatment of chronic viral infections is often fraught with problems regarding adverse effects of drugs, patient compliance and resistance development. Favorable safety profiles, simplified regimens and a low pill burden will favor good compliance, which, in turn, delays the emergence of drug resistance. The emergence of clinically-relevant resistant variants is associated with factors such as selective pressure under the treatment drug, the generic barrier of the drug to resistance and the replication fitness of the resistant variant. The resistant variants will be selected by treatment with drug concentrations insufficient to suppress completely the replication of moderately resistant viruses. Replication under drug selection pressure can then result in the accumulation of further adaptive mutations, conferring a higher degree of resistance and a higher level of fitness. When viruses become resistant to one drug, they are often automatically cross-resistant to the drugs in the same class. The treatment failure of an active treatment regimen is often derived from the emergence of drug-resistant variants and viral rebound, which is more often than not a result of poor compliance with the drug regimens due to their challenging dose schedules and/or poor side-effect profiles.

13.2.2 HIV and HIV-1 Protease Inhibitor Ritonavir

HIV was first isolated in 1983 and was soon identified as the etiological agent of acquired immunodeficiency syndrome (AIDS), responsible for more than 25 million deaths worldwide over the past 30 years. Through the collaborative work of the US Food and Drug Administration (FDA) and scientists from both academia and the pharmaceutical industry, more than 20 drugs covering six mechanistic classes have been approved to date for treating HIV infection/AIDS. Unfortunately, eradication of HIV from humans with anti-retroviral therapy has not been achieved. The current goal of antiretroviral therapy is to achieve maximum and durable suppression of plasma viremia, with preservation of CD4+ T cells and minimum development of drug resistance.[1]

Until mid-1995, chemotherapies for the treatment of AIDS and/or HIV infection were limited to only nucleoside reverse transcriptase inhibitors (NRTIs). Although these antiretroviral agents provided long-awaited relief for those suffering from this devastating disease by delaying the progression to AIDS, a death sentence at the time, the clinical benefit was overshadowed by the rapid emergence of resistant strains of HIV.[2]

HIV-1 protease, which is responsible for the processing of the Gag and Gag–Pol polyprotein gene products into mature and functional proteins, plays a critical role in virus replication *in vitro*[3] and has been pursued as a target for inhibiting HIV replication. Part of the family of aspartic proteases, HIV-1 protease is a symmetrically assembled homodimer. The active site is formed at the dimer interface, with the two aspartic acids at positions 25 and 25' located at the base of the active site. During the proteolytic processing of the substrate, a tetrahedral transition state between the enzyme and substrate bridged by water exists transiently at the active site. HIV-1 protease inhibitors (PIs) bind to the active site of protease in a way that mimics the transition state, preventing the enzyme from cleaving the post-translational polyproteins necessary for the maturation of infectious virions.[4] Treatment with potent PIs produces a rapid fall in plasma HIV RNA, with a concomitant increase in CD4+ T cells.

Three HIV-1 PIs were approved during 1995 and 1996. Ritonavir was the second PI licensed in the USA for the treatment of HIV, with a recommended dosage of 600 mg twice daily (bid). However, the clinical efficacy of PI mono-therapy is eventually lost in many patients owing to the development of PI-resistant mutations. It was discovered that the durability of virologic response was greater with higher drug exposure and the rate of appearance of resistance mutations during monotherapy was found to correlate inversely with the trough plasma drug concentrations.[5] This observation, in addition to later studies demonstrating the important role of protein binding in the determination of necessary drug exposure,[6] laid the foundation for the design of an effective regimen of antiretroviral therapy.

When ritonavir is used at the therapeutic dose of 600 mg bid, many patients experience dose-limiting adverse effects, including gastrointestinal (GI) intolerance, headaches and circumoral numbness.[7] High-dose ritonavir can also cause serum lipid disturbances, including significant increases in serum cholesterol and marked increases in serum triglycerides.[7,8] Owing to these side effects and the subsequent availability of better tolerated PIs, the use of ritonavir as a therapeutic PI is now relatively insignificant.

The combination of ritonavir and saquinavir, the first approved PI that has distinct *in vivo* resistance mutant selections from ritonavir, was brought into clinical studies with the hope that the absence of cross-resistance would delay or prevent the emergence of resistance. These studies resulted in the discovery that co-administration of saquinavir with ritonavir significantly enhanced the plasma exposure of saquinavir and improved efficacy, thus opening a new era of PK enhancement.

13.2.3 Drug-resistant Mutations and Pharmacokinetic Profiles

As mentioned earlier, resistance development is a major obstacle to antiviral therapy and all active antiviral agents have been shown to select for resistance mutations. A successful antiviral agent should be aimed at suppressing all existing viral variants, thus preventing the selection of drug-resistant quasi-species and their subsequent evolution. This implies that the number of mutations required for the first escape of the antiviral agent should be greater than the expected number of mutations present in the viral population. In addition, the need to achieve target-site drug concentrations sufficient to suppress viral replication is another key factor in successful antiviral therapy and the prevention of resistance development. For most drugs, plasma concentration can be used as a surrogate marker for the drug concentration at the target site. For example, in HCV or HIV infection, for which the major drug target site is the liver and cells of immune system, respectively, it is the plasma drug concentration, rather than target-site drug concentration, which is monitored to be reached and maintained in order to achieve an antiviral response in the form of viral load reduction. Hence PK parameters that ensure adequate target-site drug concentration become important issues for antiviral drugs used in therapy.

Figure 13.1 helps explain the importance of drug exposure in preventing the emergence of drug-resistant variants. IC_{50} is defined as the *in vitro* drug concentration necessary to inhibit viral replication by 50%. WIC_{50} and QIC_{50} represent the IC_{50} of a drug against heterogeneous virus populations of wild-type and of highly populated quasi-species, respectively; the quasi-species has a

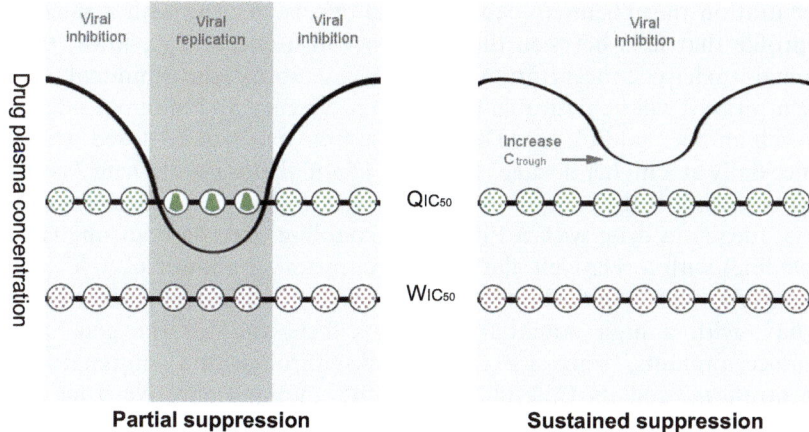

Figure 13.1   Relationship between drug concentration and viral suppression.

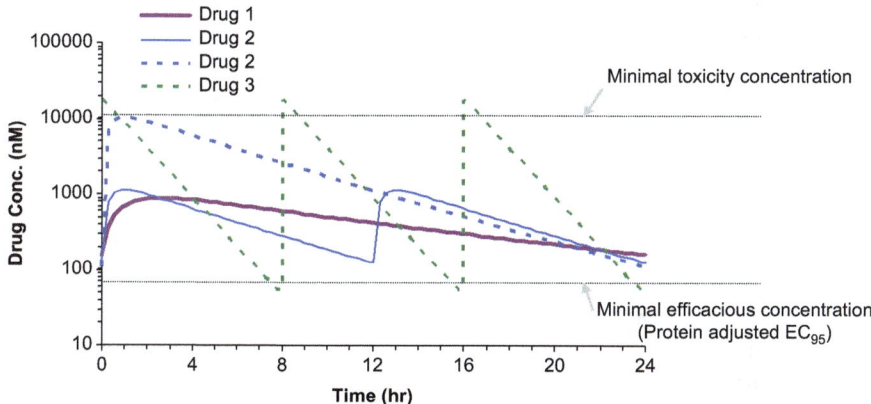

Figure 13.2   PK profiles of drugs.

reduced susceptibility to the drug. The curve is the steady-state plasma drug concentration. The steady-state minimum drug concentration (C_{trough}) in the left-hand diagram is maintained above the WIC_{50} but falls below the QIC_{50}. Thus, there is a period of time in the drug treatment cycle in which the wild-type virus is suppressed but not the quasi-species, allowing it to replicate. After an extended period of dosing, the resistance mutations will be selected under drug pressure and thus the total viral load will rebound. If the PK parameter can be improved to the degree that C_{trough} (right-hand diagram) can be adequately maintained above the IC_{50}s of the heterogeneous viral population, the viral replication will be suppressed, which will correspond to a durable viral suppression and minimal emergence of resistance.

Several scenarios of PK profiles of drugs are depicted in Figure 13.2. The 'minimal effective concentration' is the drug concentration required for durable suppression of viral replication; the 'minimal toxicity concentration' is the drug concentration that begins to cause toxicity effects. A drug with a steady-state PK profile that falls between the two concentrations, a C_{max} lower than the 'minimal toxicity concentration' and C_{trough} above the 'minimal effective concentration,' will be desirable for sufficient efficacy and minimal side effects. To reach an adequate C_{trough}, Drug 2 (blue line) needs to be dosed twice daily or once daily at a higher dosage. The C_{max} of a high dosage of Drug 2 is close to the concentration that causes toxicity, which will result in dose-related side effects. Ideally, a drug with a PK profile enabling once-daily dosing (Drug 1, purple line) with a relatively flat PK curve, meaning a lower C_{max}/C_{min} ratio, will perform most effectively. Drug 3 (black line) needs to be dosed three times per day, with a high potential for side effects and for the generation of resistance mutants, since its C_{max} breaks through the 'minimal toxicity concentration,' and its C_{min} falls below the 'minimal effective drug concentration.' More often than not, a potent drug with a good resistance profile has sufficient antiviral properties but suffers from a poor PK profile and fails to attain sufficient C_{trough} for durable viral suppression. A pharmacoenhancer can improve the PK profile of a drug and enable it to achieve adequate exposure at a lower dosage and with less frequent dosing, as exemplified by the PK profile of Drug 2, which can be improved to a level that is similar to that of Drug 1. In clinical practice, ritonavir demonstrated these desired properties and significantly enhanced the PK exposure of a co-dosed PI, enabling HIV PIs to remain a key component for HIV treatment.

13.2.4 Combination Therapy and HAART

High-level viral replication rates in conjunction with the high mutation rate of some viruses, such as HIV, cause any monotherapy treatment to fail.[9] A combination therapy using two or more drugs from different classes against the target virus forces the virus to mutate simultaneously at multiple positions in one viral genome to become resistant and has become the standard therapy for chronic diseases such as HIV-1 infection. Combination therapy results in much greater levels of viral suppression, thus reducing viral turnover, which, in turn, reduces the rate of production of mutants. In addition, the development of resistance is much more complex, as the virus must acquire mutations that induce resistance to a range of drugs, raising the genetic barrier.

The approval of first-generation HIV PIs during 1995 and 1996, followed by the approval of the non-nucleoside reverse transcriptase inhibitor (NNRTI) nevirapine, made combination therapy containing antiretroviral agents from different classes possible and brought a revolution in the treatment of HIV-infected patients. Combination therapy with three or more active agents, or highly active antiretroviral therapy (HAART), resulted in markedly reduced plasma viremia and a concomitant increase in CD4 + T cells in HIV-infected patients. The current standard of care recommended by the US Department of Health and Human Services (DHHS)[1] and the International Antiviral Society-USA (IAS)[10] for the treatment of naive HIV-infected patients is to combine two nucleoside reverse transcriptase inhibitors plus a third agent, whether a

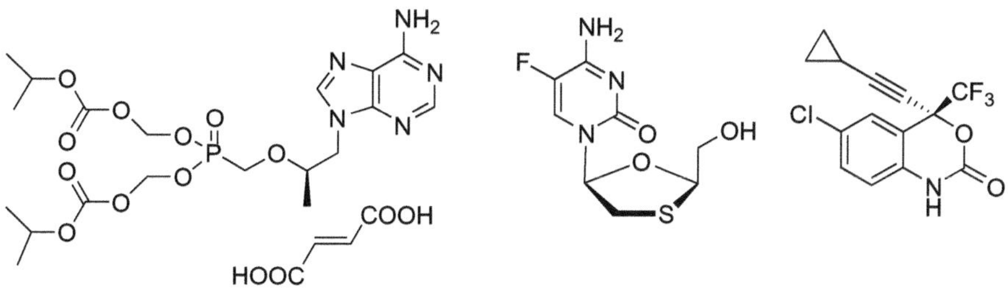

1 tenofovir disoproxil fumarate **2** emtricitavine **3** efavirenz

Figure 13.3 Structures of components of Atripla®.

non-nucleoside inhibitor, a ritonavir-boosted protease inhibitor or an integrase inhibitor. HAART has been responsible for delaying disease progression and a dramatic decline in HIV-related morbidity and mortality,[11] transforming HIV infection from a terminal condition into a chronic and manageable disease in little over a decade. Adherence, driven by a low pill burden, convenient dosing schedules and an improved tolerability and safety profile, plays a critical role in achieving long-term efficacy of HAART and preventing the emergence of drug-resistant variants.[12] Fixed-dose combination (FDC), a combination pill of two or more antiretroviral agents, was also introduced to reduce pill burden and simplify HAART regimens, with the aim of improving adherence to treatment. In 2006, the first single pill containing a complete regimen for treatment-naive individuals was approved. This pill, Atripla® (Figure 13.3), contains the NRTIs tenofovir disoproxil fumarate (**1**) and emtricitabine (**2**) in addition to the NNRTI efavirenz (**3**).

13.3 Ritonavir as a Pharmacoenhancer for HIV Therapy

The introduction of HIV-1 PIs in the mid-1990s marked the beginning of the era of HAART. HIV-1 protease inhibitors remain a mainstay of the anti-retroviral arsenal for all stages of HIV-1 infection.[13] However, poor compliance with a PI-containing antiretroviral therapy increases both the risk of incomplete viral suppression and the emergence of drug resistance.

As a class, all PIs are metabolized primarily by cytochromes P450 of the CYP3A subfamily (primarily CYP3A4 and CYP3A5) in the liver and intestine. Since they are substrates, PIs can inhibit the enzymes in a competitive manner; however, studies have shown that other mechanisms of inhibition may also play a role.[14–16] Many PIs also induce CYP3A enzymes. In addition, some HIV-1 PIs are substrates for several transport proteins, such as P-glycoprotein (P-gp) and multidrug-resistance protein 1 and 2 (MRP1 and MRP2).[14] Several PIs, including ritonavir and saquinavir, are known to inhibit P-gp activity. Most PIs have an unfavorable PK profile, including poor and/or variable oral bio-availability and rapid metabolic degradation with relatively short plasma elimination half-lives. They also have high degrees of protein binding and are

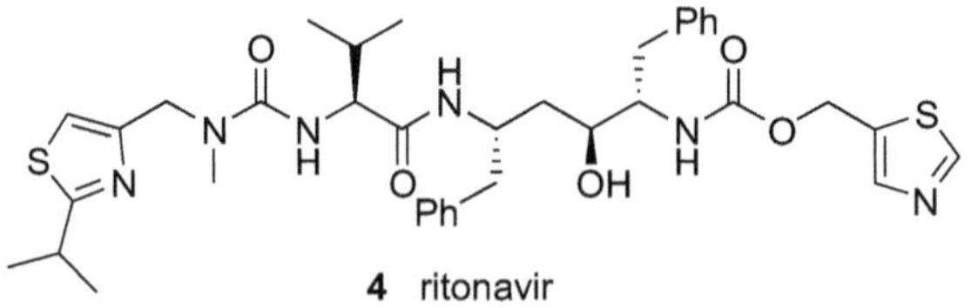

4 ritonavir

HIV EC$_{50}$: 15 nM

CYP3A (TFD) EC$_{50}$: 380 nM

Figure 13.4 Structure and activity of ritonavir. Inhibition of terfenadine hydroxylase (CYP3A) activity in human liver microsomes.

subject to efflux by P-gp. As a result, the early PI-containing regimens were typically complex and required frequent dosing, high pill burdens and strict meal and fluid requirements and were associated with significant side effects and undesirable toxicities. These effects presented adherence challenges for both clinicians and patients and, even with complete adherence to these complex regimens, substantial interpatient variability existed. Consequently, the enthusiasm for the use of regimens containing PIs began to wane. However, this changed rapidly with the discovery of the PK-enhancing effect of ritonavir (4, Figure 13.4) in the clinic.

CYP3A4 is the most abundant CYP enzyme in the liver, playing key roles in both the detoxification of xenobiotics and the metabolism of endobiotic signaling molecules.[17] Like most other PIs, the metabolism of ritonavir is mediated predominantly by CYP3A4, with a minor contribution from CYP2D6.[18] Additionally, *in vitro* metabolism studies confirmed that ritonavir is a potent inhibitor of CYP3A4, inhibiting the human liver microsomal metabolism of saquinavir (5) and indinavir (6, Figure 13.5) at low concentrations.[18,19] Subsequently, it was recognized that co-administration of ritonavir with PIs improved the PK of the latter significantly in humans, resulting in higher bioavailabilities and prolonged elimination half-lives. When co-administered with ritonavir in healthy volunteers, the plasma level of saquinavir was greatly increased, leading to a more than 50-fold increase in area under the curve (AUC) and a 22-fold increase in the maximum plasma concentration (C_{max}).[20] More importantly, the C_{12h} level of saquinavir (400 mg) increased from undetectable when dosed alone to more than 1 µM when co-dosed with ritonavir (600 mg). In addition, ritonavir reduced intersubject variability of the saquinavir AUC from 60 to 28%.[20] A substantial increase in plasma drug levels of indinavir was also observed when co administered with ritonavir.[21]

Clinical studies with combined PI therapy were first conducted in HIV-infected patients using ritonavir and saquinavir (400 mg each bid).[22] This combination, with both PIs in a therapeutic dose, proved to be safe and effective. The dramatic effect of ritonavir on the PK profile of saquinavir is consistent with a large reduction in the first-pass metabolism and post-absorption clearance of saquinavir. In addition, the GI side effects related to

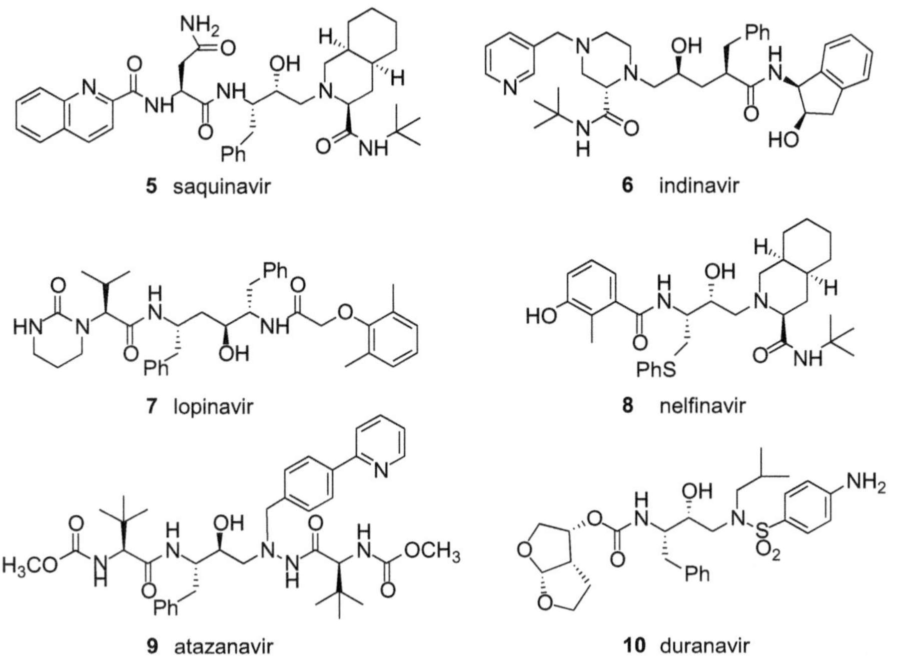

Figure 13.5 Structures of HIV-1 PIs.

dosing with ritonavir at 400 mg bid were reduced compared with those seen with ritonavir at 600 mg bid, although they remained a concern. Consequently, to improve tolerability, lower doses of ritonavir co administered with a PI were then investigated in the clinic. This clearly demonstrated that the low dose (100 or 200 mg daily) of ritonavir was sufficient to inhibit CYP3A4 metabolism and greatly enhance the PK of the co administered PIs to allow for once- or twice-daily dosing. Ritonavir-boosted PIs thus provide simplified regimens, reducing the pill burden, improving PK and treatment response, diminishing interpatient variability and obviating food restrictions, thereby enhancing adherence to therapy and slowing the emergence of resistance. As a result, co-administration of a PI with a low dose of ritonavir, often called 'a boosted PI,' has been adopted as standard practice. All currently prescribed PIs, apart from nelfinavir[23] (**8**, Figure 13.5), are typically boosted with a low dose of ritonavir;[24] whereas lopinavir (**7**) is coformulated with a low dose of ritonavir in the products Kaletra and Aluvia. Ritonavir-boosted atazanavir (**9**) or darunavir (**10**) are the preferred PIs as the third agents on treatment guidelines from the DHHS[1] and also IAS.[10]

Although boosting doses of ritonavir (100 to 200 mg once to twice daily) are better tolerated, even these lower doses are associated with safety and tolerability issues in some patients. Recent studies in healthy volunteers showed that 100 mg bid of ritonavir is still associated with adverse effects on the serum lipid profile, characterized by an increase in the concentration of total cholesterol, low-density lipoprotein (LDL) cholesterol and triglycerides and the total/ high-density lipoprotein (HDL) cholesterol ratio, and also a decrease in HDL cholesterol concentration.[25] The increase in triglyceride levels must sometimes be controlled with medication or may even necessitate switching to other drugs. GI side effects, however, are significantly reduced with low-dose ritonavir.

As ritonavir is also a potent HIV-1 PI, the principal liability of a subtherapeutic dose of ritonavir is its potential to select PI-resistant virus in drug regimens that are not fully suppressive and/or do not contain an additional HIV-1 PI. This possibility remains hypothetical, since there are insufficient clinical data to conclude that doses of ritonavir used for boosting can lead to the emergence of PI-resistant viruses.

Despite the expanded understanding of the role of the CYP450 enzyme system in drug interactions, drug interactions remain unpredictable.[26] In addition to inhibiting CYP3A, ritonavir inhibits CYP2D6 and activates xenobiotic-sensing receptors, such as the aryl hydrocarbon receptor (AhR) and the pregnane X receptor (PXR). At boosting doses, the impact of ritonavir on CYP2D6 and AhR appears to be negligible.[27] PXR is a predominant regulator of CYP3A expression. It also controls the expression of CYP2B6 and regulates multidrug-resistant gene 1 (*MDR1*) and other genes.[28] Activating PXR results in the induction of CYP3A and drug transporters.[29] Although the net effect of ritonavir on human CYP3A in the clinic is inhibitory, the inhibitor potency is reduced upon chronic dosing, as the rate of CYP3A re-synthesis is increased *via* induction. Other proteins induced by ritonavir in the clinic include CYP2B6, CYP2C9, CYP2C19, UGT1A4 and MDR1, which further complicate the drug–drug interactions.[30,31] Consequently, the dosage of therapeutic drugs co-dosed with ritonavir needs to be carefully monitored and refined, depending on their metabolism and/or route of clearance.

The poor physicochemical properties of ritonavir generate challenges in its production and formulation. Ritonavir is not bioavailable from the solid state, so it was formulated as either an oral solution or semi-solid capsules, both in an ethanol–water-based solution. The solvent system is believed to be the cause of some of the poor tolerability issues associated with the drug. After 2 years on the market, a new and extremely insoluble crystal form, polymorphism form II, appeared. Unfortunately, studies demonstrated that form II is the thermodynamically more stable form and exhibits an extremely low dissolution rate, resulting in unacceptable bioavailability. Studies of the crystal form of form II showed that it formed strong hydrogen bonds between the two molecules in a transform *via* the hydroxyl group.[32] The two ritonavir molecules depicted in the box in Figure 13.6 is the repeating unit in the crystal packing. To prevent the formation of form II, the semi-solid capsule or oral formulation requires refrigeration prior to use. Therefore, ritonavir is not easily amenable to coformulation with other PIs. Recently, the FDA has approved commercialization of a ritonavir tablet and a combined lopinavir–ritonavir tablet using melt extrusion technology (Meltrex). Overall, the poor physicochemical properties of ritonavir hinder its utility to be combined with other antiretroviral (ARV) agents in the development of a fixed-dose combination (FDC) or a fixed-dose regimen (FDR).

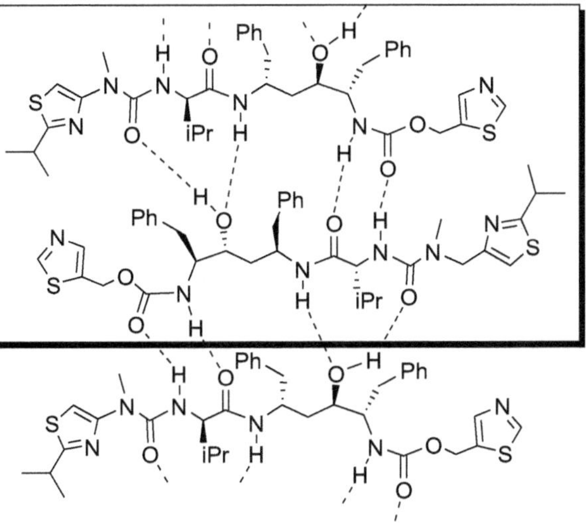

Figure 13.6 Hydrogen bonding network for ritonavir polymorph form II.

13.4 Mechanism of CYP3A Inhibition of Ritonavir

There have been many studies on the mechanism of action (MOA) of ritonavir as a CYP3A inhibitor, but no definite conclusions have been reached. The available data are inconsistent and suggest that ritonavir acts as a competitive, mixed competitive–non-competitive, quasi-irreversible or mechanism-based CYP3A inhibitor. Through kinetic and equilibrium analysis, the most recent studies[33] on the MOA of ritonavir concluded that ritonavir is an irreversible type II (Figure 13.7) inhibitor that inhibits CYP3A4 turnover both by replacing substrates in the active site and binding irreversibly to the heme iron and also through changes in the protein redox potential which preclude reduction by cytochrome P450 reductase (CPR). From the crystal structure of ritonavir and the CYP3A4 enzyme complex (Figure 13.8), it is evident that ritonavir fits perfectly into the active-site cavity with extensive hydrophobic enclosure by the enzyme and the nitrogen of the unsubstituted thiazole ligates to the heme iron.

The conclusion that ritonavir exerts its PK-enhancing effect *via* a type II binding with CYP3A through the 5-thiazolyl group is consistent with the first reported mechanism study by Kempf *et al.*[19] They also examined the structural features of ritonavir responsible for CYP binding and inhibition, showing the important contribution of the two terminal thiazoles to inhibiting the oxidizing capability of the enzyme (Table 13.1). Although compounds **4** and **11–14** potently bind to CYP3A, only ritonavir (**1**) inhibits the CYP3A-mediated oxidation of terfenadine. Compounds **11** and **12** differ from ritonavir only in a change of the terminal 5-thiazolyl to 4-thiazolyl and pyridinyl, respectively, but the inhibitory capacity was significantly reduced. Compound **13** incorporated a pyridyl carbonate instead of the isopropyl-4-thiazolylurea of ritonavir and compound **5** truncated the left portion; both lost their potency as

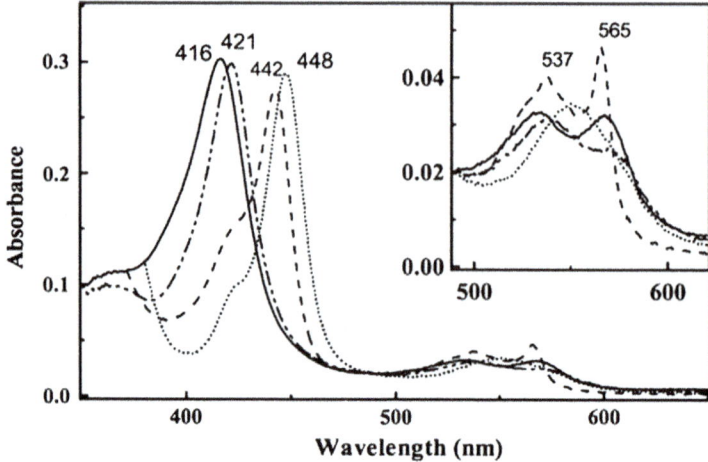

Figure 13.7 Spectral changes induced by ritonavir in CYP3A4. Absorbance spectra of ferric ligand-free (—), ferric ritonavir-bound (–·–), ferrous ritonavir-bound (- - -) and ferrous-CO adduct (····) of 3 μM CYP3A4 were recorded in buffer (50 mM phosphate, pH 7.5, 20% glycerol, and 1 mM dithiothreitol).

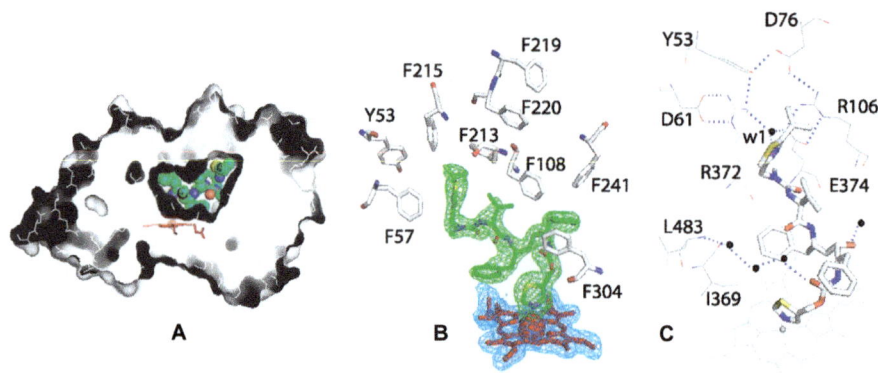

Figure 13.8 Crystal structure of the CYP3A4-ritonavir complex. (A) The active site cavity of ritonavir-bound CYP3A4. Ritonavir is green and in CPK representation; the heme is red. (B) Aromatic residues surrounding ritonavir. $2F_o - F_c$ (blue) and $F_o - F_c$ (green) electron density maps around the heme and ritonavir are contoured at 1σ and 3σ, respectively. (C) An umbrella-like charge–charge/H-bonding network connected to the isopropylthiazole moiety of ritonavir *via* a highly ordered water molecule (w1) (2.0 Å).

CYP3A inhibitors, reflecting the importance of the 5-thiazolyl moiety. These structure–activity relationship data suggest that the high CYP3A-inhibitory potency associated with ritonavir is due both to hydrophobic interactions with CYP3A and the direct ligation of the heme iron by the thiazole moieties.

Table 13.1 CYP binding and CYP3A inhibition by analogs of ritonavir.[19]

Compound	Structure	CYP binding[a] [ΔA (nmol CYP)$^{-1}$]	CYP3A inhibition[b]: IC$_{50}$ (μM)
4		0.033	0.38
11		0.024	3.0
12		0.032	1.5
13		0.033	3.8
14		0.045	2.3

[a] ΔA is a measure of the extent of binding to CYP. Microsomes were used without differentiation of CYP subtypes.
[b] Inhibition of terfenadine hydroxylase (primarily CYP3A4/5) activity in human liver microsomes.

In contrast to the results of the investigations described above, however, several *in vitro* kinetic studies have shown that ritonavir is actually a potent mechanism-based CYP3A4 inhibitor (MBI).[34] This mode of inactivation of CYP3A4 by ritonavir involves time- and cofactor-dependent metabolism of ritonavir to form a metabolite (or metabolic intermediate) that binds tightly to CYP3A4 and inactivates it. The precise identity of the active metabolite(s) is unknown, as are the details of the mechanism by which the inactivation takes place.[35] The addition of reduced glutathione, ascorbic acid, *N*-acetylcysteine or superoxide dismutase failed to protect CYPs from inactivation by ritonavir, indicating that the metabolite or metabolic intermediate is efficiently modifying the active site rather than escaping.[27,34] One of the metabolites, M1, was reported to be potentially involved in the inactivation of the CYP3A.[36] The K_I and k_{inact} values have been measured for both hepatic microsomal fractions and recombinant CYP3A4 and are listed in Table 13.2.

Reviewing the results from both the type II binding and MBI proposals, it is more likely that the overall inhibitory effects of ritonavir arise from a combination of both type II and MBI, as illustrated in Figure 13.9. Ritonavir binds to the CYP3A4/5, forming an enzyme–inhibitor (RTV/CYP3A) complex.

Table 13.2 Mechanism-based inhibition of CYP3A enzymes by ritonavir.

Enzyme source	Substrate	Comments	k_{inact} (min^{-1})	K_I (μM)	k_{inact}/K_I	Partition ratio
Intestinal microsomes	Indinavir[34a]	Different k_{inact} values measured for intestinal microsomes and recombinant CYP3A4	0.078			10
Recombinant CYP3A4			0.135			
Liver microsomes	Triazolam[34b]	Values of k_{inact} and K_I not available				
Hepatic microsomes	Testosterone[34c]		0.29	0.038	7.63	
Hepatic microsomes	Testosterone[34d]	Metabolic intermediate complex formation on CYP3A4	0.40	0.17	2.35	
Recombinant CYP3A4			0.32	0.10	3.20	
Recombinant CYP3A5			0.08	0.12	0.67	
Hepatic microsomes	Midazolam[34e]		0.45	0.38	1.18	
	Testosterone[34e]		0.28	0.18	1.56	

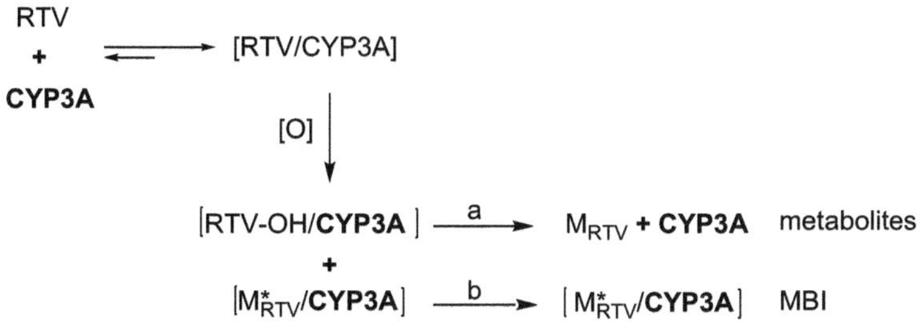

Figure 13.9 (a) Ritonavir is metabolized to form a more hydrophilic metabolite, prior to excretion from the body. (b) Ritonavir undergoes a biotransformation into a metabolic intermediate or a reactive metabolite that causes mechanism-based inhibition.

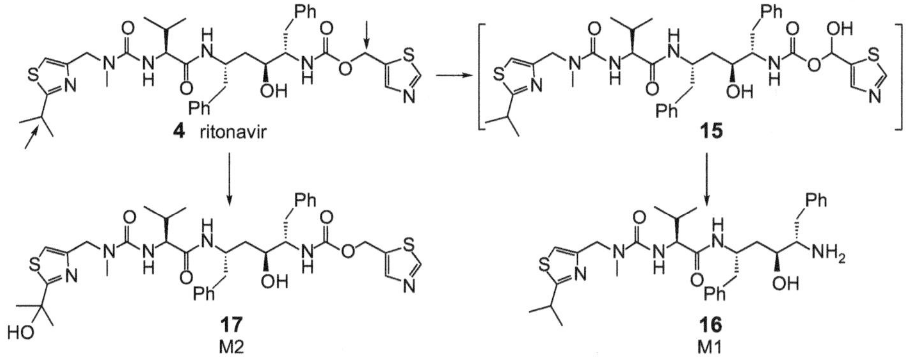

Figure 13.10 CYP3A oxidation of ritonavir and oxidative metabolites.

The type II binding capability enables ritonavir to bind to CYP3A with fast kinetics and tight binding and the disassociation rate is very slow. The RTV–CYP3A complex will either disassociate to give CYP3A and unchanged ritonavir or ritonavir will be oxidized by CYP3A. Two different types of oxidation products can be formed. The oxidation product RTV-OH, a benign intermediate, can disassociate from CYP3A (route a). RTV-OH can be a final metabolite or it will further degrade to form additional metabolites (oxidation and cleavage process), which are then excreted from the body. Two of the major ritonavir metabolites,[18] M1 and M2 (Figure 13.10), are examples of this route. M2 is a hydroxylated RTV at the isopropyl site, which is also a final metabolite, and M1 is the degradation product of RTV-OH in which hydroxylation has occurred at the α-site of thiazolyl. The other oxidation product, M*$_{RTV}$, a metabolic intermediate or a reactive metabolite (route b), either binds CYP3A tightly to form a metabolic intermediate complex (MIC) or reacts with CYP3A, thus inactivating the enzyme.

Although the direct interaction between ritonavir and the heme of cytochrome P450 as a type II ligand may play a role in the inhibition of CYP3A *in vivo*, the mechanism-based inactivation of CYP3A4 is consistent with the clinical observation of the sustained pharmacoenhancing effect of ritonavir, with the duration of effect longer than the persistence of ritonavir in plasma. The 4- and 5-thiazolyl groups of ritonavir are believed to play a key role in this mechanism-based inhibition of CYP3A4.[34a] Modifications of both moieties caused a significant decrease in ritonavir's inhibitory activity.[19]

In the clinic, ritonavir-boosted regimens favorably alter the PK of co administered PIs. Ritonavir inhibits CYP3A4 and exerts most of its effects through two mechanisms: (1) it attenuates first-pass metabolism in the gut and liver and thereby increases the amount of drug reaching the systemic circulation and/or (2) it decreases hepatic metabolism, resulting in a prolonged terminal half-life. In addition, there may be a minor contribution from its inhibition of P-gp. Ritonavir may increase the C_{min}, AUC and C_{max} of co administered PIs; however, the degree to which ritonavir affects these parameters varies with the targeted PI. For PIs that undergo extensive first-pass metabolism, such as saquinavir, ritonavir boosting increases all three parameters in addition to prolonging the elimination half-life. For the PIs indinavir and amprenavir, which have reasonable unboosted bioavailability but a short half-life, ritonavir boosting primarily increases their C_{min} and AUC.

The long-term toxicities associated with chronic inhibition of CYP3A4 are unknown. In general, a major concern for mechanism-based CYP-inhibition is that it may cause hepatotoxicity.[37] However, there is no evidence to support the association of the hepatotoxicity observed with high-dose ritonavir with its CYP inhibition. Low-dose ritonavir-related hepatotoxicity seems uncommon and does not appear to be associated with a significant increase in the risk of hepatotoxicity relative to the other PIs.[35,38] In addition, ritonavir has been marketed and prescribed to millions of patients for more than 15 years and hypersensitivity to ritonavir has not been reported. The mechanism of liver injury observed with PIs, including high doses of ritonavir, is not clear; it is likely that viral hepatitis co-infection may be a risk factor for the observations of increased ALT.[39–41]

13.5 Discovery of the Pharmacoenhancer Cobicistat

Despite advances in antiretroviral therapy, many patients experience suboptimal virologic, immunologic or clinical benefit from currently available treatment options due to development of resistance. New drugs, particular novel drug classes, are needed against highly resistant strains of HIV. Elvitegravir (EVG, **18**, Figure 13.11) is an HIV integrase strand transfer inhibitor with potent antiretroviral activity against wild-type and NRTI, NNRTI and PI-resistant laboratory strains.[42] However, it is extensively metabolized primarily by CYP3A4 in the liver and intestine *in vivo*, excluding its use as a once-daily drug.

In addition to affecting the PK of HIV PI drugs, ritonavir has been shown in clinical studies to enhance the PK of important antiviral drugs that are CYP3A substrates, including elvitegravir.[43] When dosed alone, elvitegravir needs to be administered at 400 mg bid to achieve a plasma trough concentration that is

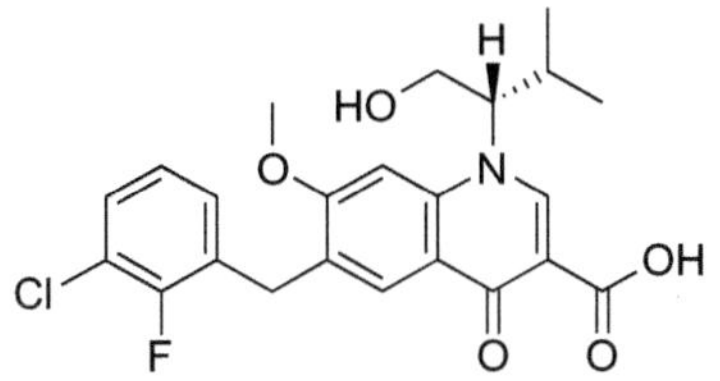

18

Figure 13.11   Structure of elvitegravir.

adequately above its protein binding-adjusted EC_{95}. When co-administered with 100 mg of ritonavir, the steady-state systemic area under the plasma concentration–time curve exposure of elvitegravir (150 mg) increased 20-fold. The trough concentration increased significantly to more than 10-fold above its protein binding-adjusted EC_{95}, achieving the targeted plasma level for durable viral suppression; therefore, it can be dosed once daily. Phase 3 studies showed that in previously treated patients, once-daily elvitegravir was non-inferior to the licensed integrase strand transfer inhibitor raltegravir when combined with regimens that included a boosted protease inhibitor. However, the use of a subtherapeutic dose of ritonavir may have the potential to select PI-resistant virus in a non-PI-containing regimen, thus limiting the use of ritonavir-boosted elvitegravir to only treatment-experienced patients or a PI-containing regimen. In addition, as discussed in previous sections, ritonavir has other disad-vantages, including causing lipid disorders and triggering undesired drug interactions as an inducer of drug-metabolizing enzymes such as CYP, P-gp and UDP glucuronosyltransferases (UGTs).

A new pharmacoenhancer to allow the broader use of once-daily elvitegravir needs to have the following targeted properties: (1) no or minimal anti-HIV activity; (2) a mechanism of CYP inhibition similar to that of ritonavir; (3) higher aqueous solubility and better physicochemical properties; (4) reduced off-target drug interactions; and (5) improved tolerability with minimal side effects due to lipid disorder and gastrointestinal disturbance.

Since ritonavir (**4**) is a unique pharmacoenhancer that exerts sustained pharmacological effects with a record of long-term safety demonstrated in clinical settings, it was used as a starting point, with the goal of eliminating its anti-HIV activity while maintaining its potent inhibitory activity and MOA on CYP3A enzymes. The initial attempts to eliminate the antiviral activity of ritonavir were focused on removal of the key hydroxyl group that mimics the transition state of amide hydrolysis through the formation of hydrogen bonds to the oxygen atoms of the catalytic Asp25 and Asp25′ residues at the active site of the HIV protease.[44] Desoxyritonavir (**19**, Figure 13.12) was about 20 times less potent than ritonavir in a cell-based antiviral assay but retained full inhibitory activity against CYP3A. After extensive structure–activity rela-tionship (SAR) studies with compound **19**, cobicistat (**20**, Figure 13.12) was identified as a potent, selective and orally bioavailable inhibitor of CYP3A.[45]

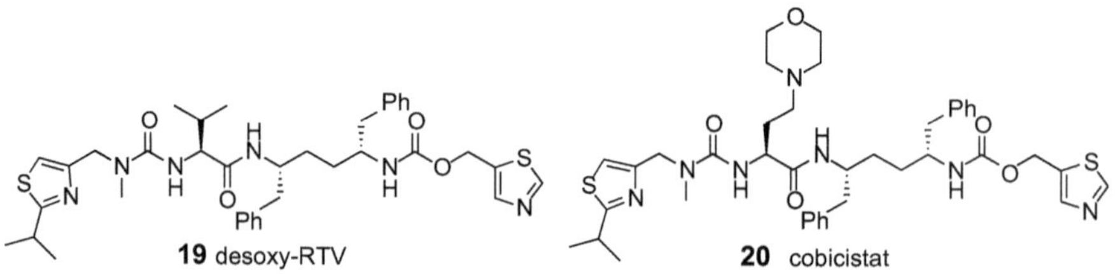

19 desoxy-RTV
HIV EC_{50}: 290 nM
CYP3A (MDZ): 104 nM

20 cobicistat
HIV EC_{50}: >30,000 nM
CYP3A (MDZ): 154 nM

Figure 13.12 Desoxy-RTV and cobicistat.

Table 13.3 Kinetics of the inactivation of human hepatic microsomal CYP3A-dependent midazolam 1′-hydroxylase activity.

Parameter	*Ritonavir*	*Cobicistat*
k_{inact} (min^{-1})	0.23 ± 0.06	0.44 ± 0.09
K_I (nM)	256 ± 90	939 ± 353
k_{inact}/K_I (min^{-1} μM^{-1})	0.90	0.57

Cobicistat's mechanism of CYP3A inhibition was compared extensively with that of ritonavir. Similarly to ritonavir, in addition to interacting directly at the heme iron of the CYP3A enzyme, cobicistat is also a potent inhibitor of CYP3A enzymes and is an effective mechanism-based inhibitor. A protocol similar to that described by Ernest *et al.*[34d] was used to measure the parameters of inactivation kinetics, k_{inact} and K_I, using midazolam as the CYP3A substrate.[46] Importantly, cobicistat and ritonavir inactivate CYP3A similarly at both low and high concentrations and in a time- and concentration-dependent manner. The corresponding estimates of k_{inact} and K_I are presented in Table 13.3. These results suggest that ritonavir and cobicistat share the same mechanism of action for the inhibition of CYP3A.

The mechanism-based inactivation of human CYP3A4 by cobicistat implies that, in the clinic, sustained CYP3A inhibition will likely be achieved even after cobicistat is cleared from the body and before new CYP enzymes are synthesized by the hepatocytes, as the half-life of CYP turnover has been estimated to be around 12–48 h.[47] *In vitro* data also suggest that cobicistat will inhibit CYP3A with similar potency to ritonavir in humans. This has been demonstrated in clinical studies, where cobicistat showed similar potency to ritonavir in affecting the PK of the model CYP3A substrate midazolam. In addition, cobicistat also showed non-linear dose-dependent and time-dependent PK, consistent with it being a mechanism-based inhibitor.[48]

CYP3A enzymes are known to display substrate dependence in their susceptibility to inhibition. To confirm that cobicistat also retains broad substrate specificity similar to that of ritonavir in inhibiting CYP3A, the abilities of ritonavir and cobicistat to inhibit the metabolism of a diverse set of substrates using assay protocols was investigated based on current industry and regulatory guidelines.[49] The potency of cobicistat as an inhibitor of human hepatic microsomal CYP3A was compared with that of ritonavir, using substrates that include terfenadine, midazolam, testosterone, atazanavir, telaprevir and elvitegravir. The inhibitory potency was either measured with marker activities for the enzymes or determined by monitoring substrate depletion. As shown in Table 13.4, the broad spectrum of inhibitory activity of cobicistat against human CYP3A was confirmed.

The data in Tables 13.3 and 13.4 indicate that cobicistat shares a similar spectrum of CYP3A substrate specificity to that of ritonavir. It retains the characteristic of mechanism-based inhibition of CYP3A and is equipotent to ritonavir for the substrates tested. Inhibition studies with the most important human CYP enzymes showed no significant inhibition at concentrations likely to be achieved clinically (Table 13.5). In addition, cobicistat is more selective than ritonavir, with much reduced inhibitory activity towards CYP2D6, CYP2C8 and CYP2C9.

As discussed in Session 13.3, ritonavir activates the pregnane X receptor (PXR) and induce metabolic proteins including CYP3A, CYP2B6, CYP2C9 and P-gp, further complicating the potential for drug–drug interactions.[29,30] It is therefore desirable to eliminate or reduce this drug interaction potential for new PK enhancers. Studies assessing the induction liability of cobicistat showed that neither cobicistat nor ritonavir showed significant stimulatory activity in the AhR-responsive assay. However, at $10\,\mu M$ in the PXR assay, ritonavir

Table 13.4 Inhibitory potencies against human hepatic microsomal CYP3A-dependent activities.

| CYP3A activity | $IC_{50}\ (nM)$ | |
	Ritonavir	Cobicistat
Midazolam 1′-hydroxylase[50]	107	154
Testosterone 6β-hydroxylase[51]	116	151
Terfenadine oxidase[52]	275	285
Elvitegravir oxidase	26	33
Atazanavir oxidation	40	44
Telaprevir oxidation	18	30

Table 13.5 Inhibitory potencies against activities catalyzed by major human hepatic microsomal cytochromes P450.

| Enzyme | Activity | Calculated $IC_{50}\ (\mu M)$ | |
		Ritonavir	Cobicistat
CYP1A2	Phenacetin O-deethylase	>25	>25
CYP2B6	Bupropion 4-hydroxylase	2.9	2.8
CYP2C8	Paclitaxel-6α-hydroxylase	2.8	>25
CYP2C9	Tolbutamide 4-hydroxylase	4.4	>25
CYP2C19	S-Mephenytoin 4′-hydroxylase	>25	>25
CYP2D6	Dextromethorphan O-demethylase	2.8	9.2

exhibited 93% of E_{max} (the response seen with $10 \mu M$ rifampicin was considered to be 100% of the E_{max}) and the corresponding EC_{50} was $1.9 \mu M$. Ritonavir is thus confirmed as a potent PXR agonist with the potential to achieve significant activation in humans where the C_{max} is around $2 \mu M$. Cobicistat was considerably weaker than ritonavir in activating PXR, with 15% of E_{max} detected at a concentration of $10 \mu M$. The shapes of the curve also suggest that a 100% response with cobicistat would be difficult to achieve. Cobicistat is therefore much weaker than ritonavir as an activator of PXR and is potentially less likely to cause clinical drug–drug interactions through induction mechanisms.

Chronic treatment of HIV-infected patients with ritonavir is known to induce changes in body fat distribution (lipodystrophy), elevated cholesterol and triglycerides (hyperlipidemia) and insulin resistance, known collectively as metabolic syndrome.[8d] It is believed that some of these effects are due, at least in part, to the direct effects of ritonavir on adipocytes.[53] *In vitro*, ritonavir has been shown to affect adipocyte functions, such as lipid accumulation during differentiation and insulin-stimulated glucose uptake.[54] Therefore, cobicistat was evaluated for its effects on adipocytes in comparison with ritonavir. Atazanavir, the PI used in the clinic with the least pronounced metabolic syndrome,[55] was also evaluated as a comparator.

The lipid accumulation assay monitored normal lipid accumulation in cultured human adipocytes following induction of differentiation in the presence of tested PIs for 9 days. Ritonavir showed a clear effect, with an EC_{50} of $16 \mu M$ (Table 13.6). In contrast, both cobicistat and atazanavir exhibited no effect at concentrations up to $30 \mu M$. The glucose uptake assay monitored insulin-stimulated glucose uptake in mouse adipocytes in the presence of $10 \mu M$ ritonavir, atazanavir or cobicistat. Ritonavir showed a pronounced effect at this concentration. In contrast, the effects on glucose uptake by cobicistat and atazanavir were significantly less. The minimal adverse effects of cobicistat in these assays suggest a lower potential for toxicity related to altered lipid metabolism compared with ritonavir.

Cobicistat greatly improved aqueous solubility compared with ritonavir under both neutral (pH 7.4: 75 *versus* $\sim 2.0 \mu g \, mL^{-1}$) and acidic (pH 2.2: > 6500 *versus* $\sim 3.1 \mu g \, mL^{-1}$) conditions.

Both cobicistat and ritonavir exhibit poor metabolic stability in preclinical species *in vitro*. Their metabolic stability is concentration dependent and both can inhibit their own metabolism at high concentrations. The PK of cobicistat in preclinical animals is consistent with this, as high clearance is observed at low

Table 13.6 Inhibition of cell functions in adipocytes.

Compound	Lipid accumulation: EC_{50} (μM)	Glucose uptake (% inhibition at $10 \mu M$)
Cobicistat	>30	9.5 ± 6.4
Ritonavir	16 ± 8	55 ± 10
Atazanavir	>30	0.4 ± 0.9

doses. Volumes of distribution are moderate. Owing to mechanism-based inhibition of human CYP3A and consequent self-inhibition of their clearance, the absorption potential in humans is the more relevant PK parameter for cobicistat and ritonavir. The absorption potential was evaluated in both portal vein-cannulated dogs and rats. The results indicated that the absorption of cobicistat is above 50% in these species, comparable to that of ritonavir. Cobicistat was therefore expected to have high absorption potential in humans.

All of these studies have shown that cobicistat is a potent and selective CYP3A inhibitor that lacks significant anti-HIV activity. *In vitro* studies also suggest that cobicistat may have a lower potential for causing undesired drug–drug interactions and lipid disorders than ritonavir. Based on these results, cobicistat was selected as a clinical candidate for further development.

13.6 Cobicistat as a Pharmacoenhancer in HIV Therapy

13.6.1 Cobicistat Boosting the PK Profile of CYP3A Substrate

Cobicistat was investigated in clinical Phase 1 studies for its safety, tolerability, pharmacokinetics and pharmacodynamics with single- and multiple-escalating oral doses in healthy subjects.[48] Studies showed that cobicistat is generally safe and well tolerated, with mild-grade headache, somnolence and abnormal dreams the most frequently reported drug-related adverse events. It is well absorbed and its PK showed non-linearity, with apparent clearance approaching a nadir after single or multiple dosing at doses > 200 mg. More importantly, pharmacodynamic studies confirmed the prediction from the preclinical *in vivo* and *in vitro* investigations that cobicistat is a potent and persistent CYP3A inhibitor. In this Phase 1 study using the CYP3A substrate midazolam as a probe, the CYP3A inhibitory activities of cobicistat and ritonavir were compared. Midazolam maleate was administered at the steady state of cobicistat (50, 100, 200 mg) and ritonavir (100 mg). Cobicistat reduced the clearance of midazolam by 95% when dosed at 200 mg once daily, comparable to that achieved with ritonavir dosed at 100 mg once daily (96%). The persistence of the inhibitory effect of cobicistat was confirmed by the sustained suppression of the formation of the CYP3A-mediated metabolite of midazolam, 1′-hydroxymidazolam. In two independent Phase 1 studies in healthy subjects, the pharmacoenhancing effects of cobicistat on atazanavir and darunavir were compared with that of ritonavir. Atazanavir and darunavir were bioequivalent when administered concomitantly with 150 mg of cobicistat once daily or 100 mg of ritonavir.[56,57] These studies demonstrated that cobicistat is comparable to ritonavir in enhancing the PK profile of a CYP3A substrate.

13.6.2 Cobicistat as a Pharmacoenhancer for Anti-HIV Agents

After demonstrating favorable PK/PD results from Phase 1, cobicistat was investigated in Phase 2 and 3 clinical studies as a pharmacoenhancer for drugs metabolized by CYP3A to improve their PK exposure. The efficacy of cobicistat versus ritonavir as a pharmacoenhancer for atrazanavir was studied in a randomized, placebo-controlled, double-blind, multicenter, 48-week Phase 3 study. In the trial, anti-retroviral treatment-naive HIV-1-infected adults received either cobicistat or ritonavir for 48 weeks; all patients also received atazanavir plus Truvada. At 48 weeks, the study found that 85% of patients on the cobicistat-containing regimen, compared with 87% of patients on the ritonavir-containing regimen, achieved the primary endpoint, with HIV RNA levels less than $50\,\text{copies}\,\text{mL}^{-1}$; 7% of the patients discontinued owing to adverse events in each arm of the study. The study found that an HIV regimen containing a cobicistat-boosted PI was non-inferior to a regimen containing a ritonavir-boosted PI at 48 weeks of therapy.[58] Currently, cobicistat is under regulatory review. If approved, cobicistat may be an effective option for boosting the potency of HIV regimens that are based on PIs.

Cobicistat has favorable physicochemical properties, especially high aqueous solubility, allowing it to be formulated as a tablet and co-formulated with other drugs as a tablet. It was co-formulated with the preferred, once-daily nucleoside/nucleotide reverse transcriptase backbone tenofovir disoproxil fumarate (TDF) and emtricitabine (FTC), and also the integrase inhibitor elvitegravir (EVG 150 mg, COBI 150 mg, FTC 200 mg and TDF 300 mg) as a once-daily single-tablet regimen (STR), known as Quad. Quad is the first and only integrase inhibitor-containing single-tablet regimen. The other two single-tablet regimens available, Atripla® and Complera®, are based on a non-nucleoside reverse transcriptase inhibitor as the third agent. Two independent, fully powered Phase 3 non-inferiority trials have compared Quad with two current standard-of-care regimens for initial HIV treatment. One regimen contained the non-nucleoside reverse transcriptase inhibitor efavirenz and the other was based on the boosted protease inhibitor atazanzavir; both used Truvada as the backbone. Results at 48 weeks from these two Phase 3 studies have been published. The two Phase 3 studies show that Quad has high efficacy and a good tolerability profile, with the limitations of drug–drug interaction and a need to be taken with food. The primary endpoint for the study is to achieve HIV RNA (viral load) fewer than $50\,\text{copies}\,\text{mL}^{-1}$ at 48 weeks. The top line results showed 88% of Quad patients *versus* 84% of Atripla® patients achieved the primary endpoint. Mean CD4+ T cell count from baseline in Quad was 239 cells mm^{-1} *versus* 206 cells mm^{-1} in Atripla®.[59] In the study comparing Quad with boosted atazanavir, 90% of Quad patients *versus* 87% of boosted atazanavir patients achieved the primary endpoint at 48 weeks.[60] Quad is statistically non-inferior to the two standard-of-care regimens. The FDA approved Quad (Stribild®) to treat HIV-1 infection in treatment-naive adults on 27 August 2012. It is expected that Quad will become an important complete regimen option for adult HIV-infected subjects.

With the availability of low dose and ease of formulation, cobicistat is also being co-formulated as a fixed-dose combination tablet with atazanavir and darunavir. It is expected to help further simplify the regimen and reduce the pill burden in boosted PI drug combinations. In addition, cobicistat may help

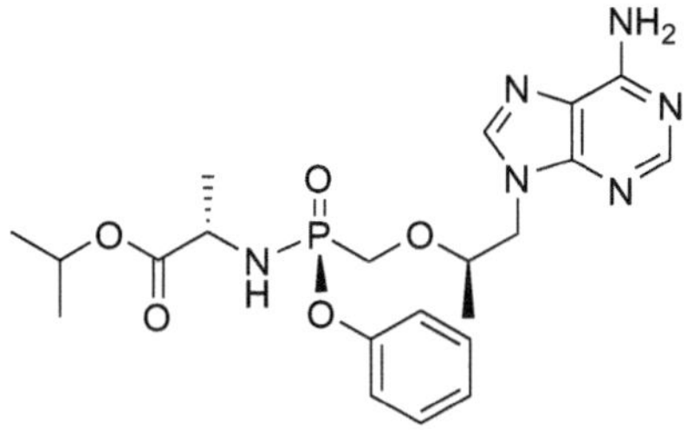

21 GS-7340

Figure 13.13 Structure of GS-7340.

generate a single-tablet regimen containing a PI. It is currently co-formulated with emtricitabine (200 mg), the PI darunavir (800 mg) and the investigational drug GS-7340 (**21**, Figure 13.13) as a single-tablet regimen. This is the first once-daily, PI-based, single-tablet regimen and is currently in clinical Phase 2 studies.

With the discovery of cobicistat as a novel and easily formulated pharmacoenhancer without antiviral activity, alternative single-tablet regimens with new drug classes or mechanisms of action that provide sustained efficacy with favorable tolerability and safety profiles for patients with HIV infection can now be realized.

13.7 Future Perspectives

13.7.1 Pharmacoenhancers in Antiviral Therapies

As mechanism-based inhibitors of CYP3A4, cobicistat and ritonavir can be expected to enhance the clinical PK of many therapeutic drugs that are CYP3A4 substrates. In addition to affecting the PK of HIV PIs and integrase inhibitors,[61] ritonavir has been shown in clinical studies to enhance the PK profile of important antiviral drugs which are CYP3A substrates, including the CCR5 antagonists aplaviroc and maraviroc[62–64] used in HIV therapy and the HCV protease inhibitors narlaprevir[65] and danoprevir.[66]

Aplaviroc is primarily metabolized by CYP3A4. Co-administration of aplaviroc with lopinavir/ritonavir (400/100 mg) increased exposure of aplaviroc by 7.7-, 6.2- and 7.1-fold for its AUC, C_{max} and C_{min} at steady state.[67] Similarly to aplaviroc, the exposure of maraviroc was increased in the presence of ritonavir in both healthy volunteers[63] and patients.[64] Maraviroc was approved in 2006 for use in treatment-experienced patients infected with CCR5-tropic HIV-1, but the development of aplaviroc has been discontinued owing to hepatotoxicity. The dosage of maraviroc requires adjustment when used in combination with PIs, with the exception of tipranavir–ritonavir combinations.[68]

Clinical studies in healthy volunteers have shown that AUC, C_{max} and $C_{12\,\text{h}}$ of a single dose of danoprevir (100 mg) before and after administration of ritonavir 100 mg bid for 10 days were increased around 5.5, 3.2 and 26.9, respectively; C_{trough} was improved to a great extent, whereas the AUC and C_{max} were less affected.[69] More importantly, the asymptomatic grade 4 ALT elevation observed with danoprevir at a dose of 900 mg bid was associated with a high AUC and C_{max}, while its efficacy corresponded with the C_{trough}. Co-administration with ritonavir allows a lower dose of danoprevir to achieve inadequate C_{trough} for therapy but with a much lower AUC and C_{max}. With the standard-of-care PEG-IFNα-2α/ribavirin, danoprevir/ritonavir 200/100 mg twice daily exhibited more potent antiviral activity than danoprevir 900 mg bid.[66] The ritonavir-boosted, low-dose danoprevir regimen is undergoing further clinical studies.

Ritonavir is also used as a pharmacoenhancer for several investigational drugs, including the HIV-1 non-nucleoside reverse transcriptase inhibitor BILR 355 and the HCV NS3 protease inhibitor ABT-405, although no detailed PK data are yet available for either of these two drugs. In the COPILOT study, ABT-450/r, plus ABT-333 and ribavirin administered for 12 weeks showed a sustained virologic response at 12 weeks post-treatment (SVR12) in 93 and 95% of treatment-naïve genotype 1 (GT1) patients. In a separate study, known as PILOT, 91% of GT1-infected, treatment-naive patients taking ABT-450/r and ABT-072 combined with ribavirin, administered for 12 weeks, achieved a sustained viral response at 24 weeks (SVR24).

13.7.2 Novel Pharmacoenhancers

Any novel pharmacoenhancers entering development that can maintain the boosting efficacy of ritonavir but overcome its liabilities, such as anti-HIV activity, adverse lipid profile and poor physical properties, are highly desirable. The development of a new chemical entity as a pharmacoenhancer was first explored with cobicistat, as the benefit of cobicistat (a pharmacoenhancer) can only be demonstrated when it is combined with a therapeutic drug, as itself it has no therapeutic effect. Ritonavir is widely used off-label as a pharmacoenhancer in the treatment of HIV infection, although it was not originally developed as a pharmacoenhancer but as an HIV PI. Cobicistat is now under regulatory review. The current challenge facing the development of a pharmacoenhancer is that a well-defined pathway is not available for advancing it through different stages of clinical studies. Also, if a pharmacoenhancer has successfully obtained regulatory approval for co-administration with a therapeutic drug, it remains unclear what type of clinical studies need to be carried out to secure indications for enhancing the same class therapeutic drugs (the 'boostees') with a similar metabolic pathway. Additionally, it is critical that the novel pharmacoenhancer has a clean chronic safety profile.

There are a number of companies engaged in the discovery and development of novel pharmacoenhancers (Figure 13.14); several are in advanced preclinical stages or in clinical studies. SPI-452 (exact structure not disclosed), a structural derivative of amprenavir, is a novel PK enhancer designed to improve the exposure of co-administered HIV medications to permit less-frequent antiviral drug dosing. Phase 1 clinical studies in combination with HIV medications

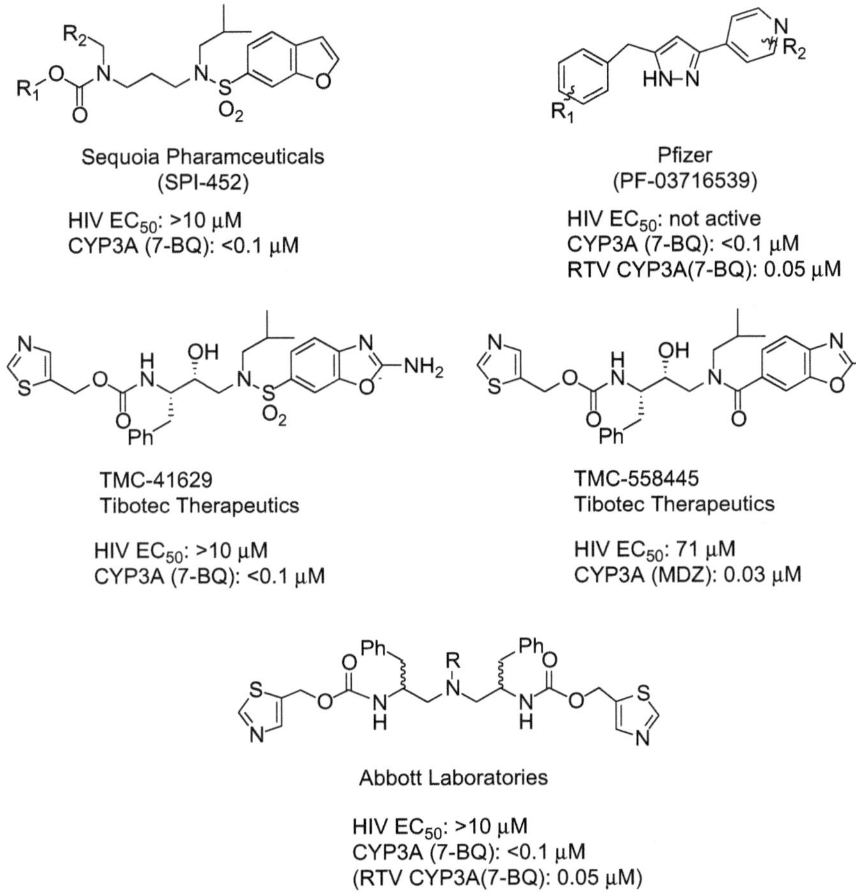

Figure 13.14 Potential novel pharmacoenhancers.

(darunavir, atazanavir or saquinavir) were reported at CROI.[70] The results indicated that SPI-452 boosted the exposure by 20- and 15-fold for darunavir and atazanavir, respectively. Another structural analog of amprenavir, TMC-41629, incorporated the 5-thiazolyl group, which is important for the CYP3A-inhibition properties of ritonavir. TMC-41629 was studied in a Phase 1 trial in healthy volunteers and the results showed that TMC-41629 enhanced the exposure of the darunavir.[71] No recent development has been reported for either SPI-452 or TMC-41629. PF-03716539 (exact structure not disclosed) was evaluated in a Phase 1 trial for safety, tolerability and PK, but the results of the study or the current status of the compound have not been disclosed. In other research, scientists from Tibotec discovered that replacing the sulfonamide in the TMC-41629 with an amide can remove its anti-HIV activity while retaining its CYP3A-inhibitory potency. TMC-558445 was identified as the newest clinical candidate and it is in Phase 1 studies.[72] Another strategy to arrive at a novel pharmacoenhancer void of anti-HIV activity was demonstrated by researchers at Abbott. A 5-thiazolyl moiety was again utilized to achieve potent CYP3A inhibitory activity and the backbone of ritonavir was modified to remove its anti-HIV activity. In preclinical studies, the lead compound demonstrated its boosting effects when co-dosed with lopinavir.[73]

The structures of these potential pharmacoenhancers deviate away from that of ritonavir. It is anticipated their long term safety profiles need to be established during clinical studies and clinical practice.

Pharmacoenhancers have demonstrated their utility in the treatment of HIV and/or HCV infections. It is expected that they will find applications in treating other life-threatening diseases, such as cancer, although close monitoring is necessary in order to avoid unfavorable drug–drug interactions.

13.8   Conclusion

Drug–drug interactions as a result of CYP enzyme inhibition are often regarded as a liability. In the case of cobicistat and ritonavir, however, associated drug–drug interactions have proven to be an asset in the treatment of life-threatening HIV infection. The co-administration of ritonavir, resulting in improvement of the PK properties of concomitant PIs, is a cornerstone of PI-containing regimens; in this respect, ritonavir plays a critical role in the development of HAART regimens and in the chronic management of HIV infection. Without the boosting effect of ritonavir, achieving convenient dosing regimens for the large peptidomimetic PIs would not have been possible. Ritonavir has been used as a pharmacoenhancer for more than 15 years. Thus far, the benefit it brings to patients with HIV infection significantly outweighs the side effects associated with its use. With the exception of predictable drug interactions, there has been no strong evidence to demonstrate significant side effects arising from sustained CYP3A4 inhibition by long-term use of ritonavir as a booster. The second-generation pharmacoenhancer cobicistat, which maintains the CYP3A4-inhibition potency without any antiviral activity and possesses significantly improved physicochemical properties, finds broader use with its improved overall profile. The first integrase inhibitor-based, once-daily single-tablet HAART regimen Quad (Stribild®) has demonstrated promising clinical results. The potent, persistent CYP3A inhibition properties and improved physicochemical properties associated with cobicistat enable elvitegravir to be a once-daily drug; its coformulation with Truvada makes a once-daily, single-tablet, complete regimen possible. Stribild® was approved to for treatment-naive HIV-infected patients and cobicistat is under regulatory review as a general pharmacoenhancer. In addition, a complete single-tablet regimen containing emtricitabine, tenofovir alafenamide fumarate (TAF - a novel prodrug of tenofovir) and cobicistat-boosted daruanavir is currently in Phase 2 studies. Cobicistat potentially makes a once-daily, PI-containing single-tablet regimen possible.

As our knowledge of CYP3A inhibition improves and our experience with pharmacoenhancers increases, it is expected that a clean and safe pharmacoenhancer will also have broad utility against other life-threatening diseases, such as cancer, albeit with close monitoring in order to avoid unfavorable drug–drug interactions.

References

1. Department of Health and Human Services (DHHS), *Guidelines for the Use of Antiretroviral Agents in HIV-1-infected Adults and Adolescents*, 2012, http://www.aidsinfo.nih.gov/guidelines/ (last accessed 28 February 2013).
2. D. D. Richman, *Rev. Infect. Dis.*, 1990, **12**, S507–S512.
3. N. E. Kohl, E. A. Emini, W. A. Schleif, L. J. Davis, J. C. Heimbach, R. A. Dixon, E. M. Scolnick and I. S. Sigal, *Proc. Natl. Acad. Sci. U. S. A.*, 1988, **85**, 4686–4690.
4. A. Patick and K. Potts, *Clin. Microbiol. Rev.*, 1998, **11**, 614–627.
5. A. Molla, M. Korneyeva, Q. Gao, S. Vasavanonda, P. J. Shipper, H.-M. Mo, M. Markowitz, T. Chernyavskiy, P. Niu, N. Lyons, A. Hsu, G. R. Granneman, D. D. Ho, C. A. B. Boucher, J. M. Leonard, D. W. Norbeck and D. J. Kempf, *Nat. Med.*, 1996, **2**, 760–766.
6. A. Molla, S. Vasavanonda, G. Kumar, H. L. Sham, M. Johnson, B. Grabowski, J. F. Denissen, W. Kohlbrenner, J. J. Plattner, J. M. Leonard, D. W. Norbeck and D. J. Kempf, *Virology*, 1998, 255–262.
7. Abbott Laboratories, *Norvir®*, *Production Information*, Abbott Laboratories, North Chicago, IL, 2008.
8. (a) D. Periard, A. Telenti and P. Sudre *et al.*, *Circulation*, 1999, **100**, 700–705; (b) S. A. Danner, A. Carr and J. M. Leonard, *N. Engl. J. Med.*, 1995, **333**, 1528–1533; (c) M. Markowitz, M. S. Saag and W. G. Powderly *et al.*, *N. Engl. J. Med.*, 1995, **333**, 1534–1539; (d) A. Carr, K. Samaras, S. Burton, M. Law, J. Freund, D. J. Chisholm and D. A. Cooper, *AIDS*, 1998, **12**, F51–F58.
9. D. Ho, *N. Engl. J. Med.*, 1995, **333**, 450–451.
10. M. A. Thompson, J. A. Aberg, J. F. Hoy, A. Telenti, C. Benson, P. Cahn, J. J. Eron, H. F. Günthard, S. M. Hammer, P. Reiss, D. D. Richman, G. Rizzardini, D. L. Thomas, D. M. Jacobsen and P. A. Volberding, *JAMA*, 2012, **300**, 387.
11. F. J. Palella, K. M. Delaney and A. C. Moorman, *N. Engl. J. Med.*, 1998, **338**, 853–860.
12. (a) E. Wood, R. S. Hogg, B. Yip, P. R. Harrigan, M. V. O'Shaughnessy and J. S. G. Montaner, *Ann. Intern. Med.*, 2003, **139**, 810–816; (b) J. A. Bartlett, M. J. Fath, R. Demasi, A. Hermes, J. Quinn, E. Mondou and F. Rousseau, *AIDS*, 2006, **20**, 2051–2064, and references therein;; (c) J. H. Willig, S. Abroms, A. O. Westfall, J. Routman, S. Adusumilli, M. Varshney, J. Allison, A. Chatham, J. L. Raper, R. A. Kaslow, M. S. Saag and M. J. Mugavero, *AIDS*, 2008, **22**, 1951–1960; (d) J.-J. Parienti, D. R. Bangsberg, R. Verdon and E. M. Gardner, *Clin. Infect. Dis.*, 2009, **48**, 484–488.

13. (a) R. S. Hogg, M. V. O'Shaughnessy, N. Gataric, B. Yip, K. Craib, M. T. Schechter and J. S. Montaner, *Lancet*, 1997, **349**, 1294; (b) R. Detels, A. Munoz, G. McFarlane, L. A. Kingsley, J. B. Margolick, J. Giorgi, L. K. Schrager and J. P. Phair, *JAMA*, 1998, **280**, 1497–1503; (c) J. A. Sterne, M. A. Hernan, B. Ledergerber, K. Tilling, R. Weber, R. Sendi, M. Rickenbach, J. M. Robins and M. Egger, *Lancet*, 2006, **366**, 378–384; (d) F. J. Palella, K. M. Delaney and A. C. Moorman *et al.*, *N. Engl. J. Med.*, 1998, **338**, 853–860.

14. J. G. Gerber, *Clin. Infect. Dis.*, 2000, **30**(Suppl. 2), S123–S129, and references therein.

15. M. Barry, F. Mulcahy, C. Merry, S. Gibbons and D. Back, *Clin. Pharmcokinet.*, 1999, **36**, 289–304, and references therein.

16. G. Moyle, *AIDS Read*, 2001, **11**, 87–98.

17. (a) D. R. Nelson, L. Koymans and T. Kamataki *et al.*, *Pharmacogenetics*, 1996, **6**, 1–42; (b) S. Rendic and F. J. Di Carlo, *Drug Metab. Rev.*, 1997, **29**, 413–580; (c) S. A. Wrighton *et al.*, *Drug Metab. Rev.*, 2000, **32**, 339–361; (d) S. Rendic, *Drug Metab. Rev.*, 2002, **34**, 83–448.

18. G. N. Kumar, A. D. Rodrigues, A. M. Buko and J. F. Denissen, *J. Pharmacol. Exp. Ther.*, 1996, **277**, 423–431.

19. D. J. Kempf, K. C. Marsh, G. Kumar, A. D. Rodrigues, J. F. Denissen, E. McDonald, M. J. Kukulka, A. Hsu, G. R. Granneman, P. A Baroldi, E. Sun, D. Pizzuti, J. J. Plattner, D. W. Norbeck and J. M. Leonard, *Antimicrob. Agents Chemother.*, 1997, **41**, 654–660.

20. A. Hsu, G. R. Granneman, G. Cao, L. Garothers, T. El-Shourbagy, P. Baroldi, K. Erdman, F. Brown, E. Sun and J. M. Leonard, *Clin. Pharmacol. Ther.*, 1998, **63**, 453–464.

21. A. Hsu, G. R. Granneman, G. Cao, L. Garothers, A. Japour, T. El-Shourbagy, S. Dennis, J. Berg, K. Erdman, J. M. Leonard and E. Sun, *Antimicrob. Agents Chemother.*, 1998, **42**, 2784–2791.

22. D. W. Cameron, A. J. Japour and Y. Xu *et al.*, *AIDS*, 1999, **13**, 213–224.

23. FDA, *Viracept US Label*, 1997, http://www.accessdata.fda.gov/scripts/cder/drugsatfda/index.cfm. (last accessed 28 March 2013), *Product Information*, Agouron, La Jolla, CA, 2007.

24. For reviews on boosted PI, see the following reviews and references therein: (a) J. G. Gerber, *Clin. Infect. Dis.*, 2000, **30**(Suppl. 2), S123–S129; (b) G. J. Moyle and D. Back, *HIV Med.*, 2001, **2**, 105–113; (c) C. L. Cooper, R. P. G. van Heeswijk, K. Gallicano and D. W. Cameron, *Clin. Infect. Dis.*, 2003, **36**(1585–1592); (d) S. L. Becker, *Expert Opin. Investig. Drugs*, 2003, **12**, 401–412; (e) J. E. Gallant, *AIDS Rev.*, 2004, **6**, 226–233; (f) M. Youle, *J. Antimicrob. Chemother.*, 2007, **60**, 1195–1205; (g) K. H. Busse and S. R. Penzak, *Expert Rev. Clin. Pharmacol.*, 2008, **1**, 533–545; (h) L. Xu and M. C. Desai, *Curr. Opin. Investig. Drugs*, 2009, **10**, 775–786.

25. S. D. Shafran, L. D. Mashinter and S. E. Roberts, *HIV Med.*, 2005, **6**, 421–425, and references therein.

26. G. K. Dresser, J. D. Spence and D. G. Bailey, *Clin. Pharmacokinet.*, 2000, **38**, 41–57.
27. V. Dixit, N. Hariparsad, F. Li, P. Desai, K. E. Thummel and J. D. Unadkat, *Drug Metab. Dispos.*, 2007, **35**, 1853–1859.
28. H. Wang and E. L. LeCluyse, *Clin. Pharmacokinet.*, 2003, **42**, 1331–1357, and references therein.
29. G. Bertilsson, J. Heidrich, K. Svensson, M. Asman, L. Jendeberg, M. Sydow-Backman, R. Ohlsson, H. Postlind, P. Blomquist and A. Berkenstam, *Proc. Natl. Acad. Sci. U. S. A.*, 1998, **95**, 12208–12213.
30. E. D. Kharasch, D. Mitchell, R. Coles and R. Blanco, *Antimicrob. Agents Chemother.*, 2008, **52**, 1663–1669.
31. M. M. Foisy, E. M. Yakiwchuk and C. A. Hughes, *Ann. Pharmacother.*, 2008, **42**, 1048–1059, and references therein.
32. J. Bauer, S. Spanton, R. Henry, J. Quick, W. Dziki, W. Porter and J. Morris, *Pharm. Res.*, 2001, **18**, 859–866.
33. I. F. Sevrioukova and T. L. Poulos, *Proc. Natl. Acad. Sci. U. S. A.*, 2010, **107**, 18422.
34. (a) T. Koudriakova, E. Iatsimirskaia, I. Utkin, E. Gangl, P. Vouros, E. Storozhuk, D. Orza, J. Marinina and N. Gerber, *Drug Metab. Dispos.*, 1998, **26**, 552–561; (b) L. L. von Moltke, A. L. B. Durol, S. X. Duan and D. J. Greenblatt, *Eur. J. Clin. Pharmacol.*, 2000, **56**, 259–261; (c) G. Luo, J. Lin, W. D. Fiske, R. Dai, T. J. Yang, S. Kim, M. Sinz, E. LeCluyse, E. Solon, J. M. Brennan, I. H. Benedek, S. Jolley, D. Gilber, L. Wang, F. W. Lee and L.S. Gan, *Drug Metab. Dispos.*, 2003, **3**, 1170–1175; (d) C. S. Ernest II, S. D. Hall and D. R. Jones, *J. Pharmacol. Exp. Ther.*, 2005, **312**, 583–591; (e) R. S. Obach, R. L. Walsky and K. Venkatakrishnan, *Drug Metab. Dispos.*, 2007, **35**, 246–255.
35. For reviews on mechanism-based inhibitors and potential issues, see: (a) S. Zhou, S. Y. Chan, B. C. Goh, W. Duan, M. Huang and H. L. McLeod, *Clin. Pharmacokinet.*, 2005, **44**, 279–304; (b) E. Fontana, P. M. Dansette and S. M. Poli, *Curr. Drug Metab.*, 2005, **6**, 413–454, and references therein.
36. B. M. VandenBrink, B. J. Kirby, J. D. Unadkat and K. L. Kunze, International Society for the Study of Xenobitics, Washington, DC, 2009.
37. M. S. Sulkowski, *Clin. Infect. Dis.*, 2004, **38**(Suppl. 2), S90–S97.
38. M. S. Sulkowski, D. L. Thomas, R. E. Chaisson and R. D. Moore, *JAMA*, 2000, **283**, 74–80.
39. M. S. Sulkowski, D. L. Thomas, S. H. Mehta, R. E. Chaisson and R. D. Moore, *Hepatology*, 2003, **38**(Suppl. 1), 698A.
40. M. S. Sulkowski, S. H. Mehta, R. E. Chaisson, D. L. Thomas and R. D. Moore, *AIDS*, 2004, **18**, 2277–2284.
41. D. J. Jevtovic, J. Ranin, D. Salemovic, I. Pesic, G. Dragovic, S. Zerjav and O. Djurkovic-Djakovic, *Biomed. Pharmacother.*, 2008, **62**, 21–25.
42. K. Shimura, E. Kodama and Y. Sakagami, *J. Virol.*, 2008, **82**, 764–774.
43. E. DeJesus, D. Berger, M. Markowitz, C. Cohen, T. Hawkins, P. Ruane, R. Elion, C. Farthing, L. Zhong, A. K. Cheng, D. McColl and B. P. Kearney, *J. Acquired Immune Defic. Syndr.*, 2006, **43**, 1–5.

44. D. J. Kempf, K. C. Marsh, J. F. Denissen, E. McDonald, S. Vasavanonda, C. A. Flentge, B. E. Green, L. Fino, C. H. Park, X.-P. Kong, N. E. Wideburg, A. Saldivar, L. Ruiz, W. M. Kati, H. L. Sham, T. Robins, K. D. Stewart, A. Hsu, J. J. Plattner, J. M. Leonard and D. W. Norbeck, *Proc. Natl. Acad. Sci. U. S. A.*, 1995, **92**, 2484–2488.
45. L. Xu, H. Liu, B. P. Murray and C. Callebaut, *ACS Med. Chem. Lett.*, 2010, **1**, 209–213.
46. (a) J. C. Gorski, S. D. Hall, D. R. Jones, M. VandenBranden and S. A. Wrighton, *Biochem. Pharmacol.*, 1994, **47**, 1643–1653; (b) J. A. Williams, B. J. Ring, V. E. Cantrell, D. R. Jones, J. Eckstein, K. Ruterbories, M. A. Hamman, S. D. Hall and S. A. Wrighton, *Drug Metab. Dispos.*, 2002, **30**, 883–891.
47. (a) M. A. Correia, *Drug Metab. Rev.*, 2003, **35**, 107–143; (b) J. Yang, M. Liao, M. Shou, M. Jamei, K. R. Yeo, G. T. Tucker and A. Rostami-Hodjegan, *Curr. Drug Metab.*, 2008, **9**, 384–393.
48. A. Mathias, P. German, B. P. Murray, L. Wei, A. Jain, S. West, D. Warren, J. Hui and B. P. Kearney, *Clin. Pharmacol. Ther.*, 2010, **87**, 322–328.
49. (a) T. D. Bjornsson, J. T. Callaghan, H. J. Einolf, V. Fischer, L. Gan, S. Grimm and J. Kao, *Drug Metab. Dispos.*, 2003, **31**, 815–32; (b) FDA, HHS, http://www.fda.gov/downloads/Drugs/GuidanceComplianceRegulatory Information/Guidances/ucm292362.pdf (last accessed 28 March 2013).
50. T. Kronbach, D. Mathys, M. Umeno, F. J. Gonzalez and U. A. Meyer, *Mol. Pharmacol.*, 1989, **36**, 89–96.
51. (a) F. P. Guengerich, M. V. Martin, P. H. Beaune, P. Kremers, T. Wolff and D. J. Waxman, *J. Biol. Chem.*, 1986, **261**, 5051–5060; (b) T. Aoyama, S. Yamano, D. J. Waxman, D. P. Lapenson, U. A. Meyer, V. Fisher, R. Tyndale, T. Inaba, W. Kalow and H. V. Gelboin, *J. Biol. Chem.*, 1989, **264**, 10388–10395.
52. K. H. Ling, G. A. Leeson, S. D. Burmaster, R. H. Hook, M. K. Reith and L. K. Chen, *Drug Metab. Dispos.*, 1995, **23**, 631–636.
53. T. M. Riddle, D. G. Kuher, L. A. Woollett, C. J. Fichtenbaum and D. Y. Hui, *J. Biol. Chem.*, 2001, **276**, 37514–37519.
54. H. Murata, P. Hruz and M. Mueckler, *J. Biol. Chem.*, 2000, **275**, 20251–20254.
55. B. Gazzard and G. Moyle, *J. HIV Ther.*, 2004, **9**, 41–44.
56. S. Ramanathan, D. Warren, L. Wei and B. P. Kearney, presented at the Interscience Conference on Antimicrobial Agents and Chemotherapy, San Francisco, 2009, Abstract 614.
57. A. Mathias, H.C. Liu, D. Warren, V. Sekar and B. P. Kearney, presented at the International Workshop on Clinical Pharmacology of HIV Therapy, Sorrento, 2010, Abstract 28.
58. J. Gallant, E. Koenig, J. Andrade-Villanueva, P. Chetchotisakd, E. DeJesus, F. Antunes, K. Arasteh, G. Moyle, G. Rizzardini, J. Fehr, Y. Liu, L. Zhong, C. Callebaut, S. Ramanathan, J. Szwarcberg, M. Rhee and A. Cheng, presented at the 19th International AIDS Conference, Washington, DC, 2012, Abstract TUAB0103.

59. P. Sax, E. DeJesus, A. Mills, A. Zolopa, C. Cohen, D. Wohl, J. Gallant, H. C. Liu, L. Zhong, K. Yale, K. White, B. P. Kearney, J. Szwarcberg, E. Quirk and A. K. Cheng, *Lancet*, 2012, **379**, 2439–2447.

60. E DeJesus, J. K. Rockstroh, K. Henry, J.-M. Molina, J. Gathe, S. Ramanathan, X. Wei, K. Yale, J. Szwarcberg, K. White, A. K. Cheng and B. P. Kearney, *Lancet*, 2012, **379**, 2429–2437.

61. E. DeJesus, D. Berger, M. Markowitx, C. Cohen, T. Hawkins, P. Ruane, R. Elion, C. Farathing, L. Zhong, A. K. Cheng, D. McColl and B. P. Kearney, *J. Acquired Immune Defic. Syndr.*, 2006, **43**, 1–5.

62. S. Ramanathan, S. Abel, S. Tweedy, S. West, J. Hui and B. P. Kearney, *J. Acquired Immune Defic. Syndr.*, 2010, **53**, 209–214.

63. S. Abel, T. M. Jenkins, L. A. Whitlock, C. E. Ridgway and G. J. Muirhead, *Br. J. Clin. Pharmacol.*, 2008, **65**(Suppl. 1), 38–46.

64. A. L. Pozniak, M. Boffito, D. Russell, C. Ridgway and G. Muirhead, *Br. J. Clin. Pharmacol.*, 2008, **65**(Suppl. 1), 54–59.

65. J. Vierling, F. Poordad and E. Lawitz, presented at the American Association for the Study of Liver Diseases (AASLD), San Francisco, 2009, Abstract LB4.

66. E. J. Gane, R. Rouzier, C. Stedman, A. Wiercinska-Drapalo, A. Horban, L. Chang, Y. Zhang, P. Sampeur, I. Nájera, P. Smith, A. S. Shulman and J. Q. Tran, *J. Hepatol.*, 2011, **55**, 972–979.

67. K. K. Adkison, A. Shachoy-Clark, L. Fang, L. Yu, V. R. Otto, M. M. Berrey and S. C. Piscitelli, *Br. J Clin. Pharmacol.*, 2006, **62**, 336–344.

68. FDA, *Selzentry US Label*, 2007, http://www.accessdata.fda.gov/scripts/cder/drugsatfda/index.cfm. (last accessed 28 February 2013).

69. J. Ö. Haznerdar, J. Fretland, G. Leong, S. Blotner, T. Hill, P. Smith and J. Q. Tran, *J. Hepatol.*, 2010, **52**, S293.

70. S. Gulnik, M. Eissenstat, E. Afonina, D. Lutke, J. Erickson, R. Dagger, B. Wynne and R. Guttendorf, presented at the 16th Conference on Retroviruses and Opportunistic Infections, Montreal, 2009, Abstract 41.

71. L. Schueller, L. Baert, S. Lachau-Durand, H. Borghys, E. Clessens, G. Van Den Mooter, E. Van Gyseghem, T. H. M. Jonckers, P. V. Van Remoortere, P. Wigerinck and J. Rosier, *Int. J. Pharm*, 2008, **355**, 45–52.

72. T. H. M. Jonckers, M.-C. Rouan, G. Haché, W. Schepens, S. Hallenberger, J. Baumeister and J. C. Sasaki, *Bioorg. Med. Chem. Lett.*, 2012, **22**, 4998–5002.

73. C. A. Flentge, J. T. Randolph, P. P. Huang, L. L. Klein, K. C. Marsh, J. E. Harlan and D. J. Kempf, *Bioorg. Med. Chem. Lett.*, 2009, **19**, 5444–5448.

CHAPTER 14

Clinical Benefits of Single-tablet Regimens

DANIELLE P. PORTER AND BILL GUYER*

Gilead Sciences, Inc., Medical Affairs, 333 Lakeside Drive, Foster City,
CA 94404, USA
*Email: bill.guyer@gilead.com

14.1 Introduction

Management of chronic diseases requiring long-term therapy with daily medications presents unique challenges to both the patient and the provider. Although the medications themselves may be highly effective in treating disease, their clinical benefits will not be fully realized if patients do not consistently take their medications as prescribed.[1] One of the most significant limitations to effective therapy is regimen complexity. Complex regimens are associated with poor patient adherence and consequently worse health outcomes.[2] Simplification of therapy aims to reduce pill burden and dosing frequency, improve medication adherence and quality of life and ultimately reduce clinical disease progression.[3]

Coformulation of two or more existing drugs into a single pill is one strategy to facilitate regimen simplification. Treatments for several chronic diseases including type 2 diabetes mellitus, cardiovascular disease (hypertension and dyslipidemia) and human immunodeficiency virus (HIV) infection and also short-term infections such as tuberculosis have benefited from the availability of coformulated medications.[2,4–11] Several different types of coformulated medications have been developed for the purposes of regimen simplification (Table 14.1). Fixed-dose combination (FDC) therapies consist of multiple

RSC Drug Discovery Series No. 32
Successful Strategies for the Discovery of Antiviral Drugs
Edited by Manoj C. Desai and Nicholas A. Meanwell
© The Royal Society of Chemistry 2013
Published by the Royal Society of Chemistry, www.rsc.org

Table 14.1 Definitions of terms for coformulated medications.

Term	Definition
Fixed-dose combination (FDC)	Two or more drugs coformulated in a single tablet, but needs to be combined with other agents to make a complete treatment regimen
Fixed-dose regimen (FDR)	Coformulation of multiple drugs into a single tablet that does not need to be combined with other agents to constitute a complete treatment regimen, but may be dosed multiple times per day
Single-tablet regimen (STR)	Coformulation of multiple drugs comprising a complete treatment regimen into a single tablet that is dosed once daily

drugs in a single tablet; however, an FDC still needs to be combined with other agents to compose a complete treatment regimen. A fixed-dose regimen (FDR) is a complete regimen that does not need to be combined with other drugs for the treatment of a chronic illness, but may be dosed multiple times per day. A single-tablet regimen (STR) combines a complete treatment regimen in a single tablet that is dosed once daily. While the benefits of FDCs, FDRs and STRs are applicable to numerous disease states, this chapter focuses on the well-documented clinical advantages of STRs for the lifelong treatment of HIV infection.

Since the advent of highly active antiretroviral therapy (HAART), considerable progress has been made in the treatment of HIV infection. Initially, combination therapy was characterized by high pill burden and multiple daily doses. STRs represent substantial improvements in the treatment of HIV infection by providing all of the components of a safe and effective antiretroviral (ARV) therapy regimen in a single pill that is dosed once daily, thereby allowing for simpler and more convenient treatment (Figure 14.1). Currently, there are three US Food and Drug Administration (FDA)-approved STRs available in the USA (Table 14.2). Two consist of combinations of two nucleoside reverse transcriptase inhibitors (NRTIs) and one non-nucleoside reverse transcriptase inhibitor (NNRTI): efavirenz/emtricitabine/tenofovir disoproxil fumarate (EFV/FTC/TDF) and emtricitabine/rilpivirine/tenofovir disoproxil fumarate (FTC/RPV/TDF). The third and newest STR consists of two NRTIs plus an integrase strand transfer inhibitor (INSTI): elvitegravir/cobicistat/emtricitabine/tenofovir disoproxil fumarate (EVG/COBI/FTC/TDF). In addition, one other STR consisting of two NRTIs plus an NNRTI is available in the developing world: efavirenz/lamivudine/tenofovir disoproxil fumarate (EFV/3TC/TDF).[9] Generic versions of the EFV/FTC/TDF STR have also been made available to enable broader use in developing countries.[12] Owing to the widely recognized benefits of STRs, several additional STRs are currently in clinical development (Table 14.2). These include INSTI-based STRs and a protease inhibitor (PI)-based STR: dolutegravir/abacavir/lamivudine (DTG/ABC/3TC), elvitegravir/cobicistat/emtricitabine/GS-7340 (EVG/COBI/FTC/7340) and darunavir/cobicistat/emtricitabine/GS-7340 (DRV/COBI/FTC/7340). Both

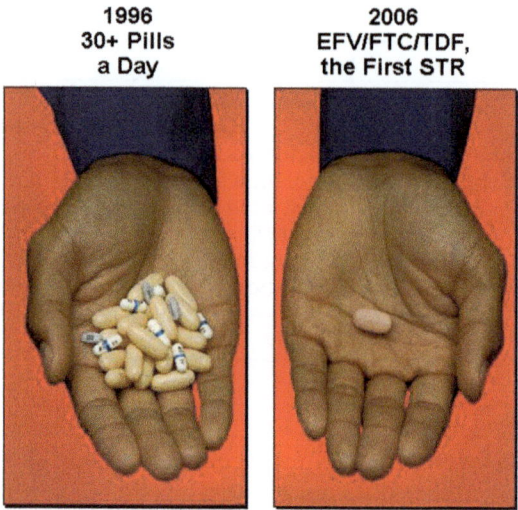

Figure 14.1 Comparison of pill burden of antiretroviral treatment regimens in 1996 and 2006.

Table 14.2 Single-tablet regimens approved and in development.

Regimen	Antiretroviral drug class
Approved single-tablet regimens	
EFV/FTC/TDF	Dual NRTI + NNRTI
FTC/RPV/TDF	Dual NRTI + NNRTI
EFV/3TC/TDF[a]	Dual NRTI + NNRTI
EVG/COBI/FTC/TDF	Dual NRTI + INSTI
Single-tablet regimens in development	
DTG/ABC/3TC	Dual NRTI + INSTI
EVG/COBI/FTC/7340	Dual NRTI + INSTI
DRV/COBI/FTC/7340	Dual NRTI + PI

[a]Only available in developing countries.

INSTI- and PI-based STRs are dual-target STRs as they are active against HIV integrase and protease, respectively, in addition to reverse transcriptase. The continued development of additional STRs involving both existing and investigational agents will provide further treatment options for patients seeking regimen simplification.

The clinical benefits of STRs are supported by the Department of Health and Human Services (DHHS) and International Antiviral Society (IAS) HIV treatment guidelines. The DHHS and IAS-USA guidelines recommend regimen simplification when possible to reduce pill burden and dosing frequency.[13,14] Fixed-dose formulations and once-daily regimens are generally preferred both for initial therapy and for convenience.[15] The major rationales behind the support for regimen simplification are to improve the patient's quality of life, maintain long-term adherence, avoid toxicities that may develop with prolonged ARV use and reduce the risk of virologic failure. Support for these recommendations includes data demonstrating that adherence is inversely related to the number of daily doses, ARV regimens with reduced dosing frequency have higher levels of adherence and patient satisfaction increases with regimens with fewer pills and reduced dosing frequency.[13]

This chapter reviews available data demonstrating the clinical benefits of STRs in the treatment of HIV disease. Large clinical trials and retrospective analyses have shown the advantages of STRs over other treatment regimens, including greater adherence and persistence, better health outcomes, improved patient preference and quality of life and reduced healthcare resource utilization. Ongoing and planned trials designed to further evaluate current and future STRs and their potential benefits in the successful management of HIV patients are also summarized.

14.2 Adherence

Adherence to antiretroviral therapy (ART), defined as the extent to which a patient complies with their prescribed treatment regimen in terms of dose, frequency and timing of administration, is essential for long-term treatment success in HIV-infected individuals.[16] High adherence rates are critical to achieve virological suppression and to minimize the emergence of drug resistance mutations, resulting in positive clinical outcomes. Patients with suboptimal adherence to ART are at increased risk for reduced treatment response and more rapid disease progression.[13,17] Studies from HIV and other disease areas have shown that medication adherence is inversely related to the number of daily doses and the number of pills per dose.[3,4,6,18–20] Results from numerous clinical trials and cohort studies have demonstrated that STRs are associated with higher adherence compared with multiple-pill ART regimens.

14.2.1 Clinical Trials

Improved adherence with the EFV/FTC/TDF STR was demonstrated in Study 934, a prospective, randomized, open-label, 144-week, non-inferiority study comparing the safety and efficacy of FTC/TDF + EFV (FTC/TDF fixed-dose combination tablets were used from week 96 to 144) *versus* zidovudine/lamivudine (AZT/3TC) twice daily + EFV once daily in ARV-naive patients. After week 144, patients could switch their ARV regimen to EFV/FTC/TDF STR. A total of 160 patients in the FTC/TDF + EFV arm and 126 patients in the AZT/3TC + EFV arm switched to EFV/FTC/TDF STR. In an *ad hoc* analysis through 240 weeks of follow-up, mean adherence rates by pill count were improved when patients had smaller daily pill burdens of one pill (97.9%; $p = 0.0005$) or two pills (97.0%; $p = 0.0262$) daily *versus* three pills daily (95.6%).[21]

Switching to an STR significantly improved adherence in the ADONE (ADherence to ONE pill) study. This was a prospective, open-label, comparative, multicenter, non-inferiority, 6-month simplification study in which virologically suppressed (HIV-1 RNA <50 copies mL^{-1} for ≥ 3 months) HIV-1-infected patients with no previous documented VF switched to EFV/FTC/TDF STR. Patients ($N = 212$) were previously on a stable regimen of EFV + TDF + either FTC or 3TC; 47% of patients were on their first ARV regimen. The primary endpoint of the study was adherence as measured by the visual analog scale (VAS). The results showed that switching to an STR improved adherence significantly at 1 month (93.8% at baseline to 96.1% at month 1; $p<0.01$) and adherence was maintained throughout the study (96.2% at 6 months).[22]

14.2.2   Retrospective and Observational Studies

ART consisting of a single pill per day (STR) was associated with increased adherence to therapy in a large retrospective analysis of the LifeLink database of managed care enrollees in the USA who received treatment for HIV or AIDS ($N = 7073$). All patients were on a complete ARV regimen (two NRTIs + a third agent) for ≥ 3 months. Data were reported for three cohorts: STR cohort ($n = 2365$), two pills per day cohort ($n = 411$) and three or more pills per day cohort ($n = 4297$). Adherence was assessed using the medication possession ratio (MPR), which was calculated as the number of prescription days supplied for all regimen components divided by the number of days from the first observed prescription in the regimen through the earliest of either the exhaustion of the days supplied of the last observed prescription or the end of follow-up. Patients on the STR consistently achieved higher adherence levels than patients on two or three or more pills per day regimens. Patients receiving a single pill per day had significantly better adherence than patients receiving multiple pills per day. Approximately 47% of patients receiving a single pill per day achieved $\geq 95\%$ adherence, compared with 41% of patients receiving two pills per day and 34% of patients receiving three or more pills per day ($p = 0.019$ for single pill *versus* two pills; $p <0.001$ for single pill *versus* three or more pills; Figure 14.2). The mean (standard deviation) MPR was 0.92 (0.09) among patients receiving a single pill per day, 0.90 (0.10) among patients receiving two pills per day and 0.90 (0.09) among patients receiving three or more pills per day ($p<0.01$ for single pill *versus* two pills and for single pill *versus* three or more pills) Multivariate logistic regression models showed that receiving a single pill per day was associated with a 59% greater likelihood of achieving a 95% adherence threshold, compared with receiving three or more pills per day [odds ratio (OR) 1.587; 95% confidence interval (CI) 1.415–1.780; $p<0.001$].[17]

Adherence was higher with EFV/FTC/TDF STR compared with non-one pill once daily ARV regimens in the REACH cohort. This prospective, observational, 6-month study assessed adherence and virologic response among a cohort of homeless and marginally housed individuals ($N = 118$) compared with historical controls in the same cohort. ARV regimens included EFV/FTC/TDF STR ($n = 47$), two NRTIs + RTV-boosted PI (PI/r; $n = 57$) or two NRTIs + NNRTI ($n = 14$). The primary endpoint was adherence by

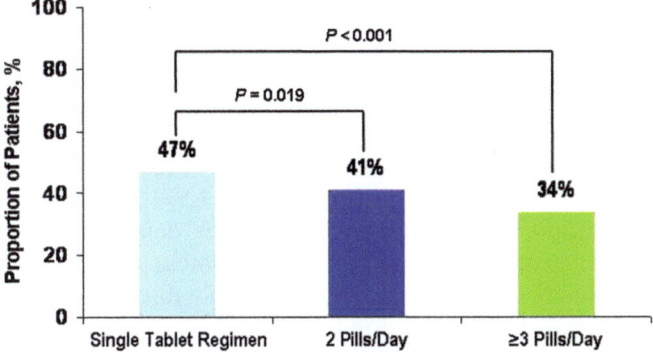

Figure 14.2 Proportion of patients achieving 95% or greater adherence by pill burden in a retrospective analysis of the LifeLink database.

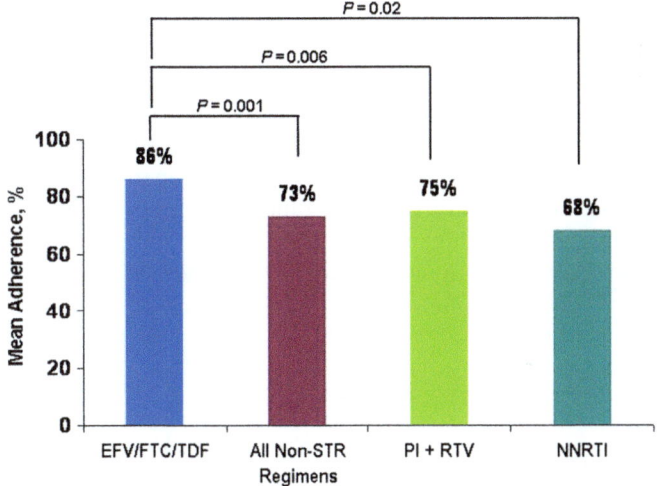

Figure 14.3 Mean adherence by antiretroviral treatment regimen in the REACH cohort.

unannounced pill counts over 6 months. Adherence in this challenging patient population was higher with EFV/FTC/TDF STR than other regimens (Figure 14.3) and after controlling for multiple confounders ($p = 0.006$).[23]

Patients taking the EFV/FTC/TDF STR were significantly more likely to have complete adherence than those on other ART regimens in a cross-sectional survey of US adults with a self-reported diagnosis of HIV/AIDS. In this study, multiple logistic regression models were used to determine independent predictors of complete adherence (defined as never missing or skipping an antiretroviral dose) in respondents taking a regimen of at least two NRTIs plus at least one PI or NNRTI ($N = 461$). Participants who reported taking an STR were 2.1 times (OR 2.1; 95% CI 1.29–3.41; $p < 0.05$) more likely to have complete adherence than respondents taking other regimens. Additionally, higher imputed daily ART pill count was associated with a lower likelihood of complete adherence (OR 0.93; 95% CI 0.90–0.97; $p < 0.05$).[24]

Patients initiating treatment with the EFV/FTC/TDF STR were significantly more adherent than those than those starting treatment with a non-STR EFV- or nevirapine (NVP)-based regimen in a retrospective study of the PharMetrics health insurance claims database of US patients with HIV who initiated first-line therapy with an NNRTI-based regimen ($N = 2974$). Adherence was assessed using proportion of days covered (PDC), defined as the ratio of the number days 'covered' by medication to the number of calendar days, both between the index date and the date of discontinuation/switching/augmentation, end of eligibility for health benefits or the end of the study period, whichever occurred first. Compared with the EFV/FTC/TDF STR, patients were less adherent (had more uncovered days) over 12 months on an EFV-based regimen other than EFV/FTC/TDF [rate ratio (RR) = 1.57; 95% CI 1.32–1.86; $p < 0.01$] or on an NVP-based regimen (RR = 2.01; 95% CI 1.51–2.67; $p < 0.01$).[25]

Patients on an STR consistently had higher adherence than patients taking a multiple-pill regimen in a retrospective analysis of Medicaid enrollees with HIV receiving ART ($N = 7783$). Adherence was measured by MPR using pharmacy records for two separate cohorts: patients who received an STR at any point in time ($n = 1838$) and patients who received a regimen consisting of two or more tablets per day at any point in time without ever receiving an STR ($n = 5945$). Adherence was higher in the STR cohort compared with the two or more pills per day regimen cohort for every level of adherence analyzed (Figure 14.4).[26]

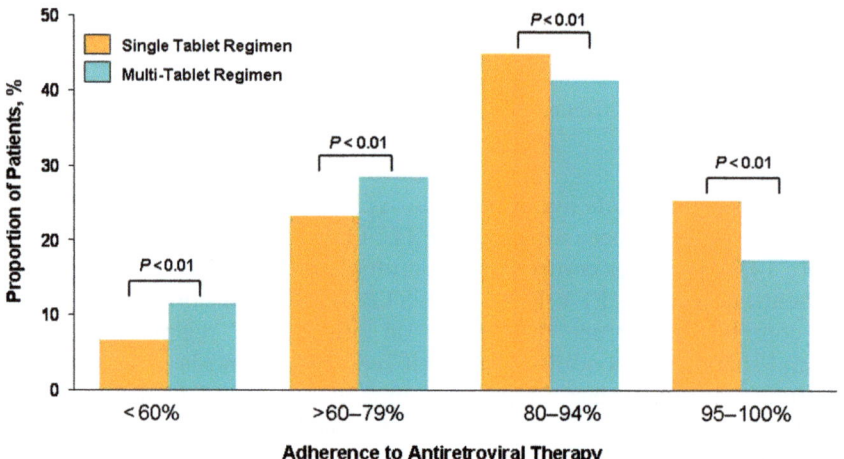

Figure 14.4   Adherence to antiretroviral therapy by number of tablets required per day in a Medicaid population analysis.
Reproduced with permission from reference 26.

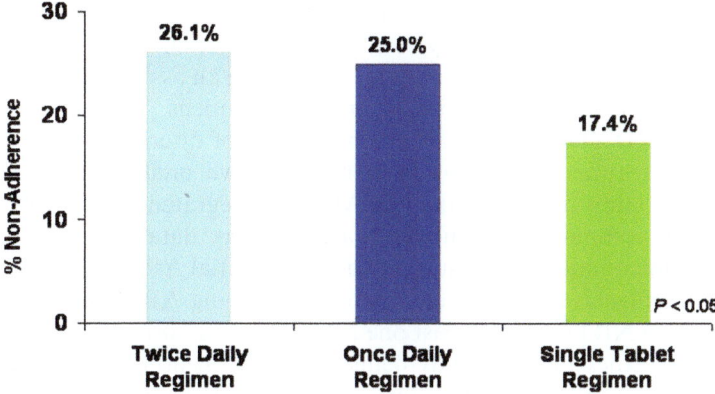

Figure 14.5 Proportion of non-adherent patients taking a twice-daily, once-daily or single-tablet regimen in an analysis of a cohort of HIV-infected individuals in Italy.

The EFV/FTC/TDF STR maintained an advantage in improving adherence compared with other ARV regimens in a study of self-reported adherence in 372 subjects attending a reference center for HIV treatment in Florence, Italy. Four measures of adherence were included in the self-administered questionnaire: (1) the proportion of ART doses taken over the preceding month, as measured by VAS; (2) any missed doses over the past week; (3) spontaneous treatment interruption of two or more days in the past 3 months; and (4) a lack of drug refill after finishing the drug in the past 3 months. Patients were defined as non-adherent if reporting any of the following: <90% of pills taken in the last month, one or more missed doses in the last week, or a spontaneous treatment interruption or refill problems in the last 3 months. Patients on the STR had the highest percentage of adherence in terms of mean percentage of pills taken in the last month (97.8%) and also a significantly lower proportion of non-adherence (17.4%; $p < 0.05$) compared with patients on non-STR regimens (Figure 14.5). Additionally, patients on the STR (OR: 0.45, 95% CI 0.22–0.42) reported lower non-adherence in a multivariable logistic regression analysis.[27]

14.3 Persistence

Persistence, defined as the duration of continuous pharmacological treatment, from initiation to discontinuation, is another important aspect of HIV management that is affected by regimen dosing. In contrast to adherence, which measures the compliance of patient behavior with a prescribed treatment regimen (as described in the previous section), persistence measures the duration or number of days during which a patient remains on a prescribed therapy without exceeding a permissible gap. Persistence is critical for HIV treatment since ARV therapy is lifelong and decreased persistence is associated with increased rates of virological failure, resistance development and less favorable clinical outcomes. Regimen complexity is one key treatment characteristic that has been associated with shorter duration of therapy or persistence.[28,29] While relatively few studies have examined ARV treatment persistence, at least two studies to date have shown an association of STRs with longer persistence compared with other ARV regimens.

The EFV/FTC/TDF STR had the lowest risk of discontinuation compared with other treatment regimens in a retrospective analysis of medical and prescription claims data from the PharMetrics Integrated Outcomes Database, a large US commercial health insurance claims database, that examined persistence in commercially insured patients on initial ARV therapy. A total of 2460 treatment-naive HIV-infected patients receiving ARV therapy consisting of at least two NRTIs plus at least one NNRTI ($n = 1388$) or PI ($n = 1072$) were followed until they met the study-defined criteria for non-persistence. Patients were considered persistent until any component of the regimen was modified (including discontinuation of or addition of an ARV medication) or there was a gap in treatment of >90 days. The EFV/FTC/TDF STR was associated with a significantly lower risk of discontinuation compared with all of the other regimens analyzed [hazard ratio (HR) 0.39; $p < 0.001$].[28]

A similar association of increased persistency with the EFV/FTC/TDF STR was seen in another retrospective study of the PharMetrics health insurance claims database of US patients with HIV who initiated first-line therapy with an NNRTI-based regimen ($N = 2974$). This study, which also evaluated adherence (as described in the previous section), assessed persistency based on the absence of evidence of therapy discontinuation, switching or augmentation of any component of the initial ART regimen over the period beginning with the index date and ending with the end of eligibility for health benefits or the end of the study period, whichever occurred first. Compared with the STR of EFV/FTC/TDF, patients were more likely to be non-persistent after initiating any other EFV-based (OR = 1.82; 95% I 1.46–2.28; $p < 0.01$) or NVP-based regimen (OR = 1.66; 95% CI 1.15–2.41; $p < 0.01$). Furthermore, patients became non-persistent more quickly after initiating another EFV-based regimen or an NVP-based regimen compared with the EFV/FTC/TDF STR.[25]

14.4   Efficacy

One of the most important clinical benefits of regimen simplification in HIV treatment is to improve virologic efficacy. Because adherence is one of the strongest predictors of efficacy, simpler regimens that improve adherence may also result in sustained virologic suppression, reduced risk of virologic failure, limited disease progression and lower risk of resistance development.[9,13] Multiple large clinical trials and cohort studies have demonstrated that the use of STRs sustains or improves efficacy compared with non-STR HAART regimens in both treatment-naive and virologically suppressed patient populations. Initiating therapy with STRs or switching stable patients from older or more complicated regimens to STRs may result in more favorable health outcomes for HIV-infected individuals and ultimately lead to greater control of the HIV epidemic.

14.4.1 Treatment-naive Studies

All three currently available STRs, EFV/FTC/TDF, FTC/RPV/TDF and EVG/COBI/FTC/TDF, have been approved for use in treatment-naive patients based on results from key clinical studies demonstrating the efficacy of these STRs in comparison with other DHHS-preferred ARV treatment regimens. Most recently, the newest STR EVG/COBI/FTC/TDF demonstrated non-inferior efficacy to two other standard-of-care regimens in two large Phase 3 studies in HIV-infected treatment-naive patients.[30,31] Study 102 was a prospective, randomized, double-blind 192-week trial comparing the efficacies of the EVG/COBI/FTC/TDF STR ($n = 348$) and the EFV/FTC/TDF STR ($n = 352$). The primary endpoint of the study was the proportion of subjects achieving HIV-1 RNA $<$ 50 copies mL^{-1} at week 48 based on intention-to-treat (ITT) FDA snapshot analysis. The predefined criterion for non-inferiority was a lower bound of a two-sided 95% CI $>-12\%$. At week 48, 87.6% of patients on EVG/COBI/FTC/TDF achieved HIV-1 RNA $<$ 50 copies mL^{-1} compared with 84.1% of patients in the EFV/FTC/TDF arm (treatment difference 3.6%; 95% CI -1.6 to 8.8%). The mean increase in CD4 cell count from baseline was significantly higher in the EVG/COBI/FTC/TDF arm than the EFV/FTC/TDF arm (239 *versus* 206 cells mm^{-3}; $p = 0.009$). In both arms, virologic failure (VF) rates were low with 2% of subjects developing primary resistance mutations.[30] Similarly, Study 103 was a prospective, randomized, double-blind 192-week trial comparing the efficacy of the EVG/COBI/FTC/TDF ($n = 353$) *versus* atazanavir (ATV) + ritonavir (RTV) + FTC/TDF ($n = 355$). The primary endpoint and criterion for non-inferiority were the same as described above for Study 102. At week 48, 89.5% of patients on EVG/COBI/FTC/TDF achieved HIV-1 RNA $<$ 50 copies mL^{-1} compared with 86.8% of patients on ATV + RTV + FTC/TDF (treatment difference 3.0%; 95% CI -1.9 to 7.8%). Mean increases in CD4 cell count from baseline were similar in both the EVG/COBI/FTC/TDF and ATV + RTV + FTC/TDF treatment arms (207 *versus* 211 cells mm^{-3}). In both arms, VF rates were low. with 1% of subjects in the EVG/COBI/FTC/TDF arm and no subjects in the ATV + RTV + FTC/TDF arm developing primary resistance mutations.[31]

Figure 14.6 summarizes the efficacy data for the EVG/COBI/FTC/TDF and EFV/FTC/TDF STRs in treatment-naive patients from Studies 102 and 103.

Patients taking the EFV/FTC/TDF STR maintained virologic suppression through week 240 in the extension phase of Study 934. Study 934 was a randomized, open-label, 144-week study evaluating the non-inferiority of FTC/TDF + EFV *versus* AZT/3TC twice daily + EFV once daily in treatment-naive patients. Through 144 weeks, 64% (146/227) of patients on FTC/TDF + EFV achieved HIV-1 RNA $<$ 50 copies mL^{-1} compared with 56% (130/231) of patients on AZT/3TC + EFV.[32] After week 144, a total of 160 patients in the FTC/TDF + EFV arm and 126 patients in the AZT/3TC + EFV arm switched to EFV/FTC/TDF STR. Among patients who switched to the STR, similar proportions had HIV-1 RNA $<$ 50 copies mL^{-1} (83% switch from FTC/TDF + EFV *versus* 82% switch from AZT/3TC + EFV) at week 240 (96 weeks post-switch to EFV/FTC/TDF STR).[21]

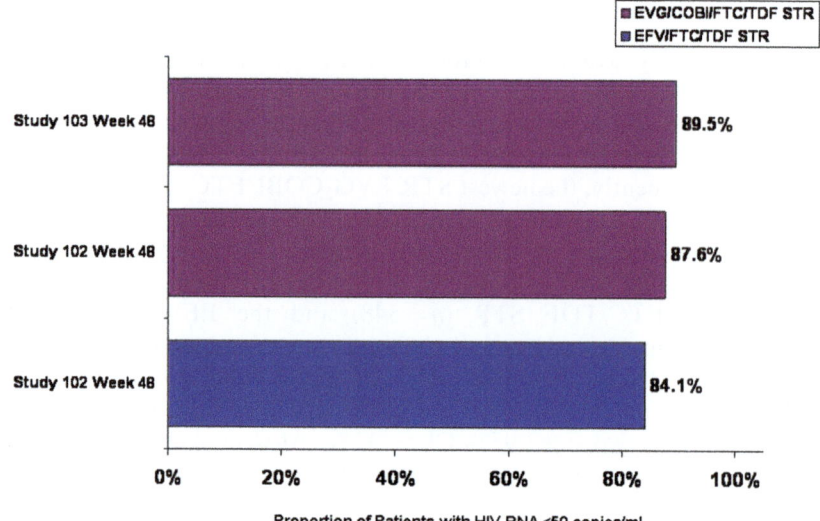

Figure 14.6   Cross-study comparison of efficacy of EVG/COBI/FTC/TDF and EFV/FTC/TDF STRs in treatment-naive patients.

14.4.2   Switch Studies

DHHS guidelines support treatment simplification for stable, virologically suppressed patients switching to newer agents, coformulated drugs, formulations with a lower pill burden, lower dosing frequency or those which are less likely to cause toxicity.[13] Several clinical trials have evaluated the safety and efficacy of switching virologically suppressed patients from their current NNRTI- or PI-based regimen to an NNRTI-containing STR.

Patients who switched to the FTC/RPV/TDF STR from the EFV/FTC/TDF STR maintained virologic suppression through week 24 in Study 111. Study 111 was a single-arm, open-label, 48-week study evaluating virologically suppressed patients on EFV/FTC/TDF who switched to FTC/RPV/TDF ($N = 49$). Patients included in this study were currently on ART consisting of only the EFV/FTC/TDF STR continuously for ≥ 3 months preceding the screening visit, had a desire to switch from EFV/FTC/TDF for tolerability issues and had no known resistance to any of the study agents at any time in the past. The primary endpoint was to evaluate the percentage of patients with HIV-1 RNA < 50 copies mL^{-1} at week 12 by the FDA snapshot analysis algorithm. Secondary endpoints included the proportion of patients with HIV-1 RNA < 50 copies mL^{-1} at weeks 24 and 48. All 49 (100%) patients in the study maintained HIV-1 RNA < 50 copies mL^{-1} at the primary endpoint at week 12 and also at the week 24 secondary endpoint. Through week 24, there were no cases of virologic failure and mean change in CD4 cell count from baseline was $+28$ cells mm^{-3}.[33]

Through 24 weeks, switching to FTC/RPV/TDF was non-inferior to remaining on a boosted PI-based regimen in the SPIRIT study. SPIRIT was a

Phase 3b randomized, open-label, multi-center, international, 48-week study to evaluate the safety and efficacy of switching from a regimen consisting of PI + RTV + two NRTIs to the FTC/RPV/TDF STR in virologically suppressed HIV-1 infected participants. Patients were randomized 2:1 to switch to FTC/RPV/TDF at baseline ($n = 317$) or maintain their current PI + RTV + two NRTIs regimen ($n = 159$) with a delayed switch to FTC/RPV/TDF at week 24. The primary endpoint was non-inferiority (12% margin) of FTC/RPV/TDF relative to PI + RTV + two NRTIs in maintaining plasma HIV-1 RNA <50 copies mL^{-1} at week 24. At week 24, 93.7% of patients who switched to FTC/RPV/TDF at baseline maintained HIV-1 RNA <50 copies mL^{-1} compared with 89.9% of patients who stayed on their PI + RTV + two NRTIs regimen (difference 3.8%, 95% CI −1.6 to 9.1). Non-inferiority was maintained regardless of viral load $\geq$ or $<100\,000$ HIV-1 RNA copies mL^{-1} prior to original ARV treatment initiation. Fewer patients who switched to FTC/RPV/TDF experienced virologic failure (0.9%) compared with the PI + RTV + two NRTIs arm (5.0%).[34]

Switching to the EFV/FTC/TDF STR was non-inferior to staying on a baseline regimen (SBR) in Study 073, a Phase 4 randomized, open-label, 48-week, non-inferiority study that evaluated switching to the EFV/FTC/TDF STR ($n = 203$) *versus* staying on a baseline regimen ($n = 97$). Patients were virologically suppressed (HIV-1 RNA <200 copies mL^{-1}) on a stable ARV regimen for ≥12 weeks with no history of virologic failure (VF). At baseline, patients were receiving an NNRTI [EFV or nevirapine (NVP)]- or a PI (most commonly ATV + RTV, lopinavir/RTV, fosempranavir + RTV or nelfinavir)-containing regimen most commonly with FTC/TDF, ABC/3TC, zidovudine (AZT)/3TC or TDF + 3TC. EFV/FTC/TDF was non-inferior to SBR for maintenance of virologic suppression through week 48 [HIV-1 RNA <200 copies mL^{-1} 89% *versus* 88% (treatment difference 1.1%; 95% CI −6.7 to 8.8%); HIV-1 RNA <50 copies mL^{-1} 87% *versus* 85% (treatment difference 2.6%; 95% CI −5.9 to 11.1%)]. Responses were similar for patients on prior NNRTI- and PI-based regimens.[35]

Switching to the STR EFV/FTC/TDF from the individual components given as a multi-pill regimen maintained virologic suppression in the ONCE study. This was a Phase 4 48-week, prospective, open-label, single-arm study in subjects stable on a first-line regimen of either TDF + FTC + EFV, FTC/TDF coformulation + EFV or TDF + 3TC + EFV for at least 24 weeks ($N = 115$). The primary endpoint was the proportion of subjects who maintained virologic suppression of HIV-1 RNA <50 copies mL^{-1} over 48 weeks. At week 48, 99.0% of subjects were virologically suppressed (HIV-1 RNA <50 copies mL^{-1}) with only one subject meeting the criteria for virologic failure (two consecutive HIV-1 RNA ≥50 copies mL^{-1} or one HIV-1 RNA ≥50 copies mL^{-1} followed by discontinuation from the study).[36]

Figure 14.7 illustrates efficacy results for the FTC/RPV/TDF and EFV/FTC/TDF STRs in virologically-suppressed patients who switched from other ARV regimens in Study 111, SPIRIT, Study 073 and ONCE.

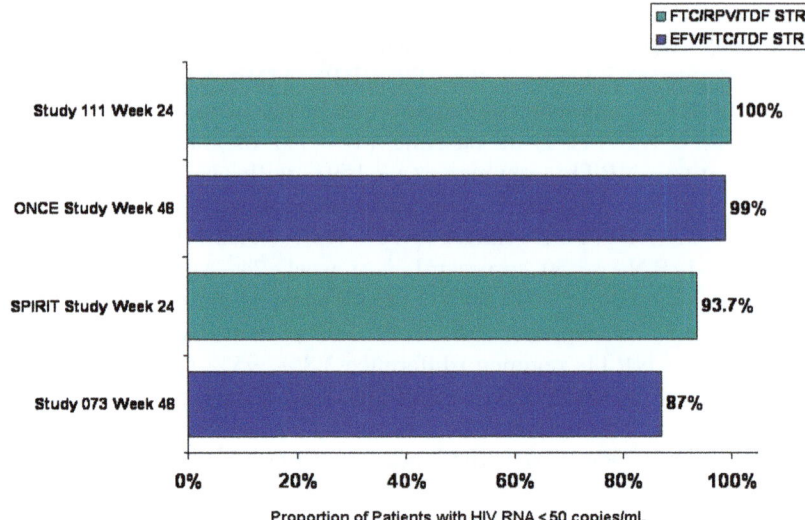

Figure 14.7 Cross-study comparison of efficacy of FTC/RPV/TDF and EFV/FTC/TDF STRs in virologically suppressed patients who switched from other ARV regimens.

14.4.3 Cohort Study

Virologic suppression rates were higher with EFV/FTC/TDF STR compared with non-one pill once-daily ARV regimens in the REACH cohort. Participants in the REACH cohort included both treatment-naive and treatment-experienced individuals with the majority of participants being nucleoside-experienced (65%) with a median of 27.6 months of prior antiretroviral therapy. More patients on EFV/FTC/TDF STR achieved HIV-1 RNA <50 copies mL^{-1} compared with non-one pill once-daily regimens (69% *versus* 46%; $p = 0.02$).[23]

14.5 Safety

Another critical objective of regimen simplification is to enhance tolerability and reduce toxicities associated with long-term ARV use. Toxicity and tolerability issues related to ARV use have been identified among the principal causes of poor adherence, treatment interruptions and regimen switching. ARV agents developed early in the HIV epidemic, particularly thymidine and adenosine analog NRTIs, have been associated with numerous side effects and adverse events (AEs), including gastrointestinal intolerance, lipodystrophy, hyperlipidemia and peripheral neuropathy.[8] ARV agents developed more recently have sought to minimize short- and long-term side effects while improving virologic efficacy and adherence. Because the treatment of HIV infection involves lifelong ARV therapy, the continued development of safe, more tolerable regimens that are easy to take is critical to successful treatment and limiting disease progression. STRs represent important advancements in the formulation of safe and effective ARVs for long-term therapy that improve adherence largely as a function of improved tolerability and better toxicity profiles. Several clinical trials have shown that initiating treatment with STRs or switching to STRs from other ARV treatment regimens resulted in improvements in safety parameters such as lipids and risk of coronary heart disease.

14.5.1 Treatment-naive Studies

The EVG/COBI/FTC/TDF STR demonstrated improvements in safety parameters compared with the EFV/FTC/TDF STR and ATV + RTV + FTC/TDF through 48 weeks in the Phase 3 Studies 102 and 103, respectively.[30,31] In Study 102, among adverse events (AEs) of any grade occurring in ≥ 10% of patients, fewer patients in the EVG/COBI/FTC/TDF arm experienced dizziness (7% *versus* 24%; $p < 0.001$), abnormal dreams (15% *versus* 27%; $p < 0.001$), insomnia (9% *versus* 14%; $p = 0.031$) and rash (6% *versus* 12%; $p = 0.009$) compared with patients in the EFV/FTC/TDF arm. Discontinuation rates due to AEs were similar between EVG/COBI/FTC/TDF and EFV/FTC/TDF (4% and 5%, respectively). Laboratory abnormalities in liver function tests for alanine aminotransferase (ALT) and aspartate aminotransferase (AST) increased in fewer patients in the EVG/COBI/FTC/TDF group (ALT, 15% *versus* 34%; AST, 18% *versus* 31%; $p < 0.001$ for both). In addition, median increases from baseline in lipid parameters were significantly lower for patients on EVG/COBI/FTC/TDF than EFV/FTC/TDF [total cholesterol (TC), 10 *versus* 19 mg dL^{-1}, $p < 0.001$; low-density lipoprotein (LDL) cholesterol, 10 *versus* 17 mg dL^{-1}, $p = 0.001$; high-density lipoprotein (HDL) cholesterol, 5 *versus* 8 mg dL^{-1}, $p = 0.001$]. Triglycerides (TGs) increased similarly in both arms (7 *versus* 7 mg dL^{-1}; $p = 0.44$).[30] In Study 103, rates of AEs (all grades) occurring in ≥ 10% of patients were similar for both EVG/COBI/FTC/TDF and ATV + RTV + FTC/TDF, with the exception of a significantly lower occurrence of ocular icterus in the EVG/COBI/FTC/TDF arm (1% *versus* 14%; $p < 0.001$). Discontinuation rates due to AEs were also similar between arms, 4% occurring in the EVG/COBI/FTC/TDF arm and 5% in the ATV + RTV + FTC/TDF arm. The proportion of patients experiencing laboratory abnormalities in total bilirubin was significantly lower for EVG/COBI/FTC/TDF than ATV + RTV + FTC/TDF (3.1% *versus* 96.3%; $p < 0.001$). Median increases in TGs were lower in the EVG/COBI/FTC/TDF arm (8 *versus* 23 mg dL^{-1}; $p = 0.006$) whereas all other lipid parameters increased similarly in both arms.[31]

14.5.2 Switch Studies

Switching to EFV/FTC/TDF led to a significant improvement in lipid parameters in the ROCKET 1 study. This Phase 4 prospective, randomized, open-label, 24-week study in virologically suppressed HIV-1-infected patients with hypercholesterolemia (total cholesterol ≥ 200 mg dL^{-1}) was designed to determine the impact of switching from ABC/3TC + EFV to EFV/FTC/TDF ($n = 79$) *versus* continuing on ABC/3TC + EFV ($n = 78$). At week 12, patients receiving EFV/FTC/TDF continued for an additional 12 weeks of treatment, whereas those receiving ABC/3TC + EFV switched to EFV/FTC/TDF. Patients who switched to EFV/FTC/TDF experienced significant median reductions in fasting lipid parameters after 12 weeks than patients continuing on ABC/3TC + EFV, including decreased TC (–33.6 *versus* +0.4 mg dL^{-1}; primary endpoint), LDL (–22.0 *versus* –0.8 mg dL^{-1}), HDL (–5.0 *versus* +1.5 mg dL^{-1}) and TG (–24.8 *versus* –2.7 mg dL^{-1}); $p < 0.001$ for all results. Decreases in lipid parameters at week 12 were similarly maintained at week 24 for patients who switched to EFV/FTC/TDF at baseline. The TC:HDL ratio was similar for the two treatment arms at both 12 and 24 weeks. Patients who switched to EFV/FTC/TDF at week 12 experienced median reductions in fasting lipid parameters from week 12 to week 24 similar to those observed in patients receiving EFV/FTC/TDF for 24 weeks. Renal parameters were similar between arms and within the normal range through week 24.[37,38]

The FTC/RPV/TDF STR was well tolerated through week 24 in Study 111, the single-arm study of patients who switched from EFV/FTC/TDF to FTC/RPV/TDF for tolerability issues. No subjects had AEs leading to study drug discontinuation through 24 weeks. The majority of study drug-related AEs were grade 1 in severity and there were no grade 3 or 4 AEs. Median changes from baseline in fasting lipid parameters were significantly improved for TC (–24 mg dL^{-1}; $p < 0.001$) and LDL cholesterol (–17 mg dL^{-1}; $p < 0.001$) through week 24.[33]

Switching to the FTC/RPV/TDF STR resulted in improvements in lipid parameters and cardiovascular risk compared with staying on a PI-based regimen through 24 weeks in the SPIRIT study, the Phase 3b study evaluating switching to FTC/RPV/TDF from PI + RTV + two NRTIs. At week 24, rates of grade 3 or 4 AEs and grade 3 or 4 laboratory abnormalities were similar in patients who switched to FTC/RPV/TDF at baseline and those who stayed on PI + RTV + two NRTIs (AEs, 5.0% *versus* 6.9%; laboratory abnormalities, 6.3% *versus* 11.3%). Reductions in lipid parameters were significantly greater for patients who switched to FTC/RPV/TDF including TC (–25 *versus* –1 mg dL^{-1}), LDL (–16 *versus* 0 mg dL^{-1}), HDL (–4 *versus* –1 mg dL^{-1}), TG (–53 *versus* +3 mg dL^{-1}) and TC:HDL ratio (–0.27 *versus* +0.08); $p < 0.001$ for all comparisons (Figure 14.8). Switching to FTC/RPV/TDF also resulted in a significant improvement in 10 year Framingham Risk Score for cardiovascular disease ($p = 0.001$).[39]

14.6   Patient-reported Outcomes

With an increasing number of safe and effective ARV options, maintaining or improving patient satisfaction and quality of life has become an increasingly important goal of ARV therapy.[40] Patient-reported outcomes have been incorporated into clinical trials more frequently to assess patient perceptions

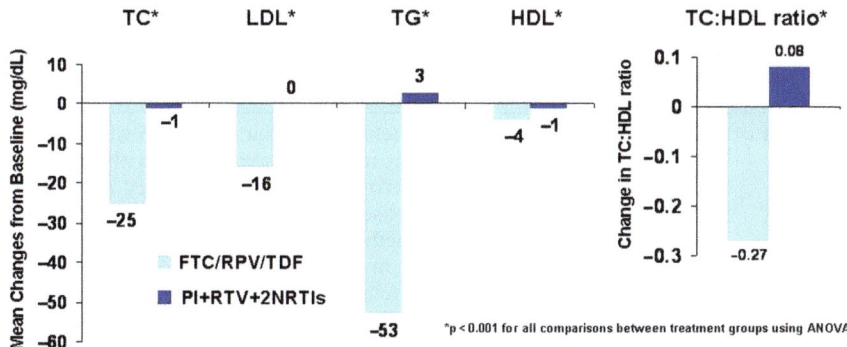

Figure 14.8 Comparison of lipid parameters between patients who switched to FTC/RPV/TDF at baseline and those who maintained a PI + RTV + two NRTIs regimen in the SPIRIT study at week 24.

and other considerations that are not captured by standard laboratory evaluations. Several validated tools are available to quantify outcomes such as quality of life, preference of medication, ease of regimen, treatment satisfaction, treatment intrusiveness and symptoms associated with HIV or its treatment.[40–44] Table 14.3 describes some of the common tools used to measure patient-reported outcomes in clinical trials. Several studies that have utilized these tools have shown significant improvements in patient-reported outcomes for STRs compared with other ARV treatment regimens.

Patients who switched to EFV/FTC/TDF reported better outcomes than patients who stayed on NNRTI- or PI-based regimens in Study 073. Results from the Preference of Medication (POM) survey for the EFV/FTC/TDF arm showed that more patients preferred EFV/FTC/TDF over their previous regimen at all post-baseline visits ($p < 0.001$) and at week 48, 85% of patients stated that EFV/FTC/TDF was 'much better' than their previous regimen. According to the Perceived Ease of Regimen for Condition (PERC) questionnaire, 68% of EFV/FTC/TDF patients *versus* 75% of SBR patients considered their regimen 'very easy to take' at baseline compared with 97% and 81% at week 48 ($p < 0.001$). Patients in the EFV/FTC/TDF arm also had a statistically significant improvement from baseline in the physical component summary of the health-related quality of life [Medical Outcomes Study 36-item Short Form (SF-36)] survey at week 48 ($p = 0.010$). Finally, self-reported symptoms from the HIV Symptom Index through week 48 demonstrated improvements in diarrhea/loose bowel movements ($p = 0.007$), bloating, pain or gas in the stomach ($p = 0.001$), changes in body appearance ($p = 0.002$) and problems having sex ($p = 0.032$) with EFV/FTC/TDF whereas patients who stayed on their NNRTI- or PI-based regimen had no significant changes.[35,40]

Patients who switched to the FTC/RPV/TDF STR experienced improvements in self-reported outcomes compared with remaining on a PI + RTV + two NRTIs regimen after 24 weeks in the SPIRIT study. Results from the HIV Symptom Index demonstrated that fewer patients who switched

Table 14.3 Tools used to assess patient-reported outcomes in HIV clinical trials.

Outcome measure	Summary description
Preference of Medication (POM) survey	Single-item survey that asks patients how the current medication compares with previous antiretroviral regimens that they have taken. Responses range from 1 (much better, I prefer this medication) to 5 (much worse, I much prefer my previous medication)
Perceived Ease of Regimen for Condition (PERC) questionnaire	Single-item questionnaire designed to assess a patient's perception of the ease or difficulty of use of their current HIV medication. Responses are scored on a scale from 1 (very easy) to 4 (very difficult)
Medical Outcomes Study 36-item Short Form (SF-36) survey	Quality of life questionnaire that yields an 8-scale profile of functional health and wellbeing scores and also physical and mental health summary measures. Physical health scales include physical functioning, role – physical, bodily pain and general health. Mental health scales include vitality, social functioning, role – emotional and mental health
HIV Symptom Index	20-item self-completed measure that addresses the presence of symptoms commonly associated with HIV or its treatment and how bothersome each symptom is to the patient
HIV Treatment Satisfaction Questionnaire (HIVTSQ)	10-item questionnaire concerning a patient's experience with their medical treatment for HIV including satisfaction with treatment, control of HIV, side effects, demands of treatment, convenience, flexibility, understanding of HIV, how well treatment fits into lifestyle, recommending treatment to others and continuing the treatment
HAART Intrusiveness Scale (HIS)	12-item scale that addresses the degree to which ART is perceived to interfere with 10 aspects of daily life such as social life, ability to work and relationships. Responses are scored on a scale from 1 to 5, where 1 indicates low interference and 5 indicates high interference

[a]Adapted with permission from reference 40.

to FTC/RPV/TDF reported having diarrhea (17.4% *versus* 45.3%; $p < 0.001$) and stomach pain or bloating (18.3% *versus* 32.1%; $p < 0.001$) compared with patients who stayed on PI + RTV + two NRTIs (Figure 14.9). Patients who switched to FTC/RPV/TDF were less likely to report fatigue ($p = 0.002$), memory loss ($p = 0.022$), headache ($p = 0.003$) and depression ($p < 0.001$) at week 24 compared with baseline. At week 24, patients who switched to FTC/RPV/TDF reported higher satisfaction with their treatment regimen by HIV Treatment Satisfaction Questionnaire (HIVTSQ) than those who stayed on PI + RTV + two NRTIs ($p < 0.001$).[39]

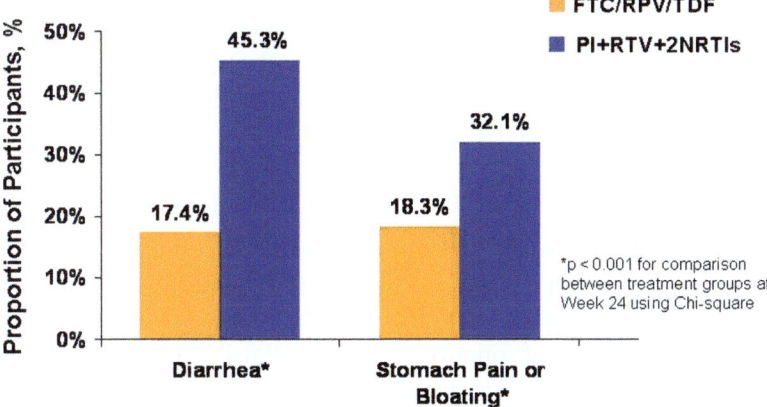

Figure 14.9 Differences in patient-reported symptoms from HIV Symptom Index between patients who switched to FTC/RPV/TDF at baseline and those who maintained a PI + RTV + two NRTIs regimen in the SPIRIT study at week 24.

Switching from ABC/3TC + EFV to EFV/FTC/TDF resulted in improvements in several patient-reported outcomes in the Rocket 1 study. At week 12, 55% of patients stated that EFV/FTC/TDF was 'much better' than their previous regimen by POM questionnaire. Switching to the EFV/FTC/TDF STR also led to a reduction in perceived treatment intrusiveness from baseline to week 12 compared with staying on ABC/3TC + EFV ($p = 0.037$) as measured by the HAART Intrusiveness Scale (HIS). Furthermore, patients who switched to EFV/FTC/TDF reported fewer symptoms (median symptom score 4 *versus* 10.5; $p < 0.0001$) and were less bothered by symptoms (median symptom score 6 *versus* 19.5; $p < 0.0001$) at week 12 compared with patients who continued ABC/3TC + EFV according to the Treatment Satisfaction and Symptoms Scale.[45]

Quality of life measures improved in patients who switched to the EFV/FTC/TDF STR from the separate components of EFV + TDF + either FTC or 3TC in the ADONE study. Patients preferred a one pill once-daily regimen over their previous regimen in terms of self-perceived tolerability, convenience, simplicity and efficacy/potency ($p \leq 0.05$ for all variables from months 1–6) Overall evaluation of quality of life (modified SF-36) improved after switching to the STR (68.8% at baseline to 72.7% at 6 months; $p = 0.042$), as did the proportion of patients not reporting any symptoms related to ARV agents (9.9% at baseline to 30.7% at 6 months; $p < 0.0001$). The number of symptoms that each patient reported also decreased significantly ($p = 0.018$).[22]

Switching to the EFV/FTC/TDF STR from its components was associated with higher treatment satisfaction over 48 weeks in the ONCE study. Treatment satisfaction was measured by treatment intrusiveness, perceived ease of regimen and regimen preference. Participants reported significantly less treatment intrusiveness by HIS 48 weeks after switching to EFV/FTC/TDF compared with baseline ($p = 0.006$). The proportion of patients reporting that their HAART regimen was 'very easy to take' on the PERC scale increased from 70.2% at baseline to 91.8% at week 48 after switching to EFV/FTC/TDF ($p < 0.0001$). At week 48, 68.4% of patients stated that EFV/FTC/TDF was 'much better' than their previous regimen according to the POM questionnaire.[46]

14.7 Healthcare Resource Utilization

The management of HIV disease involves the use of numerous healthcare services not only for long-term ART for HIV infection, but also to treat AIDS-associated symptoms and opportunistic infections and other HIV-related comorbidities. Effective ARV treatment has modified the clinical course of HIV infection by reducing disease progression, the incidence of AIDS-related complications and mortality, thereby lowering the burden of HIV disease on the healthcare system. Owing to this reduction in healthcare resource utilization, effective ART represents a cost-effective healthcare strategy.[7] STRs are of particular importance in this strategy since they have been associated with greater adherence and therefore better health outcomes. More complex regimens consisting of multiple ART components raise adherence concerns due to the risk of selective non-compliance of some of the separately administered drugs and therefore a potentially increased risk of treatment failure and worse health outcomes. In addition, ART regimens with multiple separate components may increase out-of-pocket costs for patients due to an increased number of co-payments for each prescription.[14] Pharmacoeconomic considerations such as drug cost and cost-effectiveness have implications not only for individual patients but also for healthcare policy, guidelines and global implementation of ART, especially in the context of limited healthcare resources.[8,47] Several studies have specifically investigated the association between regimen complexity and healthcare resource utilization as well as economic burden and demonstrated that STRs have significant advantages over multiple-pill regimens for each of these components.

ART consisting of a single pill per day (STR) was associated with a lower risk of hospitalization in a retrospective analysis of the LifeLink database. Hospitalizations were identified from the claims database using service codes. In a logistic regression analysis controlling for the number of pills per day and clinical and demographic characteristics, the adjusted rate of hospitalization was found to be significantly lower for patients receiving a one pill per day regimen compared with patients receiving a three or more pills per day regimen, as shown in Figure 14.10 (7.7% *versus* 9.9%; $p = 0.003$). Furthermore, there was a 24% lower risk of hospitalization among patients receiving a one pill per day regimen compared with patients receiving a three or more pills per day regimen (OR 0.764; 95% CI 0.638–0.915). Patients who achieved a 95% adherence threshold had a significantly lower rate of hospitalization, regardless of pill burden, than patients who did not achieve 95% adherence. Specifically, among patients receiving one pill per day, 6.6% of patients who achieved a

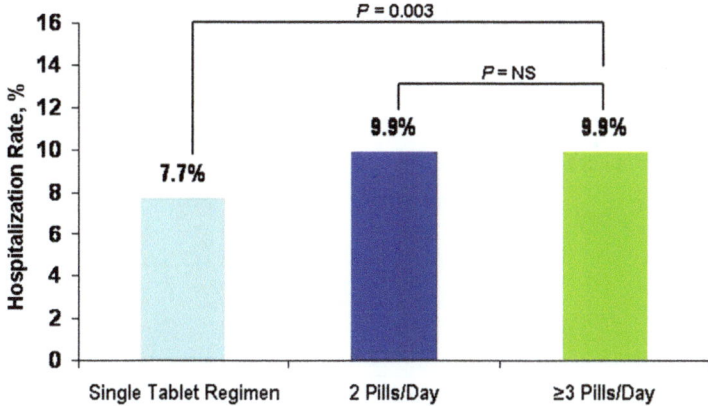

Figure 14.10 Adjusted rate of hospitalization by cohort in an analysis of the LifeLink database. NS = not significant. Reproduced with permission from reference 17.

95% adherence threshold had at least one hospitalization compared with 11.4% of patients who did not achieve 95% adherence.[17]

In a separate retrospective analysis of HIV-infected individuals receiving ART in the MarketScan Medicaid Multi-State Database, STR use was associated with a reduced risk of hospitalization and lower healthcare costs compared with multiple-pill regimens. After controlling for treatment experience and other available confounders, patients taking an STR had a 25% reduced risk of hospitalization compared with patients taking two or more pills per day (HR 0.753; $p < 0.0001$). This was true regardless of whether patients were treatment-naive or treatment-experienced. Analysis of a range of specific patient demographic and clinical profiles showed there were between 14 and 20 fewer hospitalizations per 100 patients on an STR compared with patients taking two or more pills per day.[26] STR use was also associated with a 17% reduction in total healthcare cost, due in part to significantly reduced hospitalization costs (Figure 14.11). Patients taking an STR had significantly lower monthly healthcare costs than patients receiving two or more pills per day ($605 per patient per month savings; $p < 0.001$). When the analysis was limited to only treatment-naive patients, monthly healthcare costs remained significantly lower by $922 per patient per month ($p < 0.001$) for patients on an STR compared with a multiple-pill regimen.[48]

Two economic modeling studies from European countries showed that STRs demonstrate considerable cost savings and cost-effectiveness compared with multiple-pill ART regimens. A budget analysis of antiretroviral costs from Spain and Portugal demonstrated that switching patients to the EFV/FTC/TDF STR would result in significant ART cost savings. Switching 2398 patients (95%) who are currently on the individual components of EFV and FTC/TDF to the EFV/FTC/TDF STR combined with switching 2672 patients (17%) currently receiving a PI-based regimen to the EFV/FTC/TDF

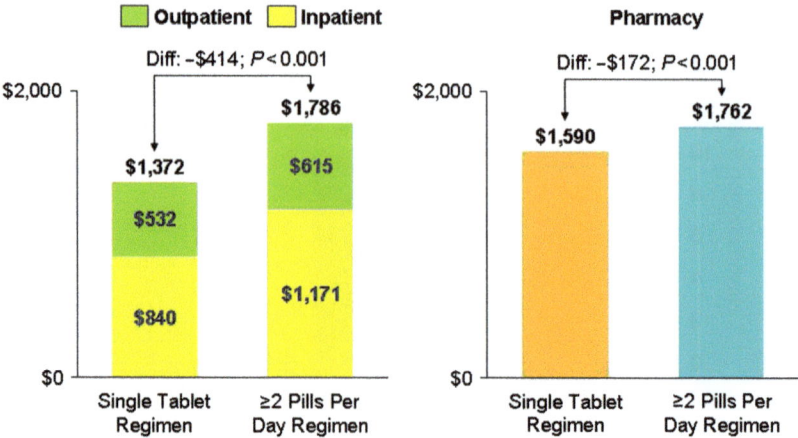

Figure 14.11   Medical and pharmacy costs per patient, per month (all-cause, unad-
justed) for patients on an STR compared with patients on a two or
more pills per day regimen in the US Medicaid analysis (based on 2010
US dollars). Reproduced with permission from reference 26.

STR would result in a savings of €7 million per year, allowing treatment with
STRs for an additional 777 patients without any ART cost increase.[49] In an
Italian simulation model, the EFV/FTC/TDF STR was the most cost-effective
treatment strategy in terms of direct health cost (including drugs, medical
examinations, hospitalizations and laboratory tests) per quality-adjusted life-
year gained compared with the other therapeutic regimens recommended by the
Italian guidelines for the first-line treatment of patients with HIV. This was
largely driven by results of the model demonstrating that patients treated with
an STR have a better quality of life with a higher number of quality-adjusted
life-years than patients on other regimens combined with the relatively low cost
of the STR relative to multiple-pill regimens.[47]

14.8   Ongoing Studies

Several studies are ongoing to evaluate further the safety, efficacy and patient-
reported outcomes of currently approved STRs compared with other antiretroviral treatment regimens for HIV. These studies are designed to answer
important questions that may have arisen during earlier studies or to evaluate
the use of a particular STR in patient populations outside the current approved
indication, such as virologically suppressed patients switching from other
regimens. The results from these studies will provide additional critical safety
and efficacy information on the use of these treatments in various patient
groups and help guide practitioners when choosing the appropriate ARV
treatment for both treatment-naive and treatment-experienced patients.

The STAR study is an ongoing Phase 3b randomized, international, open-
label, 96-week trial that will provide the first head-to-head comparison of the

two STRs FTC/RPV/TDF and EFV/FTC/TDF in ARV-naive adult patients ($N = 700$). The primary endpoint of the study is non-inferiority (12% margin) of FTC/RPV/TDF relative to EFV/FTC/TDF in achieving HIV-1 RNA <50 copies mL^{-1} at week 48. Secondary objectives of the study include evaluating the efficacy, safety and tolerability of the two regimens through 96 weeks, assessing the change from baseline in CD4 count in each treatment arm at weeks 48 and 96, assessing resistance at virologic failure and evaluating the change from baseline in fasting lipid parameters at weeks 48 and 96. Patient-reported outcome data will also be captured during the study. Primary endpoint results from STAR were anticipated in late 2012.[52]

Study 115 is an ongoing Phase 3b randomized, international, open-label, 96-week trial evaluating switching from a PI + RTV-based regimen to the EVG/COBI/FTC/TDF STR in virologically suppressed patients ($N = 420$). Patients will be randomized 2 : 1 to switch to the EVG/COBI/FTC/TDF STR at baseline or continue on their PI + RTV-based regimen for 96 weeks. The primary endpoint of the study is non-inferiority (12% margin) of switching to EVG/COBI/FTC/TDF relative to remaining on a PI + RTV-based regimen in maintaining HIV-1 RNA <50 copies mL^{-1} at week 48. Secondary objectives of the study include assessing the safety and tolerability of each treatment arm through 96 weeks and evaluating the change from baseline in CD4 cell count at weeks 48 and 96. Primary endpoint results are planned for 2013.[52]

Study 121 is an ongoing Phase 3b randomized, international, open-label, 96-week trial evaluating switching from an NNRTI-based regimen to the EVG/COBI/FTC/TDF STR in virologically suppressed patients ($N = 420$). Patients will be randomized 2 : 1 to switch to the EVG/COBI/FTC/TDF STR at baseline or continue on their NNRTI-based regimen for 96 weeks. The primary endpoint of the study is non-inferiority (12% margin) of switching to EVG/COBI/FTC/TDF relative to remaining on an NNRTI-based regimen in maintaining HIV-1 RNA <50 copies mL^{-1} at week 48. Secondary objectives of the study include assessing the safety, efficacy and tolerability of each treatment arm over 96 weeks and evaluating the change from baseline in lipid parameters at weeks 48 and 96. Patient-reported outcomes including adherence and quality of life outcomes will also be captured in the study. Primary endpoint results are anticipated in 2013.[52]

Study 123 is an ongoing Phase 3b single-arm, open-label, 48-week study evaluating virologically suppressed patients who switch from raltegravir (RAL) + FTC/TDF to the EVG/COBI/FTC/TDF STR ($N = 50$). The primary endpoint is to evaluate the proportion of patients with HIV-1 RNA <50 copies mL^{-1} at week 12. Secondary endpoints will include safety, efficacy and tolerability of EVG/COBI/FTC/TDF STR through 24 and 48 weeks after switching. Primary endpoint results are expected in 2013.[52]

14.9 Pipeline

Owing to the success of currently available STRs, several new STRs for the treatment of HIV are currently being pursued in clinical development to provide patients with even more safe and effective treatment options with one pill once-daily dosing. The STRs in the pipeline generally contain novel coformulations of two or three previously-approved drugs combined with a new investigational ARV agent. The STRs in development include a tablet based on the newest member of the INSTI class, dolutegravir (DTG), and also the first PI-based STR. Other investigational agents currently in development such as non-catalytic site integrase inhibitors (NCINIs) may provide further opportunities for the formulation of future STRs. The expansion of STRs into almost all classes of commonly prescribed ARV agents opens up the potential for a simple and convenient treatment regimen to an even wider population of HIV-infected individuals.

SINGLE is an ongoing Phase 3 randomized, double-blind, double-dummy, 96-week, non-inferiority study comparing the DTG/ABC/3TC STR and the EFV/FTC/TDF STR in treatment-naive adults ($N = 800$). The primary endpoint of the study is the proportion of patients with HIV-1 RNA <50 copies mL^{-1} at week 48. Secondary endpoint analyses will include safety, tolerability, immunologic activity, viral resistance and patient-reported outcomes. Primary endpoint results from this study were expected in late 2012 and will be used to support a future regulatory filing for DTG approval.[52]

Two of the newest STRs currently in clinical development, including the first PI-based STR, contain the investigational NNRTI GS-7340. GS-7340 is a novel amidate tenofovir prodrug that has the potential to improve upon the safety and efficacy of TDF by targeting delivery of high concentrations of tenofovir to lymphoid cells.[50,51] This increased specificity and potency allow for a lower dose of GS-7340 relative to TDF and may permit the development of new STRs not previously possible.

A Phase 2 study is currently under way evaluating the next-generation EVG/COBI/FTC/7340 STR compared with the EVG/COBI/FTC/TDF STR in ARV-naive adult patients ($N = 150$). This is a double-blind, placebo-controlled, 48-week safety and efficacy study where patients are randomized (2:1) to receive either EVG/COBI/FTC/7340 or EVG/COBI/FTC/TDF. The primary endpoint of the study is the proportion of patients with HIV-1 RNA <50 copies mL^{-1} in each treatment arm at week 24. Secondary objectives include efficacy through week 48 and change from baseline in CD4 cell count from baseline at weeks 24 and 48. Primary endpoint results are expected in 2013.[52]

Another Phase 2 study is ongoing to evaluate the GS-7340-based PI-containing STR (DRV/COBI/FTC/7340) compared with the EVG/COBI/FTC/TDF STR in ARV-naive adult patients. This is a double-blind, placebo-controlled, 48-week safety and efficacy study where patients are randomized (2:1) to receive either DRV/COBI/FTC/7340 or EVG/COBI/FTC/TDF. The primary endpoint of the study is the proportion of patients with HIV-1 RNA <50 copies mL^{-1} in each treatment arm at week 24. Secondary objectives include efficacy through week 48 and change from baseline in CD4 cell count from baseline at weeks 24 and 48. Primary endpoint results are expected in 2013.[52]

14.10 Conclusion

The development of single-tablet regimens has changed the landscape of HIV disease management. Not only is ARV therapy now simpler and more convenient for HIV-infected patients, but STRs have also demonstrated significant clinical advantages over multiple-pill regimens. One of the most critical benefits of STRs is increased treatment adherence, largely as a function of the improved tolerability and patient preference for STRs. Because medication adherence is such a critical driver of disease management and associated healthcare costs, improvements in adherence due to STRs result in better health outcomes and cost-effective treatment of HIV as a long-term chronic disease. Additional benefits of STRs include increased treatment persistence, sustained safety and efficacy and greater patient satisfaction and quality of life. The advantages of simplified regimens such as STRs are supported by the DHHS and IAS HIV treatment guidelines and data from many large clinical trials and retrospective analyses.

Because of the demonstrated advantages of STR therapies in the management of HIV and successes in other disease areas using coformulated medications, it may be beneficial to develop future STRs for the treatment of other chronic diseases such as hepatitis C virus infection. Applying the knowledge and experience gained in the development of optimal HIV treatments will hopefully advance the care of patients suffering from life-threatening diseases across therapeutic areas and around the world.

References

1. M. T. Brown and J. K. Bussell, *Mayo Clin. Proc.*, 2011, **86**, 304–314.
2. F. Aslam, A. Haque, V. Lee and J. Foody, *Patient Pref. Adher*, 2009, **3**, 61–66.
3. W. H. Frishman, *Cardiol. Rev.*, 2007, **15**, 257–263.
4. C. J. Bailey and C. Day, *Diabetes Obesity Metab.*, 2009, **11**, 527–533.
5. M. P. Curran, *Drugs*, 2010, **70**, 191–213.
6. S. A. Kamat, M. F. Bullano, C. L. Chang, S. K. Gandhi and M. J. Cziraky, *Curr. Med. Res. Opin.*, 2011, **27**, 961–968.
7. J. M. Llibre, J. R. Arribas, P. Domingo, J. M. Gatell, F. Lozano, J. R. Santos, A. Rivero, S. Moreno and B. Clotet, *AIDS*, 2011, **25**, 1683–1690.
8. J. B. Nachega, M. J. Mugavero, M. Zeier, M. Vitoria and J. E. Gallant, *Patient Pref. Adher.*, 2011, **5**, 357–367.
9. J. B. Nachega, B. Rosenkranz and P. A. Pham, *Patient Pref. Adher.*, 2011, **5**, 645–651.
10. I. Monedero and J. A. Caminero, *Int. J. Tuberc. Lung Dis.*, 2011, **15**, 433–439.
11. D. Desai, J. Wang, H. Wen, X. Li and P. Timmins, *Pharm. Dev. Technol.*, 2012, DOI: 10.3109/10837450.2012.660699.

12. S. Pujari, A. Dravid, N. Gupte, K. Joshi and V. Bele, *Medscape J. Med.*, 2008, **10**(8), 196.

13. Department of Health and Human Services (DHHS), *Guidelines for the Use of Antiretroviral Agents in HIV-1-infected Adults and Adolescents*, DHSS, Washington, DC, 2012.

14. M. A. Thompson, J. A. Aberg, J. F. Hoy, A. Telenti, C. Benson, P. Cahn, J. J. Eron, H. F. Gunthard, S. M. Hammer, P. Reiss, D. D. Richman, G. Rizzardini, D. L. Thomas, D. M. Jacobsen and P. A. Volberding, *JAMA*, 2012, **308**, 387–402.

15. M. A. Thompson, J. A. Aberg, P. Cahn, J. S. Montaner, G. Rizzardini, A. Telenti, J. M. Gatell, H. F. Gunthard, S. M. Hammer, M. S. Hirsch, D. M. Jacobsen, P. Reiss, D. D. Richman, P. A. Volberding, P. Yeni and R. T. Schooley, *JAMA*, 2010, **304**, 321–333.

16. J. A. Cramer, A. Roy, A. Burrell, C. J. Fairchild, M. J. Fuldeore, D. A. Ollendorf and P. K. Wong, *Value Health*, 2008, **11**, 44–47.

17. P. E. Sax, J. L. Meyers, M. Mugavero and K. L. Davis, *PLoS ONE*, 2012, **7**, e31591.

18. A. J. Claxton, J. Cramer and C. Pierce, *Clin. Ther.*, 2001, **23**, 1296–1310.

19. J. E. Gallant, E. DeJesus, J. R. Arribas, A. L. Pozniak, B. Gazzard, R. E. Campo, B. Lu, D. McColl, S. Chuck, J. Enejosa, J. J. Toole and A. K. Cheng, *N. Engl. J. Med.*, 2006, **354**, 251–260.

20. J. M. Molina, T. J. Podsadecki, M. A. Johnson, A. Wilkin, P. Domingo, R. Myers, J. M. Hairrell, R. A. Rode, M. S. King and G. J. Hanna, *AIDS Res. Hum. Retroviruses*, 2007, **23**, 1505–1514.

21. E. DeJesus, A. Pozniak, J. Gallant, J. Arribas, Y. Zhou, D. Warren and J. Enejosa, presented at the 49th Interscience Conference on Antimicrobial Agents and Chemotherapy (ICAAC), San Francisco, 2009.

22. M. Airoldi, M. Zaccarelli, L. Bisi, T. Bini, A. Antinori, C. Mussini, F. Bai, G. Orofino, L. Sighinolfi, A. Gori, F. Suter and F. Maggiolo, *Patient Pref. Adher*, 2010, **4**, 115–125.

23. D. R. Bangsberg, K. Ragland, A. Monk and S. G. Deeks, *AIDS*, 2010, **24**, 2835–2840.

24. T. Juday, S. Gupta, K. Grimm, S. Wagner and E. Kim, *J. HIV Clin. Trials*, 2011, **12**, 71–78.

25. C. Taneja, T. Juday, L. Gertzog, T. Correll, T. Hebden and G. Oster, presented at the 13th European AIDS Conference (EACS), Belgrade, 2011.

26. C. Cohen, K. L. Davis and J. Meyers, presented at the 51st Interscience Conference on Antimicrobial Agents and Chemotherapy (ICAAC), Chicago, 2011.

27. G. Sterrantino, L. Santoro, D. Bartolozzi, M. Trotta and M. Zaccarelli, *Patient Pref. Adher*, 2012, **6**, 427–433.

28. T. Juday, K. Grimm, A. Zoe-Powers, J. Willig and E. Kim, *J. AIDS Care*, 2011, 1–9.

29. J. W. Bae, W. Guyer, K. Grimm and F. L. Altice, *AIDS*, 2011, **25**, 279–290.

30. P. E. Sax, E. DeJesus, A. Mills, A. Zolopa, C. Cohen, D. Wohl, J. E. Gallant, H. C. Liu, L. Zhong, K. Yale, K. White, B. P. Kearney, J. Szwarcberg, E. Quirk and A. K. Cheng, *Lancet*, 2012, **379**, 2439–2448.
31. E. DeJesus, J. Rockstroh, K. Henry, J.-M. Molina, J. Gathe and S. Ramanathan *et al.*, *Lancet*, 2012, **379**, 2429–2438.
32. J. R. Arribas, A. L. Pozniak, J. E. Gallant, E. Dejesus, B. Gazzard, R. E. Campo, S. S. Chen, D. McColl, C. B. Holmes, J. Enejosa, J. J. Toole and A. K. Cheng, *J. Acquired Immune Defic. Syndr.*, 2008, **47**, 74–78.
33. A. Mills, C. Cohen, E. DeJesus, C. Brinson, K. Yale, S. Ramanathan, R. Ebrahim, S. K. Chuck and A. Cheng, presented at the 18th Annual Conference of the British HIV Association, Birmingham, 2012.
34. F. Palella, P. Tebas, B. Gazzard, P. Ruane, D. Shamblaw, J. Flamm, M. Fisher, J. van Lunzen, R. Ebrahimi, K. White, B. Guyer, H. Graham and T. Fralich, presented at the XIX International Aids Conference, Washington, DC, 2012.
35. E. DeJesus, B. Young, J. O. Morales-Ramirez, L. Sloan, D. J. Ward, J. F. Flaherty, R. Ebrahimi, J. Maa, K. Reilly, J. Ecker, D. McColl, D. Seekins and A. Farajallah, *J. Aquired Immune Defic. Syndr.*, 2009, **51**, 163–174.
36. G. Moyle, C. Orkin, M. Fisher, S. Taylor, J. Ross, H. Wang and C. Simcock, presented at the 17th Annual Conference of the British HIV Association, Bournemouth, 2011.
37. G. Moyle, C. Orkin, M. Fisher, J. Dhar, J. Anderson, J. Ewan and H. Wang, presented at the Interscience Conference on Antimicrobial Agents and Chemotherapy (ICAAC), Boston, 2010.
38. G. Moyle, C. Orkin, M. Fisher, J. Dhar, J. Anderson, J. Ewan and H. Wang, presented at the 10th International Congress on Drug Therapy in HIV Infection, Glasgow, 2010.
39. P. Tebas, F. Palella, B. Gazzard, P. Ruane, D. Shamblaw, J. Flamm, M. Fisher, J. van Lunzen, R. Ebrahimi, K. White, B. Guyer, J. Goodgame, T. Fralich, H. Graham and E. Elbert, presented at the 14th International Workshop on Co-morbidities and Adverse Drug Reactions in HIV, Washington, DC, 2012.
40. S. L. Hodder, K. Mounzer, E. Dejesus, R. Ebrahimi, K. Grimm, S. Esker, J. Ecker, A. Farajallah and J. F. Flaherty, *AIDS Patient Care STDS*, 2010, **24**, 87–96.
41. A. C. Justice, W. Holmes, A. L. Gifford, L. Rabeneck, R. Zackin, G. Sinclair, S. Weissman, J. Neidig, C. Marcus, M. Chesney, S. E. Cohn and A. W. Wu, *J. Clin. Epidemiol.*, 2001, **54**(Suppl. 1), S77–S90.
42. A. Woodcock and C. Bradley, *Qual. Life Res.*, 2001, **10**, 517–531.
43. A. Woodcock and C. Bradley, *Value Health*, 2006, **9**, 320–333.
44. V. Cooper, G. J. Moyle, M. Fisher, G. Reilly, J. Ewan, H. C. Liu and R. Horne, *AIDS Care*, 2011, **23**, 705–713.
45. V. Cooper, R. Horne and J. Ewan, presented at the 10th International Congress on Drug Therapy in HIV Infection, Glasgow, 2010.

46. V. Cooper, R. Horne, C. Simcock and D. Herath, presented at the 13th European AIDS Conference (EACS), Belgrade, 2011.
47. G. L. Colombo, V. Colangeli, A. Di Biagio, S. Di Matteo, C. Viscoli and P. Viale, *ClinicoEcon. Outcomes Res.: CEOR*, 2011, **3**, 197–205.
48. C. Cohen, K. Davis and J. Meyers, presented at the 13th European AIDS Conference (EACS), Belgrade, 2011.
49. F. Aragao, presented at the 13th European AIDS Conference (EACS), Belgrade, 2011.
50. M. Markowitz, A. Zolopa, P. Ruane, K. Squires, L. Zhong, B. P. Kearney and W. Lee, presented at the 18th Conference on Retroviruses and Opportunistic Infections (CROI), Boston, 2011.
51. P. Ruane, E. DeJeesus, D. Berger, M. Markowitz, U. F. Bredeek, C. Callebaut, L. Zhong, S. Ramanathan, M. S. Rhee and K. Yale, presented at the 19th Conference on Retroviruses and Opportunistic Infections (CROI), Seattle, 2012.
52. TrialTrove database, available at http://www.citeline.com/products/trialtrove (last accessed 27 July 2012).

Subject Index